Spring

March 1, 12:30 a.m. Local Standard Time
April 1, 10:30 p.m. Local Standard Time
May 1, 9:30 p.m. Local Standard Time

NORTH

δ Cephei
CASSIOPEIA
Double Cluster
CEPHEUS
Algol
Vega
PERSEUS
Eltanin
Polaris
CAMELOPARDALIS
DRACO
URSA MINOR
Kochab
Capella
HERCULES
M 13
AURIGA
Aldebaran
Alcor
Dubhe
CORONA BOREALIS
TAURUS
Merak
Mizar
M 35
BOOTES
URSA MAJOR
Castor
LYNX
SERPENS CAPUT
CANES VENATICI
GEMINI
Pollux
Arcturus
Betelgeuse
ORION
COMA BERENICES
LEO MINOR
M 44
Ecliptic
CANIS MINOR
M 5
Denebola
LEO
CANCER
M 42
Regulus
Procyon
MONOCEROS
LIBRA
VIRGO
Equator
SEXTANS
Sirius
Spica
Alphard
HYDRA
CANIS MAJOR
CRATER
CORVUS
PUPPIS

SOUTH
SPRING

ASTRONOMY Magazine Star Dome Map by Adolf Schaller.

Contemporary Astronomy

JAY M. PASACHOFF

Williams College – Hopkins Observatory
Williamstown, Massachusetts

SAUNDERS GOLDEN SUNBURST SERIES

1977
W. B. SAUNDERS COMPANY / PHILADELPHIA / LONDON / TORONTO

W. B. Saunders Company: West Washington Square
Philadelphia, PA 19105

1 St. Anne's Road
Eastbourne, East Sussex BN21 3UN, England

1 Goldthorne Avenue
Toronto, Ontario M8Z 5T9, Canada

Library of Congress Cataloging in Publication Data

Pasachoff, Jay M

Contemporary astronomy.

Includes bibliographies.

1. Astronomy. I. Title.

QB45.P29 520 76-014683

ISBN 0-7216-7101-2

Front cover illustration: Mars, a photograph from the Viking 1 spacecraft as it approached in June 1976. The giant canyon in the upper hemisphere has a larger diameter than the United States.

Back cover illustration: Jupiter, from the Pioneer 10 spacecraft, showing the Great Red Spot.

Photographs courtesy of NASA.

Contemporary Astronomy ISBN 0-7216-7101-2

Last digit is the print number: 9 8 7 6 5 4 3 2 1

Preface

Astronomy, as a science, combines the best of new and old traditions. Every year we find something new or unexpected; quasars, molecules in interstellar space, pulsars, coronal holes, x-ray bursts, black holes, and the nature of the surface of Mars are but a few examples from recent times. At the same time, astronomy has a rich and fascinating history.

I have attempted, in this book, to describe the state of astronomy as it is now. I have tried to fit the latest discoveries in their places alongside established results in order to give a contemporary picture of the state of our science.

CONTEMPORARY ASTRONOMY is written for students with no background in mathematics or physics. It discusses astronomy in non-mathematical terms. Students with mathematics or physics backgrounds, however, will still find the book valuable because of its unique coverage of a wide range of modern topics.

The book emphasizes topics that are of current interest to astronomers and in which current research is very active. In order to gain space in the book and time in the academic year for the contemporary material, some traditional material that is discussed at length in other texts has been minimized. But I have attempted to cover all fields of astronomy, without major gaps.

A major theme of the book has been the expansion of our senses to all parts of the spectrum. We see time and again through the entire book how observations of gamma rays, x-rays, ultraviolet light, infrared light, and radio waves are fit together with observations of visible light and with theory to improve our knowledge of the universe and the objects in it. Such new results appear throughout the entire book, rather than being segregated in a few chapters. Similarly, the historical side of astronomy is also treated as the various topics come up, rather than having historical discussions segregated.

A major goal of the book has been to describe astronomy in clear, thorough, understandable and colloquial terms. Many discussions of especially important topics have been expanded in order to have space to describe various aspects. All the steps in chains of reasoning have been included in order to make certain that the major points are clear.

An unusual aspect of CONTEMPORARY ASTRONOMY is that background discussions of physics and astronomy appear as they are needed in the text rather than as lengthy introductory chapters. This has the advantage of making more apparent the connection between the background discussions and the discussions of modern research. It has the further advantage of allowing the student to begin with real astronomy at the very start of the course, instead of having to wait until the introductory chapters are completed. Certain background material may be briefly recapitulated in later chapters even though it may have been dealt with more thoroughly earlier. Once again, the reason for this is to stress the connection between underlying concepts and their applications.

In order to depict the widespread activity in astronomy, and to demonstrate the human aspect of astronomical research, I have sometimes named individual astronomers and institutions. Further, details that may not be important in themselves are sometimes used to embellish the central point. The ability to expand in this way has enabled me to set many matters in especially interesting contexts. But at no time should students find themselves memorizing material instead of understanding concepts. As an aid in determining just what is most important to remember, I have provided summaries of each chapter in this book, and a detailed Student Study Guide (Pasachoff, Kutner, and Pasachoff, Student Study Guide to CONTEMPORARY ASTRONOMY, W. B. Saunders Company, 1977) is also available.

My stress has continually been on how to understand what is going on, and why astronomers think or act as they do. My criterion has been to include mostly material that can be remembered for years and that provides a basic understanding of the topic, rather than facts that are merely learned for examinations and then soon forgotten. In order to give a feel of what it is like to be an astronomer, I have at several places in the text described just what an astronomer does in different circumstances, such as observing at a large optical or radio telescope, observing with a satellite, or working at a solar eclipse. I have myself been fortunate to be able to observe with a variety of optical, radio, and space telescopes in this country and abroad, and I am glad to have the opportunity to describe the excitement an observer feels.

It also seems to me that after completing a course in astronomy a student should be able to carry on a reasonable conversation with an astronomer. This has led me to continue the usage of the somewhat strange units that are retained by astronomers for historical reasons rather than for reasons of logic. For example, magnitudes are used to describe the brightnesses of stars, though I try to make clear the relation of the magnitude scale to a direct scale of brightness. Anyway, I find it easier to remember magnitudes, which are simple numbers like 2 or 16, rather than luminosities like 2×10^4. If these units are madness, there is method in it.

I have included material meant to develop basic skills like conversion of units; many students already have these skills and will be able to skip these parts. (This particular skill applies as well to changing money in foreign countries as it does to changing units in an astronomical context.) I have also tried to bring across the importance of estimation and rough calculation rather than over-exact calculation based on numbers that are themselves only roughly given or known. Of course, one is free to think that a 3rd magnitude is $(2.511887\ldots)^2 = 6.309573\ldots$ times brighter than a 5th magnitude star. I prefer to say that it is about 6 times brighter—not because I can't do the exact calculation (on an electronic calculator it is no harder to do than the estimate) but because I think that the string of digits obscures the reasoning process. The understanding of estimation and the limits of accuracy can be carried by students into their daily lives.

In view of the importance to scientific research of governmental support, and the need for everyone to be informed about the value of such research, I have described from time to time the benefits that might accrue to society from certain types of astronomical research.

Arrangement of the book

Since stars are the main point of reference that most individuals have to astronomy, I have begun my discussion with them. After discussing basic definitions and terms, I discuss the sun as a special case of a star, and then proceed to stellar evolution, including its fascinating end-products: white dwarfs, neutron stars (which we observe as pulsars), and black holes. I then discuss another special case of objects closely associated with stars that we happen to be able to study because they are so near: the planets. After the discussions of stars and planets, we continue outward through the universe from small to large, with interstellar matter, our own galaxy, galaxies in general, quasars, and then cosmological consideration of the universe as a whole. The book thus describes astronomy from the inside out, beginning with the stars.

CONTEMPORARY ASTRONOMY can be used for either a one- or two-semester course. One topic that many professors treat in different manners is Section III on the planets. I have placed it in this book adjacent to the discussion of stars; one could give a two-semester course with stars and planets in the first half and the Milky Way, galaxies, and cosmology in the second half. The material in Section III on the planets could, alternatively, be taught at other places in the syllabus or even omitted; it is not dependent on the discussions that preceded it, and neither do the sections that follow depend on it. Some professors might want to teach the planets first; others might want to defer the study of planets until the very end.

For a one-semester or one-trimester survey course of all of astronomy, one should omit all the sections in smaller print and most of the boxes. I have tried to construct each chapter so that the most important material appears at the beginning, thus making it more natural to assign only the first few sections of the chapter and omit some of the less vital and more advanced material that appears at the end. One possible one-semester or one-trimester course could omit Sections 3.7 to 3.14 on observing methods and techniques; 4.2 and 4.3 on stellar masses and sizes, 4.5c on black holes in globular clusters; 5.2c-e on opacity effects, 5.8 to 5.12 on secondary uses of the sun; 6.6 on the neutrino experiments, 7.5 to 7.7 on relativity and novae; 8.8 to 8.12 on secondary pulsar phenomena; 9.4 on non-stellar black holes and 9.6 on gravitational waves; 10.4 on Bode's "law"; 12.1 on the rotation of Mercury; 13.3 and 13.5 on Venus; 16.2 and 16.4 on Jupiter's moons; any or all of Chapter 18 on comets, meteoroids, and asteroids; any or all of Chapter 19 on life in the universe; 20.5 on spiral structure; any or all of Chapter 21 on the interstellar medium and radio astronomy; any or all of Chapter 22 on infrared, x-ray, and gamma-ray studies of our galaxy; 23.2 on the origin of galactic structure, 23.3 on clusters of galaxies, 23.5 on radio galaxies, and 23.6 on the formation of galaxies; 25.5 on the hierarchical universe, 25.6 and 25.7 on the creation of the elements and their implications for the future of the universe; and any or all of Chapter 26 on how astronomical research works and on future plans and projects. This list, obviously, can be modified substantially by individual professors. Several alternate plans for semester or trimester courses are listed in the Teacher's Guide.

The *Student Study Guide to CONTEMPORARY ASTRONOMY*, written in collaboration with Marc L. Kutner and Naomi Pasachoff, provides a detailed glossary for the terms in this book, discussions of the material in CONTEMPORARY ASTRONOMY chapter by chapter, worked problems, quizzes and questions, and chapter by chapter lists of books and articles. In some cases, it will be adopted for class use along with the main text; in other cases, individual students might choose to use the book on their own.

A *Teacher's Guide to CONTEMPORARY ASTRONOMY* is also available. It contains problems, tests, and answers, and lists films, tapes, and other audio and visual aids for use as supplementary material.

The publishers and I are considering how best to make available the artwork and photographs that are included in this book in a format that can be projected in class. If you are interested in such material, please contact Mr. John J. Vondeling, Editor-in-Chief, College Department, W. B. Saunders Company, West Washington Square, Philadelphia, Pa 19105.

Acknowledgments

The publishers and I have placed a heavy premium on accuracy, and have made certain that the manuscript and proof have been read not only by students for clarity and style but also by several astronomers for their professional comment. I would like to thank, in particular, Thomas T. Arny (University of Massachusetts at Amherst), Marc L. Kutner (Rensselaer Polytechnic Institute), Richard L. Sears (University of Michigan), and Joseph F. Veverka (Cornell University) for reading the entire manuscript.

I would also like to thank other astronomers and physicists who have commented on parts of the manuscript in their particular areas of expertise. They include (in the approximate order in which the topics are treated in the book):

Jean Pierre Swings (Institut d'Astrophysique, Liège, Belgium), stars and quasars; Jeffrey L. Linsky (Joint Institute for Laboratory Astrophysics, Boulder), radiative transfer; David Park (Williams College), atomic physics; James G. Baker (Harvard College Observatory), optics; Jonathan E. Grindlay (Harvard-Smithsonian Center for Astrophysics), black holes and x-ray bursts; Peter V. Foukal (Harvard-Smithsonian Center for Astrophysics), the sun; Robert F. Howard (Hale Observatories), the sun; Lawrence Cram (Fraunhofer Institute, Freiburg, Germany), opacity effects; J. Craig Wheeler (University of Texas), stellar evolution; Raymond Davis, Jr. (Brookhaven National Laboratory), solar neutrinos; Joseph H. Taylor, Jr. (University of Massachusetts, Amherst), pulsars; John A. Wheeler (University of Texas), black holes; Kenneth Brecher (Massachusetts Institute of Technology), black holes; Peter Conti (University of Colorado), stars; Bruce E. Bohannan (University of Colorado), stars;

Owen Gingerich (Harvard-Smithsonian Center for Astrophysics), history of astronomy; David D. Morrison (University of Hawaii and NASA Headquarters), the solar system; James P. Pollack (NASA Ames Research Center), the solar system; Farouk El-Baz (NASA Headquarters and National Air and Space Museum, Smithsonian Institution), the moon; Ewen A. Whitaker (Lunar and Planetary Laboratory, University of Arizona), the moon; Gerald J. Wasserburg (California Institute of Technology), lunar chronology; William R. Moomaw (Williams College), ozone; Ian Halliday (Herzberg Institute of Astrophysics, Ottawa), Neptune and Pluto; J. Donald Fernie (David Dunlap Observatory, Toronto), the Adams-Leverrier affair; P. Kenneth Seidelmann (U.S. Naval Observatory), ephemerides and planetary parameters; Brian Marsden (Harvard-Smithsonian Center for Astrophysics), comets, asteroids, and meteoroids; George D. Gatewood (Allegheny Observatory), Barnard's Star;

Agris Kalnajs (Mt. Stromlo Observatory, Australia), spiral structure; Barry Turner (National Radio Astronomy Observatory), the latest interstellar molecules; Juri Toomre (Joint Institute for Laboratory Astrophysics), gravitational effects on galaxies; Leonid Weliachew (Observatoire de Paris, France), galaxies; George K. Miley (Leiden Observatory, the Netherlands), interferometry; Bernard J. T. Jones (Institute for Astronomy, Cambridge, England), galaxy formation; Jerome Kristian (Hale Observatories), quasars, Seyfert and N galaxies; and John Lathrop (Williams College), cosmology.

Many others have read shorter bits of manuscript, cleared up particular points, or provided special illustrations, and I thank them as well. In particular, let me acknowledge the assistance of Maarten Schmidt (Hale Observatories), Hyron Spinrad (University of California at Berkeley), Richard B. Dunn (Sacramento Peak Observatory), Freeman D. Miller (University of Michigan), Peter van de Kamp (Sproul Observatory), Kenneth D. Tucker (Fordham University), Edwin C. Krupp (Griffith Observatory), and R. Newton Mayall (American Association of Variable Star Observers). I am grateful to Elske v. P. Smith (University of Maryland) for comments early on in the preparation of this book. The many institutions and individuals who provided photographs or other illustrations are acknowledged separately; I am grateful to all of them.

Marc L. Kutner has worked closely with me on the questions.

I thank many people at the W. B. Saunders Company for their efforts on this book. John J. Vondeling, Editor-in-Chief of the College Division, merits special thanks for his continued support. He has made it possible for me to include so many illustrations and photographs. Lloyd Black of the Manuscript Editorial Department has worked long and hard on the manuscript and figures. John Hackmaster, Susan O'Neill, Celeste Brennan, and Linda Downham of the Art Department, Ray Kersey and Peg Shaw of the Illustration Department, Jay Freedman of the Manuscript Editorial Department, and Suzanne Rommel, assistant to John Vondeling, have also contributed to many phases of the book. Lorraine Battista designed the book and the cover. The detailed layout is by Joan De Lucia. The support and efforts of George Laurie of the Art Department and of Tom O'Connor and Frank Polizzano of the Production Department helped keep the book on schedule.

Most of the drawings were executed by George Kelvin of Science Graphics from my sketches. Other drawings were executed by Grant Lashbrook and John Hackmaster at Saunders, and by Ann Pierson in Williamstown, Mass.

I am grateful to Zadig Mouradian of the Observatoire de Paris and to Jean-Claude Pecker of the Collège de France for their hospitality during part of the period on which I was working on this book.

Carla A. Weiss has worked with me in Williamstown on many aspects of the manuscript, illustrations, and proof, and has been of tremendous assistance over a lengthy period of time.

Various members of my family have provided a vital and valuable editorial service. In particular, my wife Naomi Pasachoff, my father Dr. Samuel S. Pasachoff, and my mother Anne T. Pasachoff have worked many hours on the manuscript and proof. My daughter Eloise came in to help type from time to time; as she is not yet two, this did not necessarily speed things up.

The index was put together by Nancy P. Kutner and Ann Pierson, who have also aided with the page proof. Mark Pogue, a Williams College student, has suggested clarifications and read much of the page proof. Susan K. Waldron typed early drafts of the manuscript.

I am extremely grateful to all of the above individuals for their assistance. Of course, it is I who have put this all together, and I alone am responsible for any errors that have crept through our sieve. I would appreciate hearing from readers, not just about typographical or other errors, but also with suggestions for presentation of topics or even with comments about certain points that need clarifying. I invite readers to write me c/o Williams College, Hopkins Observatory, Williamstown, Mass. 01267. I promise a personal response to each writer.

JAY M. PASACHOFF

Williamstown, Massachusetts

Contents

LIST OF TABLES

Part I

The Universe: An Overview

Chapter 1

The universe is a place of great variety—after all, it has everything in it! Thus astronomers have much to study. Some of the things astronomers study are of a size and scale that we humans can easily comprehend: the planets, for instance. Most astronomical objects, however, are so large and so far away that our minds have trouble grasping the sizes and distances involved.

Moreover, astronomers study the very small in addition to the very large. The radiation we receive from distant bodies is emitted by atoms, which are much too small to see with the unaided eye. Also, the properties of the large astronomical objects are often determined by changes that take place on a minuscule atomic scale. Thus the astronomer must be an expert in the study of objects the size of atoms as well as in the study of objects the size of galaxies.

It should come as no surprise to hear that the variety of objects at very different distances from us or with very different properties often must be studied with widely differing techniques. Clearly, different tests are required to analyze the properties of Martian soil from a spacecraft on the Martian surface than are required to test the composition of a quasar deep in space. There is one unifying method, however, that links much of astronomy. This method is *spectroscopy,* which involves, as we shall see, analyzing components of the light or other radiation that we receive from distant objects and studying the components in detail. Throughout this book, we shall return to spectroscopic methods time and again, to study not only visible light but also other types of radiation. Not all of astronomy is spectroscopy, of course; much important information is gained by making images—pictures—of planets, the sun, and galaxies, for example, or by studying the variation of the amount of light coming from a star or a galaxy over time.

The explosion of astronomical research in the last few decades has been fueled by our new ability to study radiation other than light—gamma rays, x-rays, ultraviolet radiation, infrared radiation, and radio waves. Astronomers' use of their new abilities to study such radiation is a major theme of this book. All the kinds of radiation together make up the *electro-*

The Whirlpool Galaxy; M51, in the constellation Canes Venatici.

magnetic spectrum, which will be discussed in Chapter 2. As we shall see there, we can think of radiation as waves, and all the types of radiation have similar properties except for the length of the waves. Still, although to a scientist x-rays and visible light may be similar, our normal experiences tell us that very different techniques are necessary to study them.

The earth's atmosphere shields us from most kinds of radiation. Not only light waves but also radio waves penetrate the atmosphere, though most other types of radiation are blocked. Over the last forty years, radio astronomy has become a major foundation of our astronomical knowledge. For the last twenty years, we have been able to send satellites into orbit outside the earth's atmosphere, and we are no longer limited to the study of radio and visible (light) radiation. Many of the fascinating discoveries of this decade—the probable observation of a black hole, for example—were made because of our newly extended senses. We shall be discussing how all parts of the spectrum are used to help us understand the universe. And besides the information that the spectrum provides, we can also get information from direct sampling of bodies in our solar system (as we shall discuss in Part III), from cosmic rays (which are particles whizzing through space and which we shall discuss in Sections 8.3 and 22.2), and perhaps even from gravitational waves (which, as we shall discuss in Section 9.6, may or may not have been observed).

1.1 A SENSE OF SCALE

Let us try to get a sense of scale of the universe, starting with sizes that are part of our experience and then expanding toward the infinitely large. In the margin, we can keep track of the size of our field of view as we expand in powers of 100: each diagram will show a square 100 times greater on a side and thus 10,000 times greater in area than the previous diagram.

We shall use the metric system, which is in common use by scientists. The basic unit of length is the meter, which is equivalent to 39.37 inches, slightly more than a yard. Prefixes are commonly used in conjunction with the word "meter" (see Appendix 2) to define new units. The most commonly used are "milli-," meaning 1/1000, "centi-," meaning 1/100, and "kilo-," meaning 1000 times. Thus 1 millimeter is 1/1000 of a meter, or about .04 inch, and a kilometer is 1000 meters, or about 5/8 mile. We will keep track of the powers of 10 by which we multiply 1 cm by writing the number of tens we multiply together as an exponent; 1000 cm, for example, is 10^3 cm.

Literary scholars could say that a "milli-Helen" is a unit of beauty sufficient to launch one ship.

We can also keep track of distance in units that are based on the length of time that it takes light to travel. The speed of light is, according to Einstein's *special theory of relativity* (published in 1905), the greatest speed that is physically attainable. Light travels at 300,000 km/sec (186,000 miles/sec), fast enough to circle the earth 7 times in a single second. Even at that fantastic speed, we shall see that it would take years for us to reach the stars. Similarly, it has taken years for the light we see from stars to reach us, and we are thus really seeing the stars as they were years ago. In a sense, we are looking backward in time. The distance that light travels in a year is called a *light year*; note that the light year is a unit of length rather than a unit of time even though the term "year" appears in it.

Let us begin our journey through space with a view of a circle one millimeter in diameter (Fig. 1–1).

1 mm = 0.1 cm

10 cm = 100 mm

A square 100 times larger on each side is 10 centimeters × 10 centimeters (since the area of a square is the length of a side squared, the area of a 10 cm square is 10,000 times the area of a 1 mm square). The area encloses a flower (Fig. 1–2).

10 m = 1000 cm

As we move far enough away to see an area 10 meters on a side, we are seeing an area approximately that taken up by half a tennis court (Fig. 1–3).

1 km = 10^3 m

A square 100 times larger on each side is now 1 kilometer square, about 250 acres. An aerial view of several square blocks in New York City shows how big an area this is (Fig. 1–4).

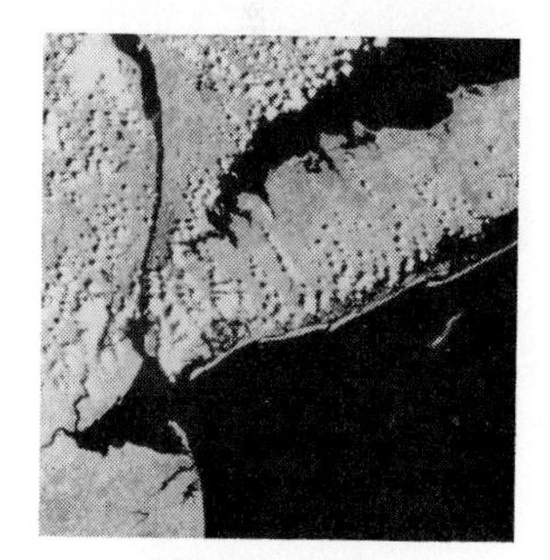

100 km = 10^5 m

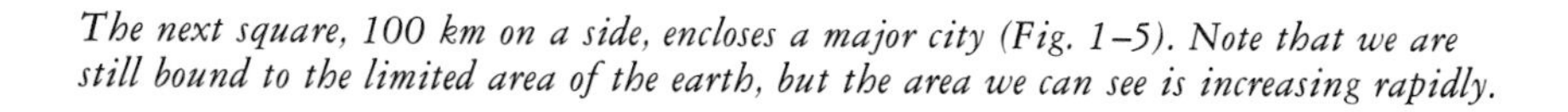

The next square, 100 km on a side, encloses a major city (Fig. 1–5). Note that we are still bound to the limited area of the earth, but the area we can see is increasing rapidly.

10,000 km = 10^7 m

A square 10,000 km on a side covers nearly the entire earth (Fig. 1–6).

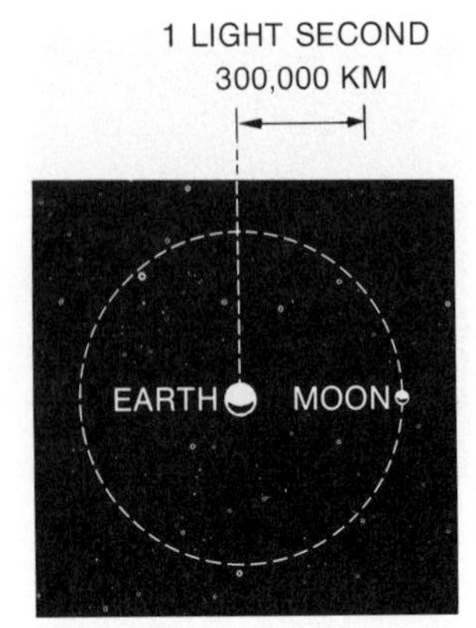

1,000,000 km = 10^9 m = 3 lt sec

When we have receded 100 times farther, we see a square 100 times larger in diameter: 1 million kilometers across. It encloses the orbit of the moon around the earth (Fig. 1–7). We can measure with our wristwatches the amount of time that it takes light to travel this distance. If we were carrying on a conversation by radio with someone at this distance, there would be pauses of noticeable length after we finished speaking before we heard an answer. This is because radio waves, even at the speed of light, take that amount of time to travel. Astronauts on the moon have to get used to these pauses when speaking to earth.

When we look on from 100 times farther away still, we see an area 100 million kilometers across, 2/3 the distance from the earth to the sun. We can now see the sun and the inner planets in our field of view (Fig. 1–8). This is the scale representing approximately the limit of our current ability to send spacecraft to sample the planets directly. We have now sent spacecraft to the moon, Mercury, Venus, Mars, and Jupiter; the results will be discussed at length in Part III.

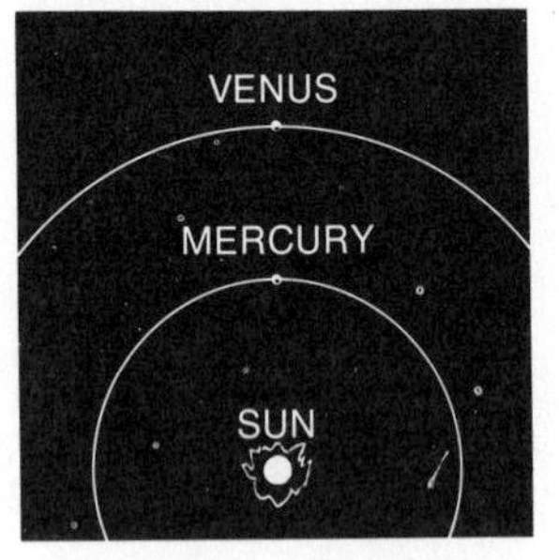

10^{11} m = 5 lt min

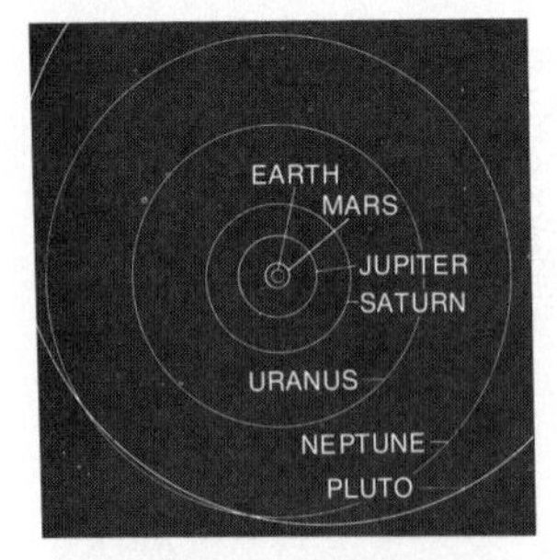

10^{13} m = 8 lt hrs

An area 10 billion kilometers across shows us the entire solar system in good perspective. It now takes light 8 hours to travel this distance. The outer planets have become visible and are receding into the distance as our journey outward continues (Fig. 1–9).

From 100 times farther away, we see little that is new. The solar system seems smaller and we see how vast is the empty space around us. We have not yet reached the scale at which another star besides the sun is in our field of view, although, of course, many stars are visible in the background (Fig. 1–10).

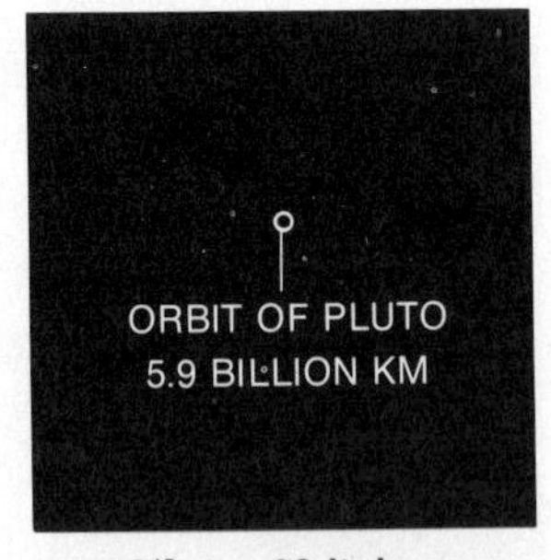

10^{15} m = 38 lt days

As we continue to recede from the solar system, the nearest stars finally come into view. We are seeing an area 10 light years across, which contains only one other star (Fig. 1–11). This star and many others turn out to be systems in which several stars revolve around each other. Part II of this book discusses the properties of stars.

10^{17} m = 10 ly

10^{19} m = 10^3 ly

By the time we are 100 times farther away, we can see a fragment of our galaxy (Fig. 1–12). We see not only many individual stars but also many clusters of stars and many areas of glowing, reflecting or opaque gas or dust called nebulae. There is even a lot of material between the stars that is mostly invisible to our eyes but that can be studied with radio telescopes. Part IV of this book is devoted to the study of our galaxy and its contents.

In a field of view 100 times larger in diameter, we can now see an entire galaxy. The photograph (Fig. 1–13) shows a galaxy called M74, located in the direction that we call the constellation Pisces. This galaxy shows arms wound in spiral form. Our galaxy also has spiral arms, though they are wound more tightly.

10^{21} m = 10^5 ly

10^{23} m = 10^7 ly

Next we move sufficiently far away so that we can see an area 10 million light years across at a glance (Fig. 1–14). There are 10^{25} centimeters in 10 million light years, about as many centimeters as there are grains of sand in all the beaches of the earth. Our galaxy is in a cluster of galaxies, called the Local Group, that would take up only 1/3 of our angle of vision. In this group are all types of galaxies, which we will discuss in Chapter 23. The photograph shows part of a cluster of galaxies in the constellation Leo.

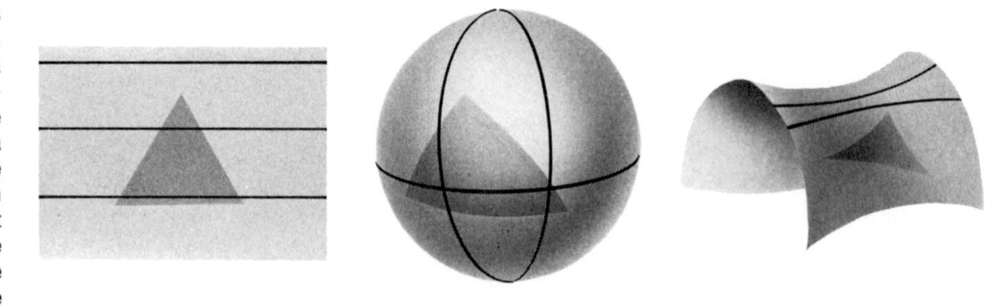

Figure 1–15 Straight lines and triangles are shown on a plane, on the surface of a sphere, and on a saddle-shaped surface. These types of surfaces have a property called curvature that is different for each type. How can we tell what the curvature is? Even if we were limited to the same two dimensions that these surfaces have, we could test which type of surface we were on by adding up the angles of a triangle. The sum of the angles of a triangle drawn on a plane is 180°. On the surface of a sphere, the sum of the angles of a triangle is always greater than 180°. (You can see this by considering a triangle with two vertices on the equator one-fourth of the way around the sphere from each other and the third vertex at the north pole. That triangle contains three right angles, which add up to 270°.) On a saddle-shaped surface, the sum of the angles of a triangle is less than 180°.

We live not in two dimensions as on these surfaces, but rather in three dimensions or, in some sense, in four dimensions (three of space plus one of time). Yet in principle we could test the curvature of our space, though in practice this has not yet been possible to do.

If we could see a field of view 1 billion light years across, our Local Group of galaxies would appear as but one of many clusters. We are not able to observe this scale.

Before we could enlarge our field of view another 100 times we might see a supercluster—a cluster of clusters of galaxies. We would be seeing almost to the distance of the quasars, which are the topic of Chapter 24. Quasars may be explosive events in the cores of normal galaxies, or may involve phenomena very different from those astronomers normally consider. We do not know what is involved. Light from the farther quasars may have taken ten billion years to reach us on earth, and if this is so we are thus looking back to epochs billions of years ago. Since we think that the big bang that began the universe took place approximately 13 billion years ago, we are looking back almost to the beginning of time.

At this scale, we cannot accurately predict what the universe would look like. Einstein's general theory of relativity predicts that space itself could show properties of being curved, just as the surface of a saddle or the surface of a sphere is fundamentally curved (Fig. 1–15). On none of these surfaces, for example, can we draw straight lines that remain parallel out to infinity, that is, never crossing and always remaining the same distance from each other. On a saddle the lines diverge; on a sphere the lines converge. Is space in the universe curved positively like a sphere, curved negatively like a saddle, or flat? We do not know. Studies of the depths of the universe may tell us about the curvature of space. In any case, we can understand certain things about the universe as a whole, and we even think that we have detected the radiation given off at the time of the big bang itself. A combination of radio, ultraviolet, and optical studies is allowing us to explore the past and predict the future of the universe. All this is discussed in Chapter 25.

1.2 THE VALUE OF ASTRONOMY

Throughout history, observations of the heavens have led to discoveries that have had major impact on people. Even the dawn of mathematics may have followed ancient observations of the sky, made in order to keep track of seasons and seasonal floods in the fertile areas of the earth. Observations of the motions of the moon and the planets, which are free of such complicating terrestrial forces as friction and which are massive enough so that gravity dominates their motions, led to an understanding of gravity and of the forces that govern all motion.

We can consider the regions of space studied by astronomers as a cosmic laboratory for the study of matter or radiation under a variety of conditions, often under conditions that we cannot duplicate on earth.

Many of the discoveries of tomorrow—perhaps the control of nuclear fusion or the discovery of new sources of energy, or perhaps something so revolutionary that it cannot now be predicted—will undoubtedly also be based on discoveries made through such basic research as the study of astronomical systems. Considered in this sense, astronomy is an investment in our future.

The impact of astronomy on our conception of the universe has been strong through the years. Discoveries that the earth is not in the center of the universe, or that the universe has been expanding for billions of years, affect our philosophical conceptions of ourselves and our relations to space and time.

Yet most of us study astronomy not for its technological and philosophical benefits but for its grandeur and inherent interest. We must stretch our minds to understand the strange objects and events that take place in the far reaches of space. The effort broadens us and continually fascinates us all. Ultimately, we study astronomy because of its fascination and mystery.

QUESTIONS

1. What is "spectroscopy"?
2. The speed of light is 3×10^5 km/sec. Express this number in meters/sec and in cm/sec.
3. During the Apollo explorations of the moon, we had a direct demonstration of the finite speed of light when we heard ground controllers speak to the astronauts. The sound from the astronauts' earpieces was sometimes picked up by the astronauts' microphones and retransmitted to earth. We then heard the words repeated. What is the time delay between the original and the "echo"?
4. The distance to the Andromeda galaxy is 2×10^6 light years. If we could travel at one-tenth the speed of light, how long would a round trip take?
5. How long would it take to travel to Andromeda at 1000 km/hr, the speed of a jet plane?
6. List the following in order of increasing size: (a) light year, (b) distance from earth to sun, (c) size of Local Group, (d) size of a football stadium, (e) size of our galaxy, (f) distance to a quasar.
7. Of the examples of scale in this chapter, which would you characterize as part of "everyday" experience? What range of scale does this encompass? How does this range compare with the total range covered in the chapter?
8. What is the largest of the scales discussed in this chapter that could reasonably be explored in person by humans with current technology?
9. How can we tell a flat space from a curved space? How could we tell if the surface of the earth is flat or curved?
10. What is the value of astronomy to you? Save your answer for comparison with your answer to a similar question in Chapter 26 or your answer when you have completed this course.

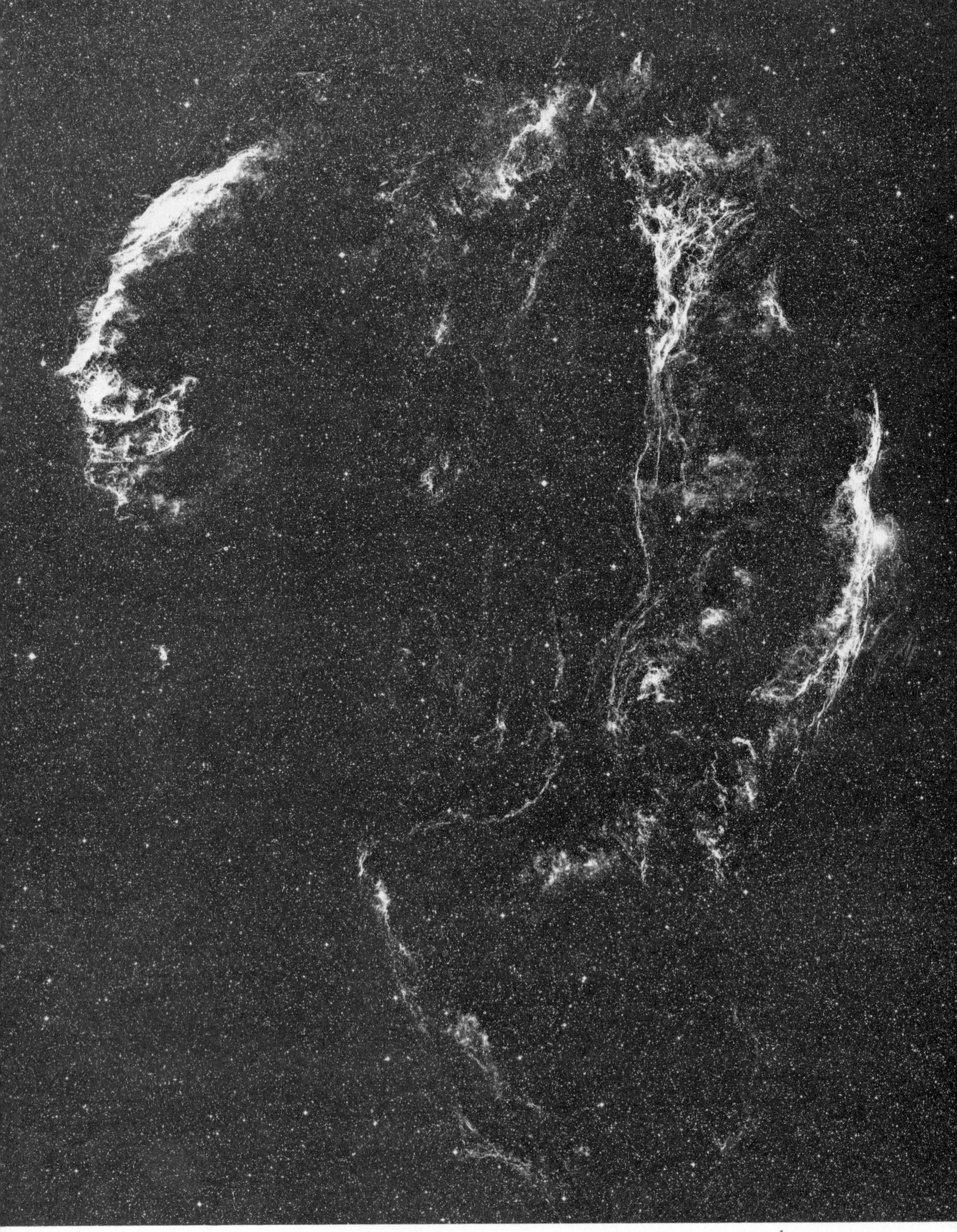

Left: The Cygnus loop, the remnant of a supernova. The remainder of the star that explodes as a supernova contracts to become either a pulsar or a black hole, depending on how much mass is left behind. *Right:* A solar prominence.

Part II

The Stars

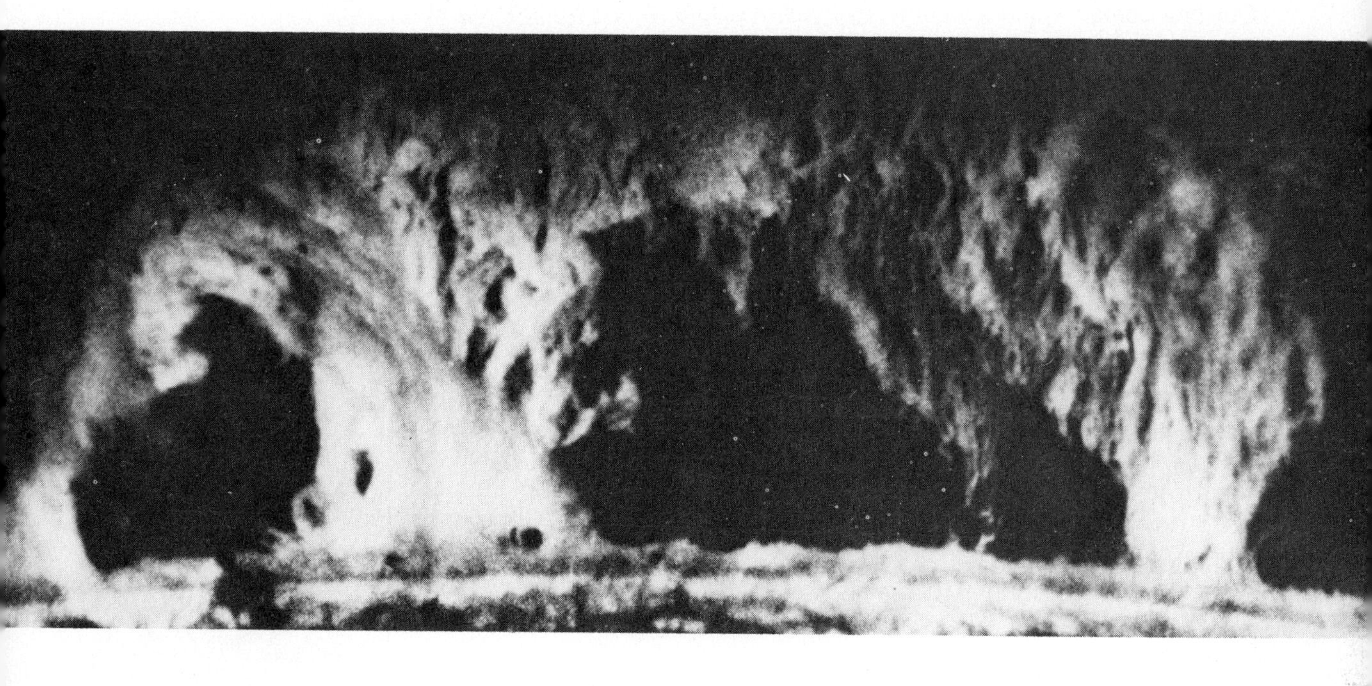

II The Stars . . .

Our first contact with astronomy involves the stars. When we look up at the sky at night, most of the objects we see are stars. In the daytime, we see the sun, which is itself a star. The moon, the planets, even a comet may give a beautiful show, but, however spectacular, they are only minor actors on the stage of observational astronomy.

From the center of a city, we may not be able to see very many stars, because city light scattered by the earth's atmosphere makes the light level of the sky brighter than most stars. All together, there are about 6000 stars in the sky bright enough to be seen with the naked eye under good observing conditions.

The first High Energy Astrophysical Observatory, HEAO-A.

Even a cursory glance at the sky shows that there are patterns in the distribution of stars. The different areas of sky are called *the constellations,* and the names that the ancients gave to these areas, associating them with figures and objects from their myths and religions, are still used down to this day. It takes a great imagination to see Orion (the Hunter) or Draco (the Dragon) in the sky above, but it is nice to be like St. Exupéry's Little Prince (Fig. 2–1), and not let our detailed knowledge cause us to lose our sense of wonder at the sky above and the objects in space.

The stars we see defining a constellation may be at very different distances from us. One star may appear bright even though it is naturally faint because it is relatively close to us; another star may appear bright even though it is far away because it is intrinsically very luminous. When we look at the constellations, we are observing only the directions of the stars and not whether the stars are physically close together (Fig. 2–3). Nor do the constellations tell us very much about the nature of stars. Astronomers tend not to be interested in the constellations because the constellations do not give us information about how the universe works. After all, the constellations merely tell the directions in which the stars lie. Most astronomers want to know *why* and *how*. Why is there a star? How does it shine? Such studies of the workings of the universe are called *astrophysics*. Almost all modern day astronomers are astrophysicists as well, for they always not only make observations but also think about their meaning.

Some properties of the stars that can be seen with the naked eye tell us about the natures of the stars themselves. For example, some stars seem blue-white in the sky, while others appear slightly reddish. From information of this nature, astronomers are able to determine the temperatures of the stars. Chapter 2 is devoted to the basic properties of stars and Chapter 3 to various methods of making astronomical observations. Some stars are not constant in brightness. In some cases, as we shall see in Chapter 4, the variations can tell us how far away these stars are, and thus give the distances to the astronomical groupings in which these stars are located. In Chapter 5, we study an average star in detail. It is the only star we can see close up—the sun. The phenomena we observe on the sun presumably take place on other stars as well.

We can't follow an individual star from cradle to grave, for that would take billions of years, but by studying many different stars we can study enough stages of development to write out a stellar biography. Chapters 6 through 9 are devoted to such life stories. Similarly, we could study the stages of human life not by watching someone's aging, but rather by studying people of all ages who are present in, say, a city on a given day.

In Chapter 6, we shall study the birth of stars, and observe some places in the sky where we expect stars to be born very soon, possibly even in our own lifetimes. We shall then consider the properties of stars during the long, stable phase in which they spend most of their histories. Then we go on to consider the ways in which stars end their lives. Stars like the sun sometimes eject shells of gas that glow beautifully; the remainders then collapse until they are as small as the earth (see Chapter 7). Sometimes when a star is newly visible in a location where no star was previously known to exist, we may be seeing the spectacular death throe of a star that had earlier been too faint to observe (see Chapter 8). The study of what happens to stars after that explosive event is currently a topic of tremendous interest to many astronomers. Some of the stars become pulsars (also discussed in Chapter 8). Other stars, we think, may even wind up as black holes, objects that are invisible and therefore difficult—though not impossible—to observe (see Chapter 9).

To do their research, astronomers still often use optical telescopes to gather light from the stars. Following how a stellar biography can be written from the clues that the light carries is like reading a wonderful detective story. But in recent years, the heavens have been studied in other ways besides observing the ordinary light that is given off by many astronomical objects. Radio waves, x-rays, gamma rays, ultraviolet light, infrared light, and cosmic rays are increasingly studied from the earth's surface or from space. Many a contemporary astronomer—even many who consider themselves observers (who mainly carry out observations) rather than theoreticians (who do not make observations, but rather construct theories)—has never looked through an optical telescope. So we mustn't grow to think that the appearance of the stars in ordinary light is the only part of astronomy. Still, studying the light from the stars has been very fruitful for astronomers, and our story begins there.

2

Ordinary Stars

What are these stars that we see as twinkling dots of light in the night-time sky? The stars are luminous balls of gas scattered throughout space. All the stars we see in the sky are among the 100 billion stellar members of a collection of stars and other matter called the Milky Way Galaxy (see Part IV). The sun, which is the controlling star of our planet, is so close to us that we can see detail on its surface, but it is just an ordinary star like the rest.

An important property of stars is that they generate their own energy and light. Deep inside them they generate energy by a process that we think we understand even though we cannot see it directly—nuclear fusion. It has been only fifty years since the realization that the stars shine in this way, and we are now looking to the stars to provide sources of energy for the earth in the future, either by harnessing the energy from the sun that falls upon the earth or by learning how to tame the nuclear fusion that takes place inside the sun and the other stars and re-creating it on earth.

We see only the outer layers of the gas of the stars; the interiors are hidden from our view by all the gas surrounding them. The outer layers are not generating energy by themselves, but are merely glowing because of the effects of energy that is transported outward from the stellar interiors. So when we study the light that we get from the stars we are not observing the processes of energy generation directly. We must use our studies of the exteriors to tell us what is going on inside.

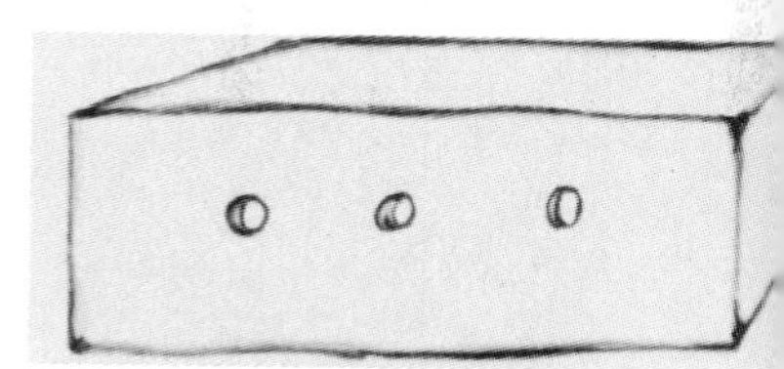

Figure 2–1 When asked to draw a sheep, the Little Prince drew the above. "This is only his box," said the Little Prince. "The sheep you asked for is inside."
From *The Little Prince* by Antoine de Saint Exupéry.

What about the twinkling? It is not a property of the stars themselves, but merely an effect of our earth's atmosphere. The starlight is always being bent by moving volumes of air in our atmosphere, making the images of the stars appear to be larger than points, to dance around slightly, and to change rapidly in intensity when observed with the naked eye (Fig. 2–2). The planets, on the other hand, do not usually seem to twinkle. This is because they are close enough to earth that they appear as tiny disks large enough to be seen through telescopes though not by the naked eye. When the disk of light representing a planet moves slightly in the sky by the same

Cassiopeia, from Johann Bayer's *Uranometria*, first published in 1603.

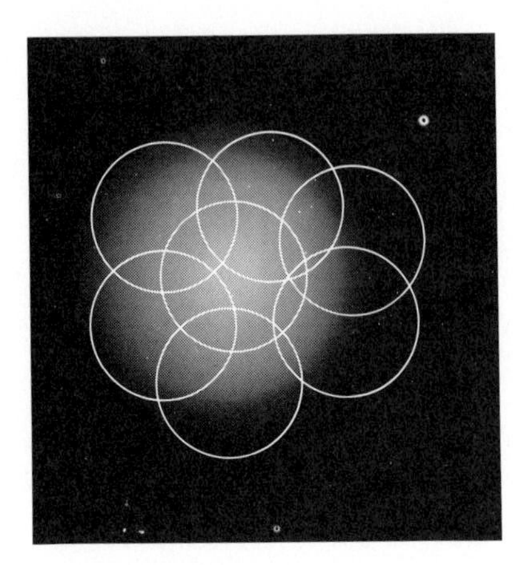

Figure 2–2 Stars appear to twinkle because they are point images (left). Because planets appear as disks (right), they do not usually twinkle, but planets can twinkle too under conditions in which the atmosphere is sufficiently turbulent.

amount as does a neighboring star, the new position of the disk still overlaps much of the former position. The visual effect to the naked eye is that the planets appear to shine more steadily, even though the earth's atmosphere is affecting each bit of their light in just the same manner as it affects the light from the stars.

The stars are balls of gas that are held together by the force of gravity, the same force that keeps us on the ground with our feet toward the center of the earth, no matter whether we are in the United States or half way around the globe in Australia. Long ago, gravity compressed large amounts of gas and dust into dense spheres, which became stars. Energy from this original contraction has heated up the interiors of the spheres until they were hot enough for nuclear fusion to begin (see Chapter 6).

Fifty years ago, Sir Arthur Eddington, a British astronomer, imagined the inhabitants of a cloud-bound planet where perennial dense clouds had prevented scientists from ever knowing that there were stars. On the basis of the laws of physics, they could nonetheless deduce that balls of gas with masses in a certain range would be shaped by gravity and caused to contract until they began to shine. Imagine their pleasure, Eddington wrote, if one day their clouds could be rolled back and these theoreticians could see the beautiful stars that they had predicted. I hasten to point out that this story requires that the laws of physics be known to the scientists of this cloud-bound planet. On our own planet, the study of the stars has been a tremendous aid to the development of these physical laws, for the astronomical objects in the sky provide an observational laboratory freed of petty annoyances like friction and air currents that complicate experiments in laboratories on earth.

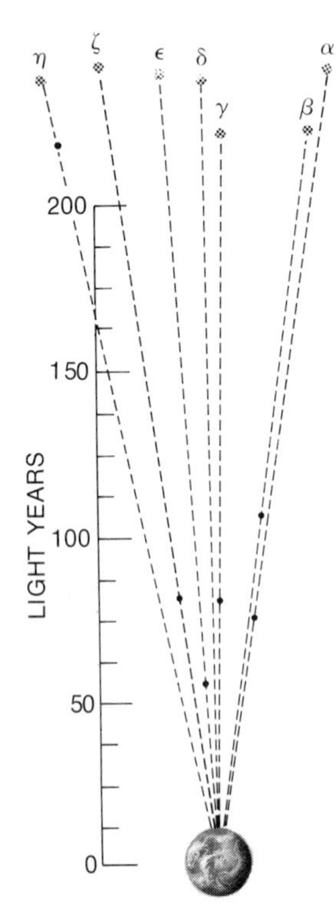

Figure 2–3 The stars we see as a constellation are actually differing distances from us. In this case we see the true distances of the stars in the Big Dipper, part of the constellation Ursa Major.

One difference between astronomers and scientists of most other fields is that many astronomers are usually *observers* rather than *experimentalists* in that they can **observe** light from distant objects but cannot ordinarily **experiment** with the objects directly. Even though we can now sample the moon and some of the planets, we cannot seriously hope to sample the distant stars.

2.1 LIGHT FROM THE STARS

The most obvious thing we notice about the stars is that they have different brightnesses. Over two thousand years ago, the Greek astronomer

Hipparchus divided the stars that he could see into classes of brightness. His measurements were extended by Claudius Ptolemy, a Greek/Egyptian astronomer in Alexandria in about A.D. 150. In their classification, the brightest stars were said to be of the first magnitude, somewhat fainter stars were of the second magnitude, and so on down to the sixth magnitude, which represented the faintest stars that could be seen with the naked eye.

In the 19th century, when astronomers became able to make quantitative measurements of the brightnesses of stars, the magnitude scale was placed on a numerical basis. It was discovered that the brightest stars that could be seen with the naked eye were about 100 times brighter than the faintest stars. This corresponded to a difference of 5 magnitudes on Hipparchus's scale, a number that was used to set up the present magnitude system.

The new system is based on the definition that a difference of five magnitudes means a ratio of intensity of exactly 100 times. Thus a star of first magnitude is exactly 100 times brighter than a star of the sixth magnitude. The brightnesses of the stars that had been classified by the Greeks were placed on this new magnitude scale (Fig. 2–4). Many of the stars that had been of the first magnitude in the old system were indeed first magnitude in the new system. But a few stars were much brighter, and thus corresponded to "zeroth" magnitude, that is, a number less than one. A couple, like Sirius, the brightest star in the sky, were brighter still. The

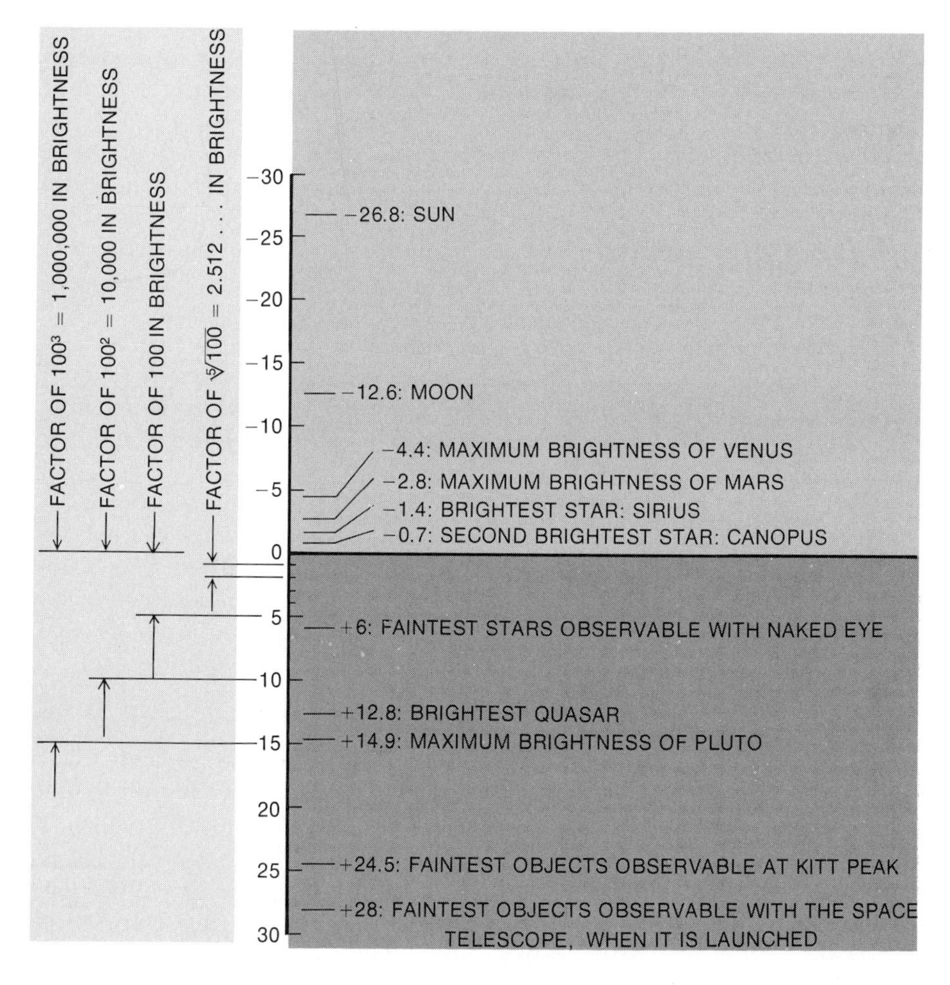

Figure 2–4 The magnitude scale.

numerical magnitude scale could easily be extended to negative numbers. On this scale, Sirius is brighter than magnitude −1. The new scale, being numerical, admits fractional magnitudes. Sirius is actually magnitude −1.4, for example. Physiologically, this type of measurement makes sense because the human eye perceives equal **ratios** of luminosity as roughly equal **intervals** of brightness (on the magnitude scale).

The scale was extended not only to stars brighter than first magnitude, but also to stars fainter than 6th magnitude. Since 5 magnitudes corresponds to a hundred times in brightness, 11th magnitude is 100 times fainter than 6th magnitude (that is, 1/100 times as bright); 16th magnitude, in turn, is 100 times fainter than 11th magnitude. The faintest stars that can be photographed with the 5-meter (200-inch) telescope on Palomar Mountain in California, the largest optical telescope in the United States and until recently (1975) the largest optical telescope in the world, are fainter than 23rd magnitude. The interval from magnitude −1.4 to magnitude 23.5 is nearly 25 magnitudes. Since each 5 magnitudes represents a multiplication of 100 times in brightness, 25 magnitudes is a factor of $100 \times 100 \times 100 \times 100 \times 100$ (since it is 100 multiplied by itself 5 times, we abbreviate this by writing 100^5), or 10,000,000,000.

½ mag = 1.585 times
1 mag = 2.512 times
2 mag = 6.310 times
3 mag = 15.85 times
4 mag = 39.81 times
5 mag = 100 times
6 mag = 251.2 times
7 mag = 631.0 times
8 mag = 1585 times
9 mag = 3981 times
10 mag = 10^4 times
15 mag = 10^6 times
20 mag = 10^8 times

In astronomy we often find ourselves writing numbers that have strings of zeros attached, so we use what is called scientific notation. *In scientific notation, which we have already employed in Chapter 1, we merely count the number of zeros, and write it as a superscript to the number 10. Thus the number 10,000,000,000, a 1 followed by 10 zeros, is written 10^{10}. If the number had been 30,000,000,000, we would have written 3×10^{10}. This method of writing numbers simplifies our writing chores considerably, and helps prevent making mistakes in copying long strings of numbers.*

We can handle numbers much less than one with negative exponents. A minus sign in the exponent of a number means that the number is actually one divided by what the quantity would be if the exponent were positive. Thus $10^{-2} = 1/10^2$. One can count the number of places by which the decimal point has to be moved until it is at the right of the first non-zero digit. Thus, for example, $.000001435 = 1.435 \times 10^{-6}$.

Scientific notation also allows simplification in calculation. The 5 in $(100)^5$ is called the exponent. *When a number with an exponent is itself raised to an exponent, as we say, then one simply multiplies the exponents. Thus, since $100 = 10^2$, we see that $(100)^5 = (10^2)^5$ and by merely multiplying 2×5 to get 10, we have 10^{10}.*

We have seen how a difference of 5 magnitudes corresponds to a factor of 100 in brightness. What does that indicate about a difference of 1 magnitude? Since an increase of 1 in the magnitude scale corresponds to a decrease in brightness by a certain factor, we need a number that, when multiplied by itself 5 times, will equal 100. This number is just $\sqrt[5]{100}$. Its value is 2.512 . . . , with the dots representing an infinite string of other digits. For many practical purposes, it is sufficient to know that it is nearly 2.5.

Thus a second magnitude star is about 2.512 times fainter than a first magnitude star. A third magnitude star is about 2.512 times fainter than a second magnitude star. Thus, a third magnitude star is approximately $(2.5)^2$, which is about 6 times fainter than a first magnitude star. (We could easily give more decimal places, writing 6.3 or 6.31, but the additional figures would not be meaningful. We started with only one digit being significant when we wrote down "2nd magnitude"—we say that the number was given to an accuracy of *one significant figure*—and we must be careful not to fool ourselves that we have gained in accuracy in the multiplication process.)

In similar fashion, using a factor of 2.512 for the single magnitudes and a factor of 100 for each group of 5 magnitudes, we can very simply find the ratio of intensities of stars of any brightness.

Example: What is the difference between magnitude -1 and magnitude 6, the ratio between the brightest and faintest stars that are seen with the naked eye?

Answer: The difference is 7 magnitudes, and the factor is thus (100)(2.5)(2.5), which is about 600. The star of -1st magnitude is 600 times brighter than the star of 6th magnitude.

One unfortunate thing about the magnitude scale is that it operates in the opposite sense from a direct measure of brightness. Thus the brighter the star, the lower the magnitude (a second magnitude star is brighter than a third magnitude star), which is certainly a confusing convention. This kind of problem comes up occasionally in an old science like astronomy, for at each stage astronomers have made their new definitions so as to have continuity with the past.

2.2 THE SPECTRUM

It was discovered over three hundred years ago that when ordinary light is passed through a prism, a band of color like the rainbow comes out the other side. Thus "white light" is composed of all the colors of the rainbow (Fig. 2–5).

These colors are always spread out in a certain order, which has traditionally been remembered by the initials of the friendly fellow ROY G. BIV, which stands for Red, Orange, Yellow, Green, Blue, Indigo, Violet. No matter what rainbow you watch, or what prism you use, the order of the colors never changes.

Modern sensibility, though, no longer sees "indigo" as a separate color of the same importance as the others listed.

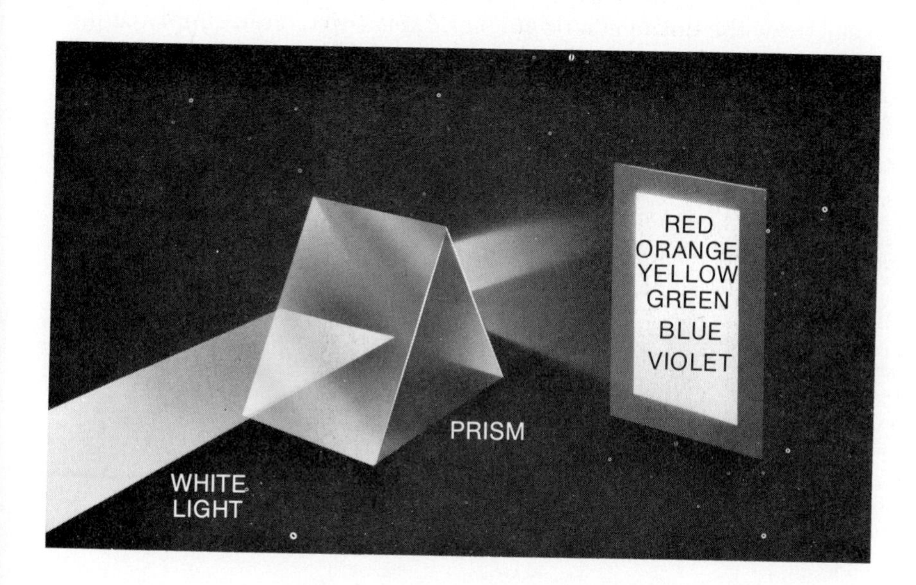

Figure 2–5 A prism disperses white light into its component colors.

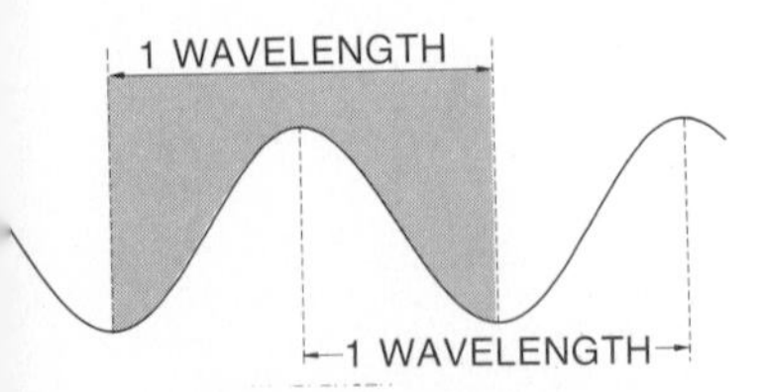

Figure 2–6 The definition of the wavelength of electromagnetic radiation.

We can understand why light contains the different colors if we think of light as waves of radiation. These waves are all traveling at the same speed, 3×10^{10} cm/sec (186,000 miles/sec). This speed is the speed that light would have if it were in a vacuum; it is normally called the *speed of light* (even though light normally travels at a slightly lower speed because the light is not usually in a perfect vacuum).

The distance between one crest of the wave to the next or one trough to the next, or in fact between any point on a wave and the similar point on the next wave, is called the *wavelength* (Fig. 2–6). Light of different wavelengths appears as different colors. It is as simple as that. Red light has approximately 1½ times the wavelength of blue light. Yellow light has a wavelength in between the two. Actually there is a continuous distribution of wavelengths, and one color blends subtly into the next.

The wavelengths of light are very short: just a few hundred-thousandths of a centimeter. Astronomers use a unit of length called an angstrom, named after the Swedish physicist A. J. Ångström. One angstrom (1 Å) is 10^{-8} cm. (Remember that this is 1 divided by 10^8, which is .000 000 01 cm.) The wavelength of blue light is approximately 4000 angstrom units, usually written 4000 angstroms (4000 Å). Yellow light is approximately 6000 Å, and red light is approximately 6500 Å in wavelength.

The human eye is not sensitive to radiation whose wavelength is much shorter than 4000 Å or much longer than 6600 Å, but other devices exist that can measure light at shorter and longer wavelengths. At wavelengths shorter than violet, the radiation is called *ultraviolet*; at wavelengths longer than red, the radiation is called *infrared.* (Note that it is not "infared," a common misspelling; "infra" is a Latin prefix that means "below.")

All these types of radiation—visible or invisible—have much in common. Many of us are familiar with the field of force produced by a magnet—a magnetic field—and the field of force produced by a charged object like a hair comb on a dry day—an electric field. It has been known for over a century that fields that vary rapidly behave in a way that no one would guess from the study of magnets and combs. In fact, light, x-rays, and radio waves are all examples of electric fields and magnetic fields that have become detached from their sources and move rapidly through space. For this reason, these radiations are referred to as *electromagnetic radiation.*

We can draw the entire *electromagnetic spectrum,* often simply called the *spectrum* (plural: *spectra*), ranging from radiation of wavelength shorter than 1 Å to radiation of wavelength many meters long and more (Fig. 2–7).

Figure 2–7 The electromagnetic spectrum.

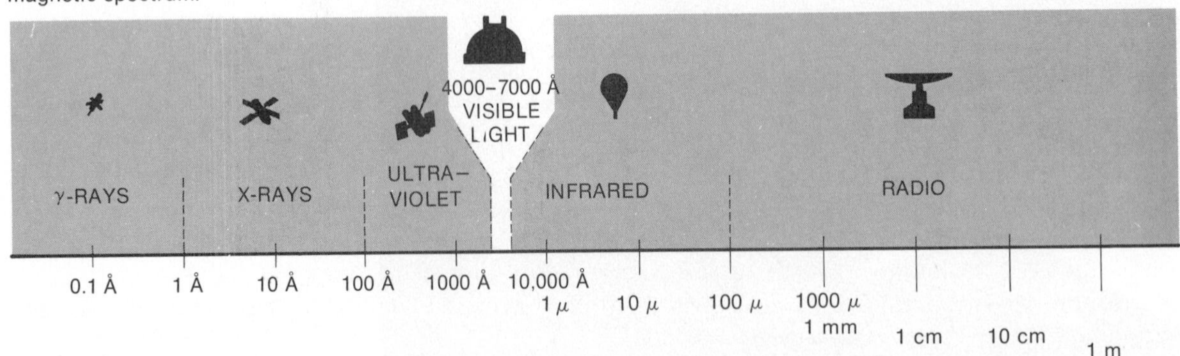

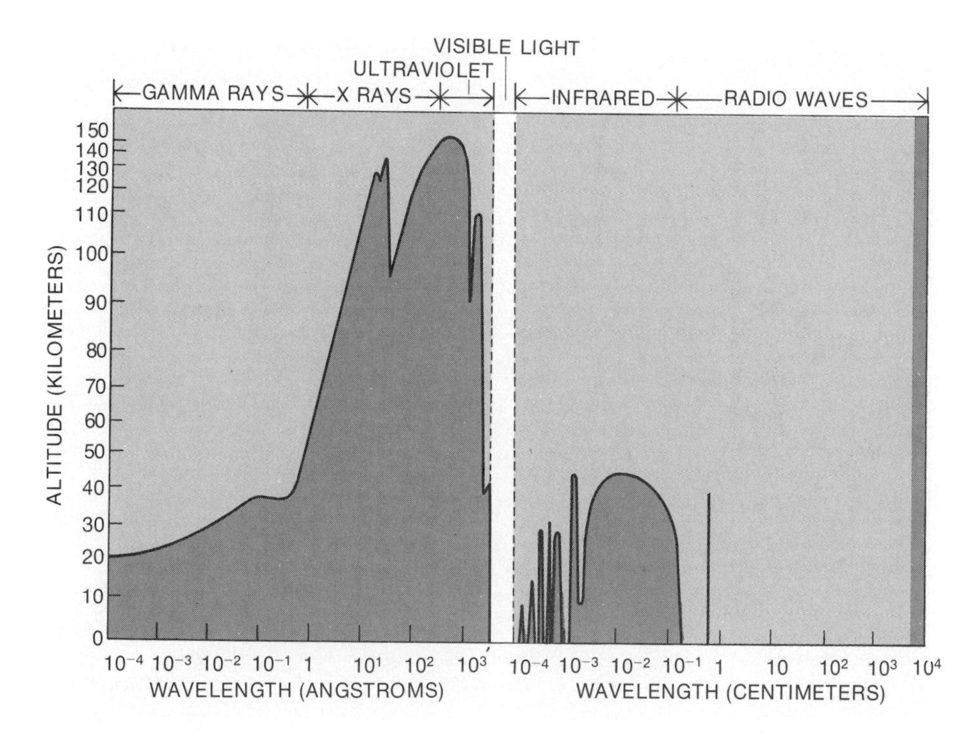

Figure 2–8 Windows of transparency in the terrestrial atmosphere allow only part of the solar spectrum to penetrate to the earth's surface. The curve specifies the altitude where the intensity of arriving radiation is reduced to half its original value.

Note that from a scientific point of view there is no real qualitative difference among types of radiation at different wavelengths. They can all be thought of as *electromagnetic waves,* and light waves comprise but one limited range. When an electromagnetic wave has a wavelength of 1 Å, we call it an x-ray. When it has a wavelength of 5000 Å, we call it light. When it has a wavelength of 1 cm (which is 10^8 Å), we call it a radio wave. Of course, there are obvious practical differences in the methods by which we detect x-rays, light, and radio waves, but the principles that govern their existence are the same.

Note also that light occupies only a very small portion of the entire electromagnetic spectrum. It is obvious that the new ability that astronomers have to study parts of the electromagnetic spectrum other than light waves enables us to increase our knowledge of celestial objects manyfold.

Only certain parts of the electromagnetic spectrum come through the earth's atmosphere. We call these parts of the spectrum at which the earth's atmosphere is transparent by the name "windows." One window passes what we call "light," and what astronomers more technically call *visible light,* or *the visible,* or *the optical part of the spectrum.* Another "window" falls in the radio part of the spectrum, and modern astronomy uses "radio telescopes" to detect that radiation (Fig. 2–8).

But we of earth are no longer bound to our planet's surface; balloons, rockets, and satellites carry telescopes above the atmosphere to observe in parts of the electromagnetic spectrum that do not reach the earth's surface. By now, we have made at least some observations in each of the named parts of the spectrum (of course, there is really no limit to how short or how long a wavelength can be). It seems strange, in view of the long identification of astronomy with visible observations, to realize that optical astronomers are perhaps in a minority nowadays.

Figure 2–9 When sunlight is dispersed by a prism, we see not only a continuous spectrum but also dark Fraunhofer lines.

2.3 SPECTRAL LINES

As early as 1666, Isaac Newton showed that sunlight was composed of all the colors of the rainbow. William Wollaston in 1804 and Joseph Fraunhofer in 1811 also studied sunlight as it was dispersed (spread out) into its rainbow of component colors (Fig. 2–9). Wollaston and Fraunhofer were able to see that at certain colors there were gaps that looked like dark lines across the spectrum at those colors. These gaps are thus called *spectral lines*; the continuous radiation in which the gaps appear is called the *continuum* (pl.: *continua*). The dark lines in the spectrum of the sun, and in the spectra of most stars, are mostly gaps representing the diminution of electromagnetic radiation at those particular wavelengths. We interpret them by supposing that originally a continuous band of color was radiated, and that something between the source and us absorbed the radiation at those particular wavelengths. These dark lines are thus called *absorption lines*; they are also known (for the sun, in particular) as *Fraunhofer lines*. It is also possible to have wavelengths at which there is some-

Figure 2–10 When we view a source that emits a continuum through a vapor that emits emission lines, we may see absorption lines at wavelengths that correspond to the emission lines.

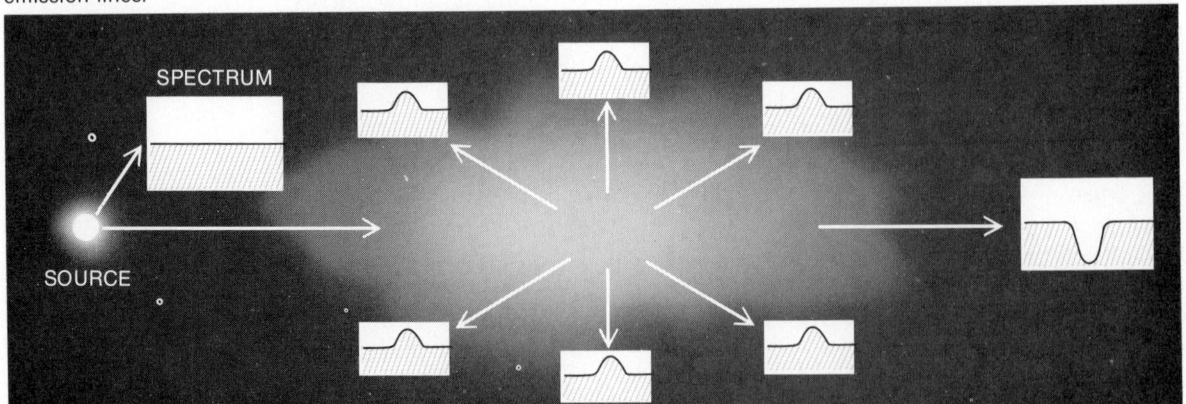

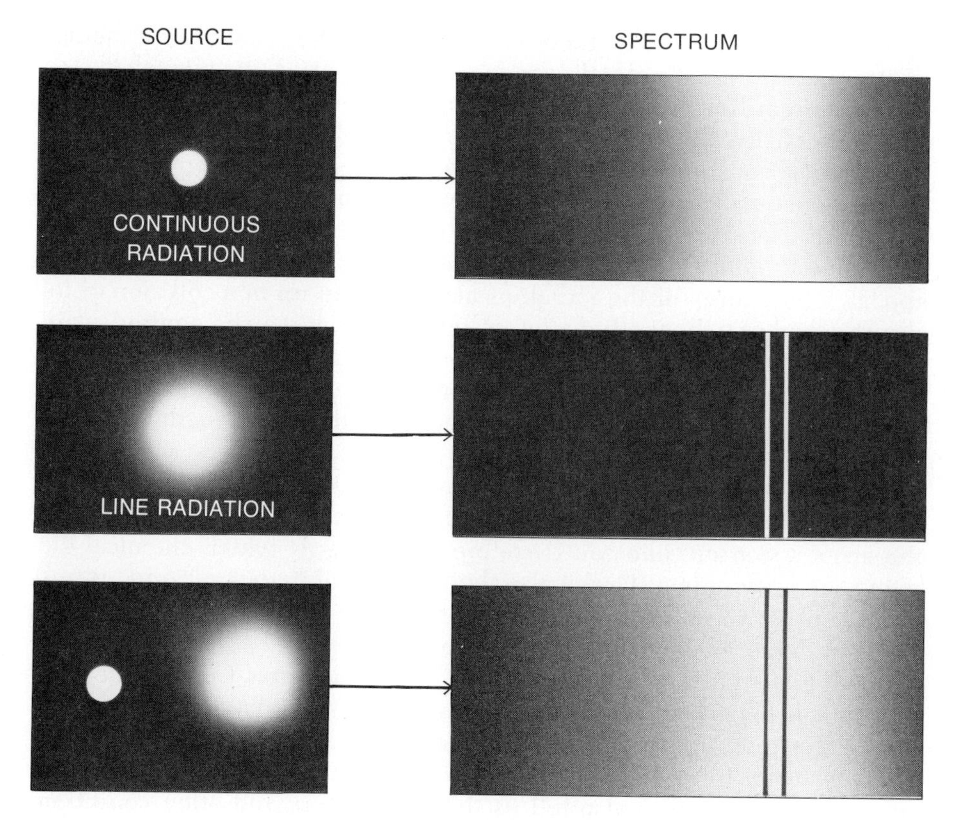

Figure 2–11 When continuous radiation from a source (*top*) shines through a gas that by itself has an emission line spectrum (*center*) then an absorption line spectrum results (*bottom*) whenever the gas in the middle is cooler than the source of the continuous radiation.

what more radiation than at neighboring wavelengths; these are called *emission lines.* We shall see that the nature of a spectrum, and whether we see emission or absorption lines, can provide considerable information about the nature of the body that was the source of the light. We say that the lines are "in emission" or "in absorption."

It was discovered in laboratories on earth that patterns of spectral lines can be explained as the absorption or emission of energy at particular wavelengths by atoms of chemical elements in gaseous form. If a vapor (the gaseous form) of any specific element is heated, it gives off a characteristic set of **emission** lines. That element, and only that element, has that specific set of lines. If on the other hand, a continuous spectrum radiated by a source of energy at a high temperature is permitted to pass through cooler vapor of any specific element, a set of **absorption** lines (Fig. 2–10) appears in the continuous spectrum at the same characteristic wavelengths of the emission lines of that element. Thus the vapor of an element through which light has passed has subtracted energy from the continuous spectrum at the set of wavelengths that is characteristic of that element (Fig. 2–11 and Color Plate 4). This was discovered by the German chemist Gustav Kirchhoff in 1859. If a continuous spectrum is directed first through the vapor of one element and then through the vapor of a second element, or through a mixture of the two gases, then the absorption spectrum that results will show the characteristic spectral absorption lines of both elements.

Kirchhoff is pronounced with a "hard" ch (that is, like k): kirk'-hoff. The name of the author of this book is also pronounced with a hard ch: pa'-sa-koff.

The same characteristic patterns of spectral lines are observed in the spectra of stars, and we conclude basically that the same elements are in the outermost layers of the stars. They absorb radiation from a continuous

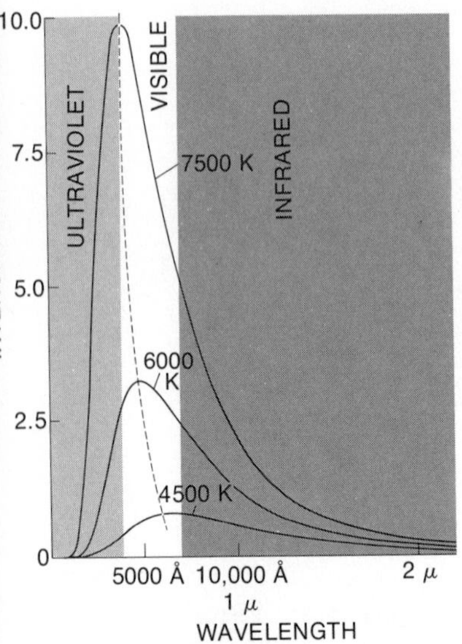

Figure 2–12 The intensity of radiation for different stellar temperatures, according to Planck's law. The wavelength scale is linear, i.e., equal spaces signify equal wavelength intervals. The dotted line shows how the peak of the curves shifts to shorter wavelengths as temperature increases; this is known as Wien's displacement law.

spectrum generated below them in the star, and thus cause the formation of absorption lines. Evidently, since each element has its own characteristic pattern of lines when it is in a certain range of conditions of temperature and density, the absorption spectrum of a star can be used to identify the chemical constituents of the star's atmosphere, in other words, the types of atoms that make up the gaseous outer layers of the star.

Where does the radiation that is absorbed go? A very basic physical law called the *law of conservation of energy* says that it cannot simply disappear. The energy of the radiation may be taken up in a collision of the absorbing atom with another atom. Alternatively, it may be emitted again, sometimes at the same wavelength, but even so it would go off in any random direction and not necessarily toward the observer. Thus, fewer bits of energy proceed straight ahead at that particular wavelength than were originally heading in that direction.

Moreover, if one element is present in relatively great abundance, then its characteristic spectral lines will be especially strong. By observing the star's spectrum, one can therefore tell not only which chemical elements are present in a star but also their relative abundances (Appendix 4).

The relative intensities of lines in the characteristic spectrum of each element depend not only on abundance but also on the temperature of the element at the location at which it is absorbing radiation. Certain lines may be stronger than other lines at a temperature of, say, 6000 to 7000 degrees, but be less intense at either lower or higher temperatures. (We shall study the reasons for this behavior later in this chapter.) Observation of a stellar spectrum can therefore also tell us the temperature and other conditions of the stellar atmosphere.

The method of spectral analysis is a powerful tool that can be used to explore the universe from our vantage point on earth. It has many uses other than those in astronomy. For example, by analyzing the spectrum, one can determine the presence of impurities in an alloy deep inside a blast furnace in a steel plant on earth. One can measure from afar the constituents of lava erupting from a volcano. Sensitive methods developed by astronomers trying to advance our knowledge of the universe often are put to practical uses in fields unrelated to astronomy.

2.4 THE COLORS OF STARS

λ, the Greek letter "lambda," is the usual symbol for wavelength.

Once we understand that the different colors merely correspond to different wavelengths, as stated in Section 2.2, we proceed to measure the colors of the stars. We can graph the intensity of light from a star at a given wavelength on one axis and the wavelength on the other axis. The measurements of intensity can be obtained either by passing the light through a spectrograph (a device that measures the spectrum, as described in Chapter 3), and measuring the intensity at each wavelength, or simply by measuring the amount of light that emerges from a set of filters that pass only certain groups of wavelengths and block others. (For example, we might first measure the magnitude of a star observed through a blue filter that passes only light of wavelengths between, say, 4000 and 4400 Å, and then through a green-yellow filter that passes only light of wavelengths between, say, 5200 and 5800 Å.)

When we examine a set of these graphs, we can see that, spectral lines aside, the radiation follows a fairly smooth curve (Fig. 2–12). Closer analysis reveals more complicated structure, which we will discuss later on. The radiation peaks in intensity at a certain wavelength, and decreases in amount more slowly on the long wavelength side of the peak than it does on the short wavelength side. (By convention, in optical astronomy longer wavelengths are always graphed to the right.)

The nature of these graphs can best be understood by first considering the radiation that represents different temperatures. When you first put an iron poker into a fire, it glows faintly red. Then it gets a brighter red, and as it gets hotter becomes yellower and then whiter. (Because the spectrum the poker emits is at its maximum intensity in the middle of the visible spectrum—we say "it peaks" there—the emission decreases to both longer and shorter wavelengths than the wavelength of the peak. The amount of radiation that is emitted at any visible wavelength is not that different from the amount emitted at any other visible wavelength. This makes the radiation seem white.) If the poker could be even hotter without melting, it would turn blue-white. White hot is hotter than red hot (Fig. 2–13).

A set of physical laws governs the sequence of events when material is heated. As the material gets hotter, the peak of the radiation shifts toward the blue. This is known as *Wien's displacement law* (Fig. 2–14).

Wien's displacement law states not only that the peak of the radiation shifts, but also specifically that the product of the wavelength where a curve peaks and the temperature of the substance is constant, $\lambda_{max} \times T = 29{,}000{,}000$, where λ is in angstroms and the temperature, T, is in the Kelvin scale (Table 2–1).

Also, as the material gets hotter, the total energy of the radiation grows rapidly. The energy follows the *Stefan-Boltzmann law,* also shown in Fig. 2–14, whereby the total energy emitted from a source of given area per second grows with the fourth power of the temperature. That is, a gas at 10,000 degrees gives off $2^4 = 2 \times 2 \times 2 \times 2 = 16$ times more energy than does the same amount of gas at 5000 degrees. (We must measure our scale of degrees from absolute zero, the minimum temperature possible. But aside from this it doesn't matter what scale of temperature we use.) We are discussing, for the moment, only the continuous part of the spectrum; it is possible to observe both a continuous spectrum and superimposed spectral lines in a single spectrum.

From Wien's displacement law, we can see that the colors of stars in the sky are telling us something about their temperatures. The reddish star Betelgeuse, thus, is a comparatively cool star, while the blue-white star Sirius is very hot. By simply measuring the intensity of a star with a set of filters at different wavelengths (see Fig. 2–43), we can assess the stellar temperature (Fig. 2–15).

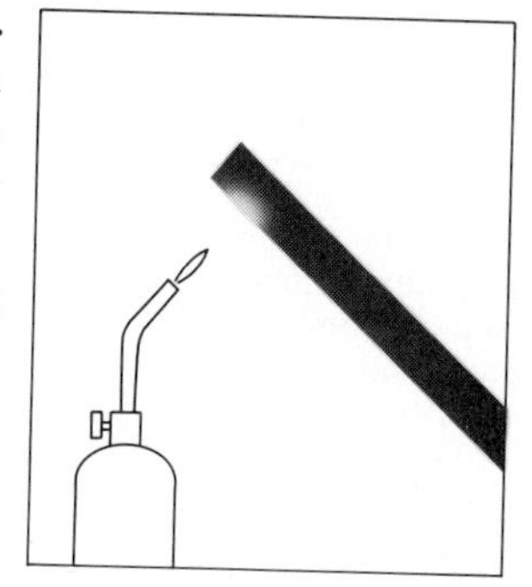

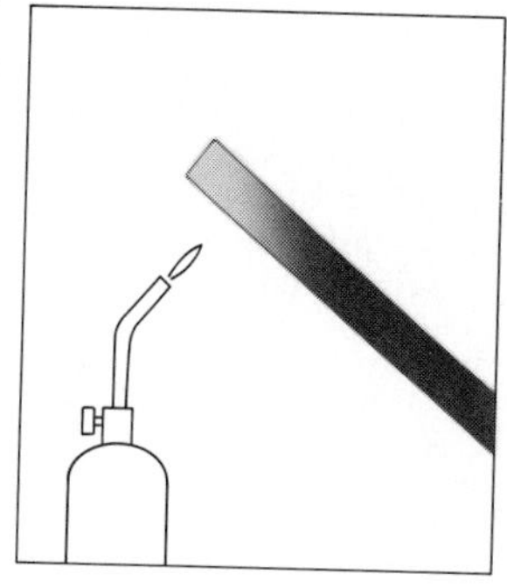

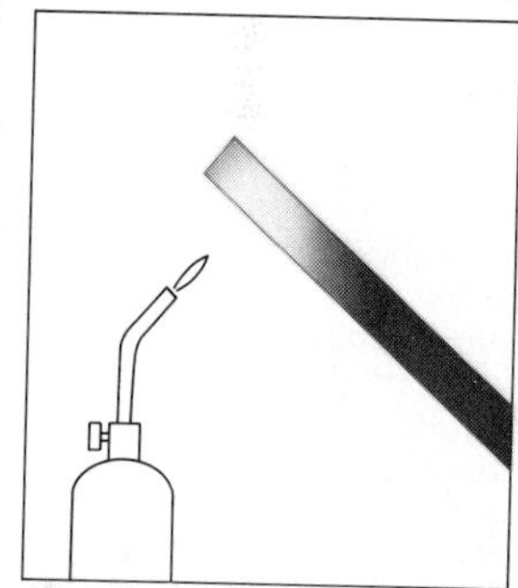

Figure 2–13 As a rod or poker (perhaps made of iron or carbon) is heated, it first glows red hot and eventually, when it is hotter, appears white hot.

TABLE 2–1 WIEN'S DISPLACEMENT LAW

Temperature	Wavelength of Peak	Spectral region of wavelength of peak
3 K	.97 mm	infrared-radio
3000 K	9660 Å	infrared
6000 K	4830 Å	green
12,000 K	2415 Å	ultraviolet
24,000 K	1207 Å	ultraviolet

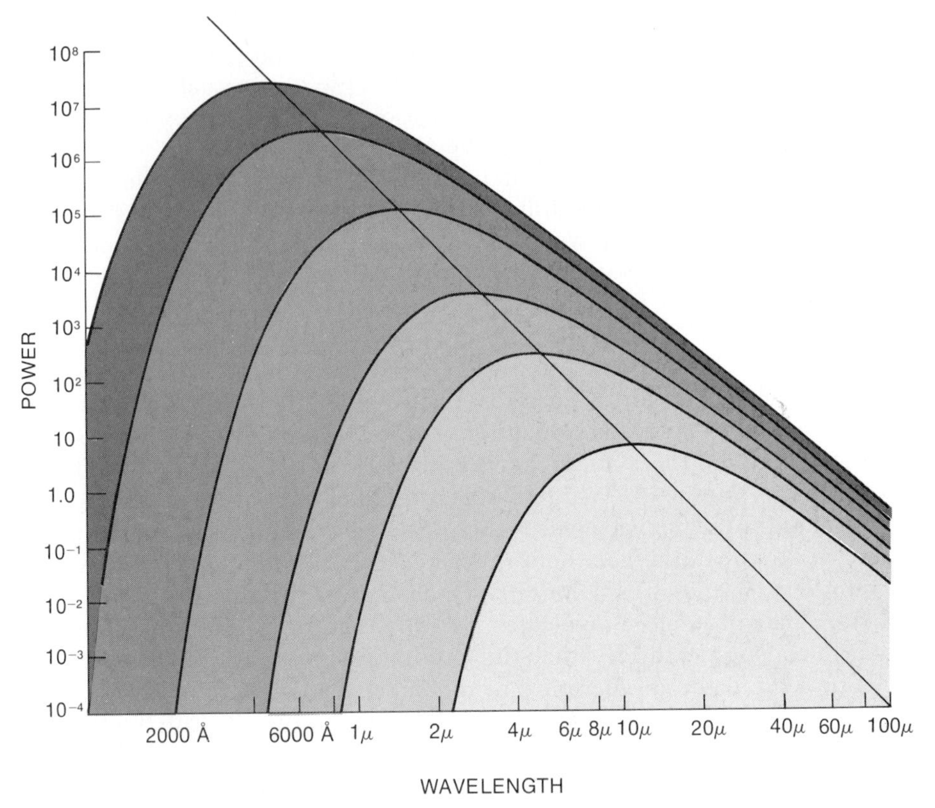

Figure 2–14 Curves for different temperatures representing the intensity of radiation according to Planck's law. The wavelength scale is logarithmic, i.e., equal spaces on the graph signify equal factors multiplying the wavelength or intensity. The peak of the curves shifts to shorter wavelengths, following Wien's displacement law. The total energy being radiated is represented by the area under each curve; it grows rapidly as the temperature increases. The Stefan-Boltzmann law states that the energy grows as the fourth power of the temperature.

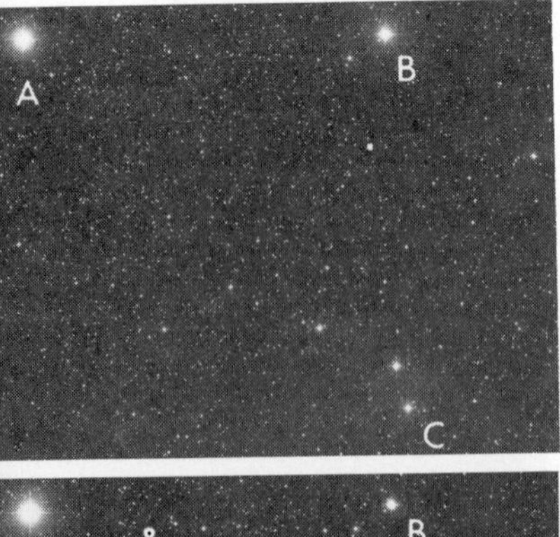

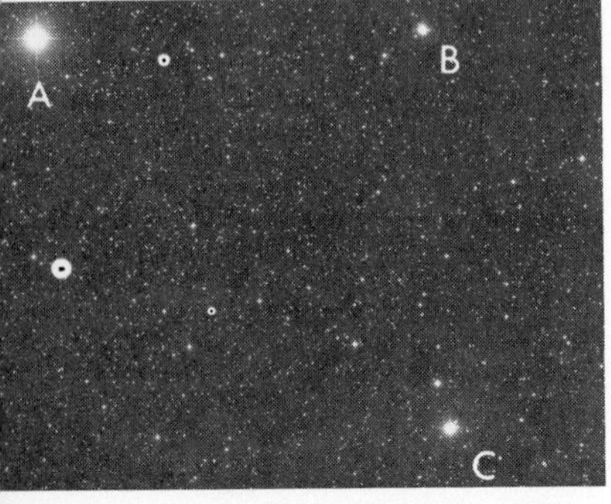

Figure 2–15 The top photograph is a view of several stars in blue light, and the bottom photograph is a view of the same stars in red light. Star A appears about the same brightness in both. Star B, a relatively hot star, appears brighter in the blue photograph. Star C, a relatively cool star, appears brighter in the red photograph.

2.5 PLANCK'S LAW AND BLACK BODIES

These laws of the distribution of energy from a heated gas are much more general than their particular application to the stars. In principle, if a gas is heated up, its continuous emission follows a certain peaked distribution called *Planck's law*, suggested in 1900 by the German physicist Max Planck (Fig. 2–16). Wien's displacement law and the Stefan-Boltzmann law, which had been discovered earlier, can be derived from Planck's law. Real gases may not, and usually don't, follow Planck's law exactly. For example, Planck's law governs only the continuous spectrum; any line emission or absorption is already a deviation from Planck's law.

Because of these difficulties, scientists choose to consider an idealization of the actual case. We call this fictional body a *black body*, since we define it as something that absorbs all radiation that falls on it. We choose the name since black things in our everyday experience do not reflect light. Of course, for a black body we are talking of an idealized black. Ordinary black paint, for example, does not absorb 100 per cent of the light that hits it. Further, something that looks black in visible light might not seem very black at all in the infrared.

An important physical law that governs the radiation from a black body says that anything that is a good absorber is a good emitter too. **The radiation from a black body follows Planck's law.** As a black body is heated, the peak of the radiation shifts towards the short wavelength end of the spectrum (Wien's displacement law) and the total amount of radiation grows rapidly (Stefan-Boltzmann law). The radiation from a black body at a higher temperature is greater at every wavelength than the radiation from a cooler black body, but as the temperature increases, the intensity of radiation at shorter wavelengths increases more rapidly than the intensity at longer wavelengths increases. Note that the curves of black bodies at different temperatures in Figure 2–14 do not cross.

In sum, black bodies are fictional, idealized objects that absorb all the radiation that hits them. They emit radiation according to Planck's law. If the temperature of a black body is sufficiently high, then it can appear quite bright. The atmospheres of stars appear, to a certain extent, as black bodies in that over a broad region of the spectrum that includes the visible, the continuous radiation from stars follows Planck's law fairly well.

Figure 2–16 Max Planck.

2.6 SPECTRAL TYPES

In the last decades of the 19th century, spectra of thousands of stars were photographed. There were many differences among spectra from different stars, and classifications of the different types of spectra were developed. The most famous worker at the vital task of classifying spectra was Annie Jump Cannon (Fig. 2–17) who, working at the Harvard College Observatory in the decades following 1896, classified over 500,000 spectra. Her catalogue, called the Henry Draper catalogue after the benefactor who made the investigation possible, is still in use today. Many stars are still known by their HD (Henry Draper catalogue) numbers, e.g., HD 176,387.

At first, the stellar spectra were classified only by the strengths of certain absorption lines from hydrogen, and were lettered alphabetically: A for stars with the strongest hydrogen lines, B for stars with slightly weaker lines, and so on. These categories are called *spectral types* or *spectral classes* (Fig. 2–18). But it was later realized that the types of spectra varied primarily because of differing temperatures of the stellar atmospheres. The hydrogen lines were strongest in stars at a certain temperature and were weaker at both higher and lower temperatures. When we now list the spectral types of stars in order of decreasing temperature, they are no longer in alphabetical order. From hottest to coolest, the spectral types are O B A F G K M.

O stars (that is, stars of type O) are the hottest, and the temperatures of M stars are more than ten times lower. Generations of American students and teachers have remembered the spectral types by the mnemonic: Oh, Be A Fine Girl, Kiss Me. The spectral types have been subdivided into 10 subcategories each. For example, the hottest B stars are B0, followed by B1, B2, B3, and so on. Spectral type B9 is followed by spectral type A0. It

Figure 2–17 The staff of the Harvard College Observatory circa 1917. Henrietta S. Leavitt, who studied variable stars (see Section 4.4b), is fifth from the left and Annie Jump Cannon, who classified hundreds of thousands of stellar spectra, is fifth from the right.

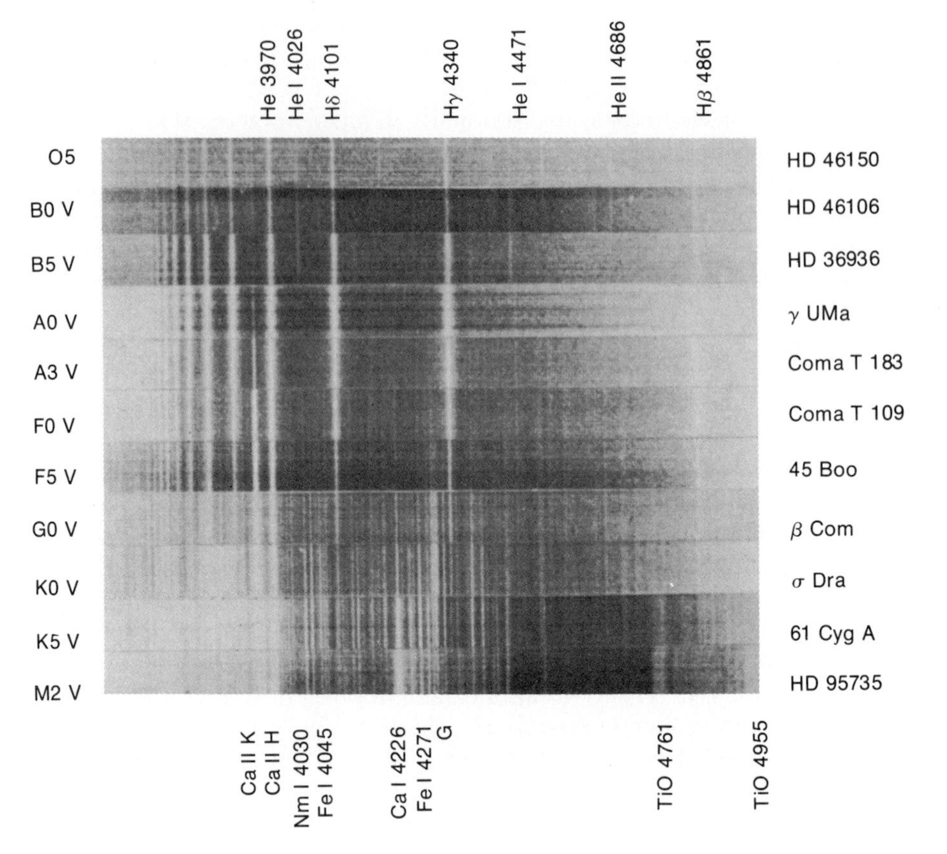

Figure 2–18 Spectral types. Note that most stellar spectra show absorption lines.

The dispersion *is the measure of how spread out the spectrum is, that is, how wide the rainbow of color is. Even a low dispersion spectrum, that is, one that is not spread out very much (we sometimes call this "classification dispersion"), is adequate to tell the spectral type to an accuracy of one half a spectral class or better (that is, differentiate F2 from F8).*

An easy way to convert from degrees Celsius to degrees Fahrenheit is to take the Celsius number, double it, subtract 10% of the result, and add 32. Example: *take 30°C, double it to get 60°, subtract 10% of 60° = 6° from 60° to get 54°, and add 32° to get the answer of 86°F.*

is fairly easy to tell the spectral type of a star by mere inspection of its spectrum. Historically, the hotter stars are called *early types* and the cooler stars are called *late types.* Thus, a B star is "earlier" than an F star, although no evolutionary sequence is now believed to exist, that is, we do not think that stars change from one spectral type to another.

We must choose the units in which we will discuss temperature for the rest of this book. In America we have used the Fahrenheit temperature scale, in which water freezes at 32°F and boils at 212°F, a span of 180°F between freezing and boiling. On this scale, absolute zero (the minimum temperature theoretically possible for any material thing) is −459.4°F (Fig. 2–19). This scale is not used in any scientific work.

Most of the rest of the world uses degrees Centigrade; a modern minor adjustment of the scale is called degrees Celsius. (The Centigrade and Celsius scales differ by less than 0.1° Centigrade or Celsius.) The United States is changing over gradually to this system; most scientists already have for their professional work. On the Celsius scale, water freezes at 0°C and boils at 100°C, a span of 100°C. Thus 180 Fahrenheit degrees are equivalent to 100 Celsius degrees; simplifying the ratio shows that 9/5°F = 1°C.

On the Celsius scale, absolute zero is −273°C. For terrestrial purposes, temperatures conveniently range around zero or a few tens. But the freezing and boiling points of water don't have much relevance to the stars. Astronomers choose to use the Kelvin temperature scale, which has its zero point at absolute zero instead of at the freezing point of water. The size of a Kelvin degree (now officially called one kelvin and abbreviated

K instead of °K) is the same as the size of a Celsius degree. Thus absolute zero is 0 K, water freezes at +273 K, and water boils at +373 K.

It is simple to change from Celsius degrees to kelvins; simply add 273 to the Celsius temperature. To change from Fahrenheit to Celsius is more complicated. One must first subtract 32°F, to align the freezing points, and then multiply by 5/9 to transform the smaller Fahrenheit degrees into the larger Celsius degrees. The formula is °C = 5/9 (°F − 32).

Stellar temperatures, whether expressed in Fahrenheit, Celsius, or Kelvin degrees, are very high—in the thousands or tens of thousands (Fig. 2–20).

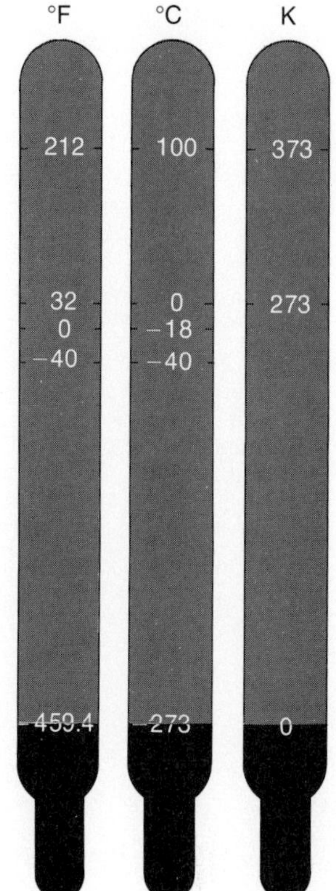

Figure 2–19 Temperature scales.

2.7 THE FORMATION OF SPECTRAL LINES

Spectral lines arise when there is a change in the amount of energy that any given atom has. The amount of energy that an atom can have is governed by the laws of *quantum mechanics,* a field of physics that was developed in the 1920's and whose applications to spectra were further worked out in the 1930's. According to the laws of quantum mechanics, light (and other electromagnetic radiation) has the properties of waves in some circumstances and the properties of particles under other circumstances. One cannot understand this dual set of properties of light intuitively and some of the greatest physicists of the time had difficulty adjusting to the idea.

An atom (Fig. 2–21) can be thought of as consisting of particles in its core, which is called the *nucleus,* surrounded by orbiting particles called *electrons.* Examples of nuclear particles are the *proton,* which carries one unit of positive electric charge, and the *neutron,* which has no electric charge. Electrons, which each have one unit of negative electric charge, are 1/1800 as massive as protons or neutrons.

According to quantum mechanics, an atom can exist only in a specific set of energy states, as opposed to a whole continuum (continuous range) of energy states. An atom can have only discrete values of energy; the energy cannot vary continuously. For example, the energy corresponding to one of the states might be 10.2 electron volts (eV). The next energy state allowable by the quantum mechanical rules might be 12.0 eV. We say that these are *allowed states.* The atom **simply could not have** an energy intermediate between 10.2 and 12.0 eV. In sum, the energy states are discrete; we say that they are *quantized.* We call them *energy levels.*

An electron volt is one of the several units in which energy can be expressed. It takes 6×10^{11} eV to equal even a small unit of energy called an erg, and 10^9 ergs per second to light a 100 watt light bulb. Obviously, one eV is pretty small.

When an atom drops from a higher energy state to a lower energy state, without colliding with another atom, the difference in energy is sent off as a bundle of radiation. This bundle of energy, a *quantum,* is also called a *photon,* which may be thought of as a particle of electromagnetic radiation (Fig. 2–22). Photons always travel at the speed of light. Each photon has a specific energy, which does not vary.

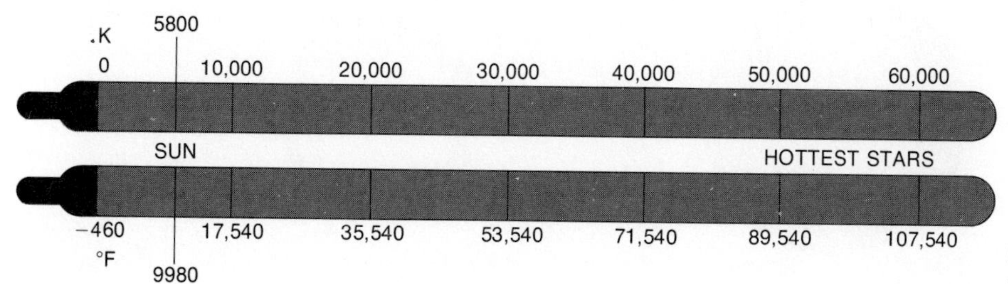

Figure 2–20 Stellar temperature scales.

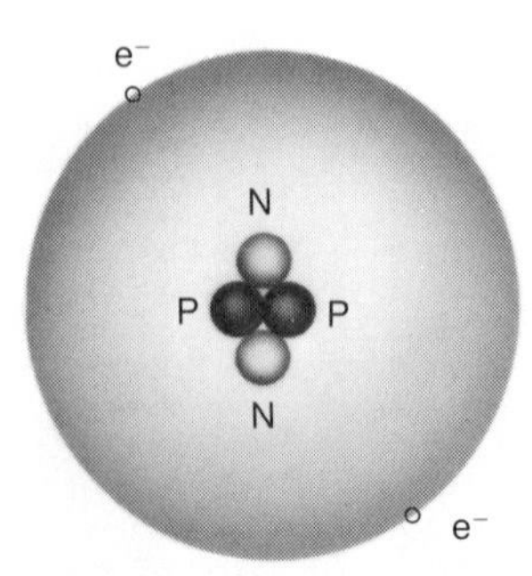

Figure 2–21 A helium atom contains two protons and two neutrons in its nucleus and two orbiting electrons.

A link between the particle version of light and the wave version is seen in the equation that relates the energy of the photon with the wavelength it has: $E = hc/\lambda$. In this equation, E is the energy, h is a constant named after Planck, c is the speed of light (in a vacuum, as always), and λ (lambda) is the wavelength. Since λ is in the denominator on the right hand side of the equation, a small λ corresponds to a photon of great energy. When, on the other hand, λ is large, the photon has less energy. An x-ray photon, for example, of wavelength 1 Å, has much more energy than does a photon of visible light of wavelength 5000 Å.

When a gas is heated up so that many of its atoms are in energy states other than the lowest possible, atoms not only are being raised to higher energy states but also spontaneously drop back to their lowest energy levels, emitting photons as they do so. These photons represent energies at certain wavelengths. As there were not necessarily any photons already existing at those wavelengths, the new photons appear, when they correspond to the visible part of the spectrum, as bright wavelengths. These are the **emission** lines (Fig. 2–23).

As we saw, when we allow continuous radiation from a body at some relatively high temperature to pass through a cooler gas, the atoms in the gas can take energy out of the continuous radiation. We now see that the atoms in the gas are being changed into higher energy states. Less radiation remains at certain wavelengths to come out of the gas than went in. These wavelengths are those of the **absorption** lines (Fig. 2–23). In the visible, these wavelengths appear dark.

Note that emission lines can appear with or without a continuum, since they are merely the addition of energy to the radiation field. But absorption lines must be absorbed **from** something, namely the continuum. A normal stellar spectrum is a continuum with absorption lines. This implies that the temperature in the outer layers of the star is decreasing with height. (See Section 2.9 for a discussion of the formation of a normal stellar spectrum.) As a result of this temperature trend with height, emission lines, which result from the presence of gas hotter than any background continuum, are rarely found in stellar spectra. Emission lines can appear, however, in spectra of stars surrounded by hot shells and in some other types of astronomical objects.

We can categorize the energy state of an atom by an amount of energy, and think of a change from one energy state to another as the result of a change in energy of the electrons orbiting the nucleus of the atom. (The production of spectral lines does not involve changes in the nucleus itself.) If an atom were at a temperature of absolute zero, then its electrons would be in the lowest possible energy levels. The lowest energy level that an

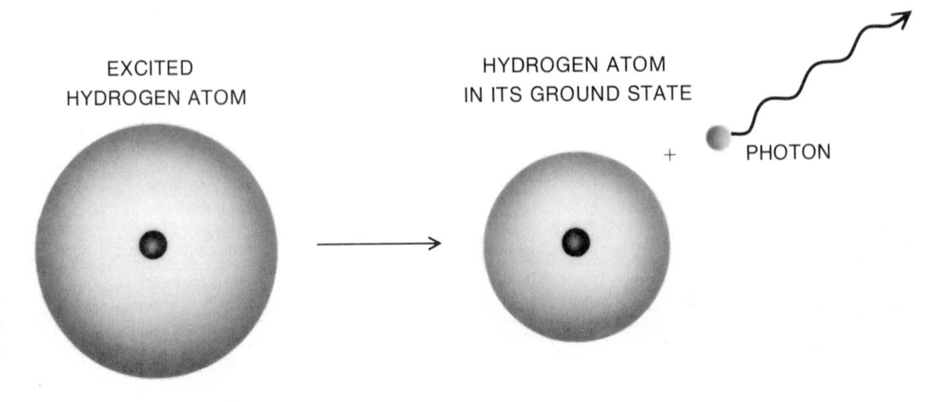

Figure 2–22 When an atom in an excited state gives off a photon, it drops back to a lower energy state, perhaps even the ground state. We see the photons as an emission line.

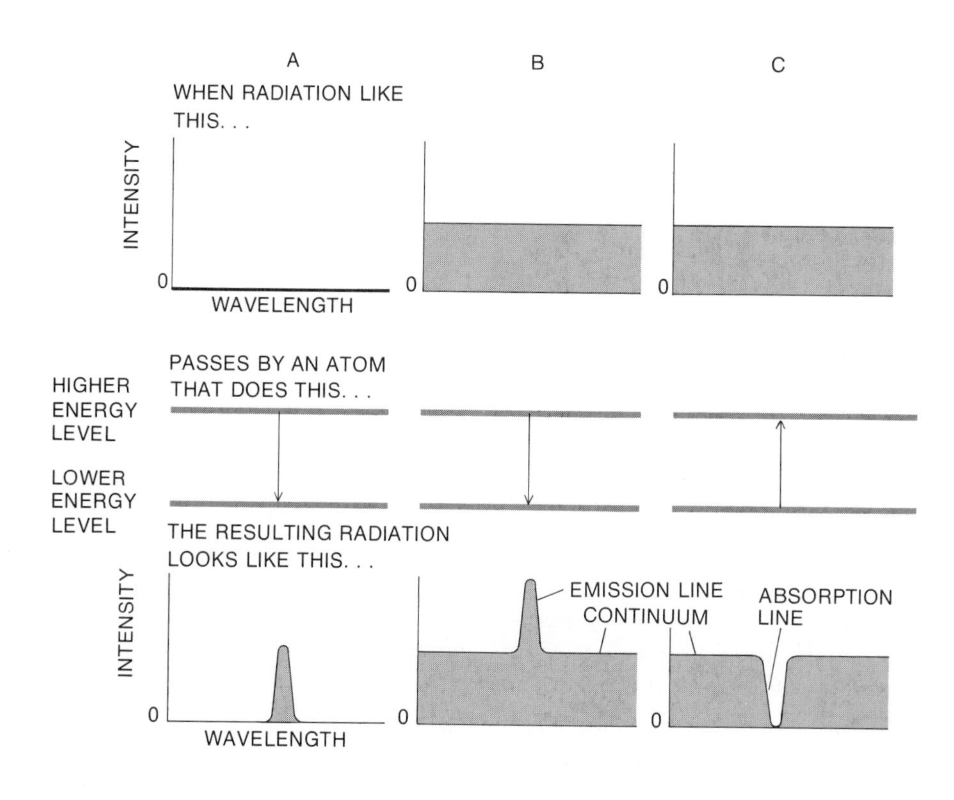

Figure 2–23 When photons are emitted, we see an emission line; a continuum may or may not also be present. An absorption line must absorb radiation from something. Hence an absorption line necessarily appears in a continuum.

electron can be on is called the *ground state* of that atom. As we add energy to the atom, we can "excite" one or more electrons to higher energy levels; that is, the electrons are pushed to higher energy levels. If we were to add even more energy, some of the electrons would be given not only sufficient energy to be excited but also enough energy to escape entirely from the atom. We then say that the atom is *ionized*, and the remnant with less than its quota of electrons is called an *ion* (Fig. 2–24). As the remnant has more positive charges in its nucleus than it has negative charges on its electrons, its net charge is positive and it is sometimes called a *positive ion*. An atom is *singly ionized* when it has lost one electron, *doubly ionized* when it has lost two electrons, and so on.

Atoms can be excited or ionized either by collisions or by radiation. In a star the former can be the case when the density is sufficiently high that collisions of the atoms with electrons are frequent. When the gas is "heated," this simply means that the constituent atoms are given more energy. Thus in a "hotter" gas, more atoms will be excited or ionized (Fig. 2–25).

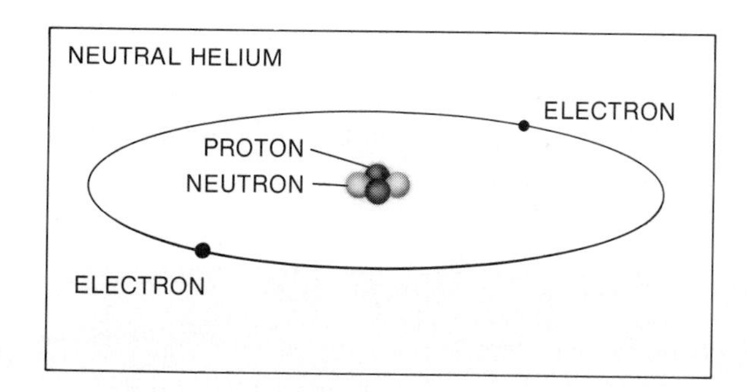

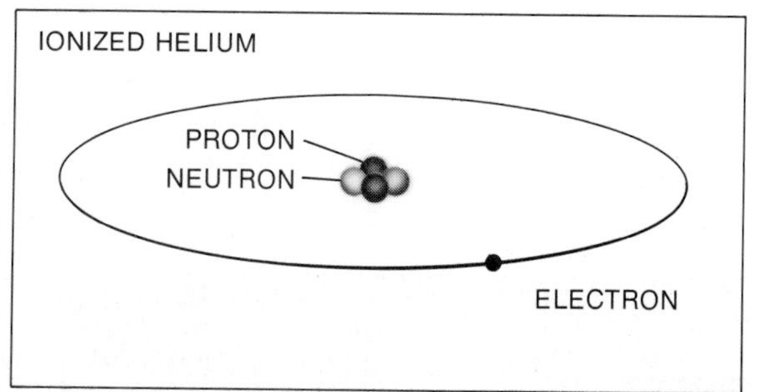

Figure 2–24 An atom missing one or more electrons is *ionized*.

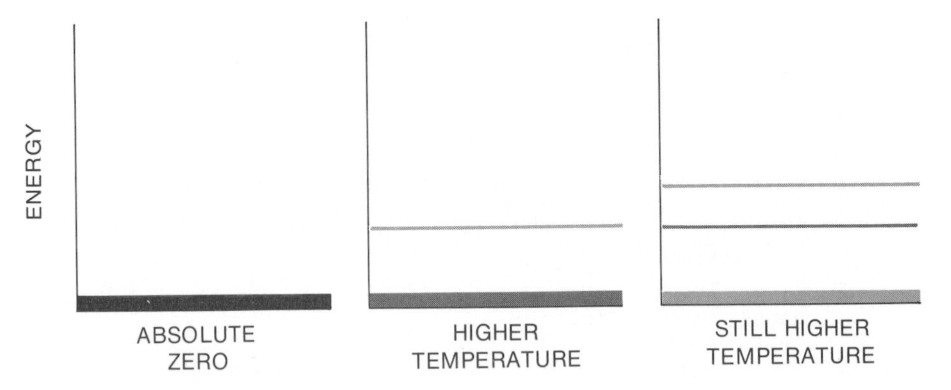

Figure 2–25 As the temperature of a cloud of gas rises, more and more electrons are excited to higher and higher energy levels. The *Boltzmann distribution* tells us how many atoms will be in each of its energy states for a given temperature.

2.8 THE HYDROGEN SPECTRUM

Hydrogen emits, and absorbs, a set of spectral lines that fall across the visible spectrum in a distinctive pattern (Fig. 2–26). The strongest line is in the red, the second strongest is in the blue, the third strongest is farther in the blue, and the other lines continue this series, with the spacing between the lines getting less and less.

The Swiss physicist J. J. Balmer noticed in 1885 that the spacing between lines follows the formula $1/\lambda = \text{constant} \times \left(\frac{1}{2^2} - \frac{1}{n^2}\right)$ where n = 3, 4, 5, and so on, is a number assigned to each energy level. (Another way of saying this is that "one over lambda is proportional to one over two squared minus one over n squared.") Since spectral lines correspond to transitions between different energy levels, we see that if the energy levels of the hydrogen atom are proportional to $1/n^2$ then the spectrum can be accounted for. Thus, since the red line of hydrogen, the first line in the Balmer series, arises from a transition from level 2 to level 3, its wavelengths would indeed correspond to $1/2^2 - 1/3^2$. The lines are labeled with the Greek alphabet, the first being alpha (α), the second beta (β), the third gamma (γ), and so on. Thus the red line is Balmer α or, since the Balmer series is simply the hydrogen (H) series that falls in the visible spectrum, simply Hα. Hβ corresponds to $1/2^2 - 1/4^2$.

It is easy to visualize the hydrogen atom in the picture (Fig. 2–27) that Niels Bohr (Fig. 2–28), the Danish physicist, laid out in 1913. In the *Bohr atom* electrons can have orbits of different sizes that correspond to different energy levels. Only certain orbits were allowable, hence the quantization of energy levels. This simple picture of the atom was superseded by the development of quantum mechanics, but the notion that the energy levels are quantized remains. The letter *n*, which labels the energy levels and which appears in the formulae for their energies, is called the *principal quantum number*.

As we saw in the previous section, the energy difference between two levels, E, is proportional to $1/\lambda$. The energy is emitted or absorbed as a photon of wavelength λ such that $E = hc/\lambda$.

Figure 2–26 The Balmer series, representing transitions down to or up from the second energy state of hydrogen.

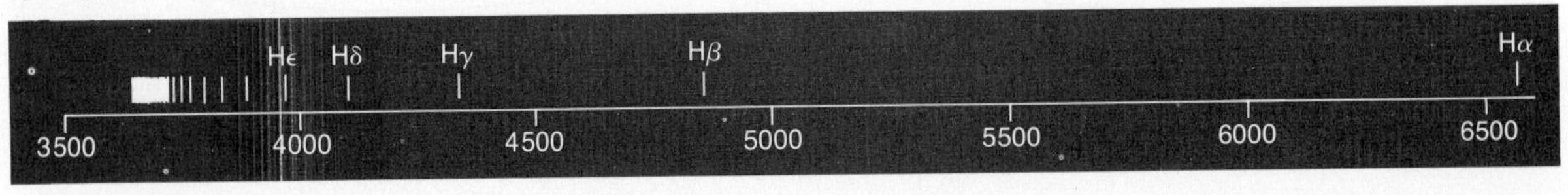

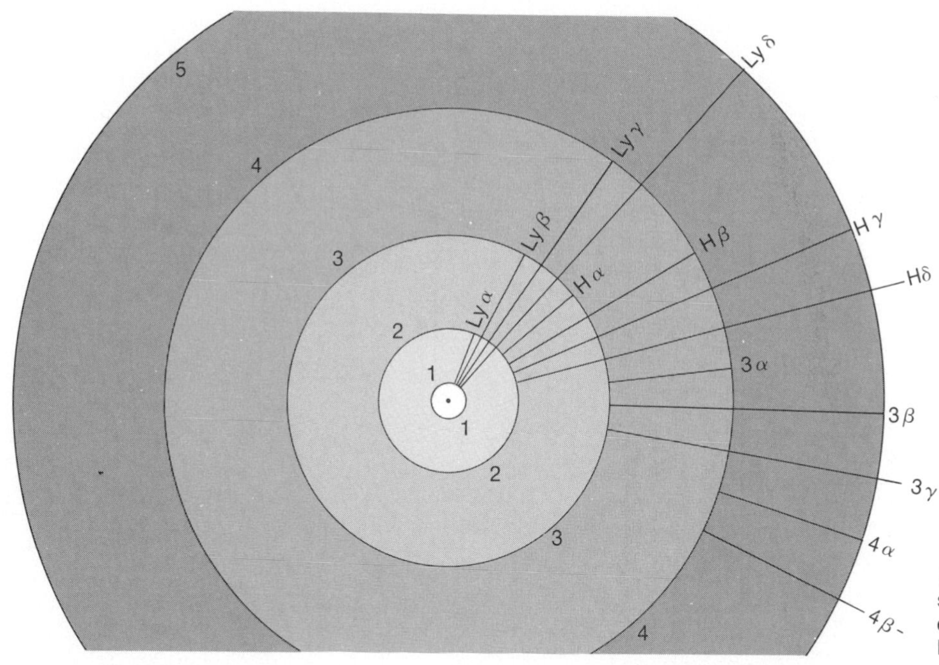

Figure 2–27 The representation of hydrogen energy levels known as the Bohr atom.

We call the energy level for $n = 1$ the *ground level* or *ground state*, as it is the lowest possible energy state (Fig. 2–29). The transitions between the ground level and higher energy levels are at shorter wavelengths than those of the Balmer series, for $(1/1^2 - 1/n^2) = (1 - 1/n^2)$ is greater than $(1/2^2 - 1/n^2) = (1/4 - 1/n^2)$. The series of transitions from or to the ground level is called the Lyman series, after the American physicist Theodore Lyman. The lines fall in the ultraviolet, at wavelengths far too short to pass through the earth's atmosphere. Lyman's observations of what we now call the Lyman lines were made in a laboratory in a vacuum tank. At present, we can observe Lyman lines from the sun and the other stars with telescopes aboard earth satellites.

Lyman went so far as to climb Mount McKinley in 1915, hoping that from that great altitude there might happen to be a window of transparency that would pass Lyman alpha. Unfortunately, there is no such window, and his photographs of the spectrum were blank in the region where ultraviolet radiation would have appeared.

There are many series of lines, each corresponding to transitions to a lower level, n^ℓ, from an upper level, n^u. We can think of the energy of the upper level as proportional to $1/n^{u^2}$ and the energy of the lower level as proportional to $1/n^{\ell^2}$. The energy difference is thus proportional to $1/n^{\ell^2} - 1/n^{u^2}$, where n^u can take on any value (integers only) greater than n^ℓ. The Balmer series, for example, corresponds to $n^\ell = 2$.

The series with $n = 3$ as the lowest term (the Paschen series) falls in the near infrared. Series with $n = 4$ and $n = 5$ (the Brackett and Pfund series) are farther in the infrared. Recently series of hydrogen lines whose lowest levels are at quantum numbers over 100 have been discovered in the radio region of the spectrum (see Chapter 21), though these lines originate in gas located between the stars rather than in the atmospheres of the stars themselves.

Since the higher energy levels have greater energy, a spectral line caused by the transition from a higher level to a lower level yields an emission line. When, on the other hand, continuous radiation falls on cool hydrogen gas, some of the atoms in the gas can be raised to higher energy

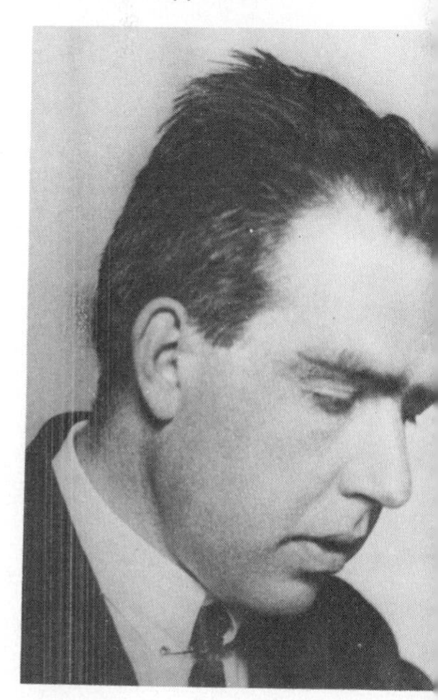

Figure 2–28 Niels Bohr.

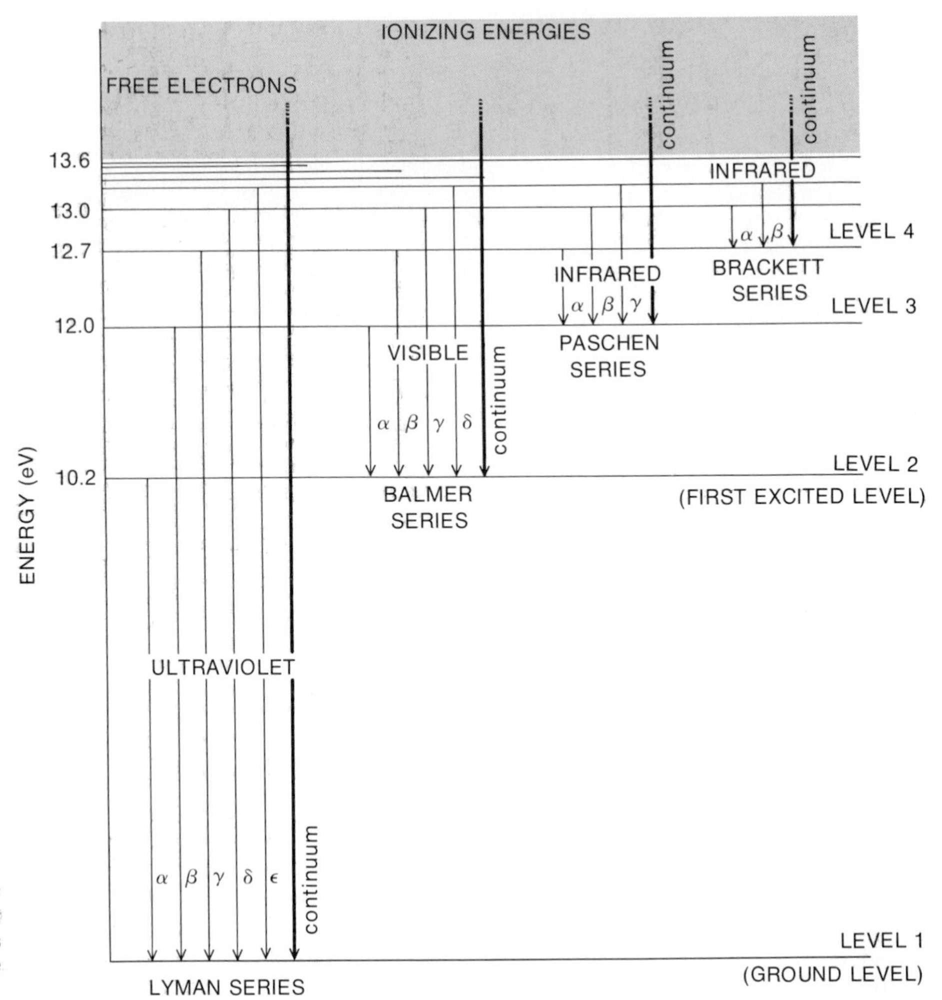

Figure 2–29 The energy levels of hydrogen and the series of transitions among the lowest of these levels.

levels. Absorption lines result in this case. The concept of the Bohr atom shows us why a given line of any element appears at the same wavelength no matter whether it is in emission or in absorption.

2.9 STELLAR SPECTRA

When we observe the light from a star, we are seeing only the radiation from a thin layer of that star's atmosphere. The interior is hidden within. As we discuss stellar spectra, let us keep this fact in mind. In parts of Chapters 5 to 9, we will be discussing the processes that go on inside stars.

As we consider stars of different spectral types (Fig. 2–18) ranging from the hottest to the coolest, we can observe the effects of temperature on the nature of the spectra. For example, the hottest stars, those of type O, have temperatures that are sufficiently high so that there is enough energy to remove the outermost electron or electrons from most of the atoms. Since most of the hydrogen is ionized, there are relatively few neutral hydrogen atoms left and the hydrogen spectral lines are weak. Helium happens not to be easily ionized—it takes a large amount of energy to do so—yet even helium appears not in its neutral state but in its singly-ionized state in an O star. Temperatures range from 30,000 K to over 60,000 K in O stars. O

Figure 2–30 The H and K lines of singly ionized calcium are the strongest lines in the visible part of the solar spectrum, and are prominent in spectra of stars of spectral types F, G, K, and M. These classes are known as *late spectral types,* while classes O, A, and B are known as *early spectral types,* although no connotations of age are meant.

stars are relatively rare since they have short lifetimes; none of the nearest stars to us are O stars. Only a handful of stars as hot as spectral types O3 or O4 are known at any distance; no O0, O1 or O2 stars have ever been discovered.

B stars are somewhat cooler, 10,000 K to 30,000 K. The hydrogen lines are stronger than they were in the O stars, and lines of neutral helium instead of ionized helium are present. Rigel and Spica are familiar B stars.

A stars are cooler still, 7500 K to 10,000 K. The lines of hydrogen are strongest in this spectral type. Lines of singly ionized elements like magnesium and calcium begin to appear. (All elements other than hydrogen or helium are known as *metals* for this purpose.) Sirius, Deneb and Vega are among the bright A stars in the sky. O, B, and A stars are all bluish in color.

F stars have temperatures of 6000 K to 7500 K. Hydrogen lines are weaker in F stars than they are in A stars, but the lines of singly ionized calcium are stronger. Singly ionized calcium has a pair of lines that are particularly conspicuous; they are easy to pick out and recognize in the spectrum. These lines are called H and K (Fig. 2–30). (Do not confuse the K in the terminology of the K line of calcium with the K in the name of the spectral type. H was the letter assigned to the strong calcium lines when Fraunhofer made out his original list and K was assigned to the shorter wavelength line later on.) Canopus, the second brightest star in the sky (not visible from the latitudes of most of the United States), and Polaris are prominent F stars.

G stars, the spectral type of our sun, are 5000 K to 6000 K. They are yellowish in color, as the peak in their spectra falls in the yellow to yellow/green part of the spectrum. The hydrogen lines are visible, but the H and K lines are the strongest in the spectrum.

K stars are relatively cool, only 3500 K to 5000 K. The spectrum is covered with many lines from neutral metals, a strong contrast to the spectra of the hottest stars, which show few spectral lines. Arcturus and Aldebaran, both visible as reddish points in the sky, are K stars.

Box 2.1 Do not confuse:
K stars: a spectral class
K: a kelvin, a unit of temperature
K: a strong line in many stellar spectra
K: the element potassium

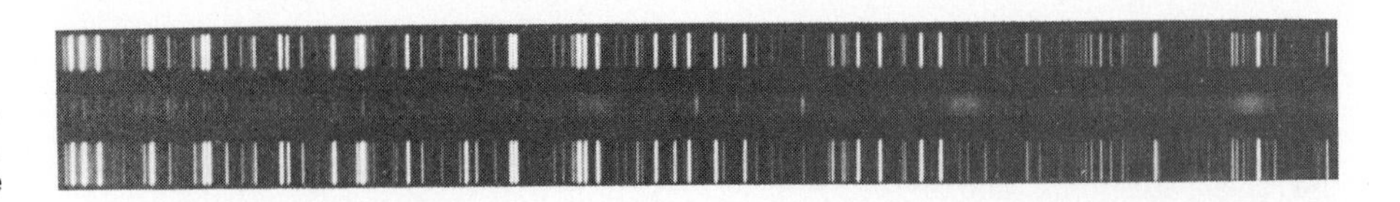

Figure 2–31 The emission line spectrum of the cataclysmic variable star R Coronae Borealis is in the center. At top and at bottom is the spectrum of iron in a special arc in the telescope dome, in order to provide a reference set of known wavelengths.

M stars are cooler yet, with temperatures less than 3500 K. Their atmospheres are so cool that molecules can exist without being torn apart, and the spectrum shows many molecular lines. (Hotter stars do not usually show molecular lines.) Lines from the molecule titanium oxide are particularly numerous. Betelgeuse is an example of such a reddish star of spectral class M.

We have listed the ordinary spectral types of stars, all of which show absorption spectra, that is, a continuum crossed with dark lines. There are many unusual stellar spectra, though, and additional spectral types. For example, R stars, N stars, or S stars have temperatures similar to those of M stars, but most of the lines in their spectra are from different atoms or molecules. (Spectral types R, N, and S are usually remembered by adding "right now, smack" to the mnemonic given in Section 2.6.) "Peculiar A stars," written Ap, show very strong lines of certain ionized metals. The strength of these lines in Ap stars varies with time. Some stars even show emission lines (Fig. 2–31): these are called Of, Oe, Be, Ae and Me stars, for example. *Wolf-Rayet stars* are O stars that not only show emission lines but also have the strange feature that the emission is particularly broad, that is, a given line may cover several angstroms of spectrum instead of a fraction of an angstrom. Astronomers think that the emission in Wolf-Rayet stars comes from shells of material that the star has ejected into the space surrounding it.

Two Be stars, γ Cassiopeiae and X Persei, have the unusual property of being sources of x-rays.

2.10 THE CONTINUOUS SPECTRUM

Where does the continuous spectrum come from? In order for there to be an absorption line, there must be some continuous radiation to be absorbed. The simplest picture, one that was believed for many years but later found to be oversimplified, is that a level just under the surface of a star emits a continuous spectrum, and the light emitted by this layer passes through a layer of gas that absorbs at certain wavelengths. These diminished intensities at certain wavelengths are, of course, the absorption lines. However, we now realize that the processes of continuous emission and spectral-line absorption take place together throughout the outer layers of the stars and not in separate layers. The detailed study of the processes of emission and absorption is called the study of *stellar atmospheres*, and nowadays large-scale digital computers are able to handle complicated equations that govern the transfer of radiation from the outer layers of a star into space. Sometimes the largest computers, such as those at the National Center for Atmospheric Research (NCAR) in Boulder, Colorado, may run for hours to calculate a model of the atmosphere of a single star.

The particular mechanism that causes the continuous emission itself is different for different spectral types. (Note that we use the word "emission" here because it is a positive contribution to the radiation at all frequencies, just as an emission line is a contribution to the radiation at a specific frequency; do not confuse continuous emission with line emission.) We have seen that most processes in atoms contribute only line radiation and not continuous radiation, because

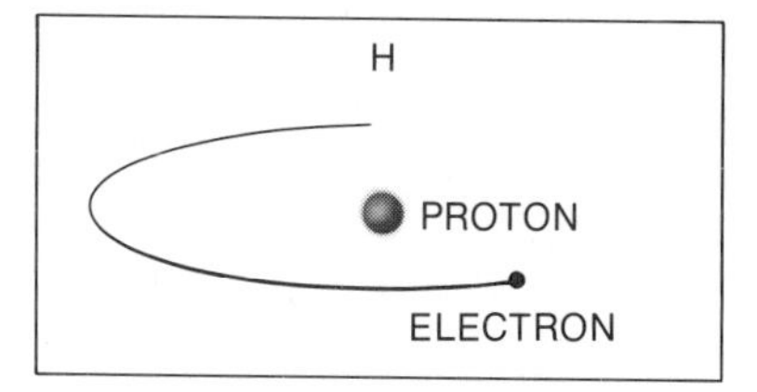

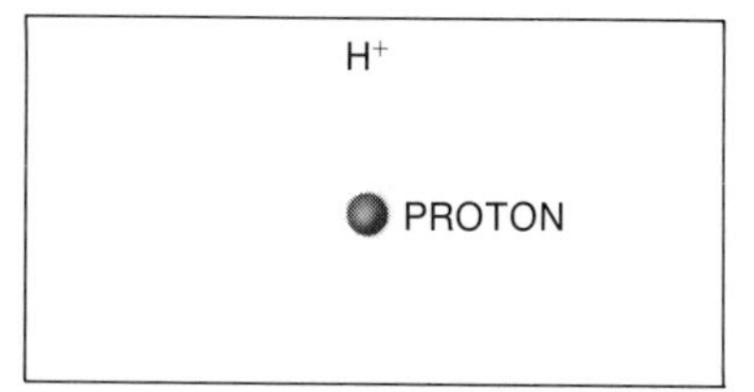

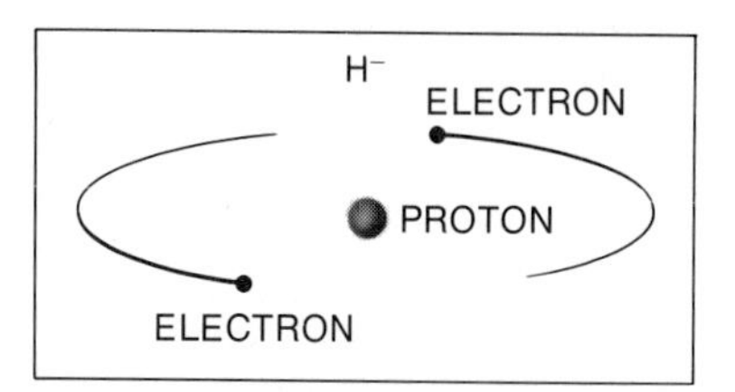

Figure 2–32 Neutral hydrogen, the hydrogen ion, and the negative hydrogen ion.

atomic energy levels are quantized. Let us consider the mechanism that causes the continuous emission in the atmosphere of the sun, a star of spectral type G, remembering that other spectral types may have different mechanisms.

In the sun, the continuous emission is caused by a strange type of ion: the negative hydrogen ion (Fig. 2–32). You will recall that when an atom has lost an electron, we say that it is an ion. These are *positive ions,* since the lost electrons leave behind a net positive charge. A negative ion is just the opposite, having gained one or more electrons. Hydrogen has one stable energy state where the single proton in its nucleus is surrounded by two electrons, rather than by its normal quota of one. This state with two electrons is called the *negative hydrogen ion,* H^-. The H^- ion is stable because each of the orbiting electrons partially shields the other electron from the nuclear charge. The other electron thus still feels some attraction, but the H^- ion is not very stable and breaks up easily. An H^{--} ion has recently been discovered.

When the H^- atom breaks up, the second electron is freed from the system and can go off with any amount of energy. To find the wavelength of radiation absorbed in this process we must find the energy difference between states before and after the breakup of the ion. The energy representing the former, negatively ionized, state is quantized, and so can take only certain values. But the electron can escape with any amount of energy, and so the energy characterizing the new state – a neutral atom plus a free electron – is not quantized. Thus the difference in energy can take any value, and the wavelength of the photon can thus be any value. One has continuous absorption instead of absorption at discrete wavelengths. This is called a *bound-free transition* (Fig. 2–33), since the electron goes from a state bound in an H^- ion to a state in which the electron is free of the atom that it leaves behind.

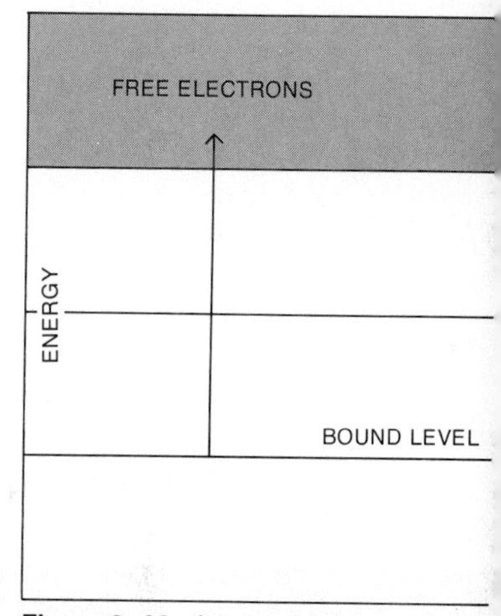

Figure 2–33 A bound-free transition in an atom that has three bound states.

This absorbing ability is called the *opacity* of the matter, which is, simply, a measure of how well the matter can absorb light. In order to have an equilibrium state in which the matter neither heats up nor cools down, clearly we must have the same amount of energy escaping as is being absorbed. Thus just as the H^- ion provides the continuous absorption, it also provides the continuous emission. When free electrons join neutral hydrogen atoms to form H^- ions, a continuous spectrum of radiation is given off.

Bound-free transitions in the negative hydrogen ion turn out to provide the continuous spectrum of the sun, but this could not have been realized intuitively. Rupert Wildt calculated in 1939 that this was the mechanism. Only one atom in 100 million in the sun is in the negative hydrogen ion state, but there are so many hydrogen atoms that even this small fraction provides a sufficient number of H^- ions to cause the observed effect.

This mechanism can provide the continuous spectrum of the sun in the visible and infrared, but different mechanisms govern other parts of the continuous spectrum of the sun. And still different mechanisms, which we also will not go into here, govern the continuous spectra of different spectral types of stars.

2.11 THE HERTZSPRUNG-RUSSELL DIAGRAM

In about 1910, Ejnar Hertzsprung in Denmark and Henry Norris Russell at Princeton University in the United States independently plotted a new kind of graph (Fig. 2–34). On one axis, the abscissa (x-axis), each

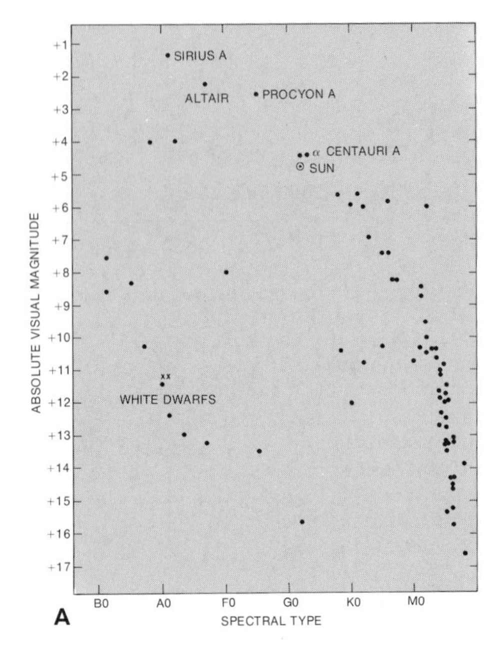

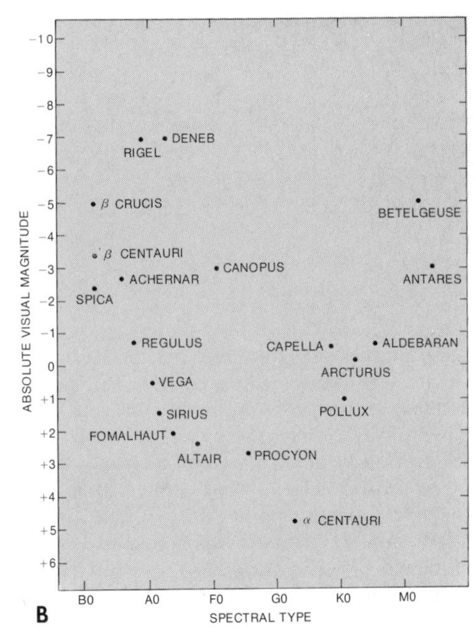

Figure 2–34 Hertzsprung-Russell diagrams (*A*) for the nearest stars in the sky and (*B*) for the brightest stars in the sky. The brightness scale is given in *absolute magnitude,* which is the magnitude the stars would have if they are all placed at the same standard distance. (This distance is 10 parsecs, equivalent to about 32.6 light years; parsecs will be discussed in the next section.) Because the effect of distance has been removed, the intrinsic properties of the stars can be compared directly on such a diagram.

graphed a measure of the temperature of stars. On the other axis, the ordinate (y-axis), each graphed a measure of the brightness of stars. They found that all the points that they plotted fell in limited regions of the graph rather than being widely distributed over the graph.

Figure 2–35 A physical grouping of stars in space, the galactic cluster NGC 5897.

There are several possible ways to plot the temperature. One doesn't need to plot the temperature itself. One can plot in its place the spectral type, for example, or even some measure of the color of a star.

To plot the brightness, there are also several possible ways. The simplest is merely to plot the magnitudes of stars, with the brightest stars (i.e., those with the most negative magnitudes) on the top. But remember that a star can appear bright either by really being intrinsically bright or alternatively by being very close to us. For the moment, we can get around this problem by plotting only stars that are at the same distance away from us. Later on we shall see what has to be done in order to determine the distances to particular stars. (Actually Hertzsprung graphed magnitudes of stars in clusters versus their colors. Russell plotted the intrinsic brightness for stars of known distance from the sun versus their spectral types. This intrinsic brightness, the amount of energy that a star emits each second, is called the *luminosity.*)

How do we find a group of stars at the same distance? Luckily, there are clusters of stars in the sky (described in more detail in Chapter 4) that are really clusters of stars in space at approximately the same distance away from us. They are not merely stars that are, even though they are in the same direction, at different distances from us. We do not have to know

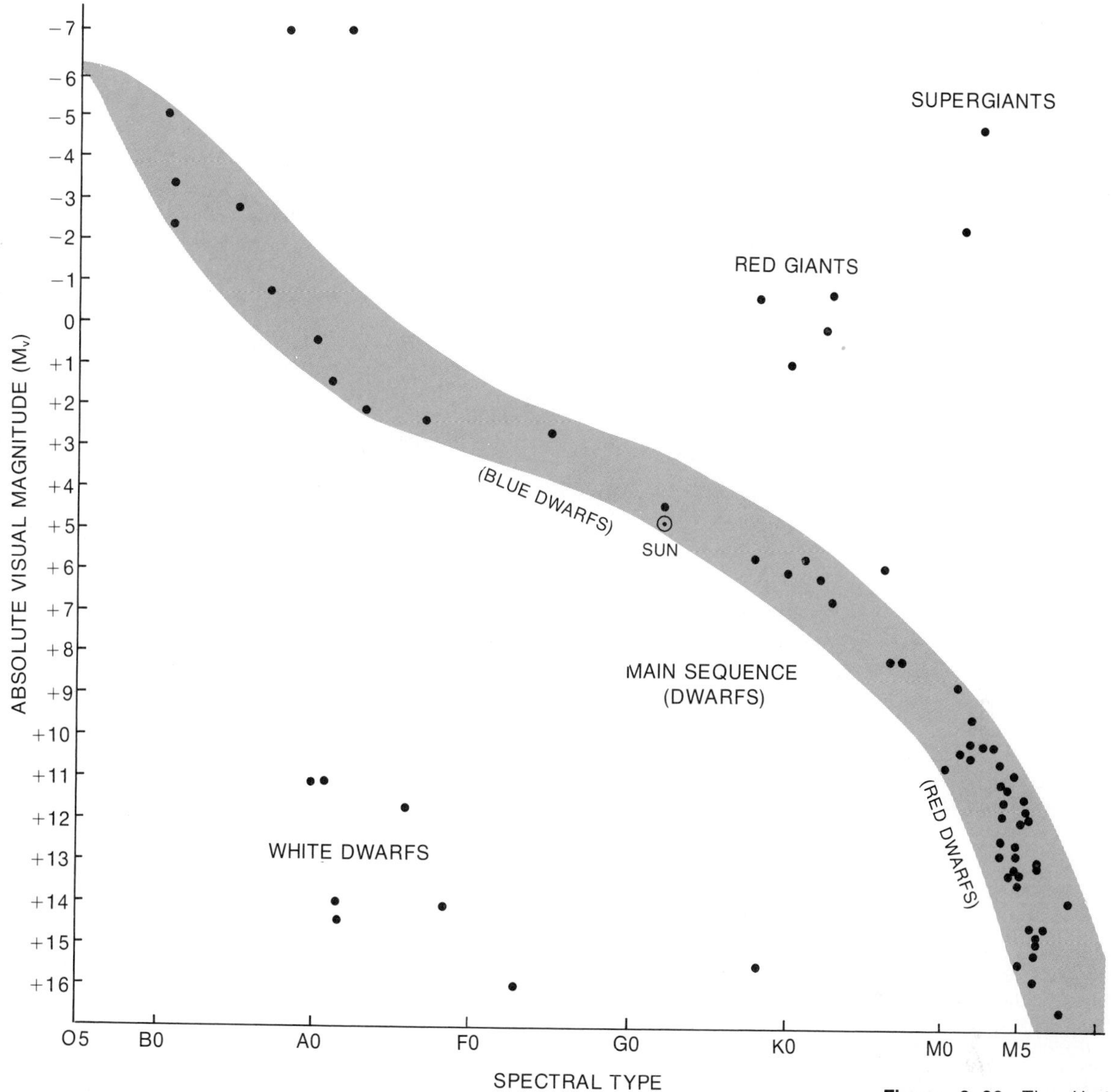

Figure 2–36 The Hertzsprung-Russell diagram, with the stars of Figure 2–34 included.

what the distance is to plot a diagram of the type plotted by Hertzsprung; all we have to know is that the stars are really clustering in space (Fig. 2–35).

A plot of temperature versus brightness is known as a *Hertzsprung-Russell diagram,* or simply as an *H-R diagram.* Note that since H-R diagrams were sometimes originally plotted by spectral type, from O to M, the hottest stars are on the left side of the graph. Thus temperature increases from right to left. Also, since the brightest stars are on the top, magnitude decreases toward the top. In some sense, thus, both axes are plotted backwards from the way a reasonable person would choose to do it if there were no historical reasons for doing it otherwise.

When plotted on a Hertzsprung-Russell diagram, the stars mainly lie on a diagonal band from upper left to lower right (Fig. 2–36). Thus the hottest stars are brighter than the cooler stars. Most stars fall very close to this band, which is called the *main sequence.* Stars on the main sequence are called *dwarf stars,* or *dwarfs.* There is nothing strange about dwarfs: they are the normal kind of stars. The sun is a type G dwarf. Some dwarfs

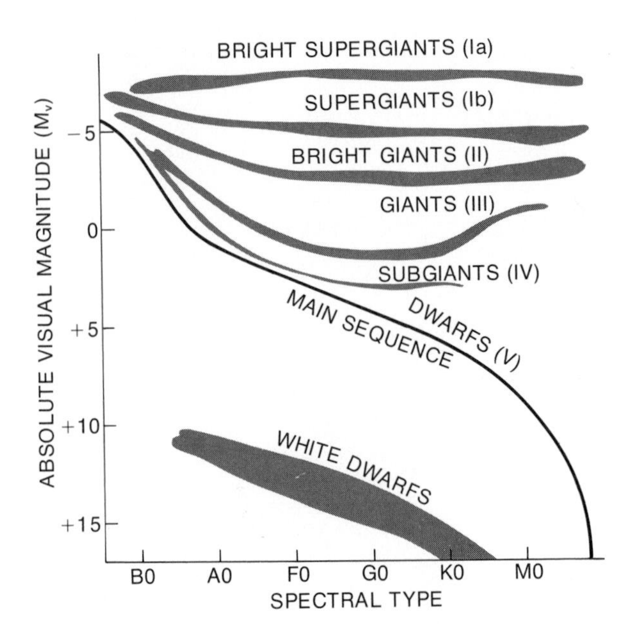

Figure 2–37 Luminosity classes are specified by numbers (shown in parentheses) or names.

are quite large and bright; the word "dwarf" is used only in the sense that these stars are not a larger brighter kind of star that we will define below as giant.

Some stars lie above and to the right of the main sequence. That is, for a given spectral type the star is intrinsically brighter than a main sequence star. These stars are called *giants,* because their luminosities are large compared to dwarfs of their spectral type. Some stars, like Betelgeuse, are even brighter than normal giants, and are called *supergiants.* Since two stars of the same spectral type have the same temperature and, according to the Stefan-Boltzmann law, the same amount of emission from each area of their surfaces, the brighter star must be bigger than the fainter star of the same spectral type. Stars in a class of faint hot objects, called *white dwarfs,* are located below and to the left of the main sequence. They are smaller and fainter than main sequence stars of the same spectral type. Distinctions among stars of the same spectral class are described as *luminosity classes* (Fig. 2–37).

The use of the Hertzsprung-Russell diagram to link the spectrum of a star with its brightness is a very important tool for stellar astronomers. We shall return to further discussion of the H-R diagram after we discuss ways of measuring the distances to stars.

Figure 2–38 "Pardon me, I thought you were further away." Without clues to indicate distance, we cannot properly estimate size. (Reproduced by special permission of PLAYBOY Magazine; copyright © 1971 by Playboy.)

2.12 STELLAR DISTANCES

We can tell a lot by looking at a star, or by examining its radiation through a spectrograph. But such observations do not tell us directly how far away the star is. Since all the stars are but points of light in the sky even when observed through the biggest telescopes, we have no reference scale to give us their distances (Fig. 2–38).

The best way to find the distance to a star is to use the principle that is also used in the rangefinders in cameras. Sight toward the star from different locations, and see how the direction toward the star changes. You can see this effect by holding out your thumb at arm's length. Examine it first with one eye closed and then with the other eye closed. Your thumb seems to change in position as projected against a distant background; it appears to move across the background by a certain angle, about 5°. This is because your eyes are a few centimeters apart from each other, and when you look through each eye you are looking from a different point of view.

Box 2.2

There are 360° to a full circle, and 90° to a right angle. Thus 1° (1 degree or, to prevent confusion with degrees of temperature, 1 degree of arc) is 1/360th of a full circle.

Each degree is divided into 60 minutes (60 minutes of arc, abbreviated 60 arc min). Each minute of arc, in turn, is divided into 60 seconds (60 seconds of arc, abbreviated to 60 arc sec).

Thus there are 60 arc sec in 1 arc min, and 3600 arc sec in 1°.

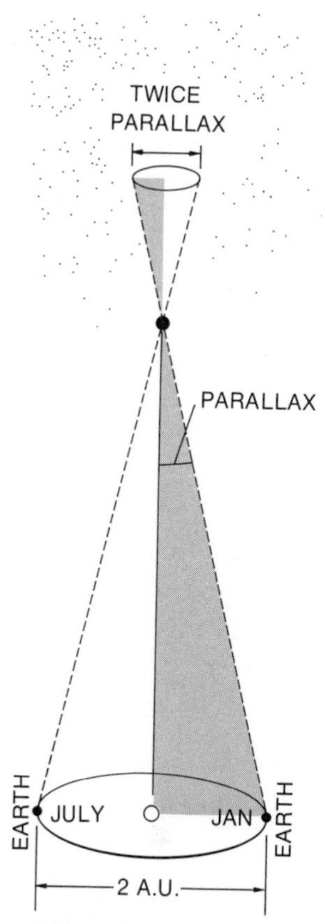

Figure 2–39 The nearer stars seem to be slightly displaced with respect to the farther stars when viewed from different locations in the earth's orbit. We call the average separation of the earth and the sun, measured in linear units like kilometers or miles, 1 Astronomical Unit (1 A.U.). 1 A.U. = 150,000,000 km (93,000,000 miles). The maximum baseline possible, 2 A. U., corresponds to observations made six months apart.

Now hold your thumb up closer to your face, just a few centimeters away. Note that the angle your thumb seems to jump across the background as you look through first one eye and then through the other is greater than it was before. Just the same effect can be used for finding the distances to nearby stars. The nearer the star is to us, the farther it will appear to move across the background of distant stars. To maximize this effect, we want to observe from two places that are separated from each other as much as possible. For us on earth, that turns out to be the position of the earth at intervals of six months. In that six-month period, the earth moves to the opposite side of the sun from where it began. In principle, we observe the position of a star in the sky very precisely, with respect to the background stars, and six months later, when the earth has moved halfway around the sun, we repeat the observation. The straight line joining the points from which we observe is called the *baseline*. In this case we have observed with a baseline twice as long as the distance from the earth to the sun, 2 A.U. in length. We can think of differences in position in the sky as an angle whose vertex is at the earth. The angle that a star seems to move between two observations made from the ends of a baseline of 1 A.U. (half the maximum possible baseline) is called the *parallax* of the star (Fig. 2–39). From the parallax, we can use simple trigonometry to calculate the distance to the star. This is called the method of *trigonometric parallax*.

The basic limitation to the method of trigonometric parallax is that the farther away the star is, the smaller is its parallax. It turns out that this method can only be used for about eight thousand of the stars closest to us; most other stars have parallaxes too small for us to measure. Even Proxima Centauri, the star nearest to us, has a parallax of only 3/4 arc sec. This is the angle subtended by a dime at a distance of 2 km. Clearly we must find other methods to measure the distances to other stars.

Since parallax measures have such an important place in the history of astronomy, a unit of distance was defined that relates to these measurements. If we were outside the solar system, and looked back at the earth and the sun, we would see a certain angle between them as measured from our vantage point. From twice as far away as Pluto, for example, when the earth and the sun were separated by the maximum amount, they would be approximately 1° apart.

The recent development of automatic measuring engines has enabled trigonometric parallaxes to be measured for stars five times more distant than could previously be studied—trigonometric parallaxes can now be measured for stars as far as 300 light years from the sun and as faint as 16th magnitude.

As we go farther and farther away from the solar system, 1 A.U. subtends a smaller and smaller angle. When we are approximately 60 times

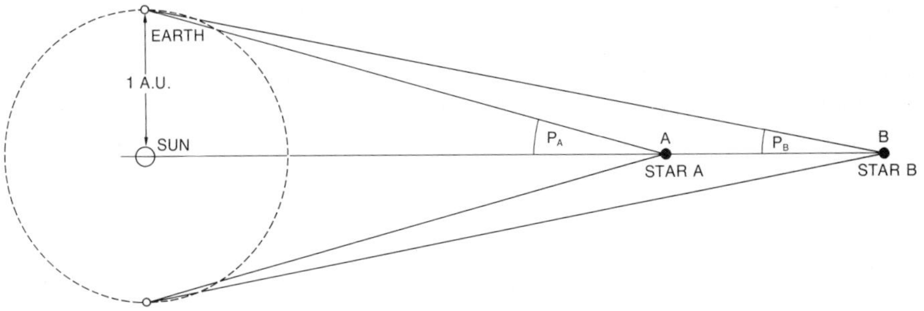

Figure 2–40 Star B is farther from the sun than star A and thus has a smaller parallax. The parallactic angles are marked P_A and P_B.

farther away, 1 A.U. will subtend only 1 arc minute, since 60 arc min = 1 arc degree. When we are approximately 60 times farther away again, 1 A.U. will subtend only 1 arc second, since 60 arc sec = 1 arc minute. Note that the angle subtended by the astronomical unit is a measure of the distance we have gone (Fig. 2–40).

When we are at the distance such that 1 A.U. subtends only 1 arc sec, we call that distance *1 parsec* (Fig. 2–41). A star that is 1 parsec from the sun has a **par**allax of one arc **sec**. 1 parsec is a long distance. Most astronomers tend to use parsecs instead of light years, the distance that light travels in a year ($=9.5 \times 10^{12}$ km), when talking about stellar distances. It takes light 3.26 years to travel 1 parsec; thus 1 parsec = 3.26 light years. Note that parsecs and light years are **distances**, just like kilometers or miles, even though their names contain references to their definitions in terms of angles or time.

Figure 2–41 A *parsec* is the distance at which 1 A.U. subtends an angle of 1 arc sec. One parsec is approximately equal to 3.26 light years.

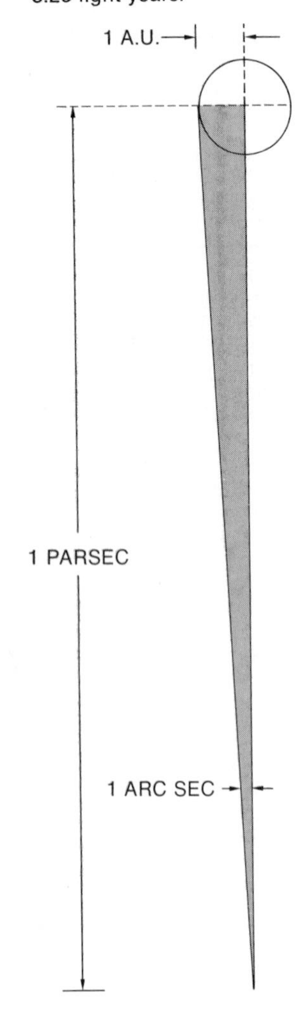

The advantage in using "parsecs" instead of "light years," "kilometers," or any other distance unit is that the distance in parsecs is equal to the inverse of the parallax angle in seconds of arc:

$$d(\text{parsecs}) = 1/p \text{ (seconds of arc)}.$$

For example, a star with a parallax angle of 0.5 arc sec is 2 parsecs away, and would thus be one of the nearest stars to the sun. Since the parallax angle is the quantity that is measured directly at the telescope, it was convenient to choose a distance unit very closely related to the parallax angle.

2.13 ABSOLUTE MAGNITUDES

We have thus far defined only the *apparent magnitudes* of stars, how bright the stars **appear** to us. However, if we chose a standard distance, we could consider how bright all stars would appear if they were at that standard distance. The standard distance that we choose is 10 parsecs. We define the *absolute magnitude* of a star to be the magnitude that the star would appear to have if it were at a distance of 10 parsecs. We normally write absolute magnitude with a capital M, and apparent magnitude with a small m.

Thus if a star happens to be exactly 10 parsecs away from us, its absolute magnitude is exactly the same as its apparent magnitude. If we were to take a star that is farther than 10 parsecs away from us, though, and moved it to be 10 parsecs away, then it would appear brighter to us than it does at its actual position. Since the star would be brighter, its absolute magnitude would be a lower number (for example, 2 rather than 6) than its apparent magnitude.

On the other hand, if we were to take a star that is closer to us than 10 parsecs, and moved it to 10 parsecs away, it would be farther away and

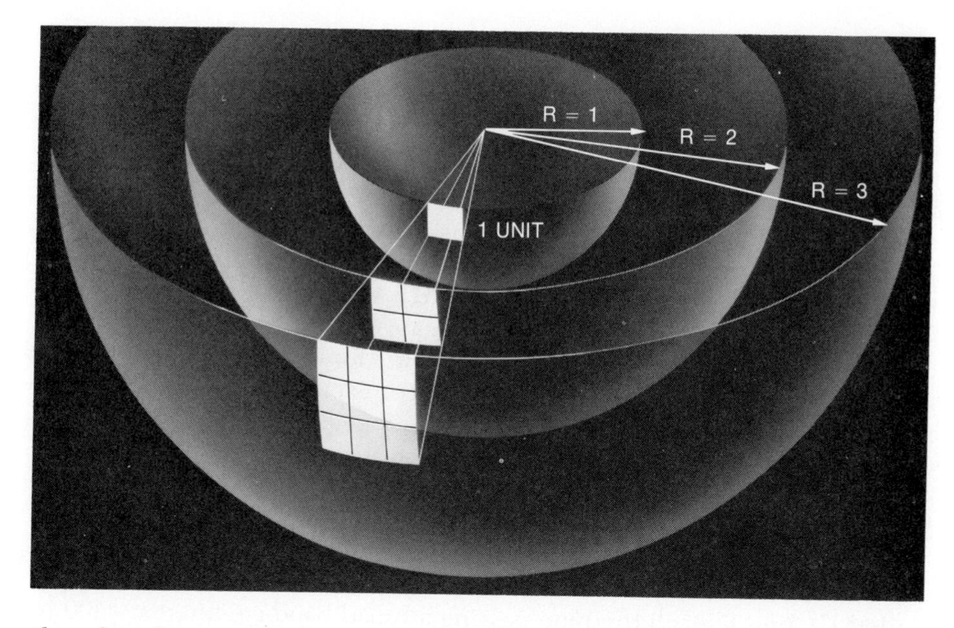

Figure 2–42 The inverse square law. Radiation passing through a sphere twice as far away as another sphere has spread out so that it covers $2^2 = 4$ times the area; n times farther away it covers n^2 times the area.

therefore fainter. Its absolute magnitude would be higher (more positive, for example, 10 instead of 6) than its apparent magnitude.

To assess just how much brighter or fainter the star would appear, we must realize that the intensity of light from a star follows the *inverse square law* (Fig. 2–42). That is, the intensity of a star varies inversely with the square of the distance of the star from us. (This law holds for all point sources of radiation, that is, sources that appear as points without length or breadth.) If we could move a star twice as far away from us, it would appear four times fainter. If we could move a star 9 times as far away from us, it would grow 81 times fainter. Of course, we are not physically moving stars (we would burn our shoulders while pushing), but merely considering how they would appear at different distances. In all this, we are assuming that there is no matter in space between the stars to absorb light. Unfortunately, this is not always a good assumption (see Chapter 20).

For many purposes, it is sufficient just to understand how the inverse square law affects brightness. Astronomers, however, normally speak in terms of magnitudes. Remember that a difference of one magnitude is a factor of approximately 2.5 in brightness. Thus, if a star is 2.5 times fainter (that is, gives off 2.5 times less light) than another star, it is 1 magnitude fainter. If it is 3 times fainter, it is slightly more than 1 magnitude fainter. If it is 6 times fainter, it is slightly less than 2 magnitudes fainter.

Example: A star is 20 parsecs away from us, and its apparent magnitude is +4. What is its absolute magnitude?

Answer: If the star were moved to the standard distance of 10 parsecs away, it would be twice as close and, therefore, by the inverse-square law, would appear four times brighter. Since 2.5 times is one magnitude, and $(2.5)^2 = 6.25$ is two magnitudes, it would be approximately 1½ magnitudes brighter. Since its apparent magnitude is 4, its absolute magnitude would be $4 - 1\frac{1}{2} = 2\frac{1}{2}$.

Astronomers have a formula they can use to carry out this calculation, but the formula merely does numerically what we have just carried out logically. The formula is $m - M = 5 \log_{10} r/10$, where m is the apparent magnitude, M is the absolute magnitude, and r is the distance in parsecs. Note that if $r = 10$, then $r/10 = 1$, log 1 = 0, and m = M. If r = 20, as in our example above, m − M = 5 log 2 = 5 × 0.3 = 1.5 magnitudes. Be careful not to get carried away using the formula without understanding the point of the manipulations.

Note that the actual value of the magnitude, either absolute or apparent, can depend on the wavelength region in which we are observing. Let us consider a blue star and a red star that give off the same amount of energy. The blue star can seem much brighter than the red star if we observe them through a blue filter, and the red star can seem much brighter than the blue if we observe them through a red filter. Thus we must specify the wavelength range in which we are observing. For example, one often sees magnitudes listed as m_{pg} and m_v for apparent photographic and visual magnitudes, respectively. The corresponding absolute magnitudes are M_{pg} and M_v. The sensitivity of photographic plates peaks farther toward the violet than does the sensitivity of the eye, which has its peak sensitivity in the green, so use of m_{pg} is equivalent to observing through a filter passing shorter wavelengths than the visual range.

A standard set of filter wavelengths has been defined with one filter in the ultraviolet passing a broad band of wavelengths centered at 3700 Å (called U), one filter in the blue passing a broad band centered at 4400 Å (called B), and one filter in the yellow passing a broad band centered at 5500 Å (called V for visual). Sets of equivalent filters exist at observatories all over the world. Thousands of stars have had their colors measured with this *UBV set of filters*; we call the process *three-color photometry*. Graphs of the sensitivities of the UBV filters appear in Figure 2–43. U, B, and V are the **apparent** ultraviolet, blue, and visual magnitudes, respectively (and are unfortunately and inconsistently written with capital letters); the corresponding **absolute** magnitudes are written M_U, M_B, and M_V. A four-color system, uvby, has ultraviolet, violet, blue, and yellow filters. This is called *four-color photometry*.

Other filters exist for other spectral ranges; a star's magnitude measured far in the infrared, for example, may be very different from magnitudes measured in the blue. R and I filters are now sometimes used together with UBV filters (Fig. 2–44).

"Bolometer," the instrument used to measure the total amount of energy arriving in all spectral regions, derives its name from "boli," Greek for "beam of light."

The physical quantity of fundamental importance is not the magnitude as observed through any given filter but rather the magnitude corresponding to the total amount of energy given off by the star summed over all spectral ranges. This is called the *bolometric magnitude*, M_{bol}. We cannot, however, measure bolometric magnitudes directly; we would have to measure the energy given off by the star over the entire spectrum and add these contributions. A major uncertainty in the past has been how much energy was contributed by the ultraviolet part of the spectrum that was blocked out by the earth's atmosphere. We could make theoretical predictions of the expected ultraviolet contribution (note that this is light further in the ultraviolet than that measured by the U magnitude), but this had to be checked

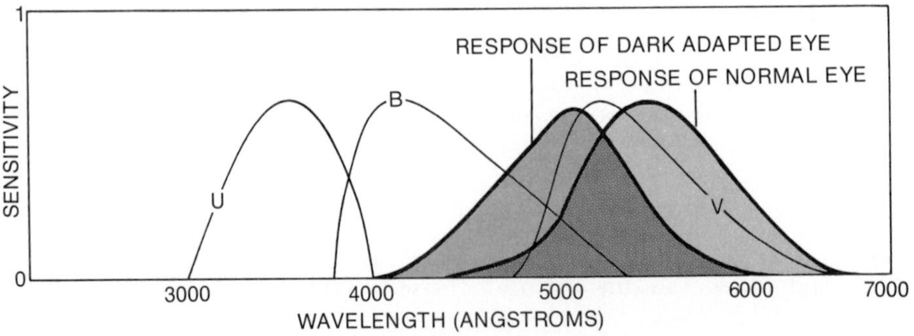

Figure 2–43 The U, B, and V curves represent the standard set of filters used by many astronomers. The response of the eye under normal conditions and the response of the dark adapted eye are also shown.

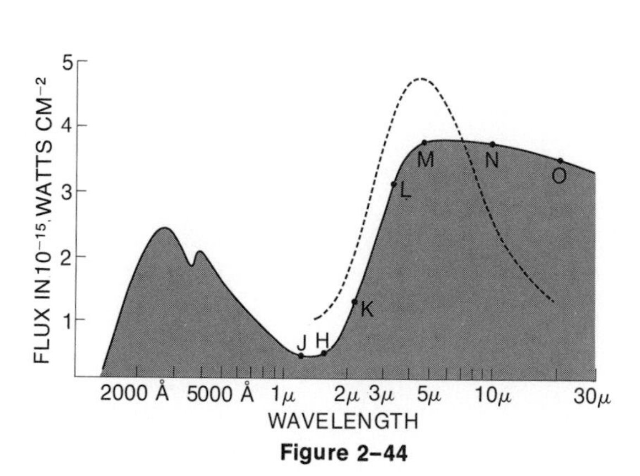

Figure 2–44

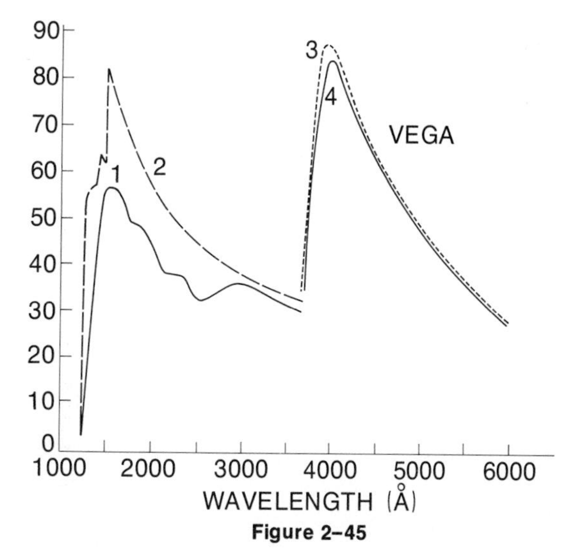

Figure 2–45

Figure 2–44 Filters other than the standard UBV set are required for use in the infrared. The H filter is out of alphabetical order as the result of a historical quirk. The wavelength of these filters is marked on the spectrum of the star HD 45677. The dotted curve at right shows the spectrum of an 800 K black body.

Figure 2–45 The spectrum of Vega, a star of spectral type A. The vertical axis is the flux—the power passing through a surface. Curve 1 shows the average of data from the second Orbiting Astronomical Observatory and from the European TD–1 satellite, while curve 2 shows a theoretical model that had been constructed for this ultraviolet part of the spectrum. Curves 3 and 4 show a model and measured values for the visible part of the spectrum, and are in better agreement with each other than are the theoretical and observational curves in the ultraviolet.

by observation. Orbiting observatories—in particular, telescopes of the Smithsonian Astrophysical Observatory and the University of Wisconsin that were aboard NASA's Orbiting Astronomical Observatory 2 (OAO-2)—measured ultraviolet intensities of many stars. Thus this is no longer a gap in our knowledge. The measured values did not always agree with theoretical predictions that had been made earlier, and improved theories have been constructed based on these data. A group of European astronomers has more recently built a spacecraft, called TD-1, that can accurately record the spectra of thousands of stars in the wavelength range from 1350 to 2550 angstroms (Fig. 2–45). As we can tell from Wien's displacement law, O and B stars are sufficiently hot to radiate most of their energy in that range, and TD-1 has surveyed all the hot stars down to 9th [apparent] magnitude. These measurements in the ultraviolet are being put together with measurements made in the visible to help us better understand the workings of the stellar atmospheres.

2.14 SPECTROSCOPIC PARALLAX

When we take a group of stars whose distances we can measure directly, by some method like that of trigonometric parallax, we can plot a Hertzsprung-Russell diagram for all these stars. We have thus established a "standard" Hertzsprung-Russell diagram, and from this standard diagram can read off the absolute magnitude that corresponds to a given spectral type (Fig. 2–46).

We can apply the standard Hertzsprung-Russell diagram to find the distance to any star whose spectrum we can observe, if we know that the star is on the main sequence. We first find the absolute magnitude that corresponds to its spectral type. We must also know the apparent magnitude of the star, but we can get that easily and directly by simply observing it. Once we know both the apparent magnitude and the absolute magnitude, we have merely to figure out how far the star has to be from the standard distance of 10 parsecs to account for the difference $m - M$.

Example: We see a G2 star on the main sequence. Its apparent magnitude is +8. How far away is it?

Answer: From the H-R diagram, we see that $M = +5$. The star appears fainter than it would be if it were at 10 parsecs. It is approximately $(2.5)^3 = 6 \times 2.5 = 15$ times fainter. (More accuracy in carrying out this cal-

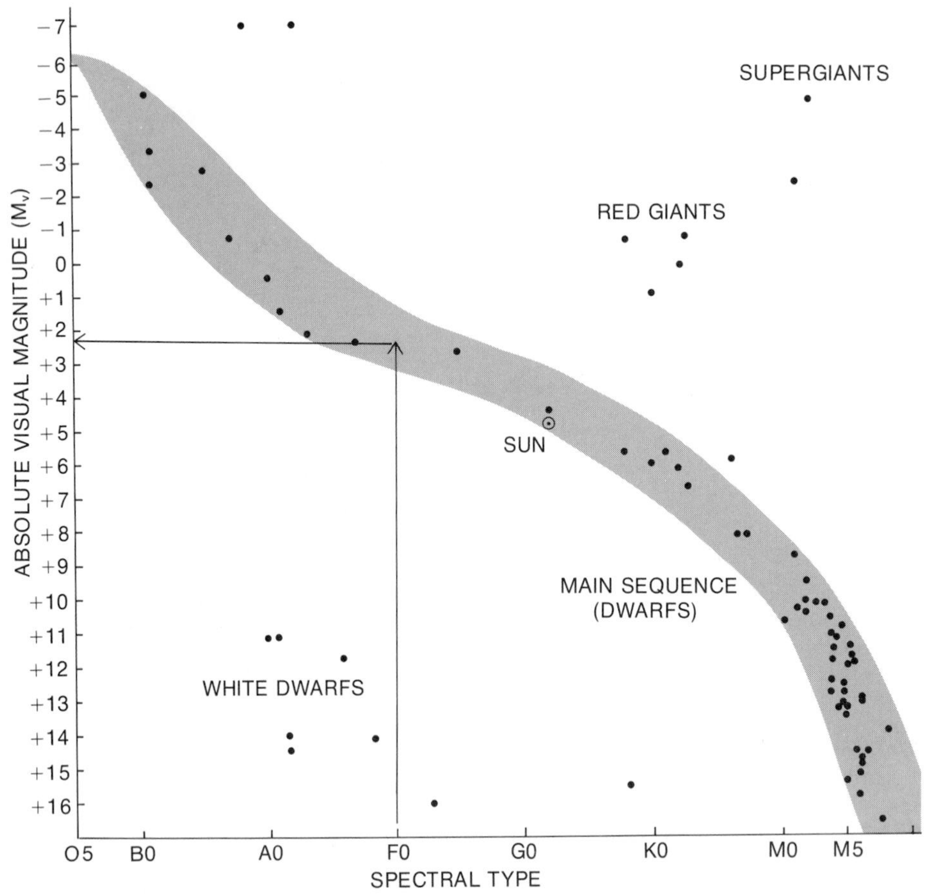

Figure 2–46 Use of the Hertzsprung-Russell diagram to measure a spectroscopic parallax.

culation would not be helpful.) By the inverse square law, this means that it is approximately 4 times farther away (exactly 4 times would be a factor of $4^2 = 16$ in brightness). Thus the star is 4 times 10 parsecs, or 40 parsecs, away.

We are measuring a distance, and not actually a parallax, but by analogy with the method of trigonometric parallax for finding distance, the method using the H-R diagram is called finding the *spectroscopic parallax.*

We can tell whether or not a star is on the main sequence by looking carefully at its spectrum. Even though a giant star and a main sequence star can have the same spectral type, the giant star is much bigger and its surface is thus farther from its center. Since gravity acts as though the mass that causes it is concentrated at the center of the star, the gravity at its surface, technically known as the *surface gravity,* will be much smaller for the giant than it is for the dwarf. Thus the atmosphere is not as compressed in the giant, and the spectral lines emitted show the subtle effects of lower atmospheric pressure. For example, lines formed under conditions of great pressure are slightly broader than lines formed under conditions of less pressure. A trained stellar spectroscopist can thus classify stars not only by spectral type but also by what their luminosity class is (see Section 2.11), whether a star is a giant, a dwarf, or in between. The distinctions between luminosity classes are easier to distinguish for cooler stars than they are for hotter stars. One must know the luminosity class of a star in order to apply the method of spectroscopic parallax to find its distance.

2.15 THE DOPPLER EFFECT

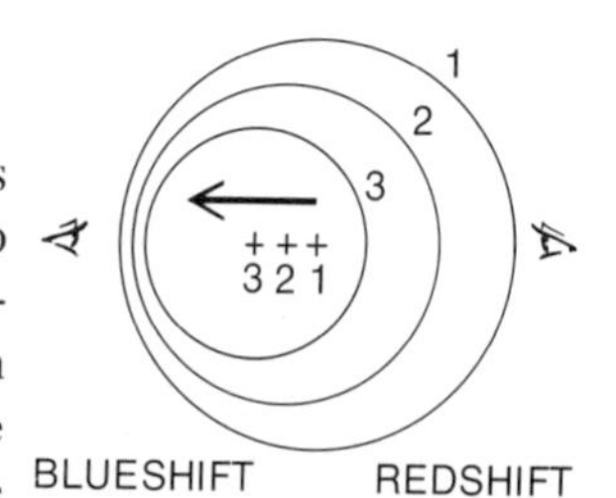

Figure 2–47 An object emits waves of radiation that can be represented by a sphere centered on the object. Observers who are being approached by the emitting source see the waves compressed. The wavelengths are shorter, and so the spectrum is blueshifted. Observers from whom the source is receding detect a redshift. We call this the Doppler effect.

The Doppler effect is one of the most important tools that astronomers can use to understand the universe. This is because without having to measure the distance to an object they can use the Doppler effect to determine its *radial velocity,* its velocity in the direction toward or away from us (the radius), measured with respect to our own velocity. Thus if we were moving at the same velocity as a source of radiation or sound, it would have no net velocity with respect to us, and no Doppler shift would be observed. The Doppler effect in sound is familiar to most of us, and its analogue in electromagnetic radiation, including light, is very similar.

In sound, you may be familiar with the change of pitch of a train whistle, or the whine of a jet engine, as the train or plane first approaches you and then passes you and begins to recede. The object that is emitting the sound waves is approaching you at first. By the time it emits a second wave, it has moved closer to you than it had been when it emitted the first wave. Thus the waves pass more frequently than they would if the source were not moving. The wavelengths seem compressed, and the pitch is higher. When the emitting source passes you, the wavelengths become stretched, and the pitch of the sound is lower (Fig. 2–47).

With light waves, the physical mechanism is not as straightforward as is the stretching or compressing of the wavelengths of sound, but the effect is the same. As a body emitting light, or other electromagnetic radiation, approaches you, the wavelengths become slightly shorter than they would be if the body were at rest. Visible radiation is thus shifted slightly in the direction of the blue (it doesn't actually have to become blue, only be shifted in that direction). We say that the radiation is *blueshifted.* Conversely, when the emitting object is receding, the radiation is said to be *redshifted* (Fig. 2–48). We generalize these terms to types of radiation other than light, and say that radiation is blueshifted whenever it changes to shorter wavelengths, and redshifted whenever it changes to longer wavelengths.

Let us first consider the wavelength when the emitter is at rest, which is called the *rest wavelength.* Now let us consider the emitter to be moving. The fraction of the rest wavelength that the wavelength is shifted is the same as the fraction of the speed of light at which the body is traveling (or, for a sound wave, the fraction of the speed of sound).

In equation form, we write

$$\frac{\Delta\lambda}{\lambda_0} = \frac{v}{c},$$

where $\Delta\lambda$ is the change in wavelength, λ_0 is the original (rest) wavelength, v is the velocity of the emitting body, and c is the velocity of light ($= 3 \times 10^{10}$ cm/sec). We define positive velocities as velocities of recession and negative velocities as velocities of approach. The new wavelength, λ, is equal to $\lambda_0 + \Delta\lambda$.

Example: A star is approaching at 30 km/sec (i.e., v = −30 km/sec). At what wavelength do we see a spectral line that was at 6000 Å when the radiation left the star?

Answer:

$$\frac{\Delta\lambda}{\lambda_0} = \frac{v}{c} = \frac{-30 \text{ km/sec} \times 10^5 \text{ cm/km}}{3 \times 10^{10} \text{ cm/sec}} = \frac{-3 \times 10^6 \text{ cm/sec}}{3 \times 10^{10} \text{ cm/sec}} = -10^{-4}.$$

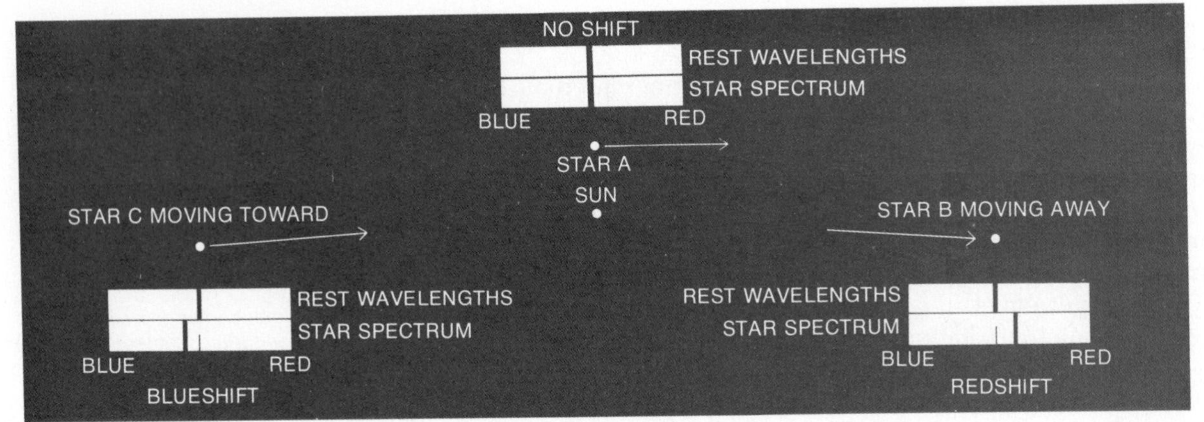

Figure 2–48 The Doppler effect. In each pair of spectra, the position of the spectral line in the laboratory is shown on top and the position observed in the spectrum of the star is shown below it. Lines from approaching stars appear blueshifted, lines from receding stars appear redshifted, and lines from stars that are moving transverse to us are not shifted because at that particular moment the star has no component of velocity toward or away from us.

Thus $\Delta\lambda = 10^{-4}\lambda_0 = -10^{-4} \times 6 \times 10^3 = -0.6$ Å. The change in wavelength, $\Delta\lambda$, is thus -0.6 Å. (Since the star is approaching, this is a blue shift, and the new wavelength is slightly shorter than the original wavelength.) The new wavelength, λ, is thus $\lambda_0 + \Delta\lambda = 6000$ Å $- 0.6$ Å $= 5999.4$ Å.

Note that since there are proportions on both sides of the equation, if we take care to use v and c in the same units (e.g., cm/sec), then $\Delta\lambda$ will be in the same units that λ is in, no matter whether that is angstroms, centimeters, or whatever.

The 30 km/sec given in the example is typical of the random velocities that stars have with respect to each other. These velocities are small on a stellar scale, but large compared to terrestrial velocities (30 km/sec = 30 km/sec × 3600 sec/hr = 108,000 km/hr). Doppler shifts resulting from these random velocities are relatively small. We have to go outside our own galaxy to find larger Doppler shifts. Nonetheless, the Doppler shifts within our own galaxy are large enough to be measured, and give us quite a lot of information about our stellar neighbors.

2.16 STELLAR MOTIONS

On the whole, the network of stars in the sky is fixed. But radial velocities can be measured for all stars, and some of the stars are seen to move slightly across the sky with respect to the more distant stars. The actual velocity of a star in 3-dimensional space, with respect to the velocity of the sun, is called its *space velocity*.

We usually deal separately with the velocity of a star toward or away from us (the radial velocity), and the angular velocity across the sky, which is called the *proper motion*. These motions are perpendicular to each other.

Proper motion is ordinarily very small, and is accordingly very difficult to measure. Usually we must compare the positions of stars taken at intervals of decades to detect a sufficient amount of proper motion to allow a measurement of any accuracy to be made. Only a few hundred stars have proper motions as large as 1 arc sec per year; proper motions of other stars are sometimes measured in seconds of arc per century!

Figure 2–49 Two photographs taken 11 months apart. In making this combined print, one of the negatives was shifted slightly. The proper motion of Barnard's star is clearly visible.

The star in the sky with the largest proper motion was discovered in 1916 by E. E. Barnard, and is known as Barnard's star (Fig. 2–49). It moves across the sky by 10¼ arc sec per year. Some call it "Barnard's runaway star" because of this proper motion, which, though the highest for any known star, is still minuscule. Barnard's star is seen to move so rapidly across the sky because it is relatively close to us. Only 1.8 parsecs away, it is the nearest star to us after the sun, Proxima Centauri and the Alpha Centauri double star. Since the moon appears half a degree across, Barnard's star moves the equivalent of the diameter of the moon across the sky in only 180 years.

The stars that show proper motions are generally the closest stars to us. If more distant stars were moving at the same velocities as measured in km/sec, their angular movement would be imperceptible (Fig. 2–50). If we know the distance to a star, from such a method as that of trigonometric parallax, and know the proper motion, then we can determine the velocity in the direction perpendicular to that of the radial velocity in linear measure such as km/sec.

From proper motion studies we can only tell the component of motion of a star across the sky, and not how rapidly it is approaching or receding from us. We can measure the velocity in our line of sight (the radial velocity) from the Doppler effect. (In order to derive it, we have to correct

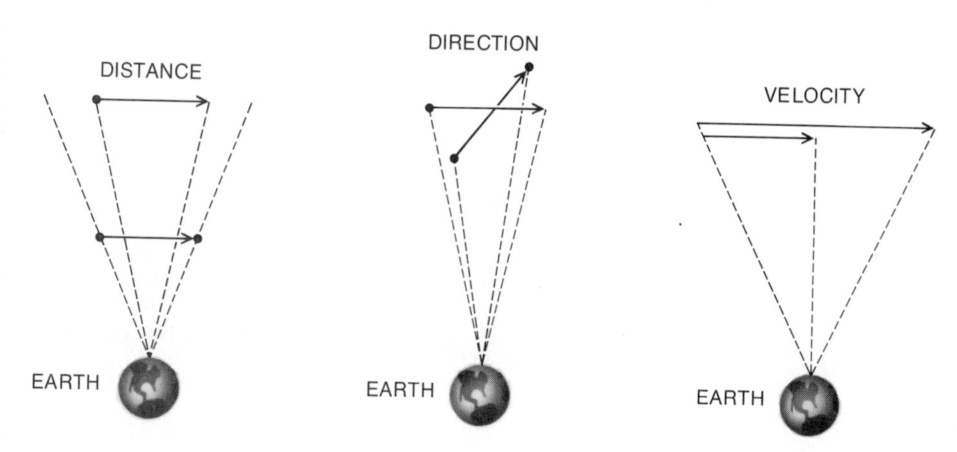

Figure 2–50 Stars can have different proper motions because (left) they are at differing distances from us even though they have the same speed through space oriented in the same direction (*middle*), their directions of motion are oriented differently even though they have the same speed through space and are at the same distance from us, or (*right*) they are actually traveling at different speeds through space.

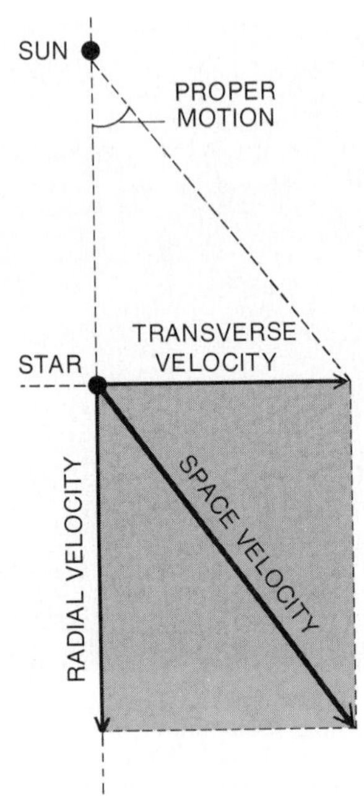

Figure 2–51 If we know a star's proper motion and its distance we can compute its linear velocity through space in the direction across our field of view. The Doppler effect gives us its linear velocity toward or away from us. These two velocities can be combined to tell us the star's actual velocity through space, its *space velocity*.

for the contribution to the radial velocity caused by the earth's orbit around the sun.) Note that since we need only measure the spectrum to find radial velocities, we are not limited to the nearby stars, as we are for proper motion studies. We can combine the values of velocity in the plane of the sky and in the radial direction to find the actual velocity of the star in space, the space velocity (Fig. 2–51).

Since the sun has its own random velocity in space, we prefer not to use the sun as a standard against which to measure velocities. We refer the velocities of stars, rather, to the average velocity of the stars in the neighborhood of the sun. If we were moving with this average velocity so that the net average velocity with respect to us were zero, we would be in a reference system called the *local standard of rest*. The actual velocities of stars with respect to the local standard of rest are called *peculiar velocities*. From the space velocities, we can calculate the peculiar velocities. Most peculiar velocities are a few tens of km/sec; very few are above 100 km/sec.

SUMMARY AND OUTLINE OF CHAPTER 2

Aims: To learn fundamentals and basic definitions pertaining to stars and spectroscopy

Cause of twinkling
Magnitude: how to convert from magnitudes to factors of brightness (Section 2.1)
The spectrum: absorption lines, emission lines, the continuum (Sections 2.2 and 2.3)
Colors of stars (Section 2.4)
Radiation laws: Planck's law, Wien's displacement law, Stefan-Boltzmann law (Section 2.5)
Spectral types (Section 2.6)
Conversion of units, temperature scales (Section 2.6)
Definition of ion (Section 2.7)
Pattern of a hydrogen spectrum (Section 2.8)
Stellar spectra (Section 2.9)
The continuous spectrum of the sun: H^- (Section 2.10)
The Hertzsprung-Russell diagram (Section 2.11)
Parallax, parsec (Section 2.12)
Inverse-square law, absolute magnitude (Section 2.13)
Spectroscopic parallax (Section 2.14)
Doppler effect, use in determining stellar motions, proper motion (Sections 2.15 and 2.16)

QUESTIONS

1. Venus can be as bright as magnitude −4. Betelgeuse is a first magnitude star (m = +1). How many times brighter is Venus at its brightest than Betelgeuse?

2. Star A has magnitude +12. Star B appears 10,000 times brighter. What is the magnitude of star B? Star C appears 10,000 times fainter than star A. What is its magnitude?

3. Star A has magnitude +10. The magnitude of star B is +8 and of star C is +3. How much brighter does star B appear than star A? How much brighter does C appear than A? How much brighter does C appear than B?

4. Why do the stars twinkle?

5. How much brighter is a -1 magnitude star than a $+1$ magnitude star?

6. What are the colors of the rainbow? What are the parts of the electromagnetic spectrum?

7. The sun's spectrum peaks at 5600 Å. The spectrum of a star whose temperature is twice that of the sun would peak at what wavelength? How much more energy than the sun would the star give off?

8. One black body peaks at 2000 Å. Another peaks at 10,000 Å. Which gives out more radiation at 2000 Å? Which gives out more radiation at 10,000 Å? What is the ratio of the total radiation given off by the two bodies?

9. Which contains more information, Wien's law or Planck's law? Explain.

10. What is the ratio of energy output for an average O star and the sun?

11. Star A appears to have the same brightness through a red and through a blue filter. Star B appears brighter in the red than in the blue. Star C appears brighter in the blue than in the red. Rank these stars in order of increasing temperature.

12. What is the difference between the continuum and an absorption line? The continuum and an emission line? Draw a continuum with absorption lines. Can you draw absorption lines without a continuum? Can you draw emission lines without a continuum?

13. Are emission, absorption, or continuous spectra given off by (a) an ordinary incandescent electric bulb, (b) a fluorescent lamp, (c) a neon sign?

14. You are driving a car, with the speedometer showing that you are going 80 km/hr. The person next to you asks why you are only going 75 km/hr. Explain why you each saw different values on the speedometer.

15. Would the parallax of a nearby star be larger or smaller than the parallax of a more distant star? Explain.

16. A star has an observed parallax of 0.1 arc second. What is its distance in parsecs? What is its distance in light years? What would its parallax be if it were ten times farther away?

17. Estimate the farthest distance for which you can detect parallax by alternately blinking your eyes. To what angle does this correspond?

18. What is the frequency of radiation whose wavelength is equal to your height?

19. What is the significance of the existence of the main sequence?

20. What factors determine spectral type? What instrument would you use to determine the spectral type of a star?

21. Why is the Balmer series the most commonly observed spectral series of atomic hydrogen?

22. Two stars have the same apparent magnitude and are the same spectral type. One is twice as far away as the other. What is the relative size of the two stars?

23. Two stars have the same absolute magnitude. One is ten times farther away than the other. What is the difference in apparent magnitudes?

24. A star has apparent magnitude of $+4$, and is 100 parsecs away from the sun. If it is a main sequence star, what is its spectral type? (Hint: refer to Fig. 2–37.)

25. A star is 20 parsecs from the sun and has apparent magnitude $+2$. What is its absolute magnitude?

26. A star has apparent magnitude $+9$ and absolute magnitude $+4$. How far away is it?

27. List spectral types of stars in order of strength, from strongest to weakest, of (a) hydrogen lines, (b) ionized calcium lines, (c) titanium oxide lines.

28. If a star is moving away from the earth at very high speed, will the star have a continuous spectrum that appears hotter or cooler than it would if the star were at rest? Explain.

29. The earth moves around the sun at about 30 kilometers a second. How much will this shift the frequency of a spectral line whose rest wavelength is 6000 Å? What is the maximum variation in observed frequency of this line from a star in the plane of the ecliptic over the course of a year?

30. If we measure the Doppler shift of a star's spectral lines, how can we calculate that star's velocity with respect to the local standard of rest? What other data do we need?

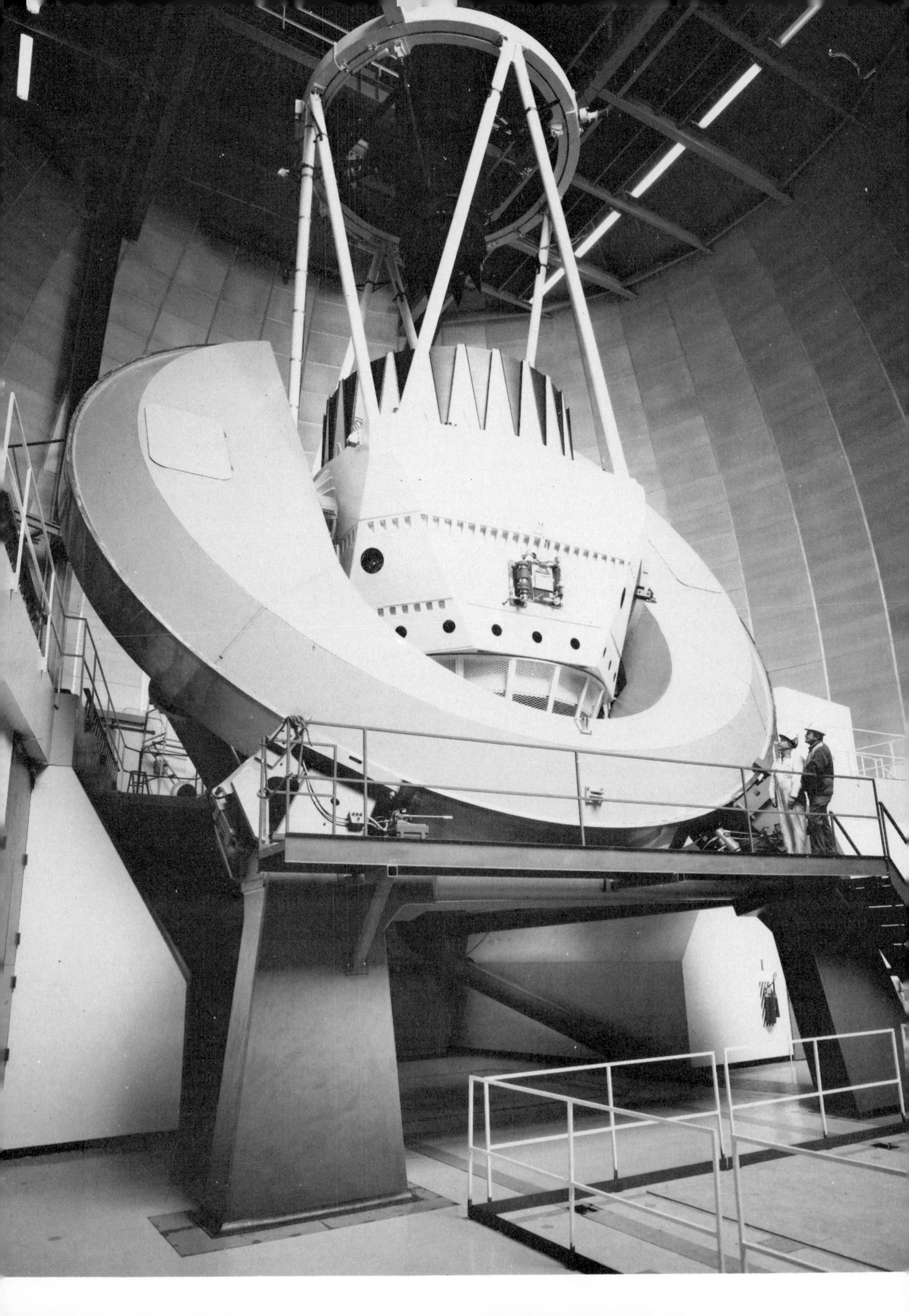

3

Observing the Stars

If you forced a group of astronomers to choose one and only one instrument to be marooned with on a desert island, they would probably choose—a computer. The notion that astronomers spend most of their time at telescopes is far from the case in modern times. Still, telescopes have been and continue to be very important to the development of astronomy, and in this chapter we will discuss telescopes and other observational devices.

Even the question of how to define a telescope has no simple answer, for a "telescope" to observe gamma rays may be a package of layers of plastic flown into the atmosphere aboard a balloon, and a radio telescope may be a large number of small aerials strewn over acres of landscape. We will begin, nevertheless, by discussing telescopes of the traditional type, which observe the radiation in the visible part of the spectrum. They are important in and of themselves because of the many things we have learned over the years by studying visible radiation. Also, the types of telescopes most used by students and by amateur astronomers (some of whom are quite professional in their approach to the subject) are most likely to be optical. Further, the principles of focusing and detecting electromagnetic radiation that were originally developed through optical observations have widespread use throughout the spectrum.

3.1 THE PURPOSE OF A TELESCOPE

Contrary to popular belief, the most important purpose for which most optical telescopes are used is usually to gather light. (For the next few sections I shall drop the qualifying word "optical.") True, telescopes can be used to magnify as well, but for the most part astronomers are interested in observing fainter and fainter objects and so must collect more light to make these objects detectable. There are certain cases where magnification is important—such as for observations of the sun (which will be discussed in

---

The 4-meter Mayall telescope at the Kitt Peak National Observatory.

Chapter 5) or for observations of the planets—but stars appear as mere points of light no matter how much magnification is applied.

When we look at the sky with our naked eyes, there are several limitations that come into play in addition to that of spectral selectivity. One limitation is that we can see only the light that passes through an opening of a certain diameter—the pupils of our eyes. In the dark, our pupils dilate so that as much light as possible can enter, but the apertures are still only a few millimeters across. A second limitation is one of time. Our brains distinguish a new image several times a second, and so we are unable to store faint images for a long time to accumulate a brighter image. Astronomers overcome both these limitations with the combination of a telescope to gather light and a recording device, such as a photographic plate, to store the light. Other equipment may also be used, such as a spectrograph to analyze the spectral content of the light, or even simple filters.

Actually, resolution depends not only on the aperture, as explained at right, but also on the wavelength of radiation being observed; the figures given in the text are for light of approximately 5000 Å. As the wavelength of radiation doubles, resolution is halved. For example, a telescope that can resolve two 5000 Å light sources that are 1 arc sec apart could only resolve two 10,000 Å sources if they were 2 arc secs apart. The limit of resolution for visible light, called Dawes' limit, is approximately 2×10^{-3} λ/d arc sec, when λ is in Å and d is in cm.

A further advantage of a telescope over the eye, or of a larger telescope over a smaller telescope, is that of *resolution*, the ability to distinguish finer details in an image. For example, a telescope can distinguish the two components of a double star from each other in many cases where the unaided eye is unable to do so. If a telescope with a collecting area ten centimeters across can resolve double stars that are separated by a certain angle, 1 arc sec, a telescope twenty centimeters across, twice the diameter, can resolve stars that are separated by half that angle, ½ arc sec. In principle, *resolution is inversely proportional to the diameter of the telescope,* that is, to the diameter of the part of the telescope that is collecting the light.

For the larger telescopes the best resolution that can be achieved is limited to about 1 arc sec on an average good night by the turbulence in the earth's atmosphere rather than by the telescope size. Thus increasing the diameter of a telescope above a certain size no longer improves the resolution, even on the best of nights, though the advantages in terms of light-gathering power remain. A telescope is capable of considerably better resolution than the eye, which is limited to a resolution of 1 arc min.

We will begin by considering the telescopes themselves, and then go on to consider the equally important devices that are used in conjunction with the telescopes.

3.2 REFRACTING TELESCOPES

Refraction is the bending of light (or other electromagnetic radiation) when the light passes from one medium (e.g., transparent object, air, interstellar space) into another. A lens uses the property of refraction to focus light, and a telescope that has a lens as its major element is called a *refracting telescope.*

Let us consider waves of light hitting a block of glass at an angle, and draw an advancing plane of radiation (for example, the plane representing the first light that would arrive if the source were to suddenly become visible). This advancing plane is called a *wave front* (Fig. 3–1). (T.V. newscasters talk of weather fronts.) The wave front is straight, hence one side meets the glass slightly before the other side. It takes the other side a very small additional amount of time to reach the glass, and in that time the first side has traveled a short distance through the glass. The speed of light is lower in glass than it is in air, so in that small additional time the part of the wave in the glass travels a shorter distance than the part of the wave still in

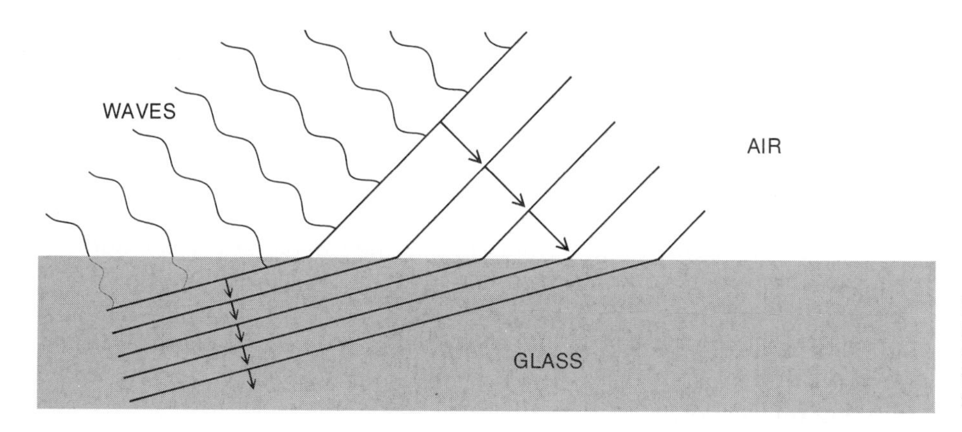

Figure 3–1 Light travels at a velocity in glass that is different from the velocity that it has in air. This leads to refraction.

the air. By the time all the wave front has reached the glass, the wave front is moving in a different direction than it was before it started through the glass. This bending of the direction of travel of the light is called *refraction.* Put a straw in a glass of water (Fig. 3–2), a branch in a lake, or a foot in a bathtub to observe this phenomenon.

A suitably curved piece of glass can be made so that all the light that travels through it is bent, bringing all the rays of each given wavelength to a focus. Such a curved piece of glass is called a lens. The lens in your eye does a similar thing to image objects of the world onto your retina. An eyeglass lens helps the lens in your eye accomplish this focusing task.

The distance of the image from the lens on the side away from the object being observed depends in part on the distance of the object from the front side of the lens. Astronomical objects are all so far away that they are, for the purpose of forming images, as though they were infinitely far away. We say that they are "at infinity." The images of objects at infinity fall at a distance called the *focal length* behind the lens (Fig. 3–3).

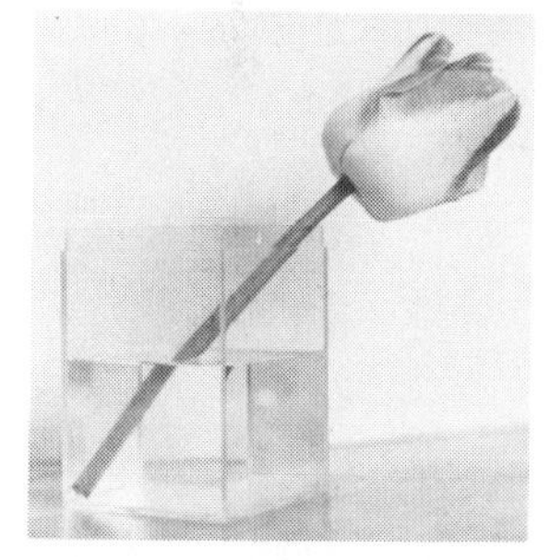

Figure 3–2 The flower stem appears both bent and displaced at the boundary between the water and the air.

The area of which an image is formed is called the *field of view.* A simple lens can only be designed to focus light from directly in front of it or from regions not too far to the side. Objects that are at angles far from the center of the field of view are not focused perfectly.

The technique of using a lens to focus faraway objects was, we think, developed in Holland in the first decade of the 17th century. It is not clear how Galileo heard of the process, but it is known that Galileo quickly bought or ground a lens and made a simple telescope that he demonstrated in 1610 to the Senate in Venice (Fig. 3–4). He put this small lens and another, smaller lens at opposite ends of a tube. The second lens, called the *eyepiece,* is used to examine and to magnify the image made by the first lens, called the *objective.* Galileo's telescope was only a few centimeters in diameter, but it magnified enough to impress the nobles of Venice, who had assembled to see the new invention.

Figure 3–3 The focal length is the distance behind a lens to the point at which objects at infinity are focused.

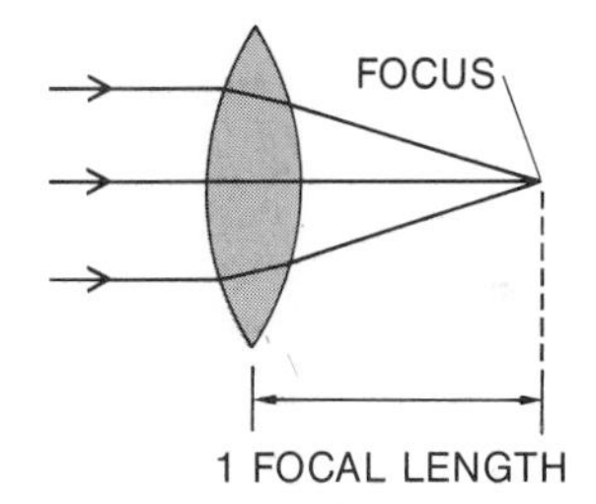

Galileo turned his simple telescope on the heavens, and what he saw revolutionized not only astronomy but also much of seventeenth century thought. He discovered, for example, that the sun had spots—blemishes—on it, that Jupiter had satellites of its own, and that Venus had phases. We shall be discussing the fundamental importance of these discoveries later on.

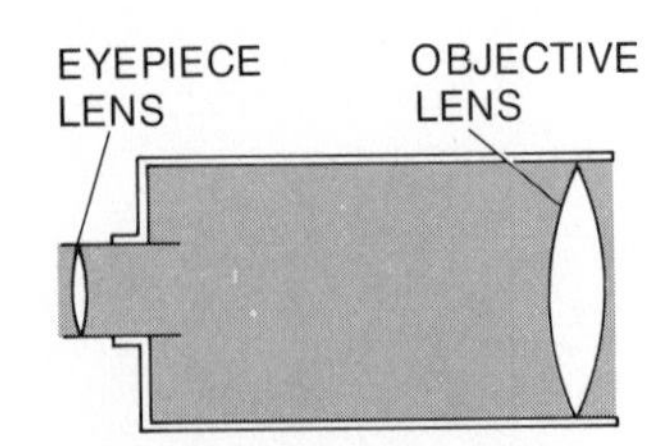

Figure 3–4 A simple refracting telescope consists of an objective lens and an eyepiece.

Larger telescopes were made, and larger still. Not only did the telescopes magnify the image but they also collected light from a larger area than the area of the pupil of a person's eye, and thus allowed objects to be seen that are too faint for the naked, unaided eye.

There are many problems in making refracting telescopes. For one thing, lenses suffer from *chromatic aberration*, the effect whereby different colors are focused at different points (Fig. 3–5). (The speed of light in a medium—in glass, for example—depends on the wavelength of the light, so not all wavelengths can be brought to a focus at the same point. Ingenious methods of making "compound" lenses of different glasses that have slightly differing properties have succeeded in reducing this problem, but the residual chromatic aberration is a fundamental problem for the use of refracting telescopes.)

In the first quarter of the nineteenth century, Joseph Fraunhofer in Germany, the same individual who had been the first to measure the lines in the solar spectrum, was one of the major lens makers in Europe. In the 1830's, after his death, the firm with which he had been associated made two 38-cm (15-inch) lenses for the Harvard College Observatory in the United States and for the Pulkovo Observatory near Leningrad in Russia.

At about that time an American, Alvan Clark, Sr., began making lenses. After a time, the quality of his work surpassed the European standard. He was pressed to make larger and larger lenses (Fig. 3–6). In 1895, he succeeded in the difficult task of making a lens 1 meter (40 inches) across for the Yerkes Observatory in Williams Bay, Wisconsin (Fig. 3–7).

With lenses that large, physical problems arise. It has been difficult to get a pure piece of glass sufficiently free of internal bubbles and sufficiently homogeneous throughout its volume to allow the construction of a lens larger than that at Yerkes. And a lens can be supported only from its rim when it is mounted in a telescope, since one cannot obstruct the aperture. Gravity then causes the lens to sag in the middle, and this effect changes as the telescope is pointed in different directions. One can try to make the lens more rigid by making it thicker, but this makes it heavier and even more difficult to get a sufficiently pure mirror blank. The percentage of light that is transmitted through the glass diminishes too. The 1-meter telescope at Yerkes, for all of these reasons, is still the largest refracting telescope in the world. Since its construction, optical astronomy has adopted reflecting telescopes, which use mirrors instead of lenses. A long time has passed, however, and modern glass-making techniques might well permit construction of a larger lens if anyone were interested enough to provide the financing.

Refracting telescopes are still the telescopes of choice in a variety of circumstances. For example, special telescopes to study the solar corona use lenses because the lenses can be carefully cleaned so that they scatter very little light and because, unlike mirrors, lenses do not have to be coated with material that scatters some of the light. Still, the largest telescopes are now all reflectors.

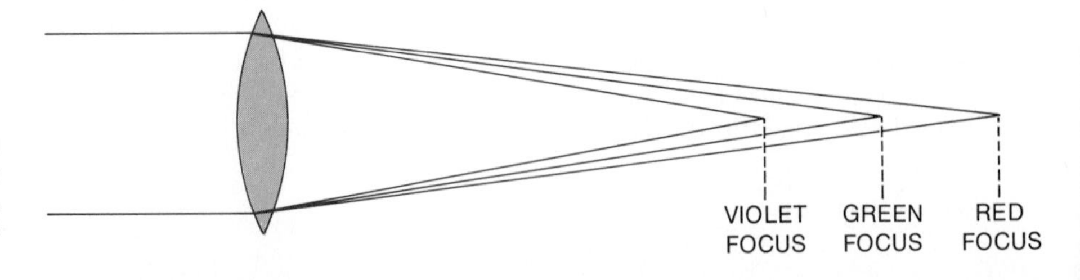

Figure 3–5 The focal length of a lens is different for different wavelengths, which leads to chromatic aberration.

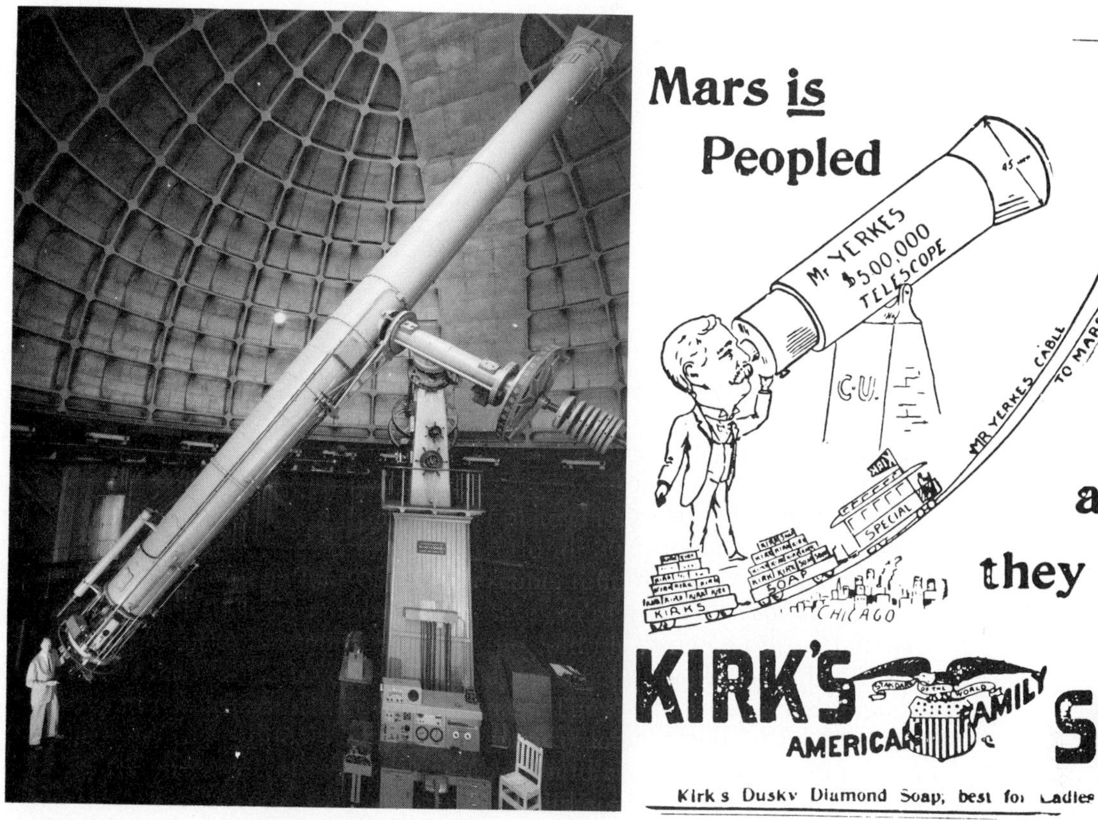

Figure 3–6

Figure 3–7

Figure 3–6 The 0.9-meter (36-inch) refractor of the Lick Observatory.

Figure 3–7 The opening of the 1-meter (40-inch) refractor of the Yerkes Observatory was the cause of much notice in the Chicago newspapers in 1893.

3.3 REFLECTING TELESCOPES

Reflecting telescopes are based on the principle of reflection with which we are so familiar from ordinary household mirrors. A mirror reflects the light that hits it so that the light bounces off at the same angle at which it approached. A flat mirror gives an image the same size as the object being reflected, but funhouse mirrors, which are not flat, cause images to be distorted.

One can construct a mirror in a shape so that it reflects incoming light to a focus. Let us consider a spherical mirror, for example. If you were at the middle of a giant spherical mirror, in whatever direction you looked you would see your own image. A sphere images whatever is at its center back

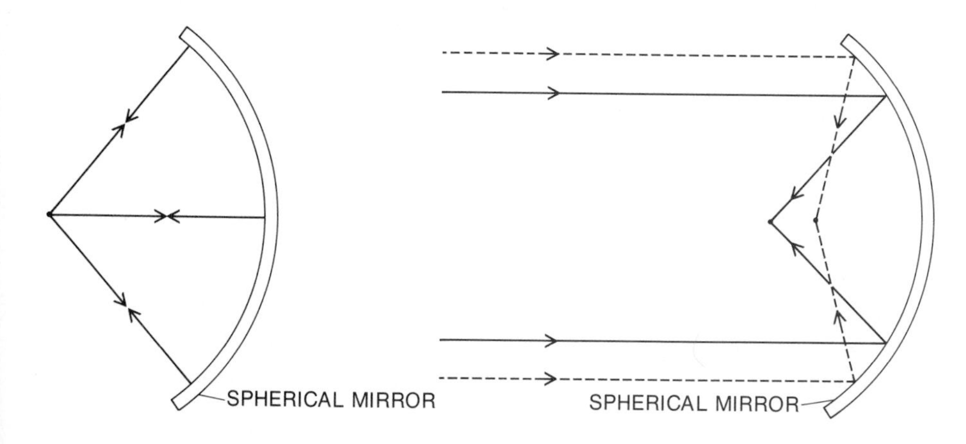

Figure 3–8 A spherical mirror focuses light that originates at its center of curvature back on itself, but suffers spherical aberration in that it does not perfectly focus light from infinity.

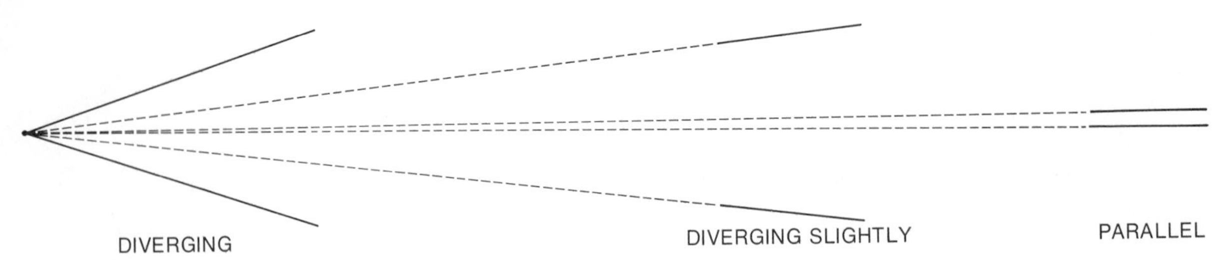

Figure 3–9 Light rays from very distant objects are diverging so slightly by the time they reach us that we speak of *parallel light.*

at its center (Fig. 3–8). If we were to use just a portion of a sphere, it would still image whatever was at the center of its curvature back to that same point.

But the stars are far away, and we must find a shape for a mirror that has the property of focusing light from celestial objects to a single point. The stars and planets are so far away that individual light rays emanating from there are diverging by such an imperceptible amount by the time they reach us on earth that they are practically parallel. We say that we are observing *parallel light* (Fig. 3–9).

A *parabola* is a two-dimensional curve that has the property of focusing parallel rays to a point. Telescope mirrors are actually *paraboloids,* which are the three-dimensional curves generated when parabolas are rotated around their axes of symmetry (Fig. 3–10). Over a small area, a paraboloid differs from a sphere only very slightly, and one can ordinarily make a parabolic mirror by first making a spherical mirror and then deepening the center slightly (Fig. 3–11). In order to have a focus that is as "sharp" as possible, one must make the shape agree with that of a paraboloid within a fraction of the wavelength of light at which one wants to observe, normally by better than one-eighth the wavelength of the light. Since the wavelength of yellow light, at 6000 Å, is about half a micron (half a millionth of a meter), it is obvious that the mirrors of optical telescopes have to be made to a very high absolute precision.

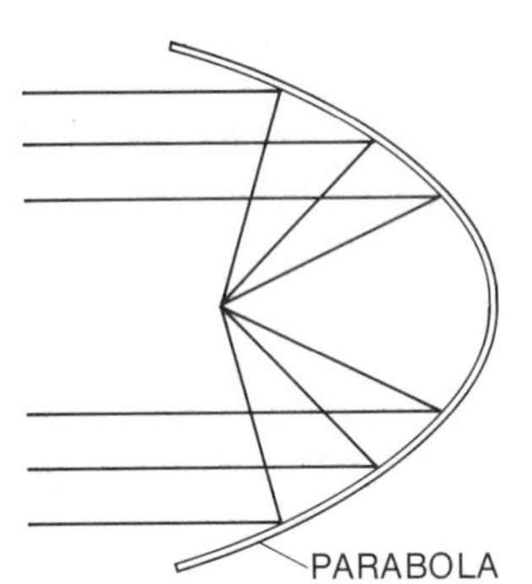

Figure 3–10 A parabola focuses parallel light to a point.

The technique of making a small telescope mirror is not excessively arduous. Many amateur astronomers have made mirrors approximately 15 cm (6 inches) across without having had previous experience; it requires perseverance and about 50 hours of time. Making a large telescope mirror is another story, of course, and usually takes years.

Figure 3–11 A parabolic mirror is usually made by deepening the center of a spherical mirror.

The advantages of reflecting telescopes over refracting telescopes are several. The reflection of light is not color-dependent, so that chromatic aberration is not a problem for reflectors. One normally deposits a thin coat of a highly reflecting material on the front of a suitably ground and polished surface. Since light does not penetrate the surface, it does not matter too much what is inside the telescope mirror. The internal material does not have to be free of bubbles or striations; it only has to have a smooth surface, though it should be free of internal stress and strain and very stable whatever the temperature is. Further, one can have a network of supports all across the back of the telescope mirror, to prevent the mirror from sagging under the force of gravity. All these advantages allow reflecting telescopes to be made much larger than refracting telescopes. (At wavelengths shorter than those of visible light, glass does not pass radiation, so only reflecting telescopes can be used in these x-ray or shorter ultraviolet regions.)

A potential problem with small reflecting telescopes could be the fact that one must gather the reflected light without impeding the incoming light. Clearly, if you put your head at the focal point of a small telescope

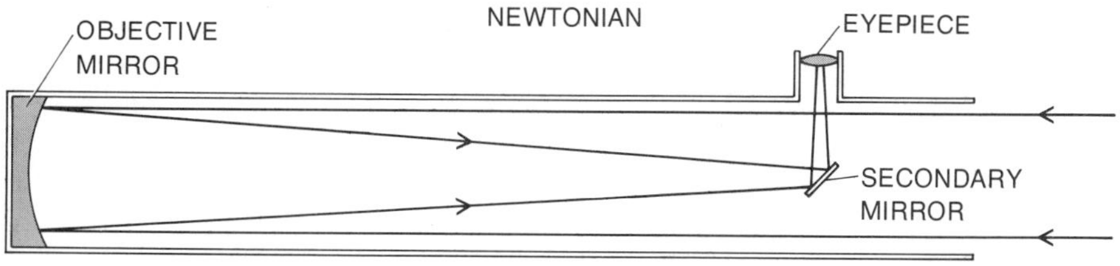

OBJECTIVE MIRROR
NEWTONIAN
EYEPIECE
SECONDARY MIRROR

Figure 3–12 A Newtonian reflector.

mirror, the back of your head would block the incoming light and you would see nothing at all!

Isaac Newton got around this problem by putting a small diagonal mirror a short distance in front of the focal point to reflect the light so that the focus was outside the tube of the telescope. This type of apparatus, in use down to this day, is known as a *Newtonian telescope* (Fig. 3–12).

In Herstmonceux, England, there is now a 2.5-meter (98-inch) reflecting telescope at the Royal Greenwich Observatory. On one side of the floor of the telescope dome, in a glass case, is a full-scale modern replica of Newton's 5-cm (2-inch) telescope from 1671. On the side of the case facing the visitor's gallery is a sign identifying the replica; on the side of the case facing inward toward the telescope is a sign that can be read only by the astronomers. The replica—and sign—is shown in Figure 3–13.

Figure 3–13 The replica of Newton's telescope in the visitor's gallery overlooking the 2.5-meter (98-inch) Isaac Newton telescope of the Royal Greenwich Observatory at Herstmonceux, England, showing the side visible only to the astronomers.

Two contemporaries of Newton invented alternative types of reflecting telescopes. N. Cassegrain, a French optician, and J. Gregory, a Scottish astronomer, invented types in which the light was reflected by the secondary mirror back toward the primary mirror and through a small hole in the center of the primary. These designs are called *Cassegrainian* (or *Cassegrain*) and *Gregorian telescopes* (Fig. 3–14).

Modern large telescopes also often have a *coudé* focus. In the coudé system, the light is bent with a series of mirrors until it travels down the axis of the telescope that is pointing at the north pole ("coudé" means "bent" or "elbowed" in French). The image never moves laterally, but only rotates. Thus the light at the coudé focus comes to the rear of the telescope in an unchanging direction. As a result, rather large spectrographs can be built into a room at the base of the telescope.

In all of these methods, a slight bit of the incoming light is blocked by the secondary mirror, and by the supports that suspend the secondary mirror in the middle of the tube. But the secondary mirror can be made small enough relative to the primary mirror that relatively little light is blocked.

The largest telescopes have several interchangeable secondary mirrors built in so that different foci (Newtonian, Cassegrainian, and so forth) can be used at different times. The very largest telescopes are so huge that the observer can even sit in a "cage" suspended at the end of the telescope

Foci *is the plural of* focus.

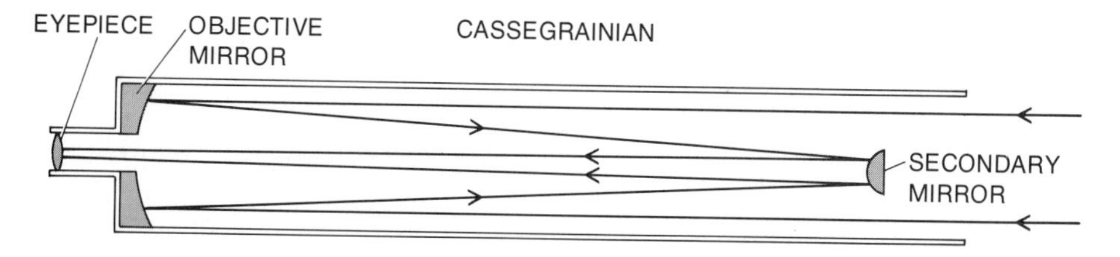

Figure 3–14 A Cassegrainian (often called Cassegrain) telescope has a convex secondary mirror.

Figure 3–15 One of Russell W. Porter's set of drawings of the 5-meter (200-inch) telescope on Palomar Mountain.

at the *prime focus,* the position of the direct reflection of the light from the primary mirror (Fig. 3–16). The observer in the prime-focus cage blocks only a small fraction of the incoming light.

Figure 3–16 The prime-focus cage of the 5-meter Palomar telescope.

Modern reflectors began with the construction of the 0.9-meter (36-inch) telescope at the Lick Observatory on Mt. Hamilton in California and then, under the aegis of George Ellery Hale, of the 1.5-meter (60-inch) and 2.5-meter (100-inch) telescopes on Mt. Wilson in California. The 2.5-

Box 3.1 Conversion of Units

In order to know what to multiply and divide when converting from one system of units to another, some people find it convenient to multiply and divide the words representing the units as though they were numbers. They set up a chain of multiplication and division of known conversion factors so that the unit names they are trying to convert **from** cancel out. Then they do the same series of multiplications and divisions with the numbers themselves. For example, 1 inch = 2.54 cm. If we divide both sides of the equation by 2.54 cm, we have the following equation: $\frac{1 \text{ in}}{2.54 \text{ cm}} = 1$. Equivalently, we can divide by 1 in to get $\frac{2.54 \text{ cm}}{1 \text{ in}} = 1$. Now, we can always multiply any number by 1 without changing its value. Thus if we have, say, 200 inches and want to put this figure in units of centimeters, we can write 200 inches $\times \frac{2.54 \text{ cm}}{1 \text{ in}}$. The "inches" in the numerator and denominator cancel, and we have 508 cm as the converted figure. Similarly, $3 \text{ m} \times \frac{100 \text{ cm}}{\text{m}} \times \frac{1 \text{ in}}{2.54 \text{ cm}} = \frac{300}{2.54}$ in. This method works just as well with units of time (months, weeks, days, hours, minutes, seconds), units of volume, and so on.

meter telescope, with a mirror made of plate glass, began operation in 1917 and was the largest telescope in the world for 30 years.

One of the limitations of any telescope, reflecting or refracting, is the length of time that the mirror or lens takes to reach its equilibrium shape when exposed to the temperature of the cold night air. In the 1930's, the Corning Glass Works invented Pyrex, a type of glass that is less sensitive to temperature variations than ordinary glass. They cast, with great difficulty, a mirror blank a full 5.08 meters (200 inches) across.

The construction of the telescope was held up by many factors, including the Second World War. Only in 1949 did active observing begin with the telescope (Fig. 3–15). It stands on Palomar Mountain in Southern California, and has only just lost its title of largest in the world. It is named the Hale telescope after George Ellery Hale, who was responsible for its construction.

This instrument has been known for years as the "200-inch telescope," but the scientific journals are now converting even this sacred name to metric units. We will round off 5.08 meters to 5 meters in this book. (It will not be amiss, I hope, to point out that the mirror is about 17 feet in diameter, the size of an ordinary room.) The hole in the mirror to allow use of the Cassegrain focus is 1 meter (40 inches) across; it blocks only $(1/5)^2 = 4$ per cent of the mirror's overall area. Even when the observing cage, which is 1.8 meters (72 inches) across, is installed at the prime focus, only $(1.8/5)^2 =$ 13 per cent of the incoming light is blocked.

Other telescopes have been built since, but with a single exception they have all been substantially smaller. The largest telescope in the world is a new 6-meter (236-inch) reflector in the Soviet Union, opened in 1976 on Mount Pastukhov in the Caucasus (Fig. 3–17).

Newer materials for telescope mirrors have been developed in recent years that are even less sensitive to heat variations than Pyrex. They include fused quartz and ceramic materials such as Cer-Vit and Ultra-Low Expansion (U.L.E.). The mirrors of newer telescopes for ground-based

A

Figure 3–17 (*A*) The Soviet 6-meter (236-inch) telescope on Mount Pastukhov in the Caucasus. *More views of this telescope appear on the following page.*

B

Figure 3–17 continued *(B)* The tube of the 6-meter telescope before installation in the dome. *(C)* The mirror blank. The completed mirror does not give images of sufficiently high quality, and will be replaced.

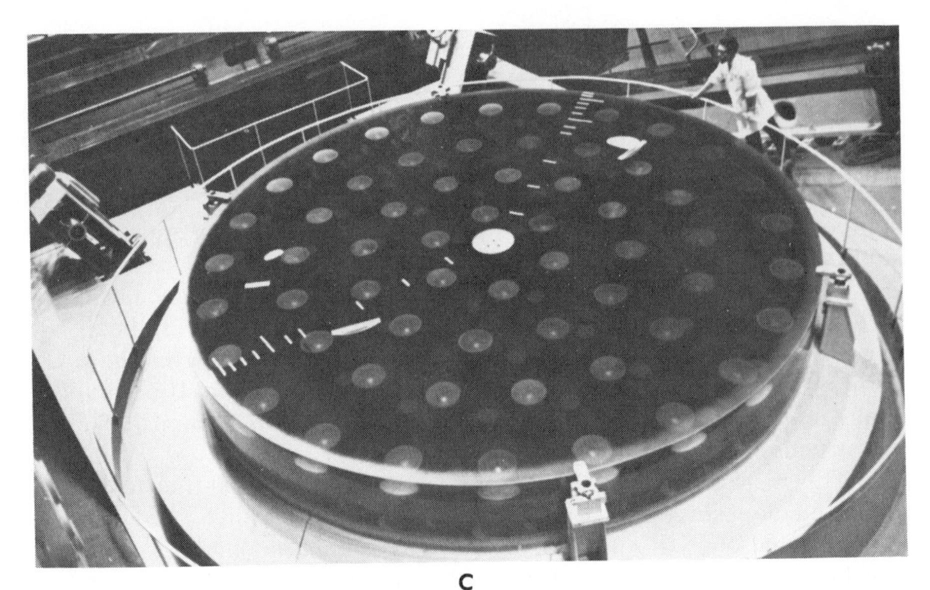

C

observatories as well as for space vehicles are made of these materials. The material of the mirror blank is itself not very reflective. The actual reflecting surface for visible light is ordinarily an ultra-thin layer of aluminum deposited on top of the mirror blank in a process called aluminizing. (To reflect ultraviolet light, mirrors are coated with gold.) A mirror usually has to be re-aluminized at intervals of approximately 5 years.

When we want to launch a telescope into space, other properties must be added to the list of desired properties. Lightness without sacrifice of strength is important, for example. Thus in recent years the ability to make small mirrors out of beryllium has been developed.

Up to now, the largest astronomical telescope that has been sent into space has been the 0.9-meter (36-inch) telescope aboard the *Copernicus* satellite, the third Orbiting Astronomical Observatory launched by NASA. Currently the Space Telescope (ST) with a mirror 2.4 meters (95 inches) in diameter is on the drawing board for possible launch in 1981 to concentrate on visible and ultraviolet studies (Fig. 3–18). Free of the twinkling effects caused by turbulence in the earth's atmosphere, the ST would be able to

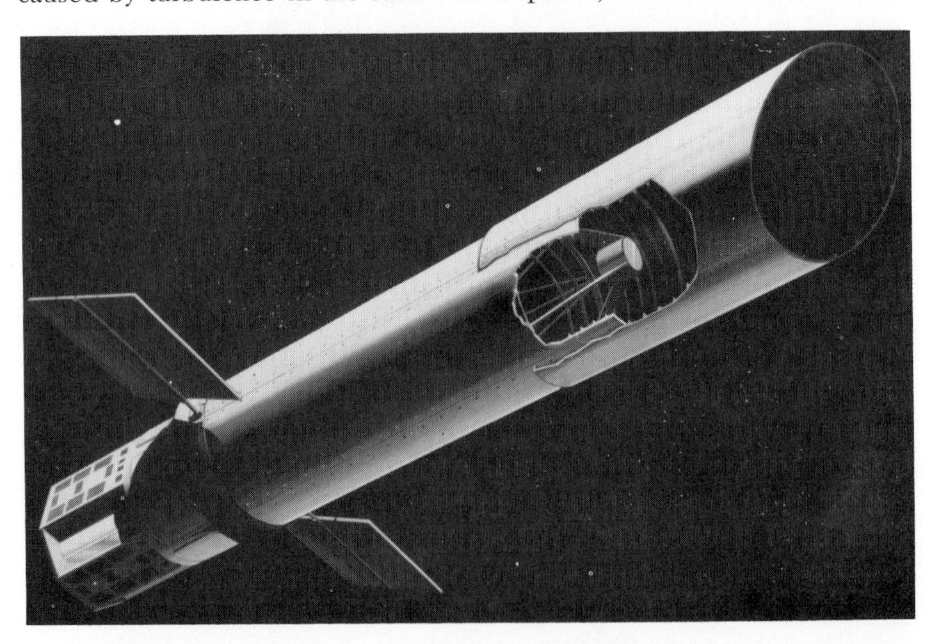

Figure 3–18 An artist's conception of the Space Telescope, with a cutaway showing the secondary mirror and its support.

observe finer details of objects in space than are visible with earth-based telescopes. Also, because it would not be limited by the brightness of the earth's atmosphere even above a remote mountain, the ST would be able to observe fainter objects than any of the ground-based equipment. A Large Infrared Telescope (LIRT) with a 2.8-meter mirror is being planned for the European Spacelab to be launched with NASA's Space Shuttle.

3.4 OPTICAL OBSERVATORIES

Once upon a time, a hundred or so years ago, telescopes were put up wherever astronomers happened to be located. Thus all across the country and all around the world, on college campuses and near cities we find old observatories. Percival Lowell, who located an observatory in Arizona in 1894, and Hale, with his decision ten years later that the clear and dark skies of Southern California were an all-important consideration, were major forces in beginning the construction of observatories at remote sites chosen for their favorable observing characteristics rather than for their proximity to a home campus. We have already discussed the reflectors for which Hale was responsible: the 1.5-meter and 2.5-meter on Mount Wilson and the 5-meter on Palomar. Mount Wilson and Palomar are now jointly operated, and called the Hale Observatories, with offices in Pasadena. Technically, the Mount Wilson Observatory is owned by the Carnegie Institution of Washington and the Palomar Observatory is owned by the California Institute of Technology. The Hale telescope on Palomar is able to undertake useful observing on about 320 nights per year. The second largest telescope in the country has, until recently, been the 3-meter (120-inch) reflector at the Lick Observatory in California. The telescope, on Mt. Hamilton, is used by astronomers of several campuses of the University of California. The headquarters of the Lick Observatory are on the campus at Santa Cruz.

The National Science Foundation has sponsored and built a United States National Observatory. Its telescopes are located on Kitt Peak, a sacred mountain of the Papago Indians, about 80 km (50 miles) southwest of Tucson, Arizona (Fig. 3–19). The headquarters of the Kitt Peak National

Figure 3–19 An aerial view of Kitt Peak. The 4-meter telescope is at the right; the solar telescope is the sloping structure at the lower left. The white dome at extreme left is the 11-meter radio telescope of the National Radio Astronomy Observatory.

Observatory are in Tucson, where the astronomers on the staff and the visiting astronomers work, except during the specific periods when they are observing with one of the telescopes.

Kitt Peak (as the National Observatory is usually called) has been set up for use by all the astronomers in the United States, with the idea in mind that it is efficient to put big telescopes at a site where the sky is clear as many nights as possible during the year, and where the atmosphere above is generally stable so that the images formed are steady and sharp. Any astronomer can apply for observing time at one of the telescopes at Kitt Peak. It is much cheaper in the long run for scientific and governmental organizations to pay for plane fares for their staffs to travel in order to use telescopes in common than it is to duplicate expensive major facilities in places of less-than-ideal climate. In fact, to equalize the expenses regardless of institutional affiliation, Kitt Peak pays the fare of the astronomers awarded observing time. A 2.1-meter (84-inch) telescope had been the workhorse of the Observatory; a 4-meter (158-inch) telescope (shown at the opening of this chapter and in Color Plate 1) was opened there in 1973. This latter telescope is named the Mayall telescope in honor of a former director of the Observatory.

Part of the sky is never visible from sites in the northern hemisphere, and many of the most interesting astronomical objects in the sky are only visible or best visible from the southern hemisphere. Therefore, considerable effort has gone into finding suitable southern hemisphere sites: telescopes in Peru, South Africa and, more recently, Australia have provided very important scientific data. In the last few years a special emphasis has been placed on constructing telescopes at southern hemisphere sites.

Many of the new sites are on coastal mountains west of the Andes range in Chile. The Kitt Peak staff has supervised the construction of the Cerro Tololo Inter-American Observatory there (Figs. 3–20 and 3–21), at which a

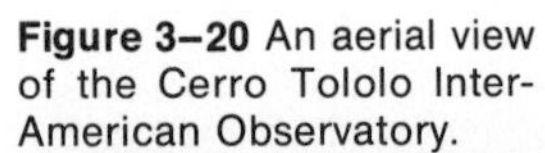

Figure 3–20 An aerial view of the Cerro Tololo Inter-American Observatory.

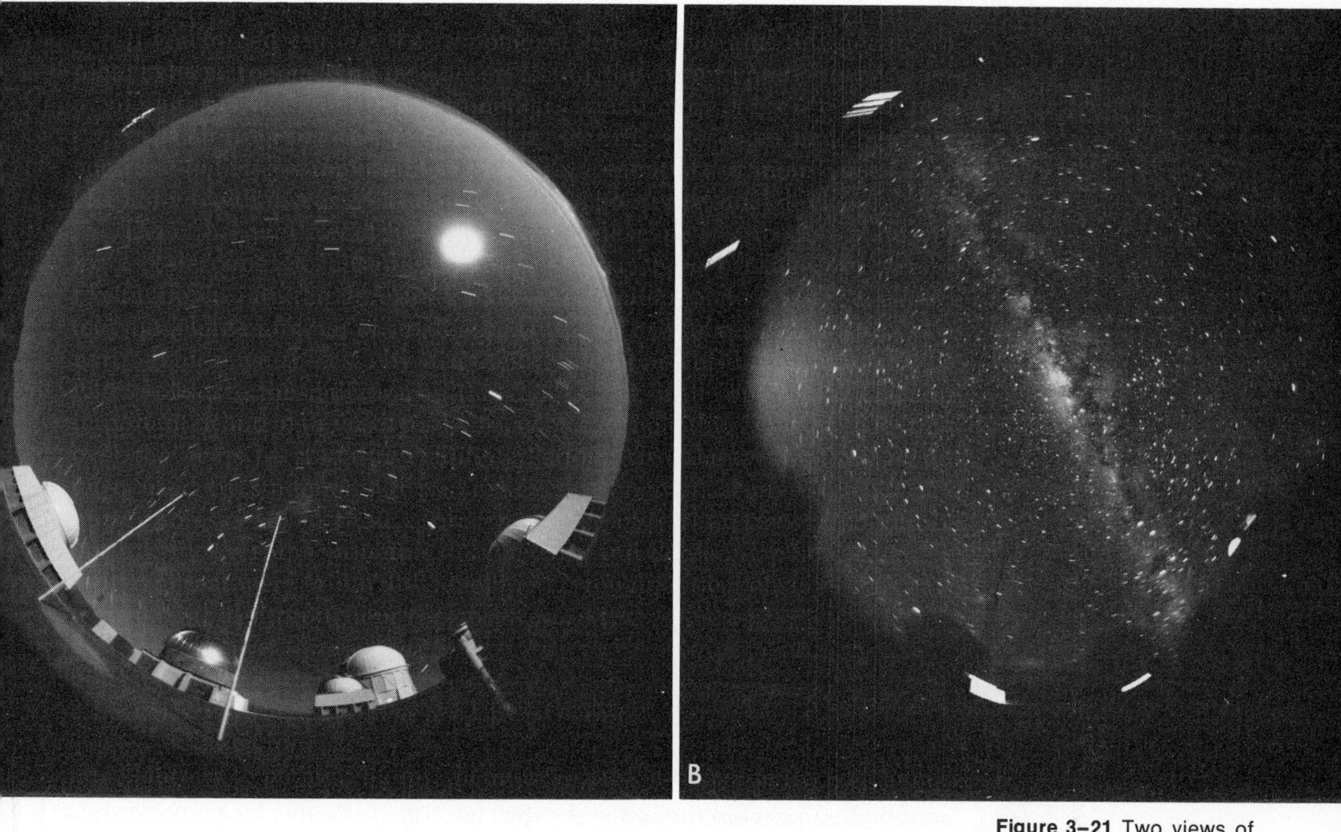

Figure 3–21 Two views of the Cerro Tololo Inter-American Observatory taken from the same location with a fish-eye lens. *(A)* The camera was held fixed, and star trails and the overexposed image of Venus show above the domes. *(B)* The camera tracked the stars; the domes are blurred but the Milky Way shows.

4-meter (158-inch) telescope has been constructed. The staff of the Hale Observatories has supervised, for the Carnegie Institution of Washington, the construction of a 2.6-meter (101-inch) telescope on Cerro las Campanas, another Chilean peak. A consortium of European observatories and universities run the European Southern Observatory, where a 3.6-meter (142-inch) telescope is being erected on still another Chilean mountain top. It is quite common in the astronomical community to go off to Chile for a while to "observe."

Another important new site is in Hawaii. Already a 2.2-meter (88-inch telescope has been installed on Mauna Kea, on the island of Hawaii. A Franco-Canadian-Hawaiian group is building a 3.6-meter (144-inch) telescope for a site there. This site is at an especially high altitude, 4145 m (13,600 ft), and the air above is relatively dry. Since water vapor blocks infrared radiation, these telescopes are especially suitable for infrared observations. A 3.2-meter (126-inch) telescope being built by NASA for Mauna Kea will concentrate on infrared work. The first mirror blank unfortunately developed a crack while it was being worked on, and a substitute blank had to be obtained, thus causing a delay.

It is not useful to get too exact in specifying the diameters of telescopes. The outer 4 cm (almost 2 in) of the 4-m mirrors are beveled, thus limiting the usable area to 3.96 m (156 in). Further, bits of the outermost part of the Kitt Peak mirror do not match the desired shape as accurately as the rest of the mirror; the outermost 19 cm (8 in) are masked off, leaving a usable area of 3.81 m (150 in). The Siding Spring telescope, with 3.89 m (153 in) of usable aperture, is actually intermediate in size between the Kitt Peak and Cerro Tololo telescopes. We shall refer to them as "4-meter telescopes."

The Australian government has long supported astronomical research. Australia remains at the forefront with an Anglo-Australian telescope of 3.9 meters (153 inches), which was opened at Siding Spring in eastern Australia in 1974. It is a joint project with Great Britain.

Other large telescopes exist at such locations as the Haute Provence Observatory in southeastern France, the Pic du Midi Observatory in the Pyrenees in southwestern France, the Royal Greenwich Observatory in southern England, the Crimean Astrophysical Observatory in the Soviet

Figure 3–22 The Paris Observatory has most of its facilities in Meudon, a suburb just to the south of the city. Few stellar observational programs are carried out from this site, though hundreds of astronomers work here.

Union, the Ondrejov Observatory in Czechoslovakia, the Purple Mountain Observatory near Peking, China, a new Spanish-German observing station in southeastern Spain, and the McDonald Observatory in Texas. The British are planning to build a 4.2-m telescope in the Canary Islands, on what they hope is an extinct volcano, and may move their 2.5-m telescope there.

The continued building of optical observatories, with the current unprecedented spate of large telescopes, shows the vitality of this type of astronomical research, even though new techniques involving space telescopes, radio astronomy, and other new technologies are now providing important data at an increasing rate.

This brief discussion mentions only optical observatories, and not the many telescopes that are now in existence on the ground and in space to observe other regions of the spectrum. Nor, perhaps, does it make it sufficiently clear that the current definition of observatory is no longer the place where the telescopes are located but rather the place where the astronomers are. Most of an astronomer's work is done by studying the data that have been recorded on film, on computer tape or otherwise, during a field trip to an observatory. Other astronomers, theoreticians, carry out their research without ever using a telescope. So it is still possible to have major observatories located in urban areas—the Center for Astrophysics of the Harvard College Observatory and the Smithsonian Astrophysical Observatory (in Cambridge, Massachusetts), and the Observatoire de Paris (in Meudon, France), shown in Figure 3–22, are among many examples. The only thing that you can't do very well at such observatories is observe.

Other laboratories have not been based on observational projects, but nonetheless provide the basic physical and spectroscopic measurements or computer facilities that are important for astrophysical research. They include the Joint Institute for Laboratory Astrophysics in Boulder, Colorado, laboratories at Berkeley, Caltech, and other universities, the Herzberg Institute in Ottawa, and many centers in Europe.

3.5 LIGHT-GATHERING POWER AND MAGNIFICATION

The principal function of large telescopes is to gather light. The amount of light collected by a telescope mirror is proportional to the area of the mirror, πr^2. (It is often convenient to work with the diameter instead of the radius, since the diameter is the number usually mentioned. The formula for the area of a circle may, of course, also be written as $\pi d^2/4$.)

The ratio of the areas of two telescopes is thus the ratio of the square of their diameters (d_1^2/d_2^2), which is usually more easily calculated as the square of the ratio of their diameters $(d_1/d_2)^2$. For example, the 200-inch telescope has an area 4 times greater than the area of the 100-inch telescope, $(200/100)^2$ (Fig. 3–23). In the metric system $(5/2.5)^2$ gives the same result. Thus during observations of equal duration the 200-inch would collect four times as much light as the 100-inch if all other things were equal, and for the same exposure time would record stars 1 or 2 magnitudes fainter.

Figure 3–23 A telescope twice the diameter of another has four times the collecting area.

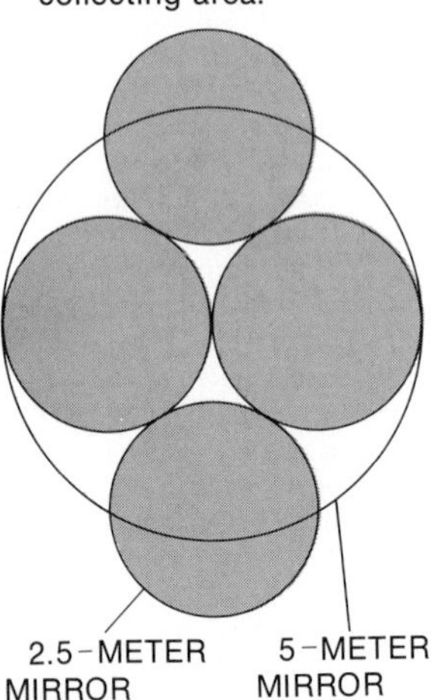

Even though the light-gathering power increases rapidly with the size of the telescope, the cost goes up even more rapidly, perhaps even with the cube of the diameter. Thus the funds have simply not been available in the United States to build a larger telescope than the 5-meter (200-inch), or even a duplicate of the overworked 5-meter.

Alternative methods of getting more aperture are now under investigation. The Harvard-Smithsonian Center for Astrophysics and the University of Arizona are now developing a multi-mirror telescope (MMT) (Fig. 3–24), which has six 1.8-meter (72-inch) paraboloids linked together and aligned by lasers, in the hope that they can succeed in focusing all the light from these several mirrors at the same point. It is much cheaper to build several 1.8 meter telescopes than it is to build one larger telescope of 4.5 meters (176 inches), the equivalent total aperture. The MMT is mounted on Mount Hopkins, south of Tucson, Arizona. A similar multi-mirror project is under way at the Pic du Midi Observatory in the Pyrenee Mountain range in France.

Figure 3–24 A model of the Multi-Mirror Telescope being installed on Mount Hopkins in Arizona.

A telescope's light-gathering power (and its ability to concentrate it in as small an area as possible) is really all that is important for the study of individual stars, since the stars are so far away that no details can be discerned no matter how much magnification is used. The main limitation on the smallest image size that can be detected is caused not in the telescope itself, but rather by turbulence in the earth's atmosphere. The limitations are connected with the twinkling effect of stars. The steadiness of the earth's atmosphere is called, technically, the *seeing*. We would say, when the atmosphere is steady, "The seeing is good tonight." "How was the seeing?" is a polite question to ask an astronomer about his most recent observing run. Another factor of importance is the *transparency*, how clear the sky is. It is quite possible for the transparency to be good but the seeing very bad, or for the transparency to be bad (a hazy sky, for example), but the seeing excellent.

For some purposes, including the study of the planets and the resolution of double stars (Section 4.1), the magnification provided by the telescope does play a role. Magnification can be important for the study of the sun, as well, and for extended objects (as distinguished from point objects) like the nebulae or galaxies.

Magnification is equal to the focal length of the telescope objective (the primary mirror or lens) divided by the focal length of the eyepiece. Thus the magnification of a telescope is only partly determined by the major part of the telescope; by simply substituting eyepieces of different focal lengths one can get different magnifications. However, there comes a point where even though using an eyepiece of short focal length will enlarge the image, it will not give you any more visible detail; when using a telescope one should not exceed the maximum useful magnification, which depends on such factors as the diameter of the objective and the quality of the seeing.

A telescope of one meter focal length will, then, give a magnification of 25 when used with a 40-millimeter eyepiece, an ordinary combination for amateur observing. If we substitute a 10-millimeter eyepiece, we will have a magnification of 100. We would say that we were observing with "100 power."

The 5-meter Hale telescope on Palomar, on the other hand, has several different focal lengths, depending on whether the Newtonian, prime, coudé, or other foci are used. The prime focus, for example, has a focal length of approximately 17 meters (55 feet). When the secondary mirrors of the 5-meter telescope are in place, they modify the focal length to 81 meters (270 feet) or to 150 meters (500 feet). An ordinary camera for snapshots has a 50 mm lens; imagine the Hale telescope as a 150,000 mm telephoto! A 100-millimeter eyepiece would thus give a magnification of 1500. Normally, of course, observational astronomy does not consist of look-

Figure 3–25 Edwin P. Hubble (see Chapter 23) at the finder of the 1.2-meter (48-inch) Schmidt camera on Palomar Mountain.

ing through a telescope eyepiece with the naked eye. One would have a photographic film or other device to record the radiation.

One of the first shocks in studying astronomy is to realize that astronomers don't spend all their time looking through eyepieces. One often then concludes that astronomers hardly ever look through the eyepiece of a giant telescope like the 5-meter. Yet that also turns out to be incorrect. One must check the alignment of the telescope to see that the image is falling on one's apparatus, and that is often done by looking through the eyepiece. Moreover, one often has to make slight corrections to the *guiding* of the giant telescope during the course of a long exposure. (The telescope tracks at a constant rate to follow the average motion of the stars in the sky, but such effects as the refraction by the earth's atmosphere prevent the procedure from being entirely automatic.) So I was very pleased to discover, when I first used the 5-meter, that I actually had to spend a large fraction of the time looking through the eyepiece at the image of the star as it fell on the entrance to the spectrograph.

3.6 OTHER TYPES OF OPTICAL TELESCOPES

A parabolic reflector gives a good focus only for light that strikes the mirror parallel to the axis of the paraboloid. Light at an angle to the axis is not focused very well. The fact that such reflectors have only a very small field in good focus limits their usefulness in certain cases. Though modern computer methods of lens design have led to advances, the fields of view of even the new 4-meter telescopes at Kitt Peak and Cerro Tololo are only 52 arc min. Refractors focus light over a much wider field.

Bernhard Schmidt, working in Germany in 1930, invented a type of telescope that combines some of the best features of both refractors and reflectors. In a *Schmidt camera* (Fig. 3–25), the main optical element is a large mirror, but it is spherical instead of being paraboloidal. Before reaching the mirror, the light passes through a thin lens, called a *correcting plate,* that distorts the incoming light in just the way that is necessary to have the spherical mirror focus it over a wide field. The image falls at a location where the eye cannot be put, and thus the image is always recorded on film (Fig. 3–26). Accordingly, this device is often called a Schmidt **camera** instead of a Schmidt telescope.

One of the largest Schmidt cameras in the world is on Palomar Mountain, in a dome near that of the 5-meter telescope. The Schmidt camera has a correcting plate 1.2 meters (48 inches) across, and a spherical mirror 1.8 meters (72 inches) in diameter. Its field of view is about 7° across, much broader than the field of view, only about 2 minutes of arc across, of the 5-meter reflector.

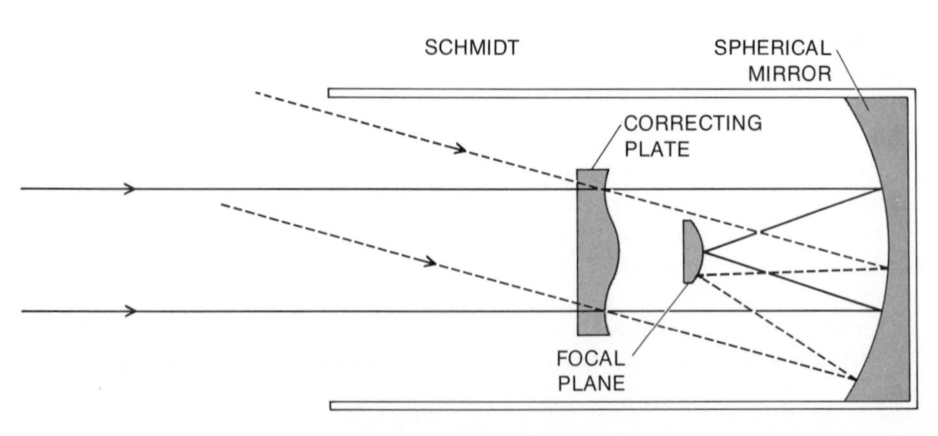

Figure 3–26 By having a non-spherical thin lens called a correcting plate, a Schmidt camera is able to focus a wide angle of sky onto a curved piece of film.

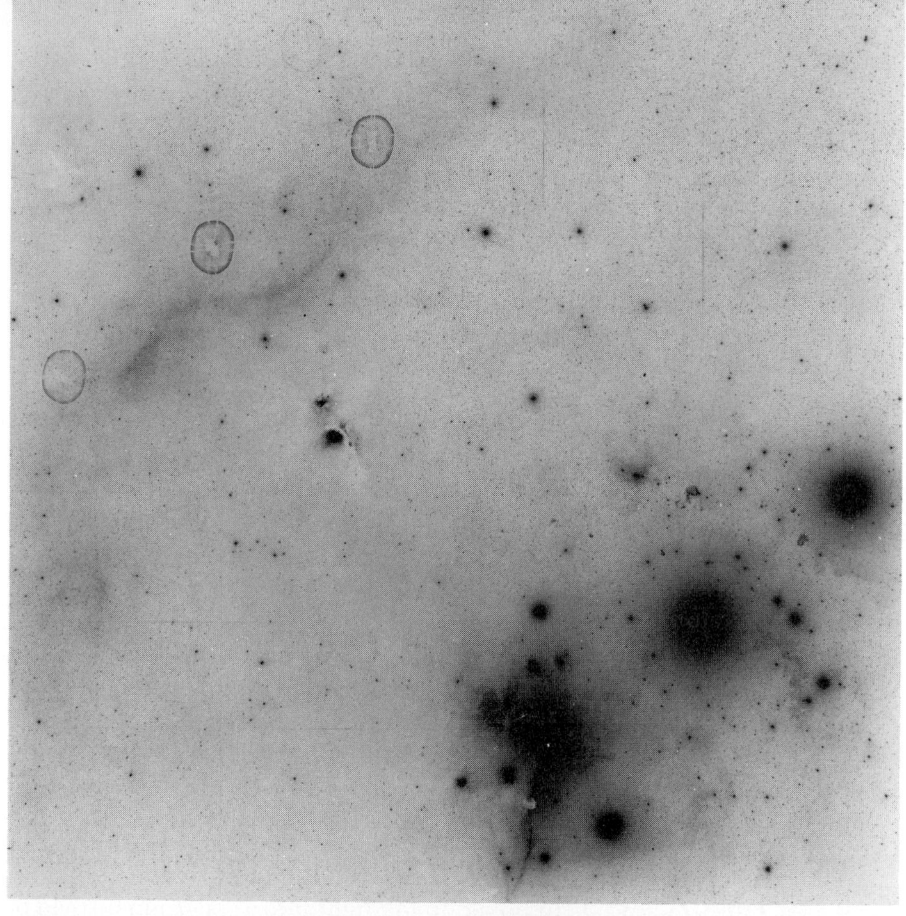

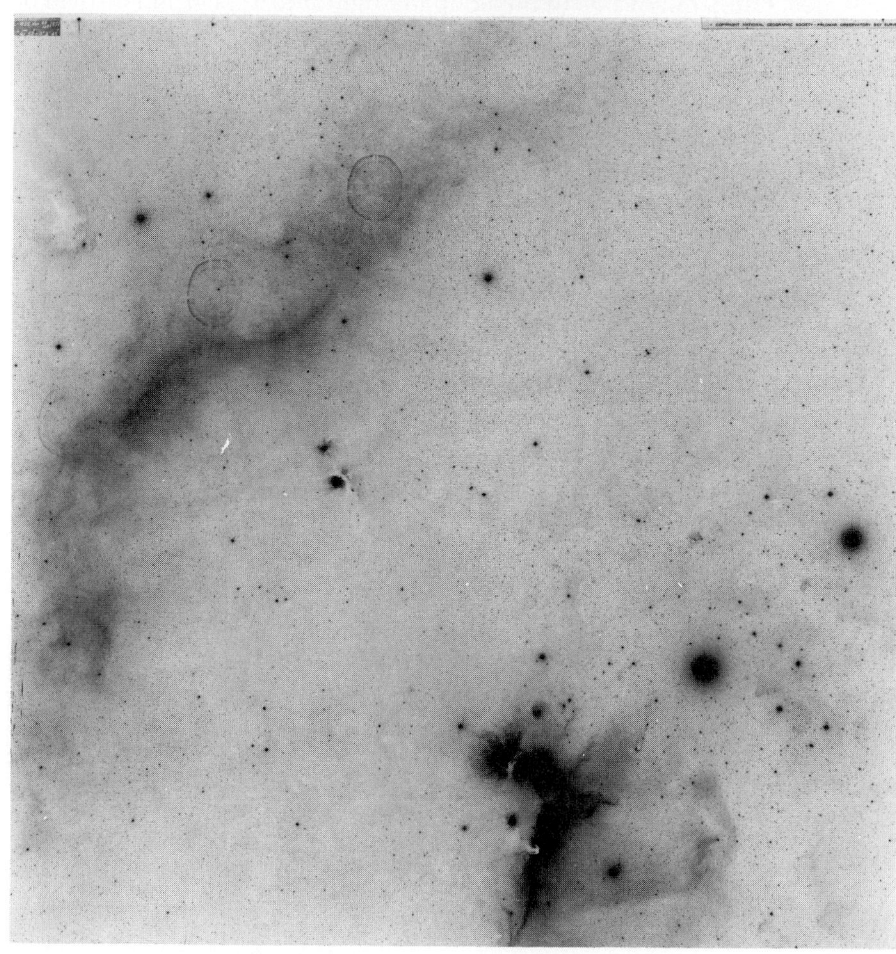

Figure 3–27 A 7° × 7° region of the sky in Orion taken with the 1.2-meter Schmidt camera as part of the National Geographic Society-Palomar Observatory Sky Survey. The faintest stars detectable are about 21st magnitude. The top photograph shows the image with a blue-sensitive plate, and the bottom photograph shows the image with a red filter and a red-sensitive plate. The print is negative, so brighter areas appear blacker. The three bright stars at the lower right are Orion's belt; they are B stars and are brighter in the blue than in the red. The bright region to their left radiates mostly H alpha, and so is brighter in the red than in the blue. The dust lanes, on the other hand, are more prominent in the blue. The Horsehead Nebula (also shown in Fig. 20–6 and in Color Plate 41) is a region of dust obscuring part of the H alpha emission that runs south of this region. Near it is a small nebula that is brighter in the blue than in the red, which leads us to conclude that we are seeing reflected starlight rather than H alpha emission.

To the upper left is the brightest section of Barnard's ring, a ring of H alpha emitting gas that surrounds most of Orion. To the lower left center a large dust cloud causes the number of stars in this region to be less than in the upper right or lower left. This is best seen on the blue print. The three blotches at the upper left are artifacts called "ghost images" caused by internal reflections in the telescope of the three brighest stars.

The field of view of the Schmidt is $(7^\circ \times 60\ \text{min}/^\circ \div 2\ \text{min})^2 = (7 \times 60/2)^2 =$ 40,000 times bigger than that of the 5-meter. So a mapping project that took 2.5 years with the Schmidt would take 100,000 years with the Hale telescope, if they exposed stars at the same rate. Even though the 5-meter can expose stars of a certain brightness $(5/1.2)^2 = 16$ times faster than the Schmidt, it would take 6000 years to make the map, too long to be practicable.

The field of view of the 5-meter is so small that it would be hopeless to try to map the whole sky with it because it would take so long. But the 1.2-meter Schmidt has been used to map the entire sky that is visible from Palomar. The survey was carried out through both red and blue filters. Sample plates from this National Geographic Society/Palomar Observatory Sky Survey are shown in Figure 3–27.

The new 1-meter Schmidt camera at the European Southern Observatory and the British 1.2-meter Schmidt camera at Siding Spring, Australia, are now being used in a joint project to extend the survey to incorporate the one-quarter of the sky that cannot be seen from Palomar.

The largest Schmidt camera in the world is at the Tautenberg Observatory, in Germany. It has a 1.4-meter (54-inch) correcting plate and a 2-meter (80-inch) spherical mirror.

Many small telescopes now commercially available are of the Maksutov design, which is similar to a Schmidt camera in that it has both a spherical mirror and a correcting plate, but the light bounces back and forth inside until it exits through a hole in the main mirror. This design allows a telescope to be made very compact and therefore portable.

3.7 SPECTROSCOPY

We have discussed the collection of quantities of light by the use of large mirrors or lenses, and the focusing of this light to a point or suitably small area. Often one wants to break up the light into a spectrum instead of merely photographing the area of sky at which the telescope is pointed. This allows us to study the spectral lines that display the excitation and ionization of atoms in astronomical sources (and thus determine the temperatures of those sources), to measure Doppler shifts, and so forth.

A prism breaks up light into its spectrum for the same reasons that light focused by a lens has chromatic aberration. A wavefront is refracted as it hits the side of the prism, but it is refracted by different amounts at different wavelengths.

When a beam of parallel light falls directly on a prism, the spectral lines that result are often indistinct because the spectrum from one place in the beam overlaps the spectrum from an adjacent spot in the beam. We can limit the blurring of spectral lines that results because of this by allowing only the light that passes through a long, thin opening called a *slit* to fall on the prism (Fig. 3–28). Then the spectral lines are sharp and relatively narrow.

Actually, astronomical spectra are no longer usually made by means of prisms. Light may be broken up into a spectrum without need for a prism if

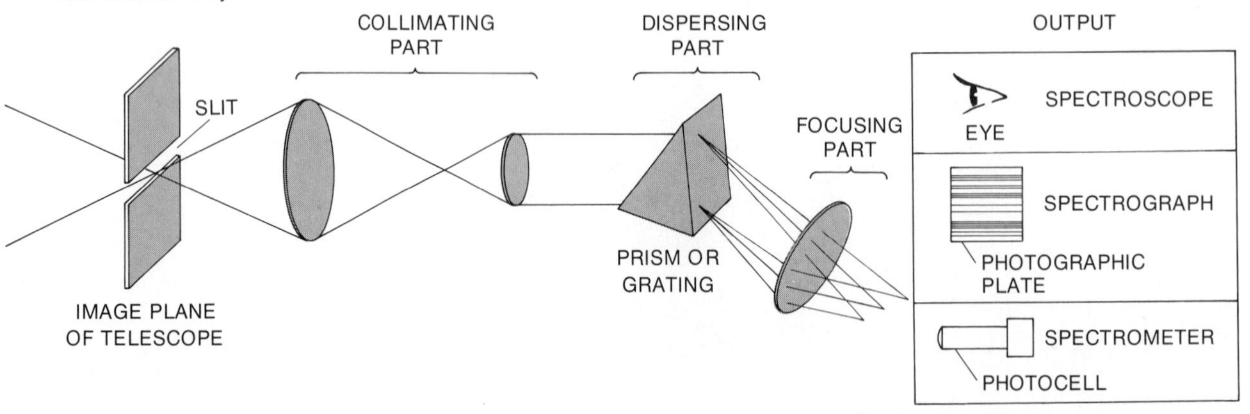

Figure 3–28 The light entering the instrument from the left passes through a slit that defines a line of radiation, is made parallel and directed on the prism (or other dispersing element), and then is focused. The device is called a *spectroscope* if we observe with the eye, a *spectrograph* if we record the data on a photographic plate, and often a *spectrometer* if we record the data electronically.

lines are ruled on a surface very close together. And I do mean close together: one can have 10,000 lines ruled in a given centimeter (25,000 lines in a given inch)! Such a ruled surface is called a *diffraction grating.* If the lines are ruled on a reflecting surface, we have a reflection diffraction grating or a *reflection grating.* Diffraction gratings can also be ruled on transparent surfaces; we then speak of a transmission diffraction grating or a *transmission grating.* The ruling of diffraction gratings is a difficult art.

One can arrange a small telescope so as to examine the spectrum that comes off a prism or grating (Fig. 3–28). A device that makes a spectrum, usually including everything between the slit at one end and the viewing eyepiece at the other, is called a *spectroscope.* When the resultant spectrum is not viewed with the eye but is rather recorded either with a photographic plate or by some other means, the device is called a *spectrograph.* Sometimes when the resultant spectrum is measured with an electronic device, rather than recorded on film, the device is called a *spectrometer.*

3.8 RECORDING THE DATA

When the desired information emerges at the end of the telescope, whether it be an image of a field of stars or a spectrum, astronomers usually want a more permanent record than is afforded by merely observing the image and more accuracy than can be guaranteed in a sketch.

Traditionally, one puts a photographic film or "plate" (short for photographic plate, a layer of light-sensitive material on a sheet of glass) instead of an eyepiece at the observer's end of the telescope, so as to get a permanent record of the image or spectrum.

This use of photographic film also significantly increases our ability to detect signals from faint sources. The eye and brain can process information for only a fraction of a second at a time and do not function cumulatively. A photographic plate, however, can be left exposed to radiation for a long period (Fig. 3–29), sometimes for hours. Just as a long exposure with an ordinary camera can record objects that are only dimly lit, a long exposure with a telescope can record fainter objects than would a shorter exposure.

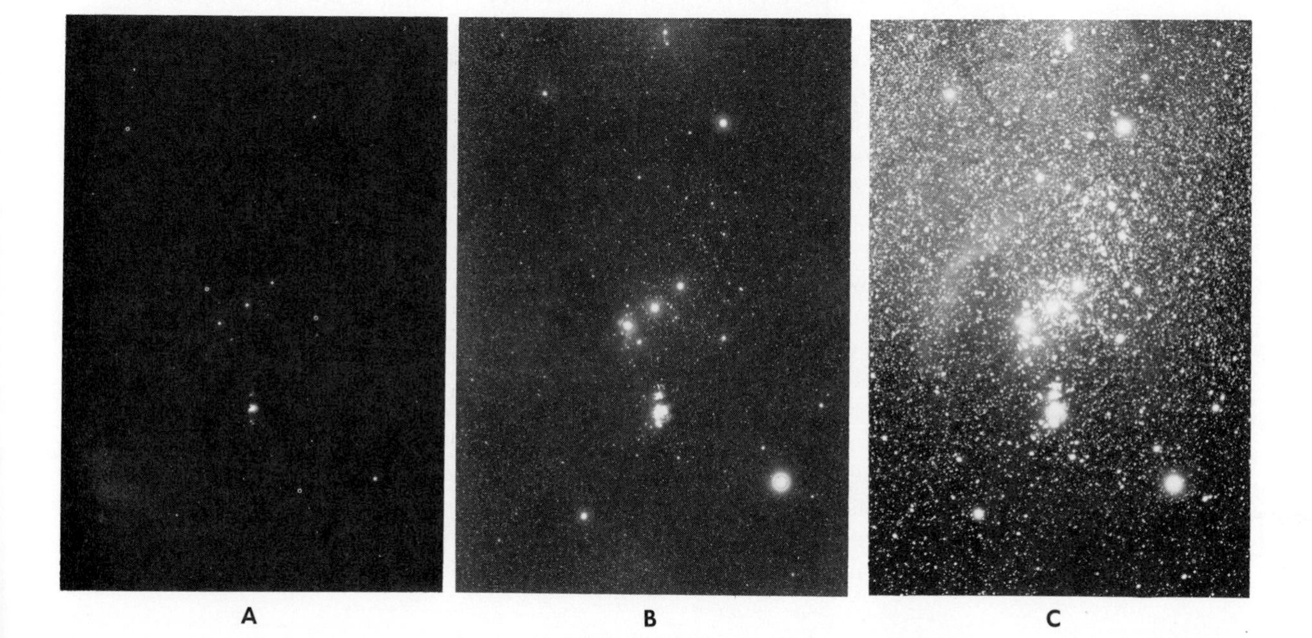

Figure 3–29 These three views of Orion (see also Fig. 8–2) show the effect of increased exposure time. The three stars representing the belt, shown in Fig. 3–27, appear at the center of the photograph. The Orion Nebula (see Color Plate 12) is below them. The red star Betelgeuse, at the upper left, does not appear very bright because these photographs were taken on blue-sensitive plates. The longest exposure shows that nebulosity covers almost the entire constellation.

Unfortunately, there are several limitations in the use of film. One of the limitations is that the density, that is, the darkness, of the film does not vary directly with the intensity of the incoming light. If two objects, one exactly twice as bright as the other, are photographed, the density ratio of the images will not be exactly 2:1. One can take a series of "calibration" test exposures so that one can calculate how the density of the photographic image on the negative varies with brightness, but this is an extra step that must always be carried out and leads to an uncertainty in the accuracy of the measures of brightness.

Another limitation of film is that merely taking a long exposure does not insure that fainter objects will be recorded. Film records images through the interaction of a photon of light with a silver compound in the emulsion. The compound is permanently affected when hit by several photons. Under a very low light level, if a second photon does not follow soon enough after a first photon, the first photon may turn out to have no effect on the silver. Essentially, the silver can "forget" that it was hit by an earlier-arriving photon. Special astronomical films have been developed to minimize this phenomenon.

The use of film has many advantages despite its disadvantages. For one thing, one can record images of everything in a whole field of view simultaneously. Film is also easily portable, and can be developed at a later, more convenient time.

Nonetheless, many other methods of recording data have been developed, and it is now being said that the day of film is past. The future seems to lie in electronic devices.

The photocell is a basic electronic device used by astronomers. Certain materials, when struck by light, give off an electric current. That current can be measured, perhaps by reading a meter or perhaps automatically, with the data recorded on punched cards or on magnetic tape. The readings can even be fed straight into a computer.

A modification of photocells has been developed whereby the electrons that become available inside the tube when light hits it can be directed to hit other special elements of the photocell. Each of the subsequent elements can be made to emit several electrons for each one hitting it. In this way, the original number of electrons can be multiplied, often by a substantial factor (perhaps 10^6), and the device is called a *photomultiplier*.

One advantage of a photocell or photomultiplier is that its response is "linear" up to some bright limit; that is, when twice as much light hits it, it gives off twice as much current. Many of the calibration problems of film are thus avoided, although electronics do have some calibration problems of their own. One of the disadvantages of photocells and photomultipliers is that they measure only the total intensity of radiation falling on them from whatever direction; they do not form an image in the way that film does. So whether photocells and photomultipliers are useful to you depends on exactly what investigation you are undertaking. For measuring the intensity of light of a star, photomultipliers are fine. Such measurements of intensity are called *photometry*.

Figure 3–30 The silicon vidicon spectrometer and data recording system of MIT, shown here, have been used in conjunction with the largest telescopes to study faint objects.

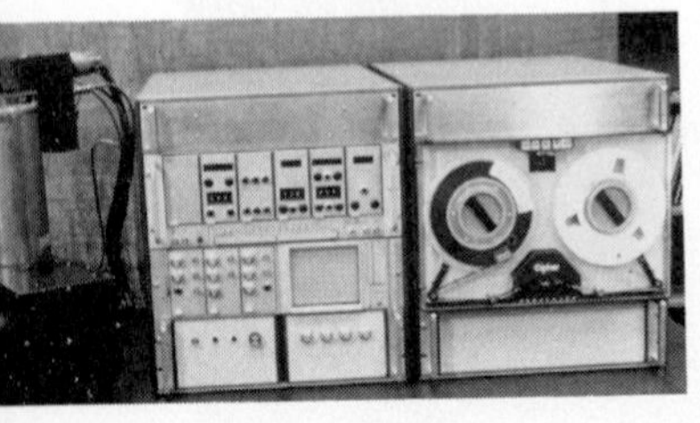

Photomultipliers are much more sensitive to faint signals than is photographic film, and also can be made to be sensitive in a wider region of the spectrum. Photometry has long been carried out with photocells and photomultipliers and can be done very precisely. In some observatories, the procedure is automated so that the signal level from the photomultiplier is monitored by a computer. When enough photons have been gathered to allow a measurement of a certain accuracy (say, to 1 per cent of the signal level), the computer ends the "exposure" and moves the telescope to the next object.

If a photocell is held at a particular location along the spectrum formed by a spectrograph, it will measure the intensity of radiation at that wavelength. By moving the photocell along the spectrum, one can electronically record the spectrum with all the photocell's advantages of high sensitivity and linearity. One disadvantage is that one must make a separate measurement at each wavelength, and it may take as long to move the photocell across a whole spectrum as it would to make a spectrum on film. However, for a short segment of spectrum, the photocell may be much faster.

Some of these problems can be overcome by having a series of photocells at different wavelengths in the spectrum. At Palomar, a multi-channel spectrometer has been built that records radiation from 33 wavelengths simultaneously. Devices like these can decrease the necessary exposure time significantly, and thus make telescopes much more efficient.

One recent device can even count the individual photons that arrive, classifying them into one of 1024 separate wavelength channels that together cover the total wavelength region under observation. It does this with very great speed, as it is a very sensitive instrument. Moreover, it is inherently linear, thus avoiding one of the major disadvantages of photographic plates.

For some years devices called *image tubes* have been used, sometimes to increase the sensitivity of film and sometimes to increase the wavelength coverage. Light from the telescope or spectrograph may fall on the faceplate of a tube and generate internal electrons or photons, but ultimately these secondary emissions still fall on a photographic plate or film. Since the photons generated need not be at the same wavelength as the incident photons, one can, say, have an infrared signal at a wavelength beyond the sensitivity of film fall on the image tube and generate green photons to fall on ordinary film or on a television-type camera.

A new series of devices is now able to provide an electronic imaging capability. The technology for making these is related to the new technology that has recently brought electronic calculators to their present versatility and price range. Among several competing imaging devices are "silicon diode vidicons" (Fig. 3–30), for which a series of small crystals, each only a few microns across, can actually be grown on a thin silicon wafer. Each point on the wafer then acts as though it were a small photocell. These devices can be used in the place of photographic plates to make either direct images of astronomical objects or of spectra. Rapid advances are being made in this field.

Figure 3–31 This enlargement of the spectrum of a distant galaxy shows clumps of grain on the photographic plate.

3.9 OBSERVING IN DIFFERENT SPECTRAL REGIONS

The most familiar detector of electromagnetic radiation is the photographic plate. Basically, a photographic plate (or film) consists of a backing covered with an *emulsion*. The backing may be glass for stability or plastic for flexibility. The emulsion is the working layer. In it are grains of a silver compound. When these grains are struck by photons of radiation, they undergo a chemical change. Then the film is placed in a chemical called the "developer," and the silver in the grains that have been hit by photons is separated out of the compound. The unaffected silver compound is later washed away by another chemical called the "fixer." The silver grains that remain appear black and form the image. This is a "negative" image, with the darkest areas corresponding to the brightest parts of the incident image. Inspection of a photographic plate under high magnification shows the grainy structure of the image (Fig. 3–31).

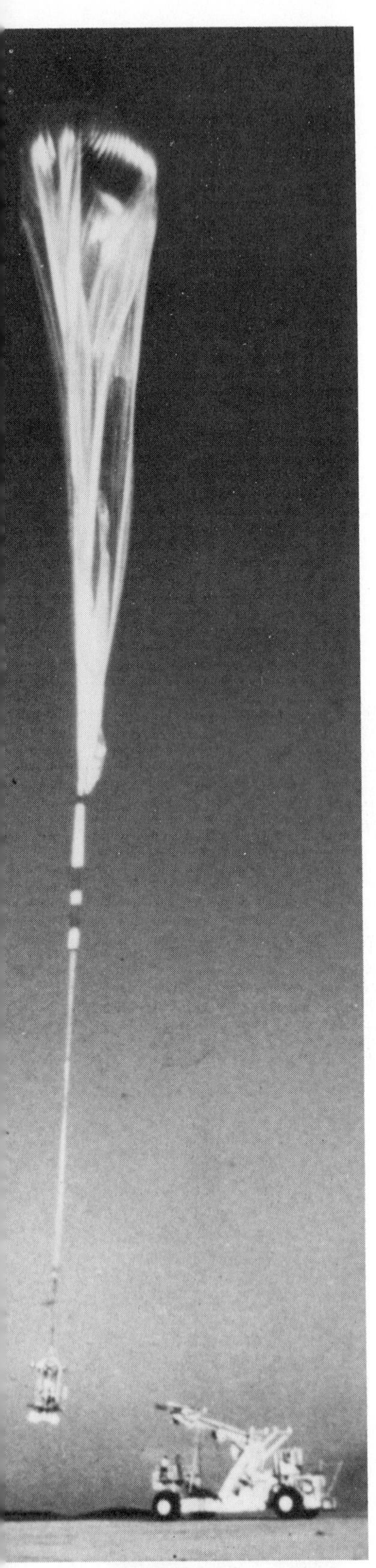
Figure 3–32 A balloon launch carries aloft a 102-cm infrared telescope.

The photons, as we have seen, have energies of hc/λ. Ultraviolet and x-ray photons have shorter wavelengths and thus greater energies than photons of visible light. Thus they also have enough energy to interact with the film grains. Photographic methods can be used, therefore, throughout the x-ray, ultraviolet, and visible parts of the spectrum.

Infrared photons have longer wavelengths and thus lower energies than visible photons. They do not have enough energy to interact with the silver grains on ordinary photographic plates. Some special types of film can be used at the very shortest infrared wavelengths, but for the most part astronomers have to employ other methods of detection in this region of the spectrum.

Devices made of special materials like lead sulfide can be sensitive in the infrared. These devices are used in the same way as photocells are used in other parts of the spectrum.

Another major limitation in the infrared is that there are very few windows of transparency in the earth's atmosphere. Most of the atmospheric absorption in the infrared is caused by water vapor, which is located at lower levels in our atmosphere than is the ozone that causes the absorption in the ultraviolet. Thus we do not have to go as high to observe infrared as we do to observe ultraviolet. It is sufficient to send up instruments attached to huge balloons (Fig. 3–32). The major balloon-launching site in the United States is in Palestine, Texas.

At even longer wavelengths, in the radio region of the spectrum, there is another window of transparency. Photocells do not respond to these photons of very low energy. We must use antenna techniques and then amplify the weak electrical signals generated by the antenna in much the same manner that we use to receive and amplify radio and television signals at home. The techniques of radio astronomy, now well established as one of the major branches of astronomy, will be discussed at length in Part IV.

At the shorter end of the spectrum, gamma rays, x-rays, and ultraviolet light do not come through the earth's atmosphere. Ozone, a molecule of three atoms of oxygen (O_3), is located in a broad layer between about 20 and 40 kilometers in altitude, and prevents all the radiation at wavelengths less than approximately 3000 Å from penetrating. Fears that the earth's ozone layer is in danger will be discussed in Section 13.5.

To get above the ozone layer, astronomers have launched telescopes and other detectors in rockets and in orbiting satellites. Often photocells are used in the orbiting spacecraft, as the signals can then be radioed back without the need to develop film. One of the advantages of the Skylab manned orbital missions of 1973 and 1974 (discussed in Section 5.4b) was that film techniques could be used to record ultraviolet spectra and images since astronauts could return the film to earth. Techniques do exist to return film from unmanned orbiting satellites, but these techniques are more often used for military surveillance reasons than for astronomy.

The ability to use film instead of photocells or television is one reason that rockets that last only a few minutes above the atmosphere and return to earth continue to be used in this age of orbiting satellites. Another reason is that we can accurately calibrate the sensitivity of the equipment both before launch and after recovery, so we can find out exactly how sensitive the instruments were at the moment when observations were made. Even when major experiments are sent up in satellites, a series of calibration rockets may be sent up every few months to compare these calibrated rocket observations with the observations being made from the satellites.

In the short wavelength regions, we cannot merely use parabolic mirrors to image the incident radiation, as the x-rays will pass right through the mirror! Fortunately, x-rays can still be bounced off a surface if they strike the surface at a very low angle. This is called *grazing incidence* (Fig. 3–33). The principle is similar to that of skipping stones across the water. If you throw a stone straight down at a lake surface, the stone will sink immediately. But if you throw a stone out at the surface some distance in front of you, the stone could bounce up and skip along a few times. By carefully choosing a variety of curved surfaces that suitably allow x-radiation to "skip" along, astronomers can now make telescopes that actually make x-ray images. But the appearances of the telescopes are very different from those of optical telescopes (Fig. 3–34).

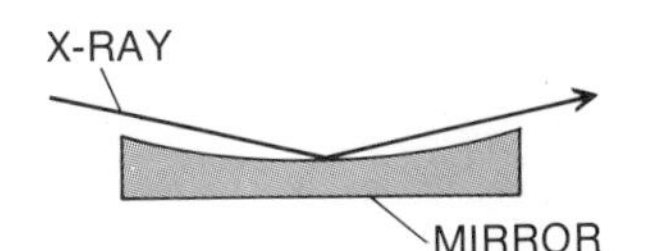

Figure 3–33 Grazing incidence.

In order to form such high-energy photons, processes must be going on out in space that involve energies very much higher than most ordinary processes that go on at the surface layers of stars. The study of the processes that bring photons or particles of matter to high energies is called *high-energy astrophysics*. The astronomical community has placed a priority on work in this field for the next few years, and the next emphasis in space observations will be in x-ray and γ-ray spectral regions that give us the most information about these processes. NASA plans the launch of two spacecraft, HEAO-A (for **H**igh-**E**nergy **A**stronomy **O**bservatory) in 1977 and HEAO-B in 1979, largely to observe in the x-ray and γ-ray spectral regions.

3.10 A NIGHT AT PALOMAR

In some sense, the opportunity to observe with what has been the largest optical telescope in the world, the 5-meter telescope on Palomar Mountain, is the ultimate reward for years of study and research in astronomy. Since Palomar is privately owned, most of the observing time goes to the staff of the Hale Observatories. A dozen individuals on the permanent staff have five nights or as many as 30 nights a year assigned to them. Younger research fellows may have two or three nights each during the years they spend at the Hale Observatories after having received their Ph.D.'s. Aside from the time allotted to the in-house staff, about one-third of the observing time goes to outside applicants from all over the world.

A

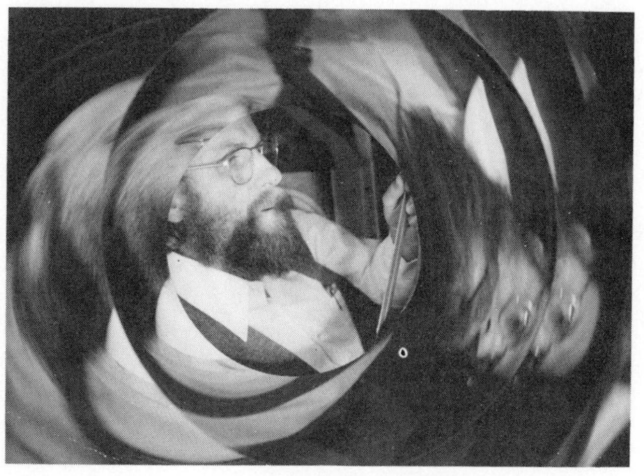

B

Figure 3–34 (*A*) The cylindrical mirror of HEAO-B does not resemble mirrors for optical telescopes. This is the outermost of four concentric mirrors. Skylab astronaut Owen Garriott is on the right in this view. (*B*) Caltech scientist Gordon Garmire looks through the mirror, polished for grazing incidence. The mirror has parabolic and hyperbolic sections, and consists of an aluminum structure coated with nickel. It has been launched in a rocket.

A

B

Figure 3–35 *(A)* An observer riding in an elevator to reach the prime-focus observing position of the 5-m telescope. *(B)* Final shaping of the 5-m mirror with a small polishing machine.

Before using the 5-meter Hale telescope, you must have qualified by acquiring experience at some other smaller telescope, probably a large telescope in its own right such as the 2.5-meter on Mount Wilson. You must then prepare a detailed proposal to a selection committee specifying what it is you want to observe and why you need the 5-meter telescope in particular to observe it. There may be 320 clear nights a year on Palomar Mountain, but many more nights a year would be needed if all the requests were to be granted.

Shortly before your observing time comes, there is a variety of last-minute preparations to make. You might have complicated new equipment to test out, or you might be using one of the standard spectroscopic, photoelectric, or photographic systems permanently installed at the telescope.

Even if you are using the standard equipment, you might have special filters to use. You will want to have prepared a list of the objects that you plan to observe, with an ample number of objects in a second category of priority in case everything goes fantastically well. You will look up in a catalogue the positions of these objects in the sky. But to take account of precession (described in Section 3.11) you have made corrections with a desk calculator or with a small computer program.

Finally, the big day comes and you set off for Palomar. If you are on the staff of the Hale Observatories, your office is in Pasadena, California, and Palomar is about a 2½ hour drive southward and slightly east. If you are at an institution farther away, you would first have to fly to Los Angeles. In any case, you have to have your suitcase packed and be prepared to spend a few days and nights away from home.

As you drive into the mountains from the desert, the foliage changes. By the time you have finished the steep last bit of the climb up Palomar, you find yourself in a pine forest. By noontime of the day you are to start observing, you check into the residence hall, which is called "the Monastery."

Figure 3–36 A moonlight view of the dome of the 5-meter Hale telescope on Palomar Mountain in California.

Meals are served at the Monastery. The noon meal is for the new arrivals and for the daytime workers; the astronomers who worked the preceding night are usually asleep. After lunch, you set off for the telescope.

The 5-meter telescope stands in a huge domed building that is 40 meters (135 ft) high, the equivalent of a 13-story building (Fig. 3–36). Let us say that you are doing photographic spectroscopy. The first thing that you would do is to go to the plate vault, a large walk-in refrigerator where special stocks of photographic emulsions are kept. The emulsions are specially chosen for their astronomical characteristics, and are not the ones that you buy in your corner camera store. In an effort to make the plates as sensitive as possible, astronomers use such methods as baking the plates in a small oven (in the dark, of course) for a few hours. This sometimes enables you to cut your exposure time by 20 per cent or more. It has recently been discovered that baking plates in an airtight container filled with nitrogen increases the sensitivity even more. In addition to preparing your plates, you would also go to the darkroom and mix the photographic chemicals you would need for developing them after they have been exposed.

Then you would make certain that the spectrograph you plan to use is in place. There are so many interchangeable pieces of equipment available for use in studying the light reflected by the 5-meter mirror that you usually have to set up your own equipment.

The rest of the afternoon might well be spent in carefully focusing the spectrograph. To do this you might use an iron arc, an arc of electricity between two iron rods. This gives off the spectrum of iron, which has very many spectral lines across the visible part of the spectrum. You vary slightly the position of optical components in the system, trying different positions, and then developing plate after plate until you are satisfied that the focus can no longer be improved. Besides the fact that you want your images to be as sharp as possible in order to study the shapes or positions of the lines very accurately, you also know that if the spectral lines are out of focus, the few photons of light that you are trying to record at that wavelength are spread over a greater area of photographic plate. Thus a bad focus will increase the exposure time that you must use.

It is finally dinner time, just before dark, and you head back to the Monastery. All the daytime personnel have left the mountain, and it is mostly the observers at this meal. There are several other telescopes on the mountain, including a 1.5-meter reflecting (60-inch) telescope and the 1.2-meter (48-inch) Schmidt telescope, so the "200-inch observer," although seated at the head of the table, is not alone at dinner.

In the twilight you walk back to the telescope. A night assistant is assigned to the telescope at all times and is responsible for all the mechanical aspects of the telescope. The night assistants know the telescope thoroughly, and also know when a strange noise might mean that the telescope must be shut down for repairs, or when high wind or blowing dust mean that the telescope dome must be closed for the safety of the mirror.

You might start first with an object in the eastern part of the sky, for that part of the sky is opposite from the setting sun and gets darker the earliest. When it is really dark, you start any very long exposures you want to make. It is not unusual for a single exposure to last one to three hours, or even all night. On infrequent occasions you might simply close the protective cover on the photographic plate at the end of the night, and resume the exposure for a second night.

The telescope is very well lubricated and very well balanced, so even though it weighs over 500 tons it turns very smoothly and needs only a moderate sized motor. But refraction effects in the earth's atmosphere may cause the light from stars to vary in position at a slightly different rate from the normal average, and you must continually check that the telescope is pointing in absolutely the right direction. This is called guiding. Especially in direct photographic work, where you are making an image, observers who are the most careful with their guiding get the best results.

Figure 3–37 The Hale telescope pointing to the zenith.

The night assistant points the telescope at the object in the sky that you are ready to observe. If it is a reasonably bright object, then you can check very simply to make sure it is the correct one: you need merely to see if the stars around the object in the field of view in the telescope have the same positions relative to each other as the stars on a star chart you have of the area around your object. If you are observing a faint object, on the other hand, you might have to point the telescope a certain fraction of the way between two objects that are bright enough to be visible. You may never see the faint object yourself, though its image will appear on the developed plate. New television-type devices that are sensitive to low levels of light sometimes enable you to see faint objects that could not previously be used for alignment or guiding.

Your exposure time depends on the quality of the seeing that night. If the seeing is very good, then the images of stars will be very small, less than 1 arc second. Most of the light from your star will then pass through the spectrograph slit and be dispersed. If the seeing is bad, however, then the images of stars will be bigger, perhaps 5 seconds of arc. Then your spectrograph slit may be much narrower than the image of the star, and much of the starlight is bounced off the slit and never reaches the emulsion. In that situation your exposures would have to be longer to record the same amount of light.

After an hour or so, your exposure is finished. You cover the slit, and close the film holder. Then you replace the exposed plate with an unexposed plate, and prepare to take another exposure, perhaps of another object. Then you will go to the darkroom and develop your plate, gently rocking it back and forth in the developer for several minutes. The night assistant might be able to help you by guiding on the second star while you are developing the plate of the spectrum of the first star. It is important to develop your first plate right away to make sure that the system is working and that there are no blockages in the light path.

In the course of the night you will probably want to go up to the floor of the dome for a short while just to appreciate the stars and the marvelous monster of an instrument that you are controlling to gather photons from afar.

All too soon the night is over. The dawn is visible in the east, and you must shut down or else the sky light will fog your plate. After developing your last plate (or perhaps leaving it to do at the beginning of the next night), you trudge home to the Monastery to bed.

After a few days on the mountain, you may have a dozen or more plates (Fig. 3–38). Your work has just begun. It may take months or years to analyze the data completely, often using measuring devices and computers at your own institution. The thrill of having yourself captured those few photons from a distant star remains.

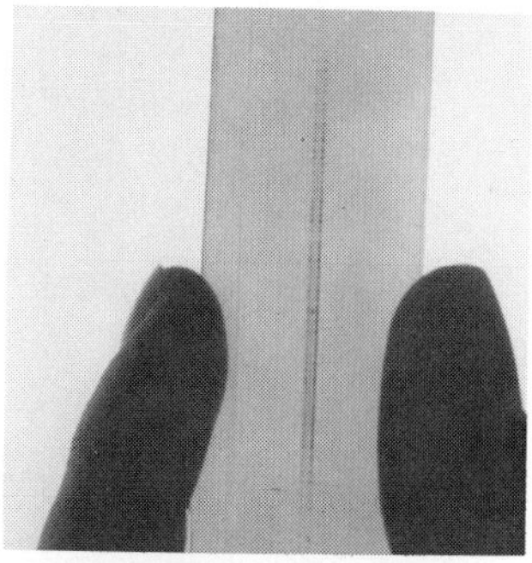

Figure 3–38 The collecting area of the Hale telescope is large, but the light is focused down to a small area on a spectrographic plate. The spectrum shown is that of the variable star R Coronae Borealis, and was taken by the author. Most of the spectrum visible is that of an iron arc operated in the telescope dome to provide known wavelengths for calibration purposes; the spectrum of the star (Fig. 2–31) is the few emission lines between the left and right halves of the calibration spectrum.

3.11 COORDINATE SYSTEMS

Astronomers and geographers share the need to set up systems of coordinates—*coordinate systems*—to designate the positions of places in the sky or on the earth.

The geographers' system is familiar to most of us: longitude and latitude. Lines (actually half-circles) of longitude called *meridians* run from the north pole to the south pole. The zero circle of longitude has been adopted, by international convention, to run through the former site of

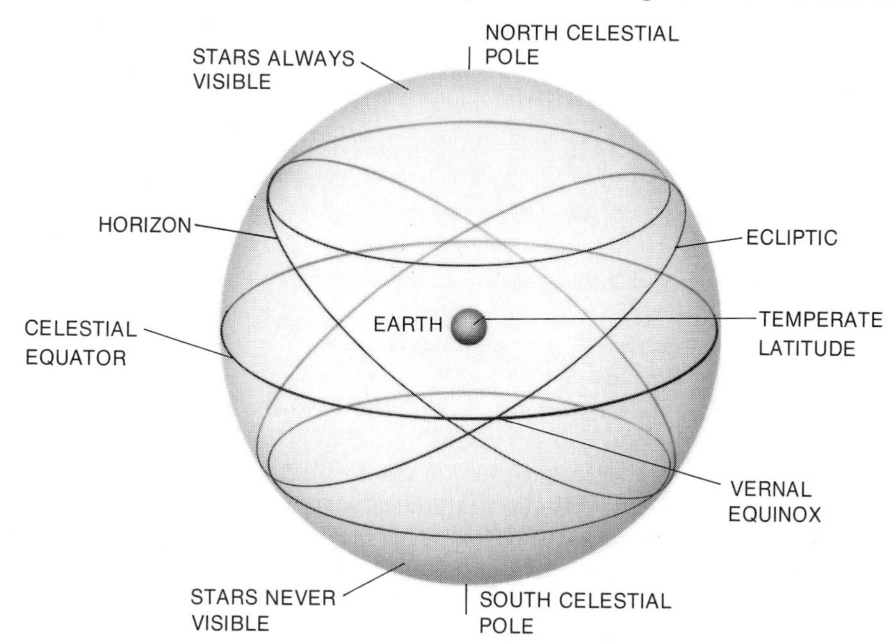

Figure 3–39 The celestial equator is the projection of the earth's equator onto the sky, and the ecliptic is the sun's yearly path through the stars. The vernal equinox is the intersection of the ecliptic and the celestial equator. From a given location at the latitude of the United States, the stars nearest the north pole never set and the stars nearest the south pole never rise above the horizon.

the Royal Greenwich Observatory in England. We measure longitude by the number of degrees east or west an object is from the meridian that passes through Greenwich.

Latitudes are defined by parallel circles that run around the earth, all parallel to the equator. Zero degrees of latitude corresponds to the equator; ninety degrees of latitude corresponds to the poles (90° of north latitude for the north pole and 90° of south latitude for the south pole).

The astronomers' system corresponds exactly to the geographers' system, except that the astronomers use the names *right ascension* for celestial longitude and *declination* for celestial latitude. They are measured with respect to a *celestial equator,* which is the extension of the earth's equator into space, and *celestial poles,* which are on the extensions of the earth's axis of spin into space and thus lie above the earth's poles (Fig. 3–39).

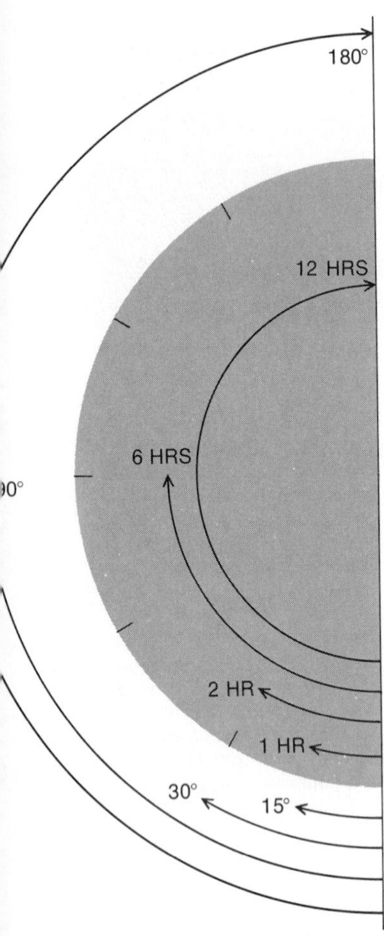

Figure 3–40 24 hours of right ascension make up a full 360° circle, so each hour of right ascension corresponds to 15° of arc.

Right ascension and declination form a coordinate system fixed to the stars. To observers on earth, the stars appear to revolve every 24 hours. The coordinate system thus appears to revolve at the same rate. Actually, of course, the earth is rotating and the stars and celestial coordinate system remain fixed.

One circle of right ascension has been chosen arbitrarily to mark the zero, just as the meridian of longitude that passes through Greenwich is the zero in the terrestrial system. Right ascension is measured eastward from that point. However, instead of measuring the number of degrees either east or west of Greenwich, astronomers measure only in the eastward direction. Moreover, instead of using degrees, they choose to use a system patterned after timekeeping. A full circle around the earth, which corresponds to 360°, is set equal to 24 hours of right ascension. Thus each hour of right ascension is equal to $360°/24 = 15°$ (Fig. 3–40). Each minute of right ascension is equal to $15°/60 = 1°/4 = 15$ minutes of arc. Each second of right ascension is equal to 15 seconds of arc. Because of this factor of 15, one must make clear whether one is talking in the system of right ascension or the system of arc measure, because the terms minutes or seconds could refer to either system.

Although this system of hours, minutes, and seconds of right ascension may seem unwieldy, it does have certain advantages for astronomers. Since the earth rotates in 24 hours, it takes approximately one hour of time (as we normally measure it, defined with respect to the sun and thus called *solar time*) for one hour of right ascension to pass overhead. Astronomers use this to set up a timekeeping system called *sidereal time* (sidereal means "by the stars") (Fig. 3–41).

By definition, it takes one hour of sidereal time for one hour of right ascension to pass overhead.

The sidereal time is the right ascension of the star that is passing directly overhead. Thus each location on the earth has a different sidereal time at each instant. Sidereal time is not divided into standard time zones the way that solar time is.

Astronomers find this system convenient because they can consult clocks that are set to run on sidereal time. Thus by merely knowing the sidereal time, they can tell whether a star is favorably placed for observing. If the right ascension of the star (which they look up in a catalogue) is the same as the sidereal time at that moment, then the star is as high in the sky as it can be. The difference between the right ascension of a star and the current sidereal time is called the star's *hour angle.* Though a star's right ascension doesn't change, its hour angle does.

Example: Let us consider Sirius, which has right ascension 6^h43^m (we can look this up in a table). When the sidereal time is 6^h43^m, Sirius will be crossing the *meridian,* the line across the sky from north to south that passes through the *zenith,* the point in the sky directly overhead. At sidereal

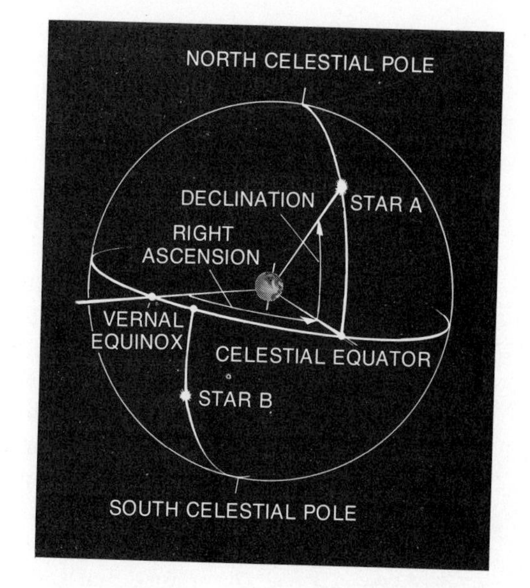

Figure 3–41 Right ascension is measured in hours of time (1 hour of time = 15° of arc) from the vernal equinox. Declination is measured in degrees north (+) or south (−) from the celestial equator. The sidereal time at any point on earth is the right ascension of a star that is crossing the meridian (the line extending from north to south and passing through the zenith). Thus sidereal time at the point on earth at which star A is due north or south is the same as the right ascension of star A (the declination doesn't matter). The hour angle is the number of hours it will take a star to reach the meridian (labeled minus) or the number of hours since a star has passed the meridian (labeled plus). Thus to find the hour angle of star B, as seen from the point at which A is on the meridian, subtract the right ascension of star B from the sidereal time.

time 5^h43^m, one hour before Sirius crosses the meridian (i.e., *transits*), the hour angle would be -1^h. At sidereal time 7^h43^m, one hour after Sirius transits, the hour angle would be $+1^h$.

Sidereal time and solar time are not exactly the same. They are both caused by the rotation of the earth on its axis. A *sidereal day* is the length of time that a given star in the sky takes to return to the same position in the sky. A *solar day* is the length of time that the sun takes to return to the same position in the sky. Since the earth is revolving around the sun every year, by the time a day has gone by, the earth has moved 1/365 of the way around the sun. Thus after the stars have returned to their same positions in the sky, it takes 1/365 of approximately 24 hours = 4 minutes for the sun to return to its position. A solar day is thus approximately 4 minutes (actually 3 minutes 56 seconds of time) longer than a sidereal day.

Every observatory has both sidereal clocks, for the astronomers to tell when to observe their stars, and solar clocks, for the astronomers to gauge when sunrise will come and to know when to go to dinner. A solar clock and a sidereal clock show the same time (on a 24-hour system) on only one day each year. The next day the sidereal clock is 4 minutes ahead, the second day afterward it is 8 minutes ahead, and so on. Six months later the two clocks differ by 12 hours and the stars that were formerly at their highest at midnight are then at their highest at noon. As a result, they may not be visible at all at that season.

The coordinate in the sky that is perpendicular to the celestial equator marks the declination. Declination, latitude in the sky, is measured in degrees north or south of the celestial equator in exactly the same manner that latitude on the earth is measured with respect to the terrestrial equator.

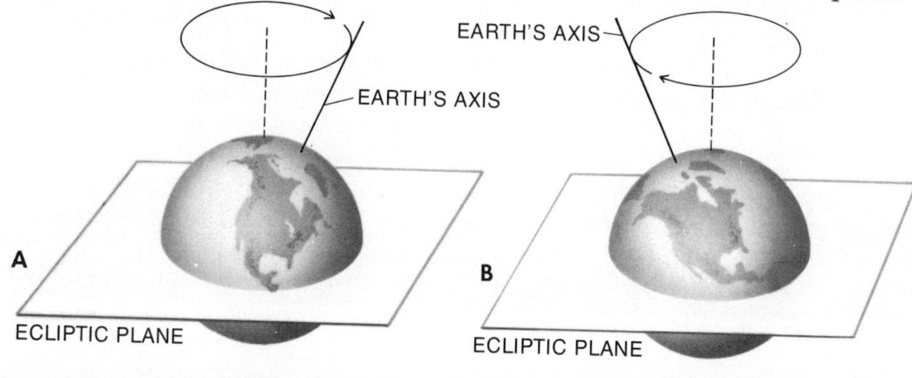

Figure 3–42 The earth's axis precesses with a period of 26,000 years.

Every star has a right ascension and declination, in just the same way that every town on earth has a longitude and latitude. For example, Williamstown, Massachusetts is always at 73°12′ western longitude (west means a longitude that is measured from Greenwich in the westward direction) and at latitude +42°43′ (a positive latitude means that the location is north of the equator). In the heavens, Sirius is always essentially fixed in the coordinate system at 6^h43^m right ascension and at declination −16°41′. The cities on earth don't change their positions on the surface of the globe, and the stars in the sky don't change their positions much either.

Actually, there are minor effects that cause the celestial coordinates of the stars to change slightly. One such effect is *precession*, which causes a slight drift in the coordinate system with a 26,000 year period (Fig. 3–42). Precession takes place because the earth's axis doesn't always point exactly at the same spot in the sky; it rather traces out a small circle and takes approximately 26,000 years to return to the same orientation. As a result of precession, one has to make small corrections in any catalogue list of celestial positions to update them to the present time. Precession is a small effect that must be taken into account only for observing with large telescopes—changes in celestial coordinates may be a few minutes in a few years. Don't worry, Polaris will be the North Star again in about 26,000 years, but in between—in, say, A.D. 14,980—the North Star will be Vega. An analogous change on earth, with which some readers may be familiar, is that of compass corrections that must be made because of the drift of the earth's magnetic poles.

Figure 3–43 The ecliptic is divided into the signs of the zodiac in this illustration from the *Epitoma In Almagestum Ptolemaei (Venice, 1496)* by Regiomontanus. This first translation into Latin of Ptolemy's *Almagest,* which contained the standard description of Ptolemy's earth-centered universe, had been formerly available only in Arabic.

Another minor effect that causes changes in the right ascensions and declinations of celestial objects is proper motion (see Section 2.16), but this effect is so minuscule that even when astronomers work at it, the proper motion is difficult to measure.

Note that although the stars are fixed in their positions in the sky, the sun's position goes through the whole range of right ascension each year. The path of the sun in the sky with respect to the stars is called the *ecliptic* (Fig. 3–43). The earth's axis is inclined by 23½° with respect to the perpendicular to the plane of its orbit (Fig. 3–57). Because of this, the ecliptic is inclined by 23½° with respect to the celestial equator. The ecliptic and the celestial equator cross at two points. The sun crosses one of those points, called the *vernal equinox,* on the first day of spring. The sun crosses the other intersection, the *autumnal equinox,* on the first day of autumn. In the northern hemisphere, these occur on approximately March 21st and

Nox *is Latin for night.*

Color Plate 1 (left): The 4-m Mayall optical telescope on Kitt Peak, near Tucson, Arizona. (Kitt Peak National Observatory photo)

Color Plate 2 (top right): The 66-m radio telescope at the Australian National Radio Observatory at Parkes, N.S.W. (Photograph by the author)

Color Plate 3 (bottom): From top to bottom, the emission spectrum of ionized calcium, the Fraunhofer spectrum of the sun, a continuous spectrum, and the emission spectrum of a fluorescent lamp. Note the H and K lines of ionized calcium at the left, which are the strongest absorption lines in the visible part of the solar spectrum. (Bausch and Lomb photos)

Color Plate 4: 47 Tucanae, a prominent globular cluster visible from the southern hemisphere. (Cerro Tololo Inter-American Observatory photograph with the 4-m telescope)

Color Plate 5: The Pleiades, M45, is a galactic cluster in the constellation Taurus, the bull. Reflection nebulae are visible around the brightest stars. (Hale Observatories photograph with the 1.2-m Schmidt camera)

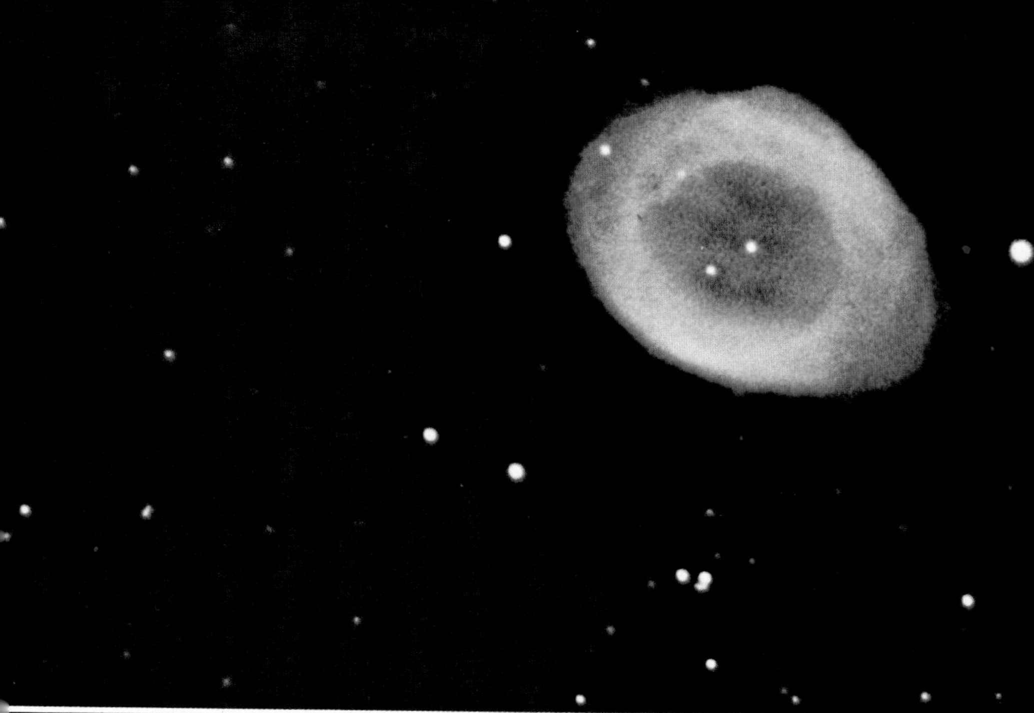

Color Plate 6 (top): Ring Nebula, M57, a planetary nebula in Lyra. Red hydrogen radiation is visible around its outer edge; green radiation from ionized oxygen shows in the center. Its central star appears distinctly. (Hale Observatories photo with the 5-m telescope)

Color Plate 7 (bottom): Dumbbell Nebula, M27, a planetary nebula in the constellation Vulpecula. (Hale Observatories photo with the 5-m telescope)

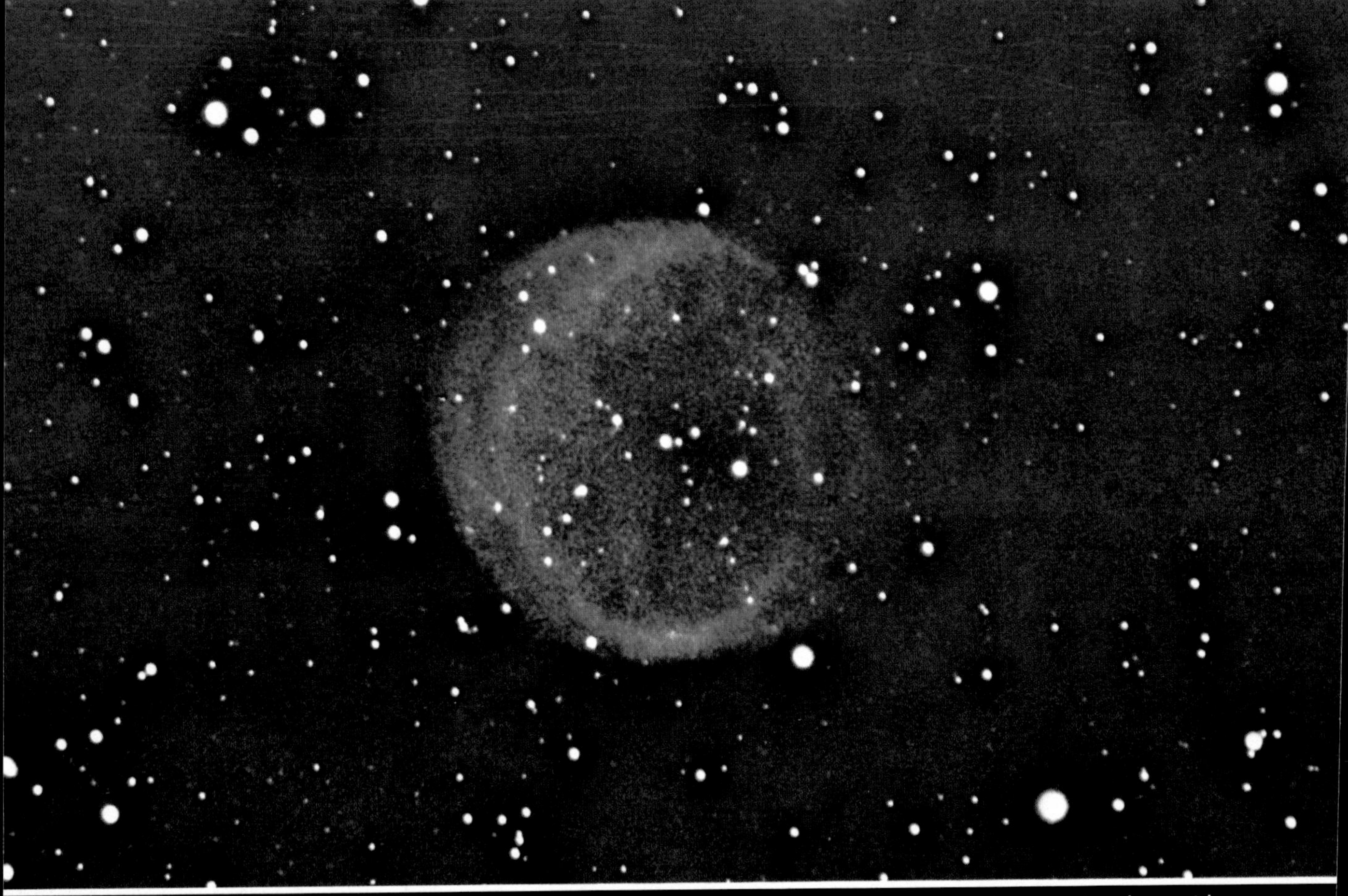

Color Plate 8 (top): A planetary nebula, NGC 6781, in the constellation Aquila. (Hale Observatories photo with the 1.2-m Schmidt camera)

Color Plate 9 (bottom): A computer display showing the distribution of x-radiation between 15 and 80 Å from the Cygnus loop, a supernova remnant. (Computer image by Richard J. Borken of the University of Wisconsin from rocket data he collected in collaboration with Saul A. Rappaport of M.I.T.)

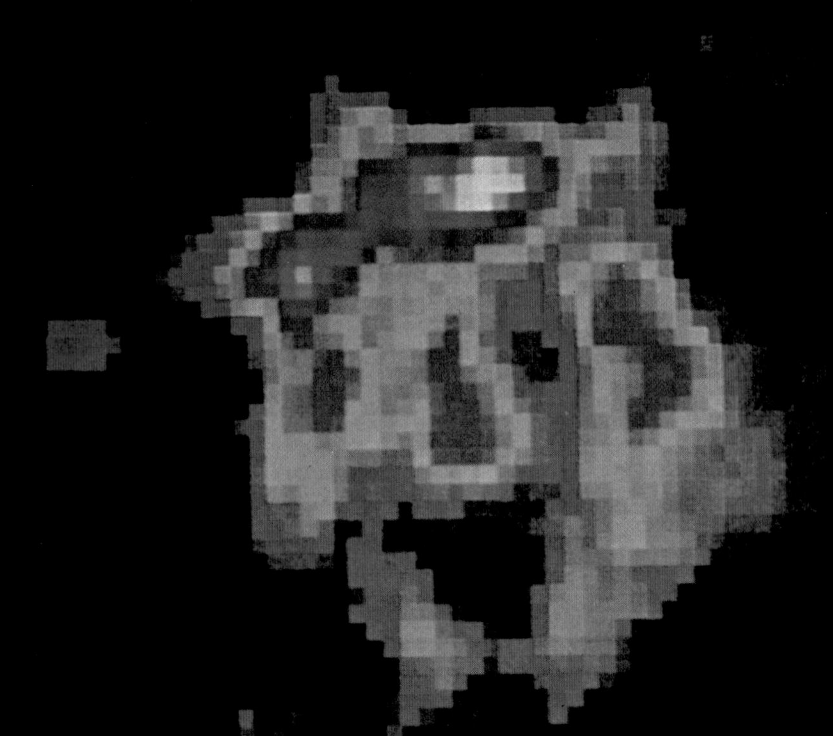

September 21st, respectively. On these days, the sun's declination is 0°; the sun's declination varies over the year from $+23\frac{1}{2}^\circ$ to $-23\frac{1}{2}^\circ$.

These points are called equinoxes because the daytime and the nighttime are supposedly equal 12-hour lengths on these days. Actually, because the refraction by the earth's atmosphere makes the sun appear to rise a little early and set a little late, and the fact that the top of the sun rises ahead of the middle of the sun, at U.S. latitudes the daytime exceeds the nighttime by about 10 minutes on the days of the equinoxes. The days of equal daytime and nighttime precede the vernal equinox and follow the autumnal equinox by a few days.

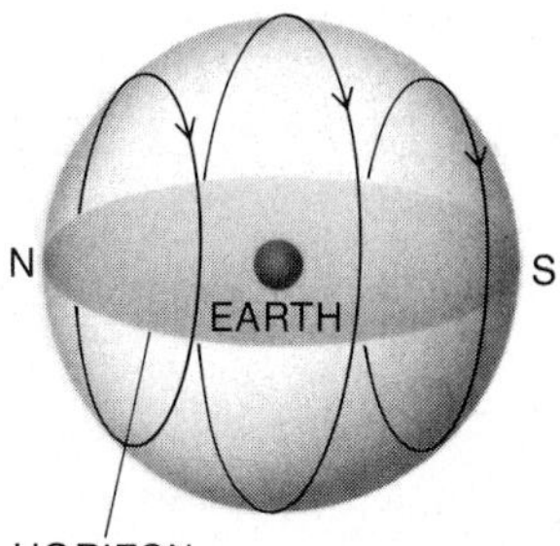

The moon goes around the earth once each month, and thus the moon's right ascension changes through the entire 24 hours of right ascension once each month. Since the moon's orbit is inclined to the celestial equator, the moon's declination also varies. The planets' motions in the sky are less easy to categorize, but they also change their right ascension and declination from day to day. Tables of the daily positions of the sun, moon, and planets are published each year by the U.S. Naval Observatory in Washington, D.C., in a book entitled *The American Ephemeris and Nautical Almanac.*

ZENITH
N
S
EARTH
HORIZON

Figure 3–44 From the equator, the stars rise straight up, pass right across the sky, and set straight down.

3.12 OBSERVING THE STARS

At the latitudes of the United States, which range from +25° for the tip of the Florida Keys to +49° for the Canadian border, or down to +19° in Hawaii and up to +67° in Alaska, the stars rise and set at angles to the horizon other than 0° or 90°. In order to understand the situation, it is best first to visualize simpler cases.

If we were standing on the equator, the stars would rise perpendicularly to the horizon (Fig. 3–44). The north celestial pole would lie exactly on the horizon in the north, and the south celestial pole would lie exactly on the horizon in the south. Each star would rise somewhere on the eastern half of the horizon; "up" would remain "up" for twelve hours, and then would set. We would be able to see all stars, no matter what their declinations, for twelve hours a day. The sun, no matter what its declination, would also rise and set twelve hours apart, so the day and night would each last twelve hours.

If, on the other hand, we were standing on the north pole, the north celestial pole would be directly overhead, and the celestial equator would be on the horizon (Fig. 3–45). All the stars would move around the sky in

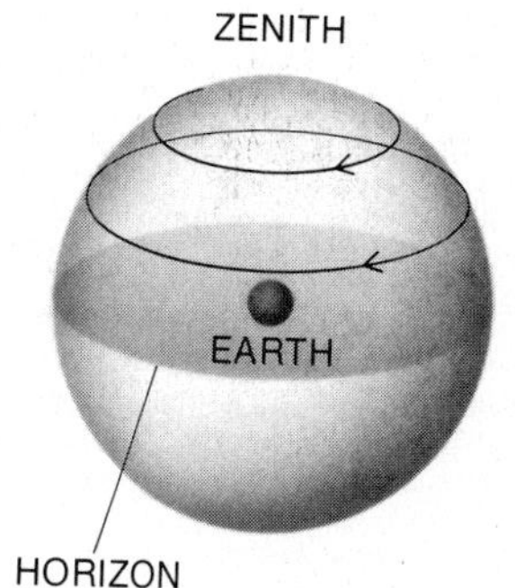

Figure 3–45 From the pole, the stars move around the sky in circles parallel to the horizon, never rising or setting.

Figure 3–46 In this series taken in June from northern Norway, above the Arctic Circle, one photograph was taken each hour for an entire day. The sun never set, a phenomenon known as the *midnight sun.* Since the site was not at the north pole, the sun and stars move somewhat higher and lower in the sky in the course of a day.

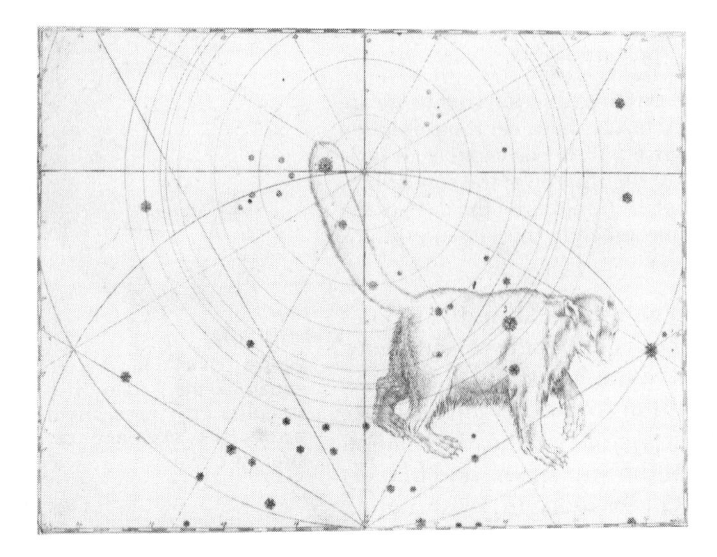

Figure 3–47 Polaris is the star at the tip of the tail of Ursa Minor, the Little Bear. The Little Dipper takes up most of the constellation. Because of precession, Polaris is somewhat closer to the north celestial pole than it was in about 1600, when this chart was drawn for Bayer's *Uranometria*. It was in this atlas of 1706 stars that Bayer began the system of assigning Greek letters to the brightest stars in each constellation. Polaris is α Ursae Minoris.

circles parallel to the horizon. Since the celestial equator is on the horizon, we could see only the stars with northern declinations. The stars with southern declinations would never be visible.

Whenever the sun had a northern declination, it would be above the horizon and we would have daytime. It would move in a circle all around us and essentially parallel to the horizon. From day to day it would appear slightly higher in the sky for 3 months, and then move gradually lower. The date when it is highest in the sky is called the *summer solstice;* it occurs approximately on June 21st each year, and on that date the sun is 23½ degrees above the horizon because its declination is +23½ degrees. The six months of the year when the sun never sets are known as the time of the

Figure 3–48 Star trails form circles around the north celestial pole; the dome of the 2.6-meter telescope of the McDonald Observatory in Texas is in the foreground.

Figure 3–49 Near the celestial equator, the star circles are so large that they appear almost straight in this view past the Yerkes Observatory in Wisconsin.

midnight sun (Fig. 3–46). The sun crosses the celestial equator and begins to have northern declination at the vernal equinox. The sun crosses the celestial equator in the other direction six months later at the autumnal equinox.

Let us consider in this paragraph a latitude between the equator and the north pole, say 40° north latitude. Then the stars seem to rise out of the horizon at oblique angles. The north celestial pole (with the north star nearby it) is always visible in the northern sky, and is at an *altitude* of 40° above the horizon. The star Polaris, a 2nd magnitude star, happens to be located within 1° of the north celestial pole, and so is called the *pole star* (Fig. 3–47). All stars within 40° of declination of the north pole, that is, all stars between +50° and +90° declination, never go below the horizon as seen from this latitude and so can be seen whenever it is dark. These stars are called *circumpolar* (Fig. 3–48). All stars below −50° declination never rise above the southern horizon, and can never be seen from this latitude. Stars with declinations just a little north of −50° (−45°, for example) rise very briefly in the southern sky. Stars of declination closer to zero (−10°, for example) stay above the horizon for longer times (Fig. 3–49).

When the sun is at the summer solstice, it is at its greatest northern declination and is above the horizon of northern hemisphere observers for the longest time each day. Thus daytimes in the summer are longer than daytimes in the winter, when the sun is at its lowest declinations. In the winter, the sun not only is above the horizon for a shorter period each day but also never rises very high in the sky. The time of its lowest declination is called the *winter solstice.* Winter in the northern hemisphere corresponds to summer in the southern hemisphere.

The *seasons* (Fig. 3–50), thus, are caused by the variation of declination of the sun, which, in turn, is caused by the fact that the earth's axis of spin is tipped by 23½° with respect to the perpendicular to the plane of the earth's orbit around the sun. This cycle, of course, repeats once each year.

When an astronomer wants to know if a star is favorably placed for observing, he must know both its right ascension and declination. By comparing the right ascension with the sidereal time to compute the hour angle, he can tell if it is the best time of year at which to observe (Fig. 3–51). But he must also know the star's declination to know how long it will be above the horizon each day.

These concepts involve spherical geometry, which many individuals

Figure 3–50 The seasons occur because the earth's axis is tipped with respect to the plane of its revolution around the sun. When the northern hemisphere is tilted toward the sun, it has its summertime; at the same time, the southern hemisphere is having winter.

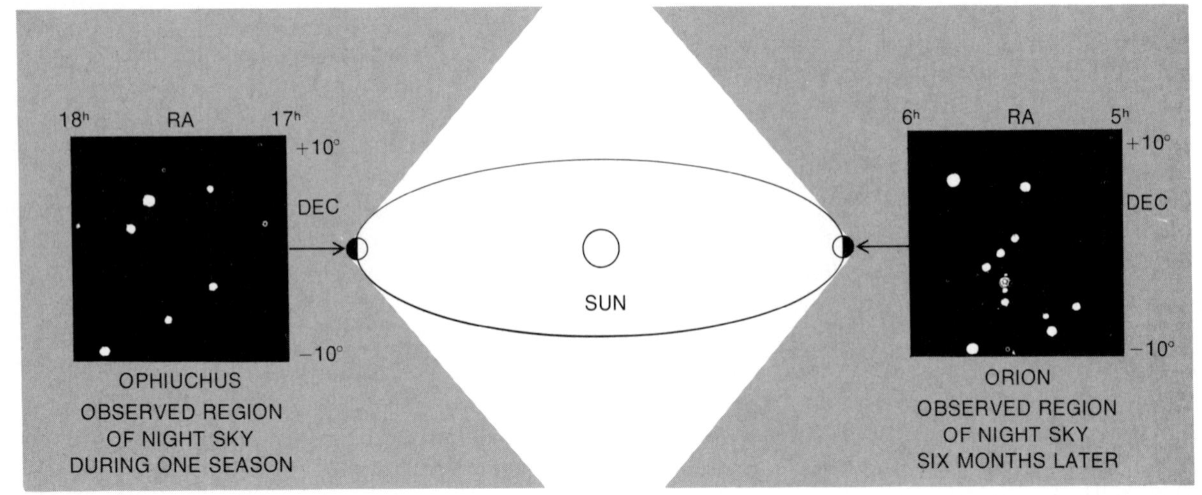

Figure 3–51 For each time of the year, we can see the constellations that are in the direction away from the sun. Because the earth rotates much more rapidly than it revolves around the sun (24 hours compared with 365 days), all observers on the earth see the same constellations in a given season, even though it is daytime for some while it is nighttime for others.

find difficult to visualize. But a little experience with a telescope can make right ascension and declination seem very easy.

Telescopes are often mounted at an angle such that one axis points directly at the north celestial pole, i.e., at a position in the sky very close to that of Polaris. (In the southern hemisphere, the axis would point at the south celestial pole.) This axis is called the *polar axis*. Thus since all stars move across the sky in circles centered nearly at Polaris, the telescope must merely turn about that axis to keep up with the stellar motions. The other axis of the telescope is used to point the telescope in declination.

Since motion around only one axis is necessary to track the stars, one need have only a small motor set to rotate once every 24 sidereal hours. This motor is linked to turn the polar axis in the direction opposite to the rotation of the earth. The principle is the same for a small telescope in your backyard as for the 5-meter telescope at Palomar. Arrangements of this type are called *equatorial mounts,* since one axis rotates perpendicularly to the celestial equator (Fig. 3–53).

The alternative to this system is to mount a telescope such that one axis rotates parallel to the horizon. This motion is called *azimuth.* A second, perpendicular, axis points the telescope up and down in *altitude* above the

Figure 3–52 Cartoons by Charles Schulz.

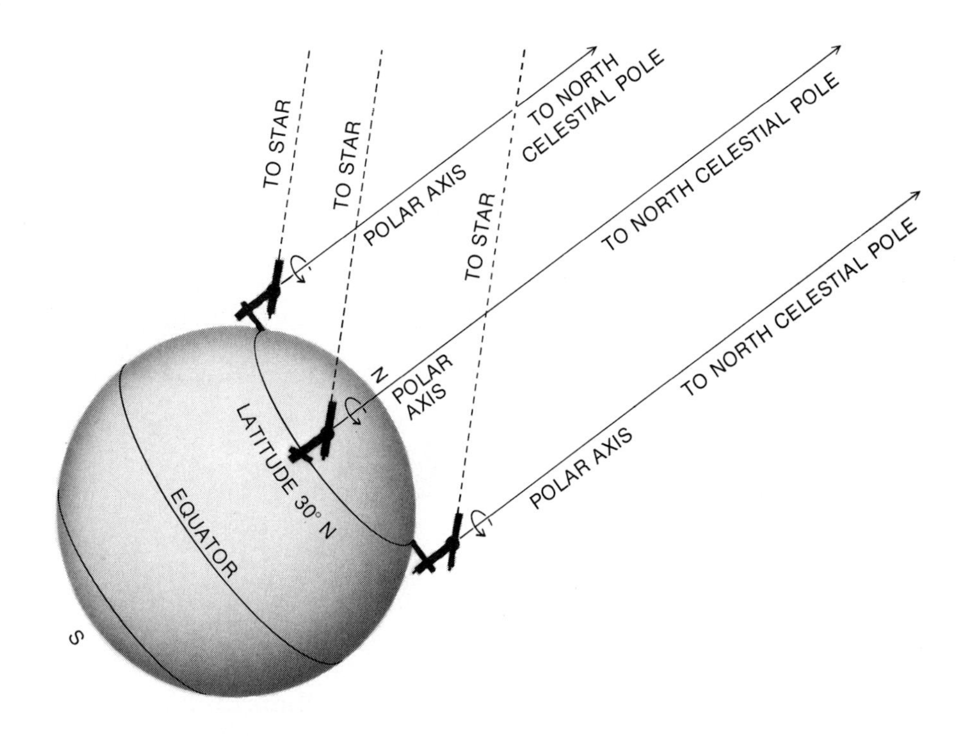

Figure 3–53 An equatorially mounted telescope need rotate on only one axis to keep pointing at a star. This axis is called the *polar axis;* it is fixed for telescopes in the northern hemisphere so that it points at the north celestial pole. Motion around the axis perpendicular to the polar axis sweeps out a circle of declination in the sky.

horizon. To track a star in this system, continual adjustments would have to be made in both axes, which was formerly very inconvenient. However, the availability of inexpensive minicomputers to make the necessary calculations of spherical geometry has allowed the mounting of large new telescopes in this system, which is called *alt-azimuth.* A large alt-azimuth mount is often less expensive to construct than an equatorial mount of the same size.

3.13 TIME AND THE INTERNATIONAL DATE LINE

Every city and town on earth used to have its own time system, based on the sun, until widespread railroad travel made this inconvenient. In 1884, an international conference agreed on a series of longitudinal time zones. Now all localities in the same zone have a *standard time* (Fig. 3–54). Since there are twenty-four hours in a day, the 360° of longitude around the earth are divided into 24 standard time zones, each 15° wide. Because the time is the same throughout each zone, the sun is not directly overhead at noon at each point in a given zone, but in principle is less than about half an hour off.

Standard time is based on a *mean solar day,* the average length of a solar day. The length of a particular day can be slightly different (up to half a minute) from the mean solar day because the earth travels at varying speeds in its elliptical orbit around the sun (as discussed in Section 10.2) and because its axis of rotation is inclined with respect to the plane of its orbit (as discussed in Section 3.11).

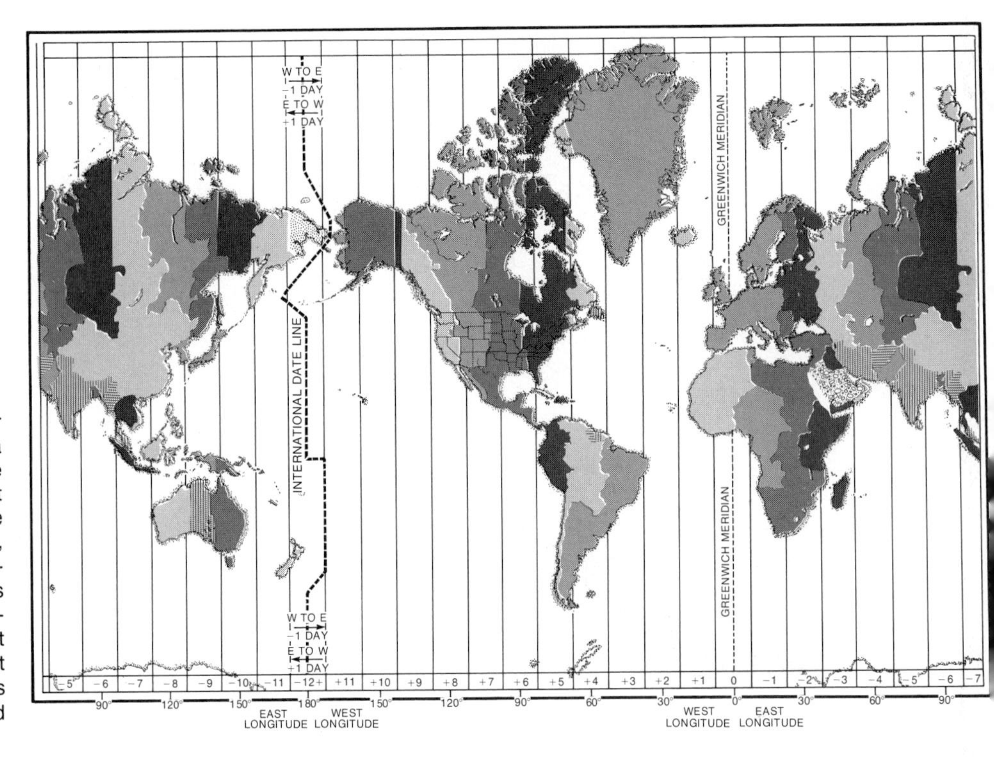

Figure 3–54 Although in principle the earth is neatly divided into 24 time zones, in practice, political and geographic boundaries have made the system much less regular. The existence of daylight-saving time in some places and not in others further confuses the time zone system. Some countries even have double daylight-saving time, adding 2 hours. Other countries, shown with cross hatching, have time zones that differ by one-half hour from a neighboring zone. At the international date line, not only the hour but also the date changes. In the U.S., most states have daylight-saving time for six months a year—from the last Sunday in April until the last Sunday in October. But Arizona, Hawaii, and parts of Indiana have standard time year round.

As the sun seems to move in the sky from east to west, the time in any one place gets later. We can visualize noon, and each hour, moving around the world from east to west, minute by minute. We would get a particular time back 24 hours later, but if the hours circled the world continuously the date would not have changed. So we specify a line of longitude and have the date change there. We call it the *international date line.* England won for Greenwich, then the site of the Royal Greenwich Observatory, the distinction of having the basic line of longitude, zero degrees. Realizing that the international date line would disrupt the calendars of those who crossed it, that line was thus put as far away from the populated areas of Europe as possible: along the 180° longitude line. The line passes from north to south through the Pacific Ocean and actually bends slightly to avoid cutting through groups of islands, thus providing them with the same date as their nearest neighbor, as shown in Figure 3–54.

In some cities, instead of paying tolls in each direction on some bridges, one pays a double toll in one direction and travels free in the other. For those who make round trips, it evens out. If one moves permanently from one side of the bridge to the other, however, one can come out slightly ahead or slightly behind.

We may think of the time along a line of longitude as starting at the international date line and gradually moving westward around the world (Fig. 3–55). Twelve hours after it began it has reached England and twenty-four hours later it has returned to the international date line. Then it disappears into the past. If we are in a ship or plane and proceed west across the date line, the time leaps twenty-four hours ahead—theoretically, it becomes the next day. If we cross the line in the other direction, the calendar will show that we arrive the day before our departure, even though the trip took only a few hours. In actuality we make up going one way what we lose going the other. But if we cross the dateline from east to west and stay on the other side, we gain a day forever.

Astronauts in space go around the earth every 90 minutes, so they go through the whole series of time zones 16 times a day. Sixteen times the

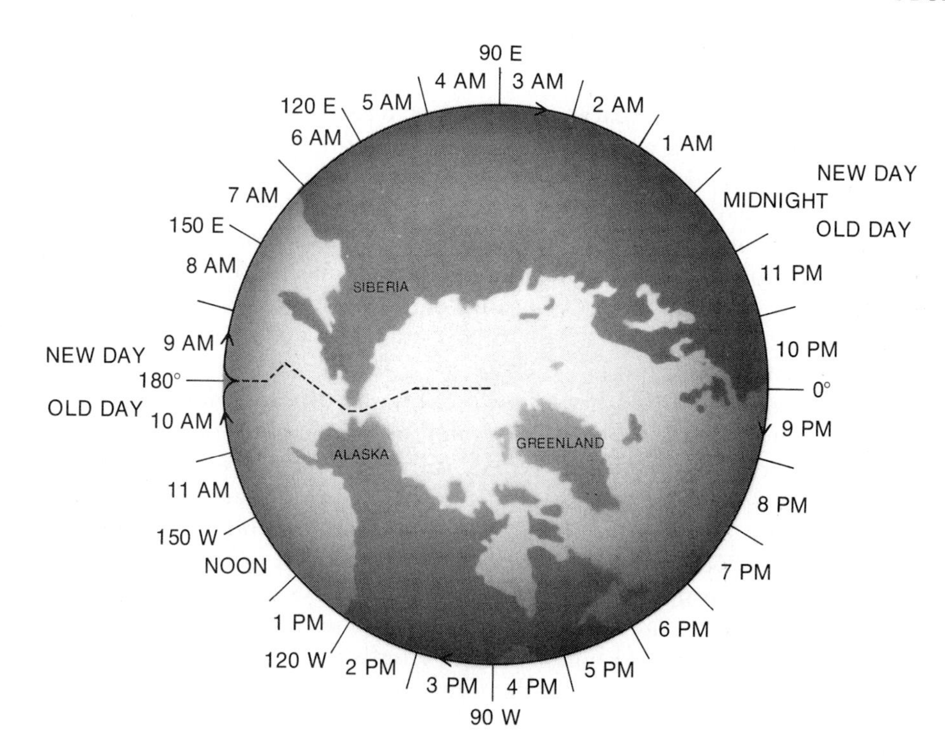

Figure 3–55 When you cross the international date line, not only does your watch change by one hour but also your calendar changes by one day. Days are born and die at the international date line.

sun rises, and sixteen times the sun sets. But after all that, it is only one day later for them, not sixteen. They have gained a day, hour by hour, sixteen times, but also crossed the international date line fifteen times, and thus set the time and date right. The Skylab 3 astronauts had sixteen New Year's Eves as 1973 turned to 1974. The astronauts themselves find it simpler to always remain on Houston time. Anyway, being an astronaut is a good profession if you like sunsets.

Astronomers often keep track of events according to the standard solar time that corresponds to the Greenwich time zone. This is called G.M.T. (Greenwich Mean Time), U.T. (Universal Time), or Z (which is colloquially called Zulu Time). One may find the time of an eclipse of the moon listed in the Ephemeris, for example, as 17:46 U.T., and must subtract (if it is in the winter, when daylight-saving time is not in effect) 5 hours to get Eastern Standard Time, 6 hours to get Central Standard Time, 7 hours to get Mountain Standard Time, and 8 hours to get Pacific Standard Time. Thus the eclipse would take place at 12:46 pm E.S.T., 11:46 am C.S.T., 10:46 am M.S.T., and 9:46 am P.S.T.

In the summer, in order to make the daylight last into later hours, many countries have adopted *daylight-saving time* (D.S.T.). Clocks are set ahead 1 hour on a certain date in the spring. Thus if darkness falls at 6 pm E.S.T., that time will be called 7 pm E.D.T., and most people will have an extra hour of daylight after work. In most places, that hour will be taken away in the fall, though some places have adopted daylight-saving time all year. The phrase to remember to help you set your clocks is "fall back, spring ahead." Of course, daylight-saving time is just a bookkeeping change in how we name the hours, and doesn't result from any astronomical changes.

Some countries have adopted daylight-saving time all year long, in order to provide more light in the evening hours. France now adds this hour all year and adds a second hour—double daylight-saving time—in the summer.

3.14 CALENDARS

The period of time that the earth takes to revolve once around the sun is called, of course, a *year*. This period is about 365¼ mean solar days. A *sidereal year* is the interval of time that it takes the sun to return to a given position with respect to the stars. A *solar year*, or *tropical year*, is the period of time that it takes the sun to return to the vernal equinox, the point where the ecliptic crosses the celestial equator. (A sidereal year differs from a solar year by 20 minutes 24 seconds because of the earth's precession, discussed in Section 3.11.) Solar years are growing shorter by 0.005 second per year (half a second per century).

The Roman calendars had, at different times, different numbers of days in a year, and so the dates rapidly drifted out of synchronization with the seasons (which follow tropical years). Julius Caesar decreed that 46 B.C. would be a 445 day year in order to catch up, and defined a calendar, the *Julian calendar*, that would be more accurate. This calendar had years that were normally 365 days in length, with an extra day inserted every fourth year in order to bring the average year to 365¼ days in length. The fourth years were, and are, called *leap years*. (The name of the fifth month, then Quintilis, was changed to honor him; in English we call it July.)

The name of the fifth month, formerly Quintilis, was changed to honor Julius Caesar; in English we call it July. The year then began in March; the last four months of the year still bear names from this system of numbering. Augustus Caesar, who carried out subsequent calendar reforms, renamed August after himself. Also he transferred a day from February in order to make August last as long as July.

The Julian calendar was much more accurate than its predecessors, but the actual solar year 1980 will be 365 days 5 hours 48 minutes 45.6 seconds long, some 11 minutes 14 seconds shorter than 365¼ days (365 days 6 hours). By 1582, the calendar was about 10 days out of phase with the date at which Easter had occurred at the time of a religious council 1250 years earlier, and Pope Gregory XIII issued a bull—a proclamation—to correct the situation. He dropped 10 days from 1582.

In the *Gregorian calendar*, years that are evenly divisible by four are leap years, except that three out of every four century years, the ones not divisible evenly by 400, have only 365 days along with all the other years. Thus 1600 was a leap year, 1700, 1800, and 1900 were not, and 2000 will again be a leap year. Although many countries adopted the Gregorian calendar as soon as it was promulgated, Great Britain (and its American colonies) did not adopt it until 1752, when 11 days were thus skipped. As a result, we celebrate George Washington's birthday on February 22nd, even though he was born on February 11, 1732. When Alaska was annexed, its calendar had to be changed over. The Gregorian calendar is the one in current use. It will be over 3000 years before this calendar is as much as one day out of step with the seasons.

Many citizens of that time objected to the supposed loss of the time from their lives and to the commercial complications. Does one pay a full month's rent for the month from which the days were omitted, for example? "Give us back our fortnight," they cried.

SUMMARY AND OUTLINE OF CHAPTER 3

Aims: To understand the different types of observing techniques and instruments used in astronomy

Optical telescopes (Sections 3.1 to 3.6)
- Resolution and twinkling of stars
- Refracting and reflecting telescopes, Schmidt cameras
- Optical observatories and their distribution across the world
- Light-gathering power and magnification

Spectroscopy and photometry (Section 3.8)

Types of astronomical detectors for different types of radiation (Section 3.9)

Coordinate systems: right ascension and declination, sidereal time, circumpolar stars, precession (Sections 3.11 and 3.12)
Standard time and the international date line (Section 3.13)
Calendars (Section 3.14)

QUESTIONS

1. What advantages does a reflecting telescope have over a refracting telescope?

2. What limits the resolving power of the 5-meter telescope?

3. List the important criteria in choosing a site for an optical observatory meant to study stars and galaxies.

4. Two reflecting telescopes have primary mirrors 2 m and 4 m in diameter. What is the difference in magnitude between the faintest stars you can observe with each telescope in a given observing time?

5. The 1.2-meter Schmidt camera has a $7° \times 7°$ field of view. Estimate how many photographs you would have to take to cover the whole sky. Explain your calculation.

6. Why can't we use ordinary photographic plates to record infrared images?

7. Why must we observe ultraviolet radiation using rockets or satellites, while balloons are sufficient for infrared observations?

8. Why might some stars appear double in blue light though they could not be resolved in red light?

9. For each of the following, identify whether it is a characteristic of a reflecting telescope, a refracting telescope, both, or neither. Give any limitations on the applicability of your answer.

   (a) Free of chromatic aberration.

   (b) Has more severe spherical aberration.

   (c) Requires aluminizing.

   (d) Can be used for photography.

   (e) Has an objective that must be supported only by its rim.

   (f) Can be made in larger sizes.

   (g) Has a coudé focus.

   (h) Has a prime focus at which a person can work without blocking the incoming light.

10. Why can radio astronomers observe during the day, while optical astronomers are (for the most part) limited to nighttime observing?

11. What are the advantages of the Multi-Mirror Telescope? How will it compare with the 5-meter telescope for studying faint objects?

12. What are the similarities and differences between making radio observations and using a reflector for optical observations. Compare the radiation path, the detection of signals, and limiting factors.

13. Why is it sometimes better to use a small telescope in orbit around the earth than it is to use a large telescope on a mountaintop?

14. A star has right ascension 5^h30^m. It is at an hour angle of 1^h. What is the local sidereal time? What will the hour angle be when the local sidereal time is 8^h?

15. Between the vernal equinox, March 21st, and the autumnal equinox, 6 months later,

    (a) by how much does the right ascension of the sun change;

    (b) by how much does the declination of the sun change;

    (c) by how much does the right ascension of Sirius change;

    (d) by how much does the declination of Sirius change?

16. On the first day of spring, refraction causes the sun to appear above the horizon for longer than 12 hours. Would you expect the day when the sun appears above the horizon for exactly 12 hours to be a few days before or after the vernal equinox? Explain. Assume that the sun is point size for this problem.

17. (a) How does the declination of the sun vary over the year?

    (b) Does its right ascension increase or decrease from day to day? Justify your answer.

18. We normally express longitude on the earth in degrees. Why would it make sense to express longitude in units of time?

19. If a planet always keeps the same side toward the sun, how many sidereal days are there in a year on that planet?

20. What is the advantage of an equatorial mount? Why did large telescopes used to be made with equatorial mounts while they are now sometimes made with altazimuth mounts?

4

Variable Stars and Stellar Groupings

We often think of stars as individual objects that shine steadily, but many stars vary in brightness and most stars actually have companions close by. Also, many stars appear as members of groupings called associations or clusters. By studying the effects that the stars have on each other, or by studying the nature of all the stars in a stellar cluster, we can derive much information that we could not discover by studying the stars one at a time. This information even leads us to a general understanding of the life history of the stars.

4.1 BINARY STARS

Most of the objects in the sky that we see as single "stars" really contain two or more component stars. The easiest such object to see with the naked eye is "one" of the stars in the handle of the Big Dipper. The brighter star is called Mizar and the companion is called Alcor. The ability to see that the system is, in fact, a double star is one of the standard tests of a person's visual acuity; only those people with the sharpest vision can "separate" Mizar from Alcor with the naked eye.

Through a telescope of any size, it is easy to see that the system is a multiple star. In fact, one can even see that one of the two major components, Mizar, is itself a double star (Fig. 4–1).

Sometimes a star appears double merely because two stars that are located at different distances from the sun appear in the same line of sight. This is the case with Alcor and Mizar. Such systems are called *optical doubles,* and will not concern us here. We are more interested in stars that are physically associated with each other.

For actual doubles, the easiest way to tell that there are two instead of

---

M13, a globular cluster in Hercules.

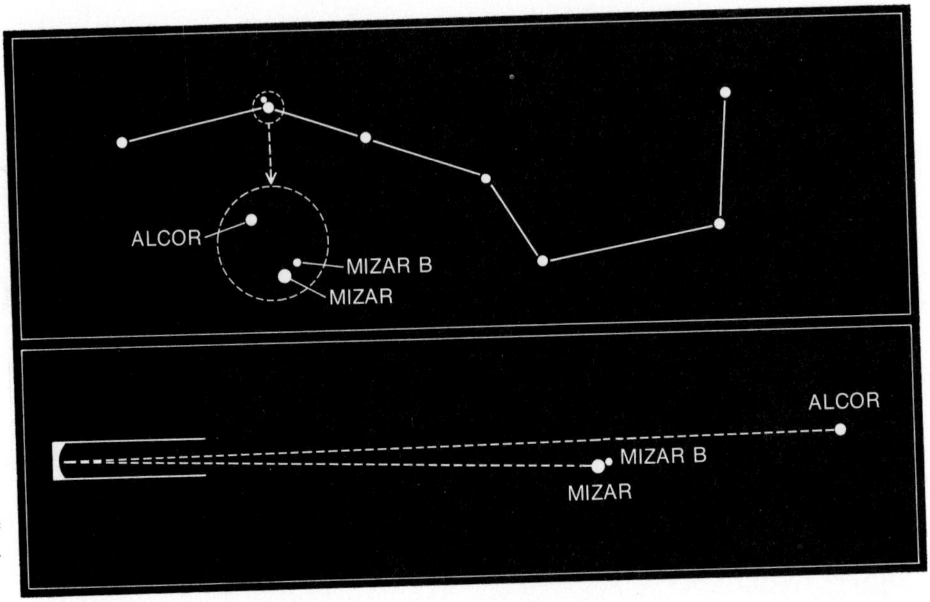

Figure 4–1 Alcor and Mizar provide examples of both visual and spectroscopic binaries.

one is by looking through a telescope of sufficiently large aperture. Stars that appear double when observed directly are called *visual binaries*. The resolution of small telescopes may not be sufficient to allow the components of a double star to be "separated" from each other, and larger telescopes can thus distinguish more double or multiple stars. When the stars have different colors, they form particularly beautiful objects to observe in even small telescopes. Five or ten per cent of the stars in the sky are visual binaries.

Albireo (β Cygni) contains a B star and a K star, which make a particularly beautiful pair because of their different colors.

We are not always fortunate enough, however, to have the two components of a double star appear far enough apart from each other to be observed visually. Sometimes a star appears as a single object through a telescope, but one can see that the spectrum of that "object" actually consists of the overlapping spectra of at least two objects. If we can detect the presence of two spectra of different types – of, say, one hot star and one cool star – then we say that the object has a *composite spectrum*. If, on the other hand, over a period of time we can actually tell that the stars are revolving around each other by the fact that the spectra are changing in wavelength with respect to each other because of the Doppler shift, then we call the "object" a *spectroscopic binary* (Fig. 4–2). As the objects orbit each other, unless we are looking straight down on the orbit from above, they each spend half of each orbit at velocities approaching us and the other half at velocities receding from us, relative to their average space motions. The velocity variations in spectroscopic binaries are periodic and, of course, the spectrum of each component varies separately in wavelength. Note that even if the spectrum of one of the stars is too faint to be seen, we can still

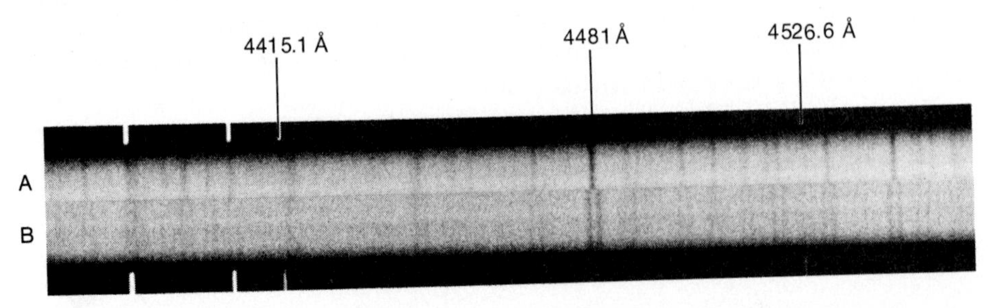

Figure 4–2 Two spectra of Mizar (ζ Ursae Majoris) taken 2 days apart show that it is a spectroscopic binary. The lines of both stars are superimposed in the upper stellar absorption spectrum, but are separated in the lower spectrum by 2 Å, which corresponds to a relative velocity of 140 km/sec. Emission lines from a laboratory source are shown at the extreme top and extreme bottom to provide a comparison with a source at rest.

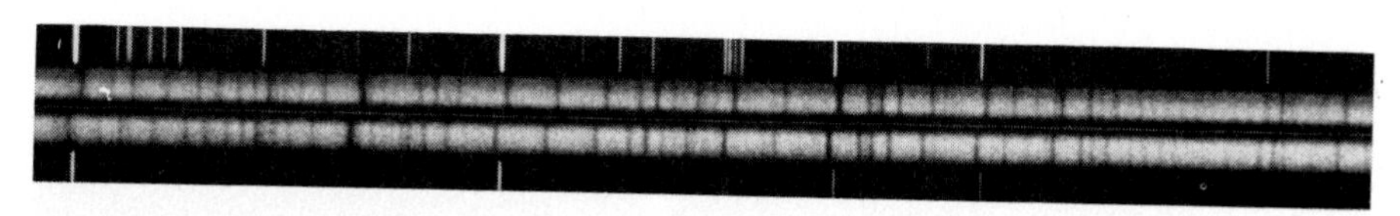

Figure 4–3 Two spectra of α Geminorum taken at different times show a Doppler shift. Thus the star is a spectroscopic binary, even though lines from only one of the components can be seen. The comparison spectrum of a laboratory source appears at the top and bottom. Note that emission lines from this laboratory source are in the same horizontal positions at extreme top and bottom, while the absorption lines of the stellar spectra are shifted laterally (that is, in wavelength) with respect to each other.

tell if the star is a spectroscopic binary from the variations in radial velocity of the visible component (Fig. 4–3). Two-thirds of all solar-type stars have stellar companions; the other third may have less massive companions (see Section 19.2). Though the presence of companions of stars of other spectral types has not been studied in such detail, we can reasonably conclude that few stars are single.

Sometimes the stars pass in front of each other, as seen from our viewpoint on the earth. The "double star" then changes in brightness periodically, as one star cuts off the light from the other. Such a pair of stars is called an *eclipsing binary* (Fig. 4–4). The easiest to observe is Algol, β Persei (beta of Perseus), in which the eclipses take place every 69 hours, dropping the total brightness of the system from magnitude 2.3 to 3.5. (A third star is present in the Algol system. It orbits the other two every 1.87 years, and does not participate in the eclipses.)

A double star may fall in more than one of these categories. For example, two stars that revolve around each other in an orbit that makes them eclipse would compose a spectroscopic binary too. Only nearby binary stars are close enough to be seen as visual binaries, but we can detect spectroscopic binaries or binaries with composite spectra at much greater distances, even in other galaxies.

Note that the way that a binary star appears to us depends on the orientation of the two stars not only with respect to each other, but also with respect to the earth. If we are looking down at the plane of their mutual orbit (Fig. 4–5) then we might see a composite spectrum; under the most favorable conditions the star might be seen as a visual binary. But in this orientation we would never be able to see the stars eclipse. We would also not be able to see the Doppler shifts typical of spectroscopic binaries, since only radial velocities contribute to the Doppler shift.

It is possible that a star could be a "double," but still not be detectable by any of the above methods. There are cases where the existence of a double star shows up only as a deviation from a straight line in the proper motion of the "star" across the sky. Such stars are called *astrometric binaries* (Fig. 4–6). We will say more of this later on, as this identification technique has been important in the discovery of white dwarfs (see Section 7.4) and of

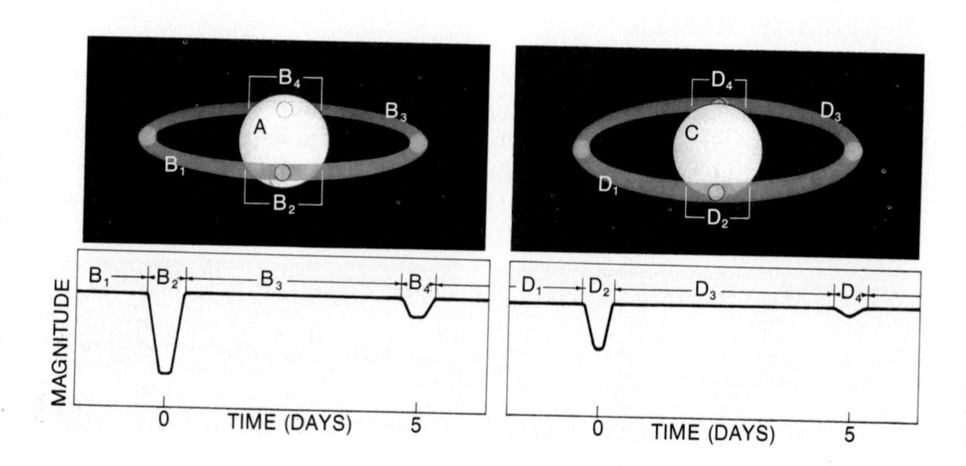

Figure 4–4 The shape of the light curve of an eclipsing binary depends on the sizes of the components and the angle from which we view them. At lower left, we see the light curve that would result for star B orbiting star A, as pictured at upper left. When star B is in the positions shown with subscripts at top, the regions of the light curve marked with the same subscripts result. At right, we see the appearance of the orbit and the light curve for star D orbiting star C, with the orbit inclined at a greater angle than at left. From earth, we observe only the light curves, and use them to determine what the binary system is really like.

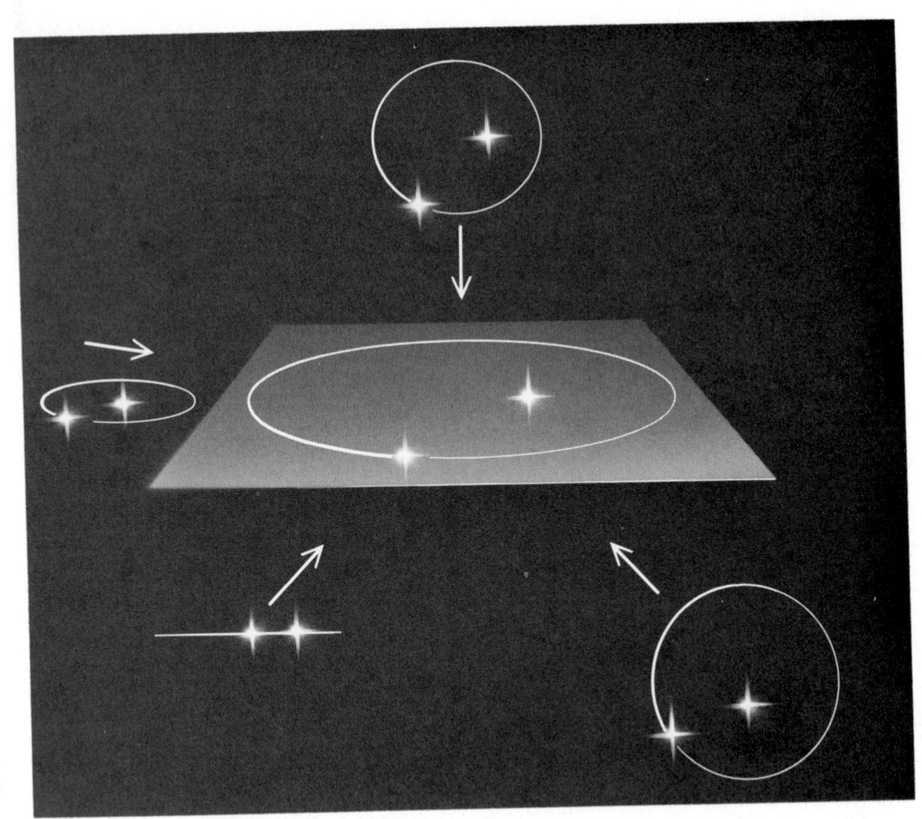

Figure 4–5 The appearance and the Doppler shift of the spectrum of a binary star depend on the angle from which we view the star.

very cool ordinary dwarfs, and is important for the discussion of the chances of finding extraterrestrial life (see Section 19.2).

Many of the celestial sources of x-rays that have been observed in recent years from orbiting telescopes turn out to be binary systems. Matter from one member of the pair falls upon the other member, heats up, and radiates x-rays. These binary x-ray sources will be discussed in Section 8.12.

Though one usually speaks of double stars, there is actually nothing to prevent stars from existing in threes, or fours, or with even more components. For example, Mizar's two visual components, called Mizar A and Mizar B, have each been found to be spectroscopic binaries. Thus Mizar is a quadruple star.

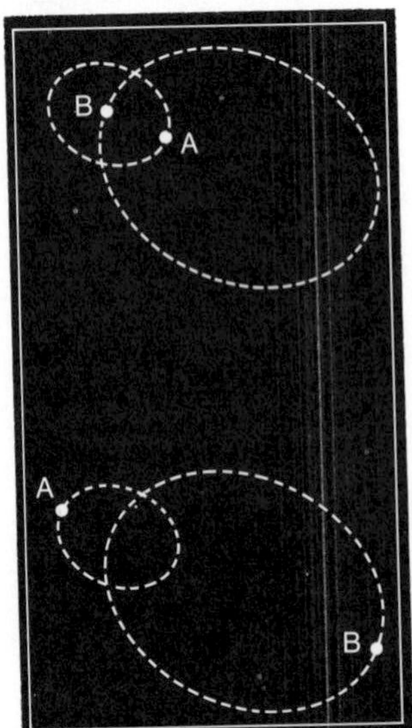

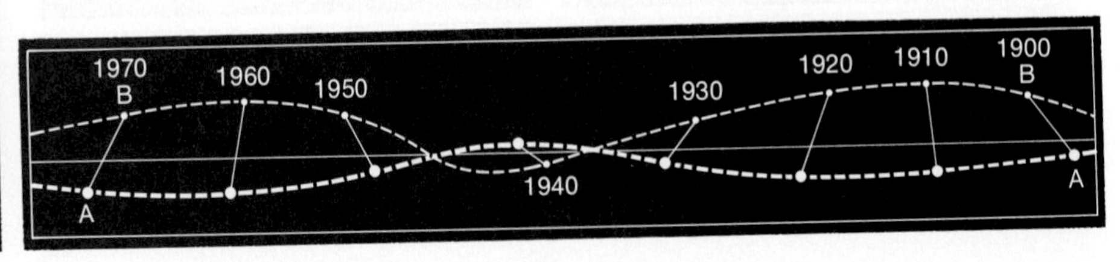

Figure 4–6 Sirius A and B, an astrometric binary. From studying the motion of Sirius A (often called, simply, Sirius) astronomers deduced the presence of Sirius B before it was seen directly. Sirius B's orbit is larger because Sirius A is a more massive star.

4.2 STELLAR MASSES

The study of binary stars is of fundamental importance in astronomy because it allows us to determine stellar masses. If we can determine the orbits of the stars around each other, we can calculate theoretically the masses of the stars necessary to produce the gravitational effects that lead to those orbits. For a star that is a visual binary with a sufficiently short period

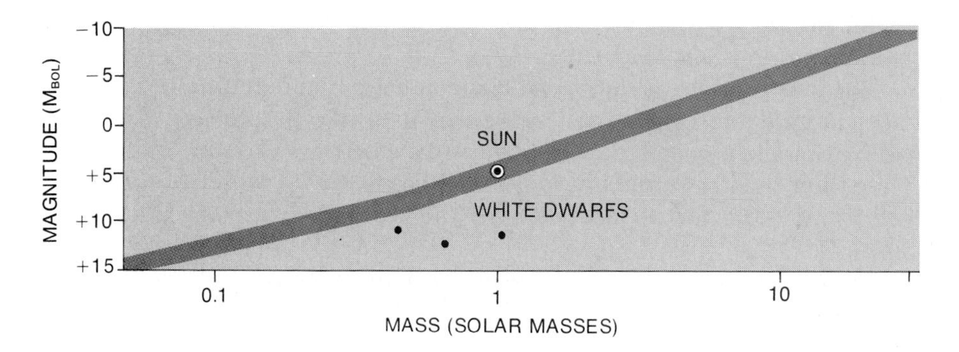

Figure 4–7 The mass-luminosity relation, measured from binary stars. Most of the stars fit within the shaded band. The relation is defined by straight lines of different slopes for stars brighter than and fainter than magnitude +7.5; we do not know if there is a real difference between the brighter and fainter stars that causes this or whether it is an effect introduced when the data are reduced, at which stage we try to take account of the energy that the stars radiate outside the visible part of the spectrum. Note that white dwarfs, which are relatively faint for their masses, lie below the mass-luminosity relation. Red giants lie above it. Thus we must realize that the mass-luminosity relation holds only for main-sequence stars.

(only 20 or even 100 years, for example), we are able to determine the masses of both of the components. If a star is a spectroscopic binary, even one with spectral lines of both stars present, we can find only lower limits for the masses—that is, we can say that the masses must be larger than certain values. This limitation occurs because we cannot usually tell the inclination of the plane of the orbit of the stars around each other. If we can detect only spectral lines from one of the components of a spectroscopic binary, we are limited to knowledge of a "mass function," a complicated term that involves the masses of both components.*

So we do not always know as much as we would like about the masses of the stars in binary systems. Of course, we are better off than we are for stars outside binary systems, for we cannot directly measure their masses at all! From studies of the several dozen binaries for which we can accurately tell the masses, astronomers have graphed the brightnesses of the stars against the masses. ("Against" in this context means that the brightnesses are plotted on one axis and the masses on the other.) Stars lie on a narrow band, the *mass-luminosity relation* (Fig. 4–7), which shows that the masses and luminosities (measured from the brightnesses) of stars are closely related. The mass-luminosity relation in this form is valid only for stars on the main sequence, that is, for normal dwarfs.

The more massive a star is, the brighter it is. In fact, the luminosity of a star is roughly proportional to the mass of the star raised to the 3.5 power; we can estimate with this formula that a star twice as massive as another is between $2^3 = 8$ and $2^4 = 16$ times more luminous, making the factor approximately 11.

The most massive stars we know are about 50 times more massive than the sun, and the least massive stars are about 25 times less massive than the sun. A dwarf star (i.e., main-sequence star) of ten solar masses is a B star. A dwarf star of two solar masses is an A star. The sun, which of course has 1 solar mass, is a G star. A dwarf of half a solar mass is a K star, and a dwarf of four-tenths a solar mass or less is an M star.

The sun has a mass of 2×10^{33} grams. (We determine this from the orbits of the planets; the physical laws we use to do this are described in Section 10.2.) The masses of the stars do not vary by much more than a factor of a thousand, so the astronomers on Eddington's hypothetical cloud-bound planet (see the introduction to Chapter 2) would have had a fairly

*If m_1 is the mass of the star that is observed, m_2 is the mass of the unseen companion, and i is the inclination of the orbit to us, the mass function is $m_2^3 \sin^3 i/(m_1 + m_2)^2$. It is displayed here to show what we mean by a "function"; it is not worth memorizing. Without observing the second star, we cannot unscramble the mass function to give the individual masses.

specific prediction to test when they went to look for gravitating spheres of gas in a certain range of masses.

The mass, in fact, is the prime characteristic that determines where on the main sequence a star will settle down to live its lifetime. The most massive stars become O stars. Less massive stars, like the sun, settle down to live their lives on a much less spectacular district of the main sequence. (See Section 6.5 for a discussion of the evolution of stars of different masses.)

About 90 per cent of stars, those on the main sequence, follow the mass-luminosity relation. The other stars, such as giants or white dwarfs, are special cases that we will discuss later on.

There are many more stars of low mass than there are of high mass. In fact, the number of stars of a given mass in any sufficiently large volume of space increases as we consider stars of lower and lower mass. Thus most of the mass in a cluster of stars, or in a galaxy, comes from stars at the lower end of the H-R diagram. Even though the many faint stars contain most of the total mass in a group of stars, most of the light that we receive from a group of stars, on the contrary, comes from the few very brightest stars. A single O star can outshine a thousand K stars. Thus the majority of the stars in a star system together give off a minor fraction of the light; the few brightest stars in that system give off most of the radiation.

4.3 STELLAR SIZES

Stars appear as points to the naked eye and as small, fuzzy disks through large telescopes. The blurring of the stellar images into disks results from distortion by the earth's atmosphere, and hides the actual size and structure of the surface of the stars. The sun is the only star for which we can measure its angular diameter easily and directly.

Astronomers use an indirect method to find the sizes of most stars. If we know the absolute magnitude of a star (from some type of parallax measurement) and the temperature of the surface of the star (from measuring its spectrum), then we can tell the amount of surface area the star must have. This, of course, depends on the radius.

One type of direct measurement works only for eclipsing binary stars. As the more distant star is hidden behind the nearer star, one can follow the rate at which the intensity of radiation from the further star declines. From this information together with Doppler measurements of the velocities of the components, one can calculate the sizes of the stars. One can sometimes even tell how the brightness varies across a star's disk.

It is much more difficult to measure the size of a single star directly. One way of measuring the diameter is by *lunar occultation*. As the moon moves with respect to the star background, it occults (hides) stars. By studying the light from a star in the fraction of a second it takes for the moon to completely block it, we can deduce the size of the stellar disk. Only a handful of stars have had their diameters measured by this technique up to the present time.

Other methods for measuring the diameters of single stars use a principle called *interferometry*, which will be discussed in the context of radio observations in Chapter 23. Even though the starlight passes through the earth's atmosphere without being occulted by either the moon or by a companion star, the method overcomes the problems of blurring by the earth's atmosphere.

The first such measurements were carried out at Mt. Wilson 50 years ago; only 7 of the stars that subtended the largest angles in the sky had their diameters measured. These were the largest of the closest stars, that is, nearby red giants and supergiants. A modern variation of this technique has been worked out at the University of Maryland by D. G. Currie, who has remeasured the diameters of 4 stars and is continuing his observations.

Over the last ten years, a new procedure called *speckle interferometry* has been developed by A. Labeyrie and his associates at the Paris Observatory in France and at the State University of New York at Stony Brook. At any one instant, the image of a star through a large telescope looks speckled because different parts of the image are affected by different small turbulent areas in the earth's atmosphere. A long exposure blurs these speckles, and gives us the fuzzy disk normally photographed. The speckle interferometry technique involves taking photographs of the speckle pattern with very short exposures—on the order of 1/100 second—or using electronic detection devices and then using mathematical techniques and computer assistance to deduce the properties of the starlight that entered the telescope. Thus, in this manner, from many short exposures of a stellar image, scientists can measure the diameter and the limb darkening. In principle, the true image can probably be reconstructed as it would appear if the atmosphere of the earth were not in the way (Fig. 4–8), though the theoretical problem involved in doing so has not yet been completely solved.

All the above interferometric methods work best for large, cool stars. For the last twenty years, R. Hanbury Brown and his associates in Australia have used another type of stellar interferometer (Fig. 4–9) that works best for hot, bright stars. They have determined the diameters of three dozen stars whose diameters cannot be measured with the other techniques. The

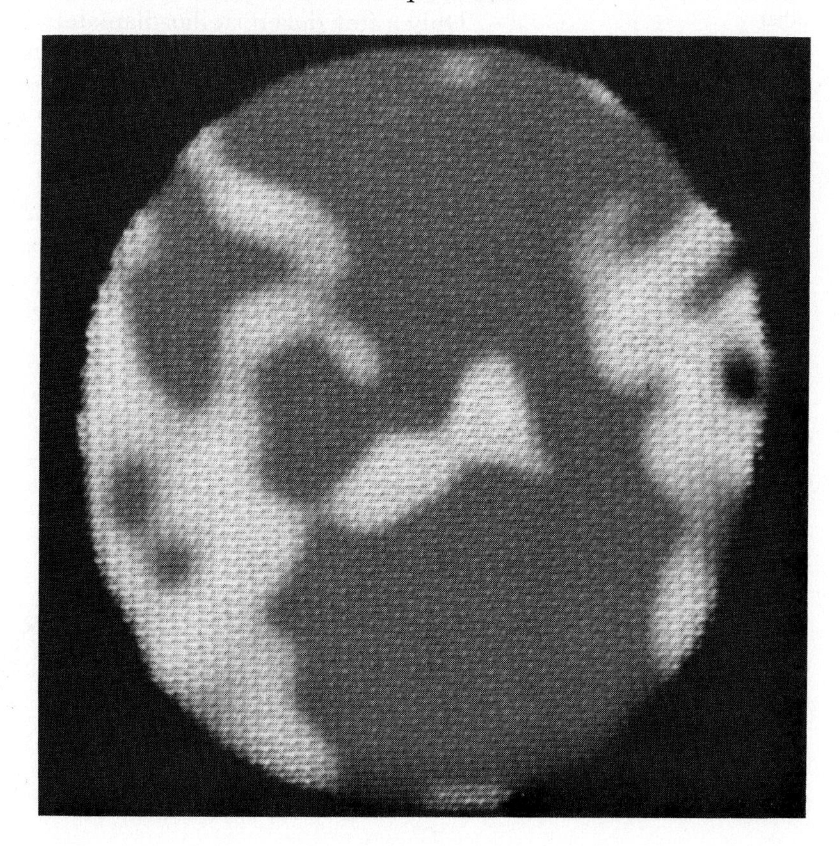

Figure 4–8 This image of the supergiant star Betelgeuse results from computer analysis of data taken with the technique of speckle interferometry. The disk is only 0.08 arc sec across. It had been thought that the large scale structures are huge regions of convection that exist on a larger scale than is observed on the sun (see Fig. 5–19), but it seems that they may unfortunately only be effects introduced in the computer analysis. Speckle analysis can easily measure the diameter of stars or limb darkening, or separate even close binary stars, but theoretical advances remain to be made before the data can be reliably interpreted to show surface structure.

Figure 4–9 Hanbury-Brown's intensity interferometer in Australia, used to measure the diameters of hot stars. The two sets of mirrors can be placed up to 188 meters apart from each other.

smallest, an O5 star called ζ Puppis (zeta of the Stern of the Ship), is only 0.00042 seconds of arc in diameter. At its distance from us, this means that its diameter is 16 times that of the sun. A cooler star, α Canis Minoris (alpha of the Little Dog), an F5 star, is 2.1 solar diameters.

Only a few dozen stellar diameters have been measured directly, and they confirm the indirect measurements (Fig. 4–10). Direct and indirect measurements show that the diameters of main-sequence stars decrease as we go from hotter to cooler; that is, O dwarfs are relatively large and F dwarfs are relatively small. (Recall that "dwarf" simply means "main-sequence star.") Indirect measurements alone show that this trend continues for K and M stars, which are smaller still. As for stars that are not on the main sequence, red giants are indeed giant in size as well as in brightness.

Interferometric techniques have been used not only to measure the

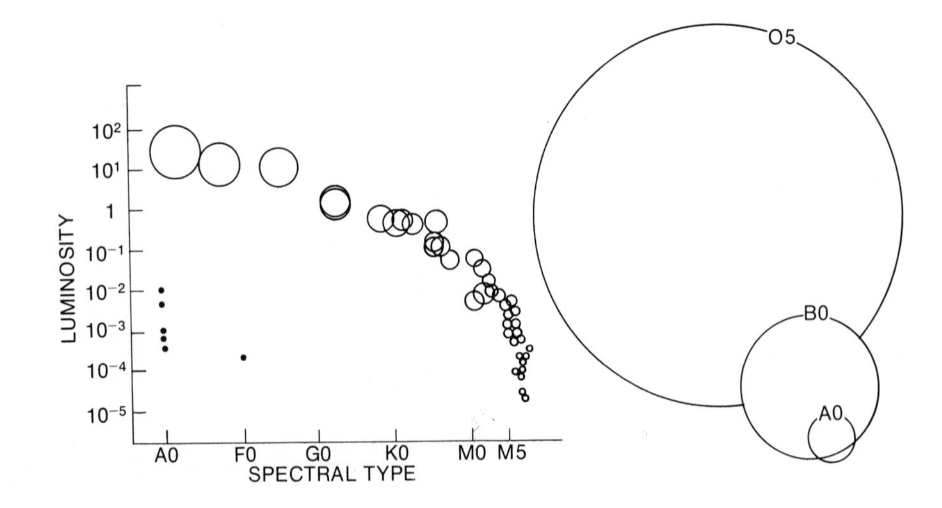

Figure 4–10 The H-R diagram for stars within 17 light years of the sun. The open circles indicate approximately the relative diameters of the main-sequence stars; the white dwarfs are shown as dots. In addition, relative sizes of O, B, and A stars are shown at right.

diameters of individual stars but also to resolve the components of close binary stars. The speckle interferometry technique, for example, has already resolved the components of 3 dozen close binary stars. Several of the systems had not been previously known to be binary.

The latest trend in stellar interferometry is the development of systems of telescopes spread over an area of ground, in order to synthesize the resolution of a single gigantic telescope covering all that area (Fig. 4–11). The technique, which is called *aperture synthesis,* has already been used in radio astronomy, as we shall describe in Section 23.5. For images in visible light, the process is still in the developmental stage, but multi-telescope interferometer systems are foreseen (Fig. 4–12).

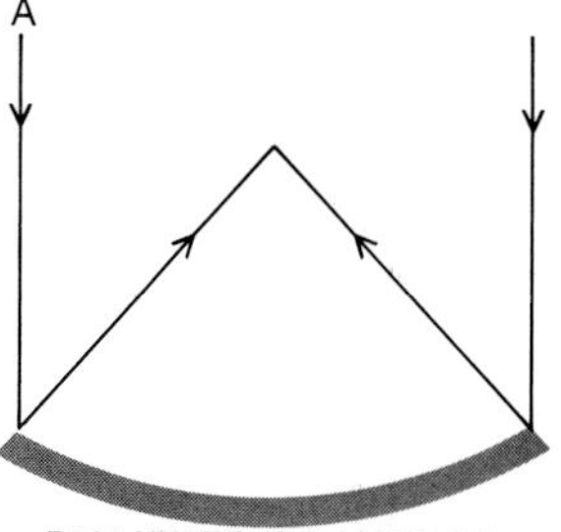

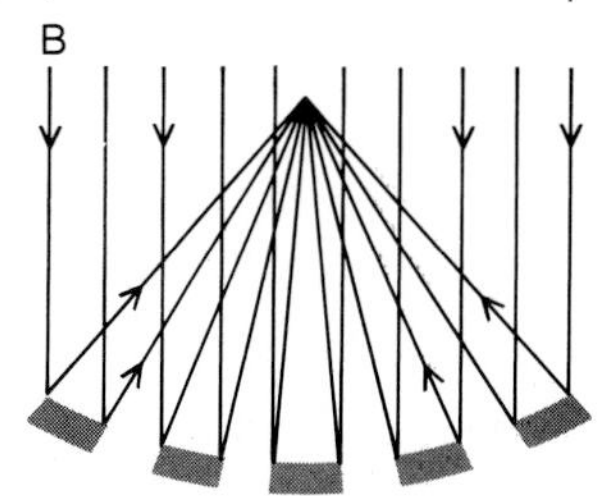

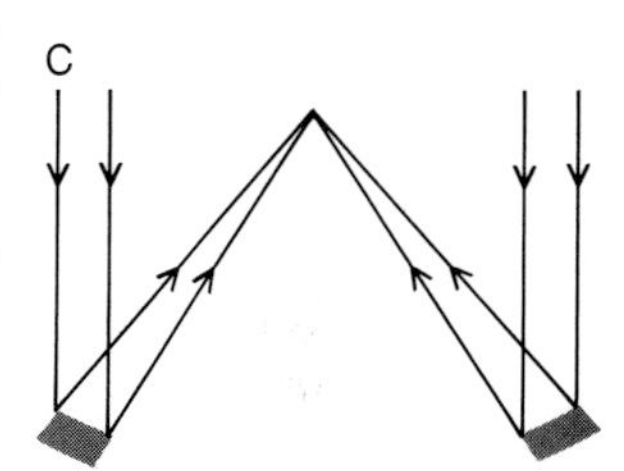

Figure 4–11 A large single mirror (*A*) can be thought of as a set of smaller mirrors (*B*). Since the resolution of a telescope for light of a certain wavelength depends only on the telescope's aperture, retaining only the outermost segments (*C*) matches the resolution of a full-aperture mirror. We can use a property of light called *interference* (see Section 23.5b) to analyze the incoming radiation. The device is then called an *interferometer.*

4.4 VARIABLE STARS

Some stars can be seen to vary in brightness with respect to time. One basic parameter that characterizes the variation of a variable star is the *period.* The period is the time it takes for a star to go through its entire cycle of variation and to return to the original degree of brightness. Thus for a periodic star we can consider the period to be the time from one maximum light phase to the next maximum light phase. We can equally well consider the period to be the time from one minimum light phase to the next minimum light phase, or the time from one point in the middle of the light curve to the next corresponding point (although it might pass through this value at some other point in the cycle as well).

One way for a star to vary in brightness, as we have seen, is for it to be an eclipsing binary. Sometimes we get the light simultaneously from both members of the binary, and sometimes the light from one of the members is at least partially blocked by the other member. But it is possible for individual stars to vary in brightness all by themselves. Thousands of such stars are known in the sky. The periods of the variations can range from seconds for some types of stars to years for others. Some types of variable stars, as we shall see, can be used to provide vital information about the distance scale.

A naming system has been adopted for variable stars (often called simply "variables") that helps us recognize them on any list of stars. The first

Figure 4–12 This proposed array of 1.5-meter telescopes could make observations with the same *resolution* as a single telescope the diameter of the array. Since each telescope could be used independently when not used as part of an interferometer, it has been suggested that each of the telescopes could be constructed and operated by a different country.

variable to be discovered in a constellation is named R, followed by the genitival form ("of the . . .") of the Latin name of the constellation (see Appendix 10), e.g., R Coronae Borealis (R of Corona Borealis). One continues with S, T, U, V, W, X, Y, Z, then RR, RS, etc., up to RZ, then SS (not SR) up to SZ, and so on up to ZZ. Then the system starts over with AA up to AZ, BB (not BA) and so on up to QZ. The letter J is omitted to avoid confusion with I. This system covers the first 334 variables; after that one numbers the stars beginning with V for variable (V335 . . .). It should be noted that variable stars with more commonly known names, such as Polaris and δ Cephei, retain their common names instead of being included in the lettering system.

Many variable stars are studied on photographic plates taken at professional observatories. In the early part of this century, visual and photographic study of variable stars was an important activity for astronomers, and major discoveries followed. Nowadays, not many professionals choose to spend their time studying variable stars, yet it takes a lot of hard work over an extended period of time to keep track of what the variables are doing. An international amateur group called the American Association of Variable Star Observers (AAVSO), with its headquarters in Cambridge, Massachusetts, is now the center of much of the observational activity on variable stars. Its thousands of members, about two-thirds from the United States and one-third from other countries around the world, send in about 150,000 visual magnitude estimates a year, which are compiled and analyzed in Cambridge with the help of computers. One member

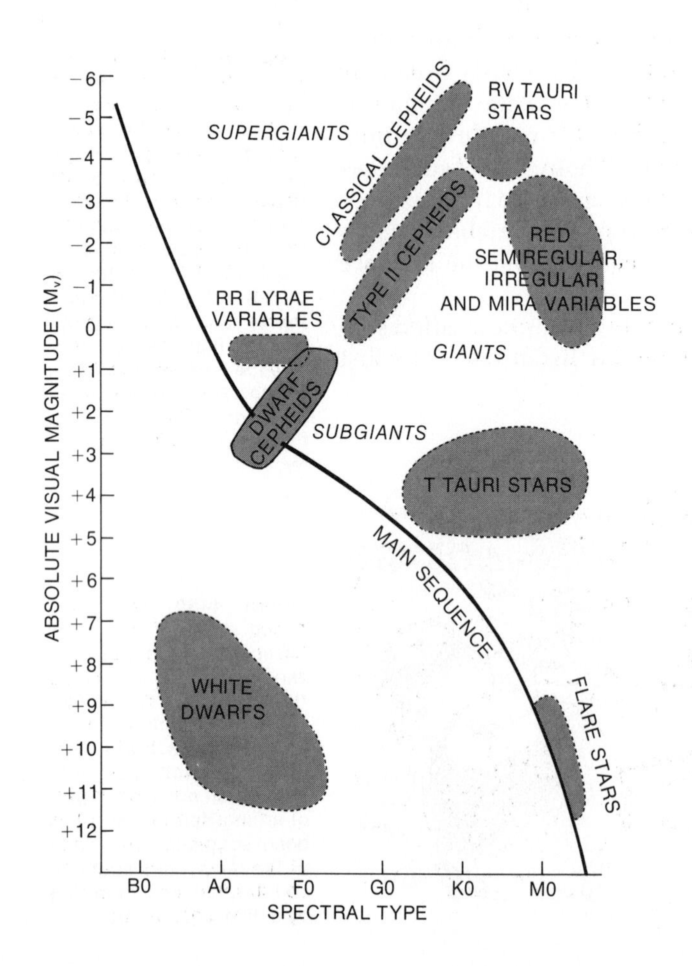

Figure 4–13 There are many types of variable stars, only some of which we discuss in the text. A given type of variable star may appear in only a very limited range of the Hertzsprung-Russell diagram.

alone sent in over 7000 observations he made in a single year. The AAVSO is an outstanding example of the interface between professional and amateur astronomers.

Members of the AAVSO observe the magnitudes of particular stars, estimating the magnitudes by comparing the stars with other stars of known, constant magnitude in the field of view of their telescopes or binoculars. The AAVSO generates charts for their members to allow them to carry out this work. With practice, one can accurately estimate brightness to a few tenths of a magnitude, if the field of view contains enough comparison stars. A very few members have even built photoelectric devices for their own telescopes to make more accurate measurements.

The headquarters staff averages the measurements of all the observers of a given star on a given day, and plots the variations of the magnitude of the observed stars with time. Such plots of magnitude versus time are called *light curves.*

In order to follow variations of stars or other astronomical objects over periods of years, we often want to find the number of days between two given dates. But it is not easy to find the number of days between, say, February 27, 1980 and July 1, 1948. Months have differing numbers of days, and even years have sometimes 365 days and sometimes 366. To eliminate these complications, astronomers often use *Julian Days,* in which days are numbered consecutively. The first Julian Day, J.D. 1, began at noon on January 1, 4713 B.C., a date at which three different cycles involving solar and lunar phenomena and intervals involved in Roman tax collection were simultaneously at their beginnings. Julian Days that correspond to the current time are shown in Table 4–1. If Julian Days are known for two observations, simple subtraction gives the time interval. Another convenience is that since Julian Days begin at noon, the Day number does not change in the course of a night's observations. Further, decimals are used instead of hours, minutes, and seconds.

The system of Julian Days was proposed by Joseph Justus Scaliger, a Renaissance scholar, in 1582. He chose the name in honor of his father, Julius Caesar Scaliger, and not directly after the Roman Emperor whose name we associate with the Julian calendar (described in Section 3.14).

TABLE 4–1 JULIAN DAYS

Noon, January 1	Julian Day	Noon, January 1	Julian Day
1901	2,415,386.0	1981	2,444,606.0
1951	2,433,648.0	1982	2,444,971.0
1976	2,442,779.0	1983	2,445,336.0
1977	2,443,145.0	1984	2,445,701.0
1978	2,443,510.0	1985	2,446,067.0
1979	2,443,875.0	1991	2,448,258.0
1980	2,444,240.0	2001	2,451,911.0

There is a varied menagerie of variable stars corresponding to different physical conditions in the stars, and thus to different locations on the H-R diagram (Fig. 4–13). We will limit our discussion to three types of variables, one of which is especially numerous, and the other two of which have provided important information about the scale of distance in the universe.

Finding the distance scale in the universe, the distances to astronomical objects in linear units like km or A.U., is one of the basic problems that astronomers attack.

4.4a Mira Variables

A type of variable star is often named after its best-known or brightest example. A star named Mira in the constellation Cetus (*o* Ceti, omicron of the Whale) fluctuates in brightness with a long period (Fig. 4–14). Red stars that share this characteristic are called *Mira variables.* The period of a given Mira-type star can be from three months up to about two years. The period of an individual star is not strictly regular; it can vary from time to time from the average period.

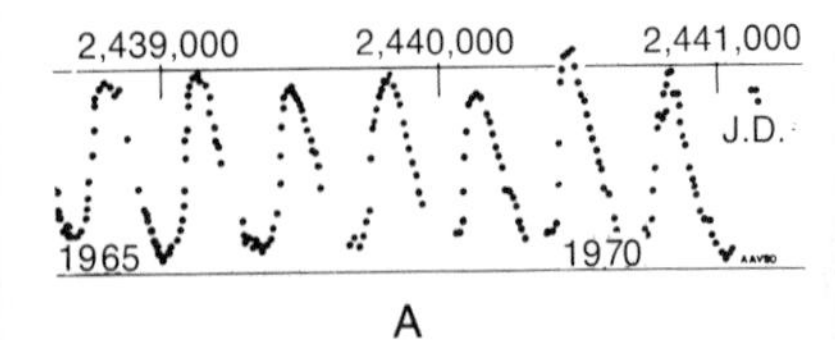

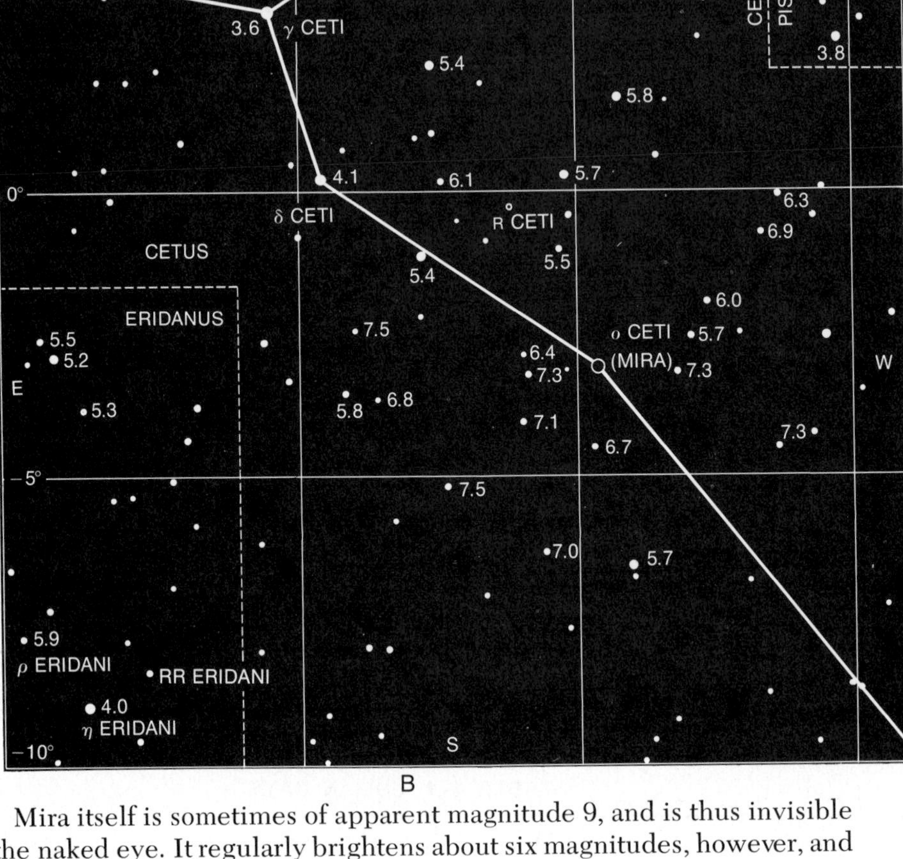

Figure 4–14 The light curve (*A*) and finding chart (*B*) for Mira, the prototype of the class of long-period variables. The magnitudes of nearby stars are marked on the finding chart.

Mira was the first variable star to be discovered. In 1596, fourteen years before the invention of the telescope, a German clergyman named Johann Goldsmit (and known by the Latinized name Fabricius) noticed that a star he had observed at second magnitude was fading from view. Within two months it became too faint for him to see. He called the star Mira (the Latin word for wonderful) because of this wonderful variation in its visibility.

Mira itself is sometimes of apparent magnitude 9, and is thus invisible to the naked eye. It regularly brightens about six magnitudes, however, and changes from fainter to brighter with a period of about 11 months. At maximum brightness it is quite noticeable in the sky. As it brightens, its spectral type changes from M5 to M9. Thus real changes are taking place at the surface of the star that result in a change of temperature.

These stars, which are the most numerous type of variable star in the sky, are also known as *long-period variables.*

4.4b Cepheid Variables

The most important type of variable star in astronomy is the *Cepheid variable* (cef′e-id). The prototype is δ Cephei (delta of Cepheus). Cepheid variables have very regular periods that, for individual Cepheids (as they are called), can be from 1 to 100 days (Fig. 4–15).

Cepheids are relatively rare stars in our galaxy; only about 700 are known. Some have been detected in other galaxies. δ Cephei itself, as an example from our own galaxy, varies between apparent magnitudes 3.6 and 4.3 with a period of 5.4 days (Fig. 4–15). Another Cepheid in our galaxy is Polaris, the north star, which is not as obviously varying since the maximum change in brightness it undergoes is only 0.1 magnitude; it varies between apparent magnitudes 2.5 and 2.6. (Polaris is actually a double star; a 9th magnitude companion is also part of the system.)

In Greek mythology, Cepheus was a king of Ethiopia, Cassiopeia's husband, and Andromeda's father.

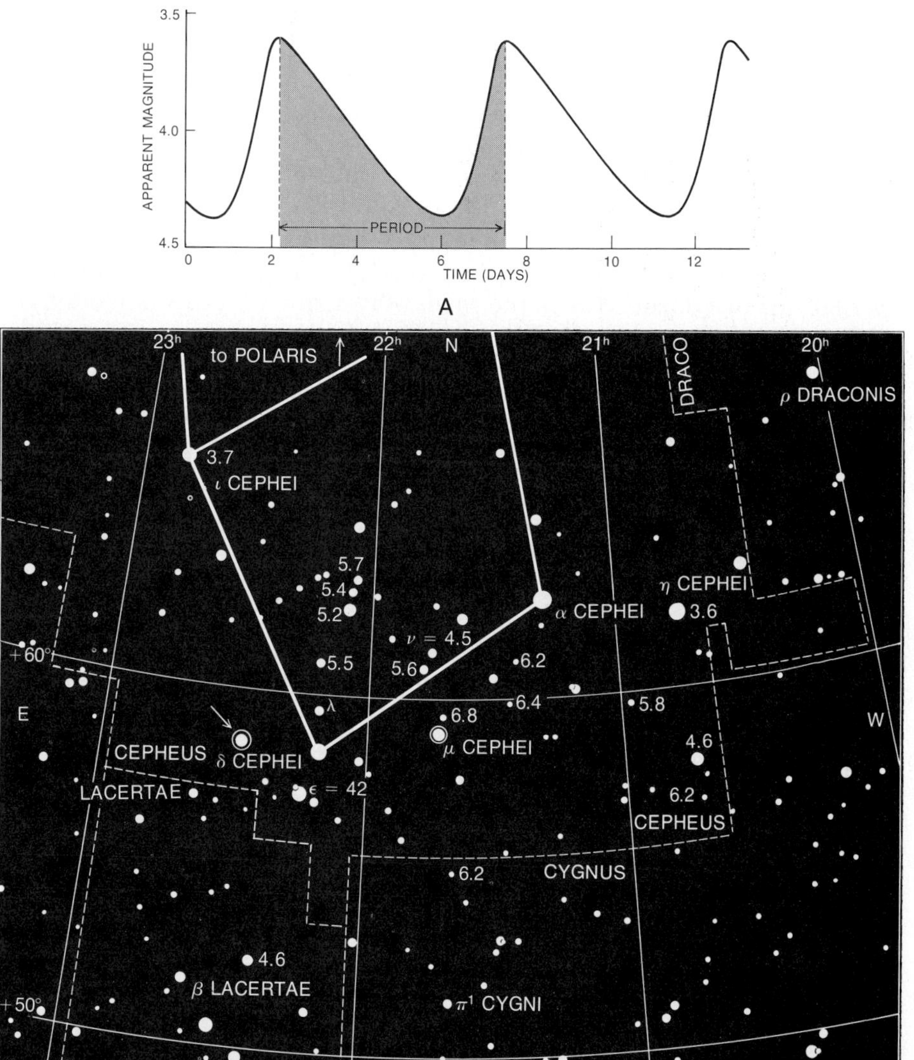

Figure 4–15 The light curve (*A*) and finding chart (*B*) for δ Cephei, the prototype of the class of Cepheid variables.

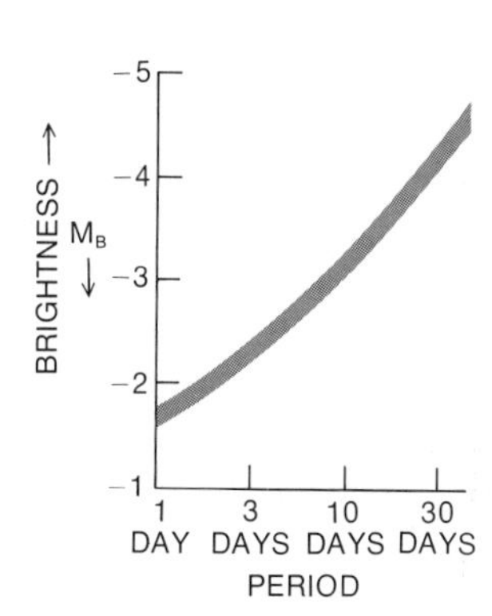

Figure 4–16 The period luminosity relation for Cepheid variables.

Cepheids are important because a relation has been found that links the periods of their light changes, which are simple to measure, with their absolute magnitudes. For example, if we measure that the period of a Cepheid is 10 days, we need only look at Figure 4–16 to see that the star is of absolute magnitude −3. From the absolute magnitudes, as we shall review later in this section, one can find their distances from the sun by using reasoning similar to that used in our earlier discussion of absolute magnitudes of stars. The study of Cepheids is the key to our current understanding of the distance scale of the universe and has allowed us to determine that the objects in the sky that we now call galaxies (discussed in Chapter 23) are giant systems comparable to that of our own galaxy (discussed in Chapter 20).

The story started 70 years ago with Henrietta Leavitt (Fig. 4–17) who, at the Harvard College Observatory, was studying the light curves of variable stars in the southern sky. In particular, she was studying the variables

Figure 4–17. Henrietta S. Leavitt, who worked out the period-luminosity relation, at her desk at the Harvard College Observatory.

in the Large and Small Magellanic Clouds (Fig. 4–18), two hazy areas in the sky that were discovered by the crew of Magellan's expedition around the world when they sailed far south. Whatever the Magellanic Clouds were—we now know them to be galaxies, but Leavitt did not know this—she knew that all the stars in each cloud were at approximately the same distance away from the earth. Thus even though she could only plot the apparent magnitudes, the relation of the absolute magnitudes to each other was exactly the same as the relation of the apparent magnitudes.

By 1912, Leavitt had established the light curves and determined the periods for two dozen stars in the small Magellanic Cloud. She plotted the magnitude of the stars (actually a median brightness, a value between the maximum and minimum brightness) against the period. She realized that there was a fairly strict relation between the two quantities, and that the Cepheids with longer periods were brighter than the Cepheids with shorter periods. By simply measuring the periods, she could determine the magnitude of one star relative to another; each period uniquely corresponds to a magnitude. If only one could find a way to determine the distance to any of the Cepheids by other methods, then one could "calibrate" the period-luminosity curve and use it to tell the absolute magnitude of all the Cepheids.

Unfortunately, not a single Cepheid is close enough to the sun to allow its distance to be determined by the method of trigonometric parallax. More complex, statistical methods had to be used to study the relationships between stellar motions and distances. These methods allowed astronomers to calibrate the period-luminosity relation. Thus Cepheids could be employed as indicators of distance, using the same method that we have already discussed (Section 2.12) to determine the distance: (1) the period-luminosity relation gives us the absolute magnitude of the Cepheid; (2) we then calculate how far a star of that absolute magnitude would have to be moved from the standard distance of 10 parsecs to appear as a star of the apparent magnitude that we observe.

When the calibration of the period-luminosity relation was worked out quantitatively by the American astronomer Harlow Shapley in 1917, the distance to the Magellanic Clouds could be calculated. They were very far away, a distance that we now know means that they are not even in our galaxy! Instead, they are galaxies by themselves, two small irregular galaxies that are companions of our own, larger, galaxy.

Cepheids can also be seen not only in our own galaxy and in the Magellanic clouds but also in more distant galaxies. They are bright stars (giants, in fact) and so can be seen at quite a distance. The use of Cepheids is the prime method of establishing the distance to all the nearer galaxies.

As the Cepheids were further studied, astronomers could make a model for their variations on the basis of their light curves and of their spectra. The absorption lines in the spectra of a Cepheid show Doppler shifts that prove that the star is actually pulsating—expanding and contracting—as it changes in brightness. The size of a Cepheid variable may change by 5 to 10 per cent as it goes through its pulsations. The surface temperature also varies. The star is brightest roughly at its hottest phase. A detailed theory of stellar pulsations that explains these effects has been worked out.

Some stars that are found in globular clusters (which are described in Section 4.5) have periods of weeks, and are about 1.5 magnitudes fainter than ordinary Cepheids for a given period of variation. At first they were thought to be ordinary Cepheids. On this assumption, distances were computed for them, and thus for the clusters in which they were located. But it was later realized that though these stars

Figure 4–18 From the southern hemisphere, the Magellanic Clouds are high in the sky. In this view from Perth, Australia, the Large Magellanic Cloud is about 20° above the horizon and the Small Magellanic Cloud is twice as high. They are not quite this obvious to the naked eye. Canopus, the second brightest star in the sky, appears at the lower left.

follow a period-luminosity diagram, they follow a slightly different period-luminosity diagram than the regular Cepheids (Fig. 4–19). In fact, they are about 1.5 magnitudes fainter for a given period than the ordinary, classical Cepheid.

These fainter stars are now called Type II Cepheids, or W Virginis stars, after the prototype of the class. The regular Cepheids are called Type I Cepheids or classical Cepheids. In this book we shall limit our use of the term "Cepheids" to mean classical Cepheids.

When the distinction between Type I and Type II Cepheids was realized, distances had to be revised for all the galaxies whose distances had been calculated on the basis of the magnitudes of their Cepheids. It had been assumed that all Cepheids followed the period-luminosity relation that we now apply only to Type II Cepheids. The distances to galaxies were correctly calculated only when a few Type I Cepheids were discovered in galactic clusters in our own galaxy (which are described in Section 4.5). The distances to these galactic clusters could be measured by fitting their H-R diagrams, as described in Section 4.5a. This work was carried out in the mid-1950's. Since the Type I Cepheids were actually intrinsically

Figure 4–19 The period-luminosity relation for Type II Cepheids is different from that for ordinary Cepheids, a discovery that led to revisions in the scale of distances to faraway objects.

Figure 4–20 The distribution of the periods of RR Lyrae stars (left peak) and Cepheid variable stars (right peak).

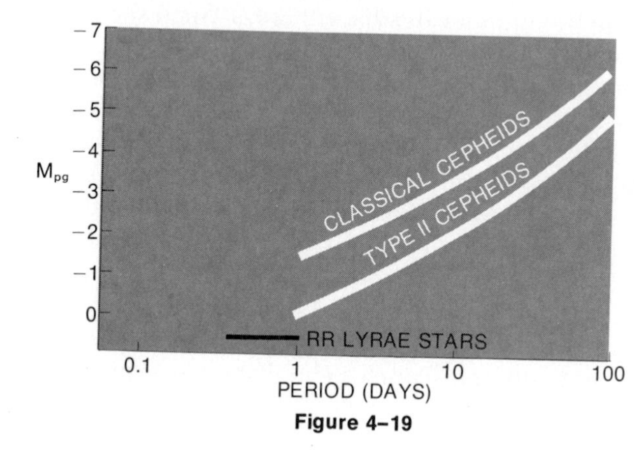

Figure 4–19

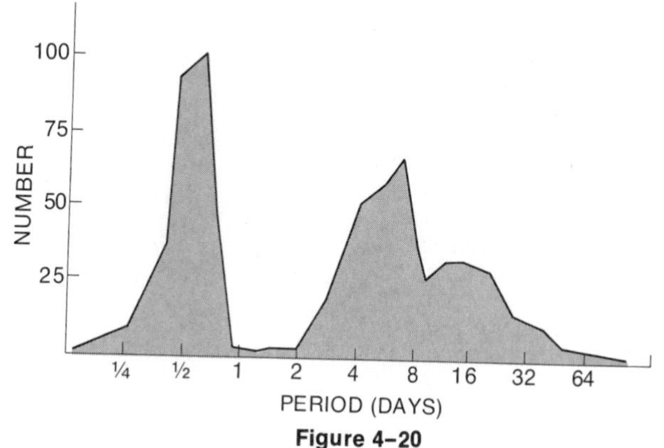

Figure 4–20

brighter than had been originally thought, the galaxies containing them were really farther away than had been previously calculated. The difference of 1.5 magnitudes corresponded to a doubling of the distance to these galaxies. The newspapers wrote that the universe had doubled in size.

4.4c *RR Lyrae Variables*

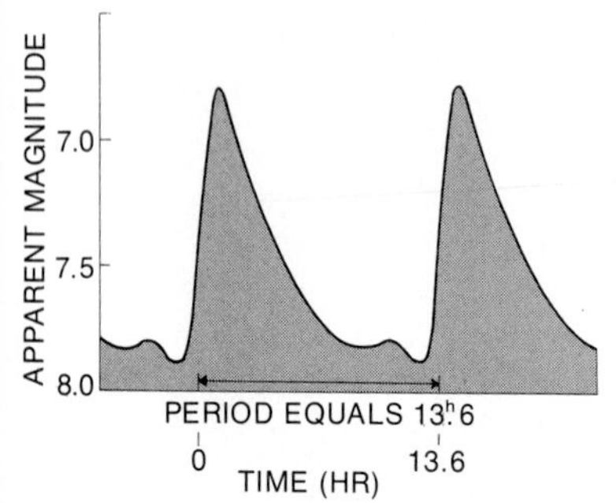

Figure 4–21 The light curve of RR Lyrae, the prototype of the class known as RR Lyrae variables or cluster variables.

Many stars are known to have short regular periods, less than one day (Fig. 4–20). Certain of these stars, no matter what their period, have the same average absolute magnitude. Again, as with Cepheids, none of these short-period variables is within range to obtain a trigonometric parallax, and the calibration of the absolute magnitude scale had to be accomplished by their membership in clusters.

Such stars are called *RR Lyrae stars* after the prototype of the class (Fig. 4–21). Since many of these stars appear in globular clusters (which will be described in Section 4.5), RR Lyrae stars are also called *cluster variables.*

In any case, it seems that all these stars are of absolute magnitude approximately 0. For a while, astronomers used the magnitude 0.0 for these stars, which seemed to be a suspiciously round number. We now think that there are several subclasses of these stars, each with a slightly different absolute magnitude, but that in any case the magnitudes are between 0 and 1 (possibly even fitting a period-luminosity relation).

Once we detect an RR Lyrae star, we immediately know its absolute magnitude, since all the absolute magnitudes are essentially the same. Just as before, we measure the star's apparent magnitude and can thus easily calculate its distance and the distance to the cluster.

4.5 CLUSTERS AND STELLAR POPULATIONS

The stars that we can observe from the earth are not distributed uniformly over the sky. Much of the lack of uniformity comes from the fact that our sun is located in the disk of a flat collection of stars, dust and gas called the Milky Way Galaxy. (We will discuss this at great length in Chapter 20). The plane of the Milky Way Galaxy appears in our sky as a hazy band that can be seen stretched from horizon to horizon when we are in a place far from city lights. From a vantage point outside our galaxy, we would see a nuclear bulge from which arms unwind in a spiral.

But even aside from the hazy band of the Milky Way, the distribution of stars in our sky is not uniform. There are certain areas where the number of stars is very much higher than the number in adjacent areas. Such sections of the sky are called *star clusters* (Color Plates 5 and 6).

The Pleiades are often known as the Seven Sisters, *after the seven daughters of Atlas who were pursued by Orion and who were given refuge in the sky. That one is missing—the Lost Pleiad—has long been noticed. Of course, a seventh star is present (and hundreds of others as well), although too faint to be plainly seen with the naked eye. The Pleiades seem to be riding on the back of Taurus.*

One type of star cluster appears as just an increase in the number of stars in that limited area of sky. Such clusters are called *open clusters,* or *galactic clusters* (Fig. 4–22). The most familiar example of a galactic cluster is the Pleiades, a group of stars visible in the winter sky. The eye sees at least six stars very close together. With binoculars or the smallest telescopes, dozens more can be seen. A larger telescope reveals hundreds of stars. Another galactic cluster is called the Hyades, which form the "v" that outlines the face of Taurus, the bull, a constellation best visible in the winter sky (Fig. 4–23). More than a thousand such clusters are known, most of them too faint to be seen except with telescopes. (The distances to many of

Figure 4–22 The galactic cluster M67 in the constellation Cancer, the crab.

these clusters can be determined, as discussed in Section 4.5a.) It has been found that all the stars in a galactic cluster are packed into a volume not more than 10 parsecs across.

Stars in galactic clusters seem to be representative of stars in the spiral

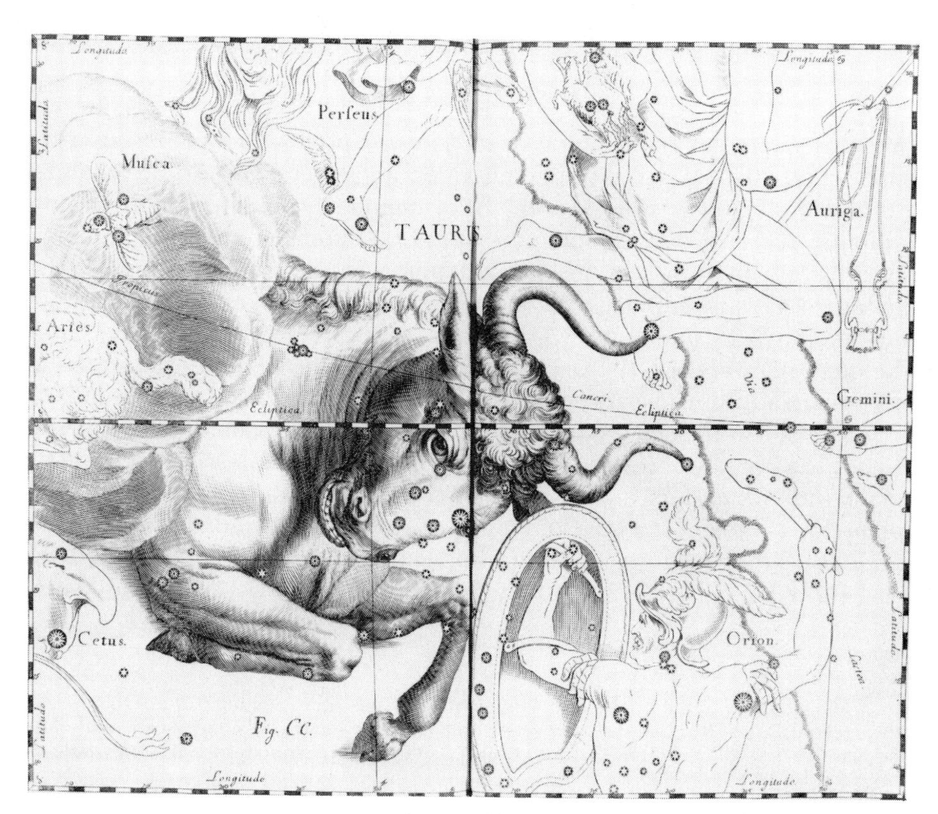

Figure 4–23 The Hyades are a galactic cluster that outlines the face of Taurus, the bull, pictured here in a plate from the atlas of Hevelius published in 1690.

A

B

Figure 4–24 (*A*) The globular cluster M3 in the constellation Canes Venatici, the hunting dogs. (*B*) The loose globular cluster NGC 6791.

arms of our galaxy and of other galaxies. (Spiral structure is described in Section 20.5.) When the spectra of stars in galactic clusters are analyzed to find the relative abundances of the chemical elements in their atmospheres, we find that over 90 per cent of the atoms are hydrogen, most of the rest are helium, and less than 1 per cent are elements heavier than helium.* This is similar to the composition of the sun (which is discussed in Section 5.2b and Table 5–1). Stars with these properties are said to belong to *Population I*. They appear in regions in which dust and gas are present in addition to stars (such dust and gas are discussed in Section 20.1).

The second major type of star cluster appears in a small telescope as a small, hazy area in the sky. Observation with larger telescopes can distinguish individual stars, and reveals that these clusters are really composed of many thousands of stars packed together in a very limited space. The clusters take spherical forms, and are known as *globular clusters* (Fig. 4–24).

Globular clusters can contain 10,000 to one million stars, in contrast

*The structure of the nuclei of the elements is discussed in Section 6.3. For the present, we need only know that hydrogen, whose dominant form contains only one nuclear particle, is the least massive element, and helium, whose dominant form contains 4 nuclear particles, is the second least massive element.

with the 20 to several hundred stars in a galactic cluster. A globular cluster can fill a volume up to 100 parsecs across. A globular cluster has fewer stars at its periphery and more closely packed stars toward the center. One can find the distance to many of these clusters by observing the RR Lyrae stars, which are found in many of them.

Globular clusters seem to avoid the plane of our galaxy and are typically found only in regions above and below the galactic plane. We say that they are in the galactic *halo*. The abundances of the elements heavier than helium in globular clusters are much lower, by a factor of 10 or more, than their abundances in the sun and in other Population I stars. Stars with these characteristics are said to belong to *Population II*. They are found in regions that do not contain any interstellar gas or dust.

Walter Baade, who originally set up this system of stellar populations on the basis of his observational work at the Mt. Wilson Observatory thirty years ago, divided all the stars into either Population I or Population II. Population I stars represent younger stars, and are typical of the spiral arms. Population II stars, which have relatively low abundances of elements heavier than helium, represent older stars. Population II stars are found in globular clusters, in the galactic halo, and in the central regions of our galaxy (Fig. 4–25). But it has since been realized that not all stars fall clearly into one division or the other, because actually there are gradations in between. Still, the division into stellar populations is useful in making broad distinctions among groups of stars and for interpreting the structure of our galaxy.

4.5a H-R Diagrams for Galactic (Open) Clusters

The Hertzsprung-Russell diagram for several galactic clusters is shown in Figure 4–26. For the purposes of this discussion it is most important to note that the horizontal axis is a measure of temperature and the vertical axis is a measure of brightness.

A measure of temperature often used for the horizontal axis in such plots is the difference between the magnitudes measured in two spectral regions, for example B-V. (This standard system of filters was defined in Section 2.12.)

The difference B-V is called the *color index* (Fig. 4–27). The blue magnitude, B, measures bluer radiation than the visual magnitude, V. A very hot star is brighter in the blue than in the visible; thus V is fainter, that is, a higher number, than B. Therefore B-V is negative. Do not be confused by the fact that the magnitude of a brighter star is a lower number (more negative) than that of a fainter star.

A

B

Figure 4–25 A photograph of the Andromeda Galaxy in ultraviolet light (*A*) shows the younger, hotter stars of Population I better than does the photograph in red light (*B*), which is better for observing the older, redder stars of Population II. Population I stars are concentrated in the spiral arms, while the nucleus consists of Population II stars.

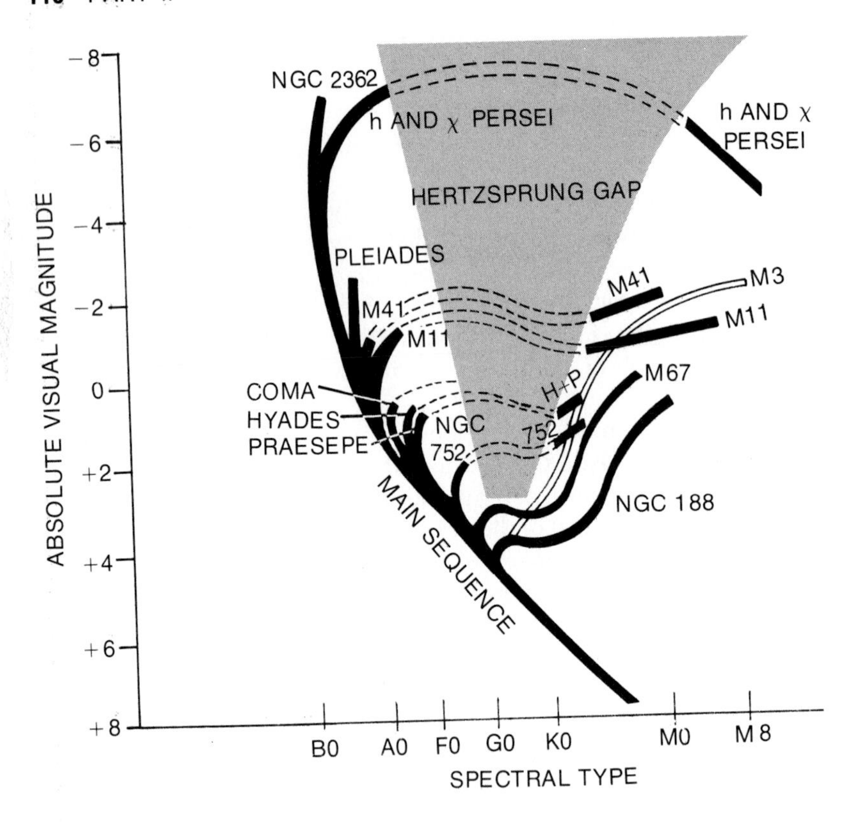

Figure 4–26 This Hertzsprung-Russell diagram of several galactic clusters shows that their fainter stars are on the main sequence, while the brighter stars are above and to the right of the main sequence. Almost all stars in the younger clusters, like h and χ Persei, follow the main sequence, while the hotter and more massive members of older clusters, like M67, have had time to evolve toward the red giant region. By observing the point where the cluster turns off the main sequence, we deduce the length of time since its stars were formed, i.e., the age of the cluster. The presence of the *Hertzsprung gap* (shaded), a region in which few stars are found, indicates that stars evolve rapidly through this part of the diagram.

Part of the H-R diagram for the globular cluster M3 (hollow bar) is graphed for comparison.

Examples: ζ Pup (blue-white) O5 B = +1.96 V = +2.25 C.I. = B-V = −0.29

Betelgeuse (red) M2 B = +2.55 V = +0.69 C.I. = B-V = +1.86

Thus labeling the horizontal axis with the color index is equivalent to labeling it with temperature in kelvins. Since the color index can be measured easily with a telescope, and does not require theoretical interpretation or analysis of spectra before graphing, it is the measure of temperature ordinarily graphed. These plots, a type of H-R diagram, are often called *color-magnitude diagrams* (Fig. 4–28).

The color index is zero for an A star of about 10,000 K. It falls in the range from −0.3 for the hottest stars to about +2.0 for the coolest. One can also compute a color index for the U and B (ultraviolet and blue) magnitudes, U-B, a color index for

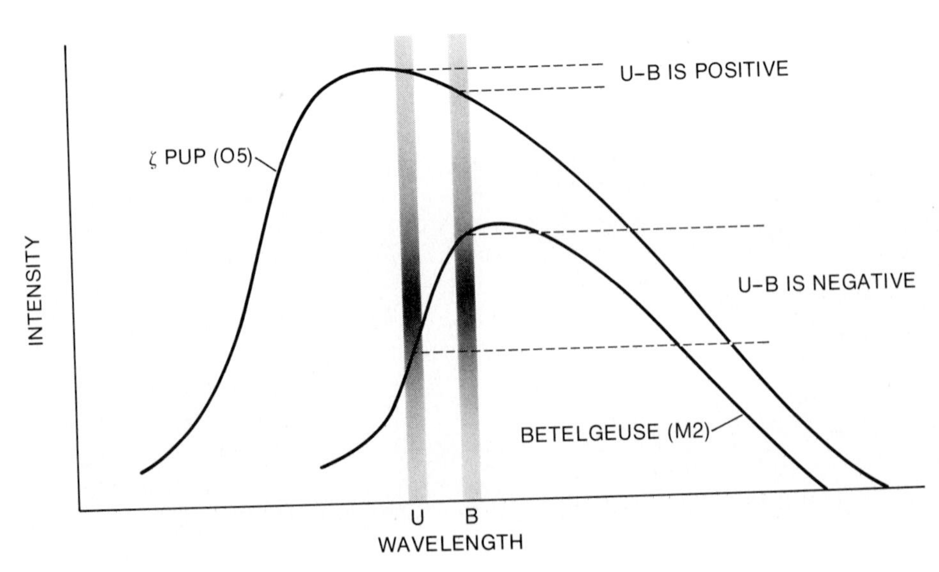

Figure 4–27 Hotter stars have negative color indices while cooler stars have positive color indices.

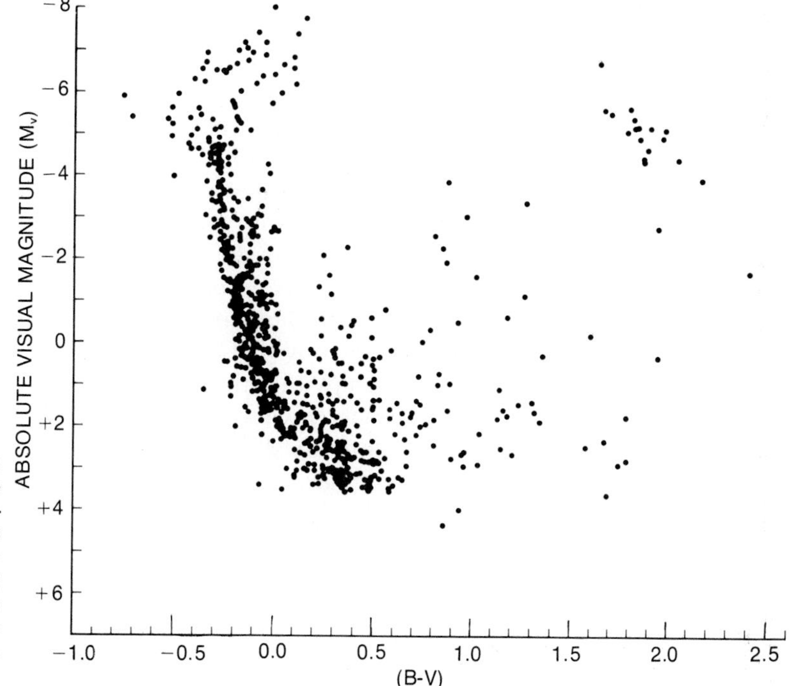

Figure 4–28 The quantities B and V can be directly measured at a telescope. The color index, B–V, is function of temperature, so plotting B or V versus B–V gives a type of H–R diagram, known as a *color-magnitude diagram,* by merely plotting these easily measured quantities. The graph shows the color-magnitude diagram for the double cluster in Perseus, h and χ Persei (h and chi Persei).

the m_{pg} and m_v (photographic and visual) magnitudes, or indeed a color index for magnitudes measured in any two spectral regions. These other color indices in the visual part of the spectrum have the same sense as the color index for B-V, that is, negative for hot stars and positive for cool ones. We must merely be certain that we are subtracting the longer wavelength measurement from the shorter wavelength measurement. Infrared astronomers use other standard systems of filters in their region of the spectrum.

The H-R diagrams for different galactic clusters appear to agree over part of the main sequence but diverge at the upper left. These stars at the upper left are the more massive ones. They are more luminous and use up their nuclear fuel at a faster rate than the cooler, more numerous, ordinary stars like the sun. The massive stars are on the main sequence in the stable, middle-age part of their lifetimes. Later in their lives they become larger and thus brighter, since they then have more surface area to radiate light. At that time they move upward on the H-R diagram from the main sequence. Their outer layers may also, after a time, become cooler, and the star will then move toward the right on the H-R diagram.

When we examine the H-R diagram of a cluster, we are observing a group of stars all of which were presumably formed at the same time in the same region of space and so have similar chemical compositions. We presume that all the stars "fell" on the main sequence during most of their lives (that is, the points representing their magnitudes and temperatures were on the main sequence). Points representing the fainter stars usually fall on the main sequence of the H-R diagram. But for some galactic clusters (Fig. 4–26) the points representing the brighter, more massive stars are to the upper right of the main sequence. We interpret this to mean that these stars have passed through the phase of their lives in which they fell on the main sequence, and have "moved off" the main sequence as they aged and evolved. Thus we deduce that a given cluster is older than the main-sequence lifetimes of those stars in that cluster that have moved off the main sequence. On the other hand, the fact that stars of a certain color and absolute magnitude, and thus a certain mass, are still on the main sequence indicates that the cluster is younger than their main-sequence lifetimes. The age of the cluster is between these two values. *Thus observing the H-R diagram for a cluster of stars by simply measuring both the B and V (apparent) magnitudes gives us enough information to determine the age of a cluster,* how long it has been since the stars in the cluster were formed. In this way we can study the evolution of stars.

If we measure apparent magnitudes for a large number of stars in a cluster, we can deduce the absolute magnitude by the following method. We already know that all the stars in the cluster are at the same distance. Since we know that the lower part of the main sequence will agree for all galactic clusters, we can take an H-R diagram measured in apparent magnitudes and merely slide it up and down along lines of constant color until we see the best agreement with respect to a standard color-magnitude diagram. Then we have found the number of magnitudes by which the absolute and apparent magnitude scales differ and thus *from the H-R diagram we can easily calculate the distance to the cluster.*

For example, let us consider just one point on the main sequence. A-type stars are known to have absolute magnitude of about +2. If we see an A star of apparent magnitude +7, we know that the cluster is farther enough away than 10 parsecs that its stars appear 5 magnitudes = 100 times fainter. Because of the inverse square law of brightness, we know that the star and its cluster are 10 times further away than 10 parsecs, which is 100 parsecs (see Section 2.13). Actually, this procedure is more precise when we *fit* (superimpose) a whole segment of a main sequence, or some other part of an H-R diagram. The procedure is thus called *fitting,* as in "fitting the main sequence." One problem, if we measure only color, is that matter located between the stars can affect the color of starlight as it passes (Section 20–4).

For a single star, study of its spectrum can tell us its spectral type and its luminosity class. Then the H-R diagram (Fig. 2–37) gives us its absolute magnitude. By comparison with the apparent magnitude, we find its distance; this is called the method of *spectroscopic parallax.*

Figure 4–26 shows that stars in the double cluster in Perseus, h and χ Persei, follow the main sequence for most of its length. The line of points

Figure 4–29 The double cluster in Perseus, h and χ Persei, a pair of galactic clusters that are readily visible in a small telescope and close enough together that they appear in the same field of view. Perseus is a northern constellation that is most prominent in the winter sky. In Greek mythology, Perseus slew the Gorgon Medusa and saved Andromeda from a sea monster.

representing the stars turns off the main sequence only at its upper left. Thus h and χ Persei are very young clusters. They may have formed only 2×10^6 years ago, a short time scale from an astronomical point of view. The Pleiades, which turns off at a somewhat more positive color index, is still a young cluster but is somewhat older than the double cluster in Perseus. It may be 10^8 years old. The Hyades, which turns off the main sequence still further down, may be much older, possibly 10^9 years. The plot of the stars representing the Pleiades on an H-R diagram gives the visual appearance of agreeing with the standard main sequence for less than half the length of the plotted curve. But there are many more of the cooler, low-mass stars than there are of the hotter, high-mass stars. Thus most of the stars in these clusters actually lie on the main sequence. Only a small fraction of the stars have evolved off the main sequence.

4.5b H-R Diagrams for Globular Clusters

The color-magnitude diagram for a globular cluster is shown in Figure 4–30. Color-magnitude diagrams for all globular clusters have similar forms.

These color-magnitude diagrams have only short main sequences, so evidently the globular clusters must be very old. Some stars even appear in a *horizontal branch* at approximately absolute magnitude zero. This horizontal branch goes between the left side of the diagram and the stars on the

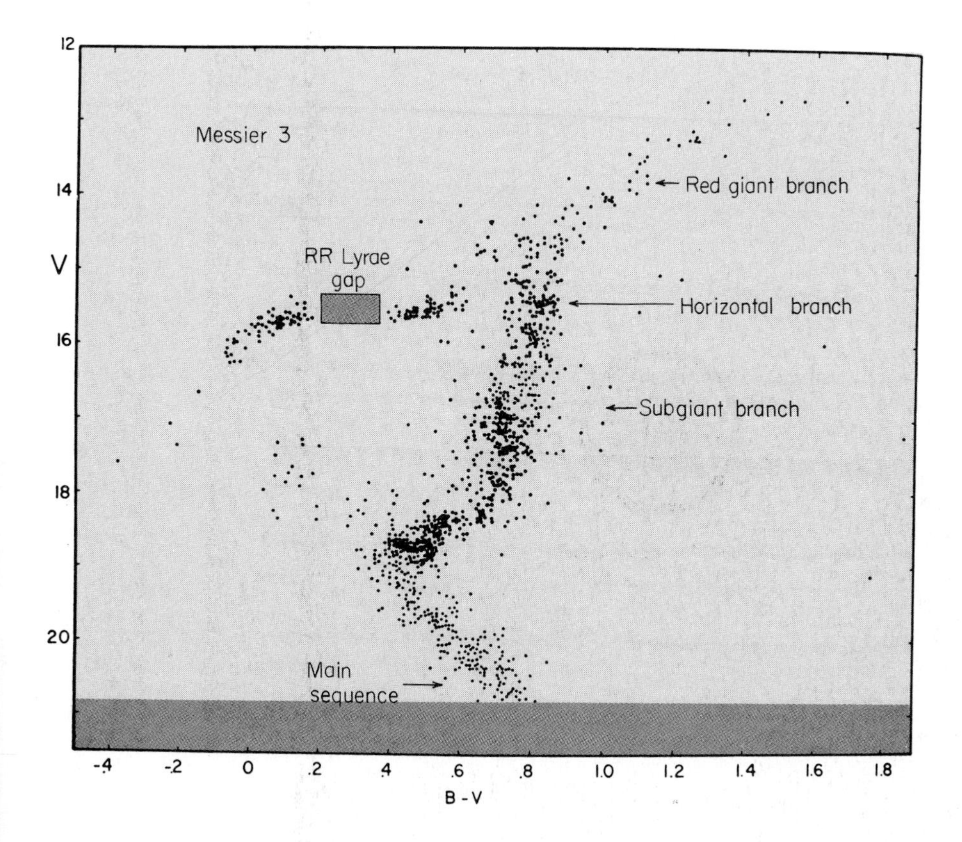

Figure 4–30 The Hertzsprung-Russell diagram for the globular cluster M3. There is a gap in the horizontal branch where no stars of constant brightness are found. The RR Lyrae variables fall here, some 200 for this cluster. We can observe only the members of the cluster that are brighter than a certain limit that depends on the size of our telescope and the darkness of the night sky. Thus no stars appear in the shaded region at the bottom.

right side of the diagram that have long since turned off the main sequence. In the horizontal branch there is a gap where stars of constant brightness are not found. Only the RR Lyrae stars fall in this gap.

The notion that the globular clusters are much older than the galactic clusters leads us to an understanding of the history of the formation of stars in our galaxy. Remember that the abundances of the metals, the elements heavier than helium, are extremely low in the Population II stars of the globular clusters. Thus there were very few heavy elements early in the history of our galaxy. The metals must have been formed later in the scheme of things, and we shall see in Chapters 6 through 8 how element formation in the interiors of stars and in the explosions of dying stars has enriched the gas from which the younger, Population I, stars have formed.

4.5c *Of Globular Clusters, X-Ray Bursts, and Black Holes*

For the last few years, since scientists have been able to place telescopes in orbit, they have been able to observe x-rays from celestial sources. (Chapter 22 contains a discussion of x-ray sources in our galaxy.) Some of the x-ray sources are associated with globular clusters; when one measures the position in the sky of one of these x-ray sources and looks at a visual image of the region, a globular cluster is right there.

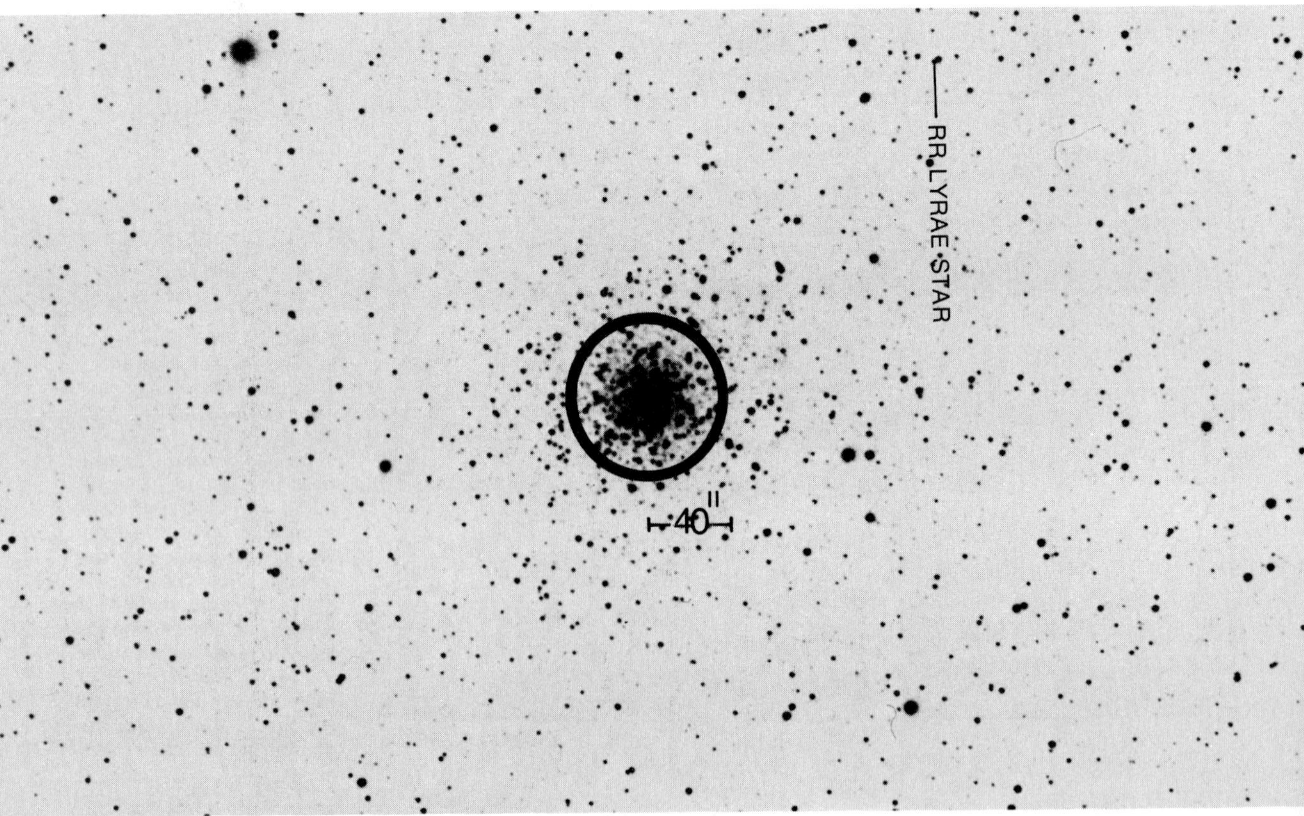

Figure 4–31 The circle shows the position of the source of x-ray pulses that are being interpreted as indicating the presence of a massive black hole. When this circle, whose size indicates the uncertainty of the position measurement from the SAS-3 satellite, is drawn on a photograph of the sky, globular cluster NGC 6624 is included in it. It is difficult to tell from inspection of the photograph whether the lone RR Lyrae star that has been detected in this field of view is indeed part of the cluster (and thus the cluster is much bigger than is apparent on the photograph), but the star's distance is the same as the 5.0/kiloparsecs distance to the cluster that has been derived from fitting its horizontal branch to a standard H-R diagram.

The strength of the x-radiation varies strongly. In one of the globular clusters, NGC 6624 (Fig. 4–31) (the 6624th object in the New General Catalogue, a standard list of visible objects in the sky compiled a hundred years ago), the x-radiation sometimes changes by a factor of 5 from one minute to the next. In 1975, telescopes aboard the Astronomical Netherlands Satellite and other satellites recorded an even stranger variation (Fig. 4–32): intense bursts of x-rays from this globular cluster, each representing an increase in intensity of 20 to 30 times in half a second, followed by a decay in the intensity for the next 8 seconds (Fig. 4–33). The bursts occur about every $4^{1}/_{2}$ hours; the interval between them varies by about 15 minutes. In each burst the cluster emits about a million times the energy radiated by the sun in the same time. At least 5 globular clusters are known to give off such bursts. Similar strange bursts also come from other directions in the sky, and it is possible that there are globular clusters there too, perhaps hidden by interstellar dust.

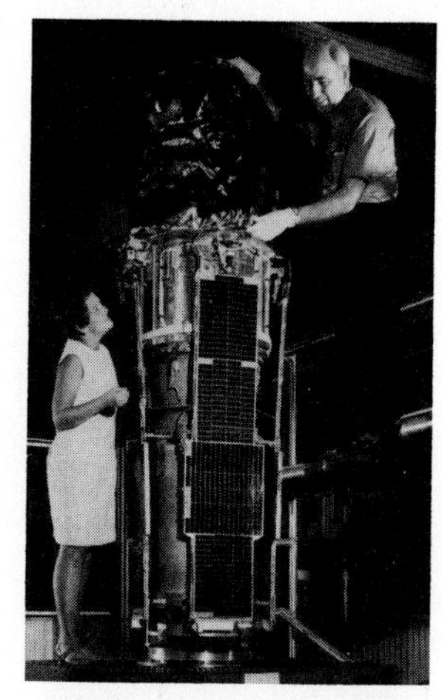

Figure 4–32 The SAS-1 satellite, one of NASA's series of Small Astronomy Satellites. SAS-3 has been observing the strange x-ray bursts.

The most likely current interpretation of these bursts involves a very strange notion that had been advanced on theoretical grounds even prior to the discovery of bursts: a very massive black hole may exist at the center of many globular clusters. (We discuss at length the properties of black holes in Chapter 9, but suffice it to say here that they are places in space where matter has collapsed in on itself to such an extent that the gravity has increased to a point where even electromagnetic radiation is prevented from escaping. Matter in the process of falling into a black hole can become very hot—after it enters the black hole itself the matter is heard from no more—and x-rays can result.)

Some theoreticians predict that the observed kinds of x-rays—with rapid variations plus specific sharp bursts—would result if a black hole containing about 1000 times the mass of the sun existed in the center of a globular cluster such as NGC 6624. Even the fact that the sharp bursts include rises in intensity over a time scale of half a second is predicted by the theoretical calculations. Strange as it sounds, the theory based on the existence of a black hole in the globular cluster is the most reasonable theory yet presented for the observed effects, though other possible explanations are also being considered. As globular clusters contain old stars, it seems reasonable that many of them could have exhausted the main-sequence phase of their lives, and, because of the force of gravity, a collection of black holes in the center of the globular cluster could have resulted. The burned-out hulks could have merged to become a single, high-mass black hole. Even some of the alternative models involve black holes, though less massive ones orbiting ordinary stars, with x-radiation resulting in the manner described in Section 9.5.

Some two dozen sources of these x-ray bursts are now known, most of which are not associated with globular clusters. They will be discussed on page 498.

The deduction that black holes may well exist in globular clusters is an interesting merger of modern and traditional astronomy. For example, to know how much energy is being emitted in the form of x-rays by the globular cluster, we must know how far away the cluster is. The distance to NGC 6624 is found by fitting the horizontal branch of its Hertzsprung-Russell diagram to a standard H-R diagram. The value comes out to be 5.0 kiloparsecs, about half the distance between the sun and the center of our galaxy. Only one RR Lyrae star can be found in the globular cluster, and even that star is so far from the visible concentration of stars that we cannot be certain that it is a member of the cluster. But it is close enough that it could be part of the outer fringe of the cluster, and its distance is also about 5 kiloparsecs from the sun.

Figure 4–33 An x-ray burst (shaded) from the globular cluster NGC 6624 observed on September 28, 1975. The burst lasted about 10 seconds.

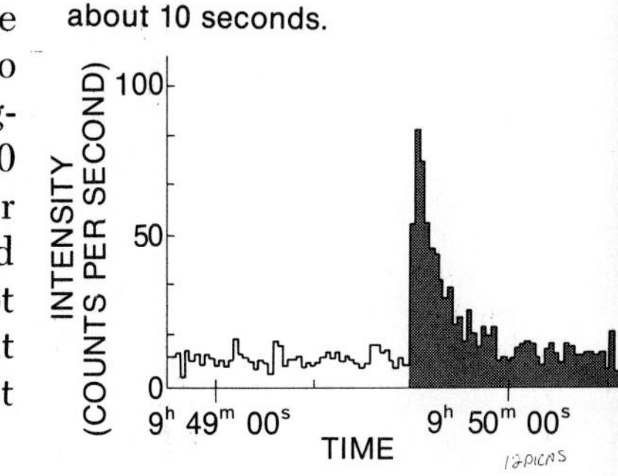

Globular clusters, far from being uninteresting relics of old stars, turn out to be fascinating places that are earning increased attention.

SUMMARY AND OUTLINE OF CHAPTER 4

Aims: To discuss multiple stars, variable stars, and stellar clusters, and to draw conclusions about the distances to stars and about the evolution of stars from their study

Binary stars (Section 4.1)
- Optical doubles
- Visual binaries
- Binaries with composite spectra
- Spectroscopic binaries
- Eclipsing binaries
- Astrometric binaries

Determination of stellar masses (Section 4.2)
- Mass-luminosity relation

Determination of stellar sizes (Section 4.3)
- Indirect: calculate from absolute magnitude or measure in eclipsing binary system
- Direct: interferometry

Variable stars (Section 4.4)
- Mira variables (Section 4.4a)
- Cepheid variables (Section 4.4b)
  - Period-luminosity relation
  - Uses for determining distances
  - Types I and II lead to different distance scale
- RR Lyrae variables (Section 4.4b)
  - All of approximately the same absolute magnitude
  - Uses for determining distances

Clusters and stellar populations (Section 4.5)
- Galactic (open) clusters
  - Population I: relatively high abundance of metals
  - Representative of spiral arms
  - 20 to several hundred members
  - Turn-off on H-R diagram gives age
- Globular clusters
  - Population II: relatively low abundance of metals
  - Representative of galactic halo
  - 10^4 to 10^6 members
  - Old enough to have H-R diagrams with horizontal branches

Color-magnitude diagrams (Section 4.5a)
- Equivalent to H-R diagrams
- U,B,V can be measured directly at the telescope
- Color index (U-B, and so on) corresponds to temperature

Spectroscopic parallax (Section 4.5a)
- Compare measured color-magnitude diagram with one calibrated in terms of absolute magnitude

X-ray bursts from globular clusters (Section 4.5c)
- Massive black holes may be the cause of the bursts

PROBLEMS

1. Sketch the orbit of a double star that is simultaneously a visual, an eclipsing, and a spectroscopic binary.
2. How much brighter than the sun is a main-sequence star whose mass is 50 times that of the sun?
3. (a) Assume that an eclipsing binary contains two identical stars. Sketch the intensity of light received as a function of time. (b) Sketch to the same scale another curve to show the result if both stars were much larger while the orbit size stayed the same.
4. A Cepheid variable has a period of 30 days. What is its absolute magnitude?
5. Explain briefly how observations of a Cepheid variable in a distant galaxy can be used to find the distance to the galaxy.
6. Briefly list the properties of Population I stars and contrast them with Population II stars.
7. Star A has color index 0.0. Star B has color index 1.0. Which has a higher surface temperature?
8. A main-sequence star is 3 times the mass of the sun. What is its luminosity relative to that of the sun?
9. (a) Use the mass-luminosity relation to determine about how many times brighter than the sun are the most massive main-sequence stars of which we know. (b) How many times fainter are the least massive main-sequence stars?
10. When we see the light from a distant galaxy, are we seeing mostly low or high mass stars?
11. When we measure the gravitational effects of a distant galaxy on its neighbors, are we measuring the effects of mostly low or high mass stars?
12. If we take two stones, one twice the diameter of the other, and put them in an oven until they are heated to the same temperature and begin glowing, what will be the relationship between the light given off by the stones?
13. (a) What is the absolute magnitude of a classical Cepheid with a 10-day period? (Refer to Fig. 4–19.) (b) What is the absolute magnitude of a Type II Cepheid with a 10-day period? (c) How many times brighter is the classical Cepheid than the Type II Cepheid? (d) If they both have the same apparent magnitude, how many times closer or farther away is the classical Cepheid?
14. Cluster X has a higher fraction of main-sequence B stars than cluster Y. Which cluster is probably older?
15. What is the advantage of studying the H-R diagram of a cluster, compared to that of stars in the general field?
16. How does the discovery of x-ray bursts from globular clusters affect your conception of astronomy, which may have been that astronomy involved a leisurely contemplation of the stars?

OBSERVING PROJECTS:

(Use the star charts on this book's end papers.)

1. (a) Try to see Alcor with your naked eye. (b) With a small telescope, observe Alcor, Mizar A, and Mizar B.
2. Observe Albiero with a small telescope.
3. Follow the brightness variation of Algol over 3 nights. (Each month, *Sky and Telescope* lists the times of minimum of Algol.)
4. Mira has a period of about 11 months. Using the finding chart in Figure 4–14, observe the brightness of Mira as often as possible for a few months.

5

The Sun

We have thus far discussed a range of individual stars of different spectral classes and have discussed groupings of stars in close physical proximity to each other. Studying these distant stars has allowed us to learn a lot about the properties of stars and how they evolve. But not all stars are far away; one is close at hand. By studying the sun, we not only learn about the properties of a particular star but also can study the details of processes that undoubtedly take place in more distant stars as well. We will first discuss the *quiet sun,* the solar phenomena that appear every day. Afterwards, we will discuss the *active sun,* solar phenomena that appear non-uniformly on the sun and vary over time.

5.1 BASIC STRUCTURE OF THE SUN

We think of the sun as the bright ball of gas that appears to travel across our sky every day. We are seeing only one layer of the sun; the properties of the solar interior below that layer and of the solar atmosphere above that layer are very different. The outermost parts of the solar atmosphere even extend through interplanetary space beyond the orbit of the earth.

The layer that we see is called the *photosphere,* which simply means the sphere from which the light comes (from the Greek *photos,* light). When we speak of the diameter of the sun, 1,400,000 km (864,000 miles), we are really speaking of the diameter of the photosphere. This is 110 times the diameter of the earth. The volume of a sphere is proportional to the cube of the radius, r, or of the diameter, d. (Since the volume of a sphere is $\frac{4}{3}\pi r^3 = \frac{4}{3}\pi\left(\frac{d}{2}\right)^3 = \frac{1}{6}\pi d^3$, the volume of the sun is $110^3 = 1$ million times the volume of the earth; one million earths could fit within the whole sun.)

The sun's mass is 2×10^{33} grams, 330,000 times the mass of the earth. As is typical of many stars, about 94 per cent of the atoms and nuclei in the outer parts are hydrogen, about 5.9 per cent are helium, and a mixture of all the other elements make up the remaining one tenth of one per cent. The overall composition of the interior is not very different.

The sun, observed in white light on July 4, 1974.

The sun is an average star, in that it is a G2 dwarf and falls right in the middle of the Hertzsprung-Russell diagram. Radiation from the photosphere peaks in the middle of the visible spectrum; after all, our eyes evolved over time to be sensitive to that region of the spectrum because the greatest amount of the solar radiation occurred there. If we lived on a planet orbiting an object that emitted mostly x-rays, we, like Superman, might have x-ray vision.

The disk of the sun takes up about one-half a degree across the sky; we say that it *subtends* one-half degree. This is large enough for us to see detail on the solar surface, and we shall describe that detail in subsequent sections.

The sun is an average of 150 million km away from the earth (93 million miles), one of the few numbers in astronomy (and in this section) that most people memorize. (The exact value of this number is one astronomical unit.) Because the earth's orbit around the sun is slightly elliptical (see Chapter 10), we are sometimes as close as 147,104,000 km (91,426,000 miles) and sometimes as far away as 152,103,000 km (94,533,000 miles). Thus sometimes the sun subtends a slightly smaller angle in the sky than at other times; it varies from 32 min 35 sec of arc to 31 min 31 sec of arc.

Beneath the photosphere is the solar *interior*. All the solar energy is generated there at the solar *core*, which is about 10 per cent of the solar diameter at this stage of the sun's life. An important theorem about stellar structure, called the Russell-Vogt theorem, states that all stars of the same mass and of the same chemical composition (if we neglect the effects of rotation and magnetic fields) evolve in the same way. This means that they follow the same evolutionary track on the H-R diagram. We will discuss the evolution of stars of 1 solar mass, including the sun, in Chapters 6 and 7.

Figure 5–1 The parts of the solar atmosphere and interior. The solar surface is depicted as it appears through unfiltered light (called *white light*), and through filters that pass only light of certain elements in certain temperature stages. These specific wavelengths, counterclockwise from the top, are the Hα line of hydrogen, which appears in the red, the K line of ionized calcium, which appears in the part of the ultraviolet that passes through the earth's atmosphere and can be seen with the naked eye, and the line of ionized helium at 304 Å in the extreme ultraviolet, which can be observed only from rockets and satellites.

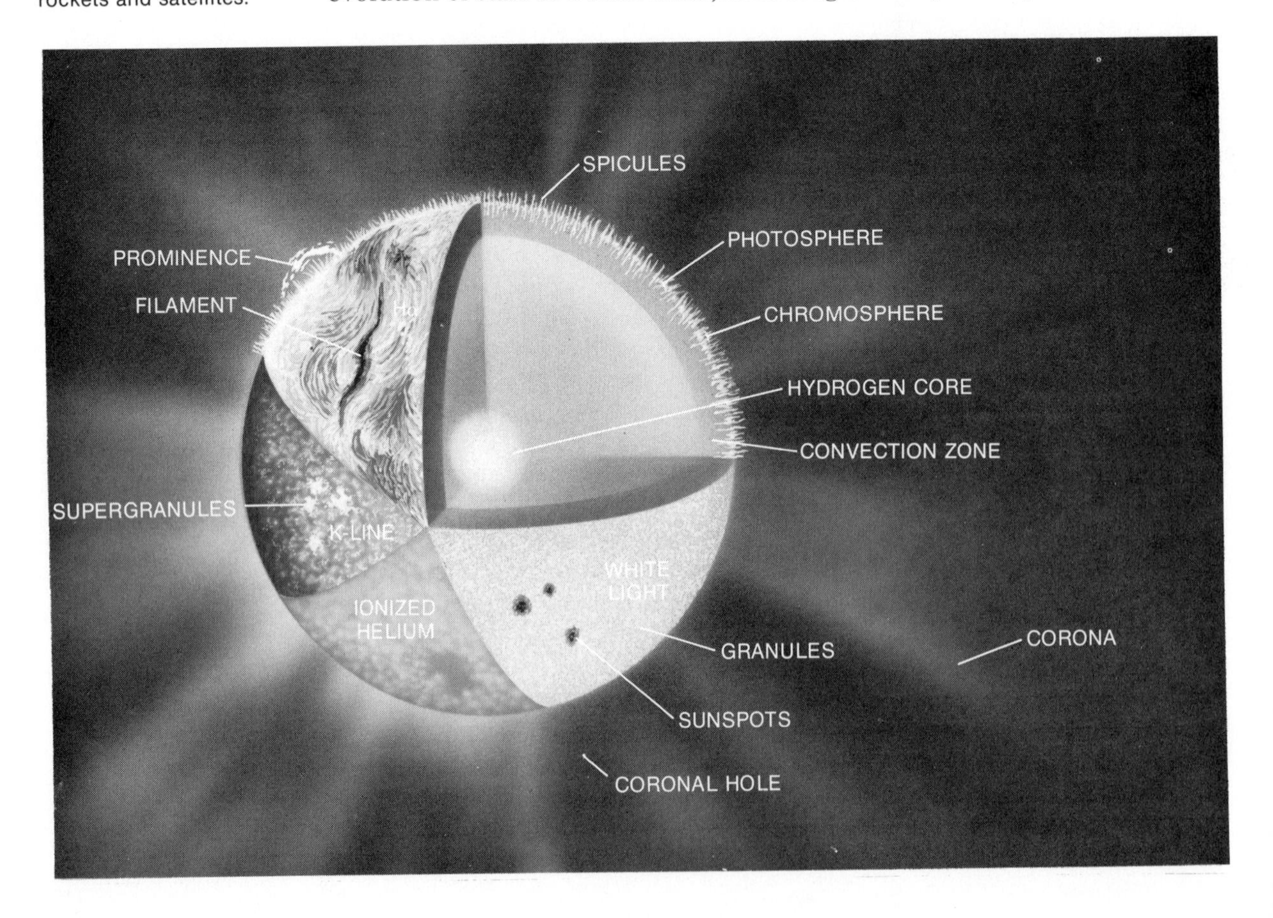

The photosphere is the lowest level of the *solar atmosphere* (Fig. 5–1). The parts of the atmosphere above the photosphere are very tenuous, and contribute only a small fraction to the total mass of the sun. These upper layers are very much fainter than the photosphere, and are best seen with the naked eye during a solar eclipse, when the moon blocks the photospheric radiation from reaching our eyes directly. Now we can also study these upper layers with special instruments on the ground and in orbit around the earth.

Just above the photosphere is a jagged, spiky layer about 10,000 km thick, only about 1.5 per cent of the solar radius. This layer glows colorfully pinkish when seen at an eclipse, and is thus called the *chromosphere* (from the Greek *chromos*, color). Above the chromosphere, a pearly-white halo called the *corona* (from the Latin, crown) extends tens of millions of kilometers into space. The corona is continually expanding into interplanetary space and in this form is called the *solar wind*. We shall discuss all these phenomena in succeeding sections.

5.2 THE PHOTOSPHERE

The sun is the only star close enough to allow us to study its surface in detail. The phenomena we discover on the sun presumably also exist on other stars.

We are discussing here only the part of the solar spectrum in the visible.

5.2a High-Resolution Observations of the Photosphere

One major limitation on observing the sun is the turbulence in the earth's atmosphere. We have seen how turbulence is a limiting factor in stellar observing, as it causes the twinkling of the stars. The problem is even more serious for studies of the sun, since the sun is up in the daytime when the atmosphere is heated by the solar radiation and so is more turbulent than it is at night.

Only at the very best observing sites, specially chosen for their steady solar observing characteristics, can one see detail on the sun subtending an angle as small as 1 second of arc. This corresponds to about 700 km on the solar surface, the distance from New York to Detroit. Occasionally, objects 1/2 arc second across can be seen. Since atmospheric turbulence causes bad "seeing" that limits our ability to observe small-scale detail (see Section 3.5), it does little good to build solar telescopes larger than about 50 cm (20 inches) in order to increase resolution, even though larger telescopes are inherently capable of resolving finer detail (see Section 3.1). The sun is so bright that we do not usually need larger apertures for collecting more light either, although sometimes solar astronomers spread out the light in its spectrum so much that larger aperture telescopes are a benefit.

When one studies the solar surface with 1 arc second resolution, one sees a salt-and-pepper texture called *granulation* (Fig. 5–2). The effect is similar to that shown by boiling liquids on earth, which are undergoing what is called *convection*. Convection is a way in which energy can be transported—conduction and radiation are other methods. Granulation on the sun seems to be an effect of convection. Each granule is only about 1000 km across, and represents a volume of gas that is rising from and falling to a shell of convection, called the *convection zone*, located below the photo-

Granulation can be seen in white light—*all the radiation from the visible part of the spectrum taken together without being filtered—as distinguished from, say, red light or light of the Hα line of hydrogen.*

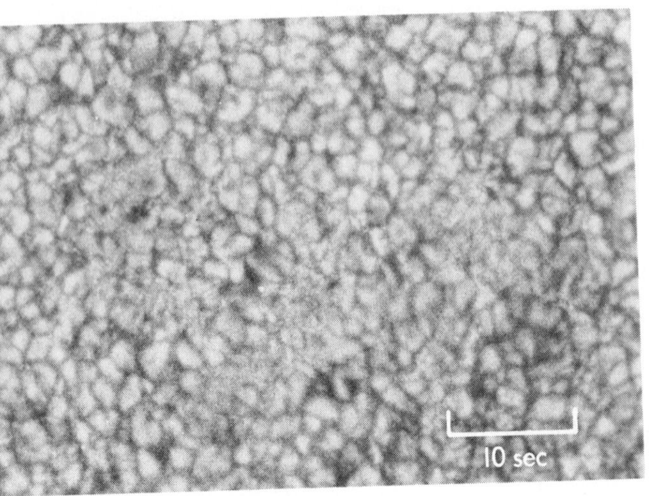

A

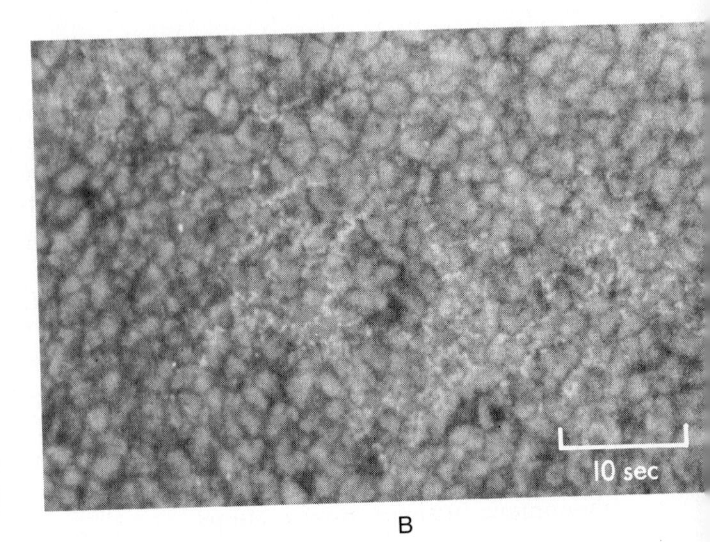

B

Figure 5–2 (*A*) Granulation on the solar surface. One arc sec is 700 km on the sun. (*B*) When we tune our filter not quite at the center of a strong spectral line (here we are 2 Å from the Hα line center), other phenomena become visible such as the bright *filigree*, seen here meandering among the granules.

sphere. The granules are convectively carrying energy from the hot solar interior to the base of the photosphere.

It was discovered in the early 1960's that areas of the upper photosphere are oscillating up and down with a 5-minute period. This result, which was totally unexpected, is now thought to be caused by waves of energy coming from the convection zone.

5.2b The Photospheric Spectrum

*We are discussing here only the part of the solar spectrum in the visible. The formation of the Fraunhofer lines was discussed in Section 2.7. Note that to absor**b** is spelled with a "**b**" but absor**p**tion is spelled with a "p."*

The spectrum of the solar photosphere, as are the spectra of all G2 dwarfs, is a continuous spectrum with absorption lines (Fig. 5–3). Tens of thousands of these absorption lines, which are also called Fraunhofer lines, have been photographed and catalogued. They come from most of the chemical elements, although some of the elements have many lines in their spectra and some have very few. The majority of the lines in the solar spectrum come from iron (Fig. 5–4) and such other elements as magnesium, aluminum, calcium, titanium, chromium, nickel, and sodium. The hydrogen Balmer lines are strong but few in number. Helium is not excited at the low temperatures of the photosphere, and so its spectral lines do not appear.

Fraunhofer, in 1814, labeled the strongest of the absorption lines in the solar spectrum with letters from A through H. His C line, in the red, is now known to be the first line in the Balmer series of hydrogen, and is called Hα (H alpha). We still use Fraunhofer's notation for some of the strong lines. The D lines, a pair of lines close together in the yellow part of the spectrum, are caused by neutral sodium (Na I). The H line and the K line, both in the part of the violet spectrum that is barely visible to the eye, are caused by singly ionized calcium (Ca II). None of the elements other than helium makes up as much as one-tenth of one per cent of the number of hydrogen atoms (Table 5–1). Nevertheless, the D, H, and K lines are among the strongest in the visible spectrum. Of the elements other than hydrogen and helium, sodium and calcium are among the most abundant. Also, most of the absorbing power of these elements at the temperatures of the photosphere and chromosphere is concentrated in just a few lines. Furthermore, these lines occur in the visible part of the spectrum. Remember, too, that

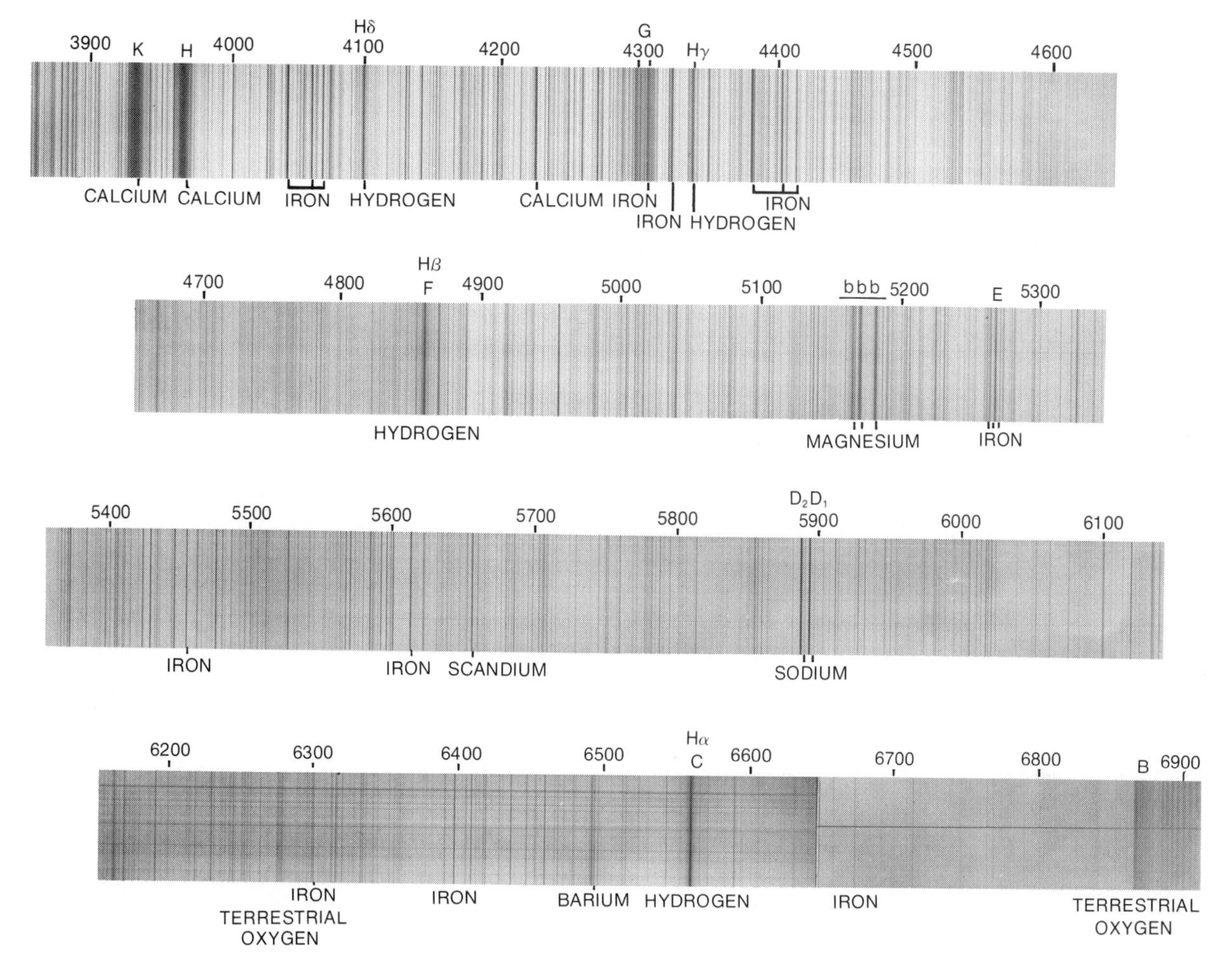

Figure 5–3 The visible part of the solar spectrum.

lines in the Balmer series of hydrogen are not transitions to the ground state, and are thus relatively weak compared to ground state (Lyman) transitions in hydrogen, whereas the D, H, and K lines arise from transitions to the ground states of sodium and calcium. At the temperatures of the photosphere and chromosphere, transitions involving the ground states of these atoms are stronger than transitions involving only higher energy levels. Lyman alpha of hydrogen, which is in the ultraviolet part of the spectrum and which cannot therefore be seen from the ground, is much stronger than Hα, and would be a fairer comparison to the D, H, and K lines.

Observations made at the highest spatial resolutions in light from differing parts of the line profiles of the Hα, H, and K lines show a variety of phenomena on a scale of 1 arc sec or smaller.

The continuous spectrum, called the *continuum*, and most of the absorption lines, are formed together throughout the photospheric layers. The

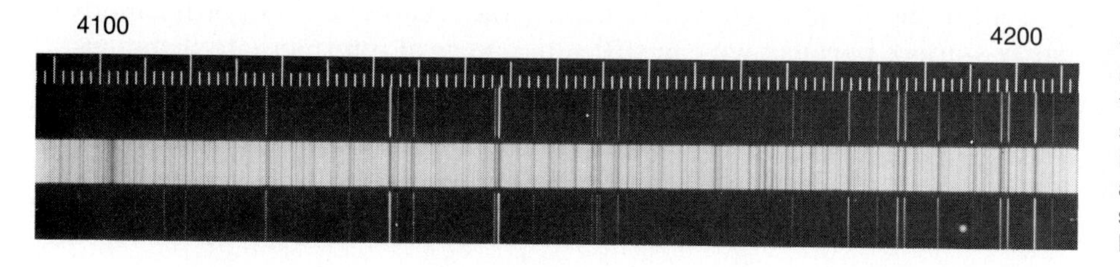

Figure 5–4 This region of the solar spectrum (center strip) is enlarged to illustrate that many of the absorption lines result from iron, whose emission lines are visible in the comparison spectrum (above and below).

TABLE 5–1 SOLAR ABUNDANCES OF THE MOST COMMON ELEMENTS

	Symbol	Atomic Number
For each 1,000,000 atoms of hydrogen, there are	H	1
63,000 atoms of helium	He	2
690 atoms of oxygen	O	8
420 atoms of carbon	C	6
87 atoms of nitrogen	N	7
45 atoms of silicon	Si	14
40 atoms of magnesium	Mg	12
37 atoms of neon	Ne	10
32 atoms of iron	Fe	26
16 atoms of sulfur	S	16
3 atoms of aluminum	Al	13
2 atoms of calcium	Ca	20
2 atoms of sodium	Na	11
2 atoms of nickel	Ni	28
1 atom of argon	Ar	18

energy emitted in them is not formed in these layers, but has rather been transported upward from the solar interior. We have seen (in Section 2.10) that the continuous part of the visible and infrared regions of the solar spectrum is formed by negative ions of hydrogen, H^-.

Now that we have discussed the solar spectrum itself, let us turn to some of the quantities that go into theoretical studies of the spectrum.

We on earth have an intuitive notion of what temperature is. But in astronomy we must be very precise about definitions of such terms. Different ways of defining the temperature for the same volume of gas may, and in fact usually do, lead to different values for the various temperatures. The temperature that we have discussed (Section 2.7) that causes electrons to be excited to high energy levels, for example, is called the *excitation temperature*.

A

B

Figure 5–5 On a day when the opacity of the atmosphere is high, we don't see very far (*A*). When the opacity of the atmosphere is lower, we can see much farther (*B*). In either case, the optical depth does not amount to much by the time our line of sight reaches nearby objects such as the foreground buildings. But in case *A* it becomes large at a distance closer to us than the Eiffel Tower. The view is from the Paris Observatory.

5.2c *The Solar Opacity and the Limb*

When we look at the sun (which we must do through special filters, for the sun is bright enough to burn our retinas if we look at the photosphere directly) it appears as a disk. But why, since the sun is just a ball of gas, does it appear to have a sharp edge? Understanding the answer to this question leads us to a fuller understanding of the properties of the photosphere and of radiating gases in general.

A gas may be transparent, partially transparent, or opaque. On a very clear day, we can see a long distance through the air. If the day is hazy, we can no longer see distant objects. The air has become opaque. Actually, it is not completely opaque, but only partially opaque or partially transparent (it is all in the point of view, like calling the weather either partly cloudy or partly sunny), because we can still see objects at intermediate distances. If the air is very foggy, then we cannot see very far at all.

The *opacity* (Fig. 5–5) is a measure of how opaque a gas is per unit of length. If the opacity of a gas is large, then we cannot see very far through it. If the opacity of a gas is very small, then the gas is fairly transparent and we can see a long way through it. However, even for a gas of low opacity (which is measured per cm), if we look through a lot of it, the opacity adds up. When the gas becomes completely opaque, we say that the *optical depth* is great. On the other extreme, when the optical depth is small, a gas is partially transparent; in other words, we can see things more or less murkily through it.

Optical depth *is a measure of how far one can see through a gas, and depends on both the opacity (per cm) and the distance.*

When we look at the sun, we see through the solar gas until the optical depth becomes so great that we cannot see any farther. We define the level to which we see when we look directly at the center of the solar disk as the *base of the photosphere.*

The sun's radius is so large that any small region of the surface seems pretty flat. Thus we can speak of levels, *of* heights *above those levels, and of* up *and* down *as though we were speaking of a flat surface.* Up *and* higher *correspond to increasing distance from the center of the sun.*

But though we can consider the solar gas in terms of its optical depth, which is related to how far we can see through it, each bit of the gas is not only absorbing radiation but also simultaneously emitting radiation. In fact, for a given temperature, the amount of emission is proportional to the amount of absorption. (This means that if gas were to absorb twice as much, perhaps because twice as much were packed into a certain volume, it would also emit twice as much, given that the temperature didn't change. For a higher temperature, however, the amount of emission would be higher even if the amount of absorption stayed the same.) The brightness that we see is a result of the emission summed up from all the gas we can see.

When we look from the earth at the center of the visible disk that is facing us, the solar gases obviously become opaque at some level; after all, we cannot see stars shining through the sun from behind it. As we look farther and farther away from the center of the disk, for a while the gas remains sufficiently opaque that we cannot see through it. Eventually, however, we reach a point (Fig. 5–6) where the gas is pretty transparent. We are still seeing through some gas, but its optical depth isn't very high.

Figure 5–6 The angle over which the solar gas changes from transparent to opaque is smaller than we can resolve with our eyes, so the solar limb appears sharp.

As we look farther and farther away from the center of the solar disk, the smallest angle between a line of sight along which the solar gas is pretty opaque and a line of sight along which the solar gas is pretty transparent is only about 1 second of arc (which, let us recall, is only 1/60 of a minute of arc and, therefore, only 1/3600 of a degree of arc). There is no jump from being transparent to being opaque; the optical depth decreases continuously over this 1 arc sec as our line of sight changes. Is this a large angle or a small angle? That depends on what we compare it to, since "large" and "small" are relative terms. If this angle—1 second of arc—were a large angle, we could see the solar opaqueness gradually diminish. In that case, the sun's edge would look fuzzy. But, as we mentioned in Section 3.1, the human eye cannot resolve angles smaller than 1 arc min. Compared to 1 arc min, 1 arc sec is a small angle. Thus to the human eye, the change from complete opacity to complete transparency appears sudden, and the solar edge (which is called the *limb*) appears sharp. Through a telescope, however, we can resolve angles that small, and the limb does not appear so sharp.

5.2d *Limb Darkening*

When we take a photograph of the sun in white light, we find that the intensity of light varies from the center of the solar disk to the solar limb. The regions near the limb are noticeably darker than the regions near the disk center. This phenomenon, which shows clearly on the photograph opening this chapter, is called *limb darkening.*

The fact that the sun looks darker near the limb, which is an **observa-**

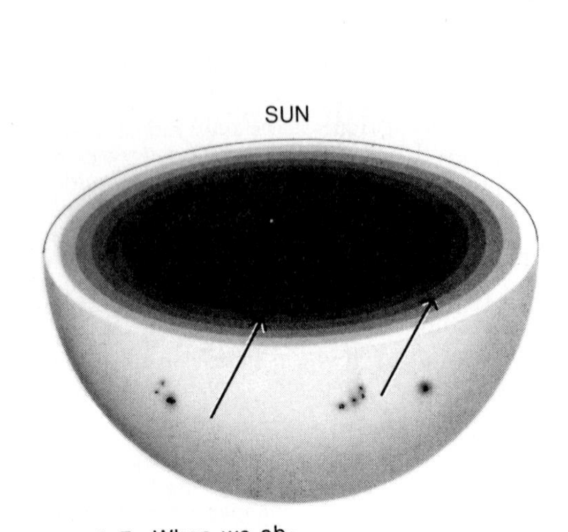

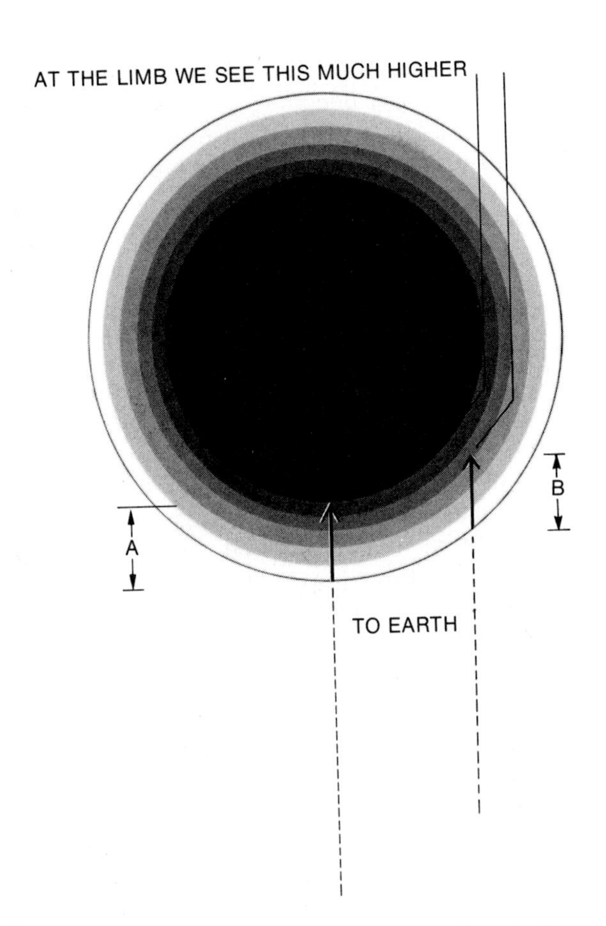

Figure 5–7 When we observe the sun, we see in until the solar gas becomes opaque (distances A or B). The illustration on the right is the top view of the hemisphere on the left. When we observe near the limb, we see obliquely through the gas. At the limb, the atmosphere becomes opaque to us at a higher, cooler level in the solar atmosphere than it does when we look at the center of the solar disk, even though the optical depth along arrow A equals that along arrow B. The cooler level radiates less than hotter levels (according to the Stefan-Boltzmann law), so we observe limb darkening. There is no significance to the shades of gray shown in the layers; the tones merely differentiate layers.

When we speak of the center of the sun, *we usually mean the center of the 3-dimensional object we call the sun. When we speak of the* center of the disk of the sun, *we mean the center of the apparent disk that we observe.*

tional result, has been **interpreted** by astronomers in terms of variation of the temperature through the sun's outer layers. Thus our knowledge of the temperature of the solar atmosphere is derived, in large part, from studying the solar limb darkening, using the reasoning that we shall now describe. The knowledge we gain in this manner, and the methods that have been developed through study of the sun, also apply to other stars.

When we look at any opaque mass of gas, we see in through the gas for a certain distance depending on the opacity of the gas. For example, when we look at the center of the sun's disk, we see to a level that we defined in Section 5.2c as the base of the photosphere. Looking through the solar atmosphere can be thought of as looking through concentric spherical shells of gas. When we look at the center of the disk, we are looking perpendicularly to the surfaces of these shells (Fig. 5–7). The gas is getting denser and denser, and thus each bit of the same size is more and more opaque (has more opacity) as we go into the sun.

We must now realize that we essentially see only the gas at the level of the base of the photosphere (this "level" may be 200 km or so thick). Why do we see essentially only the gas from this level? First, the gas above this level is fairly transparent and so we see through it. Second, since the gas becomes opaque by the level of the base of the photosphere, we obviously can't see any radiation from gas still farther down. This leaves essentially only the radiation from in between, that is, the level where the gas becomes opaque, which we have defined as the base of the photosphere. The amount of radiation we receive at each wavelength is the amount that corresponds to the Planck curve for the temperature of the gas at that level. (Planck curves were described in Section 2.5.) This gas is at a temperature of about

5800 K. If it were hotter, we would get more radiation (that is, it would be brighter), and if it were cooler, we would get less (that is, it would be fainter).

There is no way that we can see any farther into the sun than the base of the photosphere, so our knowledge of the interior of the sun is all based on very indirect studies. But for levels in the few thousand kilometers above the base of the photosphere—which corresponds to the upper part of the photosphere and the lower part of the chromosphere—we can in fact directly deduce what the temperature is from studying limb darkening.

We can do this because when we look at any point on the sun other than the center of its disk, we are no longer looking perpendicularly through the shells of gas but are rather looking through the gas at an oblique angle (as shown on Fig. 5–7). The gas still becomes opaque when we have looked through a certain amount of it, but the place at which this happens is now farther out from the center of the sun than it was when we looked at the center of the solar disk. Thus the gas becomes opaque at a level that is above the base of the photosphere. We now see gas the intensity of whose radiation is determined by the temperature at this new level. Just as we discussed above, the gas above this new level is too transparent to see, and the gas below the new level is hidden.

The gas at this higher level that we see when we look near the limb emits less radiation than the gas at the center of the disk. This is just the observational result we call limb darkening. From the fact that it emits less, we deduce by Planck's law that it is cooler. Since the sun appears darker and darker as we look closer and closer to the limb, we deduce that the temperature of the photosphere declines with height.

Note that in studying limb darkening, we are still looking on the disk of the sun. Our line of sight terminates at the level at which the gas becomes opaque. When we look off the edge, on the other hand, we are seeing different regions of the solar gas.

In this manner, by studying the solar limb darkening we can determine how the temperature of the sun changes with height in the solar atmosphere. We are thus getting information about the third, radial dimension of the sun to add to the information about the two dimensions on the solar surface of which we can take pictures. The temperature actually declines to about 4150 K at about 500 km above the base of the photosphere. This height corresponds to the lower part of the chromosphere.

Actually, it is much easier to use the limb darkening observations to map out the temperature of the sun in terms of the *optical depth* rather than in terms of kilometers. Our optical depth scale tells us whether the intensity of radiation from a certain level is diminished by a factor of 2, 10, 100, or whatever by the time it gets out. We actually measure in terms of 2.71828 . . ., a transcendental number called *e*, and specify whether the amount of radiation is diminished by 1/e, by $1/e^2 = 1/7.4$, by $1/e^3 = 1/20.1$, and so on. Technically, the optical depth is defined as the exponent of 1/e. If the amount of radiation is diminished by $1/e^1$, we say that the optical depth is 1; if the amount of radiation is diminished by $1/e^2$, the optical depth is 2; and so on. Optical depths less than 1 are considered small; optical depths larger than 1 are considered great. An optical depth of 1 is the approximate dividing line between being transparent and being opaque. The actual transfer of the temperature scale from a dependence on optical depth to a dependence on physical depth (measured in such units as kilometers above the base of the photosphere) is presently uncertain and the subject of much current research because we do not know all the causes of the opacity, especially at wavelengths in the ultraviolet.

It can also be pointed out that the most important conceptual part of the limb darkening effect is that as we look closer to the limb we see higher levels in the solar atmosphere. It happens that when we look in visible light, the temperature is decreasing outward at those levels. But when we look at some other wavelengths—observing radio waves, for example—we see higher in the solar atmosphere to heights where the temperature is increasing. Thus in the radio part of the spectrum, we see not limb darkening but rather limb brightening.

LINE CENTER (3933.7 Å)

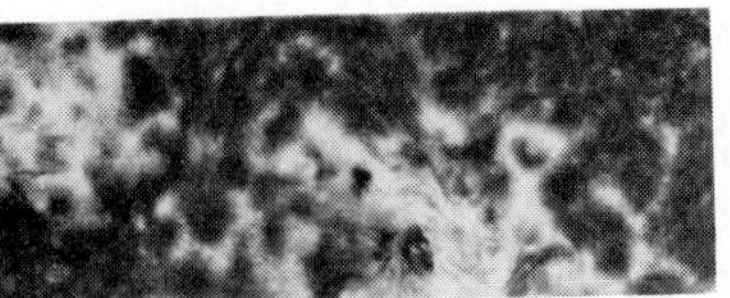

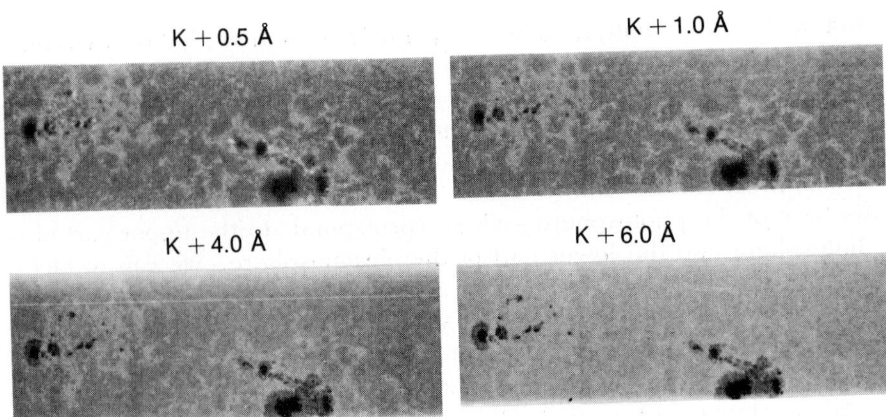

Figure 5–8 A series of photographs taken with a filter passing only light within a fraction of an angstrom of the central wavelength of the K line of Ca II. When the filter is set at the center of the K line, we see structure at higher levels than when the filter is set at longer or shorter wavelengths.

The strongest spectral lines are particularly important to study because they provide another method (besides limb darkening) of exploring the three-dimensional nature of the solar atmosphere. When we observe at a particular wavelength that corresponds to one of these strong lines, photons have a much greater probability of being absorbed than they do at some adjacent wavelength. Thus the opacity is particularly high at the wavelengths of the strong lines. This means that even when we are looking at the disk center, we do not see down as far as the base of the photosphere. Instead, we stop seeing at some higher altitude (for the strongest lines, the altitude at which we stop seeing is in the chromosphere; see Section 5.3). If we observe at the central wavelength of the calcium K line, we see a level that is approximately 1600 km above the base of the photosphere (Fig. 5–8). Of course, the K line, or any line, is not infinitely narrow. In other words, the shape of the graph of intensity vs. wavelength* varies gradually from the darkest part of the line to the continuum (Fig. 5–9). We call this shape the *line profile*. Thus by studying the line profile of the K line, we can map out the temperature and density structure at all heights between 0 and 1600 km.

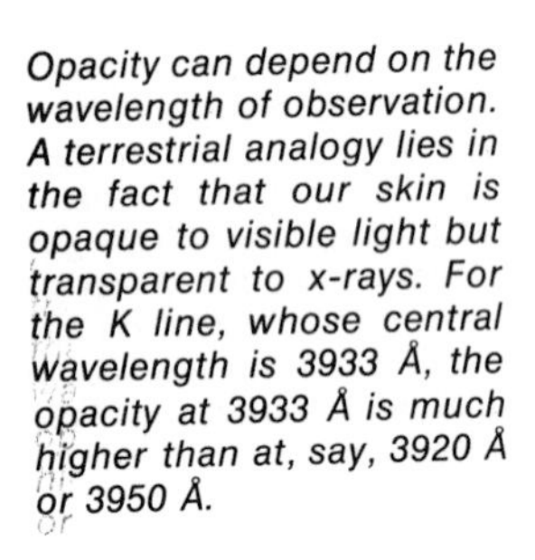

Opacity can depend on the wavelength of observation. A terrestrial analogy lies in the fact that our skin is opaque to visible light but transparent to x-rays. For the K line, whose central wavelength is 3933 Å, the opacity at 3933 Å is much higher than at, say, 3920 Å or 3950 Å.

The K line is the strongest line (the darkest absorption line) that comes through the earth's atmosphere. But slightly below the wavelength of the

*vs. = versus: that is, one axis corresponds to intensity and the other to wavelength; "against" also has this meaning.

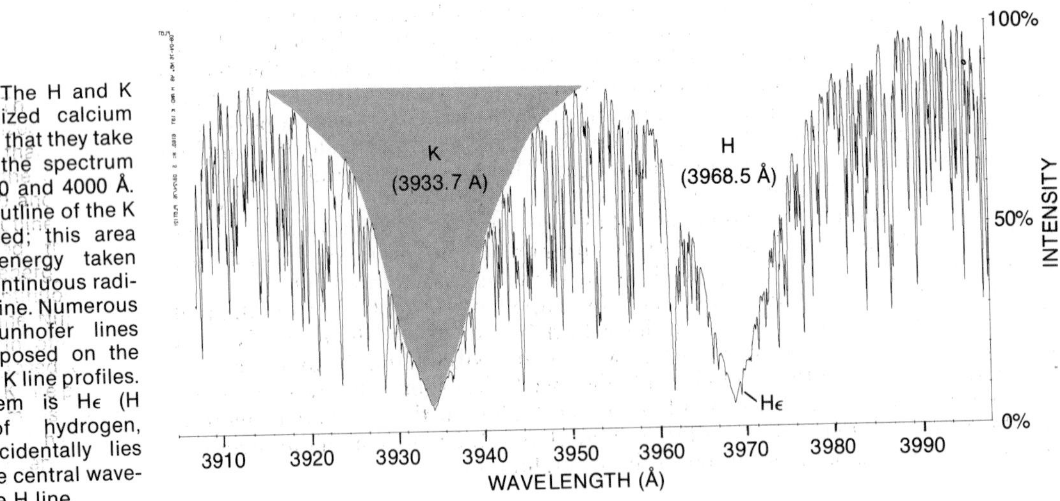

Figure 5–9 The H and K lines of ionized calcium are so strong that they take up most of the spectrum between 3900 and 4000 Å. The overall outline of the K line is shaded; this area represents energy taken out of the continuous radiation by the line. Numerous narrow Fraunhofer lines are superimposed on the broad H and K line profiles. One of them is Hε (H epsilon) of hydrogen, which coincidentally lies very near the central wavelength of the H line.

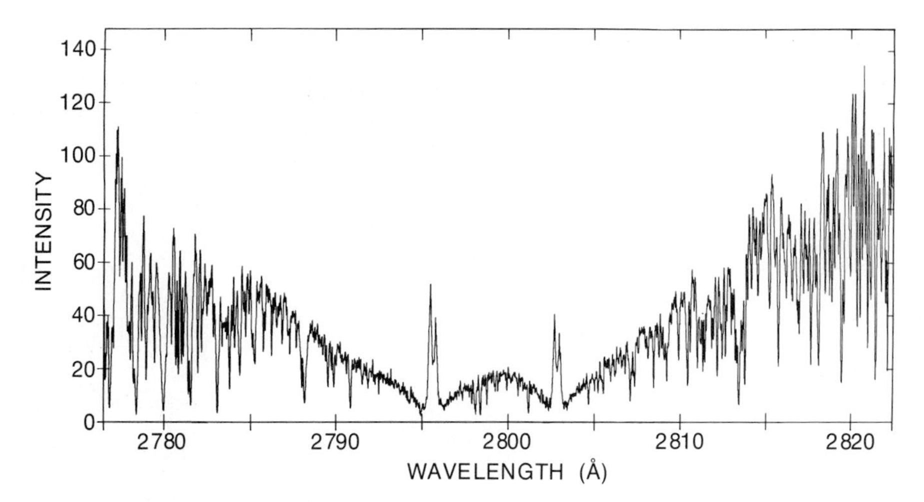

Figure 5–10 The magnesium lines shown here are at 2795 Å and 2803 Å, and so do not penetrate the terrestrial atmosphere. They were observed from a rocket. These two lines are caused by transitions in Mg II analogous to the transitions in Ca II that lead to the H and K lines, and so can be similarly interpreted. The Mg II lines are slightly stronger than the H and K lines, and the double reversal is stronger too.

cutoff from the terrestrial atmosphere lie two spectral lines that are even stronger than the K line and can thus be used both to map out structure at even slightly greater heights than the K line and to provide another way of studying heights up to 1600 km. They are two lines of ionized magnesium, Mg II, and occur at 2795 and 2803 Å (Fig. 5–10). In June 1975, NASA launched the eighth satellite in its series of Orbiting Solar Observatories. One of the two major experiments on OSO-8 studied these magnesium lines simultaneously to the H and K lines of calcium and Lyman α and Lyman β of hydrogen. The other experiment studied a larger number of ultraviolet spectral lines, though one at a time and with less spatial and spectral resolution. Observing with OSO-8 is described in Section 5.8.

The earth's atmosphere does not pass radiation of wavelengths shorter than about 3000 Å.

5.2e *Stellar Atmospheres*

In some circumstances—when the rate of collisions is relatively high and each bit of the system gains exactly as much energy as it loses in each moment of time—we have a situation called *thermodynamic equilibrium.* We have an intuitive idea of equilibrium: everything is uniform and nothing is changing. In equilibrium, the temperatures measured by different methods give the same value. Otherwise, we must specify which type of temperature we mean. In the photosphere of a star, we do not have strict equilibrium, but each small volume of gas may act as though it is in equilibrium, even though its neighboring gas may have a slightly different temperature. We call this situation *local thermodynamic equilibrium (LTE).*

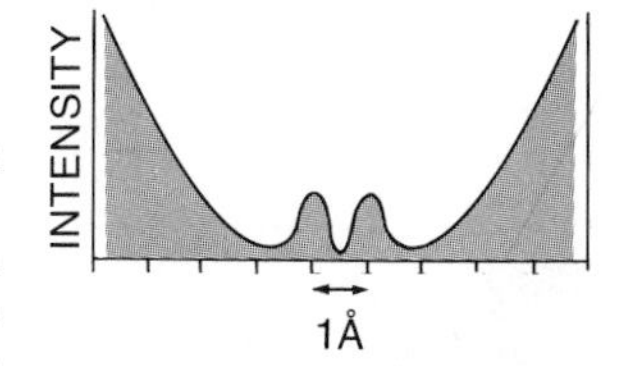

Figure 5–11 The central few angstroms of the spectrum of the K line show a peak, which reverses the normal central minimum of an absorption line, and then a reversal of the reversal, which results in a central dip. Interpretation of this *double reversal* tells us about the temperature and density structure of the upper solar photosphere and lower chromosphere.

Many theoreticians (that is, a couple of hundred) have devoted themselves to making mathematical calculations of the spectrum of the sun and the other stars. This field of research is called *stellar atmospheres.* Stellar atmospheres theoreticians usually carry out detailed calculations using sets of equations that describe the transfer of radiation through the atmosphere of a star, without necessarily assuming the existence of local thermodynamic equilibrium. They use large computers to solve the equations. In addition to the equations, they need values for the temperature (specifically defined) and density of the solar atmosphere at different heights in the atmosphere above the base of the photosphere, as well as values that describe the basic properties of the types of atoms involved.

By carrying out their calculations for different sets of assumed temperatures and densities they try to deduce which sets give predicted theoretical results that agree best with the observations. Some of the model atmospheres are quite complex, and a computer may run hours to calculate a single solution.

The study of stellar atmospheres *is applied to all stars, but meets the severest tests in explaining the sun, which we can observe in such detail.*

These astronomers try to explain not only the continuum but also the spectral lines. In particular, the strongest of the Fraunhofer lines are the H and K lines of Ca II. Careful observations of the shape of these lines show a deviation from the smooth dip characteristic of most absorption lines. There are, rather, small peaks of emission very close to the minimum value of the line (Fig. 5–11). The ability of the mathematical calculations to produce these strange shapes is a triumph for the field of stellar atmospheres.

Figure 5–12

Figure 5–13

Figure 5–14

Figure 5–15

Figure 5–12 The Sacramento Peak Observatory in Sunspot, New Mexico, has a solar tower. Light passes into a window at the top, and the entire light path inside is a vacuum in order to eliminate the effects of air currents. Support of the observatory, the nation's largest devoted solely to solar studies, was recently transferred from the Air Force to the National Science Foundation.

Figure 5–13 A large solar telescope is located at the Kitt Peak National Observatory, alongside the nighttime telescopes. A mirror at the top the size of a person reflects light down a long sloping tube shown in Fig. 3–19.

Figure 5–14 The Big Bear Solar Observatory, part of the Hale Observatories, is located on an artificial island in the middle of a lake in southern California. Because air flows smoothly over bodies of water, this leads to exceptionally good "seeing."

Figure 5–15 The Mount Wilson Observatory has a solar tower 50 meters tall, which is often used to make magnetic maps of the sun.

5.3 THE CHROMOSPHERE

Under high resolution, the chromosphere is seen not to be a spherical shell around the sun but rather to be composed of small spikes that rise and fall. They have been compared in appearance to blades of grass or burning prairies. These small spikes are called *spicules* (Fig. 5-16).

Spicules are visualized as cylinders of about 1 arc second in diameter and perhaps ten times that in height, which corresponds to about 700 km across and 7000 km tall. They seem to have lifetimes of about 5 to 15 minutes, and there may be approximately half a million of them on the surface of the sun at any given moment.

Figure 5–16 Spicules at the solar limb, observed through a filter passing a narrow band of wavelengths centered 1 Å redward of Hα.

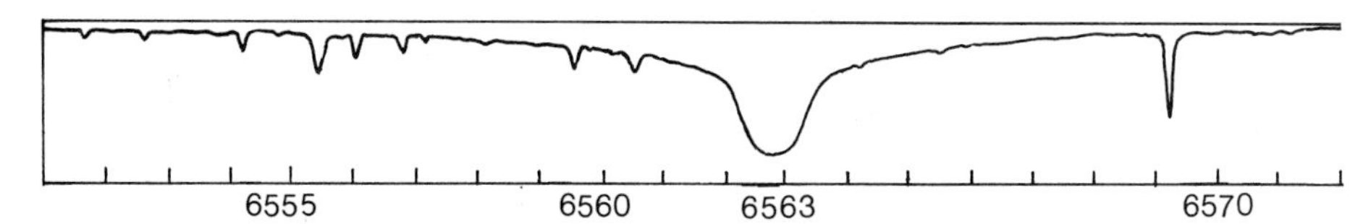

Figure 5–17 The line profile of Hα, whose central wavelength is 6563 Å.

Spicules are best seen when we are looking off the edge of the sun, beyond the limb. But we can also see the chromosphere at the center of the sun by looking through a filter that passes only the wavelength of one of the strongest absorption lines. The line most commonly used is Hα (H alpha), which is at 6563 Å in the red region of the spectrum. Spicules on the disk can be seen in Figure 5–18, taken at a wavelength not in the center of the Hα line.

Remember that the Hα line's opacity is greater than the opacity at continuum wavelengths. Accordingly, when we look only at this hydrogen radiation, the level to which we see is approximately 1500 km above the base of the photosphere.

Note that by saying that Hα is an absorption line, we mean only that it is darker than the continuum (Fig. 5–18). There is still plenty of radiation to observe at the wavelength of the Hα line center.

We are all familiar with certain kinds of filters. If you hold up a red filter to the light, for example, only red light passes through. But a regular filter that you might use for your camera may pass a band of radiation one or more thousand angstroms wide. In order to observe the hydrogen radiation, we need a special filter that passes only 1/2 Å or so of radiation. These filters are very difficult to make, and very expensive. One can also observe narrow bands of spectrum with devices called *spectroheliographs*. Observations through them, which are called *spectroheliograms*, take longer to make than observations through filters—which are called *filtergrams*. Spectroheliographs can observe narrower bands of spectrum than even so-called "narrow-band filters." Historically, spectroheliographs were developed long before narrow-band filters.

Figure 5–18 At the same wavelength as in Fig. 5–16, one can see spicules standing up as dark elements on the solar disk, although spicules cannot be distinguished on such photographs at the center of Hα. Supergranulation is also visible, as is a small active region in the foreground.

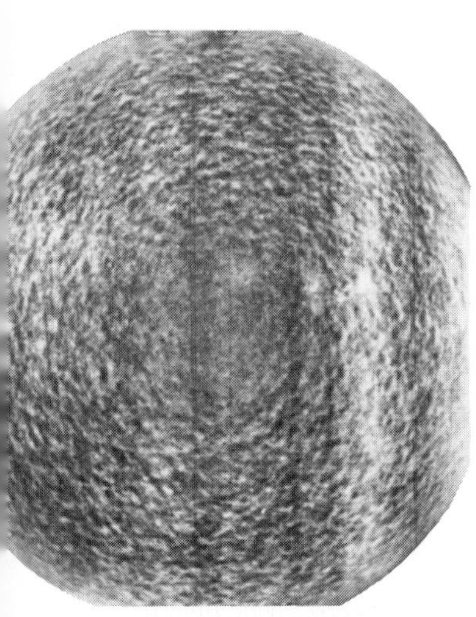

Figure 5–19 Supergranulation is best visible in an image like this one, in which the velocity field of the sun is shown. Dark areas are receding and bright areas are approaching us. We will not discuss here how such velocity images are formed other than to say that it is a composite of photographs showing different Doppler shifts.

One important discovery that was made from spectroheliograph studies, in this case studies of velocities on the sun, was the existence of large organized cells of matter on the surface of the sun called *supergranulation.* Supergranulation cells look somewhat like polygons of approximately 30,000 km diameter. Supergranulation is an entirely different phenomenon from granulation. Each supergranulation cell may contain hundreds of individual granules. Supergranules can be seen on Figure 5–19.

Matter seems to well up in the middle of a supergranule and then slowly move horizontally across the solar surface to the supergranule boundaries. The matter then sinks back down at the boundaries. This slow circulation of matter seems to be a basic process of the lower part of the solar atmosphere. The network of supergranulation boundaries, called the *chromospheric network,* is visible in the radiation of hydrogen alpha or the H and K lines of calcium.

When we examine the spectra of other, more distant stars in hydrogen or calcium light, we find unmistakable signs of chromospheres in stars of spectral types like the sun. Thus by studying the solar chromosphere we are also learning what the chromospheres of other stars are like—stars of spectral types F, G, K, and M can have chromospheres.

In 1957, Olin C. Wilson of the Hale Observatories, and M. K. V. Bappu, an astronomer from India who was then also at Mt. Wilson, discovered an important relation that links certain properties of the calcium K line in stars with the absolute magnitudes of those stars. Many attempts have been made to explain this relation theoretically, but none has been completely successful. Still, the Wilson-Bappu effect (Fig. 5–20) provides a powerful method of finding distances to F, G, K, and M stars, for from mere inspection of the spectrum one finds the intrinsic luminosity of the star.

Figure 5–20 The Wilson-Bappu effect allows us to calculate the absolute magnitude of late-type stars by merely measuring the widths of the double reversals in their K lines. (W is the line width, normally measured in wavelength, transformed for this purpose by the Doppler effect to its equivalent in km/sec.)

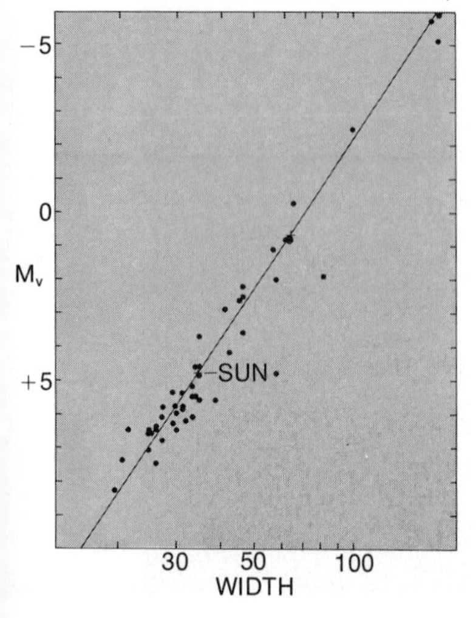

Chromospheric matter appears to be at a temperature of approximately 15,000 K, somewhat higher than the temperature of the photosphere. Thus in the upper photosphere or lower chromosphere a minimum temperature is reached. This *temperature minimum* (Fig. 5–21) of about 4150 K lies about 500 km above the base of the photosphere. Remember, though, that the continuous absorption of radiation by gas between this height and observers on earth is so low in quantity that the gas appears transparent.

We think that the temperature rises in the region above the temperature minimum because additional energy is being deposited there; after all, the temperature does not rise without reason. Waves originate in the convection zone and turn into *shock waves*—waves that compress matter in the same way that waves in a sonic boom on earth compress the air. The shock waves pass through the photosphere and dissipate their energy in the chromosphere and, even higher, in the corona. The granules and spicules, and the 5-minute oscillation, may be manifestations of these shock waves. But nobody knows exactly how the waves dissipate their energy. Several models have been proposed, but there is no agreement. This is an important field of current research.

The answers have important implications not only for astronomy but also for energy research. The gas in the sun is a mixture of ions and electrons, called a *plasma,* in a magnetic field. In developing controlled nuclear fusion on earth as a source of energy, we must learn how to contain plasmas in a magnetic field in the laboratory. For the moment, we can best study plasmas and their properties in the sun and the stars.

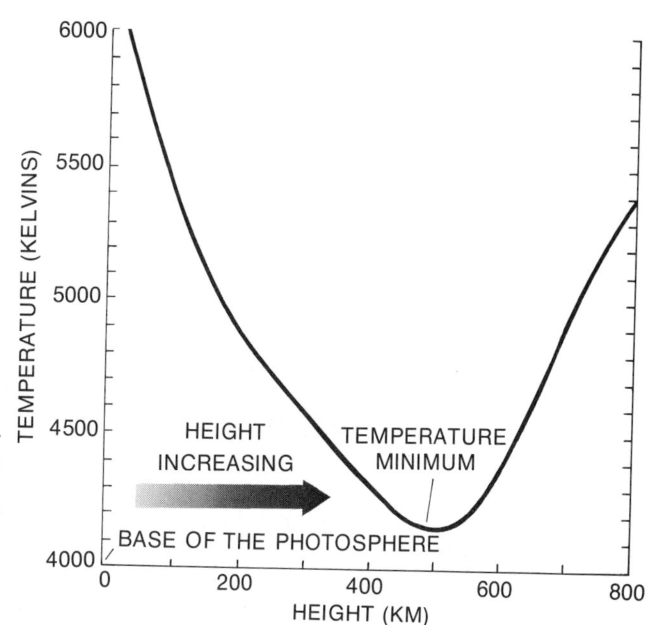

Figure 5–21 As we go out from the solar core, the solar temperature reaches a minimum in the upper photosphere and lower chromosphere, and then rises through the corona. Energy is deposited in the upper photosphere and lower chromosphere by *shock waves,* waves in which there is an abrupt jump in the pressure and in the velocities of the particles at a certain place. In regular waves, the pressure and particle velocities vary gradually.

5.3a The Chromospheric Spectrum at the Limb

During the few seconds at an eclipse of the sun that the chromosphere is visible, its spectrum can be taken. This type of observation has been performed ever since the first spectroscopes were taken to eclipses in 1868. As the photosphere is completely covered, the chromosphere becomes visible.

Since the chromosphere then appears as hot gas silhouetted against dark sky, the chromospheric spectrum consists of emission lines super-

Figure 5–22 The flash spectrum 0.6 sec after totality at the African solar eclipse of 1973. We see the chromospheric spectrum.

—Hβ 4861 Å
—Mg I 5167 Å, 5173 Å, 5184 Å
—He I D_3, 5876 Å
Na I D_2, 5890 Å
D_1, 5896 Å
—Ba II 6496 Å
—Hα 6563 Å
—He I 6678 Å

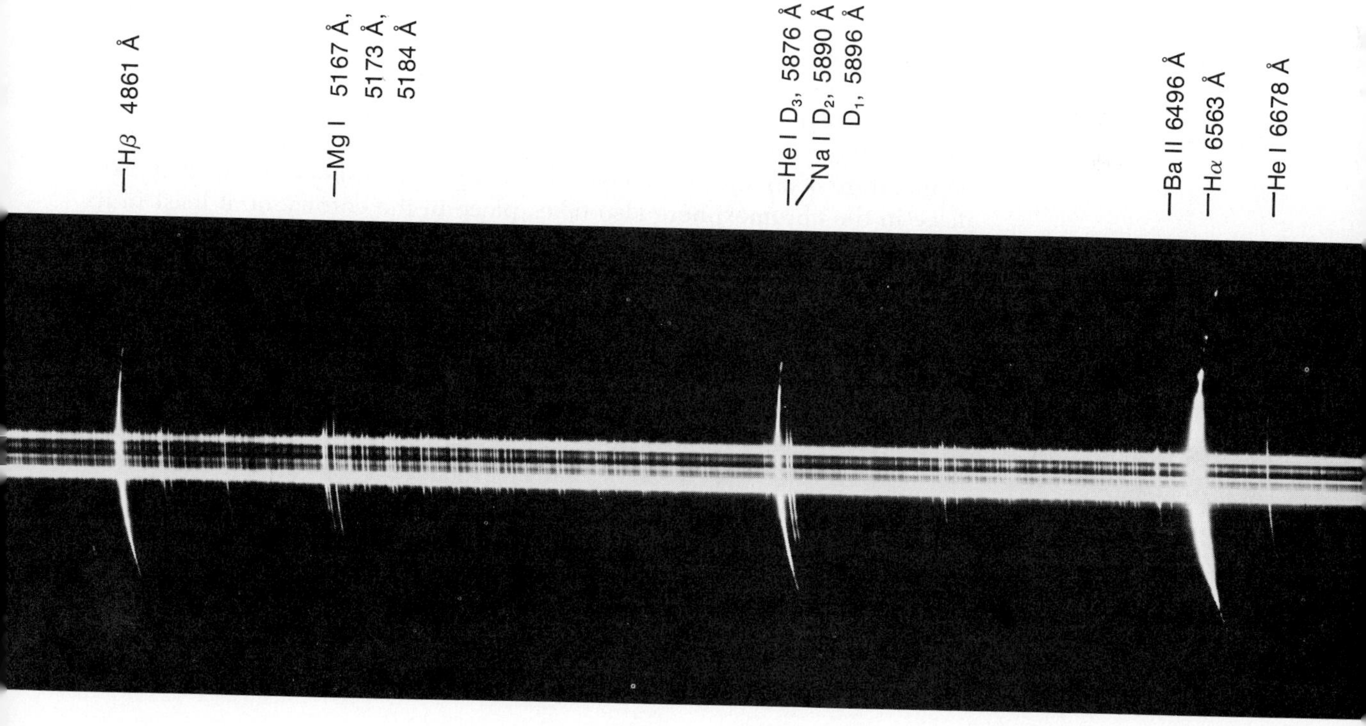

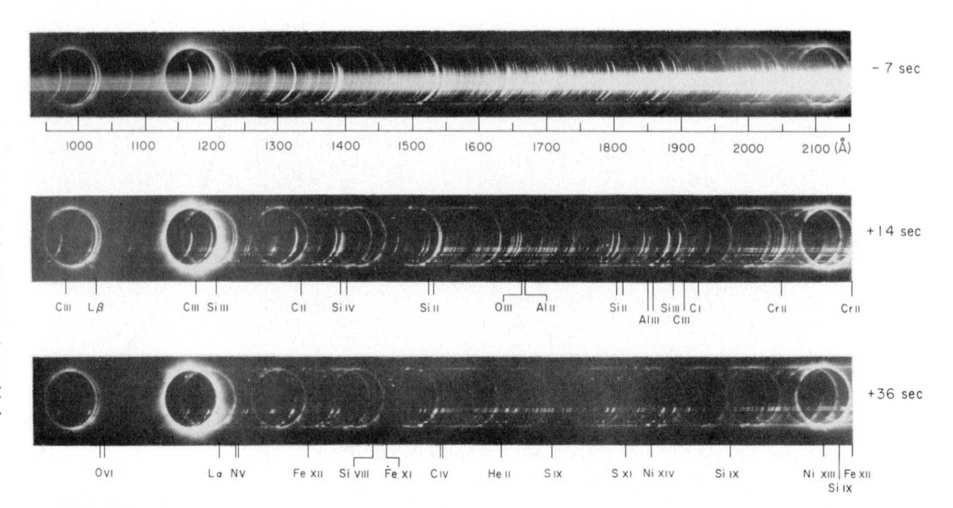

Figure 5–23 The ultraviolet spectrum of the solar chromosphere and corona, photographed from a rocket launched during the 1970 total solar eclipse. Even though the solar corona is so hot that all but a millionth of its hydrogen is ionized, enough neutral hydrogen remains to scatter photospheric light and make the corona visible at the 1216 Å Lyman α wavelength.

We are discussing the visible chromospheric spectrum in this section. In the far ultraviolet, as studied from satellites, the chromospheric and coronal lines are emission lines even when viewed at the center of the disk (Fig. 5–23 and Color Plates 17 and 18). At those wavelengths, the photosphere does not give off strong continuous radiation at these wavelengths because its gas is not hot enough.

imposed on a continuum. (The continuum is usually too faint to study well.) This situation is an excellent illustration of the relation between absorption and emission lines. When viewed with the visible spectrum of the photosphere as a background, the chromospheric gas is transparent at most wavelengths. Only for particularly opaque lines like Hα (H alpha) does it appear in absorption. But at an eclipse, when the gas is observed with no background, the spectral lines are in emission.

The chromospheric emission lines appear to flash into view at the beginning and at the end of totality, so the spectrum of the chromosphere is known as the *flash spectrum* (Fig. 5–22 and Color Plate 16). Since the chromospheric gas is hotter than the photospheric gas, most of the chromospheric emission lines are from ions.

5.4 THE CORONA

During total solar eclipses, when first the photosphere and then the chromosphere is completely hidden from view, a faint white halo around the sun becomes visible. This *corona* (Figs. 5–24 and 5–25) is the outermost part of the solar atmosphere, and is the link between the sun and interplanetary and interstellar space. It has been discovered that the corona is at a temperature of about 2,000,000 K. The dissipation of shock energy that takes place in the chromosphere also takes place in the corona, or at least in its lower part.

The temperature of the corona can be measured in many ways. One can study the excitation temperature or ionization temperature, the temperatures typical of the amount of excitation of energy levels or of the ionization of atoms in the coronal gas, respectively. One can study the kinetic temperature of the electrons—the *electron temperature*—from the effects of the resulting Doppler shifts, to be described in the discussion of the K-corona in Section 5.4a. One can also measure the electron temperature by the strength of the radio emission from the hot coronal gas. One can study the temperature of the ions by measuring the amount that the emission lines have been broadened in wavelength by the Doppler shift (Fig. 5–26). All these, and other methods, give similarly (but not identically) high values for the coronal temperature.

The actual amount of energy in the solar corona is not large, because there aren't very many coronal particles even though each particle has a

high velocity. The corona has less than one billionth the density of the earth's atmosphere, and would be considered to be a very good vacuum in a laboratory on earth. For this reason, the corona serves as a unique and valuable celestial laboratory in which we may study gaseous plasmas in a near vacuum.

Photographs of the corona show that it is very irregular in form. Beautiful long *streamers* extend away from the sun in the equatorial regions. At the poles, delicate thin *plumes* are suspended above the surface. The shape of the corona varies continuously and is thus different at each successive eclipse. The structure of the corona is maintained by the magnetic field of the sun, and studies of the corona are used to study the magnetic field. As we shall see in Section 5.6, the magnetic field is very important in shaping the processes that go on in the solar atmosphere.

There is some confusion in terms in the name "corona," for it means both the actual plasma of electrons, protons, and ions in the upper solar atmosphere and also the halo of light that we see during eclipses. The light we see during eclipses actually comes from at least three different

Figure 5–24 The 1970 total solar eclipse, photographed from Mexico by an expedition of the High Altitude Observatory. The coronal intensity falls off so rapidly with distance above the sun that photographic film cannot directly record such a large region of the corona; to take this photograph, a special filter was prepared that absorbed a large fraction of light near the center and less and less radially toward the edges. Thus the light that passed through the filter was more uniform in intensity than the incident radiation, and coronal structures are well outlined.

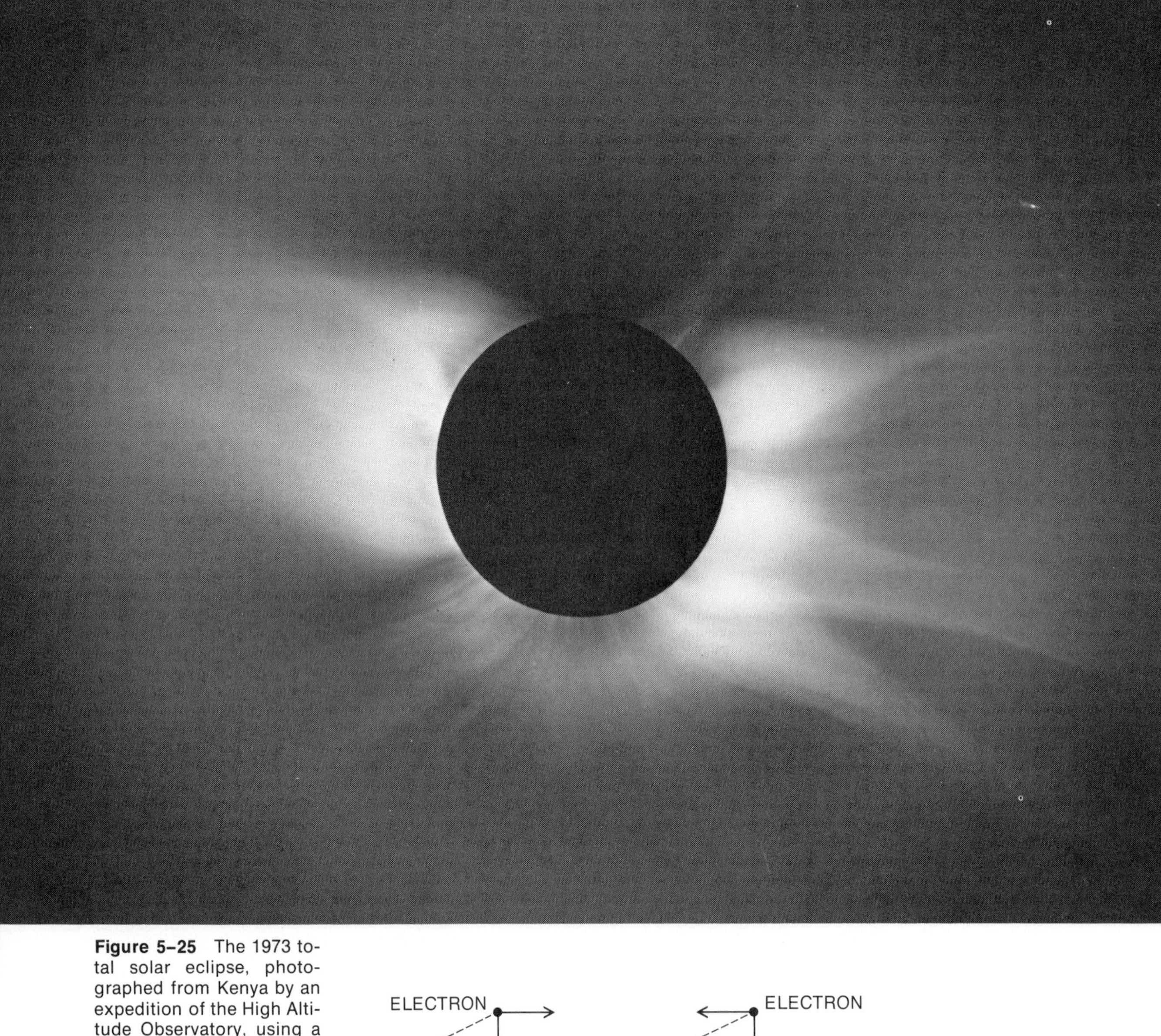

Figure 5–25 The 1973 total solar eclipse, photographed from Kenya by an expedition of the High Altitude Observatory, using a radially-graded filter as in Fig. 5–24.

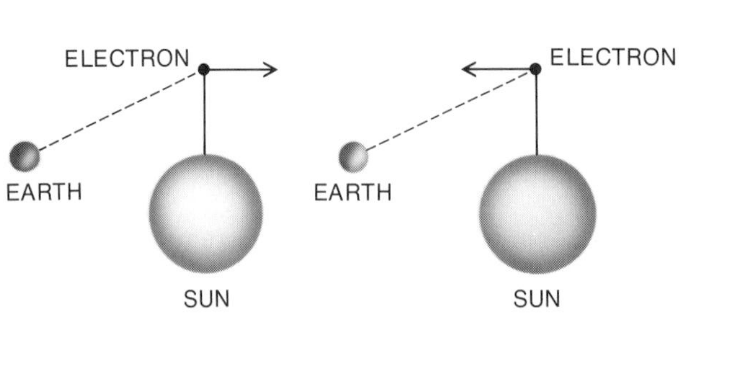

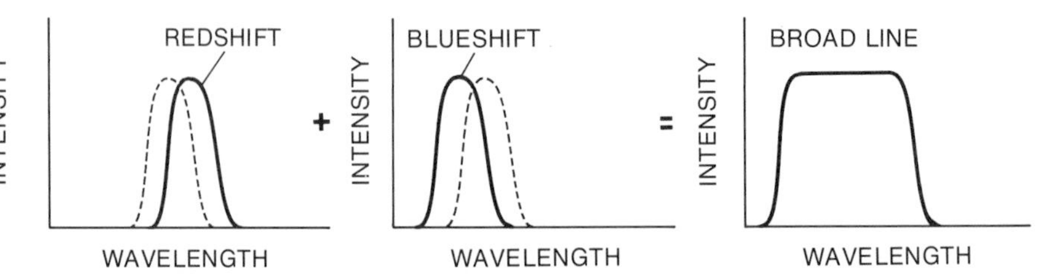

Figure 5–26 The high temperature of the corona causes the coronal emission lines to be very broad. Some individual photons are redshifted when they are scattered by electrons in the corona, as at left, and other photons are blueshifted, as at center. Since from earth we see many photons, some of which are redshifted and others of which are blueshifted, we see a broadened line, as at right.

sources, not all of them really part of the solar atmosphere. Let us discuss these sources of the coronal light as observed from earth, and their contribution to the coronal spectrum.

5.4a The Coronal Spectrum

The visible region of the coronal spectrum, when observed at eclipses, shows a continuum, and also both absorption lines and emission lines. It is easy to tell which part of the coronal spectrum comes from the emission lines: simply consider everything above the continuum level. The emission lines do not correspond to any normal spectral lines that are known in laboratories on earth or on other stars, and for many years their identification was one of the major problems in solar astronomy. In the late 1930's, it was discovered that they arose in atoms that were multiply ionized. This was the major indication that the corona was very hot. In the photosphere we find atoms that are neutral, singly ionized, or doubly ionized (Ca I, Ca II, and Ca III, for example). In the corona we find ions that are ionized approximately a dozen times (Fe XIV, for example, iron that has lost 13 of its normal quota of 26 electrons). The corona must be very hot indeed, millions of degrees, to have enough energy to strip that many electrons off atoms.

By multiply (mult-i-plē) we mean more than once, that is, twice, three times, and so on.

Approximately two dozen emission lines are known in the corona at visible wavelengths. Together, we say that they make up the *E-corona* (emission corona), which appears mixed in with the chromospheric flash spectrum on Figure 5–50 and Color Plate 16. The ions that emit the E-corona are ions located in the corona.

Once the E-corona is subtracted from the total coronal spectrum, we are left with a continuum with absorption lines. The absorption lines are almost absent when we observe in close proximity to the solar limb, and grow stronger as we look higher off the limb.

We are discussing here only the visible part of the spectrum. In the extreme ultraviolet, as for the chromosphere, the coronal spectrum consists entirely of emission lines, even in front of the disk of the sun.

There are two contributions to the continuous spectrum that we see, as we shall describe below. None of the light is emitted in the corona itself; this light is merely photospheric light being reflected toward us. The light scattered toward us by electrons in the corona has a continuous spectrum with no absorption (or emission) lines. An additional continuum with absorption lines is added to this basic coronal continuum when photospheric light is reflected by matter located in interplanetary space. The absorption lines can thus be considered as a non-coronal contaminant (Fig. 5–27). The discussion below explains these points more fully.

During an eclipse the disk of the moon prevents us from seeing the solar photosphere, photospheric light still strikes the electrons in the corona. We can look past the edge of the moon and see these electrons. Light from the photosphere strikes them, and the electrons scatter the light in all directions. Some of the light is scattered toward the earth.

Since these scattering electrons share in the high coronal temperature, they are moving at high velocities in every direction. Those that are moving in one direction may redshift the light from the photosphere, and those moving in other directions may blueshift the photospheric light. Any given absorption line thus reaches us Doppler-shifted to and fro over a range of wavelengths. This has the effect of smearing out the absorption lines; no absorption lines are left to be observed. This radiation is called the *K-corona* (K comes from *Kontinuierlich,* the German word for continuous).

If we could study the K-corona directly, we could learn about the properties

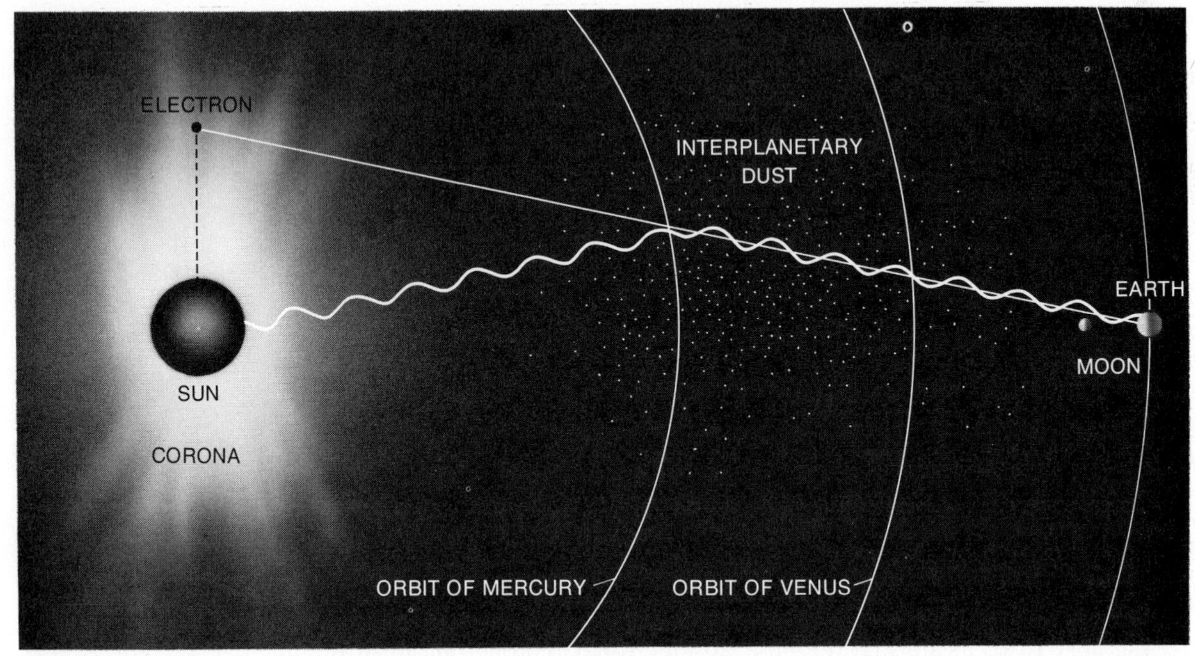

Figure 5–27 When we look at the corona from the earth, we see both light scattered to us by electrons in the corona itself (solid line), and light scattered to us by interplanetary dust (wavy line). Light that arrived via both paths comes simultaneously into our instruments or into our eyes. We must separate these two contributors to the coronal radiation in order to be able to interpret each of them properly.

of the electrons in the corona itself, the upper solar atmosphere. But dust particles in interplanetary space located approximately at the distance of the orbit of Mercury are also not shielded from the photosphere by the moon. These dust particles also scatter photospheric light so that it heads toward us, and we find that when we look along a given line of sight toward what appears to be the corona, the light scattered off coronal electrons and the light scattered off interplanetary dust particles come at us together.

The dust in interplanetary space is cool, however, compared to the solar corona, and the dust particles are much more massive than electrons. The dust particles move slowly, and so the photospheric spectral lines are not significantly Doppler shifted when scattered by the dust. The resultant spectrum, thus, shows Fraunhofer lines and is called the *F-corona* (Fraunhofer corona).

Taken together, the E-corona, K-corona, and F-corona make up the coronal light that we see. The E-corona contributes the spectral emission lines, but the total amount of energy reaching us in these lines in the visible part of the spectrum is much less than the energy reaching us in the K- and F-coronas. (The emission lines in the far ultraviolet carry much more energy than the lines in the visible.) Near the solar limb (within one or two solar radii above the limb), the K-corona is stronger than the F-corona. But the intensity of the K-corona diminishes above the limb much more rapidly than the intensity of the F-corona diminishes (Fig. 5–28), and by heights of two or so solar radii the F-corona dominates, that is, contributes most of the radiation.

Fortunately, astronomers can perform observations at eclipses to distinguish between the contributions of the F-corona and the K-corona. The strength of an absorption line that we observe in the spectrum of the corona is a mixture of the strength of the absorption line in the F-corona and of the continuum at that wavelength from the K-corona (which is entirely continuum). We can calculate just how much continuum has to be mixed in with a spectral line of a strength measured in the photospheric spectrum (which is identical to its strength in the spectrum of the F-corona) in order to give the observed strength.

The process is entirely analogous to observing a glass of chocolate milk, and deciding what the fraction of [dark] chocolate and [white] milk must be in the glass to give the mixture its observed color. If more milk is added, the mixture will be less dark; if more K-corona were contributing, the absorption lines in the observed coronal spectrum would be less dark.

Essentially, calculating the contribution of the F-corona tells us how much radiation was added to the coronal spectrum after it left the corona. We can thus subtract this added radiation, and study the remainder, which tells us about electrons in the solar atmosphere at a temperature of 2,000,000 K.

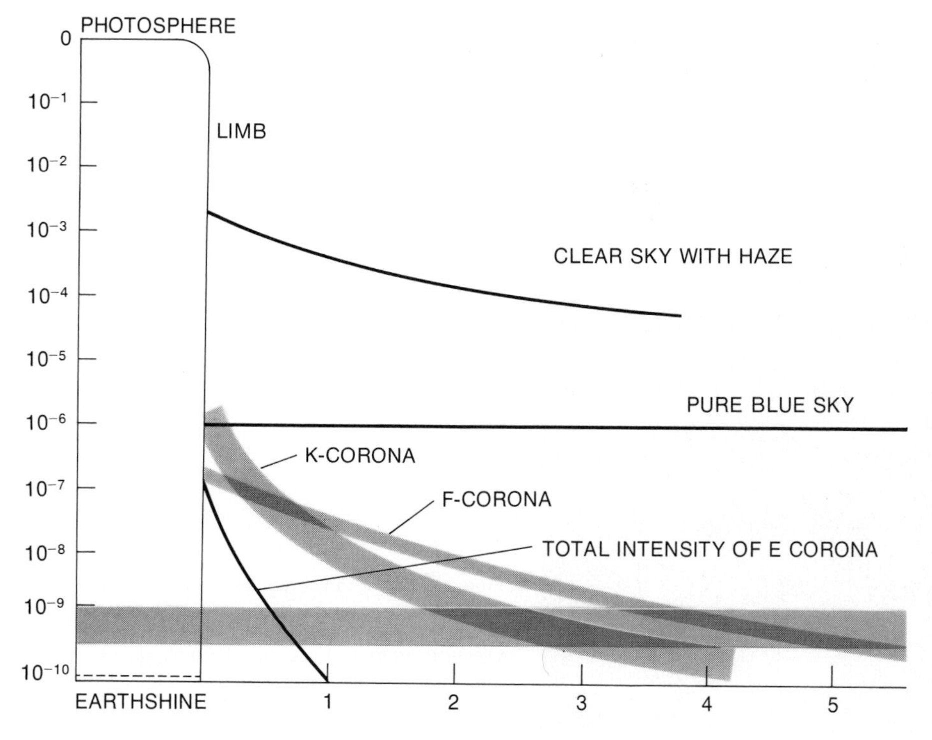

Figure 5–28 Normally, even on the clearest days, the sky is too bright to allow us to observe the corona. During an eclipse, photospheric light does not reach the terrestrial atmosphere to turn it blue, and we can observe the corona. The K-corona, from the scattering of photospheric light by electrons in the corona, falls off in intensity more rapidly than does the F-corona, from the scattering of photospheric light by interplanetary dust. Although the total energy received in all the emission lines is less than the energy received in the K-corona or F-corona, at the given wavelength of a particular emission line, the emission line is stronger than the K- and F-coronas. The intensity of the sky during an eclipse (horizontal shaded area) is about one-millionth as great as it is on an ordinary day. The amount of earthshine on the moon, sunlight reflected off the earth to the moon, is also graphed.

5.4b Non-Eclipse Observations of the Corona

The corona is normally too faint to be seen except at an eclipse of the sun because it is fainter than the everyday blue sky. But at certain locations on mountain peaks on the surface of the earth, the sky is especially clear and dust free, and the innermost part of the corona can be seen (Fig. 5–29). Special telescopes called *coronagraphs* (Fig. 5–30) have been built to study the corona from such sites. Coronagraphs are built with special attention to detail, since their object is not to gather a lot of light but rather to prevent the strong solar radiation from being scattered about within the telescope. They thus use lenses instead of mirrors, because even the best mirrors are not smooth enough to reflect light without scattering a fraction that may seem tiny but would still be sufficient to overwhelm the coronal light.

The best coronagraph site in the world is run by the University of Hawaii atop Haleakala Crater on the Hawaiian island of Maui. It is located at 3000 meters (10,000 feet) of altitude, and the sky is the purest blue overhead. Other coronagraphs are located at such sites as Sacramento Peak in New Mexico, and at the Pic du Midi Observatory in the Pyrenees in France.

Since so many of the problems in observing the faint solar corona are caused by the earth's atmosphere, it is obvious that we want to observe the corona from above the atmosphere. If we stood on the moon, which has no air, we would see the corona rising each lunar morning a bit ahead of the sun (Fig. 5–31). At any time we could stick up our hands to block out the photosphere, and see the corona around it.

But it is much less expensive to observe the corona from a satellite in orbit around the earth than it is to do so from the moon. The unmanned seventh Orbiting Solar Observatory, OSO-7 (1971–1974), and the manned Skylab missions (1973–74) used coronagraphs to photograph the corona

Figure 5–29 From a few mountain sites, the innermost corona can be photographed without need for an eclipse. These observations were taken on March 8, 1970, the day after a total eclipse, with the coronagraph of the University of Hawaii on Maui. Five exposures were superimposed to improve the image.

hour by hour in visible light (Color Plates 19 and 20). These satellites could study the corona to much greater distances from the solar surface than can be studied with coronagraphs on earth. Among the major conclusions of the research is that the corona is much more dynamic than we had thought. For example, many blobs of matter were seen to be ejected from the corona into interplanetary space, often in connection with solar flare activity (which will be discussed in Section 5.6).

One can also study the coronal spectrum from space much better than it can be studied from earth. The strongest lines of the coronal spectrum fall in the ultraviolet region below approximately 2000 Å. Since the photosphere does not radiate a strong continuum in this part of the ultraviolet, these coronal lines are strong enough to be seen whenever the sun is scanned at their specific wavelengths. Thus we can map out the solar corona across the entire solar disk from satellite observations of ultraviolet spectral lines (Color Plates 22, 23, and 24).

Many of the satellites in the OSO series had this capability. Each successive spacecraft could make observations of higher spatial resolution. (Higher resolution, of course, means that we can observe finer detail.)

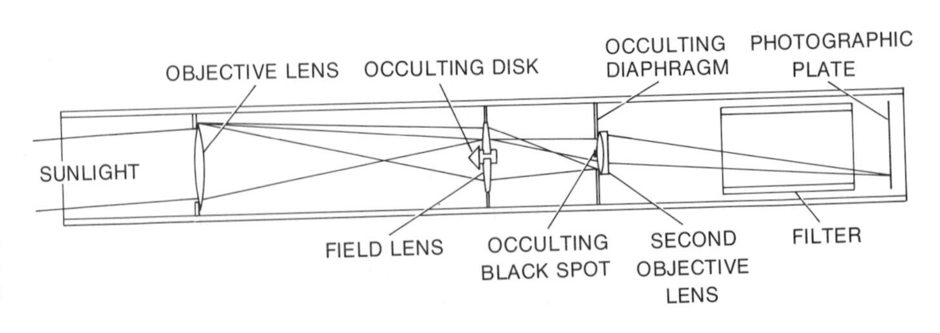

Figure 5–30 The coronagraph, invented by Bernard Lyot, enables coronal observations to be made without an eclipse by severely limiting the amount of light scattered inside the telescope through careful design, construction, and baffling (the insertion of obstructions to prevent light from bouncing around inside). The solar image is brought to a focus inside the telescope, where a small obstruction, called an *occulting disk,* blocks the image of the photosphere. Other lenses then refocus the remainder of the image farther along in the coronagraph. An occulting diaphragm and an occulting black spot on a lens block most of the solar radiation that would otherwise bounce around inside the telescope and overwhelm the radiation from the faint corona. Examining only the radiation of a coronal emission line allows the coronal structures to be better seen, because with respect to the remaining scattered photospheric light the corona is relatively intense at the wavelengths of its emission lines. Thus a filter is usually used.

Figure 5–31 The solar corona sets after the sun in the lunar evening, as photographed by the manned Apollo 15 spacecraft in orbit around the moon, where there is no atmosphere to make the sky blue. The lunar surface—a mare with mountains beyond—is illuminated by earthshine. Venus is visible just to the right of the coronal streamer.

But all these resolutions were much lower (that is, worse) than the resolution obtainable in the visible part of the spectrum from solar observatories on earth. OSO-4, for example, launched in 1967, had a resolution of 1 minute of arc (60 seconds of arc); compare this with the one second of arc size of spicules and granules.

The huge workship Skylab (see Fig. 5–78) was joined in space by three successive crews of three astronauts. A $200,000,000 battery of solar telescopes, called the Apollo Telescope Mount (ATM), was aboard, and the study of the sun with the ATM was a major scientific goal of the mission. Skylab's telescopes could attain a resolution of 5 seconds of arc and, in some cases, even 2 arc sec resolution (Fig. 5–32).

After the astronauts succeeded in repairing the Skylab module, which had been badly damaged during launch, they made a series of solar observations that are currently revolutionizing our view of the outer solar atmosphere. Astronomers would probably not have chosen to spend that

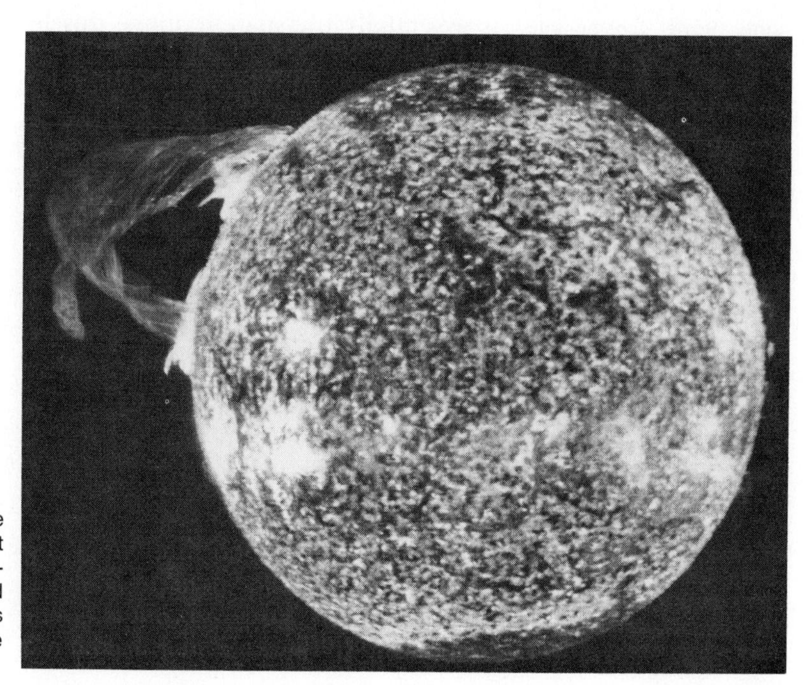

Figure 5–32 A view of the sun in the ultraviolet radiation of ionized helium at 304 Å, photographed with the Naval Research Laboratory's instrument aboard Skylab, shows a prominence that has erupted into space. It is 40 times the size of the earth.

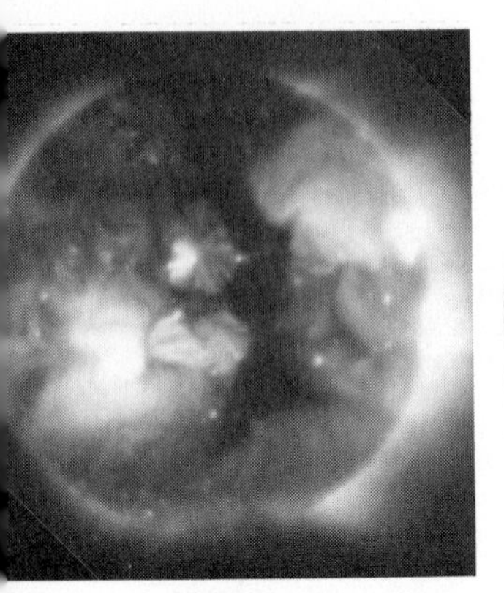

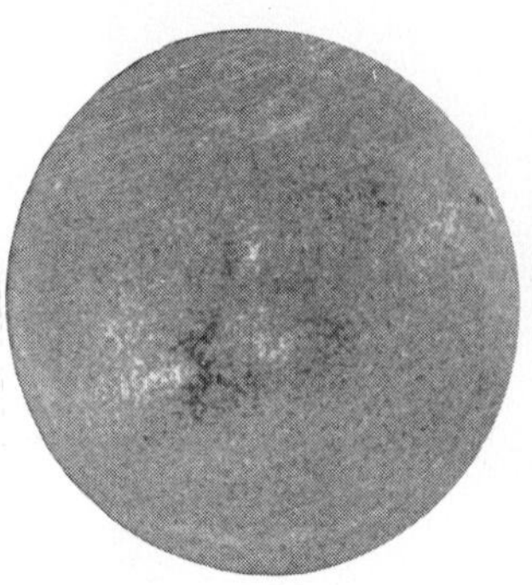

Figure 5–33 X-ray photograph of the sun (far left), made on June 1, 1973 with the American Science & Engineering Company's grazing incidence telescopes aboard Skylab. It is accompanied (left to right) by illustrations showing the sun's magnetic field, and its chromosphere as observed in Hα and in the K line, all observed simultaneously. Only the hottest regions of the sun emit x-rays.

much money in this way, but since the money was being spent as part of the ongoing space effort (which has economic and technological ramifications that overwhelm the scientific aspects), they were glad to have an opportunity to get these valuable data.

The Skylab telescopes included a coronagraph, devices to make images of the sun in different ultraviolet spectral lines, and x-ray telescopes (Fig. 5–33). These instruments observed the solar chromosphere and corona, and monitored solar activity (see succeeding Sections). The identification of coronal holes (see Section 5.7) was made on Skylab data.

The data from Skylab are still under study. The ability to make simultaneous high-resolution observations in many spectral regions has been important, as has been the ability to follow solar activity as it changed from hour to hour and day to day.

On a joint Soviet-American manned program in 1975, the Apollo spacecraft was maneuvered in between the Soyuz spacecraft and the solar disk and made a brief artificial eclipse (Fig. 5–34).

The solar corona is opaque to radio waves. Thus the radio radiation we observe from the sun originates at a level just higher than the level where we can no longer see through the corona.

We cannot receive radio waves from farther down than a certain level; the height of the level depends on the frequency of observation. The corona is essentially transparent above this level. Thus most of the radiation we detect at any frequency of observation arises almost entirely at the corresponding coronal level. Radio astronomers as well as ultraviolet astron-

Figure 5–34 An eclipse produced during the Apollo-Soyuz mission made visible long coronal streamers, Venus, and the star γ Geminorum. G. Nikolskii, the chief Soviet mission scientist, sent this picture to the author on a Christmas card. On the right, Soviet cosmonaut Alexei Leonov holds a painting he made of the event.

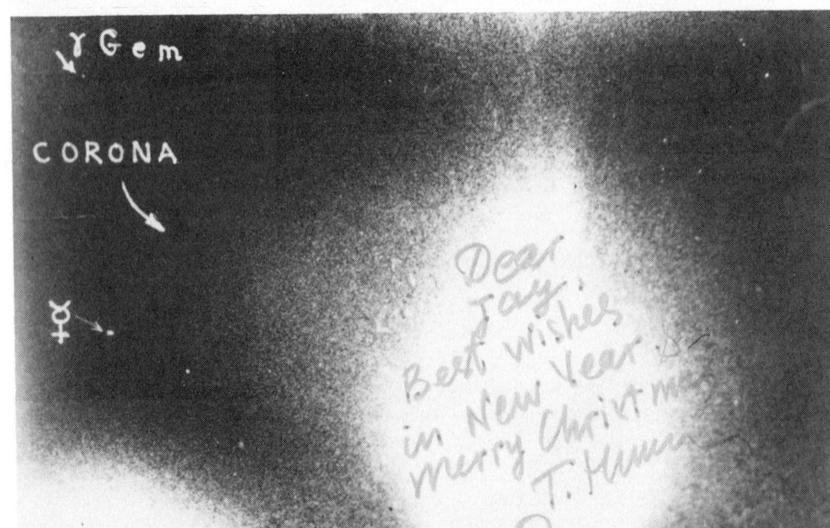

omers can study the corona every day; since radio waves pass through the earth's atmosphere, ground-based radio telescopes (Fig. 5–35) can be used.

Figure 5–35 Some of the solar radio telescopes of the U.S. Air Force's Sagamore Hill Radio Observatory in Hamilton, Massachusetts. They are small in order to provide a beam that is sufficiently large to include the whole sun. Each of the telescopes is a different size and is operated at a different wavelength chosen such that the sizes of the beams are the same. At radio wavelengths, we observe the higher levels of the solar atmosphere, including the chromosphere and corona. Thus the corona is observed best at very short or very long wavelengths.

5.5 SOLAR ECLIPSES

We have discussed the solar chromosphere and corona, parts of the solar atmosphere that are visible to the eye only at the time of a total solar eclipse. Eclipses have played a major role in solar physics, so let us discuss them in detail.

The sun (more precisely, the photosphere) is very large and very far from us. The moon is 400 times smaller in diameter than the sun, but it is also 400 times closer to earth. Because of this, the sun and the moon subtend almost exactly the same angle in the sky—about $1/2°$—which is a happy coincidence (Fig. 5–36).

The moon's position in the sky, at certain points in its orbit around the earth, comes close to the position of the sun. This happens approximately once a month at the time of the new moon, as discussed in Section 10.1. We will see there that since the lunar orbit is inclined with respect to the earth's orbit, the moon usually passes above or below the line joining the earth and the sun. But occasionally the moon passes close enough to the earth-sun line that the moon's shadow falls upon the surface of the earth (Fig. 5–37).

At a total lunar eclipse (see Section 10.1), the earth's shadow is sufficiently large that the entire moon fits inside it. Thus anyone on the surface of the earth at whose location the moon has risen sees the lunar eclipse. By contrast, at a total solar eclipse, the lunar shadow barely reaches the earth's surface. As the moon moves through space on its orbit, and as the earth rotates, this lunar shadow sweeps across the earth's surface in a band up to 300 km wide. Only observers stationed within this narrow band can see the total eclipse.

From anywhere outside this band, one sees only a partial eclipse. Sometimes the moon, sun, and earth are not precisely aligned and the darkest part of the shadow—called the *umbra*—never hits the earth. Only a partial eclipse is visible on earth under these circumstances. As long as the slightest bit of photosphere is visible, even as little as 1 per cent, one cannot see the important eclipse phenomena—the chromosphere and corona. Thus partial eclipses are of little value for most serious scientific purposes. After all, the photosphere is 1,000,000 times brighter than the corona; if one per cent is showing then we still have 10,000 times more light from the photosphere than from the corona, which is enough to ruin

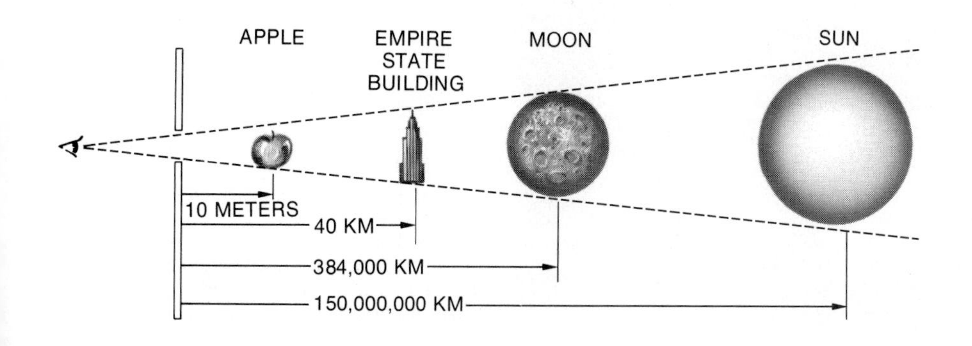

Figure 5–36 An apple, the Empire State Building, the moon and the sun are very different from each other in size, but here they subtend the same angle because they are different distances from us.

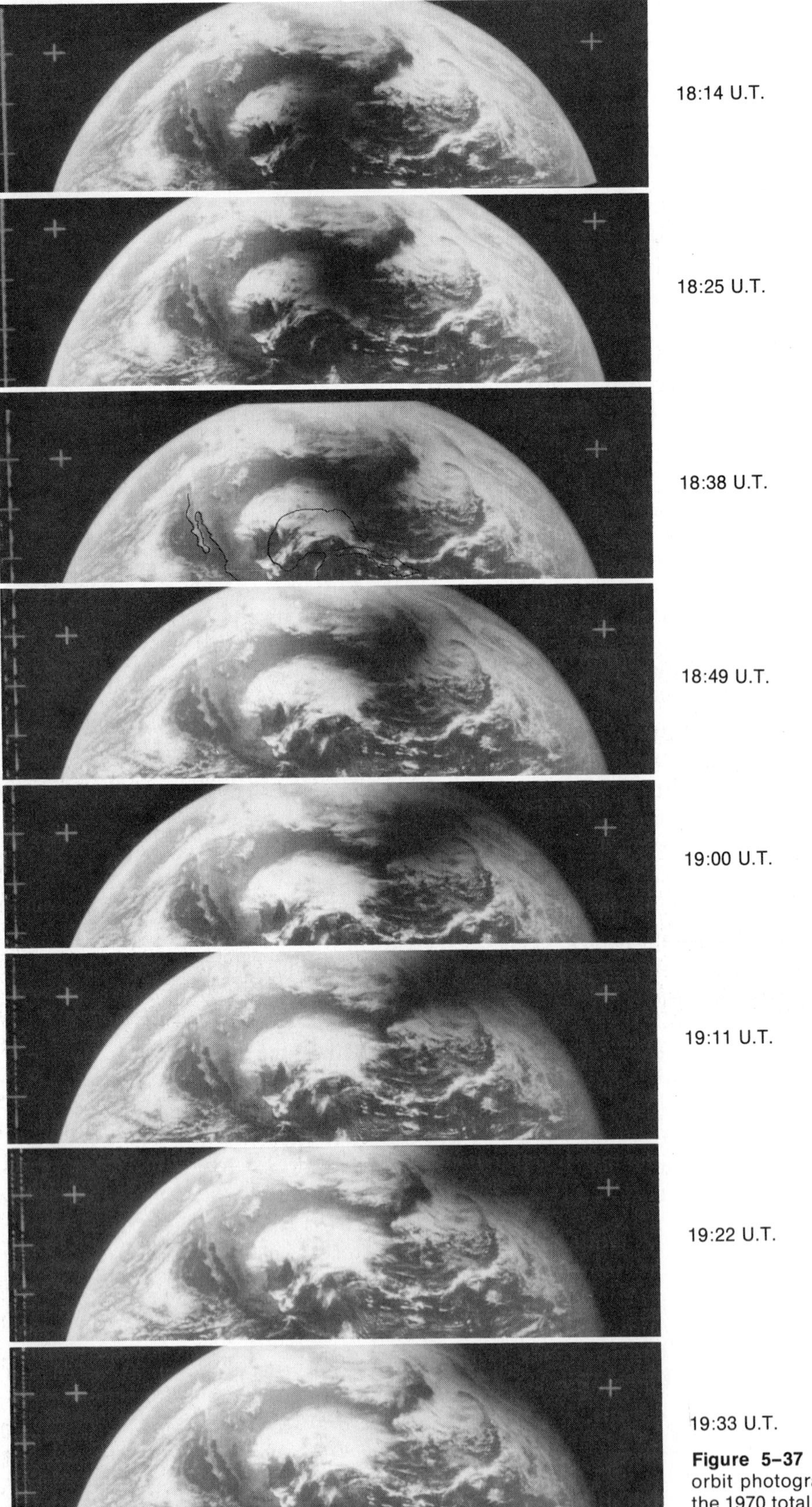

Figure 5–37 The ATS-3 satellite in earth orbit photographed the elliptical shadow of the 1970 total solar eclipse as it swept across North America.

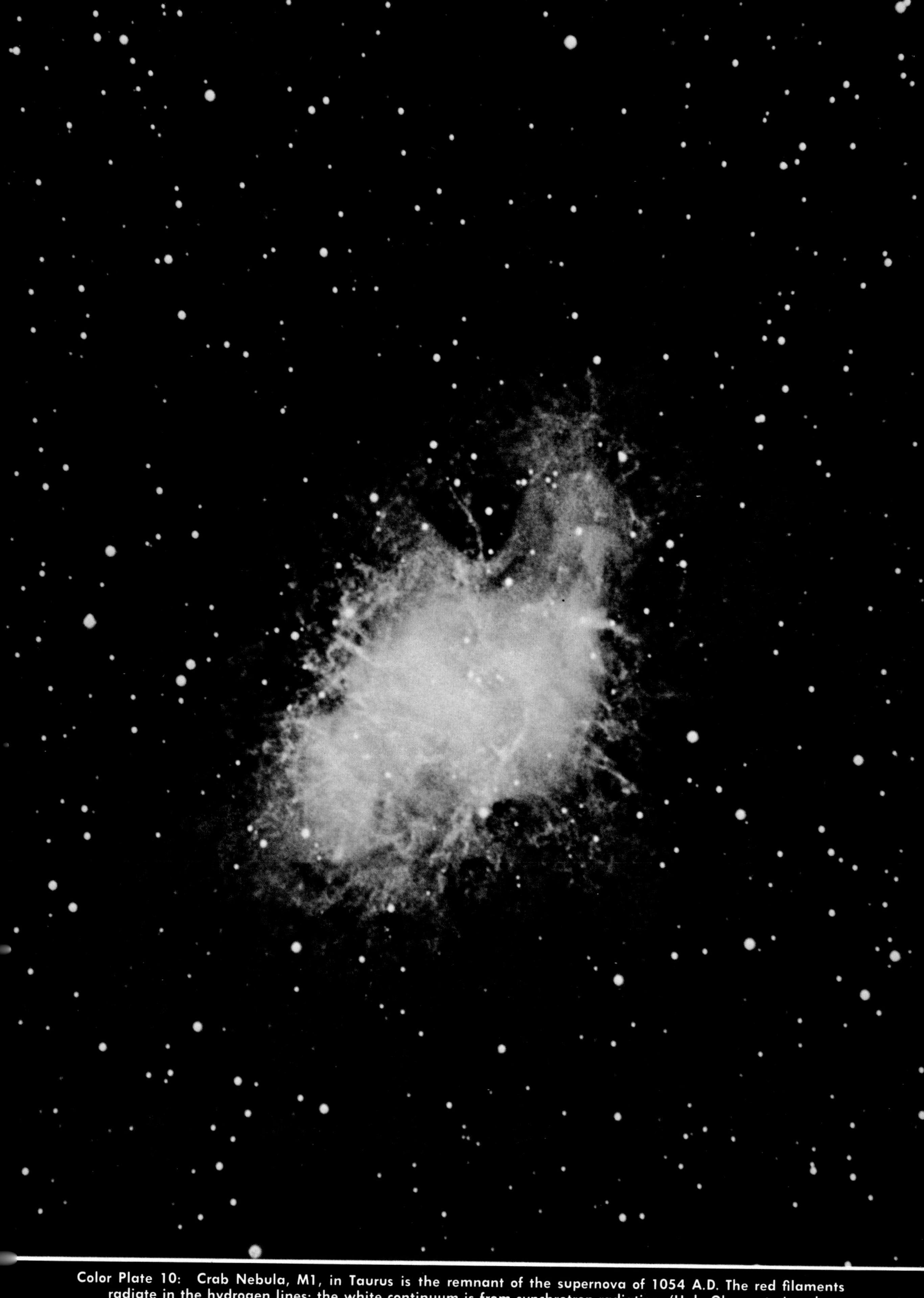

Color Plate 10: Crab Nebula, M1, in Taurus is the remnant of the supernova of 1054 A.D. The red filaments radiate in the hydrogen lines; the white continuum is from synchrotron radiation. (Hale Observatories photo with the 5-m telescope)

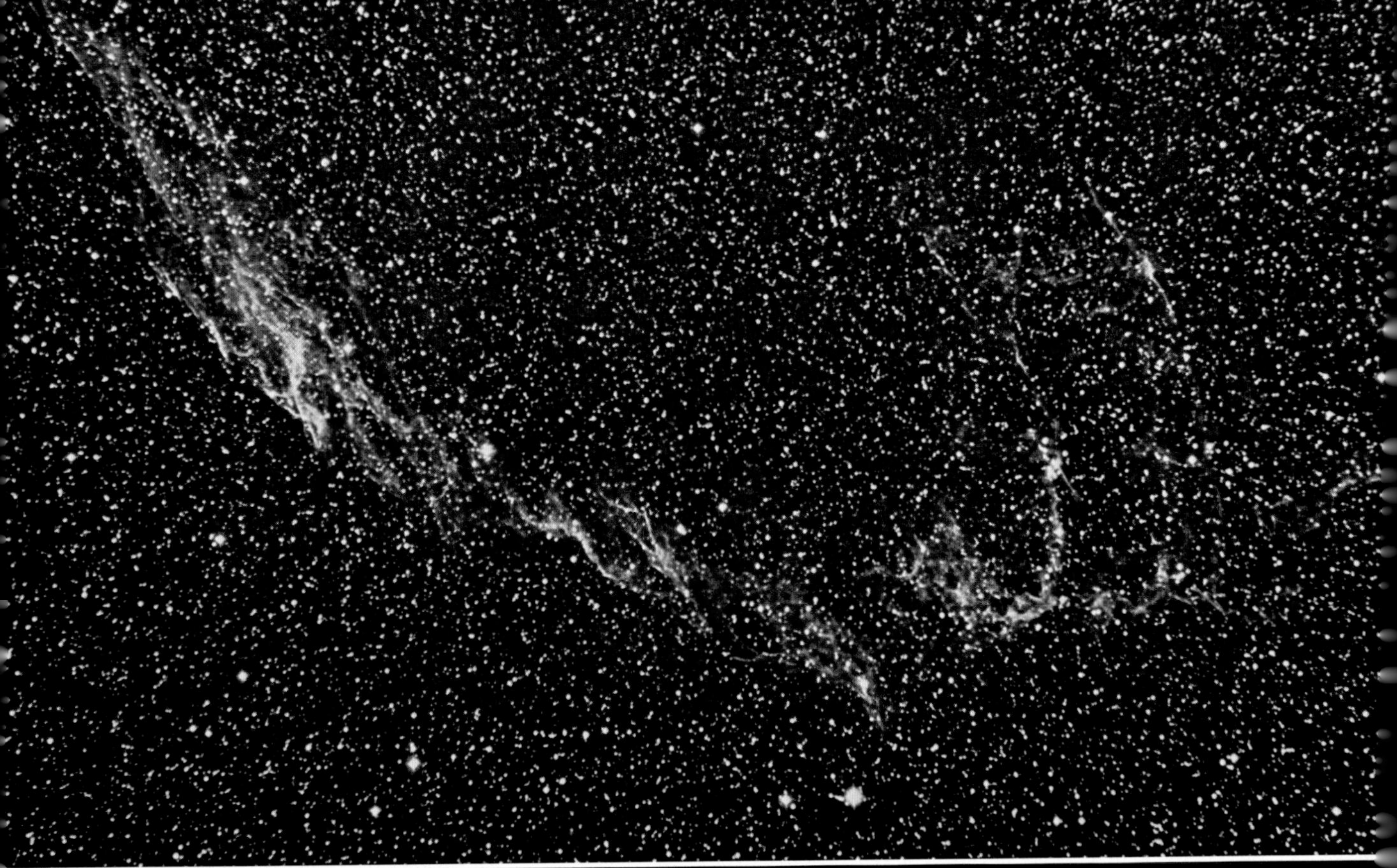

Color Plate 11: Veil Nebula, NGC 6992, part of the Cygnus loop, a supernova remnant. (Hale Observatories photo with the 1.2-m Schmidt camera)

Color Plate 12: Orion Nebula, M42, is glowing gas excited by the Trapezium, four hot stars. The nebula contains stars in formation; it is 25 ly across and 400 parsecs away. (Hale Observatories photo with the 5-m telescope)

Color Plate 13 (top): Two of the partial phases of the June 30, 1973 total solar eclipse. Note the clouds that covered part of the crescent shortly before totality.

Color Plate 14 (bottom): The diamond ring effect at the beginning of totality at the June 30, 1973 eclipse. These photographs were taken from Loiengalani, Kenya, by the author's expedition.

Color Plate 15 (top): The solar corona, photographed by the author's expedition to the 1973 solar eclipse. Note the equatorial streamers and polar plumes.

Color Plate 16 (bottom): The flash spectrum of the solar chromosphere, photographed at the 1970 total solar eclipse. (William C. Atkinson photo)

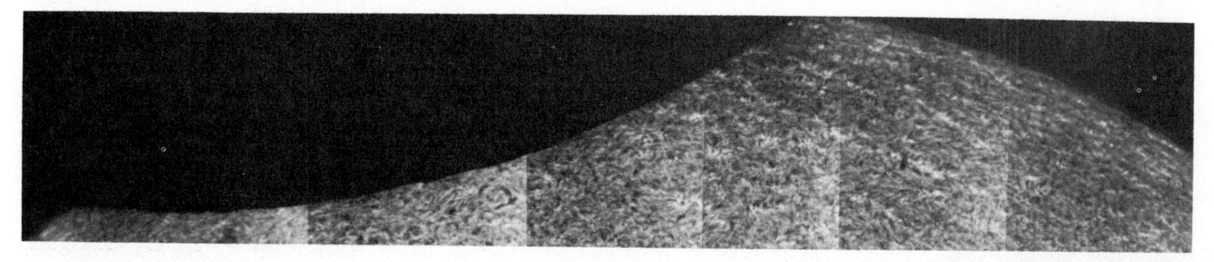

Figure 5–38 Established observatories are not usually in the path of a total solar eclipse. The partial phase, however, can be seen from a much wider area, and this closeup view in Hα from the Big Bear Solar Observatory shows the irregularities at the edge of the lunar disk projected on the solar surface. The corona could not be seen.

our opportunity to see the corona (Fig. 5–38). During the initial phase, we are in the *penumbra*.

If you are fortunate enough to be standing in the zone of a total eclipse, you will find excitement all around you as the approaching eclipse is anticipated. Even uneducated people have formed an impression of the cause of an eclipse: it immediately follows the arrival of a horde of astronomers. If the astronomers come, can the eclipse be far behind?

About an hour or an hour and a half before totality begins, the partial phase of the eclipse starts. Nothing is visible to the naked eye immediately, but if you were to look at the sun through a special filter you would see that the moon was encroaching on the sun (Fig. 5–39 and Color Plate 13). At this stage of the eclipse, it is necessary to look through a special filter (or to project the image of the sun with a telescope or a pinhole camera onto a surface such as a piece of cardboard) in order to protect your eyes, for the photosphere is visible and its direct image on your retina could cause burning and blindness (Fig. 5–40).

Figure 5–39 The partially eclipsed sun observed from Africa in 1973.

The eclipse progresses gradually, and by 15 minutes before totality the sky grows strangely dark, as though a storm were gathering. During the minute or two before totality begins, bands of shadow race across the ground. These *shadow bands* are caused in the earth's atmosphere, and as a terrestrial rather than a celestial phenomenon are not of much interest to astronomers.

You still need the special filter to watch the final seconds of the partial phase. Only a thin sliver of photosphere is visible. Then the sliver breaks up into a chain of beads along the rim of the moon. These *Baily's beads* are the photosphere shining through valleys that happen to be located at the edge of the moon and oriented so as to make the lunar rim irregular.

The last Baily's bead seems especially bright to the eye. It glistens and dazzles for a few seconds, sparkling like a diamond. This is called the *diamond ring effect* (Fig. 5–41 and Color Plate 14).

With the passage of the diamond ring effect, the total phase of the eclipse has begun. For a few seconds, a reddish glow is visible in a narrow band around the leading edge of the moon. This is the chromosphere and perhaps one or more prominences. Scientists studying the chromosphere have to work fast and have their photographic devices operating at a rapid pace in order to make their observations in these few seconds.

Then the corona is visible in all its glory (Figs. 5–25 and 5–42 and Color Plate 15). One can see the equatorial streamers and the polar plumes. At this stage, the photosphere is totally hidden so it is perfectly safe to stare at the corona with the naked eye and without filters. It has approximately the same brightness as the full moon, and is equally safe to look at. Unfortunately, many people are not adequately informed about this most important phase of the total eclipse, and miss the spectacle.

Figure 5–40 These children, who visited our eclipse site in Loiengalani, Kenya, just prior to the 1973 total solar eclipse, were practicing the use of special filters to protect their eyes when observing the partial phases. Such filters can be made by fully exposing black and white (not color) film to light, thus fogging it, and developing the film to maximum density. The black developed film then absorbs all but about 0.001% of the photospheric radiation uniformly across the visible and infrared.

The total phase may last a few seconds, or it may last as long as a little over 7 minutes. Spectrographs are operated, photographs are taken through special filters, rockets photograph the spectrum from above the earth's atmosphere, and tons of equipment are brought into operation to study the corona during this brief time of totality.

It is as though a chemist were told that he could study a certain chemical reaction only for 5 minutes a year. He might have to go to Africa to study the reaction; he could bring all his equipment and even bring all his colleagues with their own equipment, but like it or not the reaction would take place for only five minutes. Then he would have to wait a year or more for another chance.

At the end of the eclipse, the diamond ring appears on the other side of the sun, then Baily's beads, and then the final partial phases. All too soon, the eclipse is over.

On the average, a total solar eclipse occurs somewhere in the world every year and a half. There can actually be up to seven solar and lunar eclipses a year, but most of these are either lunar or only partial solar eclipses. Sometimes, too, the moon is at the far point of its orbit from earth and subtends an angle that is slightly smaller than that subtended by the photosphere. The umbra does not quite reach the earth. Even if the sun, moon, and earth are perfectly aligned, a rim of photosphere is visible all around the moon. Because of this annulus—ring—of light, the event is called an *annular eclipse* (Fig. 5–43). The corona is not visible at annular eclipses. When the umbra barely touches the earth's surface and moments of totality may or may not occur, we have a *central eclipse.*

Scientists prepare for solar eclipses years in advance. The band of totality usually does not cross populated areas of the earth (after all, the

A

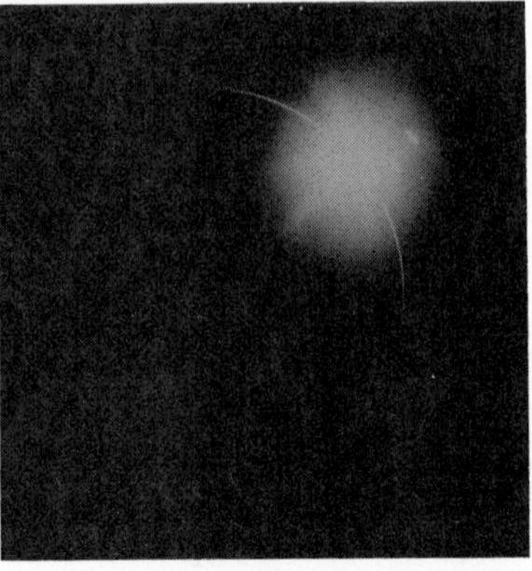

B

Figure 5–41 (*A*) The diamond ring effect at the total solar eclipse of March 7, 1970, observed from Miahuatlán, Mexico, by the author's expedition. (*B*) The diamond ring effect when Apollo 12 passed into the earth's (not the moon's) shadow while en route to the moon.

surface of the earth is largely covered with water), and astronomers usually have to travel great distances to carry out their observations. In 1965, a total eclipse crossed the South Pacific, and observations were made from tiny islands. In 1966, observations were made from the Andes in South America. In 1968, 32 seconds of totality were visible from a site in Siberia.

The last major eclipse observable from the United States occurred on March 7, 1970, when the band of totality crossed southern Mexico, and continued up the east coast of the United States and Canada. Many millions of people traveled into the band of totality to observe the eclipse from sites in Florida and Georgia (where, unfortunately, it was cloudy and the eclipse could not therefore be seen), and from sites in South Carolina, North Carolina, Virginia, and Nantucket Island off the coast of the Massachusetts mainland, at most of which the observing weather was excellent.

In 1972, an eclipse track passed over Siberia, Alaska, and Canada, but the weather was largely cloudy at the sites in eastern Canada where most observers went. The largest coordinated scientific eclipse effort ever made was carried out for the 1973 eclipse, which was visible in Africa, and which will be described in the following section. In 1974, an eclipse was visible in southwestern Australia, though the predictions of cloudy weather that kept many scientists away were borne out. In 1976 an eclipse started in Africa and was also visible in Australia, this time in the more populated southeastern regions. Again, most sites had cloudy skies.

On October 12, 1977, a total eclipse will be visible in Colombia and Venezuela in South America. The next total solar eclipse to cross the continental U.S. and Canada, and the last one to do so during the 20th century, will occur on February 26, 1979 (Fig. 5–44). The band of totality will include parts of the Pacific Northwest states of Washington and Oregon, and then continue north through Idaho, Montana, North Dakota, and central Canada. The rest of North America will see only a partial eclipse. The next total eclipse visible in the continental United States won't be until August 21, 2017. Annular eclipses will be visible in the U.S. in 1984 and 1994. Table 5–2 lists future eclipses.

Sometimes experiments are now carried out from jet planes during eclipses, but the eclipse progresses at 2000 km/hr or more across the earth. Jets fall behind. Earth satellites travel at 25,000 km/hr, and whiz through the band of totality within seconds. The greatest success ever in prolonging an eclipse was when the Anglo-French supersonic Concorde flew along with the shadow across Africa in 1973, and remained in totality for 74 minutes. Some of the astronomers aboard took advantage of the great altitude to carry out studies of infrared radiation that does not reach the ground.

In these days of orbiting satellites that can study ultraviolet coronal lines or carry coronagraphs above the earth's atmosphere, is it worth traveling to observe eclipses? The matter is controversial, but there is much to be said for the benefits of eclipse observing. Even when you consider that a certain fraction of the eclipses will be ruined by clouds, and that a certain number of experiments will fail at the last minute for a variety of reasons (ranging from the breakdown of an electronic component or a power failure to something trivial like the failure to remove a lens cap), eclipse observations are a relatively inexpensive way of observing the chromosphere and corona to find out such quantities as the temperature, density, and magnetic field structure. The whole coordinated eclipse effort that involved sending nearly a hundred individuals and a hundred tons of equipment to Africa to observe the 1973 total solar eclipse cost about a

Figure 5–42 Totality at the June 30, 1973 eclipse photographed from Loiengalani, Kenya, by the author's expedition. Taken without a filter, the photograph approximates the view that is observed by the naked eye.

Figure 5–43 A member of the author's expedition, observing the annular eclipse of December 24, 1974 in the Andes Mountains near Bogotá, Colombia.

Figure 5–44 (*A*) The 1977 and 1979 total solar eclipses. (*B*) In 1979, totality will be visible from sites in the Pacific Northwest of the U.S. and the north central U.S. and Canada.

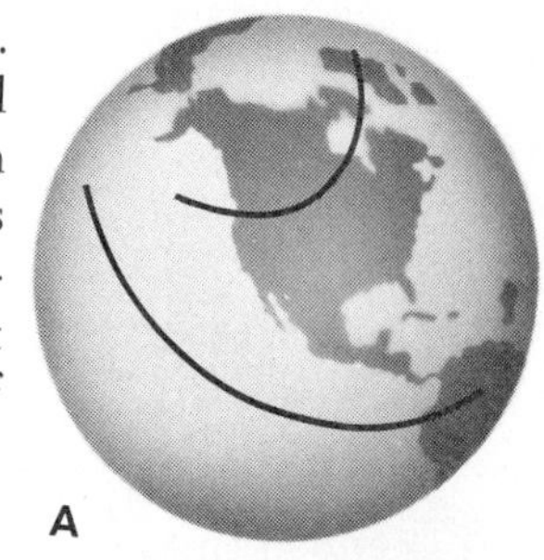

A

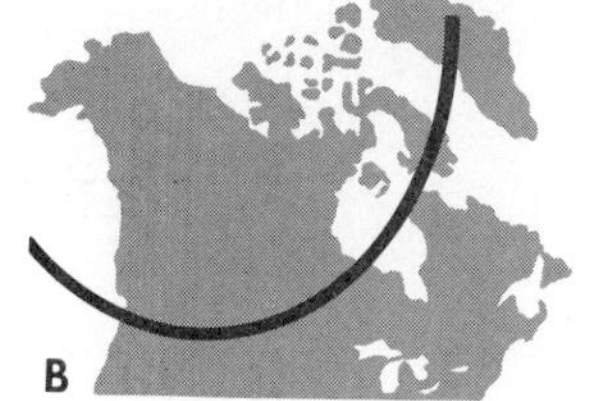

B

TABLE 5–2 TOTAL, CENTRAL, AND ANNULAR SOLAR ECLIPSES

Date	Maximum Duration	Type	Location
1977 April 18	7:05	annular	Southern Africa
1977 October 12	2:37	total	Pacific Ocean, South America (Colombia and Venezuela)
1979 February 26	2:49	total	U.S.A., Canada, Greenland
1979 August 22	6:02	annular	Antarctica
1980 February 16	4:08	total	Central Africa, India, China
1980 August 10	3:23	annular	South Pacific (Galapagos), central South America
1981 February 4	1:13	annular	Pacific Ocean and Australia
1981 July 31	2:03	total	U.S.S.R.
1983 June 11	5:11	total	Indian Ocean, Indonesia, New Guinea
1983 December 4	4:03	annular	Atlantic Ocean, central Africa
1984 May 30	1:05	annular	Mexico, southeastern U.S.A., North Africa
1984 November 22	1:59	total	New Guinea, South Pacific Ocean
1985 November 12	1:59	total	South Pacific Ocean
1986 October 3	–	central	North Atlantic Ocean
1987 March 29	–	central	Atlantic Ocean, Africa
1987 September 23	3:45	annular	China, South Pacific
1988 March 18	3:46	total	Indonesia, Philippines
1990 July 22	2:33	total	Finland, U.S.S.R., U.S.A. (Alaska)
1991 July 11	6:54	total	U.S.A. (Hawaii), Central and South America

million dollars, and about another million dollars were spent on rocket experiments. Compare this to the cost of a Skylab mission, 3 billion dollars in total (of which $200 million went to the solar telescopes), or to the $63 million cost of OSO-8. Of course, ground-based eclipse observations are limited to the visible and radio regions of the spectrum, and record the sun at only one moment of time (the corona hardly changes during the few minutes of totality nor even, usually, during the hours that the shadow takes to pass across the earth). Thus eclipses do not compete with space experiments for many types of observation. Still, it seems that eclipse observations, chancy as they are and with all their limitations, remain a cost-effective way of studying the solar atmosphere.

5.5a An Eclipse Expedition

An eclipse expedition is unlike any other kind of observing experience that an astronomer can have. A year or more of preparation is followed by a tense and pressure-packed but fascinating few weeks at the expedition site, all leading up to the inexorable deadline of the eclipse. Then, if the experiment is successful, it may take many months or several years to study the data and draw the conclusions.

On June 30, 1973, a total solar eclipse occurred that lasted over 7 minutes, close to the maximum duration theoretically possible. It was the longest eclipse of the century. Clear weather was predicted and so many teams of astronomers from countries all over the world prepared experiments to carry out during the eclipse.

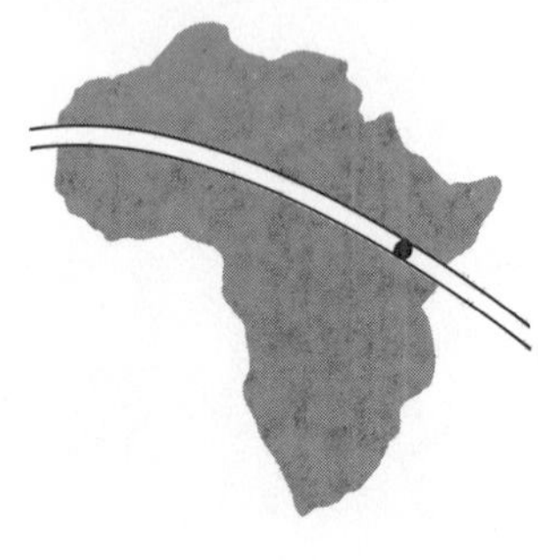

Figure 5–45 The June 30, 1973 eclipse crossed Africa from west to east at a speed that ranged from about 2000 to 3000 km/hr.

The path of the eclipse was a narrow band that crossed the continent of Africa from west to east (Fig. 5–45). The point on the map where totality was to have its maximum duration fell in the middle of the Sahara desert; the nearest towns included Timbuktu, which is in Mali, and Agadez, a similar desert outpost in Niger.

Unfortunately, a long drought in the Sahara made the air continually dusty. It did not seem that conditions would be good enough there to observe the eclipse.

The U.S. National Science Foundation decided to coordinate the American efforts and explored other sites. In view of the Saharan dust problem, they sent most of the astronomers to a site in northern Kenya. Totality was to be five minutes there, shorter than the 7-minute maximum but still a very respectable length of time. A smaller number of astronomers, who could justify specifically their need for amounts of observing time longer than 5 minutes (such as those who wanted to study the 5-minute oscillation of the solar atmosphere and needed to observe more than one complete cycle), were to go to a desert site in Mauritania, not quite as inaccessible as the sites in Mali or Niger but still in the Sahara.

Groups from about 20 U.S. observatories, research institutes, colleges, and universities were accepted as part of the overall expedition. The National Center for Atmospheric Research (NCAR), in Boulder, Colorado, was given overall responsibility for the logistics.

A month before the eclipse, the astronomers, technicians, and students who were on the expedition left for Africa in a chartered jet. The first day the plane flew from New York to Dakar, Senegal, on the western coast of Africa. The second day's flight was longer: we flew across Africa from Dakar to Nairobi, Kenya. About 70 individuals were along.

After only one day's rest in Nairobi, we set off for the eclipse site, which was in an oasis named Loiengalani, situated in the Northern Frontier District of Kenya, an area that was usually off-limits to tourists because of tribal wars. The eclipse site was a 16-hour drive north of Nairobi to the shore of Lake Turkana (then called Lake Rudolf), a huge fresh-water lake (Fig. 5–46).

At Loiengalani, we spent the next three weeks setting up equipment, checking, testing, and rechecking. Astronomers have to be very versatile at an eclipse site. They may have to construct a small building to shield their instruments from the sand and sun, they may have to pour concrete to make stable platforms for their instruments, and they may have to know details of machine-work or electronics to make last-minute repairs to balky

Figure 5–46 The eclipse site alongside Lake Rudolf (now Lake Turkana) near the oasis of Loiengalani, Kenya.

Figure 5–47 Lecturing to students at the mission school in Loiengalani.

Figure 5–46

Figure 5–47

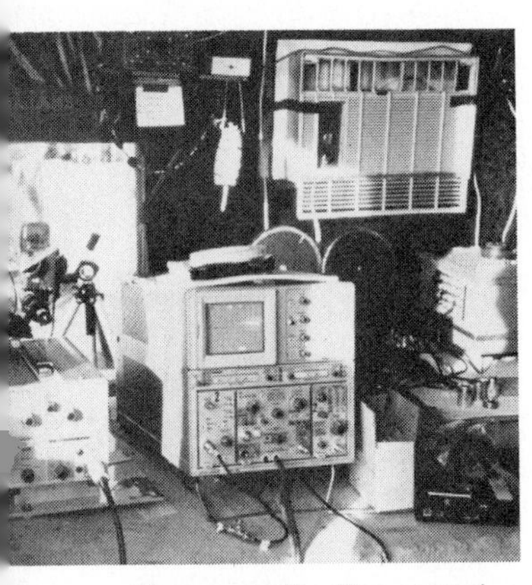

Figure 5–48 The experiment measured emission lines in the solar corona using a spectrometer that contained a silicon vidicon.

instruments. It helps to be a combination carpenter, mason, machinist, electronics engineer, optical physicist, and more.

Some of the astronomers lived in a newly-enlarged lodge while others lived in rooms in the local mission. Some of us lectured to the children at the mission school to explain what was going to happen and how to watch the eclipse safely (Fig. 5–47).

My own experiment studied the infrared spectrum of the solar corona, using an experimental silicon vidicon device (Fig. 5–48) of a type described in Section 3.8. We were observing particular strong emission lines in the E-corona that were caused by twelve-times ionized iron (Fe XIII). These lines would tell us the density at different points in the corona.

The eclipse site was situated next to one of the two villages on earth of El Molo, the smallest tribe in Africa. Their straw huts and subsistence fishing (with spears rather than with rod and line) were a contrast to our sophisticated apparatus. Also nearby were villages of Turkana and Samburu.

Tension grew as the day of the eclipse approached. One by one, almost all of the experiments were made to operate. A few days before the eclipse, reporters and tourists from Nairobi swelled the population of the oasis to a few hundred.

The morning of the eclipse, the sun rose in a perfectly clear sky. Then, within minutes, the sky completely clouded over. (Many of us were awake to watch the sunrise, because of a combination of worry and a desire to go out to check the safety of our equipment after the routinely high winds of more than 100 km/hr the previous night.)

The eclipse was to take place after noon, and the astronomers went through last-minute checks of their apparatus. The clouds gradually disappeared. At noontime we received a brief radioed report from the Mauritanian group who had just observed the eclipse, for the eclipse shadow took about 2 hours to cross Africa. They told us the shape of the corona, and mentioned special structures in the corona on which we might be interested in concentrating our attention. A sandstorm had just ended in Mauritania, and this had severely limited their ability to observe the eclipse.

The partial phase began, and the sky grew darker. We had special filters to enable us to follow the progress of the moon across the sun. A few minutes before totality, at about the time that the shadow bands occurred,

Figure 5–49 The camera was not moved between exposures in this sequence that shows the partial phases and totality of the 1973 eclipse. Note that clouds occasionally covered the sun during the partial phases.

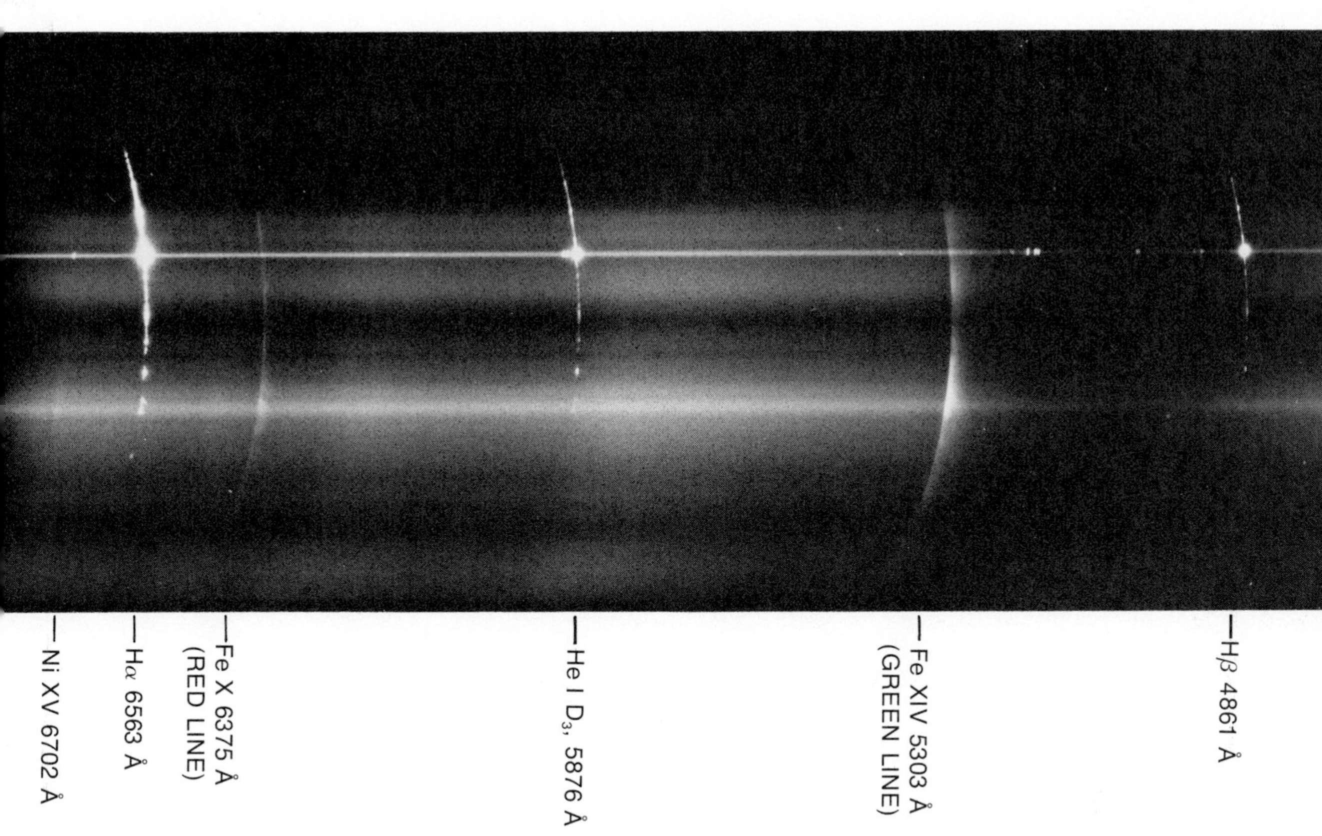

Figure 5–50 The spectrum of the solar corona, obtained by the expedition of the Tokyo Astronomical Observatory during the 1973 eclipse. Two hydrogen lines and a helium line, all from the chromospheric spectrum, also appear.

many of the experiments started recording data on instrumental tape recorders or on automatic devices with pens and moving strips of paper, or started cameras whirring. Only a few small clouds were left in the sky. Fig. 5–49 shows the progress of the eclipse phases.

Then it came. Baily's beads were visible and the diamond ring appeared. "Totality," someone shouted, for many of us were inside buildings or under tents tending our equipment and could not even see the eclipse. For five short minutes everyone worked to try to capture the data that would later on be analyzed to help us understand the solar atmosphere.

The end of the eclipse came, within a few seconds of the time predicted months and years before. The tension was over. The day had been a tremendous success.

The next day we all found that we could pack very quickly the large amount of equipment that had taken a month to unpack.

Now, years later, the results of the eclipse experiments are being published. Papers are given in meetings of the astronomical societies around the world or appear in scientific journals. Our knowledge of almost all parts of the sun has been considerably advanced.

For example, observations of the edge of the sun, made with both optical and radio telescopes during the seconds immediately preceding or following totality, allowed the limb-darkening curve to be determined with increased accuracy. From this, new measurements of the temperature and density structure of the photosphere and chromosphere were derived. Only small prominences were visible, but photographs of them that were taken through filters at special wavelengths have been analyzed to deduce the electron densities and the distribution of matter of different temperatures within the prominences. Several scientific groups made coronal spectra (Fig. 5–50) or images of the corona through suitable filters that passed only the wavelengths of coronal emission lines. These observations led to models of the distributions of coronal electrons and of the coronal temperature structure.

An especially bright and concentrated area of the corona, a *coronal condensation,* was determined to have at each height 10 times the density of the normal corona at those heights. The temperature in the condensation was determined to be 2,300,000 K. It had a cooler (1,000,000 K) dense central part that coincided with a small prominence.

Other groups of scientists observed through polarizing filters in order to study the distribution of both dust in interplanetary space and of electrons in the corona. Infrared photographs were taken to study the outer corona (the sky is darker at longer wavelengths than at shorter wavelengths, so one can observe the corona at greater distances from the sun in the infrared than in the visible). The set of observations carried out at the 1973 eclipse by scientists of many nations is noteworthy for its variety.

5.6 SUNSPOTS AND OTHER SOLAR ACTIVITY

In this section we discuss the active sun.

We have discussed the components of the sun as though it were a static object: we described the photosphere, the chromosphere, and the corona, and will describe the interiors of the sun and stars of similar mass in Chapter 6. But a host of other time-varying phenomena are superimposed on the basic structure of the sun. Many of them, notably the sunspots, vary with an 11-year cycle, which is called the *solar activity cycle.* (We shall see that the actual cycle is twice that length, 22 years in duration.)

5.6a *Sunspots*

Sunspots (Fig. 5–51) are the most obvious manifestation of solar activity. They are areas of the sun that appear relatively dark when seen in white light. They appear dark because they are giving off less radiation than

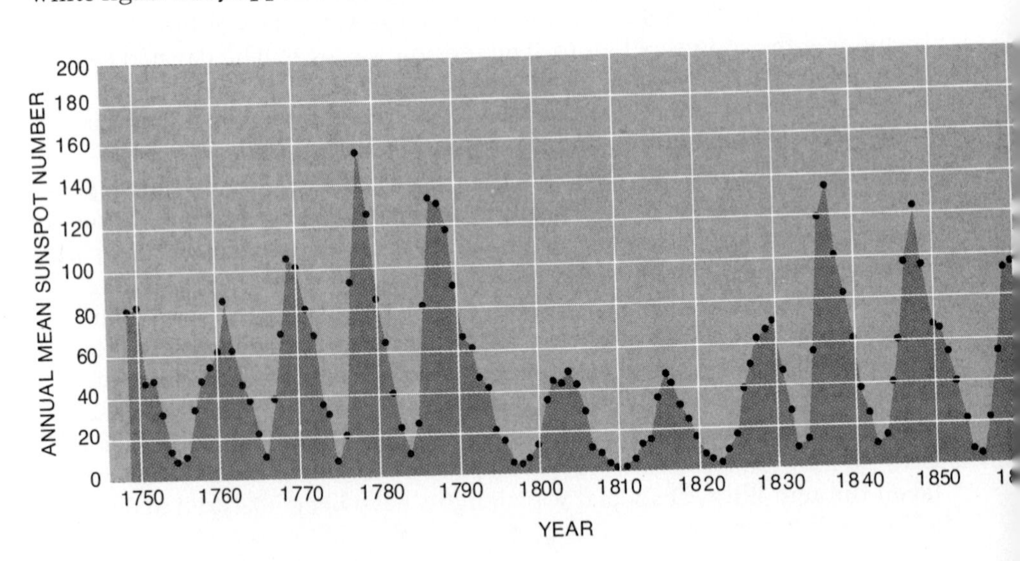

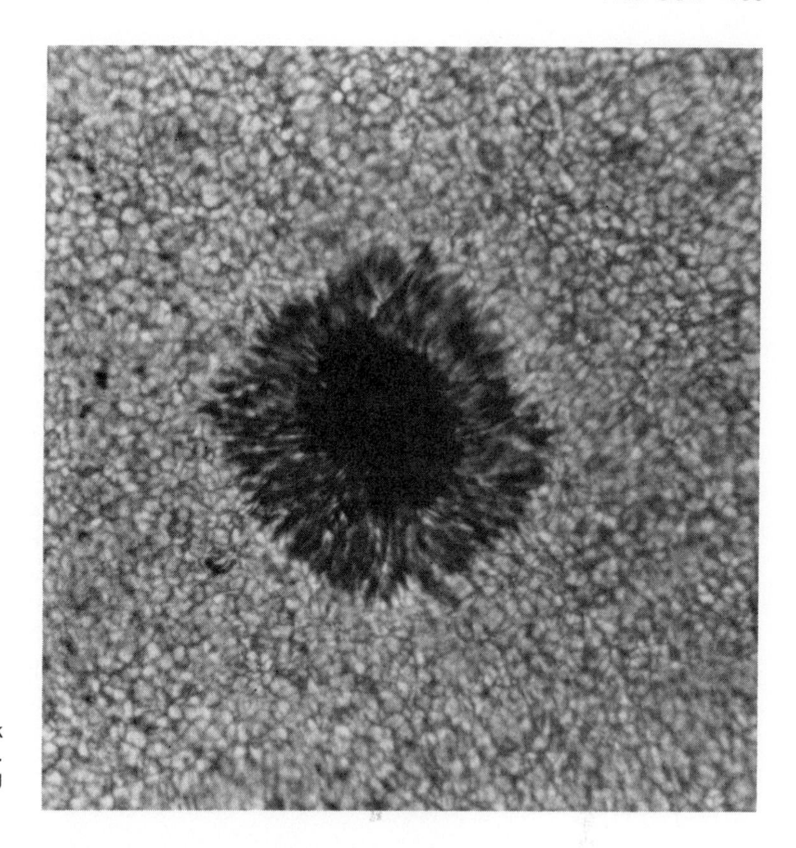

Figure 5–51 A sunspot, showing the dark *umbra* surrounded by the lighter *penumbra*. Granulation is visible in the surrounding photosphere.

the photosphere that surrounds them. This implies that they are cooler areas of the solar surface, since cooler gas radiates less than hotter gas. Actually, if we could somehow remove a sunspot from the solar surface and put it off in space, it would appear bright against the dark sky and might give off as much light as the full moon.

A sunspot includes a very dark central region, called the *umbra* from the Latin for "shadow" (pl: *umbrae*). The umbra is surrounded by a *penumbra* (pl: *penumbrae*), which is not as dark (just as during an eclipse the umbra of the shadow is the darkest part and the penumbra is less dark). Flashes of brightness that have been detected in the sunspot umbrae are presumably related to waves that have been seen running outward through

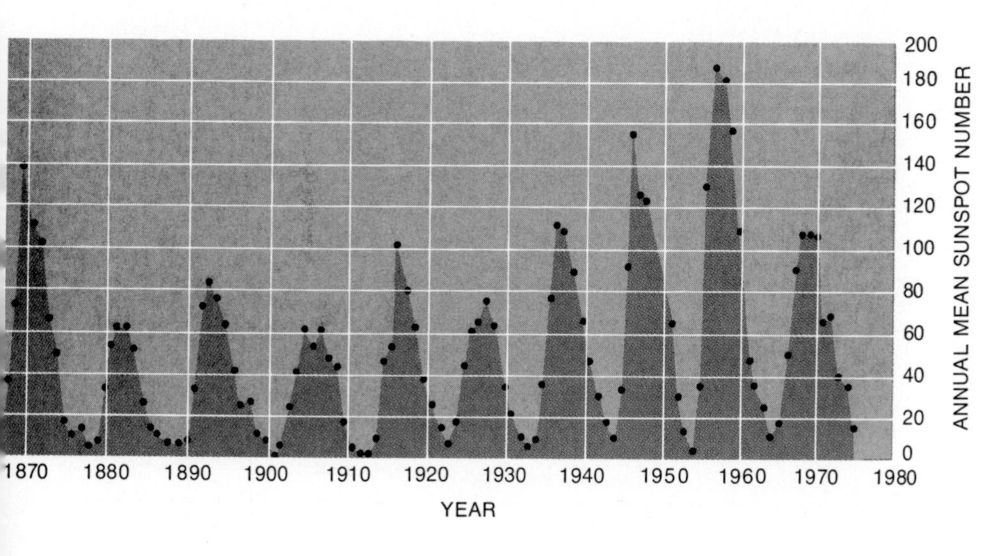

Figure 5–52 The 11-year sunspot cycle is but one manifestation of the solar activity cycle.

the penumbrae. Also, a general outflow of the matter in the penumbrae, from the umbrae to the spot boundary, has been observed (the *Evershed effect*).

Sunspots were discovered in 1610, independently by Galileo in Italy, Fabricius (who, as we noted in Section 4.4a, also discovered Mira variables), Christopher Scheiner in Germany, and Thomas Harriot in England. In about 1850, it was realized that the number of sunspots varies with an 11-year cycle, as is shown in Figure 5–52. This is called the *sunspot cycle*, although we now realize that many related signs of solar activity vary with the same period.

In 1904, E. Walter Maunder plotted, month by month, the latitude of each sunspot on the sun. He found that his diagram looked like that in Figure 5–53. Since the pattern resembles a butterfly, this is called *Maunder's butterfly diagram*. It shows that early in a sunspot cycle, new spots form close to latitude 30°. As the cycle progresses, new spots are formed closer and closer to the equator. At the end of a cycle, just before the time when the number of sunspots is at a minimum, spots of the old cycle may be appearing at the equator while the first spots of the new cycle may already be appearing at higher latitudes.

To understand sunspots we must understand magnetic fields. When iron filings are put near a simple bar magnet on earth, the filings show a

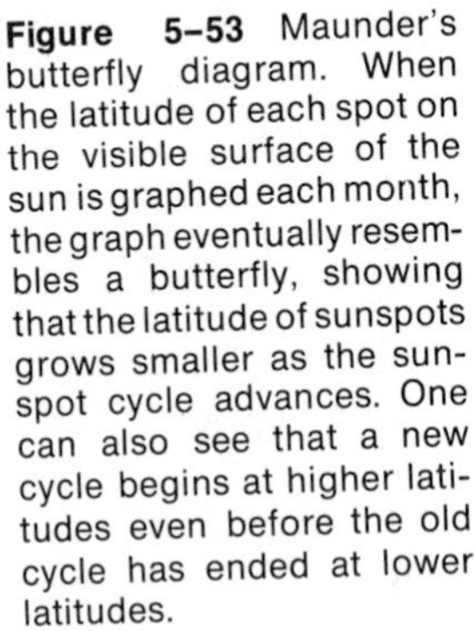

Figure 5–53 Maunder's butterfly diagram. When the latitude of each spot on the visible surface of the sun is graphed each month, the graph eventually resembles a butterfly, showing that the latitude of sunspots grows smaller as the sunspot cycle advances. One can also see that a new cycle begins at higher latitudes even before the old cycle has ended at lower latitudes.

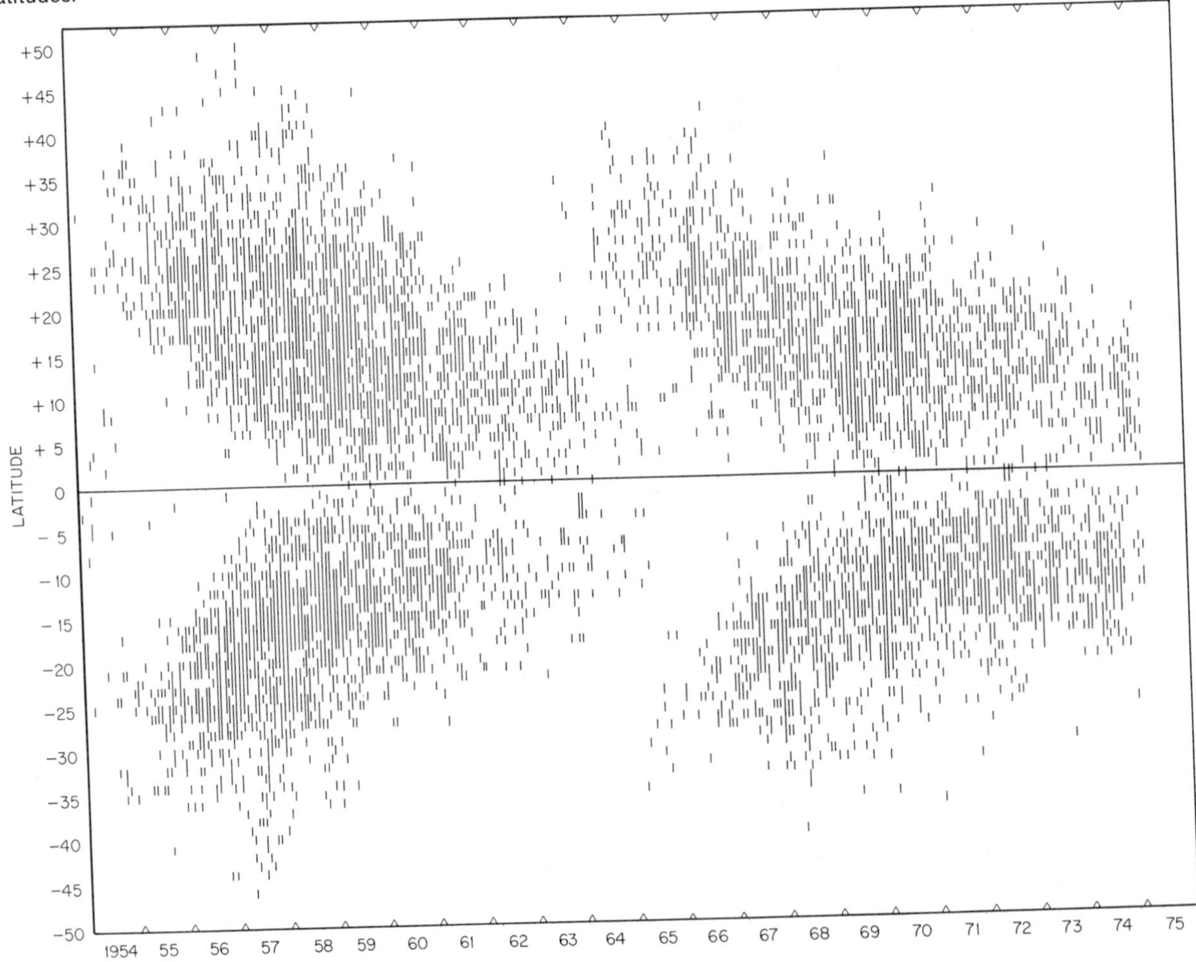

pattern that is illustrated in Figure 5–54. The magnet is said to have a north pole and a south pole, and the magnetic field linking them is characterized by what we call *magnetic lines of force,* or *magnetic field lines* (after all, the iron filings are spread out in what look like lines). The earth (as well as some other planets) has a magnetic field that has many characteristics in common with that of a bar magnet. The structure seen in the solar corona, including polar plumes and equatorial streamers, results from matter being constrained by the solar magnetic field (Fig. 5–55).

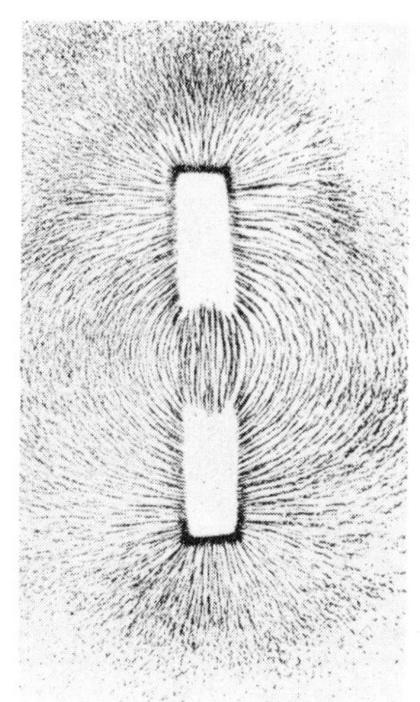

Figure 5–54 Lines of force from a bar magnet are outlined by iron filings.

We can measure magnetic fields on the sun by using a spectroscopic method. In the presence of a magnetic field, certain spectral lines are split into a number of components (Fig. 5–56) and the amount of the splitting depends on the strength of the magnetic field. This is called the *Zeeman effect.*

These measurements were first made by George Ellery Hale (who, in addition to his solar research, was responsible for the existence of the large telescopes at Mt. Wilson and Palomar). Hale showed, in 1908, that the sunspots are regions of very high magnetic field strength on the sun, thousands of times more powerful than the earth's magnetic field or than the average solar magnetic field. Sunspots usually occur in pairs, and often these pairs are part of larger groups. In each pair, one sunspot will have a polarity typical of a north magnetic pole and the other will have a polarity typical of a south magnetic pole. The shape of the structure in Hα is controlled by the magnetic fields in the solar atmosphere, so one can even get a good idea of the structure of the magnetic field merely by observing in Hα.

The spectrum of a sunspot is that typical of a gas of approximately 3800 K, cooler than the gas in the rest of the photosphere. Thus the sunspot spectrum resembles that of a star cooler than the sun, and is typical of spectral class K. In addition to the spectral lines of molecules that are typical of both sunspots and cooler stars, one sees the extreme splitting of spectral lines by the Zeeman effect. Daily maps of the solar magnetic field (Fig. 5–57) are made at the Mt. Wilson and Kitt Peak Observatories and by the National Oceanic and Atmospheric Administration's (NOAA) laboratory in Boulder, Colorado.

Figure 5–55 The configuration of the magnetic field extending through the solar corona can be calculated theoretically from ground-based observation of the magnetic fields on the solar surface. (*A*) The magnetic field calculation for September 5, 1973 is reasonably well-matched by the actual coronal configuration, as observed (*B*) in x-rays from Skylab. Theoretical work is continuing to improve the quality of the predictions.

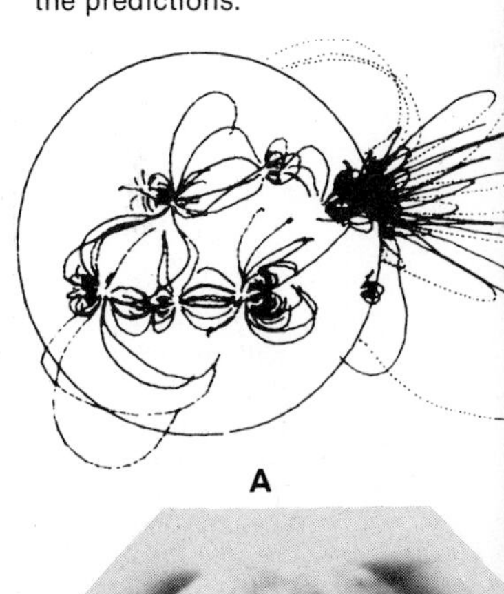

A

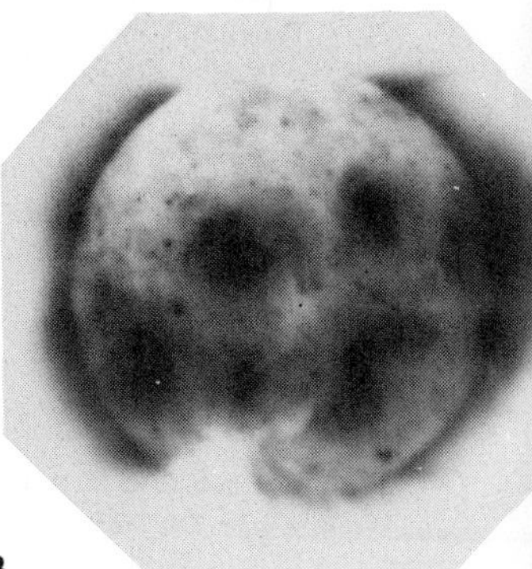

B

Besides the specific magnetic fields in sunspots, the sun seems to have a weak overall magnetic field with a north magnetic pole and a south magnetic pole, which may entirely result from the sum of the weak magnetic fields from vanished sunspots. Every 11-year cycle, the north magnetic pole and south magnetic pole on the sun reverse polarity; what had been a north magnetic pole is then a south magnetic pole and vice versa. This occurs a year or two after the number of sunspots has reached its maximum. For a time during the changeover, the sun may even have two north magnetic poles or two south magnetic poles! But the sun is not a simple bar magnet, so this strange-sounding occurrence is not prohibited. Because of the changeover, it is 22 years before the sun returns to its original configuration, so the real period of the solar activity cycle is 22 years.

The next maximum of the sunspot cycle—the time when there are the greatest number of sunspots—should take place in 1980 or 1981. NASA plans to launch a satellite called the Solar Maximum Mission to study solar activity at that time.

Although the details of the formation of sunspots are not yet understood, a general picture was suggested in 1961 by Horace W. Babcock of the Hale Observatories, elaborating on ideas that had been advanced by E. N. Parker of the University of Chicago. In the Babcock model, just under the

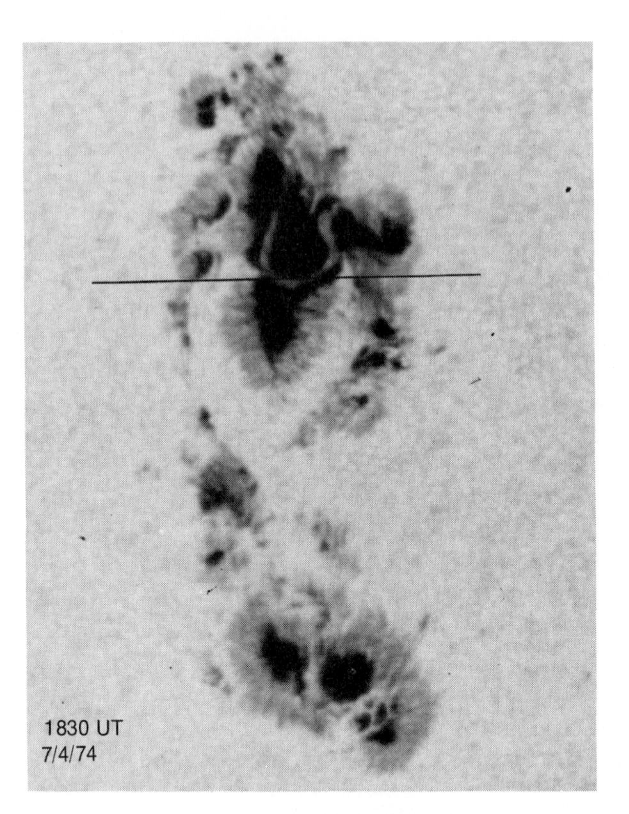

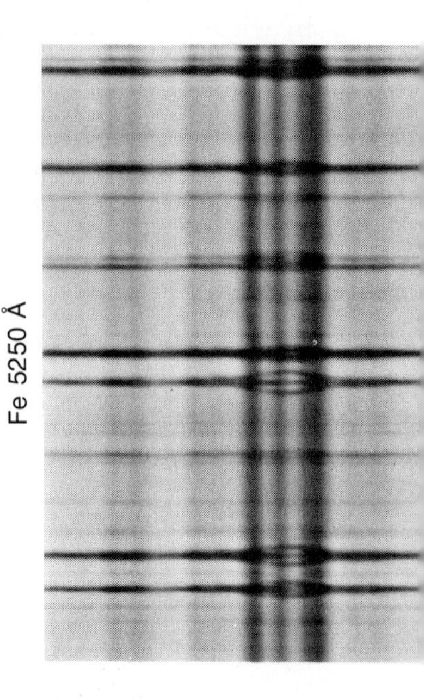

Figure 5–56 The Zeeman effect, the splitting of certain spectral lines into several components in the presence of a magnetic field, is shown at right in these Kitt Peak observations of the sunspot shown in the photograph opening this chapter. The position of the spectrograph slit on the sunspot group is shown at left. The high degree of splitting shown here of the lines, particularly the 5250 Å line of iron, indicates that the magnetic field in the spot center was thousands of gauss, compared to the earth's magnetic field of about one gauss.

solar photosphere the magnetic field lines are bunched in tubes of magnetic field that wind around the sun. These tubes are formed by a mechanism (Fig. 5–58) that will be discussed next.

The sun rotates approximately once each earth month (Fig. 5–68). Different latitudes on the sun rotate at different speeds. Though a solid ball like the earth rotates at a constant rate at all latitudes (both Tampa and New York rotate in 24 hours), a gaseous ball like the sun can rotate differentially (Fig. 5–59). Because of this *differential rotation,* if a line of sunspots started at the same longitude on the sun at a given time, the ones closest to the equator would make a full revolution faster than the ones farthest from the equator. Gas at the equator rotates in 25 days, gas at 40 degrees latitude rotates in about 28 days, and gas nearer the poles rotates even more slowly. The differential rotation can also be measured spectrographically by studying Doppler shifts. (Not only does the surface of the sun rotate differentially, but also the sun probably rotates at different speeds at different distances from its core, though we cannot directly observe the interior. Different atmospheric heights appear to rotate at slightly different speeds.)

A line of force that may have started out in the direction from north to

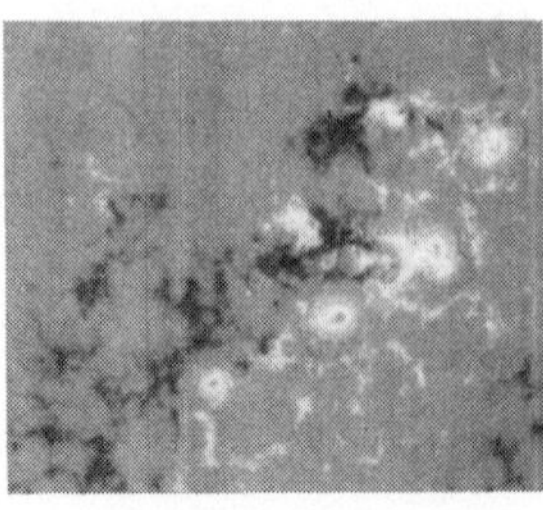

MAGNETIC FIELD

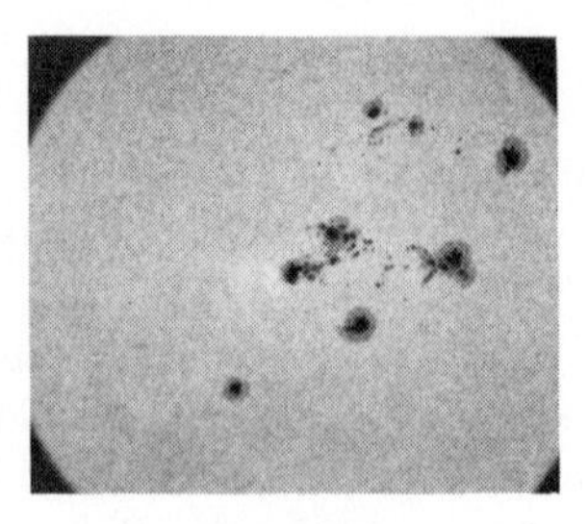

Hα +4.0 Å

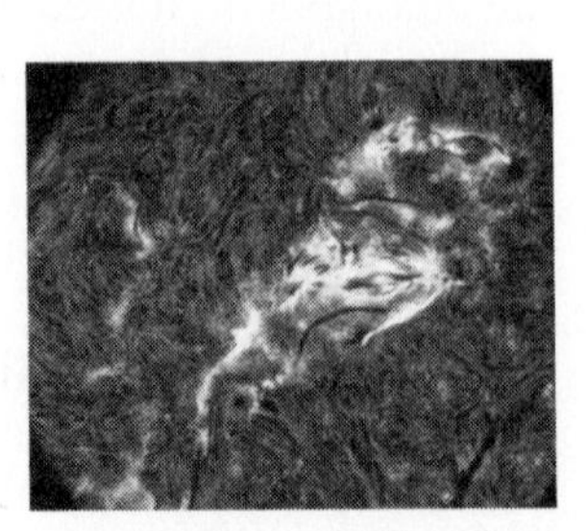

Hα LINE CENTER

Figure 5–57 From studies of the splitting of lines that are sensitive to the Zeeman effect, the magnetic field on the sun can be mapped. Black and white indicate opposite polarities. The simultaneous picture at Hα + 4 Å (in the continuum, 4 Å redward of Hα) shows the photosphere, and the right-hand picture, taken at the Hα line center, shows the chromosphere. An area of the solar surface 300,000 km square is shown; the black corners are a telescope effect and are not the edge of the sun.

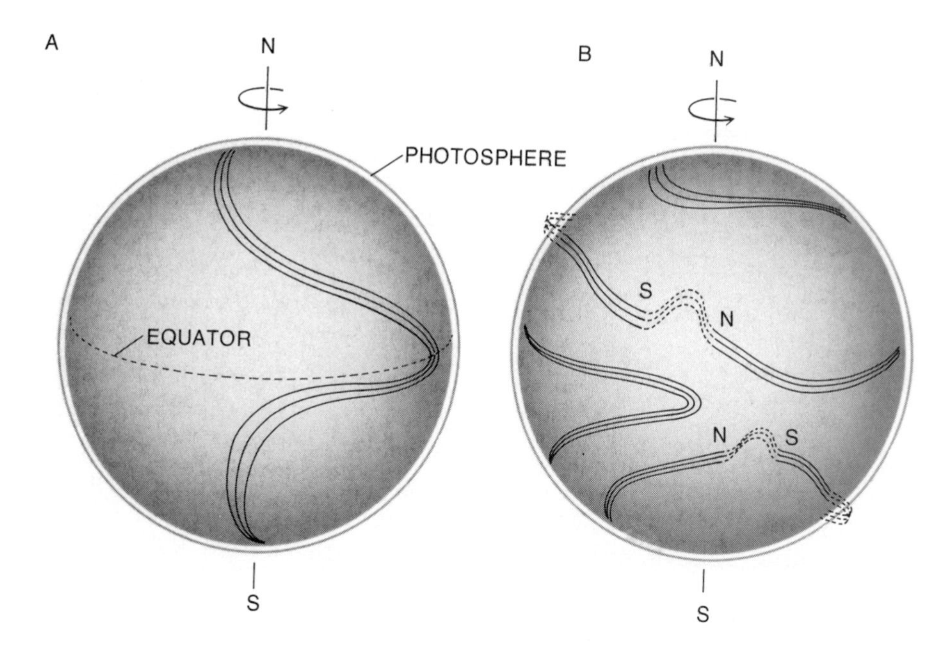

Figure 5-58 A leading model to explain sunspots suggests that the solar differential rotation winds tubes of magnetic flux around and around the sun. When the tubes break and penetrate the solar surface, we see the sunspots that occur in the areas of strong magnetic field.

south on the solar surface is wrapped around the sun by the action of the differential rotation. These lines collect in the equivalent of tubes located not far beneath the solar photosphere. Under some circumstances, buoyant forces carry part of a tube upward until the tube sticks up through the solar photosphere. Where the tube emerges we see a sunspot of one magnetic polarity, and where the tube returns through the surface we see a sunspot of the other polarity. The north and south polarities are connected by magnetic lines of force that extend above the sun.

Because of the differential rotation, the spiral winding of the magnetic lines of force is tighter at higher latitudes on the sun than at lower latitudes. Thus the instability that allows part of a tube to be carried to the surface arises first at higher solar latitudes. As the solar cycle wears on, the differential rotation continues and the tubes rise to the surface at lower and lower latitudes. This explains the latitude effect that we have seen illustrated in the butterfly diagram.

Robert Leighton of Caltech has continued the development of models of this type, and has used computers to carry out numerical calculations elaborating on the above general ideas. His calculations show how bits of magnetic field migrate over the surface of the sun as the sun rotates differentially. Hirokazu Yoshimura at Colorado further developed a theory along these lines to take into account the causes of the differential rotation. The effects that lead to the formation of sunspots arise from the interaction of the solar differential rotation with turbulence and motions in the convective zone. Dynamos in factories on earth also depend on the interaction of rotation and magnetic fields; thus these theories for the solar activity are called *dynamo theories*, and one says that the sunspots are generated by the *solar dynamo*. The dynamo theories can explain such important observations as the existence of the sunspot cycle and its 22-year period, the connection of sunspots to regions of high magnetic field, and Maunder's butterfly diagram.

Magnetic fields are able to restrain matter—this is the property we are trying to exploit on earth to contain superheated matter sufficiently long to allow nuclear fusion for energy production to take place. The strongest magnetic fields in the sun occur in sunspots. The magnetic fields in sun-

Figure 5-59 The notion that the sun rotates differentially is illustrated by this series, which shows the progress of a schematic line of sunspots month by month. The equator rotates faster than the poles by about 4 days per month.

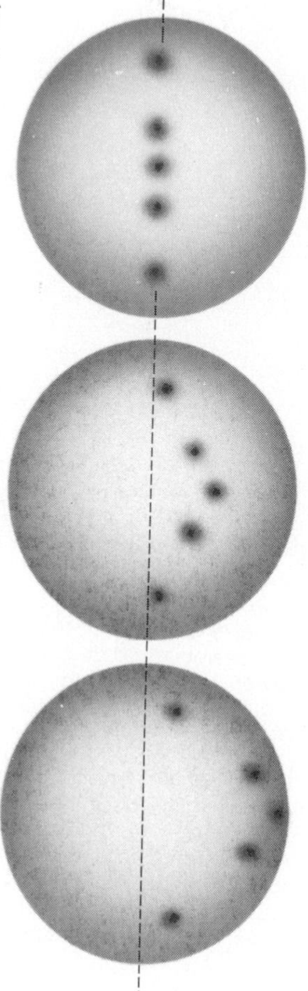

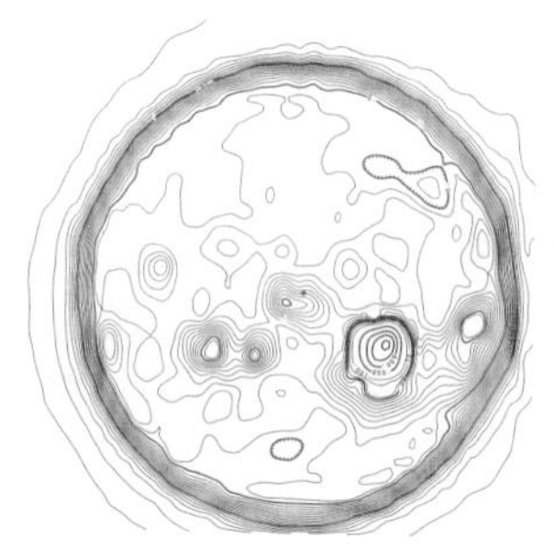

Figure 5–60 A radio map of the sun made at a wavelength of 2.8 cm with the 100-meter dish of the Max Planck Institute for Radio Astronomy of Bonn, Germany. An active region radiates in this region of the spectrum in addition to the visible.

spots restrain the motions of the matter there, and in particular limit the transport of energy to photospheric heights by convection from lower, hotter levels. This results in sunspots being cooler and darker, though exactly why they remain so for weeks is not known. The parts of the corona above active regions are hotter and denser than the normal corona. Presumably the energy is guided upward by magnetic fields. These locations are prominent in radio (Fig. 5–60) or x-ray maps of the sun.

5.6b Flares

Violent activity sometimes occurs in the regions around sunspots. Tremendous eruptions called *solar flares* (Fig. 5–61) can eject particles and emit radiation from all parts of the spectrum into space. These solar storms begin in a few seconds and can last up to four hours. A typical flare lifetime is 20 minutes. Temperatures in the flare can reach 5 million kelvins, even

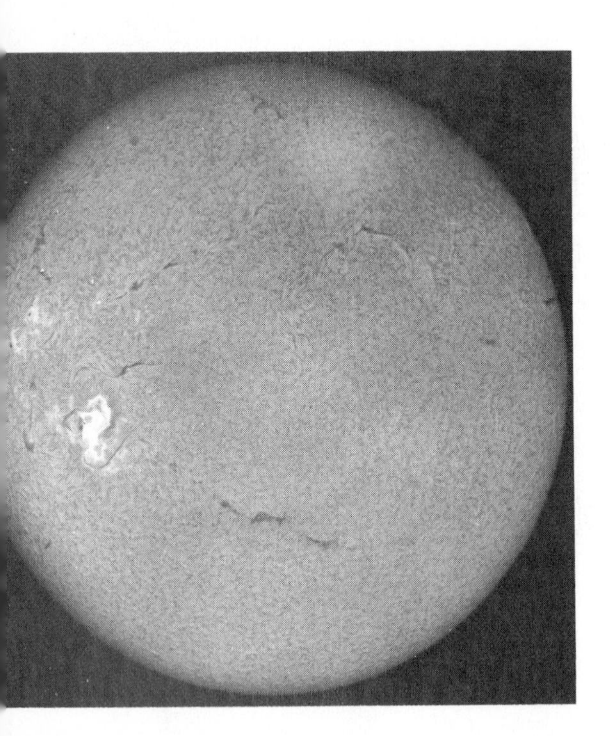

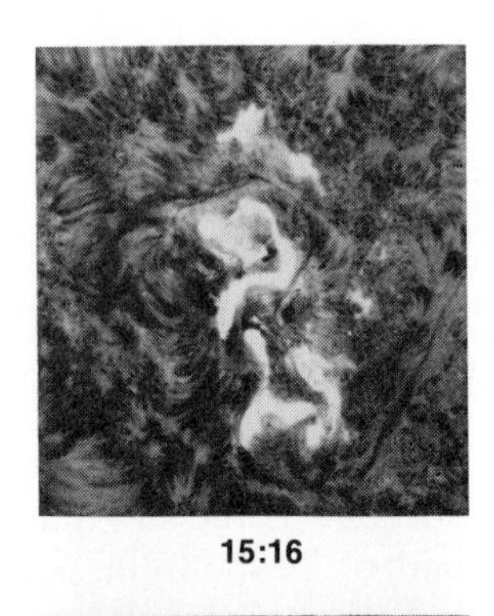

15:16

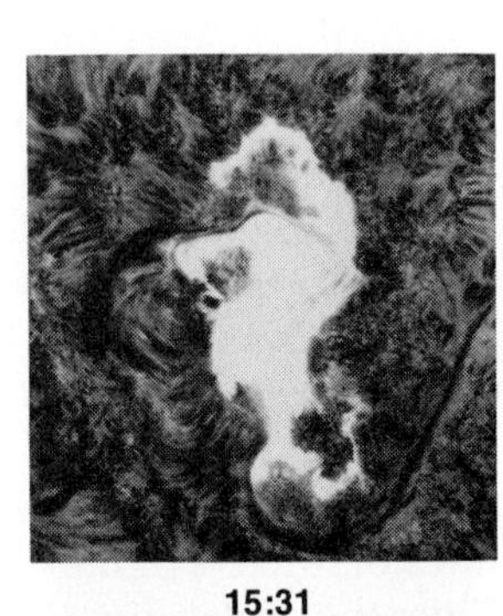

15:31

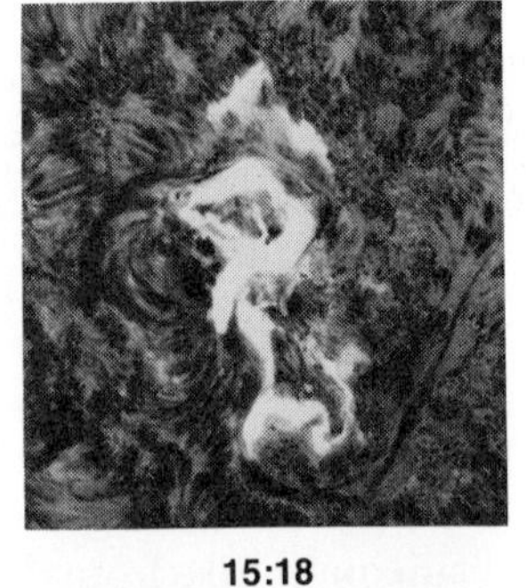

15:18

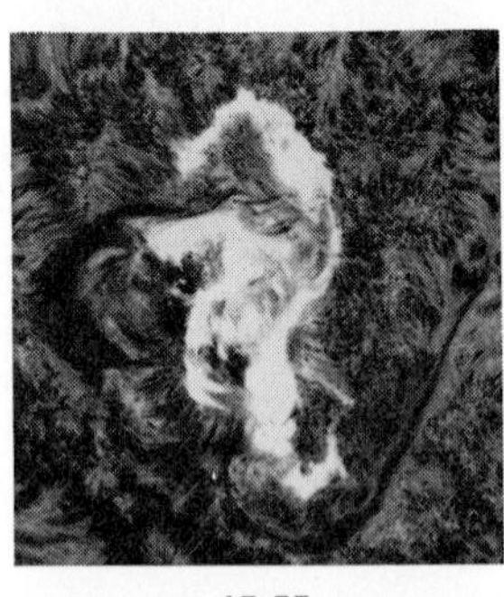

15:55

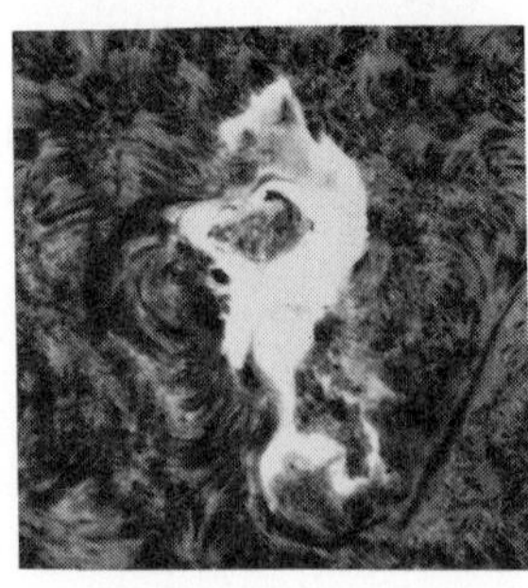

15:23

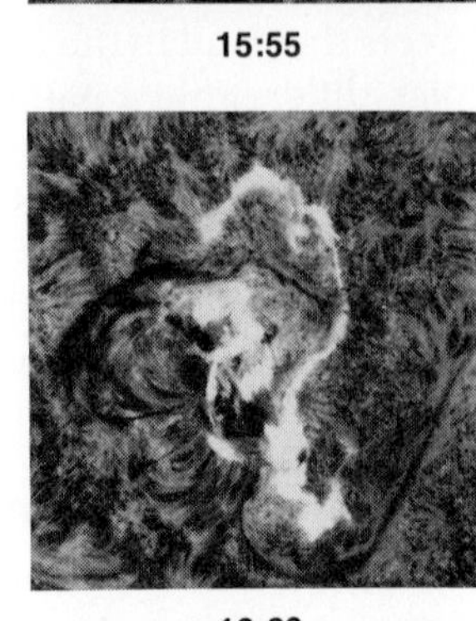

16:20

Figure 5–61 One of the largest solar flares in decades occurred on August 7, 1972, and led to power blackouts, short wave radio blackouts, and aurorae. (*A*) The whole sun is shown at the peak of the flare, 15:30 U.T. (Universal Time, approximately corresponding to Greenwich Time). (*B*) The development of the flare with time and its subsidence is shown in this sequence from the Big Bear Solar Observatory. It was hours before the flare completely died down. Other strong flares also occurred in this active region during that week. Labels show Universal Time in hours and minutes. All photographs are in Hα.

Figure 5–62 An aurora borealis, beautiful patterns of color in the sky, photographed in Alaska. The aurorae are phenomena of the earth's atmosphere, and are caused by incoming particles from the sun.

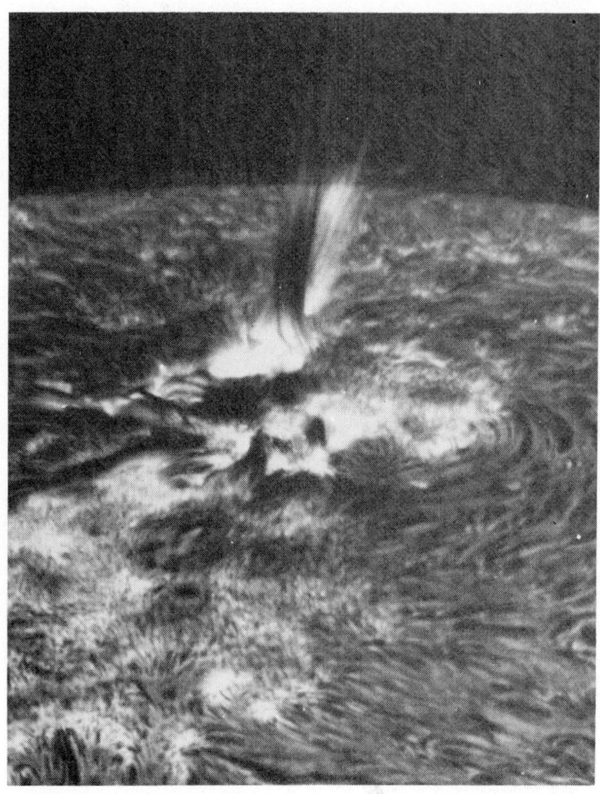

Figure 5–63 A flare near the solar limb, photographed at Big Bear on August 20, 1971.

hotter than the quiet corona. Flare particles that are ejected reach the earth in a few hours or days and can cause disruptions in radio transmission, cause the aurorae (Fig. 5–62 and Color Plate 33)—the *aurora borealis* is the "northern lights" and the *aurora australis* is the "southern lights"—and even cause surges on power lines. Because of these solar-terrestrial relationships, high priority is placed on understanding solar activity and being able to predict it. The U.S. government even has a solar weather bureau to forecast solar storms, just as it has a terrestrial weather bureau.

Solar flares are best visible when photographed through Hα filters. At the Big Bear Solar Observatory, time-lapse movies of the sun are made whenever the sun is visible, in order to catch the precursors and moments of eruption of a flare (Fig. 5–63). Other observatories, and an Air Force global network, maintain flare patrols as well. Only rarely are flares visible in white light (Fig. 5–64).

Figure 5–64 The giant flare of August 7, 1972, whose Hα images are shown in Fig. 5–61, was also visible in white light, a very unusual occurrence that happens less than once a year. The white light phase usually lasts less than 10 minutes. (Photo taken with the Solar Tower telescope, Sacramento Peak Observatory)

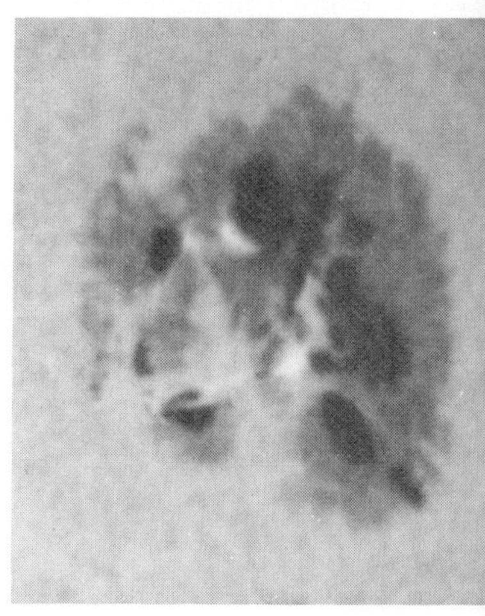

Solar flares also emit x-rays, which are studied from Skylab, from OSO's, and from meteorological satellites. The radio emission of the sun also increases at the time of a solar flare, and can be studied by special solar radio telescopes on earth. A set of 96 radio telescopes in a 3 km circle in Culgoora, Australia (Fig. 5–65) is even capable of making images of radio radiation from bursts. This radioheliograph can make images every few seconds as the bursts pass through the corona (Fig. 5–66).

No specific model is accepted as explaining the eruption of solar flares. But it seems that a tremendous amount of energy is stored in the solar magnetic fields in sunspot regions. Something unknown triggers the release of

Figure 5–65 The radioheliograph at Culgoora, Australia, consists of 96 aerials, each 14 meters in diameter, spaced around a circle 3 km in diameter.

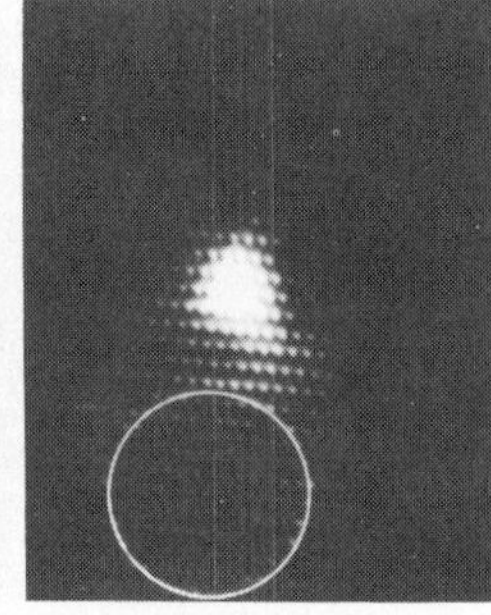

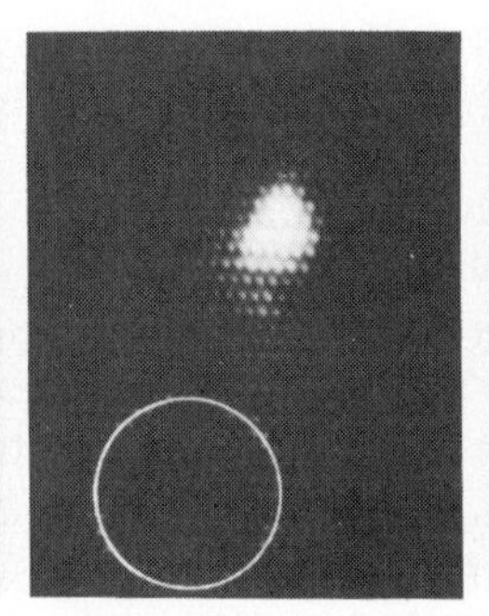

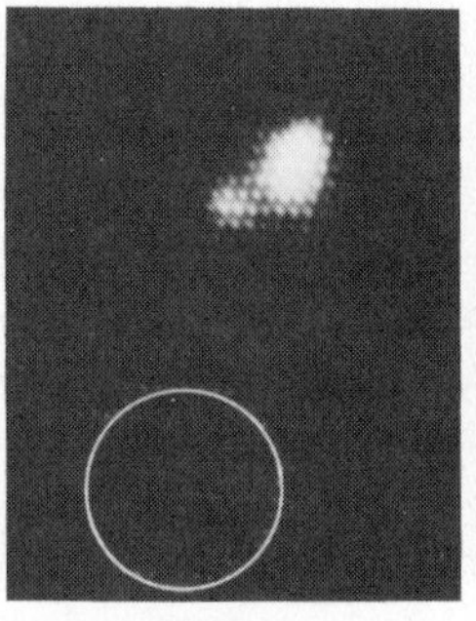

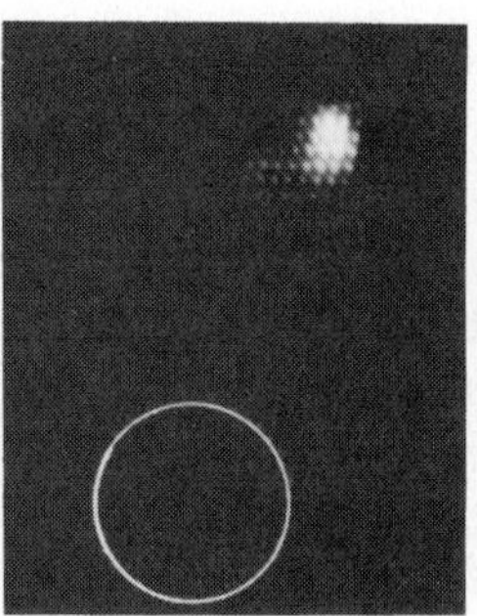

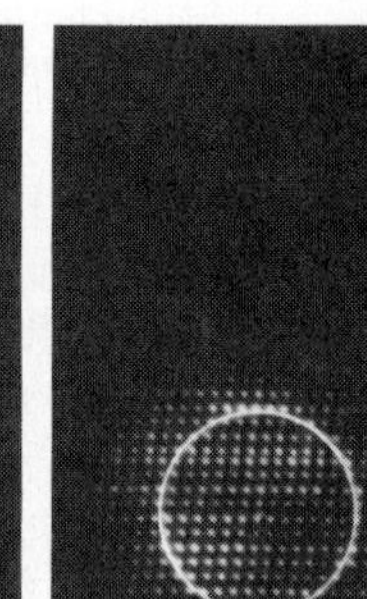

Figure 5–66 A large flare, observed by the Culgoora radioheliograph on March 2, 1969, is known colloquially as the "westward ho" event. An Hα image of the event, taken with a coronograph, appears at the left, and a series of views that shows the ejection of the radio-emitting matter over the next two hours follows. A computer calculates the radio picture on the basis of the data from the 96 aerials, used as an interferometer, and displays the picture on a screen.

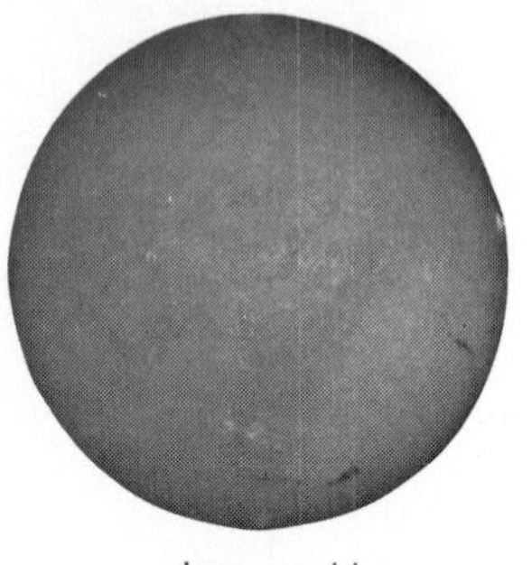

January 11
A

January 12
B

January 13
C

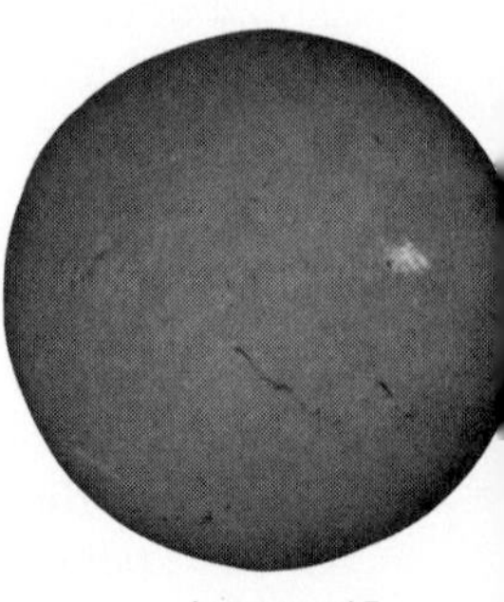

January 15
D

the energy, and matter can be ejected from the region at high velocity just as a watermelon seed can be snapped out of your fingers.

An important clue to the cause of solar flares is the fact that they usually occur in active regions with magnetic fields that are complicated and rapidly evolving. Flares do not often occur in typical sunspot groups, with their two symmetric clusters of spots of opposite magnetic polarity. Flares are much more likely to arise in complex regions, which occur when two or more sunspot groups emerge in quick succession close to one another and merge. The resulting magnetic pattern can be very complex, with north and south polarities jumbled together (Fig. 5–67). In some of the most active sunspot groups, spots seem to rotate about some common center. This pattern of rotation is often reflected in the pattern of fine structure in both the sunspots, as observed in white light, and in the overlying chromospheric structures that is observed in Hα. These white light and Hα effects help us predict when flaring is likely, though we remain unable to accurately predict individual flares.

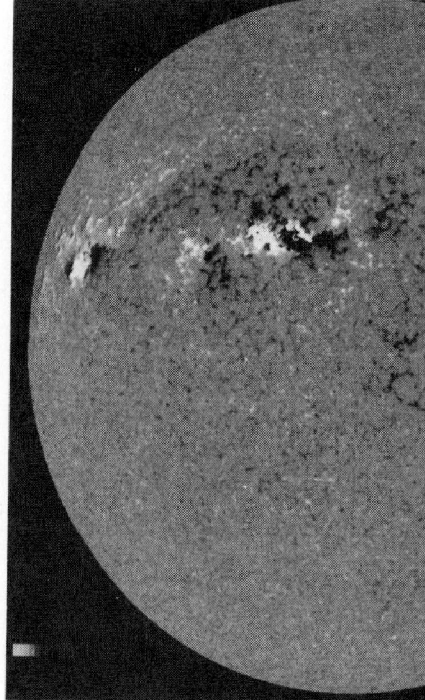

Figure 5–67 The magnetic field on July 4, 1974 was a very complex mixture of intertwined north and south polarities, and led to a large flare. The overlying sunspot group is shown in the photograph opening this chapter.

5.6c *Plages, Filaments, and Prominences*

Studies of the solar atmosphere in Hα radiation also reveal other types of solar activity. Bright areas called *plages* (from the French word for beach) surround the entire sunspot region (Fig. 5–68). The bright plage areas mark the presence of the active region long after the sunspots disappear. This is because it requires only a relatively weak magnetic field to create a chromospheric brightening compared to the very strong fields that are required to make a sunspot. Dark *filaments* are seen threading their way across the sun in the vicinity of sunspots. The longest filaments can extend for 100,000 km. They seem to represent the line of demarcation between positive and negative magnetic polarities. When these filaments happen to be on the limb of the sun, they can be seen to project into space, often in beautiful shapes. Then they are called *prominences* (Fig. 5–69). Prominences can be seen with the eye at solar eclipses, and glow pinkish at that time because of their emission in Hα and a few other spectral lines. They can be observed from the ground even without an eclipse, if an Hα filter is used. Prominences appear to be composed of matter in a similar condition of temperature and density to matter in the quiet chromosphere. Different methods that lead to deductions of the temperature of prominences lead to different values, ranging from 7000 to 30,000 K. The main point is that the temperature of a prominence is higher than that of the photosphere. Space observations can lead to maps of the temperature structure inside a prominence (Fig. 5–70 and Color Plates 22 and 23).

Figure 5–68 The rotation of the sun during the month of January 1976 can be followed by watching the motion of the bright plage area and of the filaments across the solar disk. Daily Hα photographs like this are made on a patrol basis by the National Oceanic and Atmospheric Administration (NOAA) in Boulder, Colorado.

January 16
E

January 19
F

January 20
G

January 22, 1976
H

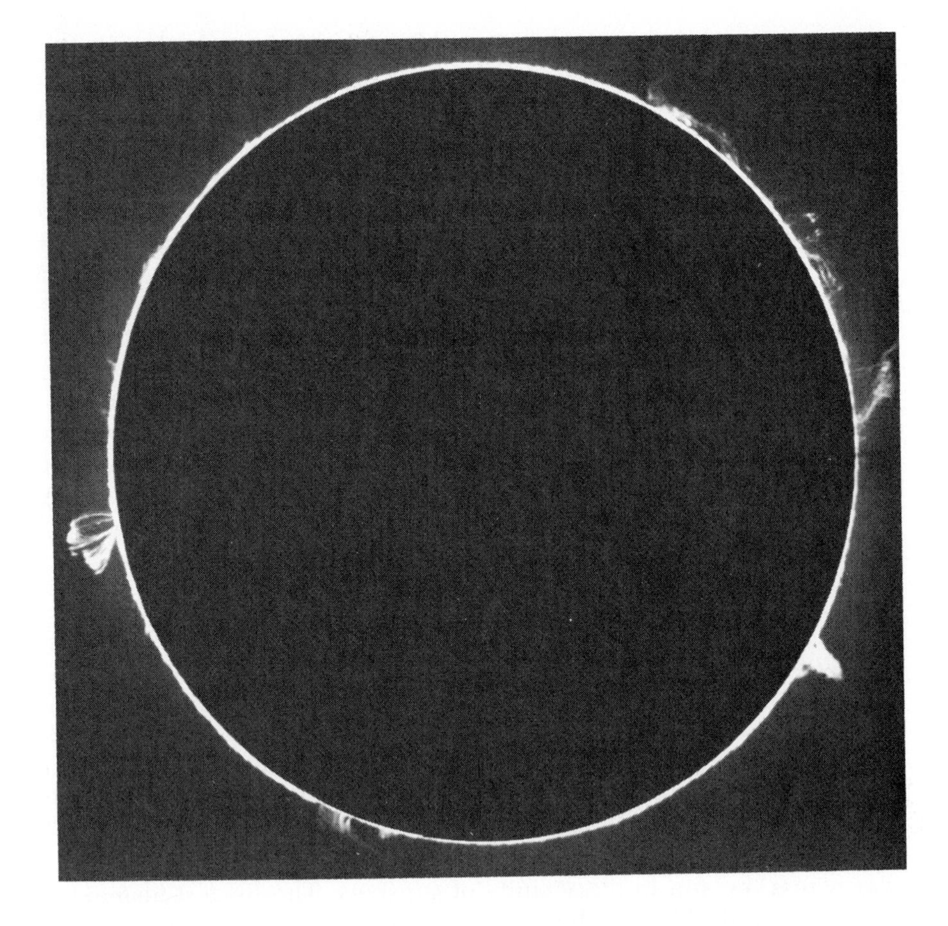

Figure 5–69 On December 9, 1929, prominences ringed the sun and were photographed in the calcium K-line. It is unusual to see this many prominences at one time.

Sometimes prominences can hover above the sun, supported by magnetic fields, for weeks or months. They are then called *quiescent prominences,* and can extend tens of thousands of kilometers above the limb. A quiescent prominence is shown on the page opening Part II. Other prominences can seem to undergo rapid changes. Prominences have been seen to erupt out to heights of a million kilometers (Fig. 5–71). On the other end of the size range, there is no clear distinction between a small prominence and a large spicule.

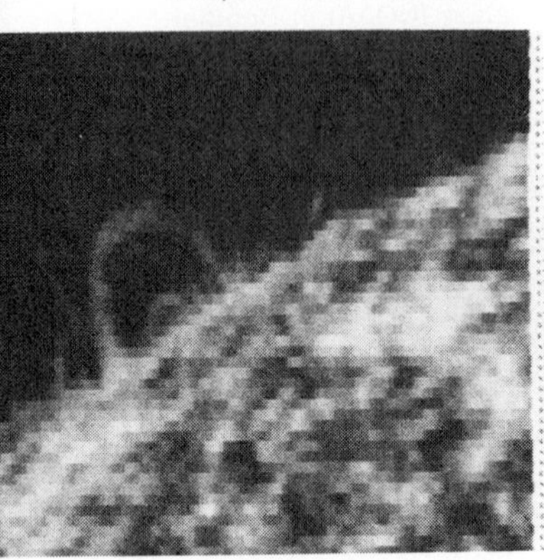

A

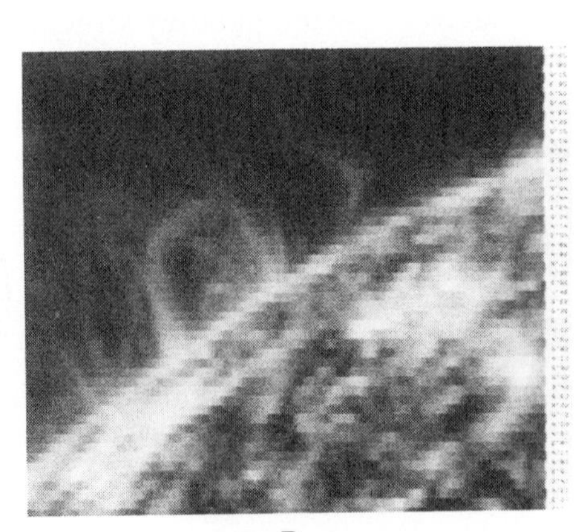

B

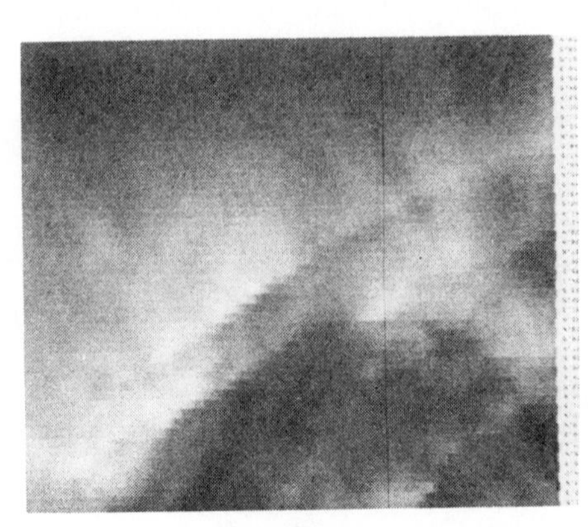

C

Figure 5–70 Observations in the ultraviolet from the Harvard College Observatory experiment aboard Skylab reveal the temperature structure in and around the prominences, which are themselves at about 15,000 K. (*A*) Triply ionized oxygen gas (O IV) is at a temperature of about 130,000 K and represents the transition zone from the prominence to the corona. (*B*) Five-times ionized oxygen gas (O VI) is somewhat hotter (300,000 K) and represents both gas in the transition zone and gas in the corona. (*C*) Nine-times ionized magnesium gas (Mg X) represents gas in the corona at a temperature of 1,500,000 K.

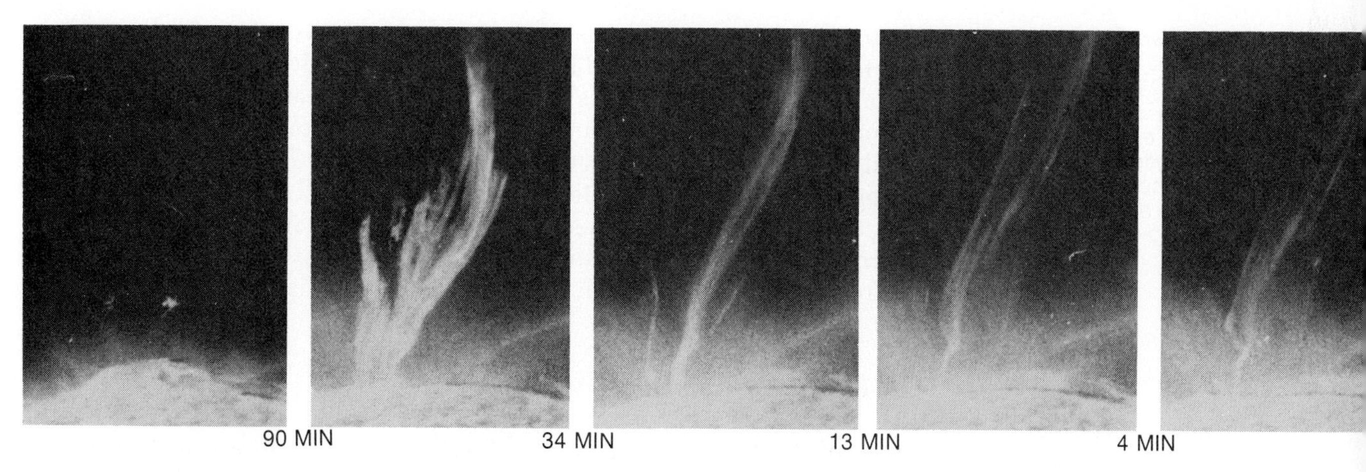

Figure 5–71 An eruptive prominence on August 21, 1973, photographed in radiation from ionized helium at 304 Å by the Naval Research Laboratory experiment on Skylab. The intervals of time between each pair of photographs are shown.

5.6d Solar-Terrestrial Relations

The solar activity cycle is now being carefully examined in order to increase our understanding of solar-terrestrial relations. Although for many years scientists were skeptical of the idea that solar activity could have a direct effect on the earth's weather, scientists presently seem to be accepting more and more the possibility of such a relationship. One possible mechanism is that particles or x-radiation from the sun interact with the earth's upper atmosphere in a way that eventually modifies the weather. The slowly-varying large-scale magnetic fields that extend from the sun to the earth also correlate with variations in terrestrial weather patterns. Many scientists are loath to accept that the sun affects the earth's weather in the absence of knowledge of the specific way in which the interaction takes place.

An extreme test of the interaction of solar activity with terrestrial weather may be provided by the interesting probability that there most likely were no sunspots at all on the sun from 1645 to 1715! The sunspot cycle may not have been operating for that 70-year period (Fig. 5–72). This was known to Maunder and others in the early years of this century but was largely forgotten until its importance was recently noted and stressed by John A. Eddy of the High Altitude Observatory. Although no counts of sunspots exist for most of that period, there is evidence that people were look-

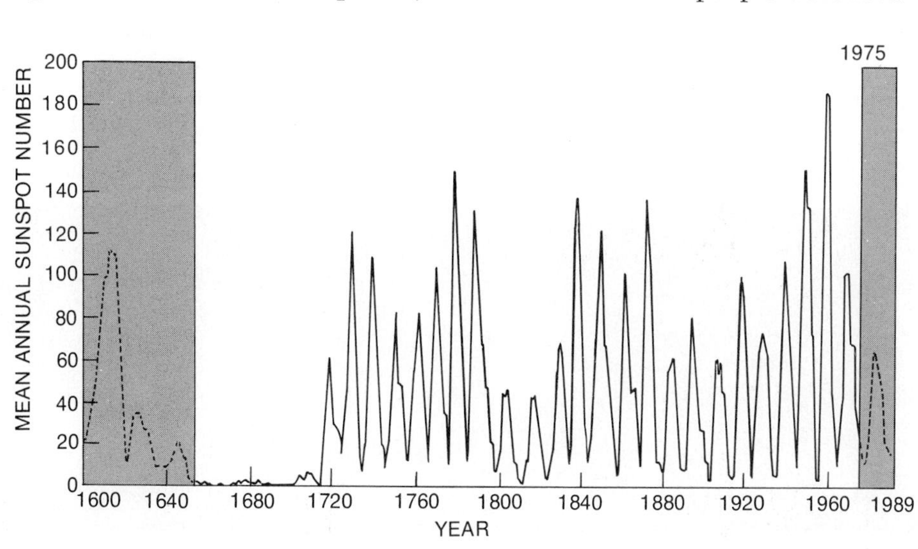

Figure 5–72 The Maunder minimum (1645–1715), when sunspot activity wqs negligible for decades, may indicate that the sun does not have as regular a cycle of activity as we had thought. Activity is extrapolated into the shaded regions.

The first white light flare ever seen, reported by Carrington in 1859, was 100 times stronger than any flare seen since. If such a strong flare were to occur nowadays, its effects could be serious.

ing for sunspots; it seems reasonable that there were no counts of sunspots because they were not there and not just because nobody was observing. A variety of indirect evidence has also been brought to bear on the question. No aurorae were reported for many years, for example, and the solar corona appeared very weak when observed at eclipses. It may be significant that the anomalous sunspotless period coincided with a "Little Ice Age" in Europe and with a drought in the southwestern United States. Another important conclusion from the existence of this sunspotless period is that the solar activity cycle may be much less regular than we had thought.

5.7 THE SOLAR WIND AND CORONAL HOLES

At about the time of the launch of the first earth satellites, in 1957, it was realized that the corona must be expanding into space. This phenomenon is called the *solar wind.* The expansion causes comet tails (Fig. 5–73) always to point away from the sun, but the existence of a solar wind had not been deduced from this fact. From spacecraft, the existence of the expansion of the corona has been documented. At the distance from the sun of the earth's orbit, particles in the corona are rushing outward at a velocity of 400 km/sec (900,000 miles per hour).

The solar wind extends into space far beyond the orbit of the earth. Calculations show that it must extend beyond the orbit of Saturn, and possibly even far beyond the orbit of Pluto. The density of particles, always low, decreases with distance from the sun.

The solar wind contains ions and electrons. The relative composition of elements and ions in the solar wind is fixed within a few solar radii of the determined base of the corona, and does not change outside of this volume. We measure the same composition even at the orbit of the earth, which is 200 solar radii from the sun.

From observations made on the Skylab manned missions, it was realized that the corona contains areas that are particularly cool and quiet. The density of gas in those areas is lower than the density in adjacent areas. It is now thought by many astronomers that the solar wind emanates from those areas, which are called *coronal holes* (Fig. 5–74). Macrospicules (shown in Color Plate 17) seem to occur only in coronal holes.

In 1975 and 1976, two joint German-American satellites named Helios were sent into the solar wind to go as close to the sun as possible, within about 44,000,000 km from the solar center. Other satellites are in orbit around the earth, and even in orbit around the sun at points quite distant from the earth, in order to give as complete a picture of the solar wind from

Figure 5–73 The solar wind causes the wavy streaming of the tails of comets, as in this view of Comet Mrkos in 1957.

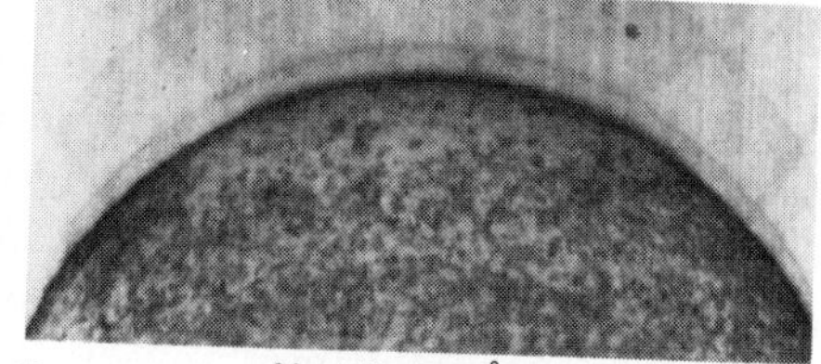

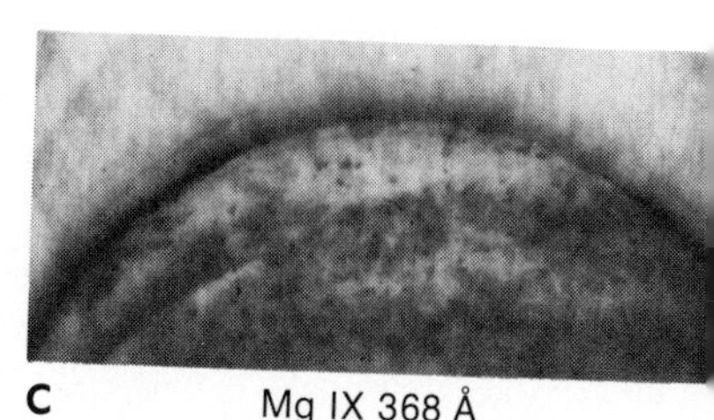

Figure 5–74 This sequence, taken on August 14, 1973, shows the coronal hole, a phenomenon that usually appears over one or both poles of the sun and often at other latitudes as well. Ionized helium (*A*) shows chromospheric temperatures, Ne VII (*B*) shows the transition zone, and Mg IX (*C*) shows the corona. A macrospicule is visible as the jet on top in the He II photograph. A coronal hole is also noticeable in visible light at the lower right on the 1970 eclipse photograph (Fig. 5–24) and in the helium ultraviolet observations shown in Fig. 5–32. The solar wind flows out of the coronal holes.

as widely spaced points of view as is possible. No observations, however, have ever been made outside the plane in which all the planetary orbits lie.

The earth's outer atmosphere is bathed in the solar wind. Thus research on the nature of the solar wind is necessary to understand our environment in the solar system.

5.8 OBSERVING WITH THE LATEST ORBITING SOLAR OBSERVATORY

Even solar astronomers who are observers rather than theoreticians need no longer go to remote mountain peaks to make their observations. They can work at their home institutions with data sent back from satellites in space.

OSO-8, the most recent of the very successful NASA series of Orbiting Solar Observatories, was launched in June 1975 (Fig. 5–75). It carries two major experiments to study the sun. Both take ultraviolet spectra of small areas (1 arc sec × 1 arc sec up to 1 arc sec × 40 arc sec) on the surface of the sun. One of these experiments is run by a group at the University of Colorado at Boulder and another by a group from the Laboratory for Stellar and Planetary Physics in France; the experiments are controlled together from computer consoles located in Boulder, Colorado.

Besides these two major experiments, which use telescopes that are kept continuously pointing at the sun, there are several additional experiments located in a part of the spacecraft that rotates every few seconds. These experiments map out swaths across the sky, and measure such things as gamma ray bursts and x-rays from objects outside the solar system.

Operating such a remote telescope requires much more planning and coordination than does simple observing with a ground-based telescope. The Colorado and French groups, along with others whose experiments were not selected, submitted their proposals to NASA in 1969 and have been working ever since on the project. They had to go through many stages of planning, design, and testing.

Since it has been realized that such an Orbiting Observatory is a major facility of the same magnitude as the observatories at Kitt Peak or Greenbank, about 50 scientists who are not on the staffs of either the French or the Colorado groups were appointed Guest Investigators. The Guest Investigators started meeting with the Principal Investigators in 1971, and had to prepare detailed proposals for their observing programs. For a space experiment, one has to work out in advance details such as how many seconds of time one should measure the intensity at a certain wavelength before moving on to an adjacent wavelength. It isn't enough just to point the telescope; every remote move of each motor on board the spacecraft must be carefully considered beforehand.

Figure 5–75 The eighth Orbiting Solar Observatory (OSO-8), launched in June 1975, being tested in the laboratory by being spun.

After launch, and a short shakedown period, the Guest Investigators came to Boulder for periods of weeks or months to help operate the spacecraft. They came from colleges, universities, and research organizations all over the United States and other parts of the world. The spacecraft operates 24 hours a day, with never a day off for cloudy weather. It orbits the earth every 90 minutes, and so goes through 16 sunrises, 16 observing "days," 16 sunsets and 16 "nighttimes" every 24 hours. Each moment has to be programmed in advance.

For the first year after launch, one astronomer for each experiment was designated as chief scientist each day. This provided a clear chain of command. Other tasks rotated on a five-day cycle. The first day, the scientist blocked out the experiments roughly. The second day the scientist fit in the details and made certain that all the different experiments were coordinated. The third day, the detailed plans were changed to computer language that could be transmitted to the satellite. These instructions were sent by the computer at Boulder to the computer at NASA's Goddard Space Flight Center at Greenbelt, Maryland, from which they were relayed to the satellite. The fourth day, the scientist supervised the general operation to see that everything was going smoothly, and may have had a chance to get a "quick look" at some of the data that were radioed back to earth by the spacecraft. And the fifth day, the scientist sat all day at a computer console checking a part of the data from each orbit to make certain that all was working well. After the first year, the observing programs were changed weekly instead of daily, but all the steps still had to be followed. Such office work is a far cry from observing at a ground-based telescope, but it carries with it the substantial reward of knowing that the data being received were hitherto unavailable and that these data could be the key to a scientific problem that the astronomer has been considering for many years.

When astronomers go home from such an expedition, they carry with them only a few graphs or sets of numbers that the computer at Boulder received from the satellite (Fig. 5–76). Most of the data is telemetered by the satellite to NASA's tracking stations around the world, where they are recorded on magnetic tape. The tape is mailed to Goddard in Maryland, and then each bit of data must be separated from other data and from informa-

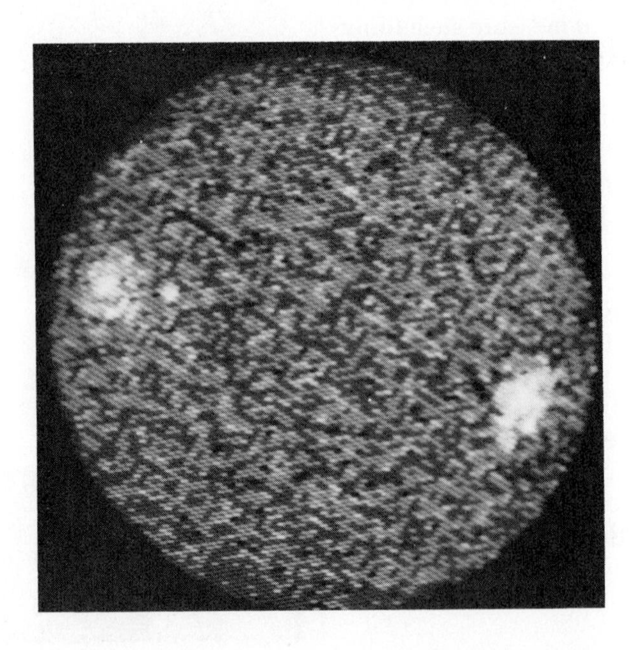

Figure 5–76 The distribution of Lyman α on the sun, as observed on July 23, 1975 by the OSO-8 experiment of the Laboratory for Atmospheric and Space Physics of the University of Colorado. Two active regions are obvious.

tion that describes the status of the spacecraft. It may be months or more before the scientists receive the data they have helped record, and only then can they go to work and study it.

Among the problems that are of specific interest to the scientists working with OSO-8 is the question of the heating of the solar chromosphere and corona. Many of the observing programs were designed to follow shock waves through the various levels of the solar atmosphere that are detectable only from above the earth's atmosphere. Some of these programs, and other investigations as well, were designed to take advantage of the fact that this was the first satellite able to make observations with such high resolution.

5.9 THE SOLAR CONSTANT

The solar constant is the amount of energy per second that would hit each square centimeter of the earth at its average distance if the earth had no atmosphere.

Every second, a certain amount of solar energy passes through each square centimeter at the average distance of the earth from the sun. This quantity is called the *solar constant.* Accurate knowledge of the solar constant is necessary to understand the terrestrial atmosphere, for to interpret the atmosphere completely we must know all the ways in which it can gain and lose energy. Further, knowledge of the solar constant enables us to calculate the amount of energy that the sun itself is giving off, and thus gives us an accurate measurement on which to base our quantitative understanding of the radiation of all the stars.

It is astonishingly difficult to measure the solar constant. Even though most of the solar energy is in the visible part of the spectrum, which passes through a window of transparency in the terrestrial atmosphere, one must also make accurate measurements of the amount of energy from the sun in other spectral regions (Fig. 5–77). Thus the capability of measuring solar

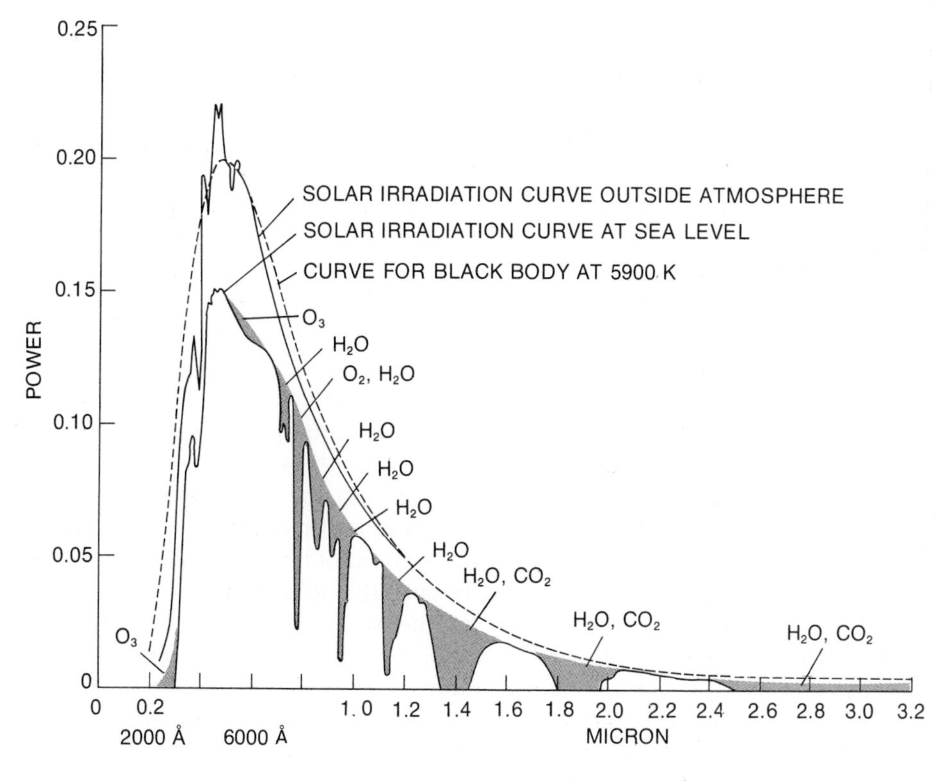

Figure 5–77 Windows of transparency in the earth's atmosphere, and the constituents of the terrestrial atmosphere (shaded) that cause the absorption in the other regions.

radiation from airplanes, balloons, rockets, and satellites above much of the earth's atmosphere has been applied to the problem of determining the solar constant. A major difficulty in all measurements—whether from the ground or from outer space—is our inability to determine the sensitivity of the measuring instruments themselves to sufficient accuracy.

The longest continuous set of measurements was made at the Smithsonian Institution in Washington from the turn of the century on. A group at the Goddard Space Flight Center of NASA in Greenbelt, Maryland, is now carrying on the work, and also considering measurements taken aloft. A German group, whose measurements were made atop the Jungfrau, a very high Swiss mountain, took advantage of the low amount of atmospheric absorption in the infrared at the great altitude to compile a table of solar radiation from 3300 to 11,000 Å. Soviet scientists have also been active in studying all the data dealing with the solar constant.

Traditionally, one recalls the simple number 2 calories/cm²/min for the solar constant. Calories are a unit of energy in the old English system of units, and watts are the unit of power in the metric system.

The latest measurements give a value of 1.94 ± 0.03 calories/cm²/min which is equivalent to 135.3 ± 2.0 milliwatts/cm². Note that the value is determined to an accuracy of 1.5 per cent; we have not yet been able to determine the solar constant to higher accuracy. Higher accuracy measurements are necessary, and an experiment for this purpose is scheduled for launch on the Solar Maximum Mission in 1980 or 1981. One interesting aim is to determine if there are long-term changes in the solar constant, perhaps arising from changes in the luminosity of the sun. Certainly our climate would be profoundly affected by small long-term changes. Of course, if the solar constant changed with time, it wouldn't really be a constant after all.

We should note that the solar constant is really not the actual amount of energy we receive from the sun, but is the amount that we would receive if the earth were at its mean distance from the sun and in the absence of the earth's atmosphere. Actually, the earth's orbit is slightly elliptical, and this leads to a variation of ± 3.5 per cent from the average in the amount of energy we receive in the course of a year.

At present, there is no firm evidence that the value changes over the solar cycle, although all observers are requesting that a careful series of the most accurate measurements be made over an entire solar cycle. One of the best ways to follow variations in the visible part of the spectrum, in which most of the energy arrives, is to study the brightness of the planets. The planets are small disks that are easy to observe while using reference stars to make certain that our instruments are not changing in sensitivity. As the planets are simply reflecting sunlight, any changes in their brightnesses might mean that similar changes in the solar brightness were taking place. Observations made intermittently over the last 27 years of Uranus, Neptune, and Saturn's satellite Titan seem to indicate such a variability, though it could be the reflecting properties of the planets that are changing.

99% of solar energy is in the range from 2760 Å to 49,600 Å (nearly 5 microns), and 99.9% is between 217 Å and 10.94 microns.

Although the solar constant itself, which is the energy received at all wavelengths, does not change appreciably with the 11-year period of the solar cycle, the amount of energy received in particular wavelength bands may change manyfold. For example, the amount of far ultraviolet or x-radiation given off by the sun varies greatly with the solar activity cycle.

From the above measurement of the solar constant, equivalent to 3.80×10^{33} ergs/sec emitted by the sun, we can determine several basic solar parameters. The sun's effective temperature is 5762 ± 21 K; that is, a black body of that temperature would give off the same amount of energy as the sun. The sun's apparent bolometric magnitude (the magnitude corresponding to the total amount of energy emitted, as defined in Section

2.13) is −26.82 and its absolute bolometric magnitude is 4.75, both figures with an uncertainty of only 0.01 magnitude.

5.10 SOLAR ENERGY FOR EARTH

Most of the energy that we use on earth can ultimately be traced back to the sun. Oil and natural gas are fossil fuels that release energy received from the sun millions of years ago. Even when we use wood in a fire we must recall that the sun supplied the energy for the growth of trees. Similarly the sun supplies the energy that makes the wind blow and heats the oceans. We are currently investigating the extraction of energy from both wind and ocean water.

The sun is also important to us as a laboratory in which we can study hot gases in a magnetic field. The knowledge we are gathering from studies of the solar gas is helping us learn to control fusion processes here on earth. If we learn to build magnetic "bottles" to contain hydrogen undergoing fusion at temperatures of millions of degrees, then we can use the hydrogen in the oceans as our energy supply and be self-sufficient in energy for millions of years.

The sun is itself a fusion reactor, properly located far from populated areas.

But progress in developing fusion for peaceful purposes has not been as rapid as had been hoped. The energy crisis probably won't allow us to wait. There is increasing interest in gathering energy directly from the sun.

Devices already exist that give off an electrical signal when struck by sunlight. Photocells (described in Section 3.8) use materials that have such properties; the type of cells under consideration for commercial generation of energy uses somewhat different processes, which we will not go into here, but the end result is the same. Whatever the details of the process, solar energy cells are still too expensive to use for general commercial and household purposes. When cost is no object, as in space satellites like Skylab or the Orbiting Solar Observatories, we can send solar cells along to generate enough electricity (Figs. 5–78 and 5–79). For some remote sites on earth where access is expensive—such as an ocean buoy—a solar cell may even be the cheapest way to provide power.

More standard and less expensive technology uses heat from solar energy to make steam, which in turn can be used to generate electricity. This may prove to be a more productive route to follow than the technology of solar cells.

There is now considerable research centering on finding ways of reducing the cost of devices that convert solar radiation into heat and electricity. Consideration is also being given to how we could best utilize such an energy conversion method. Do individuals put solar receptors on every house? Do power companies build giant farms of solar receptors that cover many square miles of countryside? Does one choose to put these farms only in regions like the southwest United States where the amount of sunshine is much higher than it is in New England? How do you store energy for a cloudy day or for the nighttime? Since the energy in sunlight falling on an area only about 100 km in diameter would meet the energy needs of the entire United States if it could be efficiently converted to electricity, these questions are obviously worth pursuing.

Solar energy can also be used directly to provide heat and hot water. One can build houses to take advantage of the solar radiation by suitably designing and orienting roofs, windows, and overhangs, and by using suit-

Figure 5–78 Skylab used solar cells to collect solar energy. One of the two paddles that contain its solar cells was crippled shortly after launch, which led to Skylab's asymmetric appearance.

Figure 5–79 Gerard O'Neill of Princeton has suggested that a self-sufficient space station like this one, permanently located in one of the points in space where the terrestrial and lunar gravity balance each other, could collect solar energy and beam it to earth in the form of microwaves. Although the initial stages would be launched from earth, much of the material used thereafter could be mined on the moon (Fig. 19–4). The concept appears promising, and is under study. (Photo from Science Year, the World Book Science Annual, © 1975 Field Enterprises Educational Corporation.)

able construction materials that tend to retain heat. Even if only part of a house's energy consumption can be supplied by such methods, the need for outside energy sources is considerably reduced.

The use of solar energy for widespread practical purposes is still in its infancy. For some purposes – providing heating for water and houses – perhaps only technological development and practical experience is necessary rather than scientific research. For other purposes – such as the generation of electricity – we have more hope that research will bring breakthroughs. The only thing that can be said for certain is that "solar energy," in the sense of heat or electricity generated from solar radiation, will play an increasingly important part in our national energy budget.

5.11 THE SUN AND THE THEORY OF RELATIVITY

The sun, as the nearest star to the earth, has been very important for testing some of the predictions of Albert Einstein's theory of gravitation, which is known as the general theory of relativity. The theory, which Einstein advanced in final form in 1916, made three predictions that depended on the presence of a large mass like the sun for experimental verification. These predictions were (1) the gravitational deflection of light, (2) the advance of the perihelion of Mercury, and (3) the gravitational redshift. We shall discuss the gravitational redshift in Section 7.5, and discuss the other two tests here.

Actually, according to the general theory of relativity, the presence of the mass warps the space. Thus the light travels in a straight line but space itself is curved, as discussed in Section 1.2 and shown in Figure 5–82. The effect is the same as it would be if space were flat and the light were bent.

As light from a distant star passes near the sun, the light is bent slightly by the presence of the large gravitational mass. An earlier, incomplete version of Einstein's theory had predicted that light would be bent by 0.84 seconds of arc (Fig. 5–81), a very small amount that is at the limit of our observing ability. (Einstein had made an arithmetic error; he should have calculated 0.87.) Luckily, this prediction was not tested at the 1914 eclipse, for Einstein had not yet developed his more complete theory. The newer version predicts a deflection of light by 1.74 seconds of arc, twice the earlier value. (The German observers who went to Russia in 1914 to observe the phenomenon were arrested as prisoners of war and thus missed the eclipse.

Figure 5–80 Albert Einstein visiting Caltech in the 1930's.

Box 5.1 The General Theory of Relativity

Einstein's *general theory of relativity* links three dimensions of space and one dimension of time to describe a four-dimensional space-time. In Einstein's own words, "If you will not take the answer too seriously, and consider it only as a kind of joke, then I can explain it as follows. It was formerly believed that if all material things disappeared out of the universe, time and space would be left. According to the relativity theory, however, time and space disappear together with the things." The general theory explains gravitation. Einstein's earlier *special theory of relativity* (1905) is a theory of relative motion. It holds that the speed of light is an absolute limit, shows how length, mass, and the rate at which time advances depend on the velocity of motion, and leads to the relation between mass and energy, $E=mc^2$.

Figure 5–81 The prediction in Einstein's own handwriting of the deflection of starlight by the sun, taken from a letter from Einstein to Hale at Mt. Wilson. "Lichtstrahl" is "light ray." Einstein asked if the effect could be measured without an eclipse, to which Hale replied negatively.

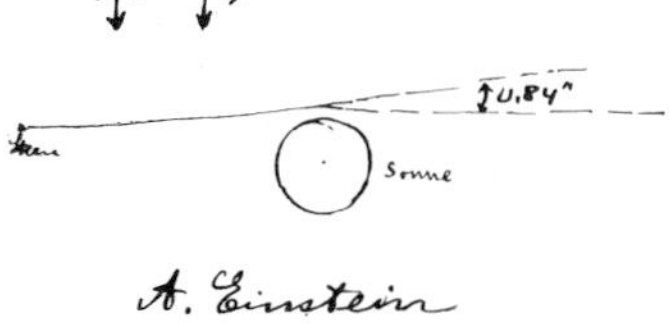

Grav. Feld
Lichtstrahl
0,84"
Sonne
A. Einstein

Had they been able to carry out their measurements, they would have indeed found a deflection, but it would have been different in quantity from that of Einstein's then current predictions. A theory is much more convincing when it predicts a hitherto unobserved result rather than merely explaining an existing observation. If the 1914 eclipse expedition had been a success, or if Hale had been able to measure the effect without an eclipse, Einstein's final version of the general theory of relativity would have

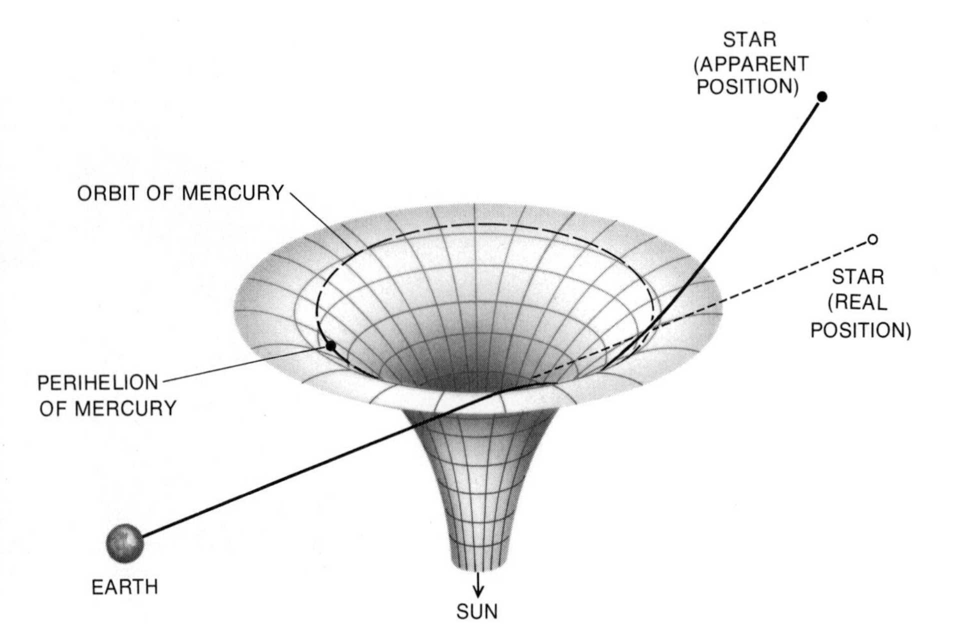

Figure 5–82 Under the general theory of relativity, the presence of a massive body essentially warps the space nearby. This can account for both the bending of light near the sun and the advance of the perihelion point of Mercury by 43 arc sec per century more than would otherwise be expected. The diagram shows how a two-dimensional surface warped into three dimensions can change the direction of a "straight" line that is constrained to its surface; the warping of space is analogous, although with a greater number of dimensions to consider.

seemed to some as an adjustment forced by lack of agreement with the experiment. In that case, the theory would not have been so readily accepted.)

Since it was predicted that the light from a star would act as though it were bent toward the sun, we on earth, looking back, would see the star from which the light was emitted as though it were shifted slightly away from the sun. Only a star whose radiation grazed the edge of the sun would seem to undergo the full 1.74 seconds of arc deflection; the effect diminishes as one considers stars farther away from the solar limb. One has to look at the sun at a time when the stars are visible, and this could be done only at a total solar eclipse.

The British astronomer Sir Arthur Eddington and other scientists observed the total solar eclipse of 1919 from a number of sites in Africa and South America. The effect for which they were looking was a very delicate one, and it was not enough merely to observe the stars at the moment of eclipse. One had to know what their positions were when the sun was not present in their midst, so the astronomers had already made photographs of the same field of stars six months earlier when the same stars were in the nighttime sky. Even though his observations had been limited by clouds, Eddington detected that light was deflected by an amount that agreed with Einstein's revised predictions. When Einstein eventually heard of the result, he was so pleased that he wrote his mother a postcard to tell her of his success. ("Dear Mother! Joyful news today," Einstein wrote. "H. A. Lorentz has telegraphed me that the English expeditions have actually proved the deflection of light near the sun.") The scientists hailed this confirmation of Einstein's theory, and from the moment of its official announcement, Einstein was recognized by scientists and the general public alike as the world's greatest scientist.

Although the experiment has been repeated at many solar eclipses, most recently in Mexico in 1970 and Africa in 1973, the precision of even the best results is only about 5 per cent (that is, we cannot tell from the data whether the actual value is as much as 5 per cent more or less than the value we specify as the "measured" value). The data agree with Einstein's prediction, but are not accurate enough to distinguish between Einstein's theory and newer, more complicated, rival theories of gravitation. Fortunately, the effect of gravitational deflection is constant through the electromagnetic spectrum, and the test can now be performed more accurately by observing the bending of radiation from radio sources, especially quasars (see Chapter 24), as the sun passes in front of them. This is done with radio interferometers (which are described in Chapter 23). The interferometer results agree with Einstein's theory to within 1 per cent, enough to make the competing theories very unlikely.

A related test involves not deflection but a delay in time of signals passing near the sun. This can now be performed with signals from interplanetary spacecraft, and these data also agree with Einstein's theory.

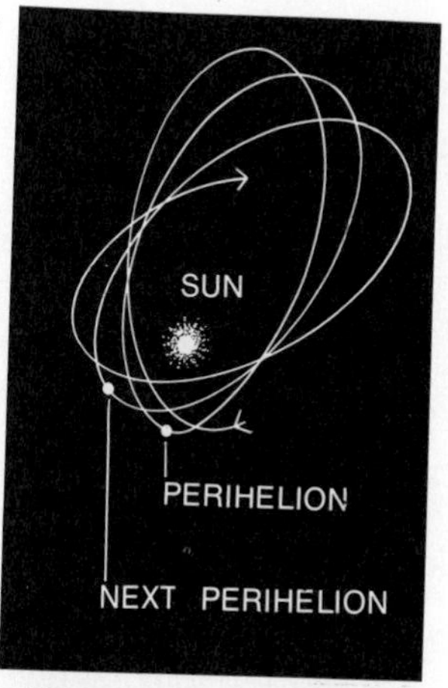

Figure 5–83 The advance of the perihelion of Mercury.

Another of the triumphs of Einstein's theory, even in its earliest versions, was that it explained the advance of the perihelion of Mercury. The orbit of Mercury, as are the orbits of all the planets, is elliptical; the point at which the orbit comes closest to the sun is called the *perihelion* (Fig. 5–83). The elliptical orbit is pulled around the sun over the years, mostly by the gravitational attraction on Mercury by the other planets, so that the perihelion point is at a different orientation in space. Each century (!) the perihelion point moves around the sun by approximately 5600 seconds of arc (which is less than 2°). Subtracting the effects of precession and of the other planets leaves 43 seconds of arc per century whose origin had not been understood before Einstein.

Einstein's general theory predicts that the presence of the sun's mass warps the space near the sun. Since Mercury is sometimes closer to and sometimes farther away from the sun, it is sometimes traveling in space that is warped more than the space it is in at other times. This should have the effect of changing the point of perihelion by 43 seconds of arc per century. The agreement of this prediction with the measured value was an important observational confirmation of Einstein's theory. More recent refinements of solar system observations have shown that the perihelions of Venus and of the earth also advance by the even smaller amounts predicted by the theory of relativity.

But in the 1960's, Robert Dicke (pronounced "Dick-ē") of Princeton suggested that there was another contribution to the perihelion advance that had been previously overlooked. He carried out detailed measurements of the shape of the sun, and found that it was out of round by a very small amount. The polar and equatorial diameters, each approximately 1,400,000 km, differ by only 35 km. But "oblateness," the deviation of the shape of a body from being round (see Section 16.1), is often caused by rotation of that body. The outer layers of the sun are rotating about once a month, not sufficiently fast to cause the oblateness that Dicke measured. Dicke therefore suggested that the inner regions of the sun, which we cannot see, were rotating very rapidly—about once every day.

If these inner regions were rotating at this velocity, then the observed solar oblateness would result, but it would have an additional observational consequence: using the latest values that Dicke reported, following further data reduction, the perihelion of Mercury would advance about 3 seconds of arc per century as a result.

When this value of 3 seconds of arc per century that would result from the rapid rotation of the solar core is subtracted from the observed 5600

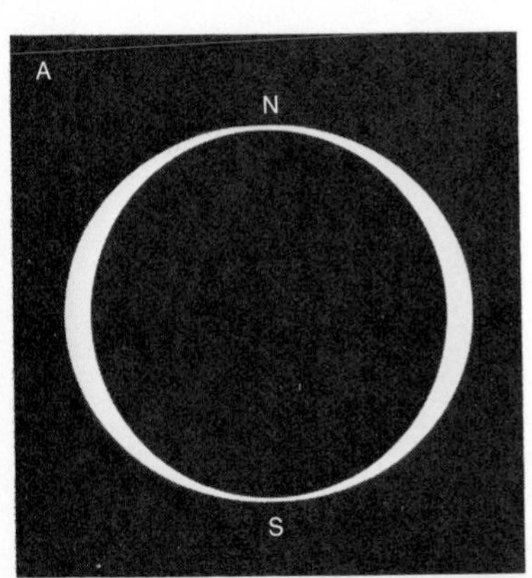

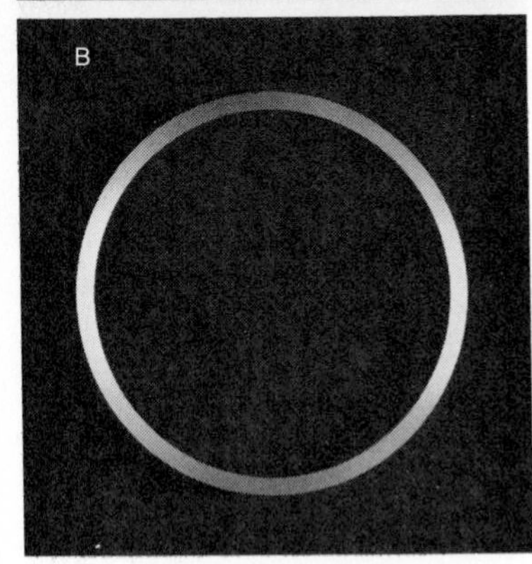

Figure 5–84 Dicke's oblateness measured the excess brightness in different parts of the solar disk, and did not directly measure the solar oblateness. The excess brightness he detected could be explained either as a real oblateness (*A*) or as the presence of brighter areas near the equator (*B*).

seconds of arc along with the precession and planetary effects, only 40 seconds of arc per century remain to be accounted for. But Einstein's theoretical prediction, as we said, was 43 seconds of arc per century. Thus Dicke concluded that Einstein's theory might be wrong.

Dicke himself, in collaboration with Carl Brans, had earlier worked out an alternative theory of gravitation. It had an adjustable parameter that could be set so that the theory predicted a perihelion advance of 40 seconds of arc per century.

This topic is extremely controversial, and has been the subject of bitter debates during the years since Dicke's measurement. Many theoreticians presented arguments why the interior of the sun could not rotate so rapidly, given the dynamics of gases. They said that turbulence inside the sun would link the inside and outside parts. Dicke, in return, rebutted their arguments with his own calculations. (The discussion centered, for a time, on why the tea in a cup stops rotating so soon after it is stirred.) Other scientists, particularly some who were knowledgeable about the outer solar atmosphere, suggested additional effects besides rotation of the core that could cause the solar equator to be slightly brighter than the solar poles. Dicke really measured the brightness of a small zone at the edge of the solar photosphere that protruded above a round disk. He interpreted excess brightness (Fig. 5–84) at the equator as the effect of an oblate sun, but it could equally well be caused if the equator was simply brighter than the poles. After all, solar plages and active regions occur preferentially along the solar equator. Dicke responded to these comments as well, and a debate in the journals and at scientific meetings followed.

Dicke's observations were never repeated, though he occasionally presented new and more complete analyses of his original data. Henry Hill of the University of Arizona, who had built much of Dicke's original apparatus at Princeton, designed a new kind of apparatus to measure the solar oblateness by directly measuring the solar equatorial and polar diameters. In 1975 he announced that his measurements showed that the sun was not as oblate as Dicke had reported. If we accept Hill's result, then the Einstein prediction remains in agreement with the observed advance of the perihelion of Mercury. Thus this experimental method remains another successful test of Einstein's theory.

Historically, in explaining the perihelion advance, general relativity was explaining an observation that had been known. On the other hand,

Figure 5–85 Einstein lecturing at the Collège de France in Paris in 1922.

in the bending of light it actually predicted a previously unobserved phenomenon.

As a general rule, scientists try to find theories that not only explain the data that are at hand, but also make predictions that can be tested. This is an important part of the *scientific method,* which is discussed further in Section 19.5. Because the bending of electromagnetic radiation by 1.74 seconds of arc was a prediction of the general theory of relativity that had not been anticipated, the verification of the prediction was a more convincing proof of the theory's validity than the theory's ability to explain the perihelion advance.

The history of these observational tests of relativity illustrates the scientific method at work. Even though the first eclipse tests of the bending of light were strong verification of the theory, astronomers have pursued the method over the years in order to make more stringent tests. They have recently generalized the test from observation of light to observation of radio waves in order to make use of new techniques that would allow more accurate measurements.

When challenges to the theory were made, they were discussed and analyzed openly both in print and at scientific meetings. A major requirement of scientific results is that they be reproducible—that any scientist working under the same conditions at any location be able to reproduce the same results. The measurements of an oblateness large enough to disprove relativity, for the moment, do not seem to be reproducible, and so we do not accept their validity. Note that this conclusion depends on the current observational situation, and that it could be modified if new results warrant it.

In sum, the theory of relativity seems to be consistent with and verified by the observational tests that concern the sun. Other observational tests will be discussed in Section 7.5.

5.12 SOLAR SEISMOLOGY

In all the years that the sun has been observed, no observer had detected an oscillation of the sun as a whole—a ringing, like a bell. But in 1975 and 1976, three groups of observers independently reported this phenomenon.

The first observations were a by-product of the careful measurements of the diameter of the sun that Hill made in order to measure the reported solar oblateness (as discussed in the previous section). Hill found that the diameter of the sun showed a periodic variation with a period of about 48 minutes (of time) and also with other longer periods.

Still more recently, British observers working in France, and Soviet observers, independently found a 2 hour 40 minute oscillation of the sun. The sun's radius would vary by about 10 kilometers with this period, a minuscule amount.

If the reality of these discoveries is confirmed, the researchers would try to tie these sets of observations together, and find out the whole range of periods of oscillation that might be detected. If a more complete set of observations is available, then the actual mechanisms for the generation of these oscillations might become apparent.

Some new observations are not finding the oscillation. More work is necessary before we can say whether or not this is a real phenomenon.

The observations are particularly valuable because the study of these oscillations might be useful for studying the solar interior. We have long studied the earth's interior by analyzing waves from distant earthquakes that travel through the earth before reaching us. Now such methods might prove applicable to the sun.

SUMMARY AND OUTLINE OF CHAPTER 5

Aims: To study the sun, which is the nearest star and an example of all the other stars whose surfaces we can not observe in such detail

Definitions: parts of the sun (interior, photosphere, chromosphere, corona, spicules, granulation, supergranulation, solar wind, coronal holes) (Sections 5.1, 5.2, 5.3, 5.4, and 5.7)

Limb darkening and its use to map the temperature over the radial direction (Section 5.2d)

Spectra:

- Photospheric spectrum: continuum with Fraunhofer lines; 5800 K; abundances of the elements (Section 5.2a)
- Chromospheric spectrum: emission at the limb during eclipses in the visible part of the spectrum (flash spectrum); the strongest lines in the visible and some of the strongest lines in the ultraviolet (observed from satellites) show us chromospheric levels; typical temperature is 15,000 K (Section 5.3a)
- Coronal spectrum: (Section 5.3a) emission lines from highly ionized elements, indicating high temperature of 2,000,000 K; continuum visible at eclipses formed by scattering of photospheric light by both electrons in the corona itself and interplanetary dust; Fraunhofer lines visible in the outer part of the corona because of the scattering from interplanetary dust; the strongest lines in the ultraviolet include lines that arise at coronal levels (Section 5.4a)

Space observations: Orbiting Solar Observatories (OSO-8 has high spatial resolution capabilities), Skylab (Sections 5.4b and 5.8)

Eclipse phenomena: band of totality, Baily's beads, diamond ring (Section 5.5)

Radio observations: radiation is from the corona and from active regions (Sections 5.4b and 5.6b)

Solar activity cycle: sunspots, magnetic fields, flares, plages, filaments, prominences (Section 5.6)

Solar wind flows from coronal holes (Section 5.7)

Solar constant and its possible variability (Section 5.9)

Solar energy (Section 5.10)

Using the solar mass to test Einstein's general theory of relativity (Section 5.11)

- Gravitational deflection
  - Predicted in advance of detection
  - First detected at eclipses
  - Now also tested in radio region of spectrum
- Advance of the perihelion of Mercury
  - Phenomenon was known in advance of Einstein's theory

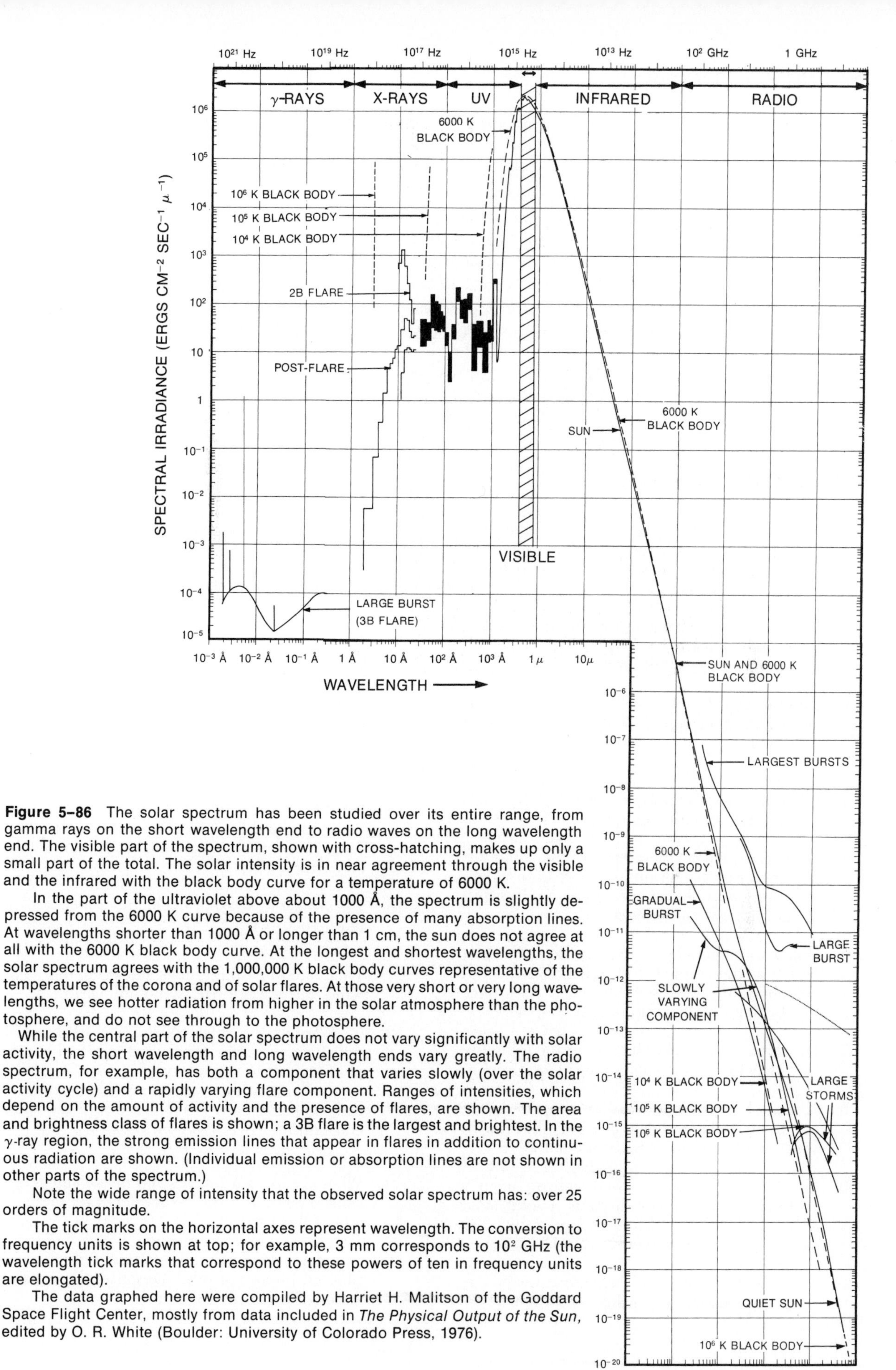

Figure 5–86 The solar spectrum has been studied over its entire range, from gamma rays on the short wavelength end to radio waves on the long wavelength end. The visible part of the spectrum, shown with cross-hatching, makes up only a small part of the total. The solar intensity is in near agreement through the visible and the infrared with the black body curve for a temperature of 6000 K.

In the part of the ultraviolet above about 1000 Å, the spectrum is slightly depressed from the 6000 K curve because of the presence of many absorption lines. At wavelengths shorter than 1000 Å or longer than 1 cm, the sun does not agree at all with the 6000 K black body curve. At the longest and shortest wavelengths, the solar spectrum agrees with the 1,000,000 K black body curves representative of the temperatures of the corona and of solar flares. At those very short or very long wavelengths, we see hotter radiation from higher in the solar atmosphere than the photosphere, and do not see through to the photosphere.

While the central part of the solar spectrum does not vary significantly with solar activity, the short wavelength and long wavelength ends vary greatly. The radio spectrum, for example, has both a component that varies slowly (over the solar activity cycle) and a rapidly varying flare component. Ranges of intensities, which depend on the amount of activity and the presence of flares, are shown. The area and brightness class of flares is shown; a 3B flare is the largest and brightest. In the γ-ray region, the strong emission lines that appear in flares in addition to continuous radiation are shown. (Individual emission or absorption lines are not shown in other parts of the spectrum.)

Note the wide range of intensity that the observed solar spectrum has: over 25 orders of magnitude.

The tick marks on the horizontal axes represent wavelength. The conversion to frequency units is shown at top; for example, 3 mm corresponds to 10^{2} GHz (the wavelength tick marks that correspond to these powers of ten in frequency units are elongated).

The data graphed here were compiled by Harriet H. Malitson of the Goddard Space Flight Center, mostly from data included in *The Physical Output of the Sun,* edited by O. R. White (Boulder: University of Colorado Press, 1976).

Einstein's theory explained the observed effect
Dicke's observations can be interpreted to suggest that the interior of the sun is rapidly rotating, which would contribute to the observed perihelion advance and thus reduce the remaining effect that must be explained. Einstein's prediction would not match the new remainder.
Hill's observations indicate that Dicke's effect not present
"Ringing" oscillations of the sun have been discovered, and may be useable to explore the solar interior just as seismic waves on earth are used to explore the earth's interior (Section 5.12)

PROBLEMS

1. Sketch the sun, labeling the interior, the photosphere, the chromosphere, the corona, sunspots, and prominences. Give the approximate temperature of each.

2. What elements make most of the lines in the Fraunhofer spectrum? What elements make the strongest lines? Why?

3. Explain why the photospheric spectrum is an absorption spectrum and the chromospheric spectrum seen at an eclipse is an emission spectrum.

4. What are two methods of observing the chromosphere without an eclipse?

5. How can the temperature of the solar atmosphere reach a minimum while both the interior on one side and the corona on the other are at higher temperatures?

6. When you pour a cup of coffee, the coffee gets darker as the cup gets full. What does this tell you about the opacity of the coffee? How does this compare with a cup of tea?

7. How does limb darkening tell us the temperature structure of the solar atmosphere?

8. Explain how at some wavelengths we can observe limb darkening while at other wavelengths we observe limb brightening.

9. (a) Why isn't there a solar eclipse once a month? (b) Whenever there is a total solar eclipse, a lunar eclipse occurs either two weeks before or two weeks after. Explain.

10. Why can't we observe the corona every day from any location on earth?

11. Describe the series of phenomena one observes at a total solar eclipse.

12. Why does the chromosphere appear pinkish at an eclipse?

13. How do we know that the corona is hot? Give two methods of determining the temperature.

14. When we see the coronal light at an eclipse, from what three mechanisms has it originated? How can we tell these apart?

15. What are three kinds of observations that have been obtained from Skylab, and how are they different from observations that we can obtain on earth?

16. Describe the sunspot cycle and a mechanism that explains it.

17. What are the differences between the solar wind and the normal "wind" on earth.

18. (a) What is the solar constant? (b) Why is it difficult to measure precisely? (c) Of what part of the spectrum is it most important to make accurate measurements in order to determine the solar constant? (d) What would the solar constant be if we lived on Mars?

19. What are the advantages and disadvantages of solar energy as a source of energy for the decade of the 1980's?

20. In what tests of general relativity does the sun play an important role? Describe the current status of these investigations.

TOPICS FOR DISCUSSION

1. Discuss the relative importance of solar observations (a) from the ground, (b) at eclipses, (c) from unmanned satellites, and (d) from manned satellites.

2. (a) Why might solar energy become a more important part of the solution to our energy problem around the year 2000 than it could be in the next few years? Discuss the advantages and disadvantages it might have relative to (b) nuclear fission, (c) nuclear fusion.

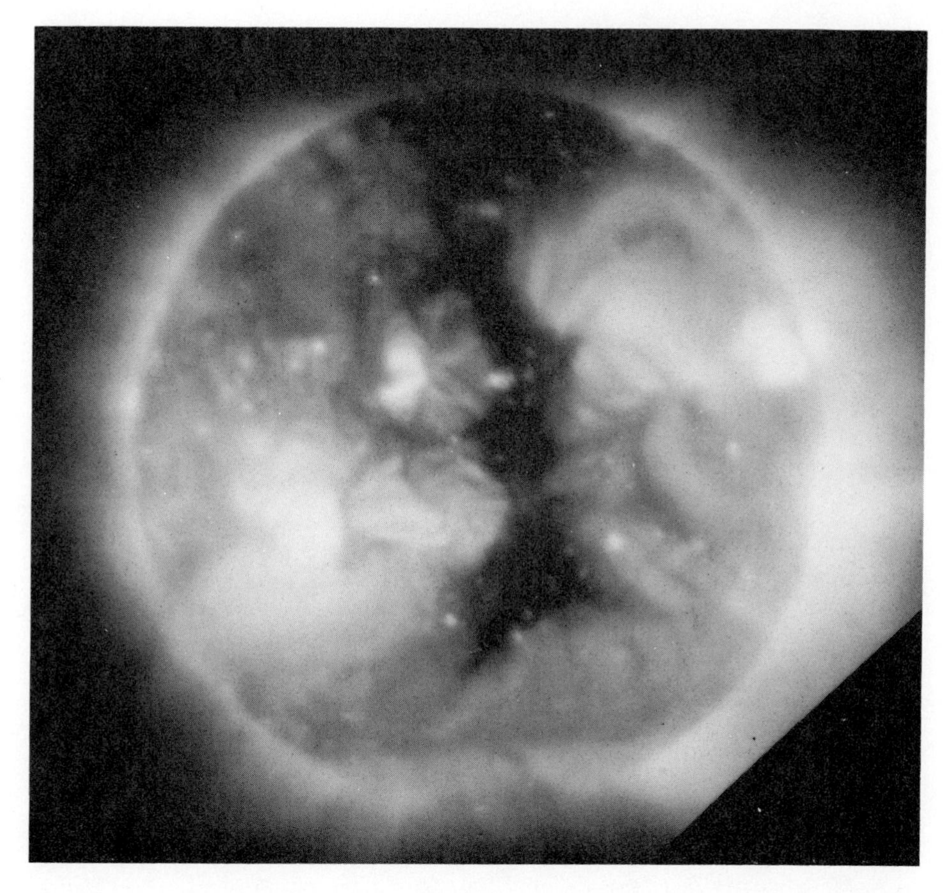

Figure 5–87 An x-ray photograph of the sun taken from Skylab on June 1, 1973. The filter passed wavelengths 2–32 Å and 44–54 Å.

6

Stellar Evolution

We have seen how the Hertzsprung-Russell diagram for a cluster of stars allows us to deduce the age of the cluster and the ages of the stars themselves (see Section 4.4). Our human lifetimes are very short compared to the billions of years that a typical star takes to form, live its life, and die. Thus our hope of understanding the life history of an individual star depends on studying large numbers of stars, for presumably we will see them at different stages of their lives. In this chapter we will discuss the birth of stars, then the processes that go on in a stellar interior during a star's life on the main sequence, and then begin the story of the evolution of stars when they finish this stage of their lives. The next three chapters will continue the story.

6.1 STARS IN FORMATION

Stars form out of the interstellar gas and dust in galaxies. In this chapter we are interested in the stars themselves; in Chapter 20 we will discuss the types of gas and dust that exist between the stars. Suffice it to say here that there are many regions of gas and dust in our galaxy in which star formation could take place.

The process of star formation starts with a region of gas and dust of slightly higher density than its surroundings. Perhaps this results only from a random fluctuation in density. If the density is high enough, the gas and dust begin to contract under the force of gravity. As they contract, energy is released, and it turns out (from a basic theorem) that half of that energy heats the matter.

As the gas and dust are heated, they begin to give off an appreciable amount of radiation. Also, as the temperature rises, the peak of the emitted radiation moves through the infrared spectrum to shorter and shorter infrared wavelengths (following Wien's displacement law, corresponding to Planck curves for higher temperature).

A photograph in red light of the region near Herbig-Haro object number 100 in the dark cloud in the constellation Corona Australis. The Herbig-Haro object is just to the lower left of the center of the photograph, near the center of the black patch that is the high-density cloud from which the H–H object, a star coming into being, was formed.

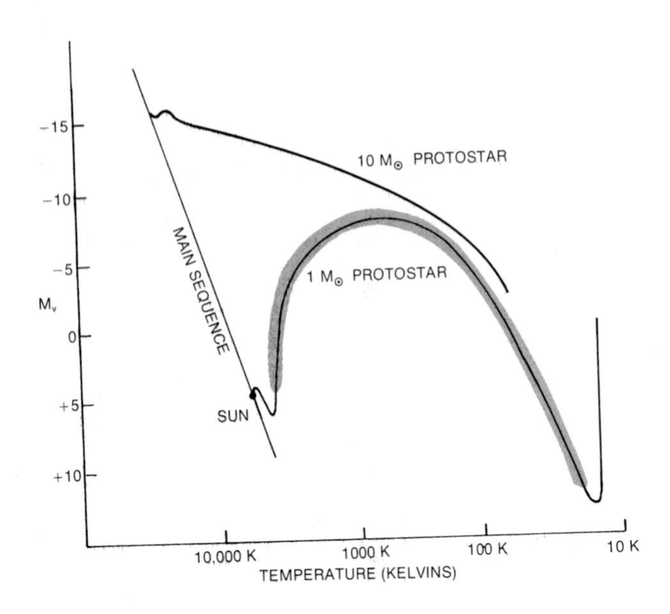

Figure 6–1 Evolutionary tracks for two protostars, one of 1 solar mass and the other of 10 solar masses. ($M_\odot$ is the symbol for the mass of the sun.) The evolution of the shaded portion of the track for the 1 solar mass star is described in the text. These stages of its life last for about 50 million years. More massive stars whip through their protostar stages more rapidly; the corresponding stages for a 10 solar mass star may last only 200,000 years. These very massive stars wind up being very luminous.

The "track" of this protostar,* the path it takes on a Hertzsprung-Russell diagram, is shown in Figure 6–1. At first, the protostar brightens, and the track moves upward and toward the left on the diagram. While this is occurring, the central part of the protostar continues to contract and the temperature rises. The higher temperature results in a higher pressure, and thus acts as a force in the outward direction. Eventually a point is reached where this outward force balances the inward force of gravity for the central region.

Box 6.1 Evolutionary Tracks

We can plot the position of a star on the Hertzsprung-Russell diagram for any particular instant of its life. As a protostar, for example, it would have a specific luminosity and temperature a million years before it began nuclear fusion, and another specific luminosity and temperature half a million years later. Each of the pairs of luminosity and temperature would correspond to a point on the H-R diagram. When hydrogen burning began, the star would also correspond to a specific point on the H-R diagram. If we connect the points representing the entire lifetime of the star, we have an *evolutionary track*. A star, of course, is only at one point of its track at any given time.

By this time the dust has vaporized and the gas is opaque, so that energy emitted from the central region does not escape directly. The outer layers continue to contract. Since the surface area is decreasing, the luminosity decreases and the track of the protostar begins to move downward on the

*"Proto-" is a prefix of Greek origin meaning "primitive."

H-R diagram. As the protostar continues to heat up, it also continues to move toward the left on the H-R diagram.

The time scale of this gravitational contraction depends on the mass of gas in the protostar. Very massive stars contract to approximately the size of our solar system in only ten thousand years or so. These massive objects become O and B stars, and are located at the top of the main sequence. They are sometimes found in groups, which are called *O and B associations.*

Less massive stars contract much more leisurely. A star of the same mass as the sun may take tens of millions of years to contract, and a less massive star may take hundreds of millions of years to pass through this stage.

Theoretical analysis shows that the dust surrounding the stellar embryo we call a protostar should absorb much of the radiation that the protostar emits. Do we detect any objects in the sky that meet the characteristics we expect from this scenario?

Theoretically, the radiation from the protostars should heat the dust to temperatures that produce primarily infrared radiation. Indeed, infrared astronomers have found several objects that are especially bright in the infrared but that have no known optical counterparts. These objects seem to be located in regions where the presence of a lot of dust and gas and other young stars indicates that star formation might be going on. Some of these infrared objects may contain stars in formation. As the technology of infrared detectors improves, astronomers are better able to map and study the distribution of energy at the wavelengths that are crucial for this study. Most astronomers think that some of the strange infrared objects being observed are the dust shells around protostars. These infrared observations are discussed in Chapter 22.

From regions that show heavy concentrations of dust, astronomers have used radio telescopes to detect many kinds of molecules and deduced temperatures and densities that are consistent with the above theory. The radio observations and other infrared observations will be discussed in Chapters 21 and 22.

In the visible part of the spectrum, several classes of stars that vary erratically are found. One of these classes, called *T Tauri stars,* includes stars that have a wide range of spectral types. Their visible radiation can vary by as much as several magnitudes, and they are known to be strong infrared emitters. Presumably these are stars that have not quite reached the main sequence and have not completely settled down to a steady and reliable existence.

The spectrum of a T Tauri star shows broad and very intense emission lines that are so strong that they overwhelm the absorption lines (Fig. 6–2). Presumably these emission lines are formed in a very extensive chromosphere surrounding the young stars.

T Tauri stars are found in close proximity to each other; these groupings are known as *T associations.* Besides the stars in the T associations one

Figure 6–2 The spectrum of T Tauri includes not only a continuum with absorption lines but also strong emission lines.

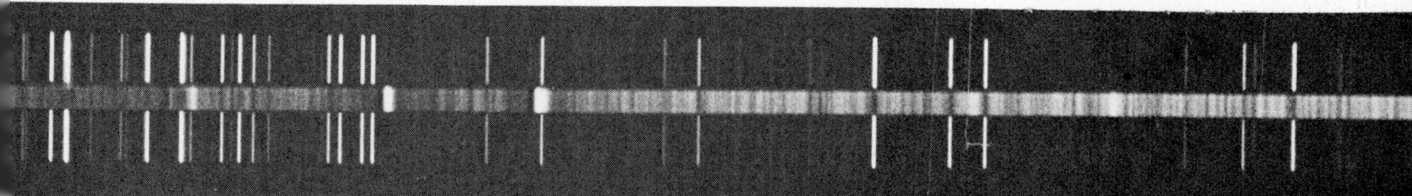

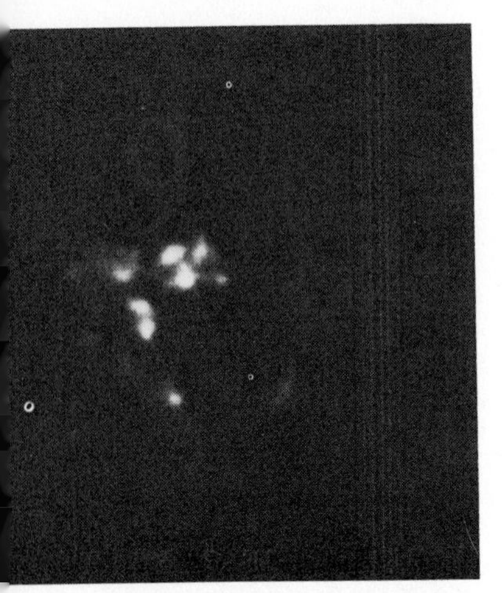

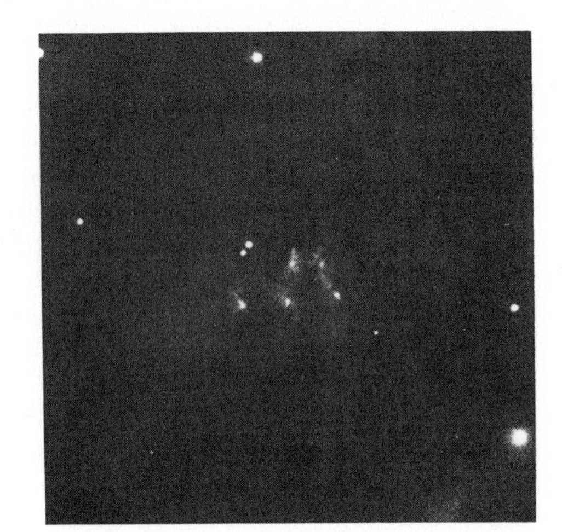

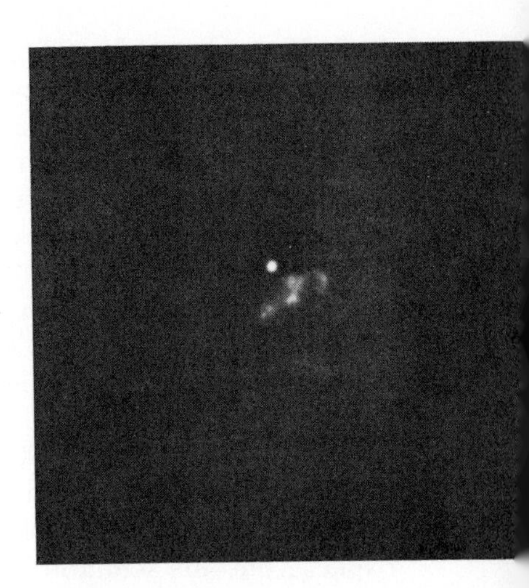

Figure 6–3 Three Herbig-Haro objects: (*left*) Haro 14a; (*middle*) Herbig 12; (*right*) Herbig 2.

finds bright nebulous regions of gas and dust called *Herbig-Haro objects*, after George Herbig of the Lick Observatory and Guillermo Haro of the Mexican National Observatory. Herbig-Haro objects may be the sites of star formation (Fig. 6–3). T Tauri itself, the prototype of the class, is embedded in the brightest Herbig-Haro object (Fig. 6–4).

Two Japanese theoreticians predict that one of the brightest of the infrared sources, an object in the Orion Nebula that was discovered by Eric Becklin and Gerald Neugebauer of Caltech, may become a T Tauri star within the next 20 years! Stay tuned. It might take only a few years for the ball of gas to shrink to the size of Mercury's orbit. If this happens, it would become visible as a cool, red star with an absolute magnitude 100 times brighter than that of the sun. Other scientists think it may take much longer – between 10^5 and 10^6 years – between the initiation of star formation and the appearance of stars.

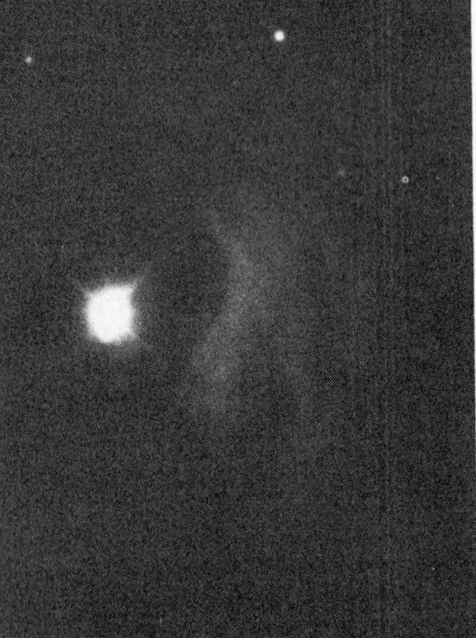

Figure 6–4 T Tauri itself is embedded in a Herbig-Haro object known as Hind's variable nebula. It has changed in brightness considerably in the last century, first fading from view and then brightening considerably. This probably results from changing illumination from T Tauri on a fixed dust cloud.

Three T Tauri stars have been observed to brighten considerably. One did so in Orion in 1936, and has been named FU Orionis (following long after ZZ and AA in the standard system of nomenclature for variable stars). Its light curve appears as Figure 6–5. Another faint variable star, V1057 in Cygnus (the 1057th variable in the Swan), brightened in 1969, and thus may also be going through the process of birth. The spectral type changed

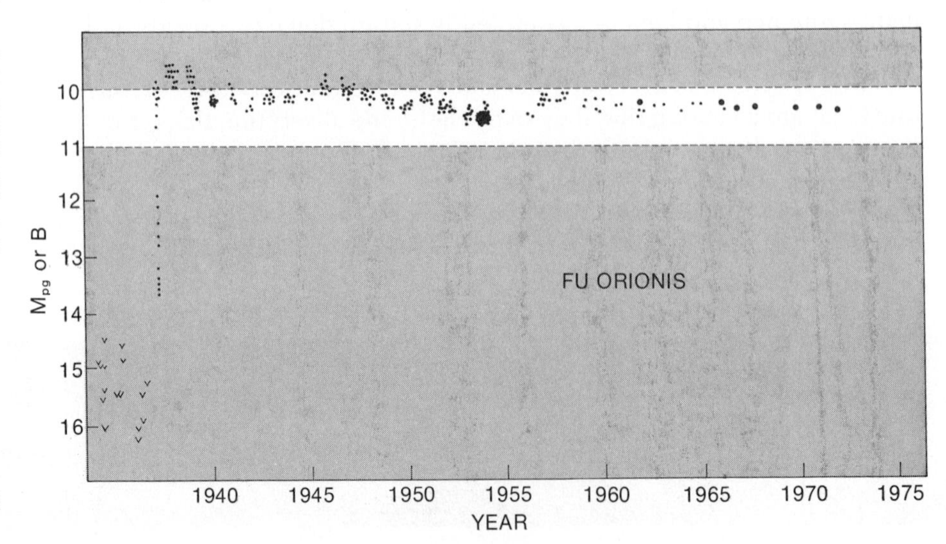

Figure 6–5 The light curve of FU Orionis.

from a cool K0 to a hot A1. Coincidentally, both stars brightened from about 16th to 10th magnitude, which resulted partly from the change in spectral type and partly from an expansion in size by a factor of 2.

Spectral lines that result from water vapor have recently been discovered in the spectra of both stars. In 1975, another T Tauri star named HL Tau was discovered to show absorption by ice—frozen water—in its infrared spectrum near a wavelength of 3 microns. It may represent an even earlier stage in stellar evolution. The brightening of FU Orionis and V1057 Cygni may have melted the ice that once surrounded them. These three stars may have advanced from the T Tauri stage to the vicinity of the main sequence before our very eyes.

6.2 STELLAR ENERGY GENERATION

All the heat energy in stars that are still contracting toward the main sequence results from the gravitational contraction itself. If this were the only source of energy, though, stars would not shine for very long on an astronomical time scale—only about 30 million years. Yet we know that even rocks on earth are older than that, since rocks billions of years old have been found. We must find some other source of energy to hold the stars up against their own gravitational pull.

The gas in a protostar will continue to heat up until the central portions become hot enough for *nuclear fusion* to take place. Using this process, which we discuss later, the star can generate enough energy inside itself to support it during its entire lifetime on the main sequence.

Nuclear fusion *is the joining of two nuclei to form a single, larger, nucleus.*

The basic fusion processes fuse four hydrogen nuclei into one helium nucleus, just as hydrogen atoms are combined into helium in a hydrogen bomb here on earth. In the process, tremendous amounts of energy are released.

In the interiors of stars, we are dealing with nuclei instead of the atoms we have discussed in stellar atmospheres, because the high temperature strips the electrons off the nuclei. The electrons are mixed in with the nuclei through the center of the star; there can be no large imbalance of positive and negative charges, or strong repulsive forces would arise inside the star.

A hydrogen nucleus is but a single proton. A helium nucleus is more complex. It consists of two protons and two neutrons (Fig. 6–6). Protons and neutrons have essentially the same mass and are similar in many of their properties, though the proton has a charge and the neutron has none.

H^1 NUCLEUS
PROTON

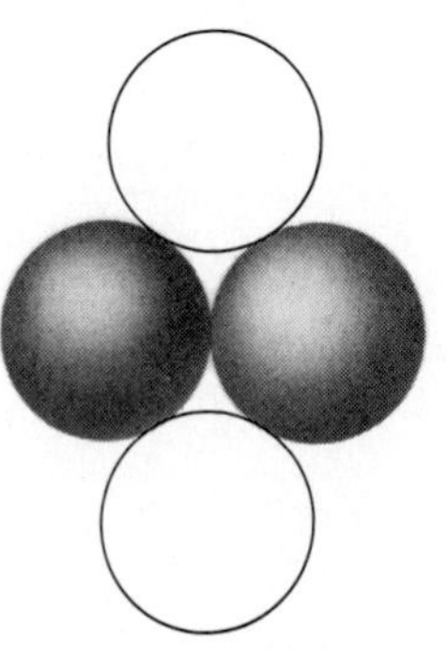

He^4 NUCLEUS

Figure 6–6 The nucleus of hydrogen's most common form is a single proton, while the nucleus of helium's most common form consists of two protons and two neutrons.

Box 6.2 Black Dwarfs

Masses of gas below about 0.07 solar mass—7 per cent of a solar mass—may never become hot enough for fusion processes to take place. Let us call these *featherweight stars.* When they radiate all the energy of their gravitational contraction, the cold, dark hulks are called *black dwarfs.* We know that there are more and more stars of lower and lower masses, so it might actually be the case that a significant fraction of the mass of a galaxy is in the form of black dwarfs. None has ever been observed.

The mass of the helium that is the final product of the fusion process is slightly less than the sum of the masses of the four hydrogen nuclei that went into it. A small amount of mass "disappears" in the process: .007 (a number that is easy to remember for James Bond aficionados) of the mass of the four hydrogen nuclei.

The mass does not really simply disappear, but is rather converted into energy according to Albert Einstein's famous formula: $E = mc^2$. Now c, the speed of light, is a large number, and c^2 is even larger. Thus even though m is only a small fraction of the original mass, the amount of energy released is prodigious. The loss of only .007 of the mass of the sun, for example, is enough to allow the sun to radiate as much as it does at its present rate for a period of at least 10^{10} years. This fact, not realized until 1920, solved the long-standing problem of where the sun and the other stars got their energy.

All the main-sequence stars are approximately 90 per cent hydrogen (that is, 90 per cent of the atoms are hydrogen), so there is lots of raw material to stoke the nuclear "fires." We speak colloquially of *nuclear burning*, although, of course, the processes are quite different from the chemical processes that are involved in the "burning" of logs or of autumn leaves. In order to be able to discuss these processes, we must first discuss the general structure of nuclei and atoms.

6.3 ATOMS

An atom consists of a small *nucleus* surrounded by *electrons*. Most of the mass of the atom is in the nucleus, which takes up a very small volume in the center of the atom. The effective size of the atom is set by the electrons, and the chemical interactions of atoms to form molecules are also determined by the electrons.

The nuclear particles with which we need be most familiar are the *proton* and *neutron*. Both these particles have nearly the same mass, only 1.6×10^{-24} gram, 1836 times greater than the mass of an electron. Many other nuclear particles have been detected—some only show up for tiny fractions of a second under extreme conditions in giant particle accelerators ("atom smashers"). Although some of these particles have important roles as the nuclear "glue" that holds nuclei together, we need not discuss the other particles in detail for most astronomical contexts. Particle physicists are currently assessing a theory that the nuclear particles are all made up of even more fundamental particles called *quarks*.

While the neutron has no electric *charge*, the proton has one unit of positive electric charge. The electrons, which surround the nucleus, have one unit each of negative electric charge, so when an atom has the same number of protons and electrons it is electrically neutral. This is called a neutral atom, and is denoted with the Roman numeral I. For example, neutral helium is denoted He I.

When an atom has lost an electron, it has a net positive charge of 1 unit. As we stated in Section 2.7, the atom is now a type of ion (Fig. 6–7); this state of the atom is denoted with a Roman numeral II. Thus an individual ion of helium that has lost one electron is in the He II state. If the atom is twice ionized, that is, has lost two electrons, it is in state III.

The definition of ions does not concern the nucleus; for the nucleus, we need keep track only of the numbers of protons and neutrons and not of orbiting electrons. The number of protons in the nucleus determines the

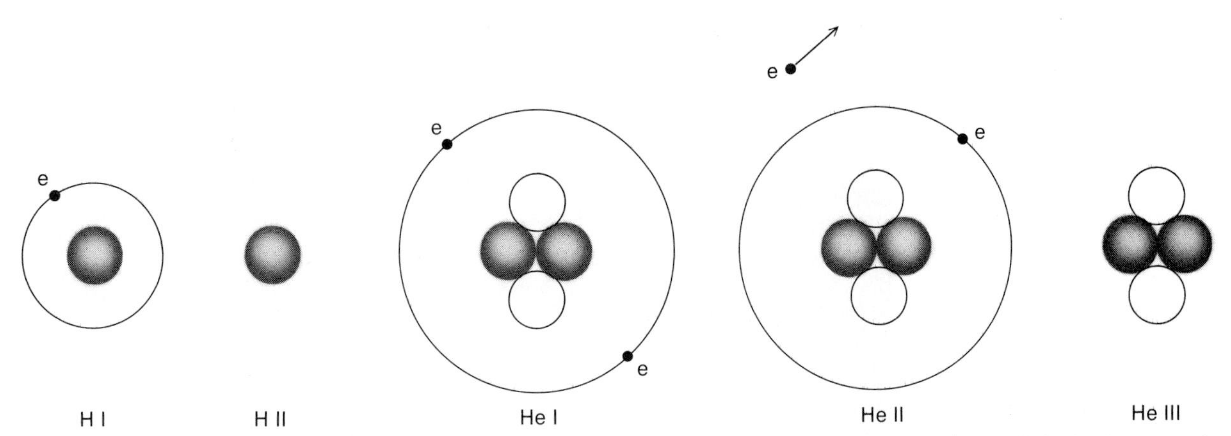

Figure 6–7 Hydrogen and helium ions. The sizes of the nuclei are greatly exaggerated with respect to the sizes of the orbits of the electrons.

quota of electrons that the neutral state of the atom must have, since it determines the charge of the nucleus. Each element is defined by the specific number of protons in its nucleus. The element with one proton is hydrogen, that with two protons is helium, that with three protons is lithium, and so on. A table of the elements appears as Appendix 4.

The number of neutrons in a nucleus is not fixed for a given element, although the number is always somewhere between 1 and 2 times the number of protons. (Hydrogen, which need have no neutrons, and helium are the only exceptions to this rule.) The possible different forms of the same element have only slightly different numbers of neutrons. These different forms are called *isotopes.*

For example, the nucleus of ordinary hydrogen contains one proton and no neutrons. An isotope of hydrogen (Fig. 6–8) called deuterium (and sometimes "heavy hydrogen") has one proton and one neutron. Another isotope of hydrogen called tritium has one proton and two neutrons.

Most isotopes do not have specific names, and one keeps track of the numbers of protons and neutrons with a system of superscripts and subscripts. Though there are several alternative ways of writing the form, we shall use the convention that a subscript before the symbol denoting the element is the number of protons, and a superscript following the symbol is the total number of protons and neutrons together. The number of protons is called the *atomic number,* and the number of protons and neutrons together is called the *mass number.* For example, ${}_1H^2$ is deuterium, since deuterium has 1 proton, which gives the subscript, and an atomic mass of 2, which gives the superscript. Deuterium has atomic number equal 1 and mass number equal 2. Similarly ${}_{92}U^{238}$ is an isotope of uranium with 92 protons (atomic number = 92) and a mass number of 238, which is divided into 92 protons and 238–92 = 146 neutrons (Fig. 6–9).

By convention, we mean a usage, agreed upon.

Each element exists only in certain isotopic forms. For example, most of naturally occurring helium is in the form ${}_2He^4$, with a lesser amount as ${}_2He^3$. Sometimes an isotope is not stable, in that after a time it will spon-

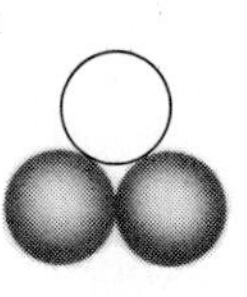
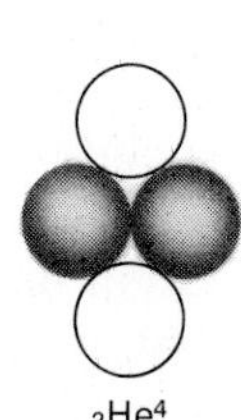

Figure 6–8 Isotopes of hydrogen and helium. ${}_1H^2$ (deuterium) and ${}_1H^3$ (tritium) are much rarer than the normal isotope, ${}_1H^1$. ${}_2He^3$ is much rarer than ${}_2He^4$.

Figure 6–9 Most atoms are much more complex than hydrogen or helium. In neutral atoms, the number of electrons is equal to the number of protons.

taneously change into another isotope or element; we say that such an isotope is *radioactive*.

While a proton is a stable particle whether it is inside or outside a nucleus, a neutron is stable only when part of a nucleus. If many "free" neutrons (neutrons outside a nucleus) are placed by themselves, after 12 minutes half of them will have changed, or "decayed," into 1 proton plus 1 electron (plus 1 neutrino, see below). We say that the *half-life* of a free neutron is 12 minutes. In a sense this is the average time it takes decay to occur: any one neutron may happen to decay in a time much shorter than or much longer than 12 minutes. Since a neutron by itself is not a nucleus, we do not say that it is radioactive; it is just an unstable particle.

During a certain type of radioactive decay, a particle called a neutrino is given off. A neutrino is a neutral particle (its name, conferred by the physicist Enrico Fermi, comes from the Italian for "little neutral one"). It has the very interesting property, for a particle, of always traveling at the speed of light. Now, ordinary matter cannot travel at the speed of light, according to Einstein's theory of relativity, because its mass gets larger and larger and approaches infinity as its speed approaches the speed of light. This problem does not arise in the case of the neutrino because the neutrino, like the photon, would have no mass if it were at rest (see Box 6.3).

Box 6.3 The Neutrino

The neutrino was first invented on theoretical grounds in 1930 by the theoretical physicist Wolfgang Pauli in order to help explain certain nuclear reactions. In a *nuclear reaction* one starts out with certain nuclear particles (protons and neutrons, for example) and possibly radiation and winds up with other particles and whatever radiation may be emitted. The particles that come out of a reaction are different from the particles that go into a reaction, just as though a Datsun and a Cadillac collided and after an x-ray zapped a passer-by, a Pinto and a Mercedes were found when the smoke cleared.

There are certain quantities that must be the same when one compares the sum of whatever comes out of with the sum of whatever goes into a reaction. For example, the total energy at the end must be the same as the total energy at the beginning. This is called *conservation of energy*. Another quantity that must be "conserved" is *spin*, a property of nuclear particles that is not unlike the spin of a top. A top (if we ignore friction) does not stop its spin abruptly or reverse its spin spontaneously, or change its spin in any other way all by itself; no isolated physical system does either. (We will discuss spin further in Section 21.4.)

Thus, when it was observed that the total energy and the total spin of the nuclear particles and radiation present after the decay of a neutron could be different from the total energy and the total spin of the particles present before this decay (namely, the spin of the neutron), Pauli and Fermi reasoned that there had to be another spin-carrying particle involved in the reactions. They were thus able to predict the existence of neutrinos long before neutrinos were detected experimentally.

Figure 6–10 Enrico Fermi, Werner Heisenberg, and Wolfgang Pauli enjoy an outing on Lake Como in 1927.

Neutrinos have a very useful property for the purposes of astronomy: they do not interact very much with matter. Thus when a neutrino is formed deep inside a star, it can usually escape to the outside without interacting with any of the matter in the star. This means that the neutrino heads right out of the star, without bumping into any of the matter in the star along the way. Electromagnetic radiation, on the other hand, does not escape from inside a star so easily. A photon of radiation can travel only about 1 cm in a stellar interior before it is absorbed, and it is millions of years before a photon zigs and zags its way to the surface.

The elusiveness of the neutrino makes it a valuable probe of the conditions inside the sun at the present time, but also makes it very difficult for us to detect on earth. Later in this chapter we shall discuss the major experiment that is being performed to study solar neutrinos.

6.4 STELLAR ENERGY CYCLES

Several chains of reactions have been proposed to account for the fusion of four hydrogen atoms into a single helium atom. Hans Bethe, now at Cornell University, suggested some of these procedures during the 1930's. The different possible chains that have been proposed are important at different temperatures, so chains that are dominant in the centers of very hot stars may be different chains from the ones that are dominant in the centers of cooler stars.

When the center of a star is at a temperature less than 15×10^6 K, the *proton-proton* chain is dominant (Fig. 6–11). The reactions of the proton-proton chain are:

$$2 \times \left\{ \begin{array}{l} H^1 + H^1 \rightarrow H^2 + e^+ + \nu \\ H^2 + H^1 \rightarrow He^3 + \gamma \end{array} \right.$$
$$He^3 + He^3 \rightarrow He^4 + H^1 + H^1$$

where e^+ stands for a positron (a particle like an electron except for having positive charge), ν (nu) is a neutrino, and γ (gamma) is radiation at a very short wavelength. In the first stage, two ordinary nuclei of hydrogen fuse to become a deuterium nucleus, a positron, and a neutrino. The neutrino immediately escapes from the star, but the positron soon collides with an electron. They annihilate each other, forming gamma rays.

Next, the deuterium nucleus fuses with yet another nucleus of ordinary hydrogen to become an isotope of helium with two protons and one neutron. High-energy radiation is released at this stage in the form of more gamma rays.

Finally, two of these helium isotopes fuse to make one nucleus of ordinary helium plus two nuclei of ordinary hydrogen. We have put in six hydrogens and wind up with one helium plus two hydrogens, a net transformation of four hydrogens into one helium. The small fraction of mass that disappears in the process is converted into energy according to the formula $E = mc^2$.

For hotter stellar interiors than that of the sun, the *carbon cycle* would dominate (Fig. 6–12). The carbon cycle begins with the fusion of a hydrogen nucleus with a carbon nucleus. After many steps, and the insertion of four hydrogen nuclei, we are left with one helium nucleus plus a carbon nucleus. Thus as much carbon remains at the end as there was at the begin-

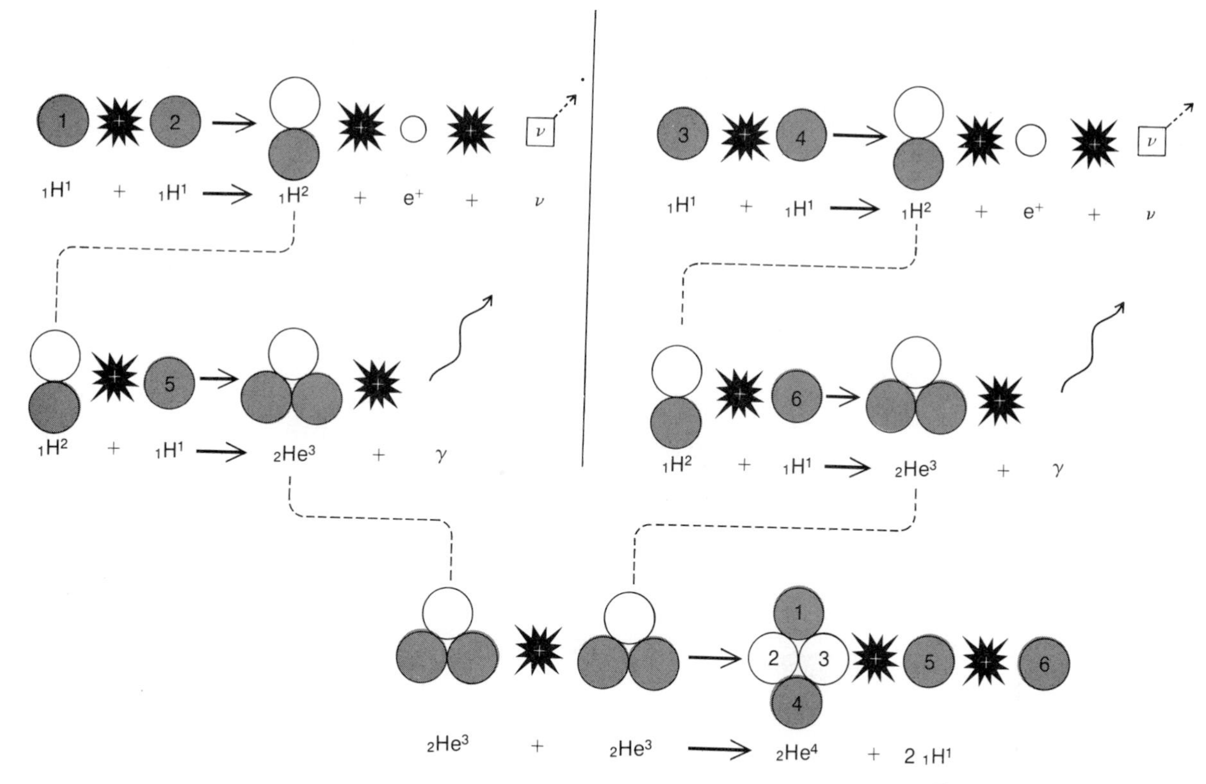

Figure 6–11 The proton-proton chain.

ning, and the carbon can start the cycle again. Again, four hydrogens have been converted into one helium, .007 of the mass has been transformed, and an equivalent amount of energy has been released according to $E = mc^2$. Since the cycle can also start with nitrogen and oxygen, the process is sometimes called the *carbon-nitrogen cycle* or the CNO cycle.

At even higher temperatures, as occur in very hot stars where the interior temperatures are above 10^8 K, we can fuse helium nuclei to make carbon nuclei. The nucleus of a helium atom is called an "alpha particle" for historical reasons, as the Greek characterizations derive from a former confusion of radiation and particles. (We now know that α particles, formerly called α rays, are helium nuclei, β particles, formerly called β rays, are electrons, and γ rays, as we have seen, are radiation of short wavelength.) Since three helium nuclei ($_2He^4$) go into making a single carbon nucleus ($_6C^{12}$) [with beryllium ($_4Be^8$) as an intermediate step], the procedure is known as the *triple-alpha process* (Fig. 6–13).

A series of other processes can build some of the heavier elements inside stars. A famous paper written in 1957 by Margaret Burbidge, Geoffrey Burbidge, William A. Fowler, and Fred Hoyle (known collectively as B^2FH) laid down the blueprint for the synthesis of heavier elements from lighter elements in the interiors of stars. This is thus the study of *nucleosynthesis*.

Their theory can account for the abundances we observe of the elements heavier than helium. Currently, we think that synthesis of isotopes of hydrogen and helium took place in the few minutes after the origin of the

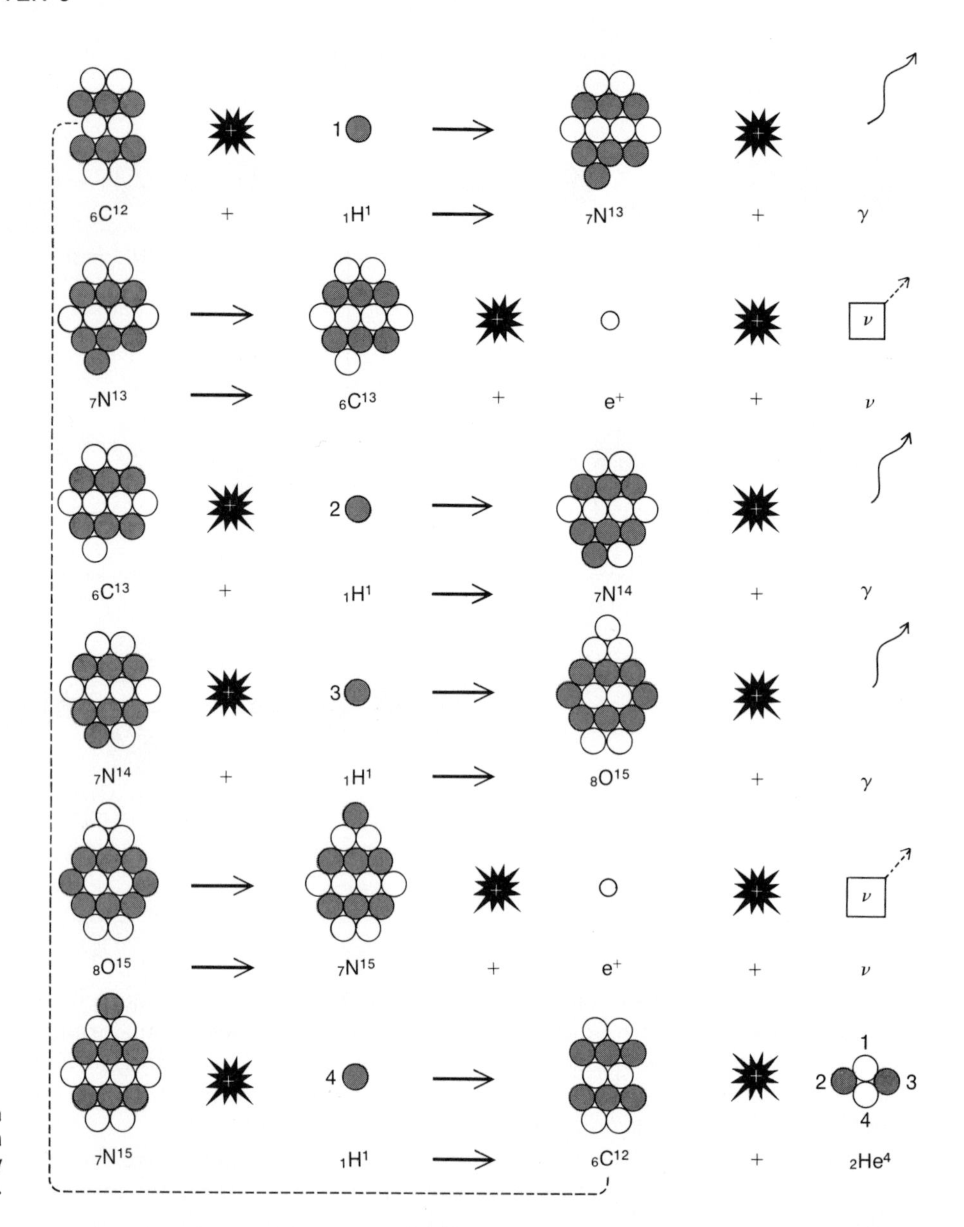

Figure 6–12 The carbon cycle. Note that the carbon is left over at the end, ready to enter into another cycle.

universe (see Section 25.6), and that the heavier elements were formed, along with additional helium, in stellar processes.

To predict the abundances of the heavy elements from other, lighter elements, one can list all the possible nuclear reactions that can build up one element from others. One needs to know the rates at which these reactions go on. (Many of the important laboratory measurements of the rates have been carried out at Caltech under the guidance of Fowler.) Then one uses a computer and calculates the resulting abundances. The theoretical calculations can satisfactorily account for many of the observational results.

6.5 THE STELLAR PRIME OF LIFE

Now that we have discussed basic nuclear processes, let us return to an astronomical situation. We last discussed a protostar in a collapsing phase, with its internal temperature rapidly rising.

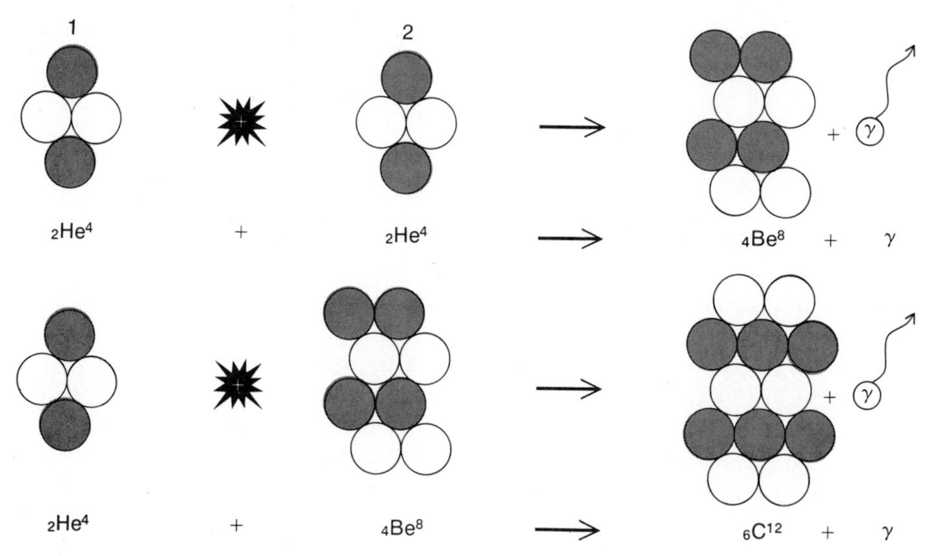

Figure 6–13 The triple-alpha process, which takes place only at temperatures above about 10^8 K.

One of the most common definitions of temperature describes the temperature in terms of the velocities of individual atoms or other particles. (See Section 5.2 for definitions of other types of temperature.) Since this type of temperature depends on the kinetics (motions) of particles, it is called the *kinetic temperature.* A higher kinetic temperature (we shall call it simply "temperature" from now on) corresponds to higher particle velocities.

Box 6.4 Types of Temperature

Excitation temperature: a number that characterizes the distribution of electrons into higher energy levels than the ground state of an atom.

Ionization temperature: a number that characterizes the relative abundances of different ionic states in a volume of gas.

Kinetic temperature, also commonly called *the temperature:* a number characterizing the velocities of molecules, atoms, or ions. The kinetic temperature of electrons, for example, is often called the electron temperature.

For a collapsing protostar, the energy from the gravitational collapse goes into giving the individual particles greater velocities; that is, the temperature rises. For nuclear fusion to begin, atomic nuclei must get close enough to each other so that the nuclear force, technically called the *strong force,* can play its part. But all nuclei have positive charges, because they are composed of protons and neutrons. The positive charges on any two nuclei cause an electrical repulsion between them, and this force tends to prevent fusion from taking place. However, at the high temperatures typical of a stellar interior, some nuclei have enough energy to overcome this electrical repulsion and to come into sufficiently close proximity for the strong force to take over. The electrical repulsion that must be overcome is the reason why hydrogen nuclei, which have net positive charges of 1, will fuse at lower temperatures than will helium nuclei, which have net positive charges of 2.

Box 6.5 Forces in the Universe

There are four known types of forces in the universe:

The *strong force,* also known as the *nuclear force,* is the strongest. It is the binding force of nuclei. Although it is very strong close up, it grows weaker rapidly with distance. It is transmitted by nuclear particles called mesons. Actually, the strong force is no longer thought to be a basic force. We now think that both mesons and nuclear particles are composed of 3 kinds of particles called *quarks* (from a line of James Joyce's *Finnegan's Wake*). The quarks are linked together by the *color force*; each kind of quark can have one of three properties called *colors.* The nuclear force is an indirect manifestation of the color force, which is carried by as yet undiscovered particles called *gluons* (because they provide the "glue").

It has recently been realized that if a fourth quark existed, then results from many experiments could be understood for the first time. This fourth quark is distinguished from the other three quarks by a quality called *charm.* (Neither *charm* nor *color* has its normal connotations in this usage.)

The *electromagnetic force,* 1/137 the strength of the strong force, leads to electromagnetic radiation in the form of photons. It is the force involved in chemical reactions.

The *weak force,* important only in the decay of certain elementary particles, is currently under careful study by particle physicists. It is very weak, only 10^{-13} the strength of the strong force, and also has a very short range. One might expect that particles would act as the transmitting agent of the weak force; such particles would be the analogy for the weak force of the mesons that we know carry the nuclear force and of the photons that we know carry the electromagnetic force. These as yet undiscovered particles for the weak force have a name waiting for them should they ever be found: either W particles or intermediate vector bosons.

The *gravitational force* is the weakest of all over short distances, only 10^{-39} the strength of the strong force. But its effect is cumulative so that on the scale of the universe it dominates the other forces. It is the only one of the four types of forces that affects all particles without exception. There is no antigravity to cancel out gravity, unlike the way positive and negative electric charges exist and can cause the electromagnetic force to cancel out. Even antimatter has the same kind of gravity as ordinary matter; antimatter and matter attract each other with the force of gravity in the same way that matter attracts matter. A search is now on for gravitational waves. It has not been proved that gravity is carried by particles analogous to the mesons that carry the strong force or the photons that carry electromagnetic radiation; if such particles are found they would be called *gravitons.*

In the center of a star, the fusion process is a self-regulating one. If the nuclear energy production rate increases, then an excess pressure is generated that would tend to make the star grow larger if it were not for the fact that this expansion in turn would cool down the gas. Thus the temperature can remain roughly constant for a very long time. This is the main-sequence phase of a stellar lifetime. Fusion processes are well regulated in stars. When we learn how to control fusion in power-generating stations on earth, which currently seems a long way off, our energy crisis will be over.

The more mass a star has, the hotter its core becomes before it generates enough pressure to counteract the gravity. Thus more massive stars use their nuclear fuel at a much higher rate than less massive stars, and even though the more massive stars have more fuel to burn, they go through it relatively quickly. This explains the mass-luminosity relation (Section 4.2).

Note that stars do not actually move along the main sequence; they stay for a long time at essentially the same place on the H-R diagram. They begin their stable hydrogen-burning phase at the lower left edge of the main sequence. This lower left edge is called the *zero-age main sequence*, often abbreviated as *ZAMS* (Fig. 6–14). We consider that a star begins its main-sequence lifetime when it reaches the ZAMS.

Actually stars do not remain exactly at the same spot on the H-R diagram while they are on the main sequence. As the hydrogen in their centers is gradually transformed into helium, they brighten slightly and thus move up on the H-R diagram, but this brightening is a very small effect. The positions representing the stars on their evolutionary tracks move only slightly off the main-sequence phase. (All this movement is upward and to the right of the ZAMS.) When the hydrogen is exhausted, stars move farther upward and to the right.

Depending on the temperature and density inside the star, and thus on how massive the star is, the helium that is formed inside the star may itself fuse into carbon via the triple-alpha process. In the interiors of the most massive stars, the carbon itself burns, and then the products of the carbon combustion burn, and so on, until eventually an iron core is built up.

Because of basic properties of nuclei, iron is the heaviest element that can form in a stellar interior. All the fusion processes that form nuclei lighter than and including iron are processes that release energy. On the other hand, one must add energy to the system in order to form nuclei heavier than iron. Since there is no source for such extra energy, iron accumulates at the center of the star.

The study of stellar interiors was for many years purely the domain of the theoretician working with equations. Sets of equations were found that accounted for the interior structure of stars, and sets of theoretical solutions were worked out. With the advent of high-speed digital computers that can perform in minutes calculations that would have taken years to do by hand, the study of stellar interiors is now carried much further. Fewer simplifying assumptions need to be made. Theoreticians now make a model of a stellar interior, considering it as a series of concentric shells. One calculates shell by shell what must be going on in the interior, making certain that the values of quantities like temperature, density, and pressure that are derived for the bottom of one layer match the values that are derived for the top of the layer immediately below (Fig. 6–15). One thus makes certain that these quantities change continuously through the stellar model in the computer, as they do in real life.

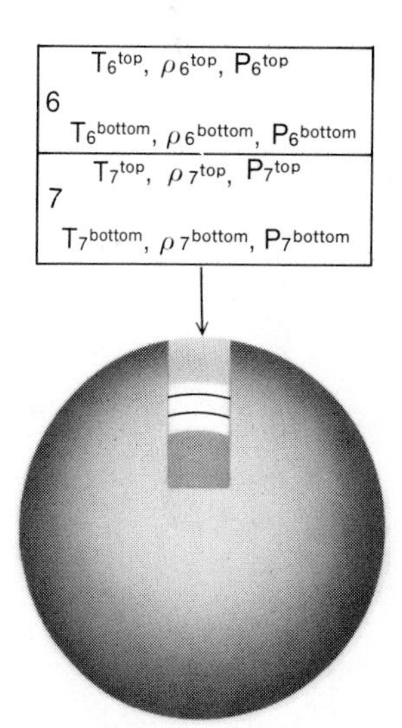

Figure 6–15 For theoretical calculations of stellar interiors, the star is often divided up into thin shells. Physical parameters such as temperature, density, and pressure must be the same at the top of one shell as they are at the bottom of the next higher shell. Two of the layers shown in white in the cutaway are enlarged at the top. T stands for temperature, ρ (the Greek letter rho) stands for density, and P stands for pressure.

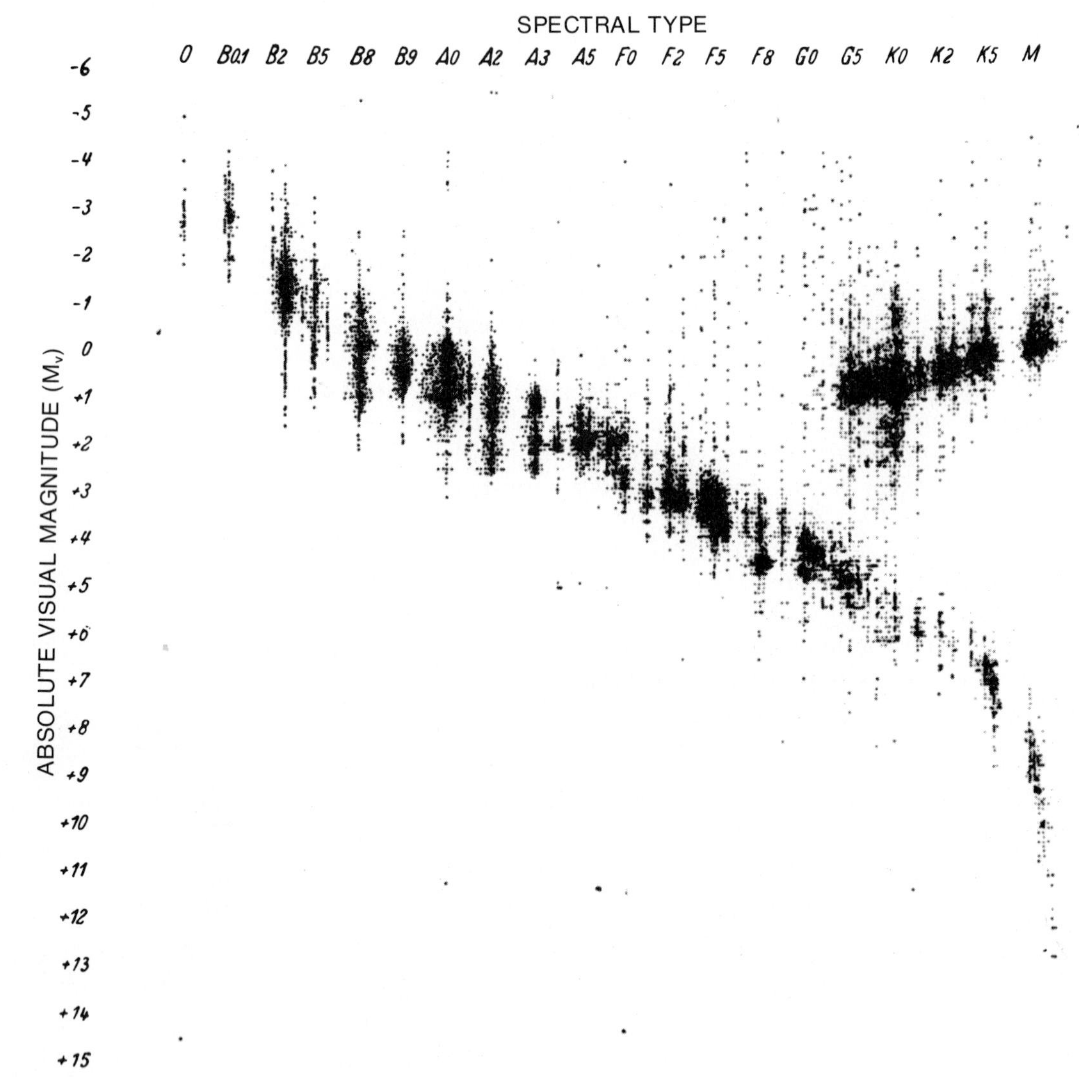

Figure 6–14 The lower left edge of the main sequence defines the *zero-age main sequence* (ZAMS). This H–R diagram includes all stars for which spectral types and distances were known at the time the diagram was compiled by W. Gyllenberg of the Lund Observatory in Sweden.

Box 6.6 Stellar Interiors

Needless to say, when one goes through the whole calculation of a model stellar interior, usually things don't exactly match; the values that one gets for half-way into a stellar interior when you calculate starting from the outside of the star and move toward the center, for example, may not quite match the values that result when you start at the center and calculate starting from the inside. You can start the calculations again, changing slightly the values of such quantities as temperature, density, and pressure, to try to improve on the previous calculation. Each time you repeat the calculation you can arrange the parameters to get better agreement. You do this over and over again until you have a reasonably consistent set of values for all the parameters. Each repeat is called an *iteration*. This method is perfectly suited for a computer—the larger, the better.

In addition to iterating over the distribution of matter in a star, one must also consider the evolution of the star with time. For example, calculations at later times will have to take into account the fact that there is a higher percentage of helium at the core than previously. One divides a stellar lifetime into time intervals for the purpose of calculation; one may, for example, construct an evolutionary sequence of models with 50-million-year intervals between them for the main-sequence stage.

6.6 THE NEUTRINO EXPERIMENT

One can make a perfectly good model for what goes on inside a star, and it can look quite satisfactory when it comes from the computer, but nonetheless one would like to find some experimental confirmation of the model. Happily, the models are consistent with the conclusions on stellar evolution that one can make from studying different types of Hertzsprung-Russell diagrams. Still, it would be nice to observe a stellar interior directly.

Since the stellar interior lies under opaque layers of gases, we cannot observe directly any electromagnetic radiation it might emit. All the gamma rays, x-rays, ultraviolet, visible, infrared, and radio radiation that we receive come from the surface layers or atmospheres of stars.

Only the neutrino, this odd, spinning, massless particle, can escape directly from a stellar interior. It interacts so weakly with matter that it is hardly affected by the presence of the rest of the solar mass, and zips right out into space at the speed of light.

Raymond Davis, Jr., a chemist at the Brookhaven National Laboratory, has spent the last ten years trying to detect the neutrinos from the solar interior that on theoretical grounds should be created by nuclear reactions there. Since the experiment is our only direct opportunity to test the theoretical work on stellar interiors, the results are exceedingly important. Because they raise serious questions about the state of our knowledge, we shall discuss them in some detail.

At best, it would be very difficult to detect a neutrino. We mentioned in the previous section that neutrinos are given off in the proton-proton chain that is typical of stars like the sun. But these neutrinos are given off with insufficient energy to allow them to be detected.

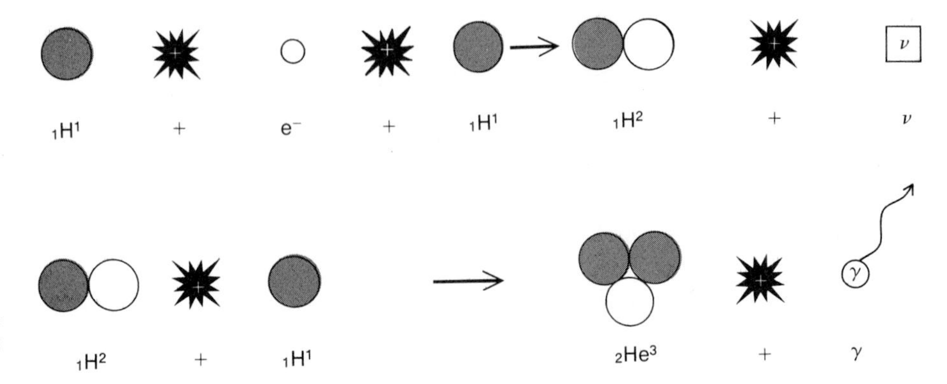

Figure 6–16 The proton-electron-proton (pep) chain.

One possible stellar fusion chain results in the formation of neutrinos of sufficiently high energy that Davis could build apparatus to search for them. This chain is a variation of the proton-proton (p-p) reaction. The normal p-p chain begins with two hydrogen nuclei (protons), and results in one deuteron, one positron, and a weak neutrino. One quarter of one per cent of the time, one time in 400, an alternative reaction occurs in which two hydrogens plus an electron (p + e + p) result in a deuteron and a neutrino. This *pep* neutrino is somewhat more energetic (has more energy) than the p-p neutrino, and can barely be detected by Davis's apparatus (Fig. 6–16).

Neutrinos that are more detectable are formed further along in the p-p and pep chains. Again, these high-energy neutrinos are not formed in the dominant branch (in which deuterium and hydrogen become He^3 and two He^3's make He^4 plus 2 H^2's), but in a branch that takes place only 0.1 per cent of the time. In the process, a boron isotope, ${}_5B^8$, decays into a radioactive beryllium isotope, ${}_4Be^8$, plus a positron plus a neutrino, and this neutrino is ten times more energetic than even the detectable pep neutrino (Fig. 6–17). (The ${}_4Be^8$ quickly decays into two ordinary heliums; we are interested here only in the neutrino.)

How do we detect such neutrinos? Neutrinos, after all, pass through the earth and sun, barely affected by their mass. At this instant, neutrinos are passing through your body. Davis makes use of the fact that very occasionally a neutrino will interact with the nucleus of an atom of chlorine, and transform it into an isotope of the noble gas argon. (The "noble" elements do not readily combine with other elements to form molecules; the atoms of argon are thus not hidden in molecules.)

This transformation takes place very rarely, so Davis had to accumulate a large number of chlorine atoms. He found it best to do this by filling a large tank with liquid cleaning fluid, C_2Cl_4, where the subscripts represent the number of atoms of carbon and chlorine in this perchloroethylene molecule. One fourth of the chlorine is the Cl^{37} isotope, which is able to interact with a neutrino. Davis now has a large tank containing 400,000 liters (100,-000 gallons) of this cleaning fluid.

Even with this huge tankful of chlorine atoms, calculations show that Davis should expect only about one neutrino-chlorine interaction every day. This experiment is surely one of the most difficult ever attempted.

When an interaction occurs, a radioactive argon atom is formed. Any such argon atoms can be removed from the tank by standard chemical techniques involving bubbling helium gas through the fluid. Then the radio-

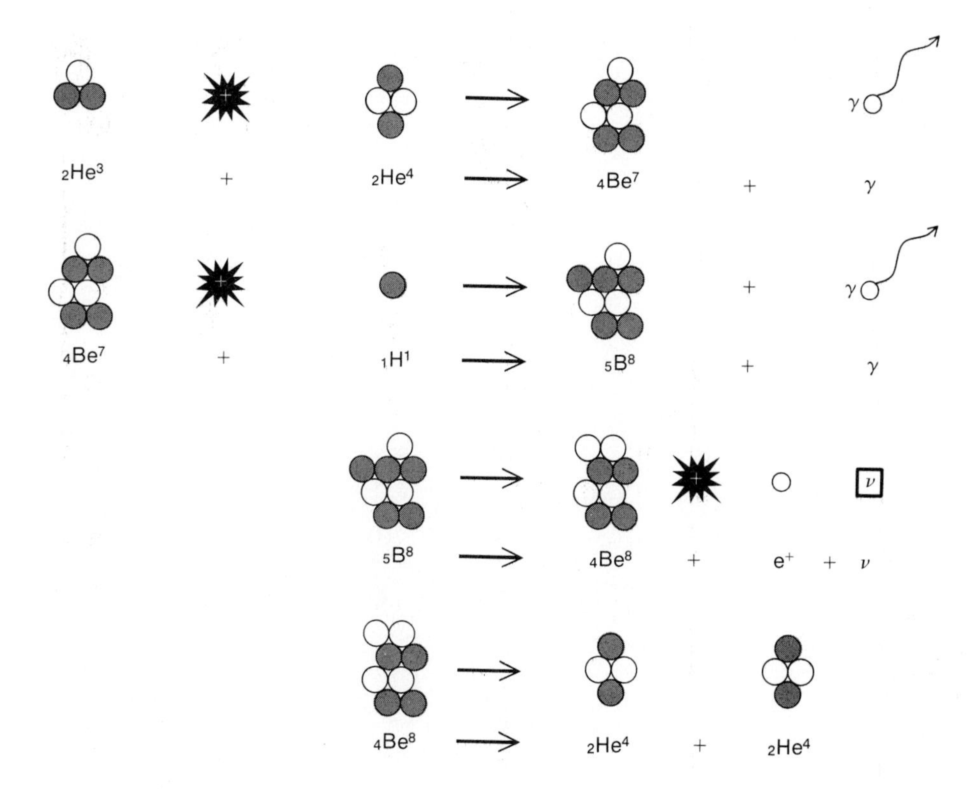

Figure 6–17 The chain of fusion via the boron-beryllium route, a subsidiary chain, leads to the ejection of neutrinos of an energy that should be detectable by Davis's apparatus.

activity of the few resulting argon atoms can be measured, also by standard means.

Davis set up his apparatus far underground, in order to shield it from other particles from space that could also interact with the chlorine. Thus 1.5 km underground, deep in the Homestake Gold Mine in Lead, South Dakota, you will find the most incredible of the world's "telescopes"—this 400,000 liter tank of cleaning fluid (Fig. 6–18). (Davis denies the story that after he bought all his cleaning fluid he was besieged by wire-coat-hanger salesmen.) The tank is even submerged in water, to further prevent contamination by outside particles (Fig. 6–19).

Davis runs the experiment for three month intervals by leaving the tank undisturbed for that length of time. Then he flushes the tank with helium to remove the argon, and measures the amount of argon. In the meantime, he prepares to "observe" for the next two or three month cycle.

It is fair to say that Davis's results have astounded the scientific community, which had confidently expected to hear reports of a small but measurable number of neutrino interactions. But Davis reported: no neutrinos.

Davis reports his results in terms of solar neutrino units, which correspond to 10^{-36} captures per second per target atom. 1 SNU (pronounced "one snew") corresponds to one capture per six days in his tank. The proton-proton chain should, it was predicted on theoretical grounds, correspond to a capture rate of about 6 SNU's. The carbon cycle would correspond to 35 SNU's. Davis's original results were negative, in that he did not definitely detect any neutrinos at all. The uncertainty in the experiment was such that there could have been as many as 3 SNU's without their having been detected.

Figure 6–18 The neutrino telescope, deep underground in the Homestake Gold Mine in Lead, South Dakota, contains a tank of 400,000 liters of carbon perchloroethylene. The tank is now surrounded by water.

This result caused some consternation in the scientific community, because it showed that the standard calculation of the capture rate gave values that were at least slightly too high. It also showed, as had already been believed, that the carbon cycle contributed less than 10 per cent of the energy of the solar interior.

Theoreticians went back to work, and tried to carry out complicated calculations including more realistic assumptions than they had previously used. They included, for example, the fact that the sun is not just a stationary ball of gas but that it is rotating. Still, the new calculations did not give predictions of appreciably lower value.

As time went on, Davis's experiment was refined and the total amount of observing time increased. His latest value is less than one-third the best theoretical prediction, which is 5.5 SNU, and his measurement is at the level that represents the minimum that standard theories are able to predict.

Several questions immediately come to mind. The first deals with Davis's apparatus, and whether there are some experimental effects that could vitiate the results. Davis has carried out a series of careful checks of his apparatus, and most astronomers and chemists are convinced that the experimental setup is not the cause of the problem.

Figure 6–19 Davis tried out the water that helps the kilometer of earth overhead shield his neutrino telescope from particles other than neutrinos.

Next, perhaps we do not understand the neutrino as well as we thought. Perhaps it disintegrates during the 8 minutes it takes to travel from the sun to the earth. That would indeed account for the failure to detect neutrinos, but in itself it would be an important result for physics. There is no experimental evidence for this situation, and not many scientists are still seriously considering it.

There are other experimental sources for possible error—for example, in the reaction rates of all the processes that lead up to the formation of the energetic neutrinos. But these have all been checked, and no error has been found.

One of the most interesting possibilities is that we really don't understand the basic physics of nuclear reactions as well as we had thought. Another interesting possibility is that our understanding of stellar interiors is not satisfactory. If, for example, stars are cooler inside than we have predicted, fewer neutrinos would be produced. Thus the neutrino experiment has clear and direct ramifications for the study of astrophysics.

Another possibility now being considered with increasing respect is the idea that maybe we are doing everything right but that the sun is simply not generating neutrinos at the moment! The neutrino experiment tells us what the sun is doing now, or really what it did eight minutes earlier. (The travel time of the neutrinos from the sun to the earth is 8 minutes.) The other way we can detect the effect of the nuclear furnace in the solar interior is to observe the solar surface, but it takes millions of years for energy generated inside to work its way up to the surface. Perhaps ice ages in the past resulted from other periods long ago when the sun was turned "off"; perhaps the current episode will lead to ice ages hundreds of thousands of years from now. We have thought of main-sequence stars as sedentary objects, regularly transforming hydrogen into helium at a steady rate. But they may be actually cyclic, and we would be merely observing the sun at a period when no nuclear fusion is going on in its interior.

The report that the sun may be oscillating with periods of 48 minutes to several hours (Section 5.12) could explain the discrepancy in the neutrino experiment. If the oscillations generate enough energy, the resulting pressure could help support the solar gas against the attraction of gravity that is tending to collapse it. Thus the amount of pressure that would have to be provided by the internal temperature of the sun is lower than had been thought, and the temperature at the center of the sun would be lower than had been thought. This would mean that the expected rate of neutrino production would be reduced below the level currently predicted.

Still another method of accounting for the neutrino experiment is the idea of Donald Clayton and his associates at Rice University that there is a mini black hole at the center of the sun. Such a mini black hole would have had to be formed in the earliest seconds of the universe, as described in Section 9.4. Even if the black hole were only a few centimeters across at present, the solar gas around it would be sufficiently heated to add substantially to the internal gas pressure. Just as with the solar oscillations, with the black hole picture the temperature at the solar core would not have to be as high as had previously been thought, and thus the predicted neutrino rate would be diminished.

These explanations may sound desperate, but the situation is not yet hopeless. Research is continuing to try to produce theoretical explanations of Davis's neutrino results. In the meantime, Davis continues to run his apparatus.

Though some of the more recent runs gave values somewhat higher than average, Davis views these as mere fluctuations of a statistical nature. Only time will tell.

Davis would like to extend his search by using a material even more sensitive to neutrinos than is chlorine. Gallium is sensitive to neutrinos of lower energy than those with which chlorine interacts. Davis calculates that he would need 50 tons of gallium, which is roughly equivalent to the entire world's production for a whole year. Nonetheless, world gallium production is increasing since this element is used in many solid state devices including the light-emitting diodes (LED's) that form the digits on most pocket calculators and digital watches. A supply might be available for purchase as the new factories are built. The gallium, of course, would not be damaged in the course of the experiment and could be used for other purposes later on. Gallium turns into germanium from an interaction with a neutrino, and Davis is now working on chemical means to separate the few germanium atoms from the gallium, which is liquid at room temperature. Another experiment, using lithium, is also being developed.

The neutrino experiment is one of the most interesting to be carried out in astronomy in recent years, and seems to be giving the most profound and unexpected results. The least that we can conclude is that until the matter is settled, we must treat all the theoretical predictions about stellar interiors with a bit of caution.

6.7 DYING STARS

We have been considering what we know and how much we may not really know about stars as they spend their middle age on the main sequence as dwarfs. After this phase of life, we have seen that the stars evolve upward on the H-R diagram. Equivalently, they become brighter. Also, their surfaces become cooler and they move to the right on the H-R diagram. On the main sequence, most stars behave similarly, regardless of their mass. Of course, their lifetimes are very different, and so are their temperatures. But once they evolve off the main sequence, mass differences play an even more important role.

We shall devote the next three chapters to the various end stages of stellar evolution. First we shall discuss the less massive stars, such as the sun. In Chapter 7, we shall see how such stars swell in size to become giants, possibly become planetary nebulae, and then end their lives as white dwarfs.

More massive stars come to more explosive ends. In Chapter 8 we shall see how some stars are blown to smithereens, and how strange objects called "pulsars" may be the remnants.

In Chapter 9, we shall discuss the strangest kind of stellar death of all. The most massive stars may become "black holes," and effectively disappear from view.

SUMMARY AND OUTLINE OF CHAPTER 6

Aims: To study the formation of stars, their main-sequence lifetimes, and the mechanisms of stellar energy generation

Protostars (Section 6.1)
- Their evolutionary tracks on the H-R diagram
- They may form in dark clouds
- Infrared observations: may be observing dust shells around protostars
- Radio observations: molecules are detected in the same regions

T Tauri stars (Section 6.1)
- Irregular variations in magnitude
- May not have reached the main sequence
- Advance of 3 objects toward the main sequence?

Stellar energy generation (Sections 6.2, 6.3, and 6.4)
- Nuclear fusion

Definition of nuclear particles, isotopes
The proton-proton chain: dominates in cooler stars, including the sun
The carbon cycle: for hotter stars
The triple-alpha process: takes place after helium has been formed and at high temperatures
Stellar nucleosynthesis
The main sequence: balance of pressure and gravity (Section 6.5)
A test of the theory: the neutrino experiment (Section 6.6)
The mechanism
The results: what SNU?
The possible explanations
The consequences of our faith in the theory
Mass as the determining factor in evolution (Section 6.7)

QUESTIONS

1. Since individual stars can live for billions of years, how can observations taken at the current time tell us about stellar evolution?

2. What is the source of energy in a protostar? At what point does a protostar become a star?

3. What is the *evolutionary track* of a star?

4. Arrange the following in order of development: OB associations; T Tauri stars; dark clouds; pulsars.

5. What is the net charge (in terms of proton charges) of Ca I, Mg II, Fe II, Fe XII. (Consult Appendix 4 for a list of the elements. The number of protons an element has is called its atomic number, and is tabulated there.)

6. (a) If you remove one neutron from helium, the remainder is what element?
   (b) Now remove one proton. What is left?
   (c) Why is He IV not observed?

7. (a) Explain why nuclear fusion takes place only in the centers of stars rather than on their surfaces as well.
   (b) What is the major fusion process that takes place in the sun?

8. If you didn't know about nuclear energy, what possible energy source would you suggest for stars? What is wrong with these alternative explanations?

9. (a) If all the hydrogen in the sun were converted to helium, what fraction of the solar mass would be lost?
   (b) How many times the mass of the earth would that be?

10. (a) How does the temperature in a stellar core determine which nuclear reactions will take place?
    (b) Why do more massive stars have shorter main-sequence lifetimes?

11. Why is the gravitational attraction between you and the earth stronger than the electrical attraction?

12. In what form is energy carried away in the proton-proton chain?

13. In the proton-proton chain, the products of some reactions serve as input for the next and therefore don't show up in the final result. Identify these intermediate products. What are the **net** input and output of the proton-proton chain?

14. What do you think would happen if nuclear reactions in the sun stopped? How long would it be before we noticed?

15. Why do neutrinos give us different information about the sun than does light?

16. Why are the results of the solar neutrino experiment so important?

7

The Death of Stars Like the Sun

The sun is just an average star, in that it falls in the middle of the main sequence. But the majority of stars have less mass than the sun. So when we discuss the end of the main-sequence lifetimes of such low-mass dwarfs, we are discussing the future of most of the stars in the universe. In this chapter we will discuss the late stages of evolution of all stars that, when they are on the main sequence, contain up to about 4 solar masses. We shall call these *lightweight stars.* Let us consider a star like the sun, in particular, remembering that all stars up to 4 solar masses go through similar stages but at different rates.

7.1 RED GIANTS

During the main-sequence phase of lightweight stars, as with all stars, hydrogen in the core gradually fuses into helium. By about 10^{10} years after a one-solar-mass star first reaches the main sequence, no hydrogen is left in its core, which is thus composed almost entirely of helium. (Some traces of heavier elements are present, at least for Population I stars, as they were present in the matter from which the star condensed.) Though there is no longer any hydrogen in the core, hydrogen is still undergoing fusion in a shell around the core.

Since no fusion is taking place in the core, there is no ongoing nuclear process to replace the heat that flows out of this hot central region. The core no longer has enough pressure to hold up both itself and the overlying layers against the force of gravity. As a result, the core then begins to contract under the force of gravity (we say it contracts *gravitationally*). This gravitational contraction not only replaces the heat lost by the core, but also in fact heats the core up further. (Half the energy from gravitational contraction always goes into kinetic motion in the interior. An example of such increased kinetic motion is the rise in temperature that we call heat. This is an important theorem that we met before in Section 6.1 in discussing the

---

The Bubble Nebula, NGC 7635, often described as a planetary nebula (although some recent work indicates that it might receive its energy from outside rather than from inside, unlike ordinary planetary nebulae).

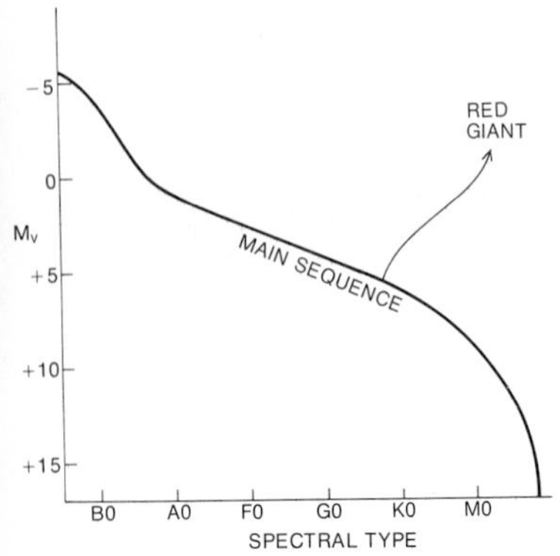

Figure 7–1 An H-R diagram showing the evolution of a 1 solar mass star toward its red giant stage.

contraction of protostars.) Thus, paradoxically, soon after the hydrogen burning in the core stops, the core becomes hotter than it was before.

As the core becomes hotter, the hydrogen-burning shell around the core becomes hotter too, and the reactions proceed at a higher rate. Part of the increasing energy production goes into expanding the outer parts of the star. In fact, the surface temperature of the star decreases. Thus, as soon as a main-sequence star forms a substantial core of helium, the outer layers grow slowly larger and redder. The star winds up giving off more energy (because the nuclear reaction rates are increasing). The total luminosity, which is the emission per unit area times the total area, is increased. The luminosity increases even though each bit of the surface is less luminous than before (as a result of its decreasing temperature) because the total surface area that is emitting increases rapidly.

The solar convection zone is discussed in Section 5.2a. Convection is the phenomenon in which hotter elements are buoyant, thus carrying their excess energy upward as they rise.

The process proceeds at an ever accelerating pace, simply because the hotter the core gets, the faster heat flows from it and the more rapidly it contracts and heats up further. The hydrogen shell gives off more and more energy. The layers outside this hydrogen-burning shell continue to expand until a zone of convection develops near their outer boundary. Through convection, energy is carried away from the core rapidly enough to limit the rate at which the core grows hotter.

The sun is presently still at the earlier stage where changes are stately and slow. But in a few billion years, when the hydrogen in its core is exhausted, the time will come for the sun to brighten and redden faster and faster until it eventually swells and engulfs Mercury and Venus. At this point, the sun will no longer be a dwarf but will rather be what is called a *red giant* (Fig. 7–1). Its great luminosity will sear and char everything on earth. When the sun is sufficiently bloated to be nearly the size of the earth's orbit (Fig. 7–2), we won't be around to admire the show.

The gravity of a spherical mass acts, to a body outside the spatial limits taken up by that mass, as though all the mass were concentrated at its center. A red giant is so bloated that its outer layers are relatively far from its center, so the force of gravity at the surface is not very great. Thus some matter can escape from the star. In particular, carbon can escape and solidify, forming grains of dust whose importance we shall study in Chapters 20 and 21. A significant fraction of a star's mass is normally lost in this red giant stage.

A red giant has a schizophrenic nature: it has a core that is very dense and hot, and outer layers that are just the opposite. The next stages of stellar evolution take place so rapidly that it is hard for even large computers to follow the development. Still, astronomers have been able to calculate the general outline of the subsequent stages.

The helium flash *is the rapid onset of the conversion of hydrogen to helium through the triple-alpha process.*

While the outer layers are expanding, the inner layers continue to heat up, and eventually reach 10^8 K. At this point, the triple-alpha process begins for all but the least massive lightweight stars, as groups of three helium nuclei fuse into single carbon nuclei. (This triple-alpha process was described in Section 6.4.) For stars about the mass of the sun, the onset of the triple-alpha process happens rapidly. The development of this *helium flash* produces a very large amount of energy in the core, but only for a few years. The consequences of this input of energy at the center are to reverse the evolution that took place prior to the helium flash: now the core expands and the outer portion of the star contracts. As the star adjusts to this situation it becomes smaller and less luminous, moving back down and to the left on the H-R diagram. When the effect of the helium flash is stabilized, the star is burning helium in the core and hydrogen in the shell. At this stage, the star may be unstable—we observe such stars as variables (see Fig. 4–13).

7.2 PLANETARY NEBULAE

Sometime after the helium burning sets in, a substantial carbon core forms. This carbon core contracts and heats up as did the helium core before it, driving up the helium reaction rates in the shell surrounding the core. The star then moves up and to the right on the H-R diagram, becoming a luminous red giant once again. This time, however, the swelling and cooling continue on and on until, after a time, the outer layers grow sufficiently cool that the nuclei and electrons combine and form neutral atoms. As electrons that are free have more energy than electrons that are bound up as parts of atoms, the system as a whole experiences an increase in energy when the previously free electrons recombine with the nuclei. (Astronomers call this process *recombination.*)

The excess energy is given off as photons, which are absorbed by the gas and blow the outer layers outward. But as the outer layers expand, they in turn become cooler, so that still more nuclei and electrons recombine and still more energy is released. In this way a vicious circle quickly sets in, which is broken in many cases when the outer layers of the star are blasted off. By the time the blown-off layers have moved a few astronomical units away from the star, they have spread into a shell thin and cool enough to be transparent. (1 astronomical unit is the average earth-sun distance.)

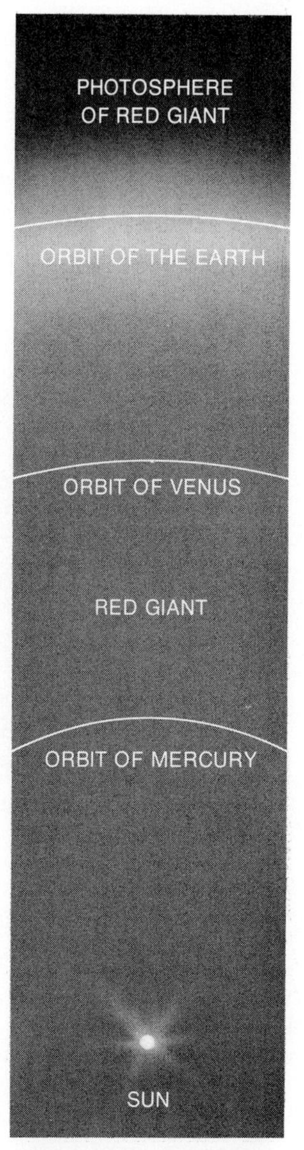

Figure 7–2 A red giant swells so much it can be the diameter of the earth's orbit.

We know of a thousand such objects in our galaxy that can be explained by the above theoretical picture, and there may be 10,000 more. They were named *planetary nebulae* because hundreds of years ago, when they were discovered, they appeared similar to the planets Uranus and Neptune when viewed in a small telescope. Both planets and "planetary nebulae" appeared as small, greenish disks. We have since discovered what planetary nebulae really are; they have nothing at all to do with planets, but they retain their old name for historical reasons. Their greenish color is caused by the presence of certain strong emission lines of multiply ionized oxygen (that is, oxygen that has lost more than one electron) and other elements. These lines happen to fall in the green region of the visible spectrum.

Planetary nebulae are exceedingly beautiful objects (Figs. 7–3, 7–4, and 7–5, and Color Plates 6, 7, and 8). These nebulae are actually semitransparent spheres or cylinders of gas. When we look at their edges, we are looking obliquely through the shells of gas (Fig. 7–6), and there is enough gas along our line of sight to be visible. But when we look through the center of the spherical shell, there is less gas along our line of sight, and it appears transparent.

Figure 7–3 NGC 2392, a planetary nebula in the constellation Gemini, the Twins. This object is sometimes known as the Eskimo Nebula.

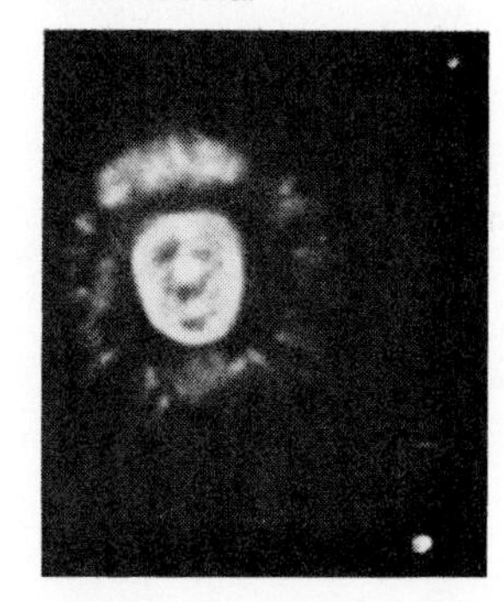

In the middle of many planetary nebulae, we can see the central star that blew off the gas. Of course, we are really looking at what we formerly considered to be the core of the star. The core is much hotter than the outside of the red giant of which it was once a part, and the central stars of planetary nebulae form a class of stars that are located on the left half of the H-R diagram (Fig. 7–7). They have temperatures over 20,000 K, and sometimes even up to 300,000 K, hotter than any main-sequence star.

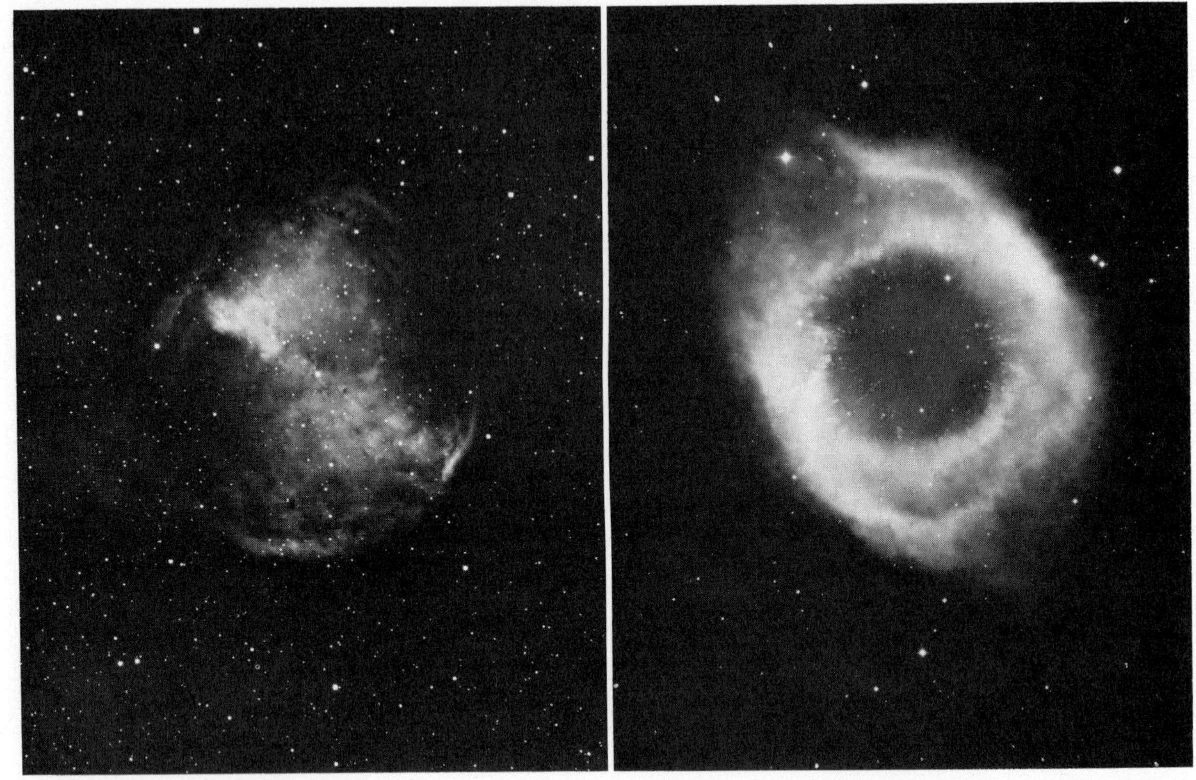

Figure 7–4

Figure 7–5

Figure 7–4 The Dumbbell Nebula, M27, a planetary nebula in the constellation Vulpecula, the Fox.

Figure 7–5 NGC 7293, a planetary nebula in the constellation Aquarius, the Water Bearer.

Astronomers are particularly interested in planetary nebulae because they want to study the various means by which stars can eject mass into interstellar space. This tells them both how a star can reduce its own mass and also about the origin of some of the interstellar matter. Each planetary nebula represents the ejection of 10 or 20 per cent of a solar mass, which is only a small fraction of the mass of the star.

We can measure the ages of planetary nebulae by tracing back the shells at their current rate of expansion and calculating when they would have been ejected from the star. Most of the planetary nebulae we can observe appear to be less than 50,000 years old. This is consistent with our picture of planetary nebulae as a transient phase in the lifetimes of stars of about one solar mass; after about 100,000 years the nebulae will have expanded so much that the gas will be invisible.

In order to calculate the ages of the planetary nebulae by this method, we must know their distances from us. We can determine the distance directly for only a few planetary nebulae: one is close enough to allow its trigonometric parallax to be measured, another is in a binary system with a star whose distance can be found by the method of spectroscopic parallax (see Section 2.14), and a third is included in a globular cluster whose distance is known. For a few planetary nebulae, we can measure the angular expansion directly on a series of photographs taken years or decades apart. If we assume that the expansion is spherically symmetric, that is, that the planetary nebula is expanding at the same rate from front to back as it is from side to side (Fig. 7–8), then we can calculate how distant it is. The front is approaching us and the back is receding from us relative to the average velocity of the nebula. We merely calculate the distance necessary to make the angular expansion from side to side correspond to the rate in kilometers/second we measure from the Doppler shift of the front relative to the back. This discussion indicates the lengths to which astronomers have to go in order to find the distances to even relatively nearby astronomical objects.

Figure 7–6 A hollow shell, when viewed from outside, can give the appearance of a ring, since when looking at the center we see perpendicularly through nearly-transparent gas. Only near the edges of the shell, where we are looking through the gas at a very oblique angle, does the gas become readily visible (that is, does its optical depth become significantly large).

A study of infrared objects carried out in 1975 turned up a strange infrared emitting object that may be a planetary nebula in the process of formation. The object is called CRL-618 (it is number 618 in a catalogue of infrared sources compiled by the Air Force **C**ambridge **R**esearch **L**aboratory in Massachusetts from the results of observations made from rockets).

CRL–618 is an object at a temperature of 32,000 degrees, which corresponds to the temperature of the central star of a planetary nebula. If it is indeed a planetary nebula the hot source would be surrounded by a dense cloud of gas and dust that emits strongly in the infrared. Presumably the dust would be at a temperature of a few hundred degrees, which would account for the observations. Most other known planetary nebulae are in more advanced stages of evolution, and this is a rare opportunity to study a new member of the group.

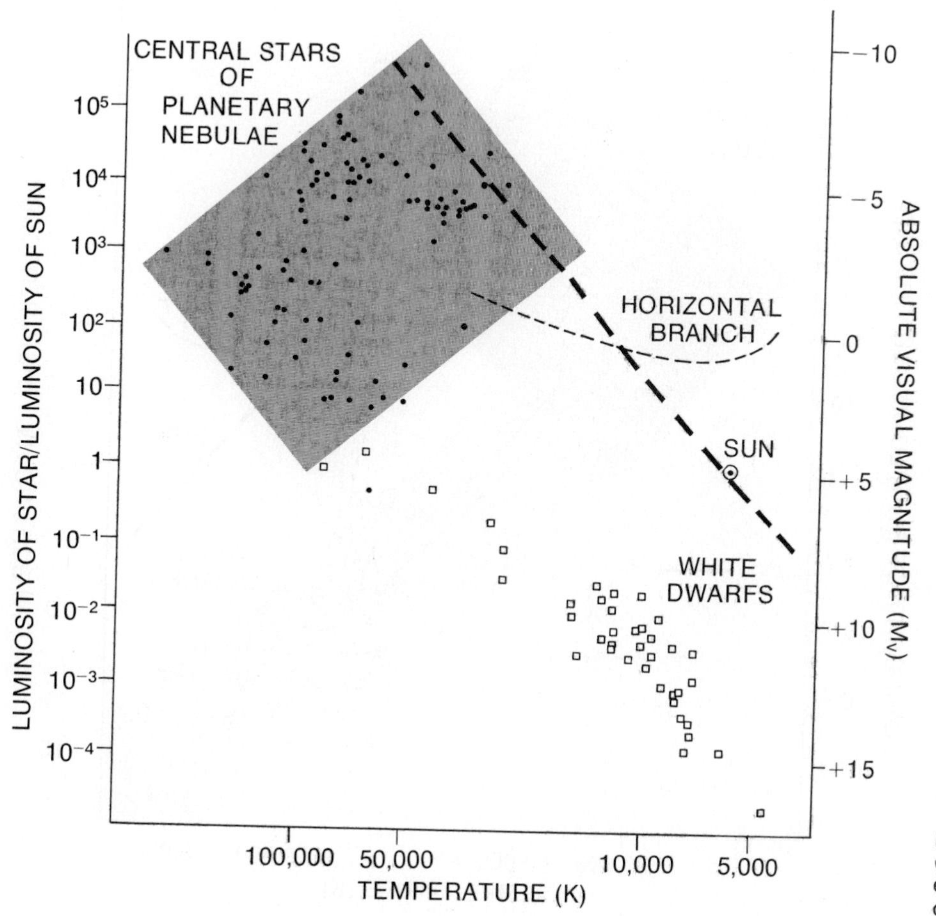

Figure 7–7 The positions of the central stars of planetary nebulae on the H-R diagram.

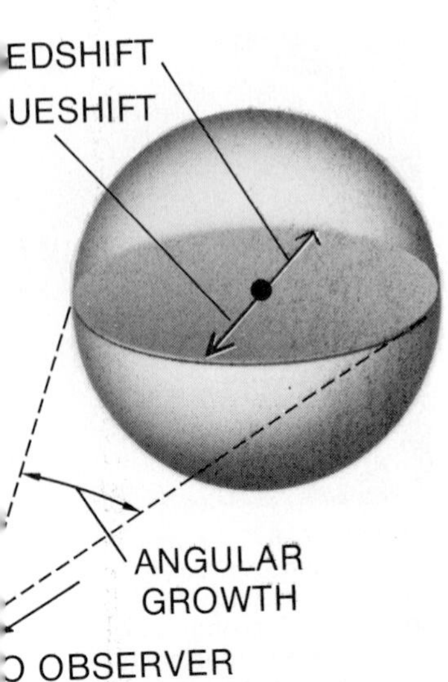

Figure 7–8 We can measure the expansion of a planetary nebula along the direction radial to us from the Doppler effect, and can sometimes measure the angular expansion of the shell of gas from observations taken decades apart. If we assume that the gas shell is expanding at the same rate in all directions, we can use the above two pieces of information to derive the distance to the planetary nebula from the earth.

7.3 WHITE DWARFS

When a star of approximately one solar mass (and other low-mass stars) has burned all the hydrogen in its core into helium and then all the helium in its core into carbon, it does not heat up sufficiently to allow the carbon to fuse into still heavier elements. Accordingly, there comes a time when nuclear reactions are no longer generating energy to maintain or increase the internal pressure that balances the force of gravity.

All stars that at this stage in their lives contain less than 1.4 solar masses have the same fate. Many or all stars in this low-mass group come from the red giant phase, and may pass through the phase of being central stars of planetary nebulae. Those that do not go through a planetary nebula stage lose their mass in some other way, perhaps during the giant stage. In any case, they lose large fractions of their original masses. Stars that contained one solar mass when they were on the main sequence now contain 0.5 solar mass. Stars that contained 4 solar masses when they were on the main sequence may now contain only 1.4 solar masses.

When their nuclear fires die out for good, the stars with less than 1.4 solar masses remaining shrink in size, and reach a stable condition that we shall describe below. As they shrink they grow very faint (in the opposite manner to that in which red giants grew brighter as they grew bigger). Because the Planck curve of radiation from many of the stars peaks to the

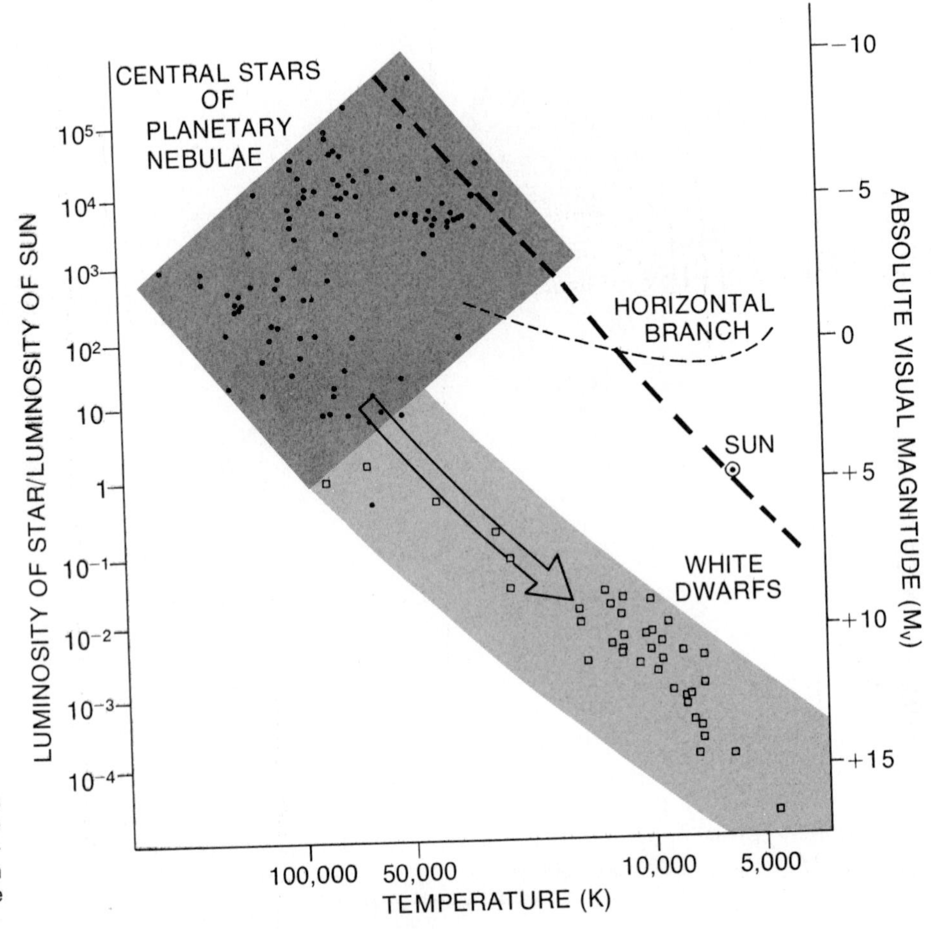

Figure 7–9 Central stars of planetary nebulae evolve into white dwarfs. The positions of white dwarfs on the H-R diagram are shown.

left of the center of the visible part of the spectrum, most appear whitish in color. (Since the intensity falls off slowly on the long wavelength side of the Planck curve, even those stars whose radiation peaks in the uv give off amounts of radiation that vary only slightly across the visible, and so also appear whitish.) Some stars of this type are cooler, though, and so appear yellow or even red. Whatever their actual color, all these stars are called *white dwarfs*. The white dwarfs occupy a region of the Hertzsprung-Russell diagram that is below and to the left of the main sequence (Fig. 7–9).

Do not confuse the term "white dwarf" with the term "dwarf." The former refers to the dead hulks of stars in the lower left of the H-R diagram, while the latter refers to normal stars on the main sequence.

White dwarfs represent a stable phase in which stars of less than 1.4 solar masses live out their old age. Obviously, something must be holding up the material in the white dwarfs against the force of gravity; nuclear reactions generating thermal pressure no longer take place in their interiors. The property that holds up the white dwarfs is a condition called *electron degeneracy;* we thus speak of "degenerate white dwarfs."

Electron degeneracy is a condition that arises in accordance with certain laws of quantum mechanics, and is not something that is intuitively obvious. As the star contracts and the electrons get closer together, there is a continued increase in their resistance to being pushed even closer. This manifests itself as a pressure. There comes a time when the pressure generated in this way exceeds the normal thermal pressure. This only happens at very great densities. When this pressure from the degenerate electrons is sufficiently great, it balances the force of gravity and the star stops contracting.

Figure 7–10 The sizes of the white dwarfs are not very different from that of the earth. A white dwarf contains about 300,000 times more mass than does the earth, however.

Electron degeneracy arises because of a principle related to quantum mechanics: the Pauli Exclusion Principle. This principle, advanced forty years ago by Wolfgang Pauli, says that electrons can be only in specific states of existence, and moreover that no two electrons can be in exactly the same state.

The Pauli Exclusion Principle limits the number of electrons that can be in the lowest energy levels. Thus some electrons must be in higher energy levels, even though the temperature is not sufficiently high that they would be excited normally to those levels. Because of the momentum that particles have, the particles exert a certain pressure. As the star contracts, electrons are forced into higher and higher energy levels by the Pauli Exclusion Principle and so the pressure increases.

A fuller, more complete description of this phenomenon is beyond the scope of this book. In any case, the star's inability to contract further because of electron degeneracy is as real and definitive to the star as our inability to pass through a brick wall is to us.

The effect of the degenerate electron pressure is to stop the white dwarf from contracting at a point when the gas is in a very compressed state. In a white dwarf, a mass approximately that of the sun is compressed into a volume only the size of the earth (Fig. 7–10). A single teaspoonful of a white dwarf weighs 5 tons; it would collapse a table if you somehow had some in your hand and tried to put it down. A white dwarf contains matter so dense that it is in a truly incredible state.

What will happen to the white dwarfs with the passage of time? Because of their electron degeneracy, they can never contract further. Still, they have some heat stored in the nuclear particles present, and that heat will be radiated away over the next billions of years. Then the star will be a burned-out hulk called a *black dwarf*, though it is probable that no white dwarfs have yet lived long enough to reach that final stage. It will be billions of years before the sun becomes a white dwarf and then many billions more before it reaches the black dwarf stage.

Some black dwarfs come from featherweight stars (Section 6.1), stars that were not massive enough to begin hydrogen burning; others are cooled white dwarfs.

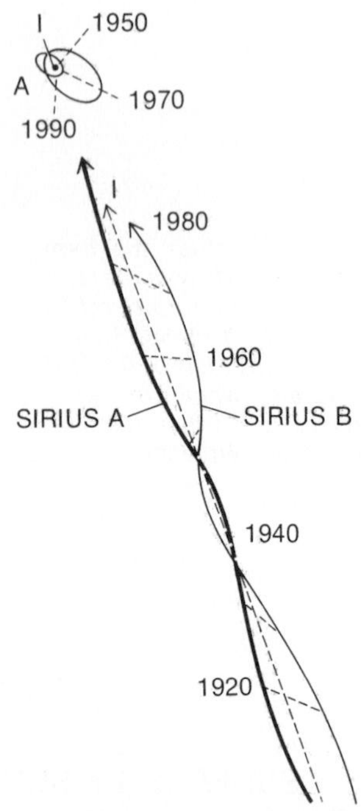

Figure 7–11 The masses of Sirius A and Sirius B can be derived by studying their paths across the sky over a period of many years. Sirius A, which is easily seen compared to its faint companion, appears to wobble in its proper motion. This reveals the presence of the companion without needing to observe it directly.

Figure 7–12 Sirius B can be seen as the faint dot alongside the much brighter Sirius A image in this photograph from the Lick Observatory. The spikes of light seemingly radiating from Sirius A are an optical effect known as diffraction. Judicious arrangement of the position of the supports can cause much of the light of Sirius A to go in non-obstructing directions.

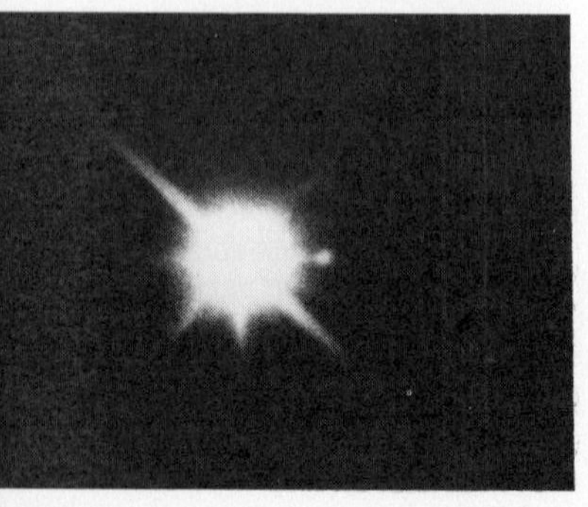

7.4 OBSERVING WHITE DWARFS

White dwarfs, as we can see from their positions on the H-R diagram, are very faint and thus should be correspondingly difficult to detect. This is indeed the case.

White dwarfs are discovered by looking for blue stars with high proper motions or by their gravitational effect on their companion stars if they are in binary systems.

In the former case, by studying proper motions we find the stars that are close to the sun. The ones that are fainter than main-sequence stars would be at their distances must be white dwarfs. To understand the latter case, we must realize that for any system of masses orbiting each other, we can define an imaginary point called the *center of mass* around which each individual mass appears to orbit. This center of mass, in its proper motion, moves in a straight line, though the path in the sky of any of the individual bodies appears wavy (Fig. 7–11).

A basic physical rule tells us that a body travels in a straight line unless affected by some outside force. This important law was put forward by Isaac Newton in his *Principia* in 1687, and is called his first law of motion. For a gravitating system of two bodies, the two bodies affect each other, but the center of mass follows a line just as though all the total mass of the two bodies were present at that point.

Thus when we look at the proper motion of a star in the sky, we can tell whether it is in a single or in a binary system. If it is a single star, it will travel at a constant rate in a straight line. If it is in a binary system, the star will appear to wobble from side to side as shown in Fig. 4–6 (unless its wobble around the straight line happens to be oriented toward and away from us instead of side to side). In Section 4.1, we defined this system as an astrometric binary. One can find the straight line followed by the center of mass of the system by drawing it down the center of the wobble. From this wobble one can infer that there is another body in the system, even though it may be invisible. Further, one can deduce limits on the mass of that other body.

In this manner, it was suggested in 1844 that Sirius, the brightest star in the sky, had a hitherto invisible companion. In 1862, astronomers were able to see the faint companion for the first time. The companion, called Sirius B, is magnitude 8.7, while neighboring Sirius A's magnitude is –1.4 (Fig. 7–12). But it was not for another 50 years that astronomers realized that the companion was not a normal red main-sequence dwarf. When it was observed in 1915 to be white, and therefore hot, the astronomers realized that it must be much smaller than had been assumed.

At least three of the 40 stars within 5 parsecs of the sun—Sirius, 40 Eridani, and Procyon—have white dwarf companions. Another nearby object, known as van Maanen's star, is a white dwarf, although not in a multiple system. So even though we are not able to detect white dwarfs at great distances from the sun, there seems to be a great number of them. Hundreds of white dwarfs have been discovered.

Even though the theory that explains white dwarfs has seemed consistent with actual observations, their size had never before been measured directly until 1975 when a set of observations was made in the ultraviolet from the Copernicus satellite. The ultraviolet part of the spectrum of Sirius B could be distinguished from that of its much brighter neighbor, and was found to peak at 1100 Å. This corresponds to a temperature of 30,000 K.

Since the absolute magnitude of Sirius B is known, its surface area and thus its radius can be found. The radius of Sirius B is indeed only 4200 km, smaller than the earth's. Even if we allow for an uncertainty of 1000 or 2000 km, that does not change the import of the result that this is direct confirmation of the picture of white dwarfs that we drew in this chapter.

As the companion to the Dog Star, Sirius B is sometimes called "the Pup."

7.5 WHITE DWARFS AND THE THEORY OF RELATIVITY

One special reason to study white dwarfs is to make use of them as a laboratory to test extreme physical conditions. It is impossible to create such strong gravity in a laboratory on earth. We have already discussed two tests of general relativity in Section 5.11: the deflection of starlight by the sun, and the advance of the perihelion of the planet Mercury. A third test is the *gravitational redshift* of light.

A gravitational redshift *is a redshift caused by gravity, as predicted by the general theory of relativity. Einstein's theory predicts that light leaving a mass would be slightly redshifted. The greater the amount of gravity at the place where the light is emitted, the greater the redshift.*

The sun is bright enough that we can spread out its spectrum and measure the wavelengths of spectral lines from the photosphere with tremendous accuracy. The redshift that Einstein predicted has been successfully detected, though the effect is minuscule. Furthermore, turbulence in the solar photosphere, such as the rise and fall of granules, distorts and confuses the results. One would like to find a body with stronger gravity than the sun, and white dwarfs are just such objects.

Unfortunately, one must know the mass and size of the white dwarf to perform the test accurately, and the masses and sizes of white dwarfs are not known very well. The observational part of the test is not difficult, because strong redshifts are found. The results agree with the predictions of the theory of relativity to within the possible error that results from our uncertain knowledge of the masses and sizes of white dwarfs. The test has best been carried out with 40 Eridani B.

In the 1960's, the sensitivity of the measuring process was improved through the application of an effect that had been discovered a few years earlier by Rudolph Mössbauer. Scientists were able to use the Mössbauer effect to measure the redshift caused by the earth's own gravity. The gravitational redshift was measured for radiation as it traveled from the basement to the top floor of a physics building. This earthbound experiment using the Mössbauer effect is now our most sensitive test of this prediction of relativity theory.

A related prediction of general relativity is that "clocks" run slower in a stronger gravity field than they do in a weaker one. Atoms can act as such clocks, because a spectral line emitted by an atom can be characterized by its frequency, and frequency is measured in cycles/**second** (now called hertz). In 1976, a group from the University of Maryland succeeded in measuring this effect of relativity by showing that atomic clocks run the expected amount (47 billionths of a second) faster than they do at sea level when carried aloft to a height of 10 km. At this altitude, the earth's gravity is slightly weaker than it is below.

Again we see that both astronomical and terrestrial tests back the validity of Einstein's theory.

7.6 NOVAE

Although the stars are generally thought to be unchanging on a human time scale, occasionally a "new star," a *nova* (pl: *novae*), becomes visible.

Figure 7–13 Nova Herculis 1934, showing its rapid fading from 3rd magnitude on March 10, 1935 to below 12th magnitude on May 6, 1935.

Such occurrences have been noted for thousands of years; ancient Oriental chronicles report many such events.

A nova is a newly visible star rather than a really new star. (It is certainly not new in the manner of the new stars detected in the infrared, as discussed in Section 6.2.) A nova, rather, represents a brightening of a star by 5 to 15 magnitudes or more, which is equivalent to a brightening by hundreds or millions of times. Often, while brightening by this large factor, a nova passes from the realm of objects that cannot be seen with the naked eye to that of objects that can be seen with the naked eye, or from objects that are too faint to be seen even with the largest telescopes to objects that can be observed.

A nova (Fig. 7–13) may brighten within a few days or weeks. It ordinarily fades drastically within months, and then continues to fade gradually over the years. Several "recurrent" novae that appear to brighten at intervals of years or decades are known.

The spectra of novae show absorption lines that are Doppler shifted to the blue. This implies that gas between the star and earth is moving toward the earth. From this we deduce that gas is being thrown off the star. Soon the photospheric material that has been ejected thins out enough to become transparent, and we begin to see emission lines from the gas that has ex-

A

B

Figure 7–14 (*A*) In 1951, a shell of gas could be seen to surround Nova Herculis 1934. (*B*) The appearance in 1949 of the shell of gas surrounding Nova Persei 1901.

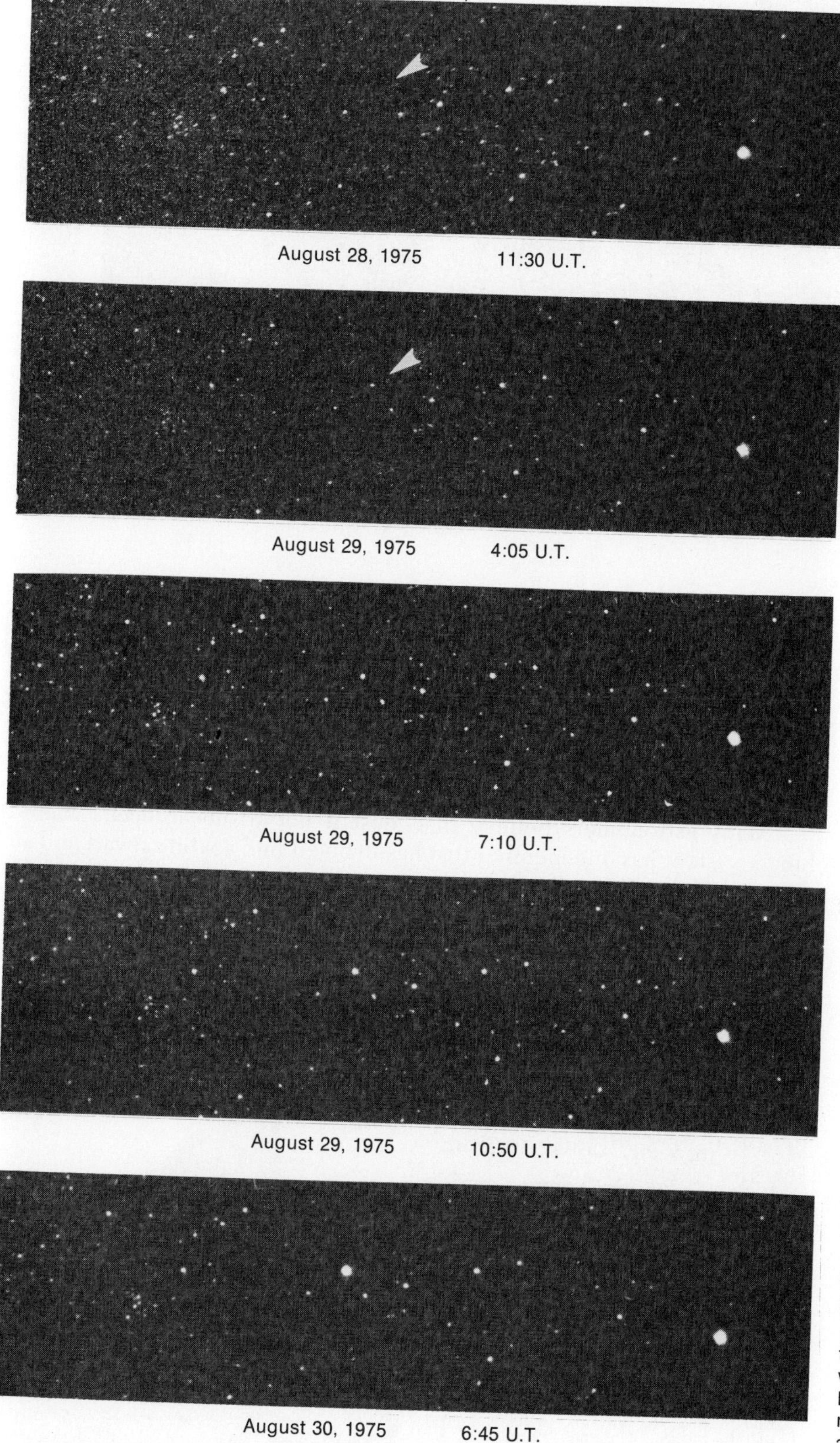

Figure 7–15 This unique series of observations covers the eruption of Nova Cygni 1975 during the period of its brightening. A Los Angeles amateur astronomer, Ben Mayer, was repeatedly photographing this area of the sky at the crucial times to search for meteors. Never before had a nova's brightening been so well observed. On August 28 (*top*), no star was visible at the arrow. The next night, the nova became visible and brightened steadily. By August 30 (*bottom*), the nova had reached 2nd magnitude, the brightness of Deneb, which is seen at the right. Each exposure was 25-min long on Tri-X film with an f/3.5 lens.

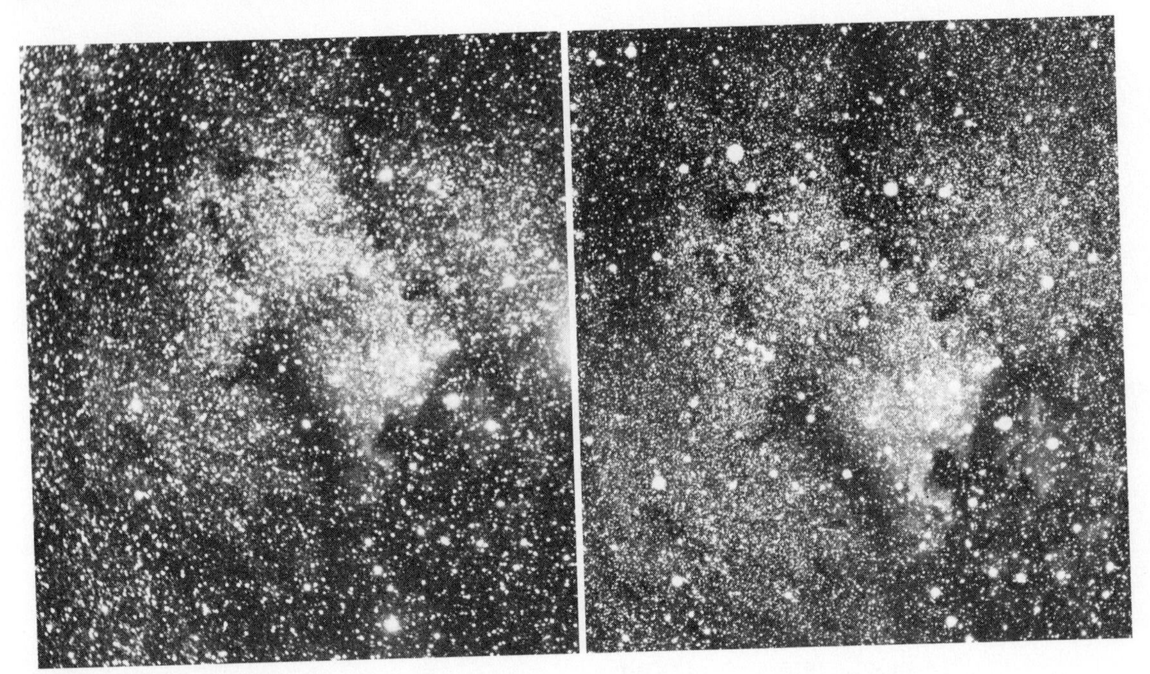

Figure 7–16 At its brightest, Nova Cygni 1975 (*top left*) was as bright as Deneb (*right of center*), and appeared as a strange new member of the Northern Cross high overhead for American observers. Long exposures, such as these taken with telescopes of the University of Colorado at Boulder, also show the North America Nebula.

panded in all directions from the star. Only 10^{-4} or so of a solar mass is thrown off in a nova outburst, so there is no reason that recurrent novae can't repeat their outbursts many times. Months or years after the outburst of light, the shell of gas sometimes becomes detectable through optical telescopes (Fig. 7–14).

Many astronomers believe that most if not all novae occur when a binary system has one member that has evolved into a white dwarf and an-

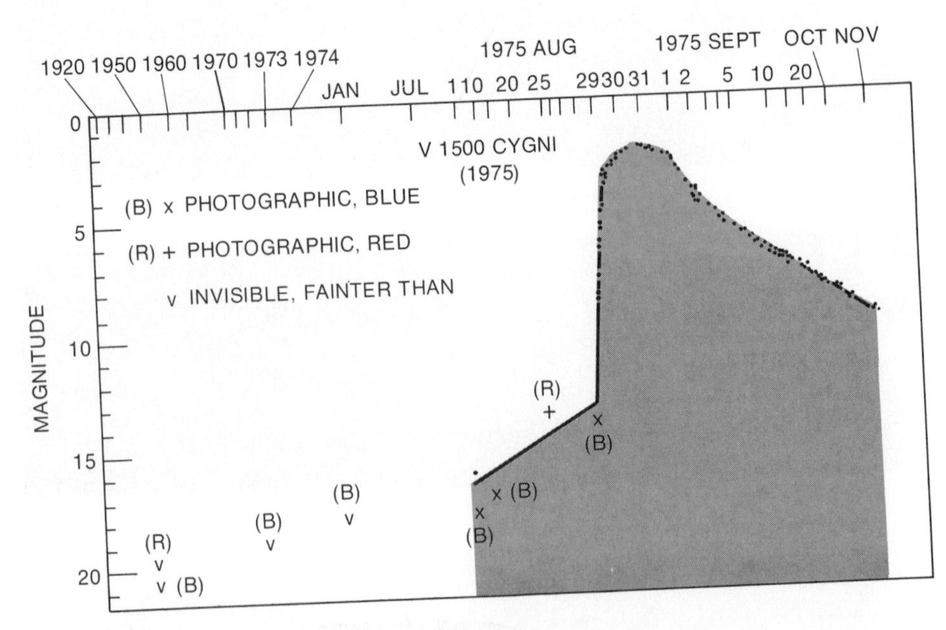

Figure 7–17 The light curve of Nova Cygni 1975.

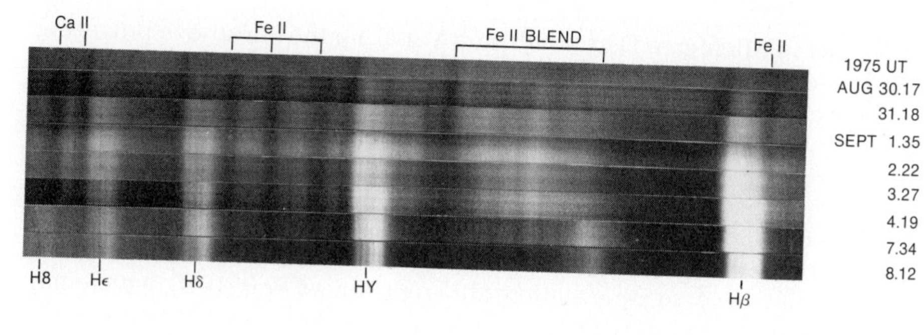

Figure 7–18 A series of spectra of Nova Cygni 1975, showing the violet at the left to the blue-green at the right. After September 1st, bright hydrogen emission lines became visible. Times are expressed in decimal parts of a day.

other member that is en route to becoming a red giant. Though the details are not certain, the nova might occur through the following processes. We have mentioned that the outer layers of a red giant are not held very strongly by the star's gravity. If a white dwarf is nearby, some of the matter originally from the red giant can surround one or both of the stars. Some may fall on the surface of the white dwarf and perhaps trigger the proton-proton chain on its surface for a brief time. The energy thus produced blows off the material. Obviously, this mechanism could recur every few years as more new matter falls on the white dwarf.

The brightest nova in thirty years appeared in the constellation Cygnus in August 1975. Within two days it brightened by perhaps as many as 15 magnitudes, and became magnitude 1.8, brighter than the star Deneb (Figs. 7–15 and 7–16) and one of the 3 dozen brightest stars in the sky. Nova Cygni 1975 underwent an unusually rapid rise in brightness and then soon became fainter—it fell below naked-eye visibility within a week (Fig. 7–17). The total range of brightening was also especially large for a nova. Yet it had a spectrum that varied over time as do the spectra of other novae, and showed the same spectral lines as do other novae, so it is accepted as an extreme example of the nova phenomenon (rather than a more catastrophic event of a type to be discussed in Section 8.2, called a supernova). Perhaps Nova Cygni 1975 was "going nova" for the first time.

Nova Cygni 1975 is called V 1500 Cygni, the 1500th variable star to be discovered in Cygnus.

Fortunately, there were several independent early discoveries of this nova—it was certainly exciting for all those individuals to discover such a bright nova—so detailed observations were made as it brightened. The intensity and spectral data (Fig. 7–18) on Nova Cygni 1975 are so abundant that their detailed analysis should significantly improve our understanding of novae. It has been suggested, for example, that Nova Cygni might even be a single star whose outburst was triggered by a sudden infall of interstellar material rather than a member of a binary system with an outburst triggered by material from its stellar companion.

In this century, exceptionally bright novae appeared in 1901, 1918, 1925, 1934, 1942, and 1975.

7.7 X-RAY NOVAE

Objects in other parts of the spectrum besides the visible can also brighten suddenly. An already strong x-ray source in the constellation Monoceros (the Unicorn), formerly known only by its catalogue number as A0620–00, brightened in August 1975 until it was more than five times as bright as any other x-ray source. Radio emission was detected from the source, and an 8th magnitude optical star was identified with the x-ray object using a telescope newly moved from an urban site to Kitt Peak by the

University of Michigan, Dartmouth, and M.I.T. for the purpose of providing coordinated visible and x-ray observations.

A search of the library of old photographic plates at the Harvard Observatory showed that the star had been an optical nova before, in 1917. Thus the object, now called Nova Monocerotis 1975, is obviously a recurrent nova. This is the first recurrent nova to erupt since x-ray astronomy developed sufficiently to allow observations to be made in that part of the spectrum. Because of this, we do not know how significant it is that this recurrent nova was an x-ray source, although Nova Cygni 1975, which may have been undergoing its first eruption, did not show any detectable x-ray emission.

SUMMARY AND OUTLINE OF CHAPTER 7

Aims: To understand what happens to stars of up to 4 solar masses when they have finished their time on the main sequence of the H-R diagram

Red giants (Section 7.1)
- They follow the end of hydrogen burning in the core
  - Core contracts gravitationally
  - Core and hydrogen-burning shell become hotter
  - The increased energy expands the outer layers
- The total luminosity is the luminosity of each bit of surface times the total surface area
  - Star is red, so each bit of surface has relatively low luminosity
  - Surface area is very greatly increased, so total luminosity is greatly increased
- Helium flash
  - Rapid onset of triple-alpha process

Planetary nebulae (Section 7.2)
- Their appearance
- Ages: less than 50,000 years old
- The gas is blown off when it absorbs photons
- Central stars: 20,000–300,000 K
- Mass loss: 0.1 or 0.2 solar mass

White dwarfs (Sections 7.3, 7.4, and 7.5)
- End result of lightweight stars
- Less than 1.4 solar masses remaining
- Electron degeneracy
- Detecting white dwarfs via proper motion or by their presence in a binary system
- Use in testing the gravitational redshift
- Einstein's general theory of relativity endorsed

Novae (Sections 7.6 and 7.7)
- Mass loss: only 10^{-4}
- Binary model: interaction of a red giant and a white dwarf
- Possible single-star model for Nova Cygni 1975
- X-ray novae

QUESTIONS

1. What event signals the end of the main-sequence life of a star?
2. When hydrogen burning in the core stops, the core contracts and heats up again. Why doesn't hydrogen burning start again?
3. When the core starts contracting, what eventually halts the collapse?
4. If you are outside a spherical mass, the force of gravity just varies inversely as the square of the distance from the center. What is the ratio of the force of gravity at the surface of the sun to what it will be when the sun has a radius of one astronomical unit?
5. Why is helium "flash" an appropriate name?
6. If you compare a photograph of a nearby planetary nebula taken 80 years ago with one taken now, how would you expect them to differ?
7. Why is a star brighter after it sheds a planetary nebula?
8. What supports a white dwarf?
9. What are the differences between the sun and a one solar mass white dwarf?
10. When the sun becomes a white dwarf, approximately how much mass will it have? Where will the rest of the mass have gone?
11. Which has a higher surface temperature, the sun or a white dwarf?
12. Sketch an H-R diagram indicating the main sequence, and the region of white dwarfs. Now indicate the position of an ordinary dwarf that appears yellowish.
13. Why does the study of the gravitational effects of a system often give a better estimate of the mass of the system than the direct observation of all the components?
14. What evidence do we have that material is being ejected from novae?
15. When the proton-proton chain starts at the center of a star, it continues for billions of years. When it starts at the surface (as in a nova) it only lasts a few weeks. How can you explain the difference?

8

Supernovae, Neutron Stars, and Pulsars

We have seen how the run-of-the-mill lightweight stars end, not with a bang but a whimper. More massive stars, stars that contain more than 4 solar masses when they are on the main sequence, put on a more dazzling display. They cook the heavy elements deep inside themselves and then blow themselves almost to bits, forming still more heavy elements in the process. Then the matter that is left behind settles down to even stranger states of existence than that of a white dwarf. In this chapter we shall discuss the death of stars of 4 to 8 solar masses, which we may call *middle-weight stars.* Let us consider one of the possible versions of the story, though we shall not attempt here to discuss all the different models that have been advanced to explain supernovae. In Chapter 9, we shall discuss the death of *heavyweight stars,* stars of more than 8 solar masses.

The division into featherweight (Section 6.2), lightweight (Chapter 7), middleweight, and heavyweight stars was suggested by Martin Schwarzschild of Princeton. Note that we are really discussing mass rather than weight.

8.1 RED SUPERGIANTS

The stars that are much more massive than the sun whip through their main-sequence lifetimes at a rapid pace. These prodigal stars use up their store of hydrogen very quickly. A star of 15 solar masses may take only 10 million years from the time it first reaches the main sequence until the time when it has exhausted all the hydrogen in its core. This is a lifetime a thousand times shorter than that of the sun. When the star exhausts the hydrogen in its core, the outer layers expand and the star becomes a red giant.

For these massive stars, the core can then gradually heat up to 100 million degrees, and the triple-alpha process begins to transform helium into carbon. However, for massive stars the helium burns steadily after the helium flash, unlike the case for less massive stars.

By the time helium burning is concluded, the outer layers have expanded even further, and the star has become much brighter than even a

Figure 8–1 An H-R diagram showing the evolution of a massive star off the main sequence to that of a red supergiant.

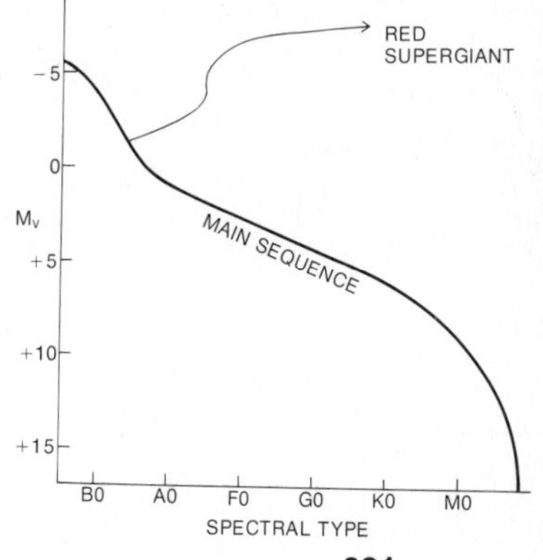

Filaments at the north end of the supernova remnant known as the Cygnus Loop.

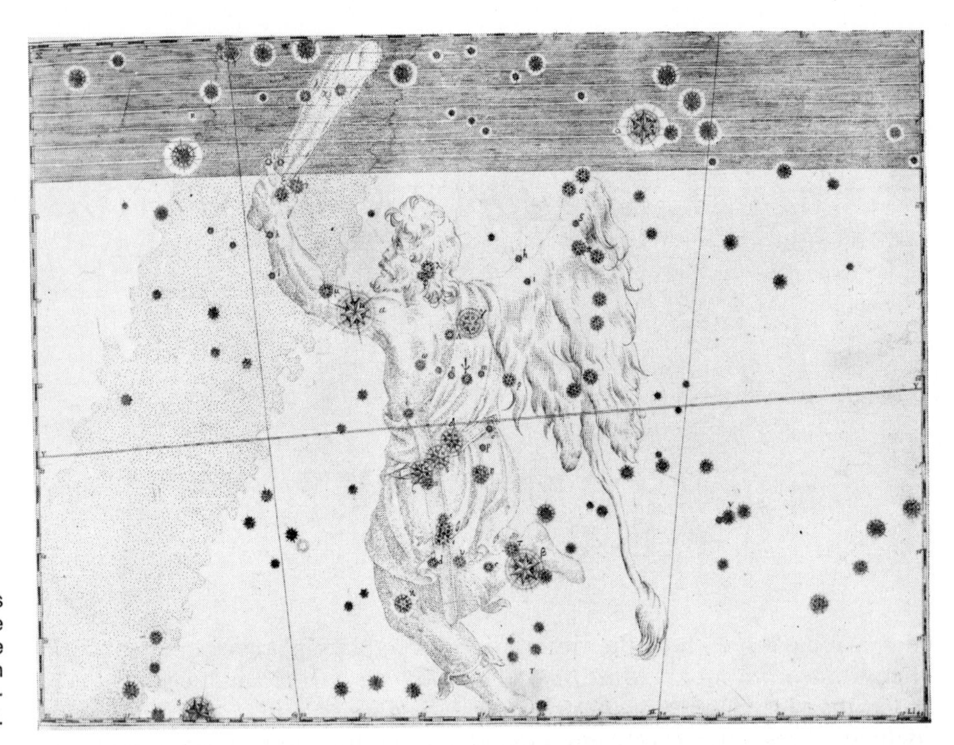

Figure 8–2 Betelgeuse is the star labeled α in the shoulder of Orion, the hunter, shown here in Johann Bayer's *Uranometria,* first published in 1603.

red giant. We call it a *red supergiant* (Fig. 8–1); Betelgeuse, the star that marks the shoulder of Orion, is the best-known example (Fig. 8–2). Supergiants are inherently very luminous stars, with absolute magnitudes of up to −10. The sun, with an absolute magnitude of +5, is only one-millionth as luminous as the most brilliant of the red supergiants. A supergiant's mass is spread out over such a tremendous volume that its average density is less than one millionth that of the sun.

Some of the supergiants are unstable and begin to oscillate. These are the Cepheid variables that we discussed in Section 4.4b. Cepheids fall in a particular region of the H-R diagram, as shown in Figure 4–13. This region of instability is near that of the RR Lyrae variables, which are somewhat less massive than the Cepheids.

Some scientists think that the carbon core can contract, heat up, and begin fusing into still heavier elements. Two C^{12} nuclei can fuse into magnesium, Mg^{24}, for example. Eventually, in some cases, even elements in the iron group (iron, cobalt, and nickel) build up. The core, containing the elements with the highest mass numbers, is surrounded by layers of elements of different mass, with the lightest toward the periphery and the heaviest toward the center.

8.2 SUPERNOVAE

After reaching supergiant status, a very massive star will eventually explode in a glorious burst called a *supernova* (pl: *supernovae*). The conditions in the center of the star change so quickly at this stage that it becomes very difficult to model them satisfactorily in sets of equations or on computers. Many different models have been advanced to describe what happens in a supernova; none is universally accepted.

Let us describe an example of a specific set of events that could lead to a supernova explosion. We present the details of the process, which most likely occurs for heavyweight rather than middleweight stars, in order to illustrate how the advance of stellar evolution can run away with itself and go out of control. Other specific processes can also explain the supernova phenomenon, and may be more important in middleweight stars.

In the case we are considering, a substantial iron core has been formed and begins to shrink and heat up. The iron represents the ashes of the previous stage of nuclear burning. Eventually the temperature becomes high enough for the iron to undergo nuclear reactions. The stage is now set for disaster because iron nuclei have a fundamentally different property from other nuclei when undergoing nuclear reactions. Unlike the other nuclei, iron absorbs rather than produces energy in order to undergo either fusion or fission. Thus iron takes up energy when it is being transformed to other elements, either heavier or lighter. This energy is no longer available to heat the core.

This model explains Type II supernovae. Type I super*novae, on the other hand, may result from the addition of mass to a white dwarf, which leads to the star's collapse. Type I and Type II supernovae can be distinguished from each other by their spectra.*

The core responds to this loss of energy by shrinking and heating up still more. The iron can be broken up into lighter nuclei by the high-energy photons of radiation that are generated, and in so doing absorbs still more energy. The process goes out of control. Within seconds—a fantastically short time for a star that has lived for millions and millions of years—the core collapses and heats up catastrophically. It was formerly thought that matter falling in upon the collapsing core from the layers above would undergo violent nuclear reactions in the intense heat of the core, but this process is now thought to be less likely and less important than the following process:

As matter falls in upon the collapsing core from the layers above, electrons and protons fuse into neutrons. In the process, huge numbers of neutrinos, the elusive particles we discussed in Sections 6.3 and 6.6, are given off. At this stage of stellar collapse, the outer layers of the star are so dense that they stop the neutrinos and in so doing are pushed away from the core with great force. Shock waves (defined in Section 5.3)—similar to the sonic booms that we experience on earth—race through these outer layers as they are blown off, possibly causing heavy elements to form. With a tremendous explosion, the star is destroyed. Only the core is left behind. This explosion of the star is called its supernova phase. To summarize, the essential part of the story is that once the core is made up of iron, the star is likely to explode.

Figure 8–3 Views of the central part of the galaxy NGC 5252, taken in 1959 and in 1972. The supernova that appeared in 1972 was nearly as bright as the rest of the galaxy.

The heavy elements formed either in the center of the star or in the supernova explosion itself are spread out into space, where they enrich the interstellar gas. This is why Population I objects, which are relatively young, have higher proportions of heavy elements than the older, Population II objects. (Stellar populations were discussed in Section 4.5.) When a star forms out of such enriched gas, these heavy elements are present to make up some of any planets that may coalesce. Thus supernovae provide the heavy elements that are necessary for life to arise. Many of the atoms in each of us have been through such supernova explosions.

In the supernova explosion, the core is compressed and becomes a neutron star, which we shall discuss in the following section. The outer layers, representing most of the mass of the star, are shredded, as is the magnetic field of the star.

In the days or weeks following the explosion, the amount of radiation emitted by the supernova can equal that emitted by the rest of its entire galaxy. Most of the supernovae that we observe (Figs. 8–3 and 8–4) are in distant galaxies, and we can see a single star outshine its galaxy for a period

Figure 8–4 Views of the galaxy NGC 4725 in the constellation Coma Berenices. There was an 8 month interval between the photographs—May 10, 1940 to January 2, 1941. The line points to the supernova.

of weeks. It may brighten by over 20 magnitudes, a factor of 10^8 in luminosity.

Optical astronomers have photographed two dozen of these stellar shreds, which are known as *supernova remnants* (Figs. 8–5 and 8–6, the photograph that opens this chapter, and Color Plates 10 and 11), in our galaxy alone. The gas gives off not only optical radiation but also strong radio radiation, and the supernova remnants are best studied by means of radio astronomy (which is described in Chapter 21). About 100 supernova remnants are known from observations made in the radio part of the spectrum (Fig. 8–7). Some of these have now also been observed in the x-ray part of the spectrum (Color Plate 9).

Supernovae were named at a time when it was thought that they were merely unusually bright novae. But now we know that novae and supernovae are very different phenomena. A supernova explosion represents the death of a star and the scattering of most of its material, while a nova uses up only a small fraction of a stellar mass and can be recurrent.

In our own galaxy, we have historical records of only five supernovae in the last thousand years, so we can conclude that many of the 24 optical remnants and over 100 radio remnants known must come from stars that exploded before such events were recorded.

In 1054 A.D., Chinese chronicles recorded the appearance of a "guest star" in the sky that was sufficiently bright that it could be seen in the daytime. No one is certain why no European records of the supernova were made, for the Bayeux tapestry illustrates a comet and thus shows that even in the Middle Ages people were aware of celestial events. Certain cave and rock paintings made by Indians in the American southwest may show the supernova (Fig. 8–8), though this interpretation is controversial.

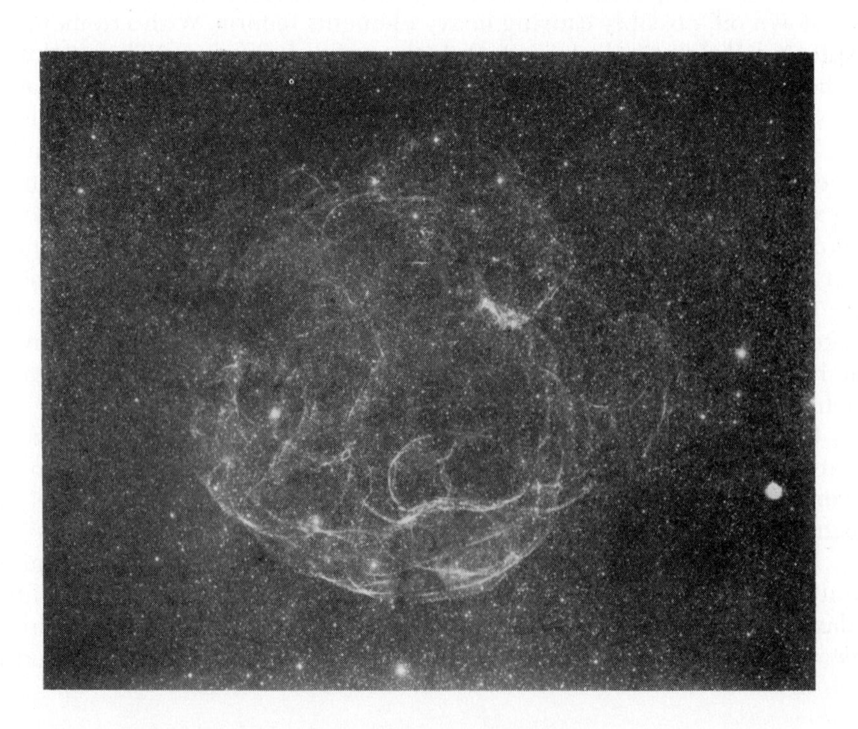

Figure 8–5 S147, the remnant of a supernova explosion in our galaxy. The long delicate filaments shown cover an area over 3° × 3°, about 40 times the area of the moon. The distance to S147 (the 147th object in a catalogue of nebulae by the Soviet astronomers Shajn and Gaze) is not known, though from comparison with the Vela remnant it might be less than 1 kpc from the sun.

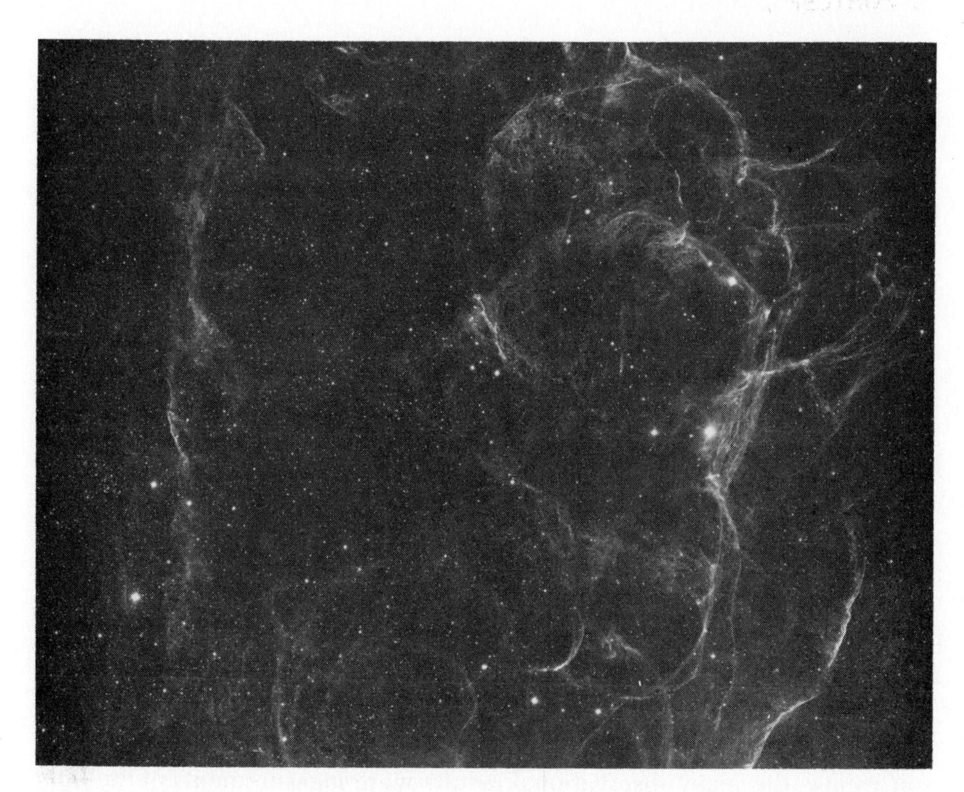

Figure 8–6 The network of nebulosity associated with the pulsar and supernova remnant in the constellation Vela. The photograph, in ultraviolet light, was taken with the University of Michigan's Schmidt camera at Cerro Tololo in Chile.

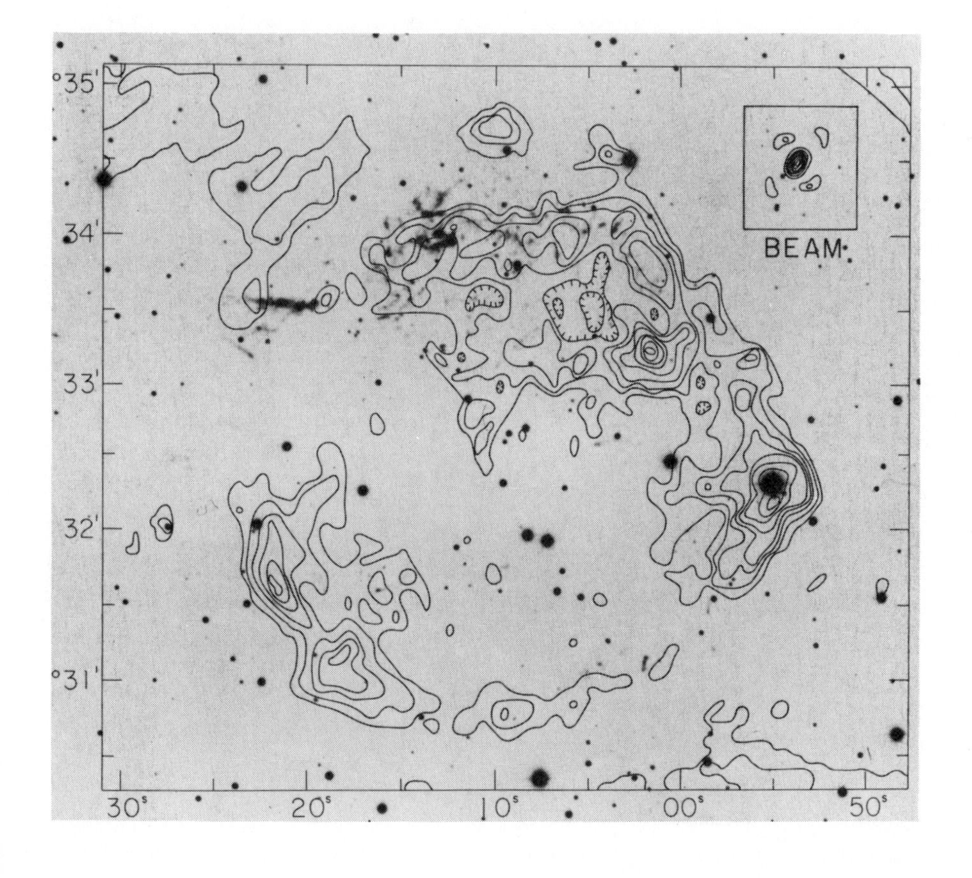

Figure 8–7 A map of the strong radio source Cassiopeia A, made at the National Radio Astronomy Observatory, shown as contours superimposed on a negative of the same region of the sky in visible light taken at the Hale Observatories. The radio source is the remnant of a supernova that exploded 310 years ago at a distance of 3 kpc from the sun. The optical image of the supernova remnant is hidden by interstellar dust, so we see little besides stars relatively nearby the sun. The supernova explosion itself was not observed.

Figure 8–8 An American Indian cave painting discovered in northern Arizona that may depict the supernova explosion 900 years ago that led to the Crab Nebula. It had been wondered why only Oriental astronomers reported the supernova, so searches have been made in other parts of the world for additional observations. Note that the drawing is obviously distorted if it is indeed of the supernova, as the star could not appear over the dark disk of the unlit part of the moon.

When we look at the reported position in the sky, in the constellation Taurus, we see an object that clearly looks as though it is a star torn to shreds (Fig. 8–18 and Color Plate 10). It is called the Crab Nebula. The Crab appears to be about 2000 parsecs from us, about 1/5 of the distance from the sun to the center of our galaxy. There is no doubt that the Crab is the remnant of the supernova whose radiation reached earth in 1054 A.D., twelve years before William the Conqueror invaded England. One check on this is to note the rate at which the filaments in the Crab Nebula are expanding. Tracing them back in time, we find that the filaments would have been at a single point approximately 900 years ago, the length of time since the reported supernova.

Mention has been found in historical texts of two other supernovae similarly long ago, in 1006 A.D. in the constellation Lupus and in 1181 A.D. in the constellation Cassiopeia. The 1006 event was observed by monasteries in Switzerland and Italy in addition to being seen by Oriental and Arabic observers.

The two best studied supernovae in our galaxy—the two most recent to be seen—were successively discovered in 1572 and 1604. They are called Tycho's supernova (Fig. 8–9) and Kepler's supernova after Tycho Brahe and Johannes Kepler, respectively, two of the most distinguished astronomers of that time and, indeed, of all times. (The planetary studies for which Tycho and Kepler are best known are discussed in Section 10.1.)

In the pre-telescopic era of the most recent supernovae observed in our galaxy, the main observational results were measurements of the light curves and of the positions. The positional results have enabled us to identify these supernovae with their optical, radio, and x-ray remnants. Of these

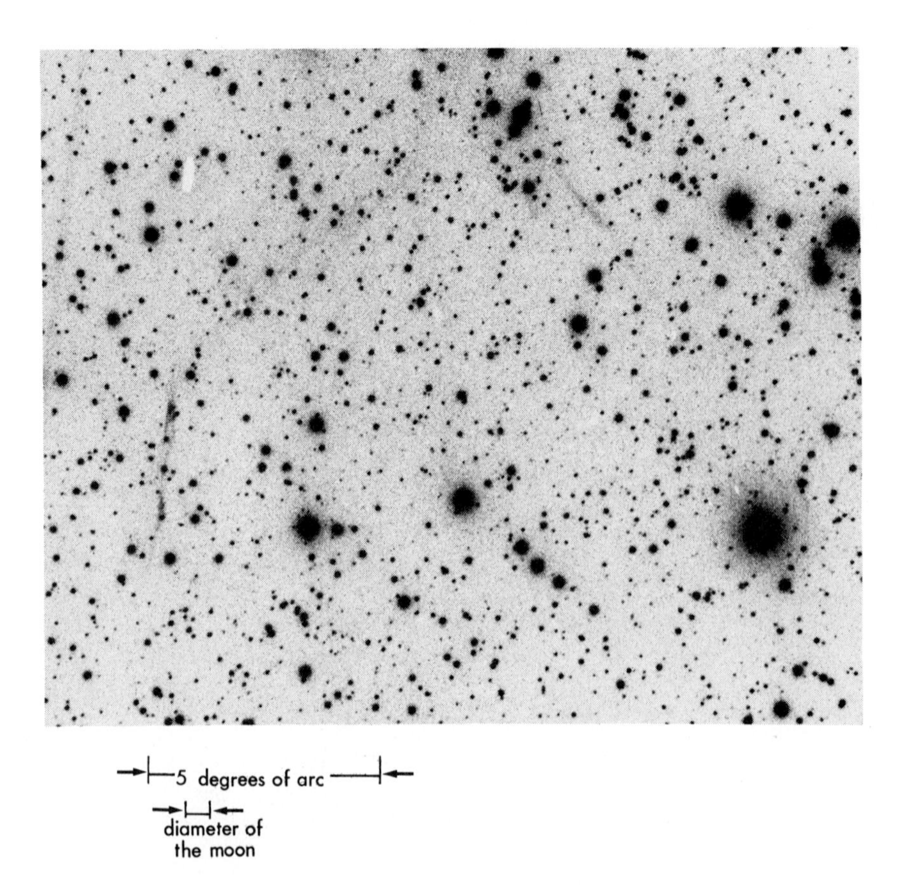

Figure 8–9 The remnant of Tycho's supernova is barely observable in this negative print of a photograph taken in the visible part of the spectrum. The event made such an impression that Bayer showed the supernova on his chart of Cassiopeia (p. 12) drawn 30 years later. Tycho's supernovae is the bright object at the left edge of the chair.

five "young" supernovae, optical and radio remnants are known for all and x-ray remnants are known for all except Kepler's.

Two much older "guest stars" recorded in Chinese chronicles were events that took place in 185 A.D. and 393 A.D. They were probably also supernovae. It has also been suggested that a supernova of 10,000 years ago is pictured in Sumerian cuneiform, and that its explosion may even have inspired the Sumerians to observe the sky and develop astronomy and mathematics. This conclusion rests on an interpretation of documents, and is an example of an interface between astronomy and history.

From a study of the rate at which supernovae seem to appear in distant galaxies, it is estimated that supernovae should appear in our galaxy about once every 30 to 50 years. There have been only a few known supernovae in our galaxy in the past 900 years, though any that occurred on the far side of our galaxy may have been hidden from our view by all the interstellar matter. There could be a supernova in our part of the galaxy any day. We've already been waiting for hundreds of years.

A supernova on our side of our galaxy will be visible to the naked eye and could be quite spectacular. In the meantime, searches go on for new supernovae in other galaxies. Each year, a dozen new supernovae may be found in other galaxies; as a consequence of their distance they are very faint, say, 15th or 20th magnitude. A computer-controlled telescope has been built in New Mexico to scan the sky constantly for supernovae, observing a programmed series of galaxies at regular intervals.

We would like to discover a supernova in the earliest possible stages of its eruption, so that we can follow its evolution from beginning to end. As soon as we detect a supernova we follow not only its brightness but also its spectrum as it fades from view. As far as we can tell, all supernovae have essentially the same maximum brightness. Thus by observing the peak **apparent** magnitude, we can use the method we described in Section 2.12 to calculate the distance to the supernova and the galaxy in which it is located. This is an important method of measuring distances to galaxies.

If a star near the sun were to explode as a supernova, it would greatly increase the amount of x-radiation that reaches the earth. On the basis of our knowledge of the number of stars within about 10 parsecs of us, and how stars become supernovae, it seems reasonable that several of these stars would have exploded in the 3 or 4 billion years that life has existed on earth. Mutations caused by the high-energy particles and radiation from such a nearby supernova may have played a significant role in the evolution of animal and plant species on earth.

8.3 COSMIC RAYS

Most of the information we have discussed thus far in this book has been gleaned from the study of electromagnetic radiation. But certain high-energy particles have been discovered to be traveling through space in addition to the photons. These particles are nuclei of atoms moving at tremendous velocities. They are called *cosmic rays.*

To best capture cosmic rays from outer space, one must travel above most of the earth's atmosphere in a rocket, satellite, or balloon. (Only a few of the most energetic cosmic rays reach the ground.) Often balloons bear stacks of photographic emulsion to altitudes of 50 kilometers; the cosmic rays leave marks as they travel through the film, and a three-dimensional picture of the track can be built up because of the three-dimensional nature of the stack of emulsion. The cosmic rays formed deep in space, called *primary cosmic rays,* often interact with atoms in the earth's atmosphere, at which time *secondary cosmic rays* are given off. It is possible to detect many such secondary rays without leaving the earth's surface.

The origin of the primary cosmic rays has long been a matter of controversy. It is known that the lower-energy cosmic rays are affected by the sun, for example, since the cosmic rays vary with the eleven-year cycle of solar activity. But it is now thought that most cosmic rays are not **formed** by the sun. In fact, it appears that the effect of increased solar activity is to increase the interplanetary magnetic field so that it shields us from cosmic rays more efficiently than we are shielded at times of low solar activity. Thus the flux of cosmic rays peaks at the minimum of the sunspot cycle.

It seems likely to many astronomers that the cosmic rays are particles accelerated to tremendous velocities in supernova explosions, and that they have been traveling through space since they were ejected.

A strong piece of evidence to back up this theory has come from a telescope that observes gamma rays. This telescope is carried on NASA's SAS-2 (Small Astronomy Satellite 2), which is still aloft. When cosmic rays interact with interstellar gas, they produce gamma rays. The paths of cosmic rays are bent by magnetic fields in space, and so come at the earth from all directions even though they may not have been produced in those directions. We thus cannot trace them back to their origins. Gamma rays, however, are not affected by interstellar magnetic fields. The new measurements show that the distribution of gamma rays matches the distribution of supernova remnants, and thus the distribution of the sources of the cosmic rays presumably does too. This evidence strongly endorses the idea that cosmic rays are a phenomenon related to supernovae.

8.4 NEUTRON STARS

We have discussed the fate of the outer layers of a star that explodes as a supernova. Now let us discuss the fate of the core.

As iron fills the core of a massive star, the temperatures are so high that the iron nuclei begin to break apart into smaller units like alpha particles (helium nuclei). The pressure is no longer high enough to counteract gravity, and the outer layers collapse.

In the supernova explosion that ensues, the matter left behind in the core has been compressed to the point where the iron has broken apart. Protons (which have positive charge) and electrons (which have negative charge) are forced together, and neutrons (which have no charge) result. There have to have been the same number of protons and electrons throughout the core; had there not been, then the imbalance of charge would have exerted a powerful force of repulsion.

The core may contain as little as a few tenths of a solar mass or possibly as much as two or three solar masses following the explosion. This remainder is at an even higher density than that at which electron degeneracy holds up a white dwarf. At this density a condition called *neutron degeneracy,* in which the neutrons cannot be packed any more tightly, appears. This condition is completely analogous to electron degeneracy, with the difference that neutron degeneracy comes from the application of quantum mechanical rules to the neutrons. The pressure caused by neutron degeneracy balances the gravitational force that tends to collapse the core, and as a result the core reaches equilibrium as a *neutron star.*

If more than two or three solar masses remain, then the force of gravity will overwhelm even neutron degeneracy. We shall discuss that case in Chapter 9.

Alternatively to the supernova picture, a neutron star could theoretically result if a white dwarf in a binary system gained mass from the other star in the system. If enough mass were gained to put the mass over the electron degeneracy limit, the star would collapse until it reached a stage of neutron degeneracy. We do not know if any stars actually become neutron stars through this process.

Figure 8–10 A neutron star may be only 20 kilometers in diameter (10 km in radius), the size of a city, even though it may contain a solar mass or more.

Whereas a white dwarf packs the mass of the sun into a volume the size of the earth, the density of a neutron star is even more extreme. A neutron star may be only 20 kilometers or so across (Fig. 8–10), in which space it may contain the mass of about two suns. In its high density, it is like a single, giant nucleus.

Since the earth is about 12,000 km in diameter, a neutron star is $(12{,}000/20)^3 = 6^3 \times (10^2)^3 \approx 200 \times 10^6 \approx 10^8$ times smaller in volume. Since a teaspoonful of a white dwarf could weigh 5 tons, a teaspoonful of a neutron star could weigh a billion tons. This state of matter may seem simply inconceivable, but on theoretical grounds it is possible for it to exist even on such a large scale.

This calculation is only to get a sense of scale, so we can ignore small factors like 2; ≈ means "approximately equal to."

What would a neutron star be like? Recall the sun, with its gaseous atmosphere many thousands of kilometers thick. A neutron star, on the other hand, might have a solid, crystalline crust about a hundred meters thick. Above these outer layers, its atmosphere—photosphere, chromosphere, and corona together—probably takes up only another few **centimeters.** Since the crust is crystalline, there may be irregular structures like mountains. Limited by the immense gravitational field (which is about 100 billion times that present on the surface of the earth), the mountains might only poke up a few centimeters through the atmosphere. This may sound like a toy model, but if you came too close you would certainly feel its gravity.

Before it collapses, the core (like the sun) has only a weak magnetic field. But as the core collapses, the magnetic field is concentrated, and grows stronger as a result. By the time the core shrinks to neutron star size, it has an extremely powerful magnetic field, much stronger than any we can produce on earth.

A neutron star may be the strangest type of star of which we can conceive. When neutron stars were discussed in theoretical analyses in the 1930's, there seemed to be no hope of actually observing one. Nobody had a good idea of how to look for a neutron star. Let us leave this story here for a moment, and jump to consider events of 1967, which will later prove to be related to the search for neutron stars.

8.5 THE DISCOVERY OF PULSARS

By 1967, radio astronomy had become a flourishing science. Dozens of large radio telescopes were in existence all around the world, and were being used to observe radio emission from objects in space. The details of radio astronomy will be described in Chapter 21, but for the current discussion all we need know is that various objects in space emit electromagnetic signals in the radio part of the spectrum, and that for the last forty years astronomers have been able to build giant antennas called *radio telescopes* to collect and focus these radio waves. Radio telescopes, just like radio receivers in our living rooms, are subject to static—rapid variations in the strength of the signal. It is difficult to measure the average intensity of a signal if the signal strength is jumping up and down many times a second, so radio astronomers usually adjust their instruments so that they do not record any variation in signal shorter than a second or so.

This should have had the effect of merely "smoothing" the incoming signal, while not distorting it. Of course it would mask any rapid variations in signal, but none were expected to arise in the source itself, as most astronomical sources were thought to be relatively steady in their emission of electromagnetic radiation. The only rapid variations expected besides terrestrial static were those that correspond to the twinkling of stars.

An astronomer at Cambridge University in England, Antony Hewish, wanted to study this "twinkling" of radio sources, which is called *scintillation*. The light from stars twinkles because of effects of our earth's atmosphere. The radio waves from radio sources scintillate not because of any terrestrial effects but rather because radio signals are affected by clouds of electrons in the solar wind. Electrons in the earth's ionosphere and in interstellar space contribute lesser amounts to the observed scintillation of radio sources.

For the same reasons that stars appear to twinkle although planets do not (see Fig. 2–2), radio sources that are so far away or so small that they appear as points will scintillate while radio sources that are somewhat extended across the sky will not. In order to study which sources appeared as points by seeing if they scintillated, Hewish adjusted his radio telescope so that it would record rapid variations in signals. (His "telescope" was actually a set of aerials set up in a large field and connected to a central receiving apparatus.) The ability to record rapid variations in signals made this array of aerials different from almost all other radio telescopes that had been operated up to that time and proved to be the key to the discovery that was forthcoming. The telescope had the additional valuable property that its total collecting area was very large, equivalent to a dish 150 meters (500 feet) in diameter. This made it very sensitive. The two properties of ability to monitor rapid variations and high sensitivity together led to the discovery of pulsars.

One day in 1967, Jocelyn Bell Burnell, then a student working on the project, noticed that a set of especially strong variations in the signal appeared in the middle of the night, when scintillations caused by the solar wind are usually weak. After a month of observation, it became clear that the position of the source of the signals remained fixed with respect to the stars (that is, according to sidereal time, which was discussed in Section 3.11), a sure sign that the object was not terrestrial or solar.

Figure 8–11 The radio telescope—actually a field of aerials—with which pulsars were discovered.

Detailed examination of the signal showed that, surprisingly, it was a rapid set of pulses, with one pulse every 1.3373011 seconds (Fig. 8–12A). The pulses were very regularly spaced, and since there are 24 hours × 60 min/hour × 60 sec/min = 86,400 seconds per day, there are approximately 60,000 pulses per day. Thus the period of pulsation could be established to great accuracy very quickly by merely counting the number of pulses per day and dividing it into the length of a day. The pulses seemed to be at least as regular as the most regular clockwork on earth.

One immediate thought was that the signal represented an interstellar

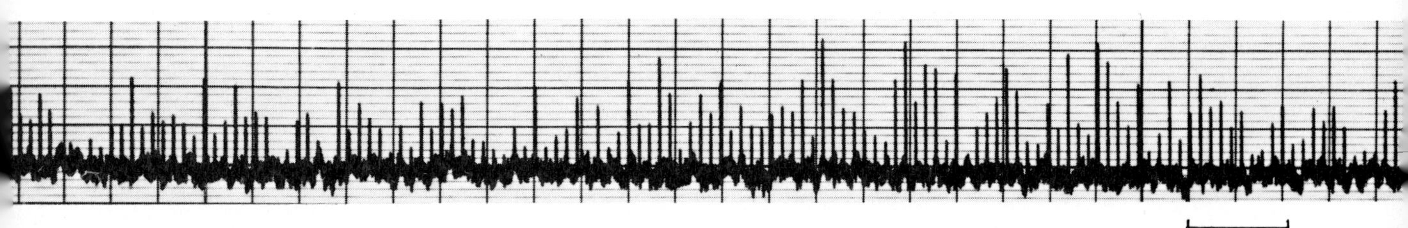

Figure 8–12A A series of pulses of radio waves observed at the National Radio Astronomy Observatory at Green Bank, West Virginia. The pulsar shown here, PSR 2021+51, has a period of 0.529 sec.

beacon sent out by extraterrestrial life on another star. For a time the source was called an LGM, for Little Green Men. Soon, Burnell located three other sources, pulsing with regular periods of .253065, 1.187911, and 1.2737635 seconds, respectively.

They were briefly called LGM 1, LGM 2, LGM 3, and LGM 4, but by this time it was obvious that the signals had not been sent out by extraterrestrial life. For one thing, it seemed unlikely that there would be four such beacons at widely spaced locations in our galaxy. Further, it was determined that the energy was being radiated over a wide range of wavelengths, and if the LGM's were smart enough to send out such precise signals, they wouldn't be stupid enough to squander their energy so. Besides, any beings would probably be on a planet orbiting a star, and no effect of a Doppler shift from any orbital motion was detected. (The Doppler effect caused by the earth's orbital motion around the sun was, of course, calculated, and its effect removed from the data.)

The discovery was kept secret at Cambridge for a few months (a controversial decision) while the signals from these four sources were analyzed, and then was announced to an astonished astronomical community. It was immediately apparent that the discovery of these pulsating radio sources, called *pulsars,* was one of the most important astronomical discoveries of the decade. But what were they?

Other observatories turned their radio telescopes to search the heavens for new pulsars. This often required the purchase of new equipment to be able to follow signals that varied in time so quickly. By this time, the four original pulsars were called by such names as CP 0950. CP stands for "**C**ambridge **P**ulsar," and 0950 is the right ascension in hours and minutes of the source (i.e., 9 hours and 50 minutes of right ascension).

Dozens of new pulsars were found by such observatories as Harvard (HP, for Harvard Pulsar), Arecibo in Puerto Rico (AP), the National Radio Astronomy Observatory (NP), Molonglo in Australia (MP), and Peking (PP). Sometimes the names are simply PSR (for **p**ul**s**a**r**) followed by both the right ascension and declination, as in PSR 2045–16 (20 hours and 45 minutes of right ascension and −16° of declination). The periods range from hundredths of a second to four seconds.

8.6 PROPERTIES OF PULSAR SIGNALS

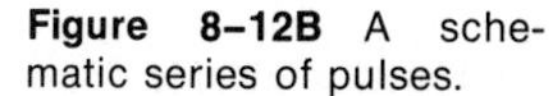

Figure 8–12B A schematic series of pulses.

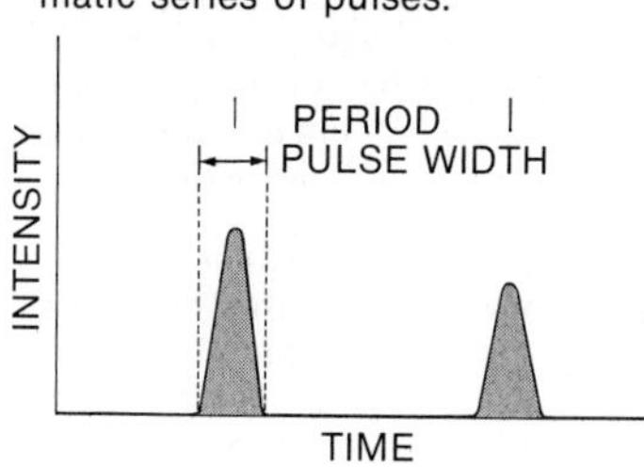

A typical pulse shape is shown in Figure 8–12B. The time from one pulse to the next pulse is called the *period.* (It doesn't matter whether one measures from the peak of one pulse to the peak of the next, or from the beginning of one pulse to the beginning of the next.) The pulse itself lasts only a *pulse width,* a small fraction of the period. The *duty cycle,* the duration of the pulse width divided by the period, is the fraction of the period taken up by the pulse.

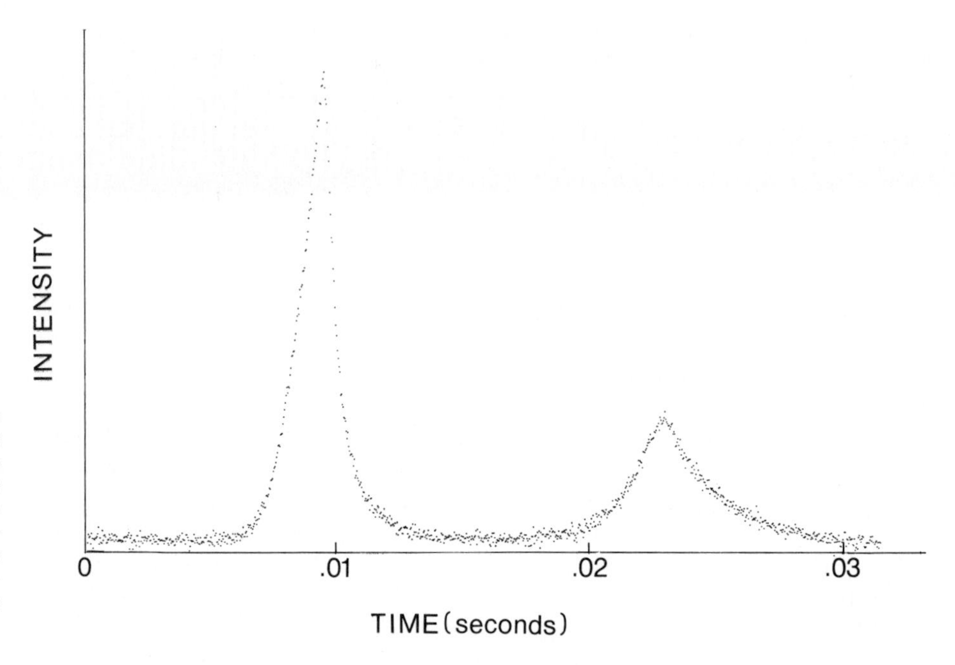

Figure 8–13 Some pulsars have not only a main pulse but also a subsidiary pulse called an *interpulse.* The pulse shown here is the result of averaging many pulses from the pulsar in the Crab Nebula, NP 0532, observed in visible light. Its period is 0.033 sec.

Pulses from a given pulsar are not of the same strength. Sometimes a few pulses are strong, and then the strength fades out and no pulses at all are detectable for a time. But when the signal returns, the pulsar is still pulsing on the same beat, as shown in Figure 8–12A.

Some pulsars have not only a main pulse, but also a weaker *interpulse,* as shown in Figure 8–13. The interpulse does not necessarily occur midway between successive pulses.

When the positions of all the known pulsars are plotted on a chart of the heavens, it can easily be seen that they are concentrated along the plane of our galaxy (Fig. 8–14) rather than being uniformly distributed over the sky. Thus they are clearly objects in our galaxy, for if they were extragalactic objects (objects located outside our galaxy), we would expect them to be distributed uniformly. (The idea of isotropy, the notion that the universe as a whole looks the same no matter in which direction we look, is one of the most fundamental notions of astronomy, and is discussed in Section 25.2.)

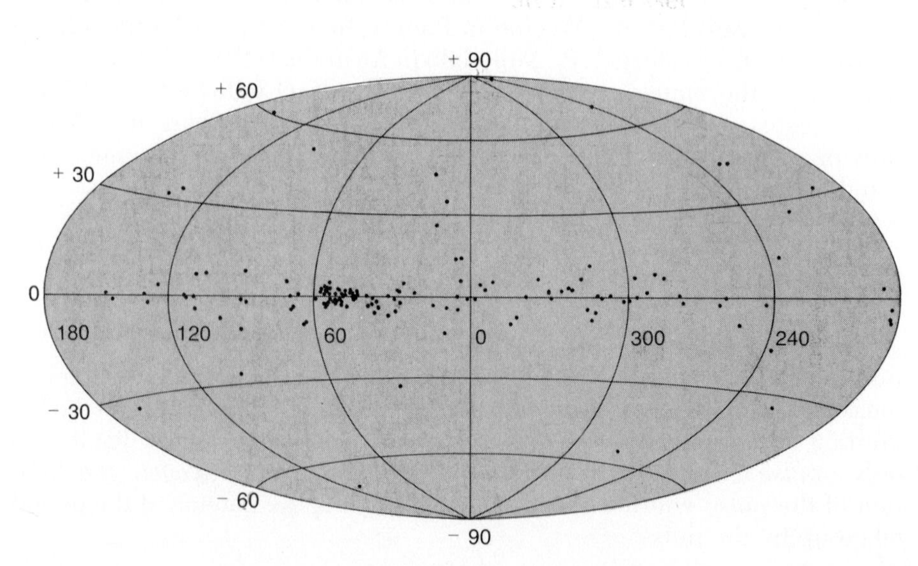

Figure 8–14 The distribution of the 154 known pulsars on a projection that maps the entire sky, with the plane of the Milky Way along the zero degree line on the map. The concentration of pulsars near 60° galactic longitude on this map merely represents the fact that this section of the sky has been especially carefully searched for pulsars because it is the area of the Milky Way best visible from the Arecibo Observatory. From the concentration of pulsars along the plane of our galaxy, we can conclude that pulsars are members of our galaxy; had they been extragalactic, we would have expected to see as many near the poles of this map. We can extrapolate that there are at least 100,000 pulsars in our galaxy.

8.7 WHAT ARE PULSARS?

Theoreticians went to work to try to explain the source of the pulsars' signal. The problem had essentially two parts. The first part was to explain what supplies the energy of the signal. The second part was to explain what causes the signal to be regularly timed, that is, what the clock mechanism is.

From studies of the second part of the problem, the choices were quickly narrowed down. The signals could not be coming from a pulsation of a whole normal main-sequence star because such a star would be too big. If the sun, for example, were to turn off all at one instant, we on earth would not see it go dark all at once. The point nearest to us would disappear first, and in the next seconds the darkening would spread farther back around the side of the sun (Fig. 8–15). It simply takes the light from the far side of the sun, which is 1,400,000 kilometers (a million miles) farther away from us than the nearest point, about 5 extra seconds to reach us since it travels at 300,000 kilometers/second (186,000 miles/second). Thus it would take the sun 5 seconds to go dark. On the other hand, pulsars had pulse widths as short as 1/40 second. Clearly, the emitting regions on pulsars had to be smaller than 1/40 light second across.

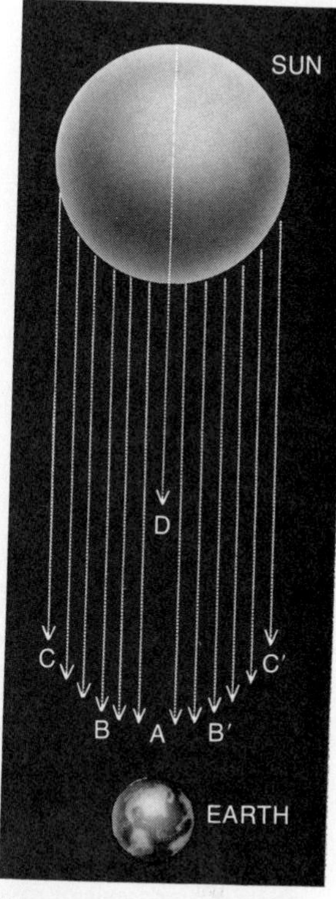

Figure 8–15 Even if a large star were to turn off all at one time, the size of the star is such that it would appear to us to darken over a measurable period of time because radiation travels at the finite speed of light. Pulsars pulse so rapidly that we know that they cannot be large objects flashing.

That left white dwarfs and neutron stars as the likely candidates. Remember, at that time neutron stars were merely objects that had been predicted theoretically but had never been detected observationally. Could the pulsars be the more ordinary objects, namely, special kinds of white dwarfs?

There were two basic mechanisms astronomers could conceive of as possibly controlling the pulsar timing. It was thought that the pulses might be coming from a star that was actually oscillating in size and brightness, or that they were coming from a star that was rotating. Another less likely alternative was that the emitting mechanism involved two stars orbiting around each other in a binary system.

The case of the two orbiting stars can be eliminated first. The mutual revolution would be more rapid the closer the two stars were together. But two white dwarfs, because of their size, could not approach close enough to each other to orbit in periods of a second or so. What about two orbiting neutron stars? More careful reasoning is necessary to rule out this possibility. It can be shown that, according to the general theory of relativity, if two neutron stars were orbiting each other once a second, then they would generate energy in the form of gravitational waves (see Section 9.6), and their orbital period would increase very rapidly (that is, the pulses would seem to slow down). Such a rapid increase of the pulsation period is not observed (though the periods do increase at a lesser rate), so the case of orbiting neutron stars can be ruled out.

The laws governing the oscillations of stars had been worked out very thoroughly, since Cepheids and other variable stars oscillate in size, continually getting larger and smaller. The period of oscillation of a star depends on the density of the star. The denser the star, the more rapidly it oscillates. So we can predict the period of oscillation of a star by examining its properties, just as we can predict the note on which a bell on earth will ring by examining the bell.

Because of its density, a white dwarf would oscillate in less than a minute. However, it could not oscillate as rapidly as once every second, as would be required if pulsars were oscillating white dwarfs. Neutron stars, denser yet, would indeed oscillate more rapidly, but they would oscillate

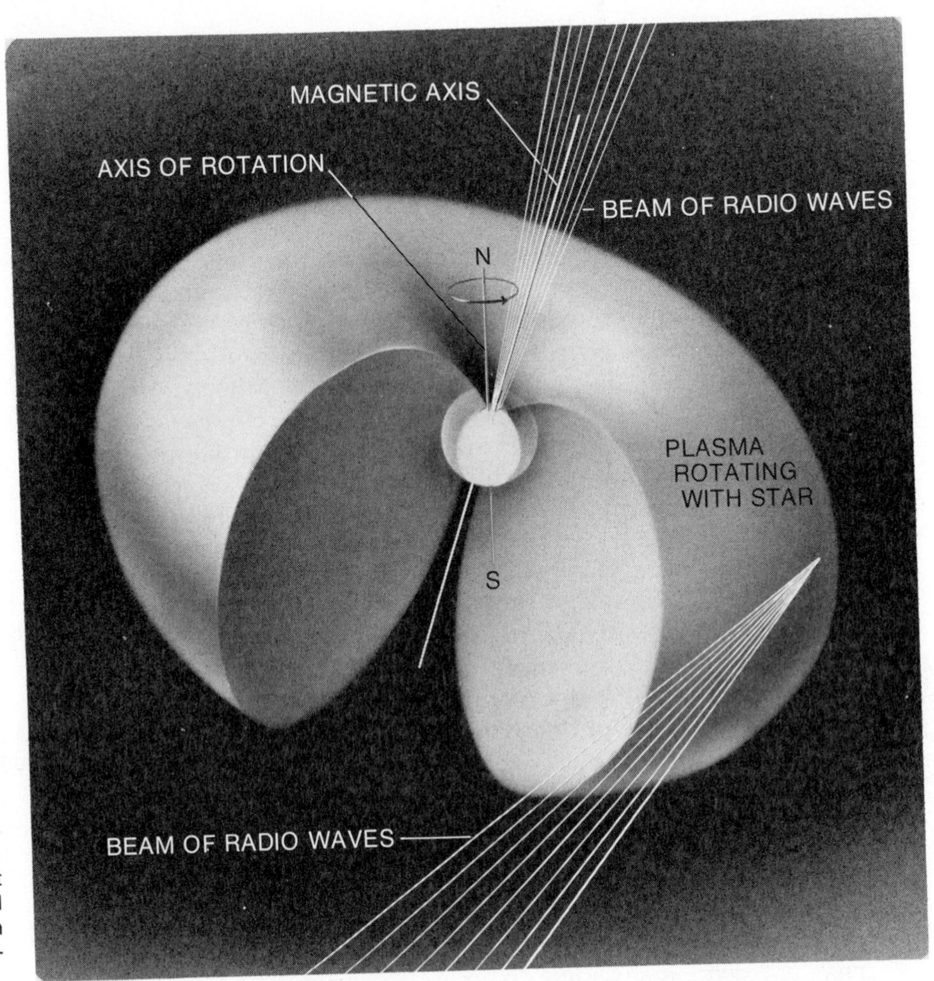

Figure 8–16 In the lighthouse model for pulsars, which is now commonly accepted, a beam of radiation flashes by us once each pulsar period, just as a lighthouse beam appears to flash by a ship at sea. The mechanism by which the beam is generated is not currently understood. Two possible variations of the lighthouse theory are shown here. The beam may be emitted along the magnetic axis, as shown in the top half of the figure. Alternatively, it is believed that the generation of a pulsar beam is related to the neutron star's magnetic axis being aligned in a different direction than the neutron star's axis of rotation. The beam would then be generated on the surface of a doughnut of magnetic field (shown cut away) that surrounds and rotates with the neutron star, as shown in the bottom half of the figure.

once every 1/100 second or so, too rapidly to account for the pulsars. Thus oscillating neutron stars are ruled out as well.

What about rotation? Consider a lighthouse whose beacon casts its powerful beam many miles out to sea. As the beacon goes around, the beam sweeps past any ship very quickly, and returns again to illuminate that ship after it has made a complete rotation. Perhaps pulsars do the same thing: they emit a beam of radio waves that sweeps out a path in space. We see a pulse each time the beam passes the earth (Fig. 8–16). The wider the beam, the longer is the duty cycle we observe.

The first pulsars discovered all had periods longer than 1/4 sec.

But what type of star could rotate at the speed required to account for the pulsations? Could a white dwarf be rotating fast enough? Recall that a white dwarf is approximately the size of the earth. If an object that size were to rotate once every 1/4 second, the centrifugal force would overcome even the immense gravitational force of a white dwarf, and the outer layers would begin to be torn off. This indicated that pulsars were probably not white dwarfs; if a pulsar with a period shorter than 1/4 second were to be found, then the white dwarf model would be completely ruled out.

Neutron stars, on the other hand, are much smaller, so the centrifugal force would be weaker and the gravitational force would be stronger. As a result, neutron stars can indeed rotate four times a second. There is nothing to rule out their identification with pulsars. Since no other reasonable possibility has been found, astronomers accept the idea that pulsars are in fact

TABLE 8–1 POSSIBLE EXPLANATIONS OF PULSARS

Hypothesis	Probability
Regular dwarf or giant	Ruled out
System of orbiting white dwarfs	Ruled out
System of orbiting neutron stars	Ruled out
Oscillating white dwarf	Ruled out
Oscillating neutron star	Ruled out
Rotating white dwarf	Ruled out
Rotating neutron star	Most likely

neutron stars that are rotating. This is called the *lighthouse model*. Thus by discovering pulsars, we have also discovered neutron stars.

Note that for this argument to be complete, and convincing, we must be as certain as we can be that there are no other possibilities to consider. Whenever we argue from elimination of other possibilities, we must be certain that we have a complete list (Table 8–1).

At the moment, about 150 pulsars have been discovered. But the lighthouse model implies that there are many more, for we can see only those pulsars whose beams happen to strike the earth. If a pulsar were spinning at a different angle, then we would not know that it was there.

We have dealt above with only the second part of the explanation of the emission from pulsars: the clock mechanism. We understand much less well the mechanism by which the radiation is actually emitted in a beam. Presumably it has something to do with the extremely powerful magnetic field of the neutron star. Energy may be generated far above the surface of the neutron star in the magnetic field, perhaps near the location where the magnetic field lines, which are swept along with the star's rotation, attain velocities that approach that of light. The radio waves may be generated by charged particles that have escaped from the neutron star near its magnetic poles. These magnetic poles may not coincide with the neutron star's poles of rotation. After all, the earth's north magnetic pole is not at the north pole but near Hudson's Bay, Canada (Fig. 8–17).

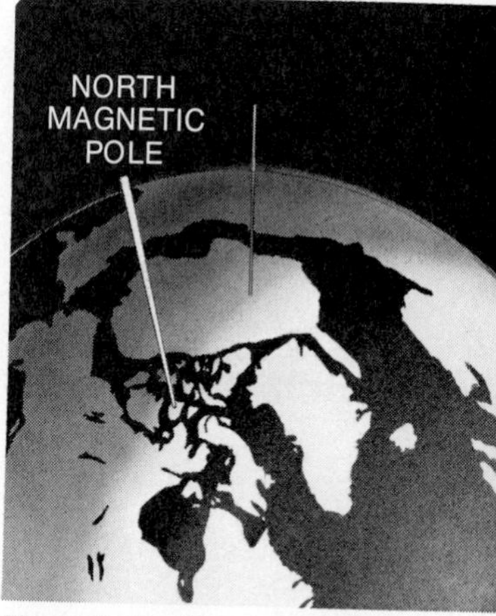

Figure 8–17 Note that the earth's magnetic axis, like that of a pulsar, is aligned in a different direction from that of the earth's axis of rotation; our magnetic north pole lies in the vicinity of Hudson's Bay in Canada. The earth rotates in 24 hours; a neutron star may rotate in 1/30 sec, which is a period 2,500,000 times shorter. Thus the effect of the magnetic field being carried around at a skew angle is intensified manyfold.

8.8 THE PULSAR IN THE CRAB NEBULA

Several months after the first pulsars had been discovered, strong bursts of radio energy were discovered coming from the direction of the Crab Nebula (Fig. 8–18). They were first observed at the National Radio Astronomy Observatory, and so the source was named **NP** 0532. The bursts were sporadic rather than periodic, but there was hope that the astronomers were seeing only the strongest pulses and that a periodicity could be found.

Within two weeks of feverish activity at several radio observatories, it was discovered at Arecibo (Fig. 19–6) that the Crab pulsar had a period of 0.033 second, the shortest period by far of all the known pulsars. The fastest pulsar previously known pulsed four times a second, while NP 0532 pulsed at the very rapid rate of thirty times a second. No white dwarf could possibly rotate that fast.

Furthermore, the Crab Nebula is a supernova remnant, and theory predicts that neutron stars should exist at the centers of supernova remnants. Thus the discovery of a pulsar there, exactly where a neutron star would be expected, was the discovery that clinched the identification of pulsars with

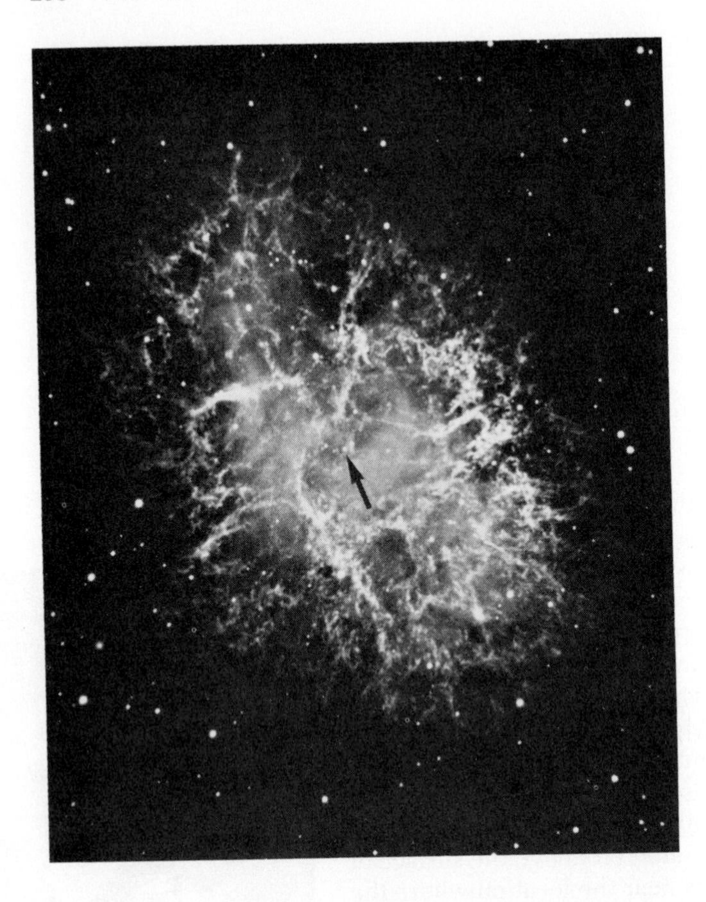

Figure 8–18 The Crab Nebula, the remnant of a supernova explosion that became visible on earth in 1054 A.D. The pulsar, the only pulsar yet detected to be blinking on and off in the visible part of the spectrum, is marked with an arrow. In the long exposure necessary to take this photograph, the star turns on and off so many times that it appears to be a normal star. Only after its radio pulsation had been detected was it observed to be blinking in optical light. It had long been suspected, however, of being the leftover core of the supernova because of its unusual spectrum, which has only a continuum and no spectral lines. Since this had been pointed out by Walter Baade, it has been known as Baade's star.

neutron stars. Presumably, we do not detect pulsars in the positions of Tycho's and Kepler's supernovae because their spin axes are oriented such that their beams of radiation do not strike the earth.

Optical astronomers, of course, had searched very carefully at the reported positions of pulsars in order to try to discover optical objects there. A substantial amount of observing time with the largest telescopes in the world had been devoted to these projects. But no optical objects had been found, and at this time the position of the Crab Nebula had not been examined since it was not yet known to contain a pulsar. After several failures, the search was abandoned.

Soon after NP 0532 was discovered, three astronomers at the University of Arizona decided to examine its position to look for an optical pulsar. They turned their telescope towards a faint star in the midst of the Crab Nebula, and it very soon became apparent that the star was pulsing! Essentially they did something that many astronomers had come to think impossible—they found an optical pulsar.

The star in question had long been suspected of being the core of the supernova. It was a 15th magnitude star, and it had been known that it did not seem to have a normal spectrum with absorption lines.

Now it was seen that the star was actually turning on and off 30 times a second. When it was on, it was substantially brighter than 15th magnitude; when it was off, it was substantially fainter. Photographic plates, of course, showed only its average brightness.

Sometime after the discovery of the optical pulsations, an astronomer at the 3-meter telescope of the Lick Observatory used a stroboscopic technique in conjunction with a television camera to examine the image. When

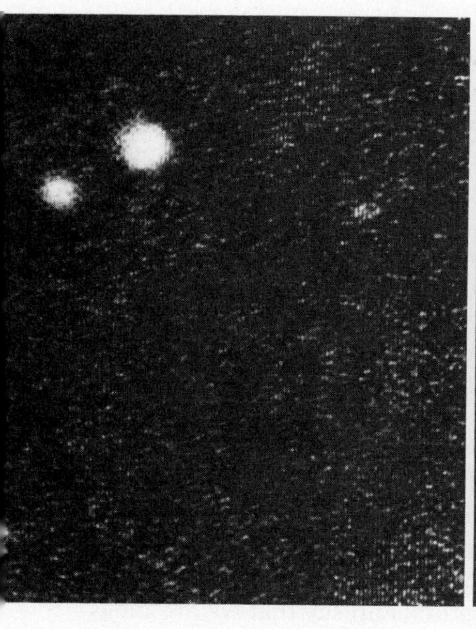

Figure 8–19 As observed on the television screen that showed the image of the pulsar observed through the Lick stroboscope, Baade's star, marked in Fig. 8–18, was observed to blink on and off while other stars in the field remained constant in intensity. This clinched the identification of this star with the pulsar.

a light is flashing stroboscopically, one can see what is going on only at certain intervals, and one cannot see what is going on in between (Fig. 8–19). For example, stroboscopic techniques are now sometimes used to illuminate a dance floor.

The Lick stroboscope (Fig. 8–20) was a rotating wheel with a slot in it. The wheel was set to rotate with the same period as NP 0532. When the slot was synchronized in position with the "on" phase of the pulsar, that is, when light was allowed to pass through the wheel whenever the star was "on," two stars were visible close together near the center of the Crab. When the slot was synchronized in position with the "off" phase of the pulsar, only one of the stars could be seen. Thus, the star that was consistently seen was radiating steadily, as do normal stars, and the other star was blinking on and off. It was conclusive proof that the blinking star was identical with the radio source. Its light curve is shown in Figure 8–13.

About ten years have passed since that discovery, and still no other optical pulsar has been found. For the moment there are 150 objects known to be pulsing in the radio spectrum, and only one—which has the shortest period yet discovered and which is thus presumably the youngest known pulsar—in the optical spectrum. Several x-ray pulsars are known, and will be discussed in a subsequent section. The Crab and one other source are also gamma-ray pulsars.

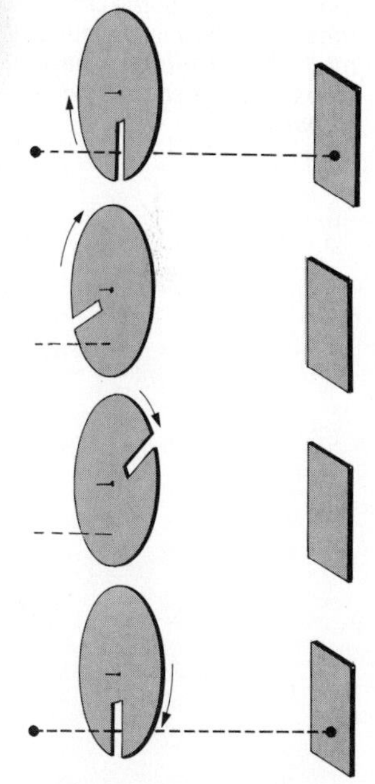

Figure 8–20 An illustration of the principle behind the stroboscope used at the Lick Observatory to detect the pulsation of the Crab Nebula pulsar 30 times per second. If the stroboscope's rotation speed is synchronized exactly with the period of the pulsar, the tendency of the human eye and brain to perceive images as persisting for a fraction of a second makes the pulsar look as though it is always on (or, if the position of the slot cut in the rotating disk is different, always off). But if the stroboscope's rotation speed is put slightly out of synchronization, the pulsar appears to blink slowly on and off at a rate that is easily visible to the eye.

8.9 DISTANCES AND DISPERSION

It is easy to tell the direction to a pulsar, but it is difficult to tell exactly how far away it is. We must know the distance in order to tell what the intrinsic luminosity of the pulsar is—that is, exactly how much energy it is giving off—just as we must know the distance to an ordinary star in order to determine its absolute magnitude.

The radio signals from a pulsar have a property that allows us to tell how far away the pulsar is. When a pulsar emits a pulse of radiation, signals of all frequencies are emitted at the same instant. But signals of different frequencies travel at different velocities through space.

One may ask, doesn't all electromagnetic radiation travel at the speed of light? Actually, the speed of light is constant in a vacuum, and signals of all frequencies travel at the same speed in a vacuum. But whenever ionized matter is present, then radiation travels somewhat more slowly than "the speed of light" and different frequencies travel at slightly different velocities. In interstellar space, it is principally the free electrons there that affect the speed of radiation.

Electrons are scattered throughout interstellar space, about 1 electron for every 30 cubic centimeters. This is an exceedingly low density of matter, for there are over 10^{19} atoms in one cubic centimeter of air in the room in which you are sitting. Yet before reaching us the radiation from pulsars has to travel large distances—thousands of parsecs in some cases—so the total amount of material it traverses adds up.

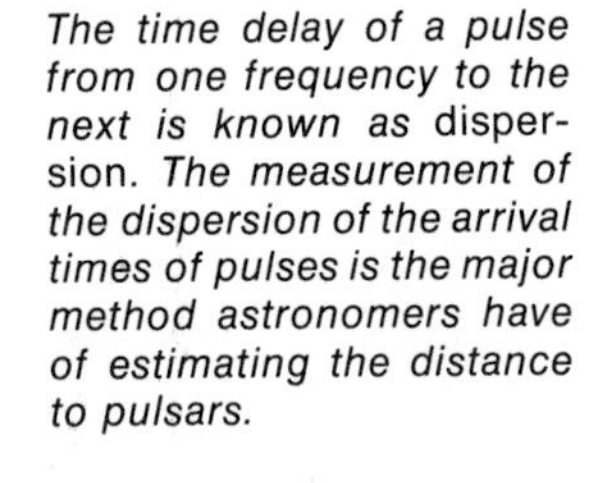

The time delay of a pulse from one frequency to the next is known as dispersion. The measurement of the dispersion of the arrival times of pulses is the major method astronomers have of estimating the distance to pulsars.

Thus by the time the signal from a pulsar has passed through interstellar space and reaches us, pulses at different frequencies arrive at different times (Fig. 8–21). By merely measuring the delay in arrival time from frequency to frequency we can determine the total number of electrons that the signal has traversed (we actually consider the number of electrons in a column with a 1 cm² cross section, and say that we are measuring the *column density*). Then, if we know from other studies the average density of electrons in interstellar space, we can calculate the approximate distance to the pulsar.

Let us consider the analogous situation of the dispersion of hurdlers arriving at the finish line of a race. All the hurdlers set off together, at the sound of the starter's gun. Let us say that we know that Leslie jumps slightly faster than Hilary by a constant amount (and so that without hurdles their speeds would be equal). If we observe the hurdlers after 50 meters of the race, Leslie will be a certain distance ahead. If we observe the hurdlers after 100 meters, Leslie will be twice as far ahead. If we go away and happen to observe them some time later, we can tell how far the hurdlers have gone by how far Leslie has pulled ahead (we must know if Leslie has lapped Hilary, that is, if Leslie has made an extra revolution of the track). We could say that the arrival time of the hurdlers has dispersed. The situation is exactly the same for pulsars.

If the distance to a pulsar happens to be known by some other method—for example, the distance to the Crab Nebula is known to be 2000 parsecs from optical studies—then one can use the observed dispersion to find the average electron density in space. This value can then be used to find the distances to other pulsars. The currently accepted value for the electron density of 1 electron per 30 cubic centimeters is somewhat lower than had been previously thought.

This method of determining distances is not foolproof, for it depends on the assumption that the electron density of space is the same every-

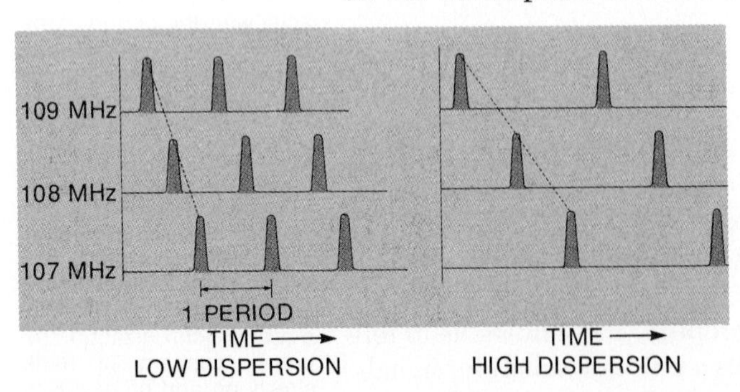

Figure 8–21 Because of the presence of electrons in the interstellar medium, the arrival time of pulses at lower frequencies are slightly delayed over the arrival time of pulses at slightly higher frequencies. The slope of the line showing the time delay as a function of frequency can be directly interpreted to give the number of electrons between the pulsar and us. Thus the pulsar whose pulses are shown on the right is farther away from us than the pulsar whose pulses are shown on the left.

where. We know that there are clouds of high-density gas scattered throughout the plane of our galaxy, and the electron density is apt to be slightly higher in the plane of our galaxy than when we look out of the plane. We can take these effects into account in part. We must simply be aware of the uncertainties in our estimates of distance based on the dispersion method.

Another line of evidence that shows that pulsars are located in our own galaxy is the measurement of proper motions. Putting together their angular velocities from proper motion studies with their distances from dispersion studies indicates that some pulsars are even moving fast enough to escape from the gravitational field of our galaxy.

8.10 THE CLOCK SLOWS DOWN

When pulsars were first discovered, their most prominent feature was the extreme regularity of the pulses. It was hoped that they could be used as precise timekeepers, perhaps even more exact than atomic clocks on earth.

After they had been observed for some time, however, it was noticed that the pulsars were slowing down very slightly. Their periods were all gradually increasing. This was an exacting measurement, for the period can be measured precisely to more than 8 decimal places, and the change in the period is often in the last decimal place. Only occasionally do astronomers, who often work with orders of magnitude (factors of ten), have reason to measure anything so precisely. The Crab Nebula pulsar, which has the shortest period, is slowing down at a faster rate than the others: its period is increasing by 3×10^{-8} seconds/day.

The filaments in the Crab Nebula still glow brightly, even though over 900 years have passed since the explosion. Furthermore, the Crab gives off tremendous amounts of energy across the spectrum from x-rays to radio waves. It had long been wondered where the Crab got this energy. Theoretical calculations show that the amount of rotational energy lost by the Crab's neutron star as its rotation slows down is just the right amount to provide the energy radiated by the entire nebula. Thus the discovery of the Crab pulsar solved a long-standing problem in astrophysics: where the energy originates that keeps the Crab Nebula shining. It has also been suggested that pulsars, rather than supernovae, are the sources of the cosmic rays.

Pulsars with the longest known periods, about 4 seconds in duration, are found to be slowing down at a lesser rate than the pulsars with shorter periods. This fits in with the theory that a pulsar is a neutron star. Astronomers think that soon after the supernova explosion that forms the neutron star, the neutron star rotates very rapidly. Over the years, it slows down as it loses energy, perhaps because its magnetic field interacts with the magnetic field of interstellar space or with charged particles. Thus we would expect the youngest pulsars to be rotating most rapidly; they also slow down at a greater rate.

Another young pulsar, with a very rapid rotation period of 0.059 second (59 milliseconds), is located in the constellation Vela (Fig. 8–22), at the site of a supernova remnant (Fig. 8–6) four times closer to us than the Crab Nebula. The Vela pulsar is slowing down at a rate slightly less than the rate at which NP 0532 is slowing down.

Imagine the surprise of astronomers when all of a sudden in 1969 this

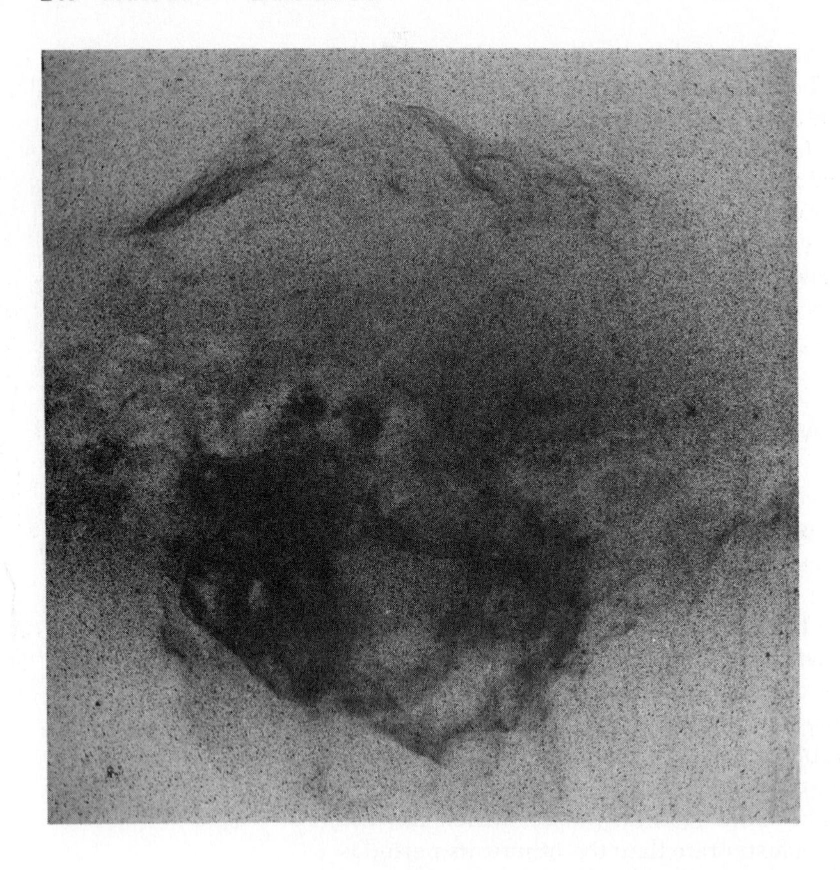

Figure 8–22 The Gum Nebula, investigated by Colin Gum from 1952 on, covers a wide area of sky in the southern hemisphere, as shown in this negative print of a photograph taken with an ordinary wide angle lens. The field of view is 40° × 40°. The Vela pulsar and supernova remnant lie within the Gum Nebula. (Photo by John C. Brandt, Robert G. Roosen, J. Thompson, and D. J. Ludden of NASA's Goddard Space Flight Center)

pulsar briefly sped up. That is, its period decreased. Then it resumed slowing down at the same rate as before (Fig. 8–23).

Scientists were astounded. Some said that it was a once-in-a-lifetime event that we were very fortunate to observe, and it would never happen again. But when, in 1971, the Vela pulsar again sped up and then again re-

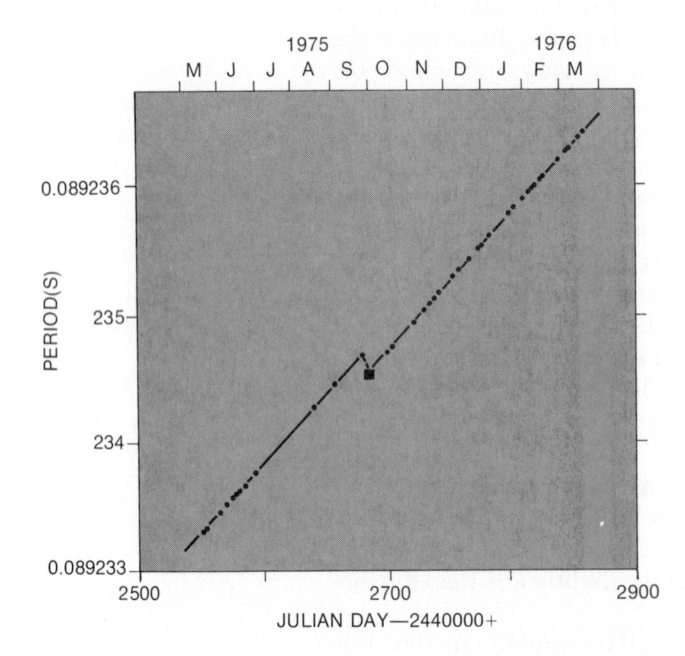

Figure 8–23 The Vela pulsar has undergone several glitches, times at which its period has decreased, which correspond to an increase in the rotation speed of the pulsar. After each glitch, the pulsar resumed slowing down (that is, its period resumed lengthening) at the same slow rate.

sumed its gradual slowdown from its newly shortened period, this idea was discarded. A third speedup occurred in 1975. These brief events at which the pulsar speeds up are called *glitches*. The pulsar in the Crab Nebula also undergoes glitches.

Let us go back to our model of a neutron star with a crystalline crust. It has been calculated that the change in period would be accounted for if the crust fractured and settled by as little as a millimeter. As a result of this "starquake," the matter would then be distributed slightly closer to the center of the star, and the star's rate of rotation would increase to compensate. (We will discuss conservation of angular momentum in Section 10.3.) Since this redistribution of material can happen sporadically, we think that Vela will continue having glitches for some time.

8.11 THE BINARY PULSAR

We recall that astronomers can only determine masses for stars that are in binary systems. All the pulsars that had been found prior to 1974 were lone objects in space. Thus although it seemed likely that pulsars would have the masses that were predicted theoretically for neutron stars, we could not confirm this directly.

Over the past few years, Joseph H. Taylor has been conducting a search for new pulsars, with the detection of unusual specimens as one of his goals. In 1974, Taylor and Russell Hulse, both of the University of Massachusetts at Amherst, found a pulsar whose period (approximately 0.059 sec) did not seem very regular. They finally discovered that its variation in pulse times could be explained if the pulsar were orbiting another star. If so, when the pulsar was approaching us the pulses would be jammed together a bit, and when it was receding the pulses would be spread apart in time in a type of Doppler effect.

No optical object has been found at the location of the binary pulsar, so it can be concluded that both objects are too faint to be seen from the earth. From dispersion measurements and the period of the orbit, the orbit of the pulsing component was determined to have a radius that is approximately the radius of the sun. Thus the accompanying object cannot be a star of normal size; it may be a neutron star too. No pulses have been detected from it, but it could simply be oriented at a different angle.

Unfortunately, when only one star of a binary pair is detectable, we cannot uniquely determine the mass of the stars, that is, determine one and only one value that is consistent with the observations. We can only calculate limits on the masses, and a quantity, also mentioned in Section 4.2, that is a mixture of sums and products of the two masses (called the *mass function*). But at least it can be concluded that the results are consistent with the interpretation of pulsars as neutron stars.

The main interest in the binary pulsar is that it may provide a powerful test of the theory of relativity. In Section 5.11, we discussed the excess advance of the perihelion of Mercury, which is only 43 seconds of arc per

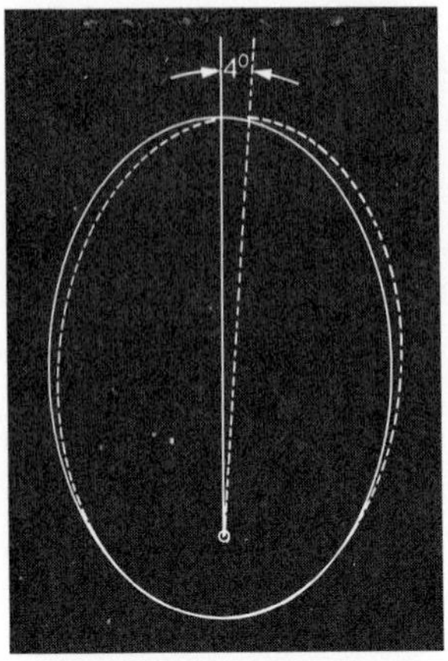

century. It can be calculated that the binary pulsar, which is in a much stronger gravitational field than is Mercury, should have its periastron, the close point of its orbit to its companion (corresponding to the "perihelion" point, the closest approach of a planet's orbit to the sun), advance by 4° per **year** (Fig. 8–24). (Converting 4° to arc sec shows that the advance is by 14,400 arc sec/year or 1,440,000 arc sec/century. We can see that 4° per year is about 35,000 times the size of the effect for Mercury.) A change of 4° is large enough to measure easily. The first year's observations of the binary pulsar seem to detect the advance and so, if the perihelion advance is caused by an effect of Einstein's general theory of relativity, provide a strong confirmation of the theory.

Figure 8–24 The periastron of the binary pulsar PSR 1913+16 advances by 4° per year, which can be interpreted as a strong endorsement of Einstein's general theory of relativity. The angle is shown for the apastron, the farthest point of the pulsar from the other object.

8.12 X-RAY PULSARS

Telescopes in orbit that are sensitive to x-rays have detected a number of strong x-ray sources. While these telescopes are not able to make the high-speed observations that are necessary to see if pulsations are taking place, x-ray telescopes sent up for brief periods in rockets have been able to make such observations.

Three x-ray pulsars have been found, that is, objects that pulse only in the x-ray region of the spectrum. Two of them, Hercules X-1 and Centaurus X-3 (the first x-ray source to be discovered in the constellation Hercules and the third in Centaurus), are located in our galaxy. The third is not in our galaxy, but rather in the Small Magellanic Cloud, a companion galaxy to our own system (and is thus called SMC X-1). It was discovered from observations made during the joint American-Soviet Apollo-Soyuz mission of July 1975. The astronauts were able to carry up equipment that was capable of higher time-resolution than equipment that had been sent up in unmanned satellites. The pulsar they discovered is a companion to a supergiant star named Sanduleak 160, which it orbits every 3.9 days. It pulses x-rays every 0.716 second (Fig. 8–25). It radiates a million times more energy than does our sun.

Figure 8–25 The x-ray pulsar SMC X-1 in the Small Magellanic Cloud has a period of 0.716 sec. These observations were taken with instrumentation of the Naval Research Laboratory during the joint Soviet-American Apollo-Soyuz project in 1975.

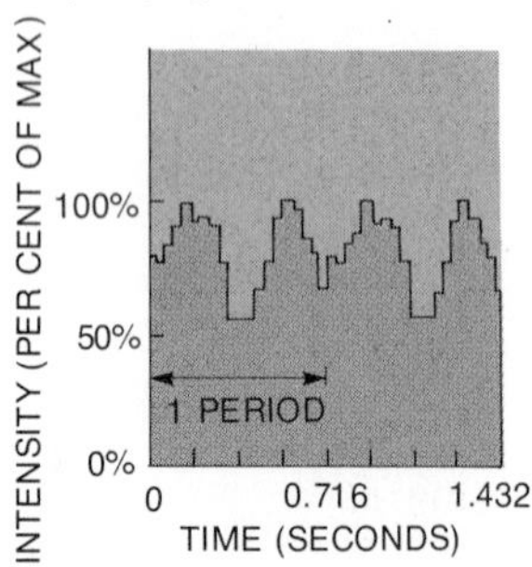

The 1.24 second period of the x-ray pulses from Hercules X-1, unlike the period of an ordinary pulsar, is growing shorter, that is, the rotation of the neutron star is speeding up. Theoreticians think that the x-ray pulsars are part of binary systems, and that they are radiating because mass from the companion star is being funneled towards the poles of the pulsar by the pulsar's strong magnetic field. Thus energy is being added to the rotating neutron star, and unlike regular (radio) pulsars, it speeds up instead of slowing down.

Hercules X-1 is the companion to a long-observed variable star (a single-line spectroscopic binary, a type of double star described in Section 4.1) named HZ Herculis (variable star HZ in the constellation Hercules); they orbit each other every 1.7 days. The mass of Hercules X-1 has been carefully evaluated. The observations indicate that the companion has 1.3 solar masses, within the limit that neutron degeneracy can support (as discussed in Section 7.3). The latest electronic techniques were used to detect individual photons emitted from the optically visible star when the beam of x-rays from the dark companion, according to the lighthouse theory, strikes it every 1.24 seconds. These observations make it plausible that Hercules X-1 is indeed a neutron star like the other pulsars.

While there are three pulsars that are pulsing only x-rays and are not detectable in the radio region of the spectrum, two other pulsars are known to pulse x-radiation and gamma radiation in addition to radio radiation. Since these two objects are the pulsars in the Crab Nebula and in Vela, two of the three fastest radio pulsars, it seems reasonable that the existence of x-radiation may be linked with the length of time that has elapsed since the supernova explosion that formed the neutron star. Further, the ability of the pulsars to generate electromagnetic radiation with energy high enough to fall in the gamma ray range indicates that it is reasonable that they should also be able to accelerate matter to high energy (see Section 8.3).

The discovery of pulsars was a complete surprise that provided great excitement for astronomers over the last decade. What will the next decade bring?

SUMMARY AND OUTLINE OF CHAPTER 8

Aims: To see how stars of medium and heavy mass become supernovae, with the cores of the stars of medium mass condensing to become neutron stars. We observe the neutron stars as pulsars.

This entire chapter concerns the evolution of stars of more than 4 solar masses.

When they exhaust their hydrogen in the core, they become red giants.

Later, heavier elements are built up in the core, and the stars become supergiants (Section 8.1).

After iron cores form, the stars explode as supernovae. Only a few optical supernovae have been seen in our galaxy in the last 900 years. Supernova remnants can be studied with techniques of radio astronomy and x-ray astronomy (Section 8.2).

Cosmic rays, high-energy particles in space, probably come from supernovae (Section 8.3).

After a star's fusion stops, neutron degeneracy can support a remaining mass of up to 2 or 3 solar masses; more massive stars will be discussed in the next chapter.

A neutron star may be only 10 km in radius, and so is fantastically dense (Section 8.4).

Pulsars were discovered when a radio telescope was built that did not mask rapid time variations in the signal (Section 8.5).

Pulsars have very regular signals, although they have been discovered to be slowing down very slightly. Pulse periods range from .033 to 4 sec (Section 8.6).

Pulsars are considered to be objects that emit in the radio spectrum, although one (the Crab) also emits light and this and one other (Vela) also pulse x-rays and γ-rays. Three sources pulse only x-rays (Section 8.12).

Scientists have identified pulsars with rotating neutron stars by a process of elimination (Section 8.7):

They can't be full-size stars, because the pulse width is too small.

They can't be a binary pair of white dwarfs, because the white dwarfs couldn't orbit each other sufficiently rapidly.

They can't be a binary pair of neutron stars, because neutron stars would give off too much gravitational radiation and their orbital period would increase too rapidly.

They can't be oscillating white dwarfs or neutron stars, because the periods of oscillation would be too long or too short, respectively.

They can't be rotating white dwarfs, because white dwarfs can barely rotate fast enough to account for the first known pulsars and surely can't rotate fast enough to account for the pulsar in the Crab Nebula.

Rotating neutron stars are all that are left, and could satisfy the observations. Thus we accept the lighthouse model.

A pulsar is observed in the center of the Crab Nebula, a supernova remnant, where we would expect to observe a neutron star (Section 8.8).

We can tell the distance to pulsars from the dispersion in the arrival time of pulses of different frequencies. If we know the distance to a pulsar from other methods, then we can measure the electron density of interstellar space (Section 8.9).

Pulsars are gradually slowing down as they grow older. Two pulsars have shown "glitches"—brief speedups—and then resumed slowing down at the same rate as before (Section 8.10).

One pulsar has been found to be in a binary system, and has led to a confirmation of the general theory of relativity (Section 8.11).

An x-ray pulsar has been observed to be speeding up gradually, opposite to the ordinary pulsars (Section 8.12).

QUESTIONS

1. Why are red supergiants so bright?
2. What are the basic differences between a nova and a supernova?
3. If we see a massive main-sequence star (a heavyweight star), what can we assume about its age, relative to most stars? Why?
4. What do we know about the core of a star when it leaves the main sequence?
5. In a supernova explosion of a 20 solar mass star, about how much material is blown away?
6. A supernova can brighten by 20 magnitudes. Verify that this is a factor of 10^8.
7. A typical galaxy has a luminosity 10^{11} times that of the sun. If a supernova equals this luminosity, what was its luminosity before it exploded.
8. Would you expect the appearance of the Crab Nebula to change in the next 500 years?
9. If the light from a supernova 2000 parsecs from us was received in 1054, when did the star actually explode?

10. (a) What are cosmic rays?
    (b) Why do we think they arise in supernova explosions?

11. A pulsar has a period of 1 sec and a duty cycle of 5 per cent. What is the pulse width?

12. In your own words, fill in Table 8–1 with the reasons why each of the first 5 explanations of pulsars was ruled out.

13. How do we find the distances to pulsars?

14. If there is 1 electron every 30 cubic cm, how many electrons are in a "tube" 1 cm in diameter and 2000 parsecs long?

15. In view of current theories about supernovae and pulsars, list the pieces of evidence that indicate that the Crab pulsar is a young one.

16. How can you explain the fact that x-ray pulsars seem to be speeding up while ordinary pulsars are slowing down?

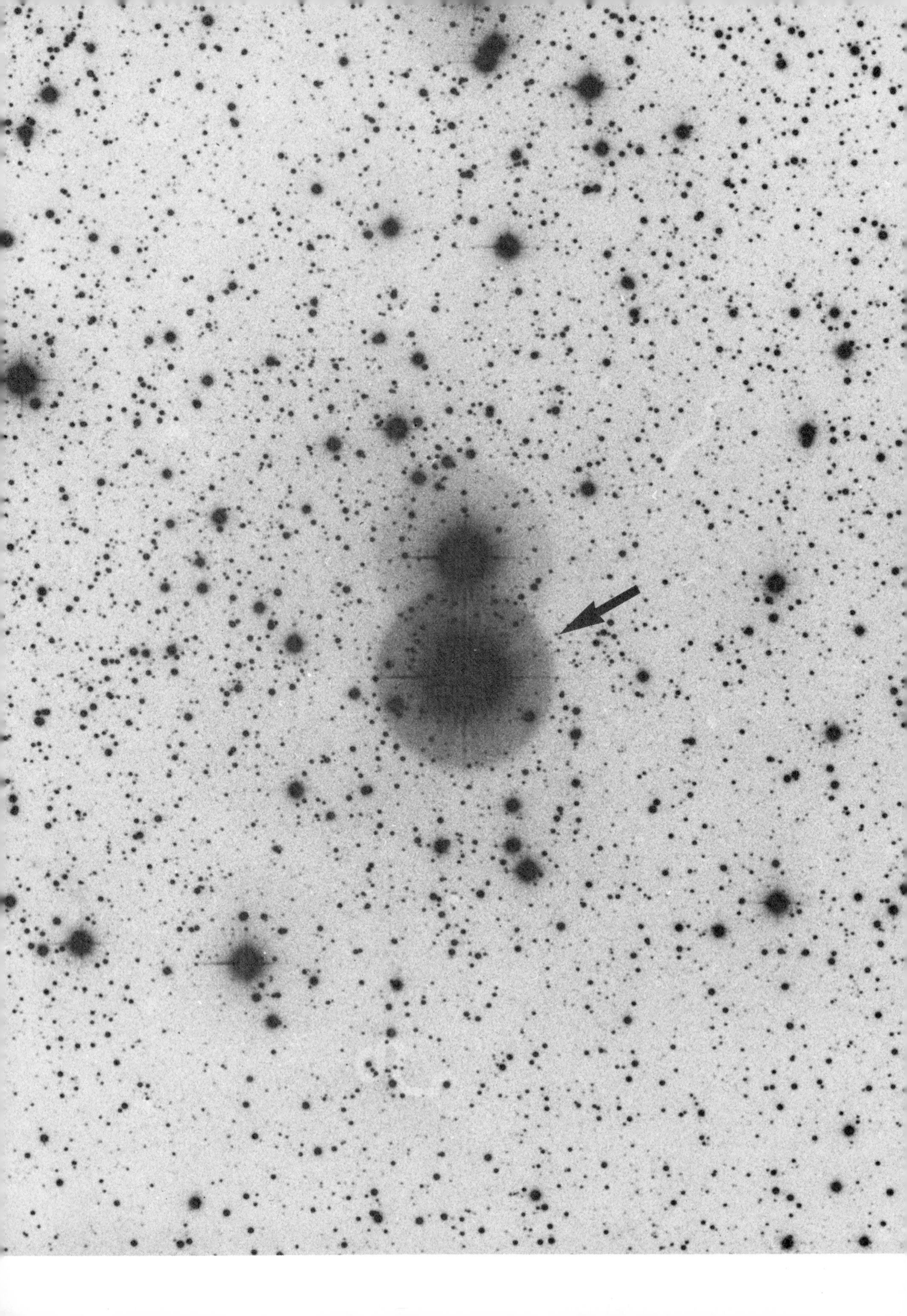

9

Black Holes

The strange forces of electron and neutron degeneracy support dying lightweight and middleweight stars against the force of gravity. The strangest case of all occurs at the death of heavyweight stars, which contained 8 to 60 or more solar masses when they were on the main sequence. These stars generally undergo supernova explosions like the middleweight stars, but some of the heavyweight stars may retain cores of over 2 or 3 solar masses. Nothing in the universe is strong enough to hold up the remaining masses against the force of gravity. The masses collapse, and continue to collapse forever. We call the result of such a collapse a *black hole.*

Do not confuse a black hole with a black body. A black body is merely a radiating body whose radiation follows Planck's law. A black hole, essentially, does not radiate (though an exception to this rule—really an alternate way of looking at the whole picture—will be discussed in Section 9.4).

9.1 THE FORMATION OF A STELLAR BLACK HOLE

When nuclear fusion ends in the core of a star, gravity causes the star to contract. We have seen in Chapter 7 that a star that has a final mass of less than 1.4 solar masses will end its life as a white dwarf. In Chapter 8 we saw that a heavier star will become a supernova, and that if the remaining mass of the core is less than 2 or 3 solar masses it will wind up as a neutron star. We are now able to observe both white dwarfs and neutron stars and so can study their properties directly.

It seems reasonable that in some cases more than 2 or 3 solar masses remain after the supernova explosion. The star collapses through the neutron star stage, and we know of no force—not even neutron degeneracy—that can stop the collapse. In some cases, the matter may have become so dense as the star collapsed that it passed through the neutron star stage before any explosion took place; in these cases, no supernova resulted.

For a long time, astronomers assumed that the most massive stars would somehow lose most of their mass before or in the process of collapsing, and end up as white dwarfs. That is one reason why astronomers have long been interested in studying the various mechanisms of *mass loss.* Several such

The overexposed dark object in the center of this negative print is a blue supergiant star HDE 226868, which is thought to be the companion of the first black hole to be discovered, Cygnus X-1.

processes are known: mass is known to flow gradually into space from the outer layers of giants and supergiants, and mass is ejected into space as the shell of a planetary nebula or in a supernova explosion. The solar wind represents such a net loss of mass from our sun. Even though stars with masses as large as sixty solar masses are known, it was long assumed that somehow almost all of the mass would be lost as the star evolved so that all stars could wind up as white dwarfs.

But astronomers no longer make that assumption. Pulsars have been identified with neutron stars, and some seem to have masses greater than 1.4 times that of the sun. Further, the image that most astronomers had decades ago of a placid universe was radically altered by the discovery of violent events that give rise to bursts of radiation in the x-ray or radio regions of the spectrum. No longer was there a preconception against the idea that very massive stars can collapse without losing most of their mass, winding up in exotic states.

The value for the maximum mass that a neutron star can have is a result of theoretical calculation. We must always be aware of the limits of accuracy of any calculation, and particularly a calculation that deals with matter in a state that is very different from states that we have been able to study experimentally. Still, the best modern calculations show that the limit of a neutron star mass is about 2 solar masses, and is probably less than 3 solar masses, though some theoreticians set higher limits.

We may then ask what happens to a 5 or 10 or 50 solar mass star as it collapses, if it retains more than 3 solar masses. It must keep collapsing, getting denser and denser. We have seen that Einstein's general theory of relativity predicts that a strong gravitational field will redshift radiation, and that this prediction has been verified both for the sun and for white dwarfs (as discussed in Section 7.5). Also, radiation will be bent by a gravitational field, or at least appear to us on earth as though it were bent. This prediction has been verified for the sun (see Section 5.11).

As the mass contracts, radiation is continuously redshifted more and more, and radiation leaving the star other than perpendicularly to the surface is bent more and more. Eventually, when the mass has been compressed to a certain size, radiation from the star can no longer escape into space. The star has withdrawn from our observable universe, in that we can no longer receive radiation from it. We say that the star has become a *black hole*.

Figure 9–1 As the star contracts, a light beam emitted other than radially outward will be bent.

Why do we call it a black hole? We think of a black surface as a surface that reflects none of the light that hits it. Similarly, any radiation that hits the surface of a black hole continues into the black hole and is not reflected. In this sense, the object is perfectly black.

Any ordinary surface is not truly "black." A black piece of paper, for example, may not reflect light and certainly doesn't emit any light of its own, but it does radiate infrared radiation corresponding to its temperature (which is presumably room temperature). A black hole, on the other hand, devours radiation at all wavelengths and emits nothing; no radiation would be seen coming from the black hole itself.

9.2 THE SPHERE OF LEAKING PHOTONS

Let us consider what happens to radiation emitted by the surface of a star as it contracts. Although what we will discuss affects radiation of all wavelengths, let us simply visualize standing on the surface of the collapsing star with a flashlight.

If we stand on the surface of a supergiant star, we note only very small effects of gravity on the light from our flashlight. We can shine the beam straight up, or at any angle, and it seems to go straight out into space.

As the star collapses, two effects begin to occur. Although we on the surface of the star cannot notice them ourselves, a friend of ours on a planet revolving around the star can detect the effects and radio back information to us about them. For one thing, our friend will see that our flashlight beam is redshifted; he could tell this either by observing a spectral line, if any, or by noting the shift of the peak of the Planck curve of the radiation toward the red. Second, our flashlight beam would be bent by the gravitational field of the star (Fig. 9–1). If we shined the beam straight up, it would continue to go straight up. But if we shined it away from the vertical, the beam would be bent slightly away from the vertical.

Figure 9–2 Light can be bent so that it falls back onto the star.

As the star contracts more and more, and the gravity gets stronger and stronger, we would begin to find that when we shined our flashlight at low angles to the horizon, the light would be bent sufficiently that it turned downward and hit the star (Fig. 9–2). This light thus never escapes into space. Still, if we shine our flashlight in the vertical direction, the light escapes.

Only if the flashlight is pointed within a certain angle of the vertical does the light continue outward. This angle forms a cone, with its apex at the flashlight, and is called the *exit cone.* As the star grows smaller yet, we find that the flashlight has to be pointed more directly upward in order for its light to escape. The exit cone grows smaller as the star continues to shrink.

When we shine our flashlight upward in the exit cone, the light escapes. When we shine our flashlight in a direction outside the exit cone, the light is bent sufficiently that it falls back to the surface of the star. When we shine our flashlight exactly along the side of the exit cone, the light goes into orbit around the star, neither escaping nor returning to the surface (Fig. 9–3). The alignment must be precise for this to happen; if the angle is the slightest bit off the exit cone, the light will either gradually escape or gradually return (though it might circle the star many times before it does so).

Figure 9–3 When the star has contracted enough, only light emitted within the exit cone escapes. Light emitted on the exit cone goes into the photon sphere.

The sphere around the star in which the light can orbit is called the *photon sphere.* Its size can be calculated theoretically. It is 13.5 km in radius for a star of 3 solar masses, and its size varies linearly with the mass; that is, it is 27 km in radius for a star of 6 solar masses, and so on.

As the star continues to contract, the exit cone gets narrower and narrower. Light emitted within the exit cone edge still escapes. The photon sphere remains at the same height even though the matter inside it has contracted further. Photons that are just outside the photon sphere leak away and can eventually reach us.

9.3 THE EVENT HORIZON

It we were to depend on our intuition, which is based on classical physics as advanced by Newton, we might think that the exit cone would simply continue to get narrower. But when we apply the general theory of relativity, we find that at a certain radius the cone vanishes. Light no longer

The Schwarzschild solutions *of equations from Einstein's general theory of relativity are the theoretical basis for the existence of black holes.*

Figure 9–4 When the star becomes smaller than its Schwarzschild radius, we can no longer observe it. We say that it has passed its *event horizon,* by analogy to the statement that we cannot see a thing on earth once it has passed our horizon.

can escape into space, even when it is traveling straight up, away from the center of the gravitational mass.

The solution to Einstein's equations that predicts this was worked out by Karl Schwarzschild in 1916, shortly after Einstein advanced his general theory. The radius of the star at the time at which light can no longer escape is called the *Schwarzschild radius* or the *gravitational radius,* and the spherical surface at that radius is called the *event horizon* (Fig. 9–4). The radius of the photon sphere is exactly 3/2 times this Schwarzschild radius.

We can visualize the event horizon in another way, by considering a classical picture (that is, one based on the Newtonian theory of gravitation). The picture is essentially that conceived in 1796 by Laplace (who was also responsible for the nebular hypothesis of planetary formation as described in Section 10.5). A body must have a certain velocity, called the *escape velocity,* to escape from the gravitational pull of another body. For example, we have to launch rockets at 11 km/sec (25,000 miles/hr) in order for them to escape from the earth's gravity. For a more massive body of the same size, the escape velocity would be higher. Now imagine that this body contracts, and we are drawn closer to the center of the mass with all the mass concentrated in a sphere below us. As this happens, the escape velocity would rise. As the body crosses its Schwarzschild radius, the escape velocity becomes equal to the speed of light. Thus even light cannot escape. If we then begin to apply the special theory of relativity, we might then reason that since nothing can go faster than the speed of light, nothing can escape. Now let us return to the picture according to the general theory of relativity.

"Now, here, you see, it takes all the running you can do, to keep in the same place. If you want to get somewhere else, you must run twice as fast as that."
The Red Queen,
Through the Looking Glass
by Lewis Carroll, 1871

The size of the Schwarzschild radius depends linearly on the amount of mass that is collapsing.* A star of three solar masses, for example, would have a Schwarzschild radius of 9 km. A star of six solar masses would have a Schwarzschild radius of twice 9 km, or 18 km. One can calculate the Schwarzschild radii for less massive stars as well, although the less massive stars would be held up in the white dwarf or neutron star stages and not

*The formula is $R = \frac{2\,GM}{c^2}$, where G is the universal gravitational constant $= 6.67 \times 10^{-8}$ when the mass (M) is expressed in grams, the radius (R) in centimeters, and the speed of light (c) in centimeters/second. This works out to mean that R is approximately the mass × 3 kilometers when the mass is expressed in terms of the mass of the sun. The radius of the photon sphere is 3/2 times larger, and equals the mass (in terms of solar masses) × 4.5 km.

Figure 9–5 This drawing by Charles Addams is reprinted with permission of *The New Yorker Magazine, Inc.,* © 1974.

collapse to their Schwarzschild radii. The sun's Schwarzschild radius is only 3 km. Note that neutron stars would not have to collapse very much before they reached their event horizons; nevertheless, they are unable to do so. The Schwarzschild radius for the earth is only 9 mm; that is, the earth would have to be compressed to a sphere only 9 mm in radius in order to form an event horizon and be a black hole.

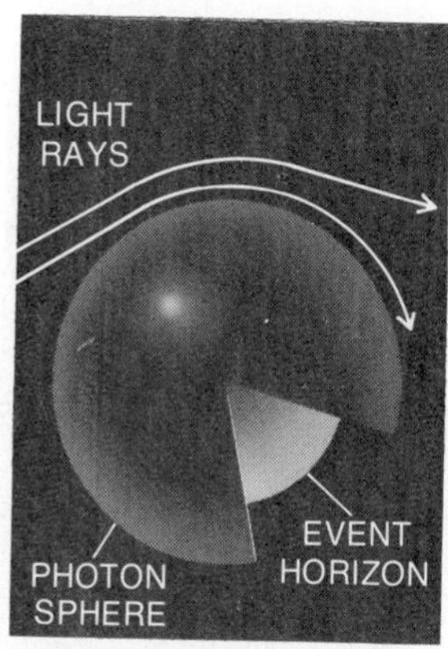

Figure 9–6 Light rays from afar can orbit a black hole in the *photon sphere.* The two light rays shown here will both escape from the black hole, though one is only slightly outside the photon sphere and so barely escapes. The light rays are, of course, deflected.

Note that anyone or anything on the surface of a star as it passed its event horizon would not be able to survive. An observer would be torn apart by the tremendous difference in gravity between head and foot. If that effect could be ignored, though, the observer on the surface of the star would not notice anything particularly wrong as the star passed its event horizon. If we chose to stay with the observer at that time, we could still be pointing our flashlights up into space trying to signal our friends on earth. But once we passed the event horizon, no answer would ever come because our signal would never get out.

What if light (or other radiation) from afar passed close by the contracting mass? If it were pointed sufficiently on line, it would be bent by the strong gravity and enter the hole. If it were pointed sufficiently off line, it would be bent by gravity, but could continue en route without hitting the hole. If light were approaching the black hole at just the specific angle in between these two situations, it could go into orbit around the hole in the photon sphere (Fig. 9–6).

Once the star passes inside its event horizon, it continues to contract. Nothing can ever stop its contraction. In fact, the mathematical theory predicts that it will contract to zero radius, a situation that seems physically impossible to conceive of. The point at which it will have zero radius (it has infinite density there) is called a *singularity.* Strange as it seems, theory predicts that a black hole contains a singularity.

Since we cannot ever hope to hold up a meter stick from the center to the event horizon of a black hole to measure its radius—after all, there is a singularity at its center—the radius of a black hole must be defined differently. We could, in principle, wrap a tape measure around a black hole to measure its circumference. We define the radius of a black hole as its circumference divided by 2π.

Even though the mass that causes the black hole has contracted further, the event horizon doesn't change. It remains at the same radius forever, as long as the amount of mass inside doesn't change. Any matter that ventures too close to the black hole will be sucked in, and the mass inside the event horizon will become larger. Thus, since the event horizon depends on the amount of mass, the event horizon will become slightly larger. As the black hole sits in space, or even moves through space as it continues the proper motion across space that the original star had, it may encounter material that it can suck up. Thus the black hole will continually grow larger. It will consume whatever it encounters, and nothing can stop it.

Once matter is inside a black hole, it loses its identity in the sense that from outside a black hole, all we can tell is the mass of the black hole, the rate at which it is spinning, and what electric charge it has. These three quantities are sufficient to completely describe the black hole. Thus, in a sense, black holes are simple objects to describe physically, because we only have to know three numbers to characterize each one. By contrast, for the earth we have to know shapes, sizes, densities, motions, and other parameters for the interior, the surface, and the atmosphere. The theorem that describes the simplicity of black holes is often colloquially stated by astronomers active in the field as "a black hole has no hair"; that is, a black hole has no basic properties at all that can be described aside from mass, spin, and charge. All other properties, such as size, can be derived from these three basic properties.

Most of the theoretical calculations about black holes, and the Schwarzschild solutions in particular, are based on the assumption that black holes do not rotate. But this assumption is only a convenience; we think, in fact, that the rotation of a black hole is one of its important properties. It was not until 1963 that Roy P. Kerr solved Einstein's equations for a situation that was later interpreted in terms of the notion that a black hole is rotating. In this more general case, there are two event horizons with somewhat different properties. Unlike the situation for a non-rotating black hole, in a rotating black hole (Fig. 9–7) it is not necessary that matter inside the event horizons reach the singularity. The matter might be able to reappear at some other point in our universe, or even reappear in our universe at some other time in the past or future! And, also unlike the case of a non-rotating black hole, for which the singularity is always unreachably hidden within an event horizon, when the rotation is very fast Kerr's solutions allow the existence of a singularity out in the open, not surrounded by an event horizon. Such a point is called a *naked singularity* and, if one exists, we would have no warning before we ran into it. Most theoreticians assume the existence of a law of "cosmic censorship," which requires all singularities to be "clothed" in event horizons, that is, not naked.

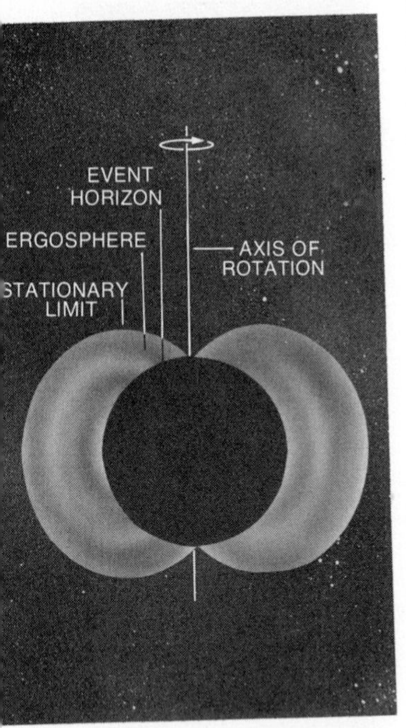

Figure 9–7 The description of a black hole so far in this chapter has assumed that the black hole is not rotating. But it seems reasonable, or even likely, that a black hole would rotate (just as a neutron star is rotating rapidly); a set of solutions to Einstein's equations discovered by Roy P. Kerr can be interpreted to describe a rotating black hole. From the top, the object looks like an ordinary non-rotating black hole. But from the side, one finds a region called the *ergosphere* because, in principle, work (which in Greek is *ergon*) can be extracted from it. The ergosphere is bounded on the outside by the *stationary limit;* within this limit nothing can be at rest. Light directed in the sense of the black hole's rotation will escape from the ergosphere, and an astronaut could too.

An object can be shot into the ergosphere of a rotating black hole at such an angle that, after it is made to split into two, one part continues into the event horizon while the other part is ejected from the ergosphere with more energy than the incoming particle had. This energy must have come from somewhere; it could only have come from the rotational energy of the black hole because there are no other possibilities. Thus by judicious use of the ergosphere, we can tap the energy of the black hole. We are always looking for new energy sources and this is the most efficient known, although it is obviously very far from being practical.

9.4 NON-STELLAR BLACK HOLES

We have discussed how black holes can form by the collapse of a star of greater than 3 solar masses. But theoretically a black hole will result if a mass of any amount is sufficiently compressed. (A neutron star, for example, could become a black hole if enough mass were added to it.) No mass less than 3 solar masses will contract sufficiently under the force of its own gravity in the course of stellar evolution. But the density of matter was so high at the time of the origin of the universe (see Chapter 26) that smaller masses may have been sufficiently compressed to form what are called mini black holes.

There is no good evidence for a mini hole, but some scientists predict that there could be some the size of pinheads (and thus masses equivalent to those of asteroids) floating around in space. Occasionally it is suggested that some untoward event that occurs on earth is caused by one. For example, in 1908 there was a tremendous explosion in Siberia that leveled trees in an area fifty miles around. All the trees were lying so that they were pointing away from the center of the explosion. No trace of a meteor (see Chapter 18) has ever been found at the site, and a collision of a mini black hole with the earth has been suggested as the source of the explosion. Attempts to search records from places near Iceland, where the mini black hole would have emerged after passing through the earth, did not work out. Other more prosaic explanations for the explosion, such as a collision with the ice, dust, and rock from a comet head, have also been suggested, so in the absence of contradictory evidence we need not, and thus should not, assume that such an odd thing as a black hole was involved. We may never really know.

But speculation continues, and it would be nice if a mini hole could be detected and corralled. If we could somehow store a spinning mini black hole by putting it in orbit around the earth, we could gradually extract energy from the black hole's rotation. We could send rockets near it, and arrange their trajectories such that the rockets were given additional energy by the black hole. We could then return that energy to earth for our use.

Stephen Hawking, the English astrophysicist who suggested the existence of mini black holes, has further deduced that small black holes can emit energy in the form of elementary particles (neutrinos, and so forth). The mini holes would thus evaporate and disappear. This seems a contradiction to the concept that mass can't escape from a black hole. But the exception is only important for the smallest mini black holes, for the amount of radiation decreases sharply as we consider black holes of greater and greater masses. Only mini black holes up to the mass of an asteroid—far short of stellar masses—would have had time to disappear since the origin of the universe. Hawking's ideas set a lower limit on the size of black holes now in existence, since we think the mini black holes were formed only at the time of the big bang (which is discussed in Chapter 26).

On the other extreme of mass, we can consider what a black hole would be like if it contained a very large number, i.e., thousands or millions, of solar masses. The density is very high when a black hole of the mass of even a large star is formed. For a black hole of larger mass, the density that it had when it went through its event horizon was smaller. If we consider a black hole of 3 hundred million solar masses (3×10^8 $M_\odot$), then the density is reduced by 10^{16} over the density of a 3 solar mass ($3M_\odot$) black hole. The density would be quite low when this mass disappeared within its event horizon, even approaching that of water. The radius of this black hole would be a hundred million times the radius of a 3 solar mass black hole, namely, 900 million km. This is only $1/_{10,000}$ of a light year, or 6 A.U., approximately the size of the orbit of Jupiter.

The Schwarzschild radius of a black hole varies linearly with the mass and so the volume contained inside the event horizon varies with the cube of the mass. The density, which is the mass divided by the volume, is thus proportional to one over the square of the mass. Density = mass ÷ volume $\propto m \div d^3 = m \div m^3 = 1/m^2$. For very large masses, therefore, the density can be very low.

Thus if we were traveling through the universe in a spaceship, we couldn't count on detecting a black hole by noticing a volume of high density. We could pass through the event horizon of a high-mass black hole without even noticing. We would never be able to get out, but it would be hours before we would notice that we were being drawn into the center at an accelerating rate.

Where could such a supermassive black hole be located? As we saw in Section 4.5c, there is some evidence that black holes of about 1000 solar masses are at the center of globular clusters. The center of our galaxy is one possibility for an even more massive black hole. Strange bursts of x-ray emission have been detected, and it is reasonable that there could be a concentration of mass there (see Chapter 22). Though we do not expect to observe radiation from the black hole itself, we hope to observe radiation from the gas surrounding it. Quasars are another kind of object we do not understand very well, and perhaps there are black holes, or even white holes, at their centers (see Chapter 24). We have a lot to learn.

Some scientists think that the universe itself may be inside a black hole. The universe is currently expanding (Chapter 23), but if it were inside a black hole it would eventually collapse. Current observations do not show sufficient density in the universe to provide enough gravity for this collapse to take place (as will be discussed in Section 25.7).

Figure 9–8 Lewis Carroll's Cheshire Cat, from *Alice's Adventures in Wonderland,* shown here in John Tenniel's drawings, is analogous to a black hole in that it left its grin behind when it disappeared while a black hole leaves its gravity behind when its mass disappears. Alice thought that the Cheshire Cat's persisting grin was "the most curious thing I ever saw in all my life!" We might say the same about the black hole and its persisting gravity.

9.5 DETECTING A BLACK HOLE

What if we were watching the black hole from the outside? Photons that were just barely inside the exit cone when emitted would leak very slowly away from the black hole, as would external radiation that hit the photon sphere almost but not quite tangent to its surface. Thus radiation from the vicinity of the black hole would never totally disappear, because there would always be a small amount of radiation spiraling out from just above the photon sphere. The closer the photons were to the exit cone, the longer they would take to reach us. Also, light emitted just before the surface reached its Schwarzschild radius would be visible although severely redshifted.

Nevertheless, these few photons would be spread out in time, and the image would be very faint. Further, as the star went through its event horizon, it would get very red and then blink out. It would do this in a fraction of a second, so the odds that we would actually be watching a collapsing star as it went through the event horizon.

But all hope is not lost for detecting a black hole, even though we can't hope to see the photons from the moments of collapse. The black hole disappears, but it leaves its gravity behind. It is a bit like the Cheshire Cat from *Alice's Adventures in Wonderland,* which fades away, leaving only its grin behind (Fig. 9–8).

The black hole attracts matter, and the matter accelerates toward it. Some of the matter will be pulled directly into the black hole, never to be seen again. But other matter will go into orbit around the black hole, and will orbit at a high velocity.

It seems likely that the gas in orbit will be heated. Theoretical calculations of the friction that will take place between adjacent filaments of gas in orbit around the black hole show that the heating will be so great that the gas will radiate strongly in the x-ray region of the spectrum. Thus, though we cannot observe the black hole itself, we can hope to observe x-rays from the gas surrounding it.

In fact, a large number of x-ray sources are known in the sky. They have been best surveyed from a NASA satellite called Uhuru. Some of these sources have been identified with galaxies or quasars. Others pulse regularly, like Hercules X-1, and are undoubtedly neutron stars (see Section 8.12). Some of the rest, which pulse sporadically, may be related to black holes.

The Uhuru satellite was launched in 1970 in an international effort by an Italian team from a site in Kenya; Uhuru means "freedom" in Swahili, and is the Kenyan national motto.

It is not enough to find an x-ray source that gives off sporadic pulses, for one can think of other mechanisms besides revolution of matter around a black hole that can lead to such pulses. One would like to show that a collapsed star of greater than 3 solar masses is present.

In Section 4.2, we discussed how masses are determined from studies of binary stars. Our hope of pinning down the mass of a collapsed object, and thus identifying it with a black hole, rests on such a study.

The study of spectroscopic binaries, which we discussed in Section 4.1, turns out to be important for our hope of discovering a black hole.

When we search the position of the x-ray sources we are interested in finding a spectroscopic binary (that is, a star whose spectrum shows a Doppler shift that indicates the presence of an invisible companion). Then, if we can show that the companion is too faint to be a normal, main-sequence star, it must be a collapsed star. If, further, the mass of the unobservable companion is greater than 3 solar masses, it must be a black hole.

Many cases of spectroscopic binaries are known, and they must all be examined carefully. Before 1970, a list of such binaries was examined to see if any met the conditions above. None did. After Uhuru compiled a catalogue of x-ray sources in 1972, the x-ray sources that emitted sporadic pulses of radiation were examined, and it was discovered with optical telescopes that some of these sources were previously unsuspected optical spectroscopic binaries.

Four of these binaries are particularly suspicious. Analysis of the spectral variations indicates that three of them have invisible stars as one member of the binary, and that the mass of the invisible star is between 2 and 3 solar masses. Thus these companions could be black holes, but the derived value is too close to the calculated limit of mass at which stars do not have to become black holes for us to have confidence in the deduction.

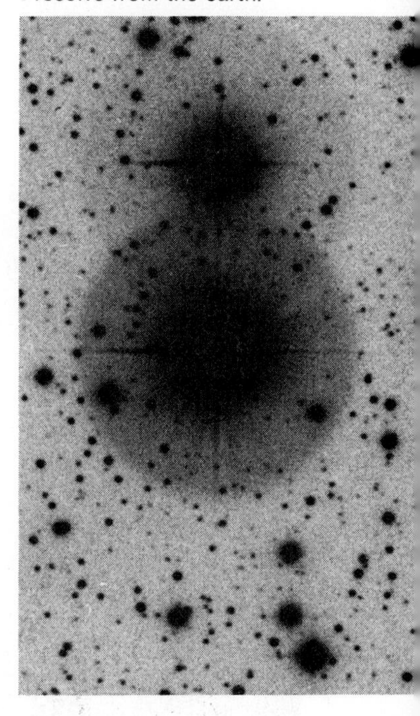

Figure 9–9 A section of the image reproduced in the photograph opening this chapter of the blue supergiant star HDE 226868. The black hole Cygnus X-1 that is thought to be orbiting the supergiant star is not visible. Note that the fact that the image of the supergiant appears so large is entirely an overexposure effect in the film and does not represent the actual angular diameter subtended by the star, which is too small to resolve from the earth.

The remaining binary, on the other hand, seems to be a much more persuasive case. This x-ray source is named Cygnus X-1, and was the first x-ray source that was found to vary in intensity on a time scale of milliseconds. In 1971, radio radiation was found to come from the same direction. For instrumental reasons (see Section 23.5b on interferometry), the position of the radio source could be measured more precisely than the position of the x-ray source. A 9th magnitude star called HDE 226868 was found at that location (Fig. 9–9 and page 246). The identification of the x-ray source with the radio source was proved conclusively when both simultaneously underwent an abrupt change in the intensity of the radiation emitted (Fig. 9–10).

HDE 226868 has the spectrum of a blue supergiant, and thus has a mass of about 15 times that of the sun. Its spectrum is observed to vary in radial velocity with a period of 5.6 days, indicating that the supergiant and the invisible companion are orbiting each other with that period. From the orbit, it is deduced that the invisible companion must certainly have a mass greater than 4 solar masses; the best estimate is 8 solar masses. Because this is so much greater than the limit of 2 or 3 solar masses above which neutron stars cannot exist, it seems that even allowing for possible errors in measurement or in the theoretical calculations, too much mass is present to allow the matter to settle down as a neutron star. Thus many, and

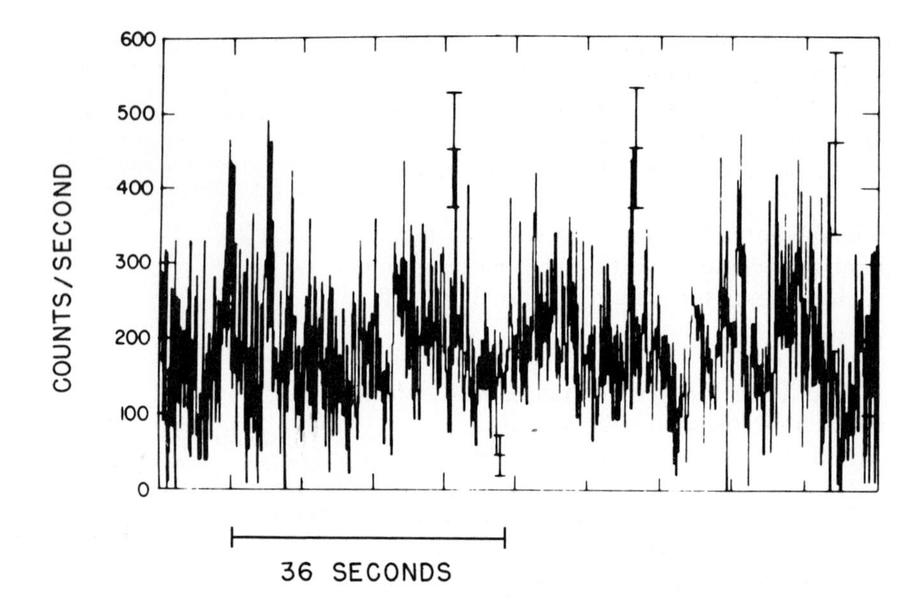

Figure 9–10 Variations in the x-ray intensity of Cygnus X-1, as observed from the Uhuru satellite.

probably most, astronomers believe that a black hole has been found in Cygnus X-1.

We must be careful to consider the uncertainties in this chain of deductions. Although HDE 226868's spectrum appears normal, the star is certainly in an abnormal situation if it is indeed orbiting mutually with a black hole. Thus we cannot be certain that it fits normally on an H-R diagram, although we have no specific information that it doesn't. Once we assume that HDE 226868 has a normal spectrum we use the method of spectroscopic parallax to find its absolute magnitude. (We can compare its absolute magnitude, in turn, with its apparent magnitude to find out that it is 10,000 light years away.) To find its mass, we also use the assumption that it is a normal blue supergiant.

Further, it has been suggested that the system could involve three and not just two stars. If this were the case, then the mass of the x-ray emitting star need not be so high. This suggestion is not generally accepted, though it would provide a way of accounting for the observed characteristics of Cygnus X-1 without requiring the presence of a black hole.

Figure 9–11 An artist's conception of the disk of swirling gas that would develop around a black hole like Cygnus X-1 (*right*) as its gravity pulled matter off the companion supergiant (*left*). The x-radiation would arise in the disk. The painting is by Lois Cohen and is used courtesy of the Griffith Observatory.

Figure 9–12 High-Energy Astronomy Observatories (HEAO's) A and B. They are each about 5 meters long and 3 tons in weight.

Calculations have been carried out, especially by Kip Thorne of Caltech, to predict how the supergiant and the black hole would interact if we accept the binary theory. Thorne finds that the binary theory predicts that the gravity of the black hole would attract matter from the outer layers of the supergiant, and that this matter would go into orbit around the black hole. Because of the orbital motion of the matter, it would form a disk around the black hole, not unlike a large-scale version of Saturn's rings (Fig. 9–11). As the matter is heated, it gives off x-rays. Thorne's calculations and computer simulations predict the shape of the disk and of other aspects of the interaction, given certain assumptions.

Matter in the disk around the black hole would orbit very quickly. If a hot spot developed somewhere on the disk, it might beam a cone of radiation into space, similar to the lighthouse model of pulsars. If the hot spot lasted for several rotations, we could detect a pulse of x-rays every time the cone swept past the earth. Thorne predicts that the period of the x-ray pulses would be extremely short, only a few milliseconds.

The Uhuru x-ray telescopes are neither large enough nor sensitive enough to variations of the signal with time to detect such periodic structure. But in 1977 NASA plans to launch HEAO-A (High-Energy Astronomy Observatory A), with a large, sensitive x-ray telescope (Fig. 9–12). Perhaps the rapid pulses will be detected. Probably the results will be inconclusive. Even if the pulses are not detected, it would not argue against Cygnus X-1 being a black hole. Only the hot spot theory would be tested.

It still looks as though Cygnus X-1 represents the first observational detection of a black hole. The new generation of x-ray telescopes should tell us more.

9.6 GRAVITATIONAL WAVES

Most of the data we have thus far discussed in this book have come from studies of the electromagnetic spectrum. But Einstein's general

Figure 9–13 Joseph Weber and his gravitational wave detector, a large aluminum cylinder, weighing 4 tons, delicately suspended so that gravitational waves would set it vibrating 1660 times per second.

theory of relativity predicts the existence of still another type of signal that we might be able to detect: gravitational waves.

Whereas electromagnetic waves affect only charged particles, we would expect gravitational waves to affect all matter. But gravitational waves, if they exist, would probably be very weak. They would be emitted only in situations where large masses were being accelerated rapidly. In the previous chapter, we mentioned one possible example: two neutron stars revolving in close orbits around each other. Another possible example would be a supernova or the collapse of a massive star to become a black hole. Even a simple close binary system containing a white dwarf or a neutron star will give out a reasonable amount of gravitational radiation.

Joseph Weber of the University of Maryland was the first to build sensitive apparatus to try to search for gravitational waves. His detector (Fig. 9–13) is a large metal cylinder, delicately suspended from wires so as to insulate it from restraint and from local motion. When a gravitational wave hits the cylinder, it should begin vibrating. Weber has sensitive apparatus to measure the vibrations.

Starting in 1969, Weber reported success in detecting gravitational waves. He even set up a second cylinder in Chicago, reasoning that the chance was negligible of anything but a gravitational wave simultaneously affecting the two cylinders in Maryland and Chicago. And he soon reported coincident "events" starting both his cylinders vibrating simultaneously.

Weber's results seemed exciting, but they were treated with skepticism by many scientists because the frequency of events is much higher than we would have expected. Nobody had predicted, for example, that black holes would form as often in our galaxy as Weber's results would indicate, or that so much mass would fall into a giant black hole even if there were one in the center of our galaxy.

Eventually other scientists built other gravitational wave apparatus, some in configurations that they had calculated would be more sensitive than Weber's. None of the other scientists has detected any gravitational waves, and most astronomers feel that Weber's detectors must have been affected by other kinds of events, perhaps of terrestrial origin.

Nonetheless, Weber gets credit for having stimulated this exciting field. Several groups continue to look for gravitational waves, and continually refine their equipment so as to allow them to detect weaker signals or waves of different wavelength. For example, a proposal has been made to put a spacecraft far into space with a fantastically accurate clock aboard to generate ultra-precisely timed radio signals. Gravity waves of very long wavelength compared to Weber's would cause fluctuations in the motion of spacecraft with respect to the earth.

Even though Weber may not have detected gravitational waves yet, many scientists hope that some will be detected soon and that this will lead us to a better understanding of gravitational events.

SUMMARY AND OUTLINE OF CHAPTER 9

Aims: To understand what black holes are, how they form, and how they might be detected

Gravitational collapse occurs if more than 2 or 3 $M_{\odot}$ remains (Section 9.1)
Critical radii of a black hole
  Photon sphere: exit cones form (Section 9.2)
  Event horizon: exit cones close (Section 9.3)
  Schwarzschild radius (gravitational radius) defines the limit of the black hole, the event horizon
  Singularity
Non-stellar black holes (Section 9.4)
  Mini black holes could have formed in the big bang
  Very massive black holes do not have high densities
Detecting a black hole (Section 9.5)
  X-radiation expected
  Detection in a spectroscopic binary, invisible high-mass companion
  Cygnus X-1: a black hole?
Gravitational waves (Section 9.6)
  They are not electromagnetic waves
  They might set matter on earth vibrating, though it is difficult to show that the detectors have been set vibrating by gravitational waves rather than by other causes
  Current situation: probably not yet detected

QUESTIONS

1. Why doesn't electron or neutron degeneracy prevent a star from becoming a black hole?

2. Why is a black hole blacker than a black piece of paper?

3. (a) Is light acting more like a particle or more like a wave when it is bent by gravity?
   (b) Can you explain the bending of light as a property of a warping of space, as discussed in Section 5.11?

4. (a) How does the escape velocity of the moon compare with the escape velocity of the earth? Would a larger rocket engine be necessary to escape from the gravity of the earth or from the gravity of the moon?
   (b) How will the velocity of escape from the surface of the sun change when the sun becomes a red giant? A white dwarf?

5. (a) What is the Schwarzschild radius for a 10 solar mass star?
   (b) What is your Schwarzschild radius?

6. What is the relation in size of the photon sphere and the event horizon? If you were an astronaut in space, could you escape from within the photon sphere of a black hole? Could you escape from within its event horizon?

7. (a) Why does the mass of a black hole that results from a collapsed star tend to increase rather than decrease with time?
   (b) What property would show up in mini black holes, if they exist, that would allow them to eventually lose mass?

8. Would we always notice when we reached a black hole by its high density? Explain.

9. Could we detect a black hole that was not part of a binary system?

10. (a) Under what circumstances does the presence of an x-ray source associated with a spectroscopic binary suggest to many astronomers the possible presence of a black hole? Why?
    (b) For what additional properties of the objects in the binary system do we search?

Part III

The Solar System

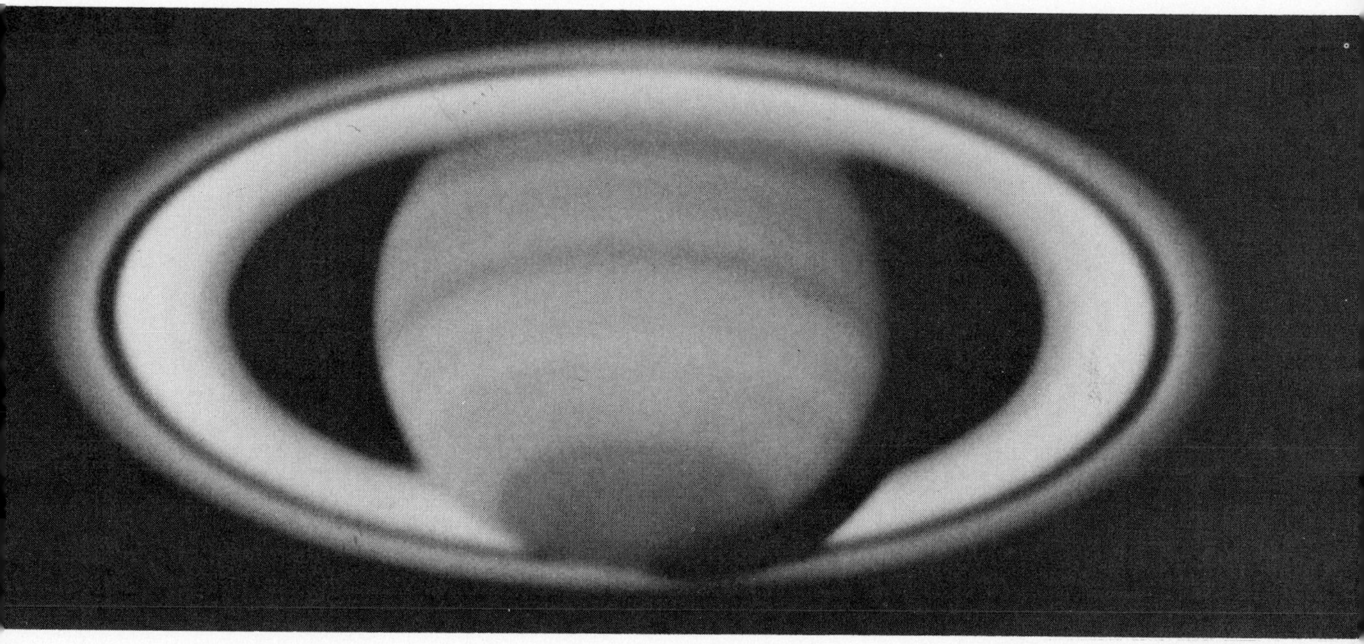

Saturn

Viking 1 pushes aside a rock on Mars, in order to sample soil that has been protected from solar ultraviolet radiation.

The Solar System . . .

The Earth and the rest of the solar system may be important to us, but they are only minor companions to the stars. In *Captain Stormfield's Visit to Heaven,* by Mark Twain, the Captain races with a comet and gets off course. He comes into heaven by a wrong gate, and finds that nobody there has heard "of the world" ("**the** world, there's billions of them!" says a gatekeeper). Finally, the gatekeepers send a man up in a balloon to try to detect "the world" on a huge map. The balloonist has to travel so far that he rises into clouds and after a day or two of searching he comes back to report that he has found it: an unimportant planet, more properly named "the Wart."

We too must learn humility as we ponder the other objects in space. And while it is no doubt the case that the heavens are filled with a vast assortment of suns and planets more spectacular than our own, still, the solar system is our own local environment. We would like to understand it and come to terms with it as best we can. Besides, in understanding our own solar system, we may even find some keys to understanding the rest of the Universe.

To get an idea of its scale, imagine that the solar system is scaled down and placed on a map of the United States. Let us say that the sun is a hot ball of gas taking up all of Rockefeller Center, more than a kilometer across, in the center of New York City.

We would then find Mercury to be a ball 4 meters across at the distance of mid–Long Island, and Venus to be a 10 meter ball one and a half times farther away. The Earth is only slightly bigger and is located at the distance of Trenton, New Jersey. Mars is half that size, 5 meters across, located past Philadelphia.

Only for the planets beyond Mars would the planets be much different in size from the Earth, and the separations become much greater. Jupiter is 100 meters across, the size of a baseball stadium, past Pittsburgh at the Ohio line. Saturn without its rings is a little smaller than Jupiter (including the rings it is a little larger), and is past Cincinnati toward the Indiana line. Uranus and Neptune are each about 30 meters across, about the size of a baseball infield, and are at the distances of Topeka and Santa Fe, respectively. And Pluto, as small as the inner planets, is as far away as Los Angeles, 40 times farther away from the sun than the Earth is. Occasionally a comet sweeps in from Alaska, or some other random direction, passes around the Sun, and returns in the general direction from which it came.

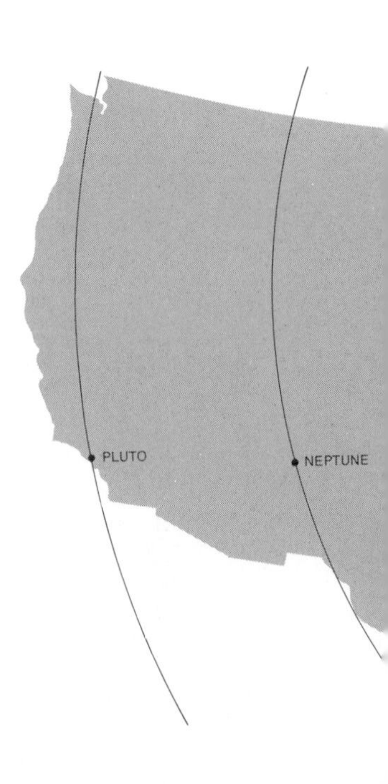

The planets fall naturally into two groups. The first group, the *terrestrial planets,* consists of Mercury, Venus, Earth, and Mars. All are rocky in nature. The terrestrial planets are not very large, and have densities about five times that of water. (When the metric system was set up, the gram was defined so that water would have a convenient density of exactly 1 gram/cubic centimeter. Thus the terrestrial planets have densities of about 5 grams/cm^3.)

The second group, the *giant planets,* consists of Jupiter, Saturn, Uranus, and Neptune. All these planets are much larger than the terrestrial planets, and also much less dense, ranging down to slightly below the density of water. Jupiter and Saturn are largely gaseous in nature, similar to the Sun in composition. Uranus and Neptune have large gaseous atmospheres but have rocky or icy surfaces and interiors. Pluto, planet number nine, is anomalous in several of its properties, as we shall discuss in Section 17.3, and so may have had a very different history from the other planets.

Between the orbits of the terrestrial planets and the orbits of the giant planets are the orbits of thousands of chunks of small "minor planets." These *asteroids* range up to 1000 kilometers across. Sometimes much smaller chunks of interplanetary rock penetrate the Earth's atmosphere and hit the ground. We shall discuss these *meteorites* and where they came from together with asteroids and comets in Chapter 18.

Many people are interested in the planets in order to study their history—how they formed, how they have evolved since, and how they will change in the future. Others are more interested in what the planets are like today. Still others are interested in the planets mainly to consider whether they are harboring intelligent life. The study of the origin of the solar system and the study of the origin of the Universe have the same name: *cosmogony.* And because the Moon, planets, and other objects in the solar system have undergone less water erosion on their surfaces than has the Earth, in observing these celestial bodies we actually see them as they appeared eons ago. In this way the study of the planets and, in particular, their surfaces, provides information about the solar system's early stages and its cosmogony. So even though the planets together contain only about one tenth of one percent of the mass of the Sun, their study may help us discover the origins of the solar system.

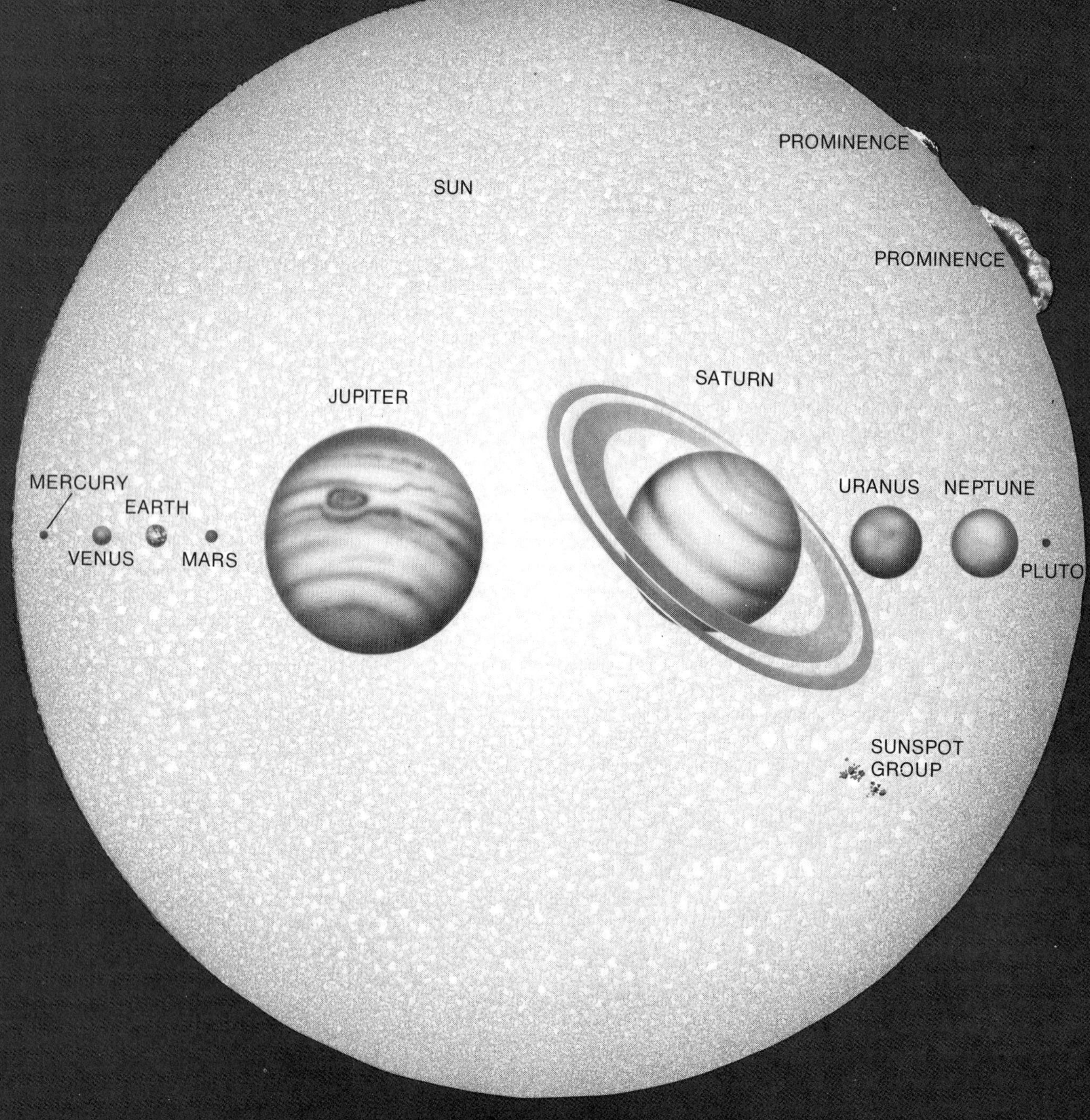

PROMINENCE

SUN

PROMINENCE

SATURN

JUPITER

MERCURY

URANUS

NEPTUNE

EARTH

VENUS

MARS

PLUTO

SUNSPOT
GROUP

10

The Planets and Their Origins

10.1 THE PHASES OF THE MOON AND PLANETS

"The Moon" is often capitalized to distinguish it from moons of other planets; in general writing, it is usually written with a small "m." In Part III, we shall capitalize "Earth" to put it on a par with the other planets and the Moon. For consistency, we shall also capitalize "Sun" and "Universe."

From the simple observation that the apparent shapes of the Moon and planets change, we can make conclusions that are important for our understanding of the mechanics of the solar system. The fact that the Moon goes through a set of *phases* approximately once every month is perhaps the most familiar everyday astronomical observation (Fig. 10–1). In fact, the name month comes from the word moon. The actual period of the phases, the interval between a particular phase of the Moon and its next repetition, is approximately 29½ Earth days (Fig. 10–2). This period can vary by as much as 13 hours. The explanation of the phases is quite simple.

The phases of moons or planets are the shapes of the sunlighted areas as seen from our vantage point.

The Moon is a sphere, and at all times the side that faces the Sun is lighted and the side that faces away from the Sun is dark. The phase of the Moon that we see from the Earth, as the Moon revolves around us, depends on the relative orientation of the three bodies: Sun, Moon, and Earth. The situation is simplified by the fact that the plane of the Moon's revolution around the Earth is nearly, although not quite, the plane of the Earth's revolution around the Sun.

Basically, when the Moon is almost exactly between the Earth and the Sun, the dark side of the Moon faces us. We call this a "new moon." A few days earlier or later we see a sliver of the lighted side of the Moon, and call this a "crescent." As the month wears on, the crescent gets bigger, and about 7 days after new moon, half the face of the Moon that is visible to us is lighted. We sometimes call this a "half moon." Since the lighted half we see is one fourth of the whole moon, this is more properly called a "first-quarter moon." (Instead of apologizing for the fact that astronomers call the same phase both "quarter moon" and "half moon," I'll just continue with a straight face and try to pretend that there is nothing strange about it.)

When over half the Moon's disk is visible, we have a "gibbous" moon.

The relative sizes of the planets and the Sun.

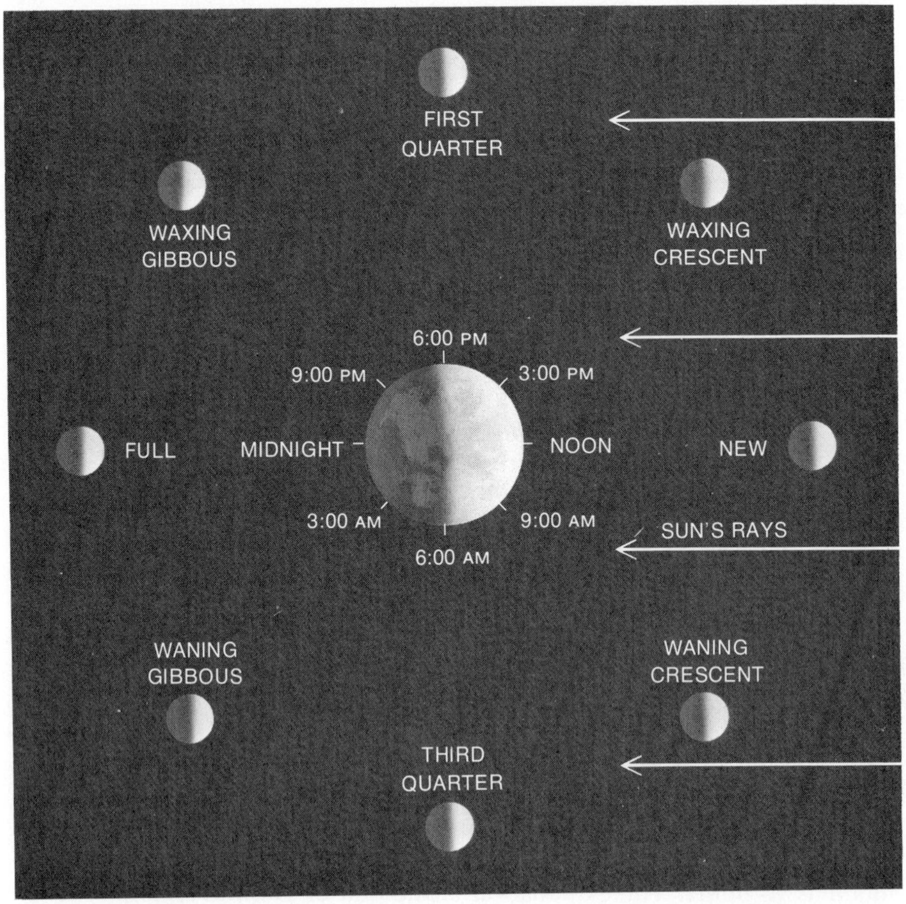

FIRST QUARTER
WAXING GIBBOUS
WAXING CRESCENT
6:00 PM
9:00 PM
3:00 PM
FULL
MIDNIGHT
NOON
NEW
3:00 AM
9:00 AM
SUN'S RAYS
6:00 AM
WANING GIBBOUS
WANING CRESCENT
THIRD QUARTER

Figure 10–1 The phases of the Moon depend on the Moon's position in its orbit around the Earth.

One week after the first-quarter moon, the Moon is on the opposite side of the Earth from the Sun, and the entire face visible to us is lighted. This is called a "full moon." One week later, we have a "third-quarter," and then we go back to "new moon" again and repeat the cycle of phases.

Note that since the phase of the Moon is related to the position of the Moon with respect to the Sun, one can tell when the Moon will rise at a certain phase. For example, since the Moon is 180° across the sky from the Sun when it is full, a full moon is always rising just as the Sun sets (Fig. 10–3). Each day thereafter, the moon rises approximately one hour later (24 hours divided by the period of revolution of the Moon around the

Figure 10–2 The phases of the Moon.

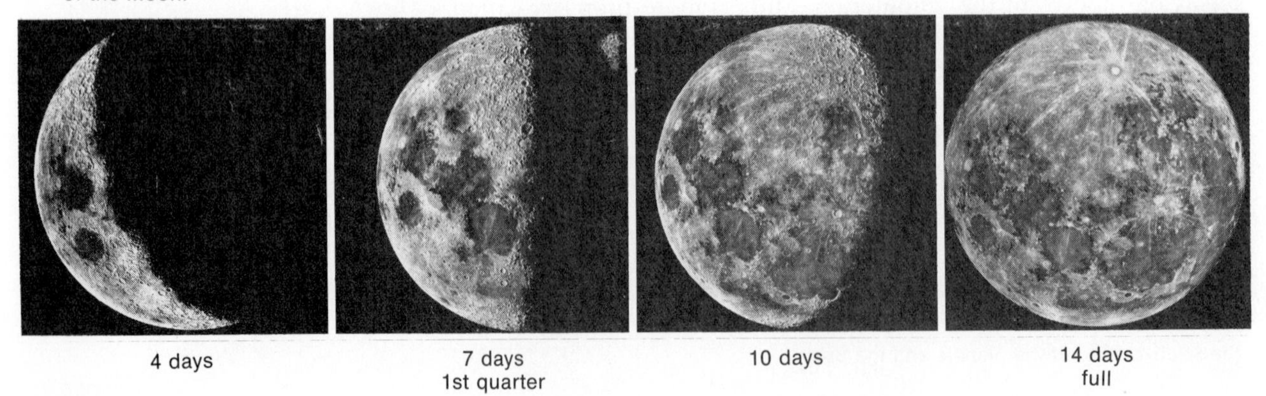

4 days | 7 days 1st quarter | 10 days | 14 days full

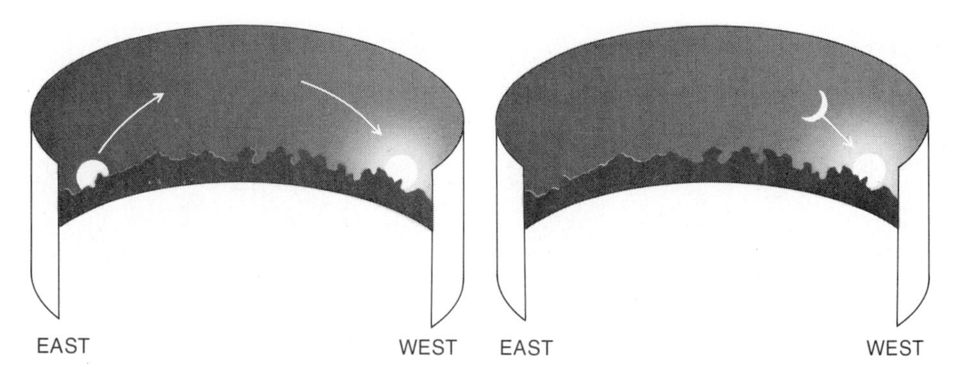

Figure 10–3 Because the phase of the Moon depends on its position in the sky with respect to the Sun, we can see why a full moon is always rising at sunset while a crescent moon is either setting at sunset, as shown here, or rising at sunrise.

Box 10.1 From *To Jane (The Keen Stars Were Twinkling)* by Percy Bysshe Shelley (written in 1822)

The stars will awaken,
Though the moon sleep a full hour later,
Tonight. . . .

Earth). The third-quarter moon, then, rises at midnight, and at temperate latitudes is high in the sky at sunrise. The new moon rises with the Sun in the east at dawn. The first-quarter moon rises at noon and is high in the sky at sunset. It is often visible in the late afternoon.

It can be amusing to find literary references to astronomical phenomena in which impossible situations are described. For example, one could never see the full moon setting just after dinner. Nor could the Ancient Mariner have seen "Above the eastern bar/the hornèd Moon, with one bright star/Within the nether tip," no matter how nicely Coleridge described it (Fig. 10–4). Moreover, he certainly didn't see a crescent in the east in the evening.

Because the Moon's orbit around the Earth, and the Earth's orbit around the Sun are not precisely in the same plane (Fig. 10–5), the Moon usually passes slightly above or below the Earth's shadow at full moon, and the Earth usually passes slightly above or below the Moon's shadow at new moon. But every once in a while, up to seven times a year, the Moon is at the part of its orbit that crosses the Earth's orbital plane at full moon or new moon. When that happens, we have a lunar or a solar eclipse.

Figure 10–4 Because the remainder of the disk of the Moon is present although dark when we see only a crescent, one can never see a star "within the nether tip."

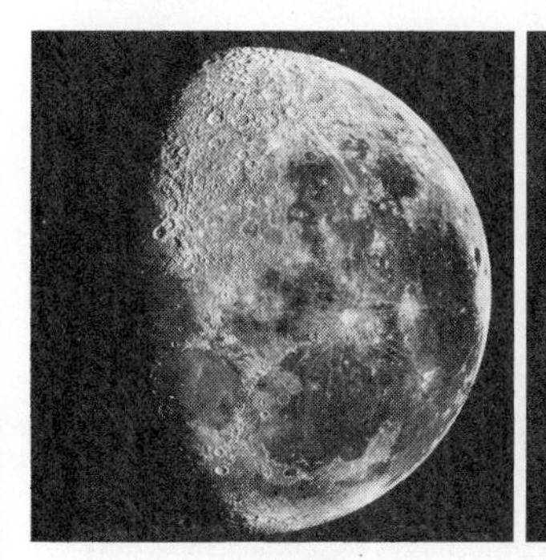

20 days

22 days
third quarter

24 days

26 days

Figure 10–5 The plane of the Moon's orbit is tipped with respect to the plane of the Earth's orbit, so the Moon usually passes above or below the Earth's shadow.

We have already discussed eclipses of the Sun, when the Moon comes directly between the Earth and the Sun. Eclipses of the Moon, when the Moon goes into the Earth's shadow, are more commonly seen.

Many more people see a lunar eclipse when one occurs, because the Moon lies entirely in the Earth's shadow and the sunlight is entirely cut off from it (Fig. 10–6). So anywhere on the Earth that the Moon has risen, the eclipse is visible. In a solar eclipse, on the other hand, the alignment of the Moon between the Sun and the Earth must be precise, and only those people in a narrow band on the surface of the Earth see the eclipse.

Of the seven eclipses of the Sun and the Moon that can occur in a single year, most are partial eclipses, where the Sun or the Moon is only partially covered. The total eclipses, which are rarer, are much more spectacular.

A lunar eclipse is a much more leisurely event to watch than a solar eclipse. The partial phase, when the Earth's shadow gradually covers the Moon, lasts for hours. And then the total phase, when the Moon is entirely

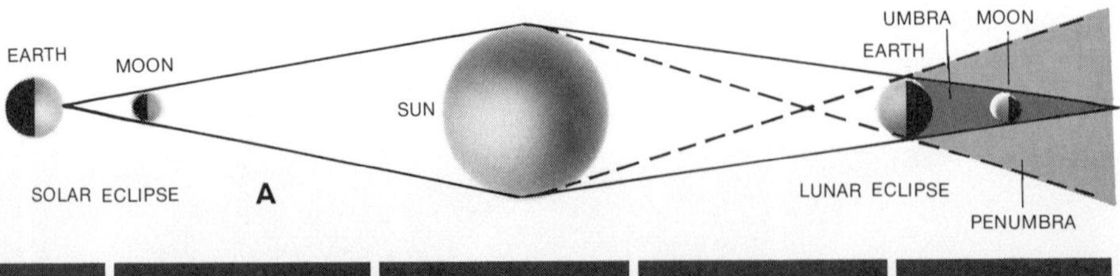

B

Figure 10–6 (*A*) When the Moon is between the Earth and the Sun, we observe an eclipse of the Sun. When the Moon is on the far side of the Earth from the Sun, we see a lunar eclipse. The part of the Earth's shadow from which the Sun is only partially shielded from the Moon's view is called the *penumbra*; the part of the Earth's shadow from which the Sun is entirely shielded from the Moon's view is called the *umbra*. (*B*) A lunar eclipse; the Moon remains visible even during totality, though it is relatively faint and longer exposures must be used.

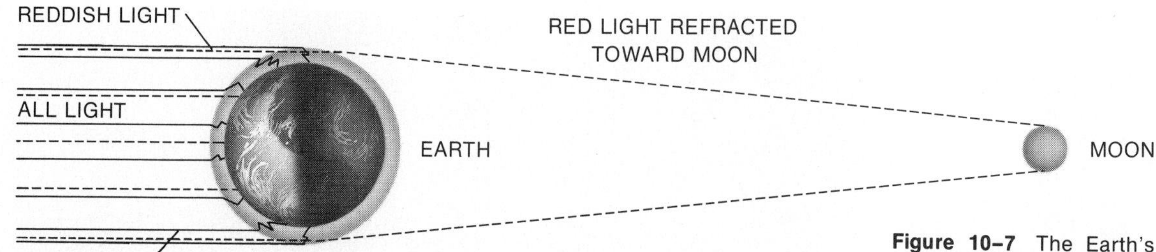

Figure 10–7 The Earth's atmosphere scatters shorter wavelengths much more than it scatters longer wavelengths, so red light is scattered much less than blue light. Some of the red light survives its passage through the Earth's atmosphere and is bent, or refracted, toward the Moon. Thus the Moon often appears reddish during the total phase of a lunar eclipse.

within the Earth's shadow, can itself last for over an hour. During this time, the sunlight is not entirely shut off from the Moon (Fig. 10–7). A small amount is refracted around the edge of the Earth by our atmosphere. Most of the blue light is taken out during the sunlight's passage through our atmosphere; this is how blue skies are made for the people one fourth of the way around the globe from the point at which the Sun is overhead. The remaining light is reddish, and this is the light that falls on the Moon. Thus, the eclipsed Moon appears reddish.

Sometimes there is a particularly dark lunar eclipse, when the Moon appears ash gray rather than red. This seems to correspond to the presence of a lot of particulate matter in the Earth's atmosphere. For example, after the explosion of Krakatoa in what was then Java in 1883, there were dark eclipses for years. Recent lunar eclipses have been fairly dark, corresponding to the eruption of Fuega in Guatemala in 1974. Thus the color and perhaps darkness of a lunar eclipse are correlated with eruptions of volcanoes and other phenomena in the Earth's atmosphere, such as overall cloud cover, rather than with celestial phenomena.

The eruptions of the volcanoes Krakatoa and Fuega were followed by dark lunar eclipses.

Scientists can make observations at lunar eclipses to study the properties of the lunar surface. For example, one can monitor the rate of cooling of areas on the lunar surface with infrared radiometers. But now we have landed on the Moon and have made direct measurements at a few specific places, so the scientific value of lunar eclipses has been greatly diminished but not eliminated. There remains little of scientific value to do during lunar eclipses, quite a different situation from that at a solar eclipse.

The Moon is not the only object in the solar system that is seen to go through phases. Mercury and Venus both orbit inside the Earth's orbit, and so sometimes we see the side that faces away from the Sun and sometimes we see the side that faces toward the Sun. Thus at times Mercury and Venus are seen as crescents, though it takes a telescope to observe their shapes.

The outer planets, though, from Mars on outward, are never between the Sun and the Earth. Thus they never appear as crescents, although sometimes they can appear gibbous.

10.2 THE MOTIONS OF THE PLANETS

When we observe the planets in the sky, we notice that their positions vary from night to night with respect to each other and with respect to the stars. The stars, on the other hand, are so far away that their positions are relatively fixed with respect to each other (stellar "proper motion" was dis-

Figure 10–8 The retrograde loop of Mars in 1977 and 1978. Opposition occurs in Cancer on January 22, 1978; Gemini and Canis Minor are among the other constellations visible in this simulation photographed at the Vanderbilt Planetarium by George Lovi. Pollux is the bright star just above and to the right side of the retrograde loop. Regulus is the bright star to the left that Mars passes close by.

It has been known for thousands of years that some of the bright dots that we see in the sky at night are unchanging (fixed) in position—the fixed stars—while others appear to move among them—the wandering stars.

cussed in Section 2.16). The fact that the planets appear as "wandering stars" was known to the ancients; the word "planet" comes from the Greek word for "wanderer." Of course, both stars and planets revolve together across our sky essentially once every 24 hours; by the wandering of the planets we mean that the planets appear to move at a very slightly different rate so that over a period of weeks or months they change position with respect to the fixed stars.

The motion of the planets in the sky is not continuously in the same direction with respect to the stars. Most of the time they appear to drift eastward, that is, move across the sky slightly slower than the stars, but sometimes they drift backwards, that is, move slightly faster than the stars and thus westward (Fig. 10–8). We call the backward motion *retrograde motion.* (The forward motion is formally called *prograde,* but the term is not commonly used; sometimes we say that the planets are revolving in the *direct* sense.)

Figure 10–9 Ptolemy. (Burndy Library, photograph by Owen Gingerich).

One of the earliest and greatest philosophers, Aristotle, who lived in Greece about 350 B.C., summarized the astronomical knowledge of his day into a qualitative cosmology that remained unchallenged for 1800 years. On the basis of what seemed to be very good evidence—what he saw—Aristotle thought, and actually believed that he **knew**, that the Earth was at the center of the universe and that the planets, Sun, and stars revolved around it. The universe was made up of a set of 55 celestial spheres that fit around each other and that had rotation as their natural motion. Each of the heavenly bodies was carried around the heavens by a sphere, whose motion was still further compounded by that of other spheres. These motions combined to account for the various observed motions, including revolution around the Earth, retrograde motion, and variation in latitude. Each planet had several spheres in order to account for its various motions. The outermost sphere was that of the fixed stars, beyond which lay the prime mover, *primum mobile,* that caused the general rotation of the stars overhead.

Aristotle's theories ranged through much of science. He held that all bodies below the spheres of the Moon were made of four basic "elements": earth, air, fire, and water. The fifth "essence"—the quintessence—was a perfect, unchanging, transparent element of which the celestial spheres around that of the Moon are formed.

Aristotle's theories dominated scientific thinking for almost two millennia, until the Renaissance. Unfortunately, most of his theories were far from what we now consider to be correct, so we tend to think that the widespread acceptance of Aristotelian physics impeded the development of science.

Almost 500 years after Aristotle, in about 140 A.D., the Greek astronomer Claudius Ptolemy (Fig. 10–9) elaborated on Aristotle's thoughts and presented a detailed theory of the universe that explained the retrograde motion. Ptolemy's model, as was Aristotle's, was Earth-centered (Fig. 10–10). To account for the retrograde motion of the planets, the planets had to be moving not just on large circles around the Earth but rather on smaller circles, called *epicycles,* whose centers moved around the Earth on larger circles, called *deferents* (Fig. 10–11). It seemed natural that the planets should follow circles in their motion since circles were thought to be "perfect" figures. Sometimes the center of the deferent was not centered at the Earth. The planets moved at a constant rate of angular motion but another complication was that the point around which the planet's angular motion moved uniformly was neither at the center of the Earth nor at the center of the deferent. Ptolemy's views were very influential in the study of astron-

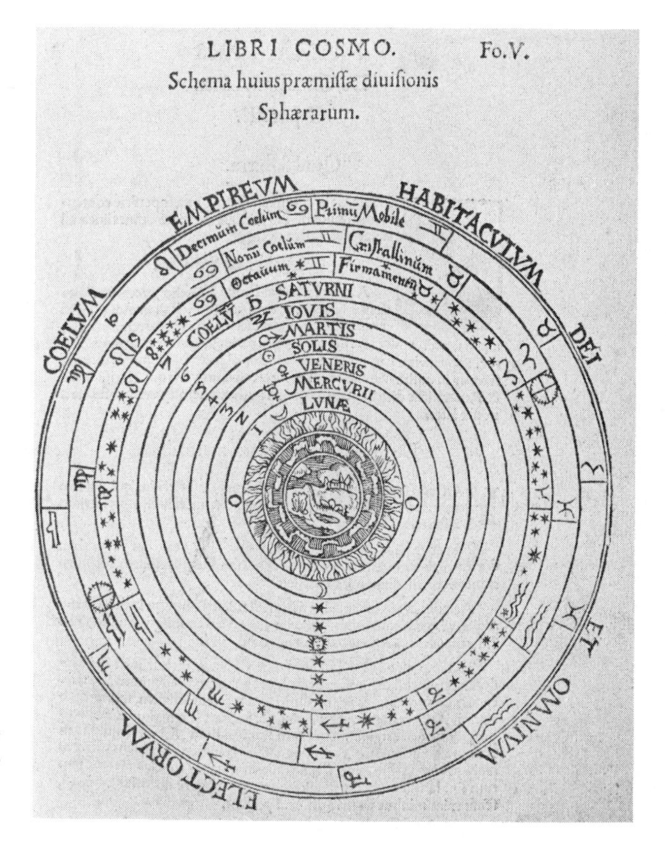

Figure 10–10 Aristotle's Earth-centered theory. The Earth is at the center, orbited in larger and larger circles by the Moon, Mercury, Venus, the Sun, Mars, Jupiter, Saturn, and the stars.

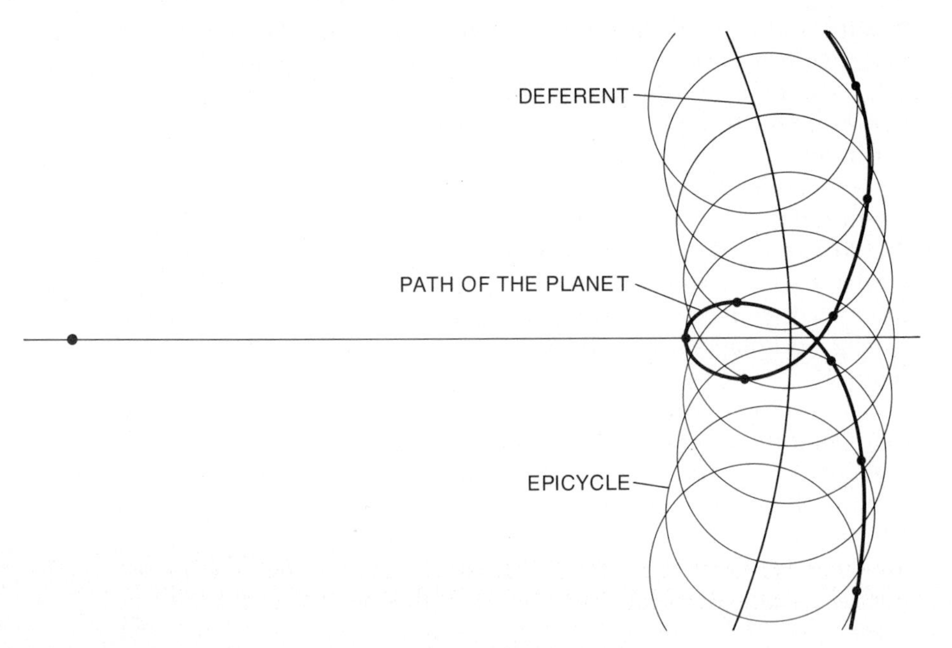

Figure 10–11 In the Ptolemaic theory, retrograde motion was caused by motion of the planets on small circles called epicycles that moved along a larger circle, centered at the Earth, called the deferent.

omy, for versions of his ideas and of the tables of planetary motions that he computed were accepted for nearly 15 centuries.

The credit for the breakthrough in our understanding of the solar system is commonly given to Nicholas Copernicus (Fig. 10–12), a 16th-century Polish astronomer whose five-hundredth birthday was celebrated by the astronomical community in 1973. Copernicus advanced a *heliocentric*—

Figure 10–12 Copernicus, in a 16th century woodcut.

Figure 10–13 The heliocentric theory, as illustrated in Copernicus's *De Revolutionibus,* published in 1543.

Figure 10–12

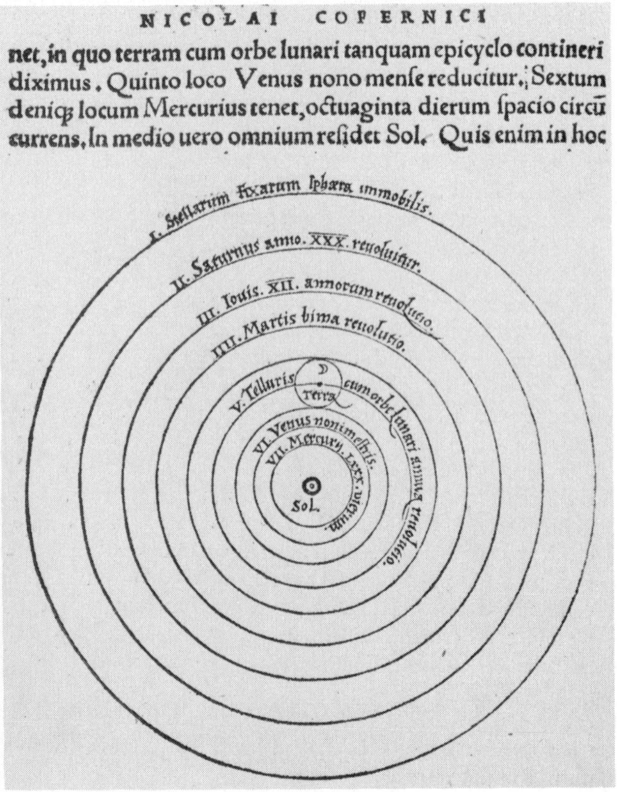

NICOLAI COPERNICI

net, in quo terram cum orbe lunari tanquam epicyclo contineri diximus. Quinto loco Venus nono menſe reducitur. Sextum deniq; locum Mercurius tenet, octuaginta dierum ſpacio circũ currens. In medio uero omnium reſidet Sol. Quis enim in hoc

Figure 10–13

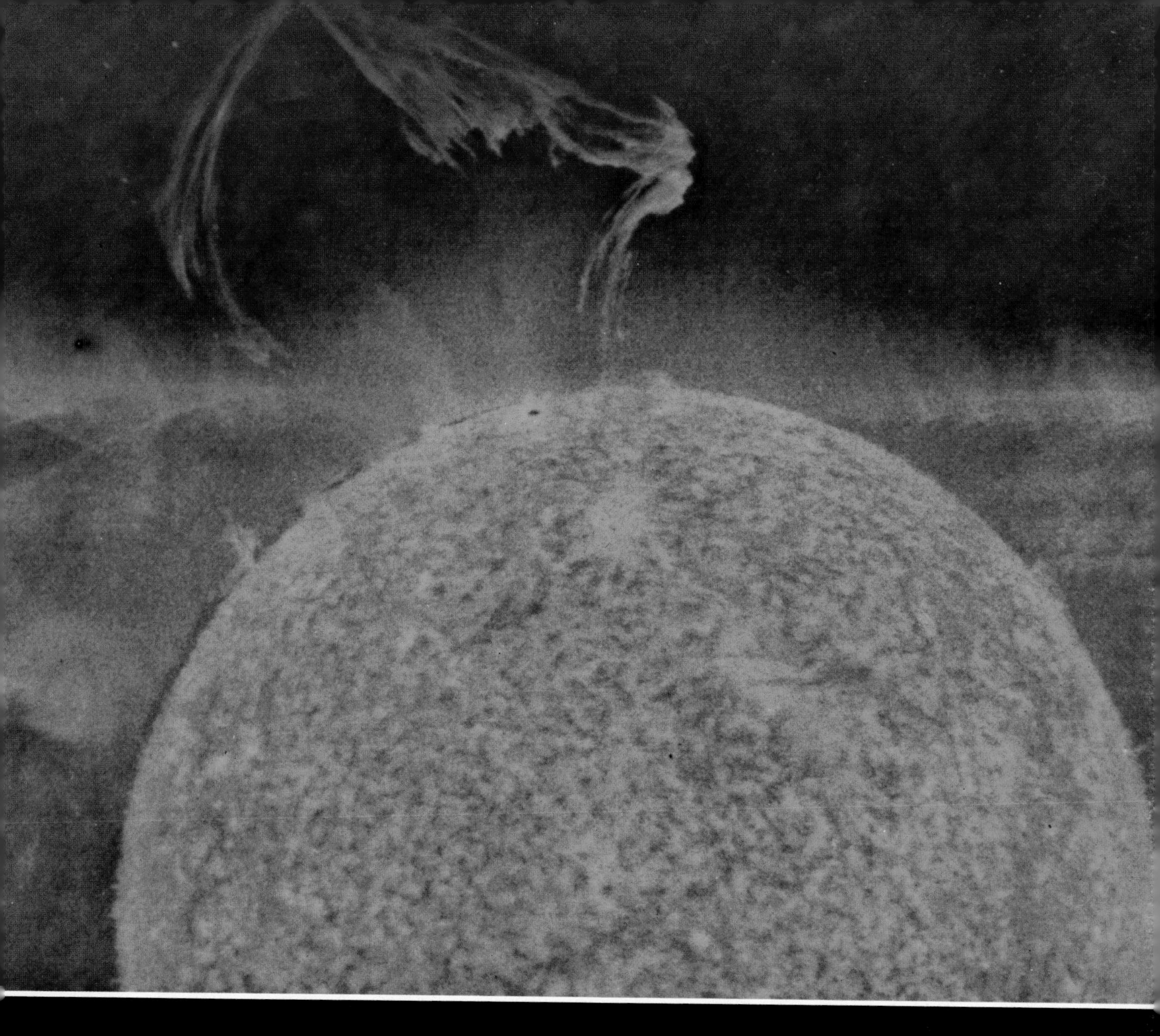

Color Plate 17 (top): An erupting prominence on the sun, photographed on August 9, 1973 in the ultraviolet radiation of ionized helium at 304 Å with the Naval Research Laboratory's slitless spectroheliograph aboard Skylab. Because of the technique used, radiation at nearby wavelengths appears as a background. A black dot visible on the solar surface slightly to the left of the base of the prominence may be the site of the flare that caused the eruption. Supergranulation is visible on the solar disk in the light of He II, and a macrospicule shows on the extreme right. (NRL/NASA photo)

Color Plate 18 (bottom): Contours of equal intensity computed for another eruptive prominence observed in the radiation of ionized helium with the NRL instrument aboard Skylab. Each color represents a different strength of emission. This view shows the prominence 90 minutes after its eruption, when the gas had moved 500,000 km from the solar surface. (NRL/NASA photo)

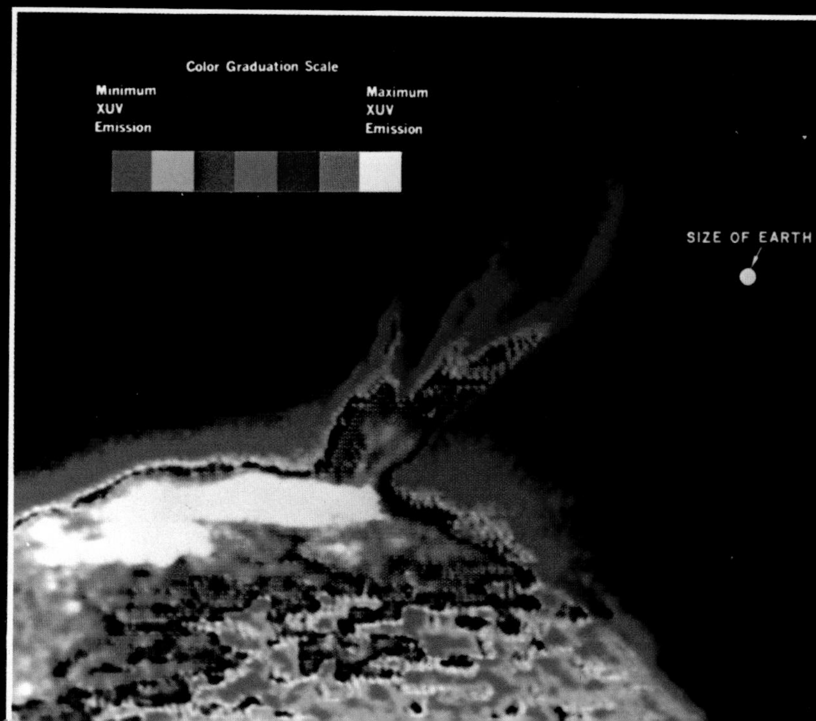

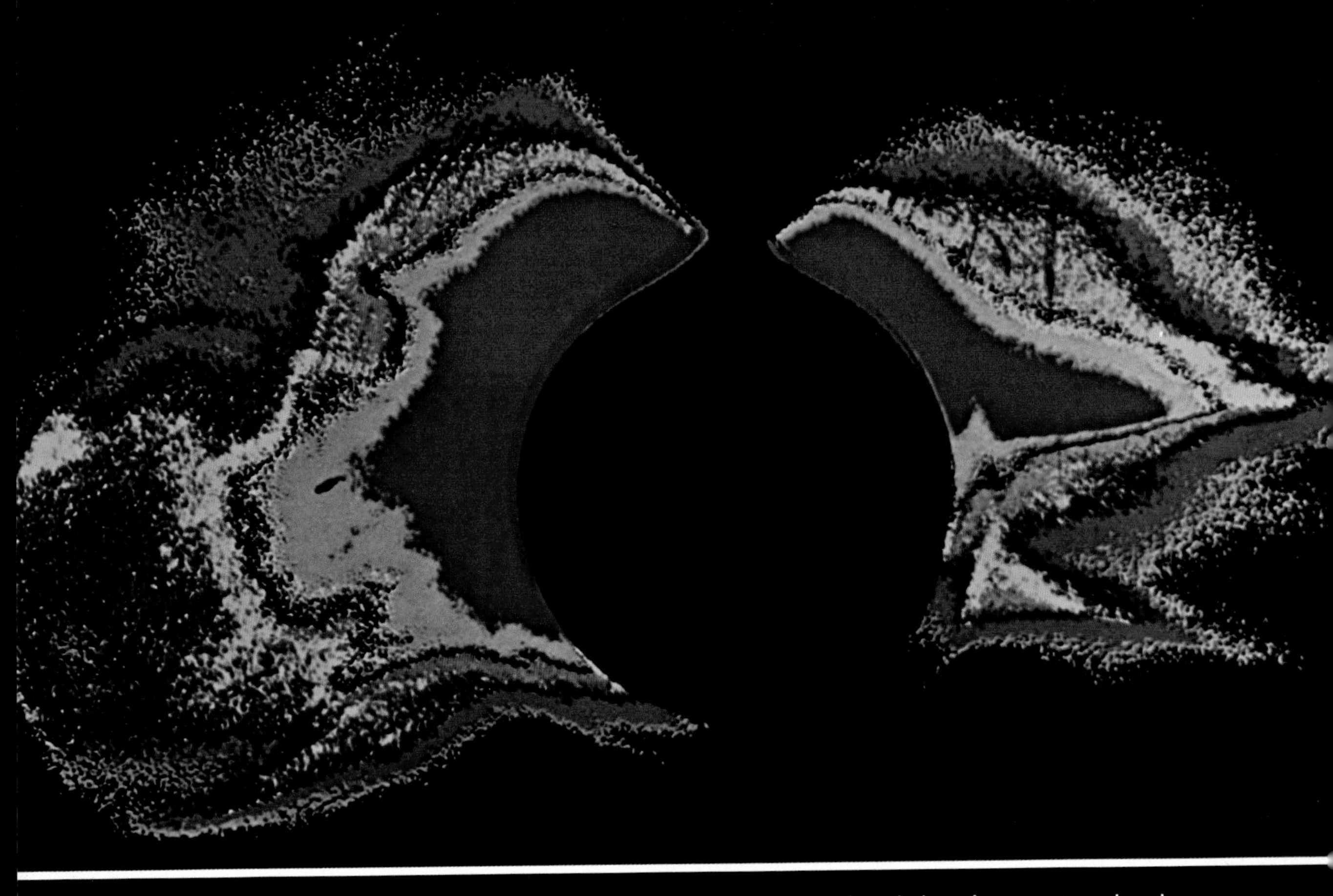

Color Plate 19 (top): Isophotes showing the intensity of the white light solar corona, reduced from observations made by the High Altitude Observatory's coronagraph aboard Skylab. The solar photosphere has been artificially occulted. (HAO/NASA photo)

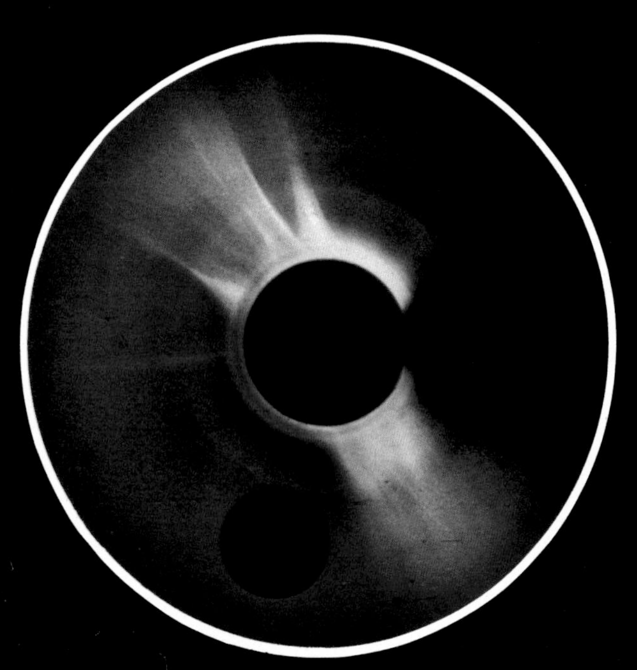

Color Plate 20 (left): Skylab did not pass into totality during the total solar eclipse of June 30, 1973, but this image shows the Moon as it crossed the solar corona as observed on that date from Skylab with the High Altitude Observatory's coronagraph. (HAO/NASA photo)

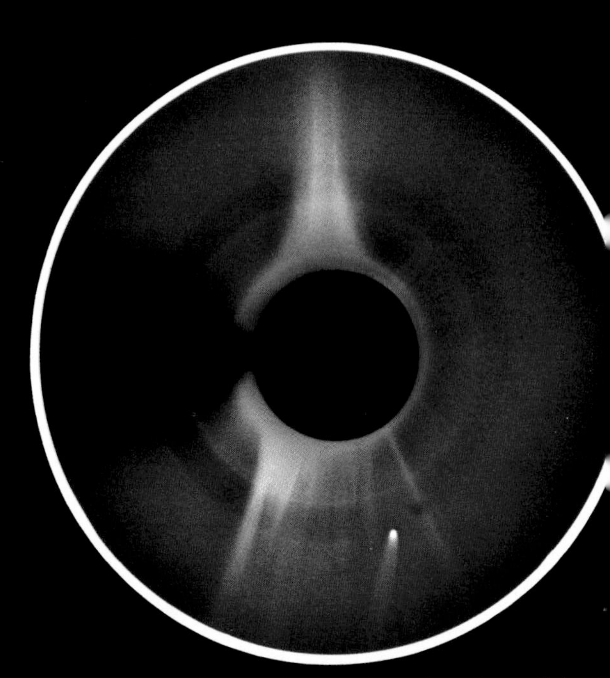

Color Plate 21 (right): On the few days when Comet Kohout was closest to its perihelion, it was invisible to observe on earth. The Skylab astronauts, fortunately, were ab to sketch and photograph it. Here we see the image Comet Kohoutek in the High Altitude Observator coronagraph on December 27, 1973. (HAO/NASA pho

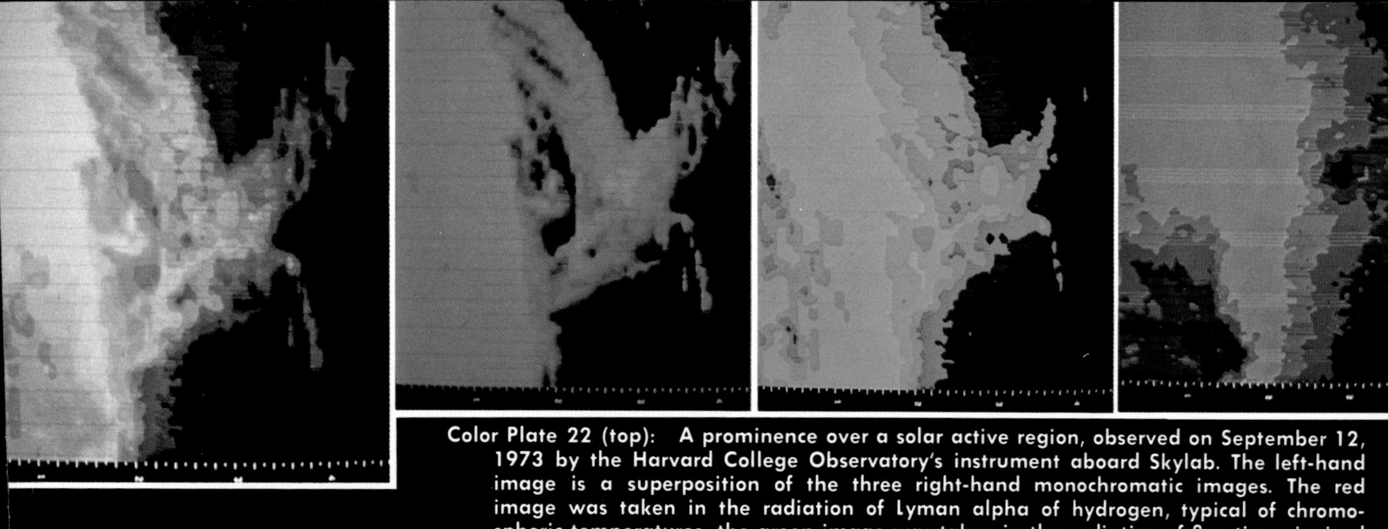

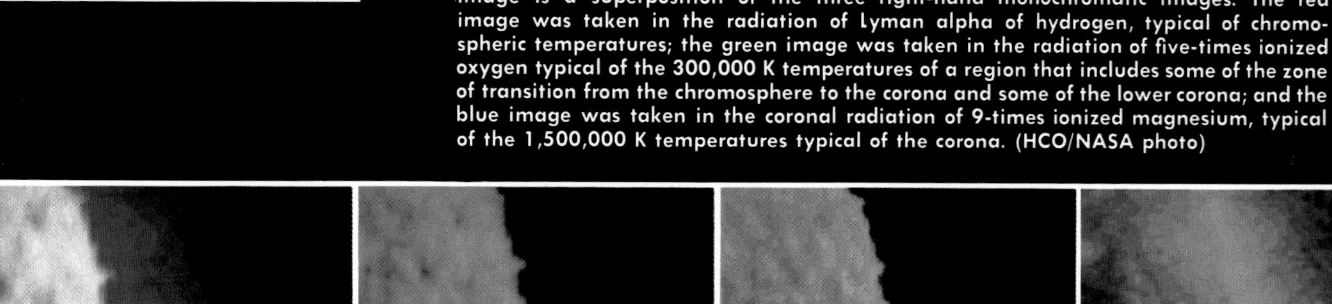

Color Plate 22 (top): A prominence over a solar active region, observed on September 12, 1973 by the Harvard College Observatory's instrument aboard Skylab. The left-hand image is a superposition of the three right-hand monochromatic images. The red image was taken in the radiation of Lyman alpha of hydrogen, typical of chromospheric temperatures; the green image was taken in the radiation of five-times ionized oxygen typical of the 300,000 K temperatures of a region that includes some of the zone of transition from the chromosphere to the corona and some of the lower corona; and the blue image was taken in the coronal radiation of 9-times ionized magnesium, typical of the 1,500,000 K temperatures typical of the corona. (HCO/NASA photo)

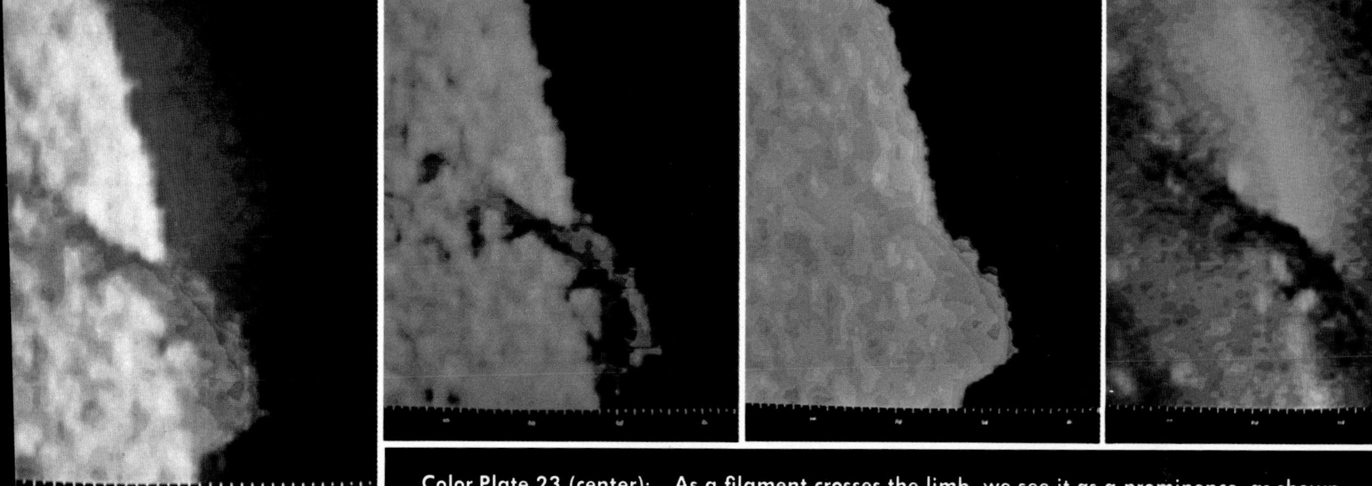

Color Plate 23 (center): As a filament crosses the limb, we see it as a prominence, as shown in Lyman alpha, the red image. Since this is relatively cool gas, the coronal image, shown in blue, demonstrates an absence of hot gas at that location. The green image, in radiation of 3-times ionized oxygen at a temperature of 130,000 K, shows the chromosphere-corona transition region. (HCO/NASA photo based on observations from Skylab)

Color Plate 24 (bottom): A time sequence, covering a 22 minute period, showing the rise and fall of an eruptive prominence on August 30, 1973. The red image shows Lyman alpha in the chromosphere, the green image shows 3-times ionized oxygen in the transition zone, and the blue image shows 9-times ionized magnesium in the corona. (HCO/NASA photo based on observations from Skylab)

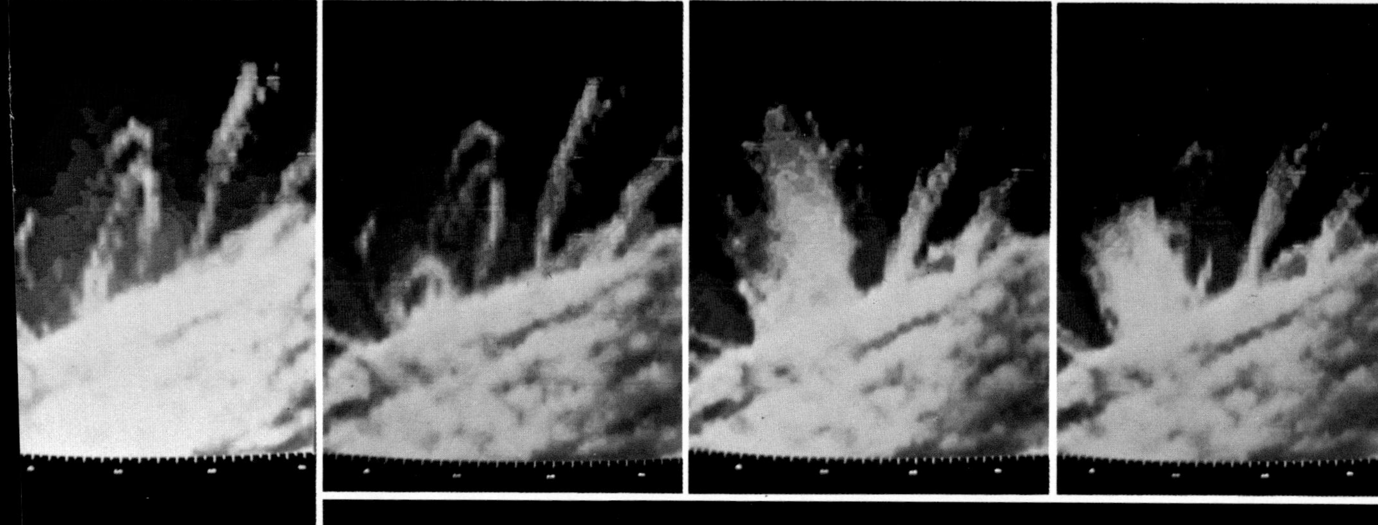

Color Plate 25: Apollo 17 view of the Taurus-Littrow Valley on the Moon. This huge, fragmented boulder had rolled a kilometer down the side of the North Massif to here. Scientist-astronaut Harrison Schmitt is at the left. The Lunar Rover is at the right. (NASA photo)
Color Plate 26: The lunar buggy carried by Apollo 17. (NASA photo)

Color Plate 27: Apollo 17 astronauts found a hint of orange in the lunar soil (picture's center), initially causing excitement since the possibility of volcanic origin was raised. (NASA photo)
Color Plate 28 (bottom, left): The blast-off of Apollo 17, a night launch. (Photo by the author)
Color Plate 29 (bottom, center): An Apollo 17 astronaut with an instrument package. (NASA photo)
Color Plate 30 (bottom, right): The crescent Earth seen from Apollo 11 in orbit around the Moon. (NASA photo)

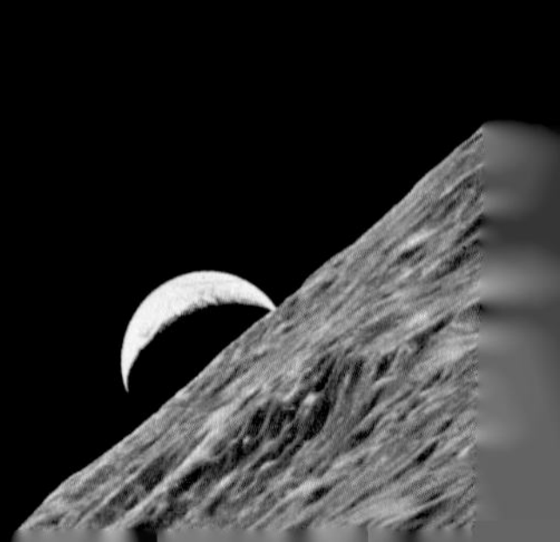

sun centered*—theory (Fig. 10–13) that the retrograde motion of the planets could be readily explained if the Sun rather than the Earth was at the center of the universe, that the Earth is a planet, and that the planets move around the Sun in circles. (Aristarchus of Samos, a Greek scientist, had also suggested a heliocentric theory 18 centuries earlier, though we do not know how detailed a picture of planetary motions he presented. In any case, his heliocentric suggestion required the apparently ridiculous notion that the Earth moved, in contradiction to our senses and to the theories of Aristotle.† Aristarchus's theory was lost in antiquity.)

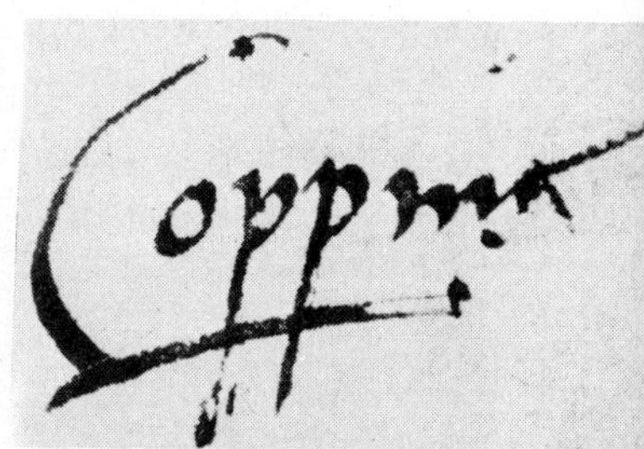

Figure 10–14 Copernicus's signature from the Uppsala University Library in Sweden, photographed by Charles Eames and reproduced courtesy of Owen Gingerich.

Copernicus used his own version of the heliocentric theory to calculate tables of planetary positions. His detailed predictions were not in much better agreement with the existing observations than tables based on Ptolemy's model, because Copernicus still used Ptolemy's observations. Also, the errors in the observations were very large. The heliocentric theory appealed to Copernicus primarily on the philosophical grounds of simplicity.

Copernicus's heliocentric theory, published in 1543 in a book known as *De Revolutionibus (On the Revolutions,* shown in Figure 10–15), explained the retrograde motion of the planets in the following manner (Fig. 10–16).

Let us consider, first, an outer planet like Mars as seen from the Earth (Fig. 10–17). As the Earth approaches the part of its orbit that is closest to Mars (which is orbiting much more slowly), the projection of the Earth-

*Helios was the Sun god in Greek mythology.

†Aristotle, Aristarchus, and Ptolemy also reasoned that if the Earth moved around the Sun, the stars, during the course of a year, would show slight displacements in the sky—parallaxes (Section 2.11)—but these parallaxes were not seen. They did not know, in that pretelescopic era, that the parallaxes would be too small to be observed. They deserve credit, at least, for the fact that their theories agreed in this sense better with observation than the heliocentric theory. It was only later that accurate observations changed the weight of the evidence on this point.

NICOLAI CO-
PERNICI TORINENSIS
DE REVOLVTIONIBVS ORBI-
um cœlestium, Libri VI.

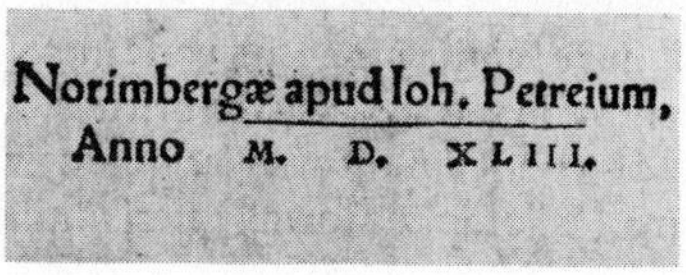

Norimbergæ apud Ioh. Petreium,
Anno M. D. XLIII.

Figure 10–15 From the title page of Copernicus's *De Revolutionibus*. About 200 copies of this work are currently known to be extant.

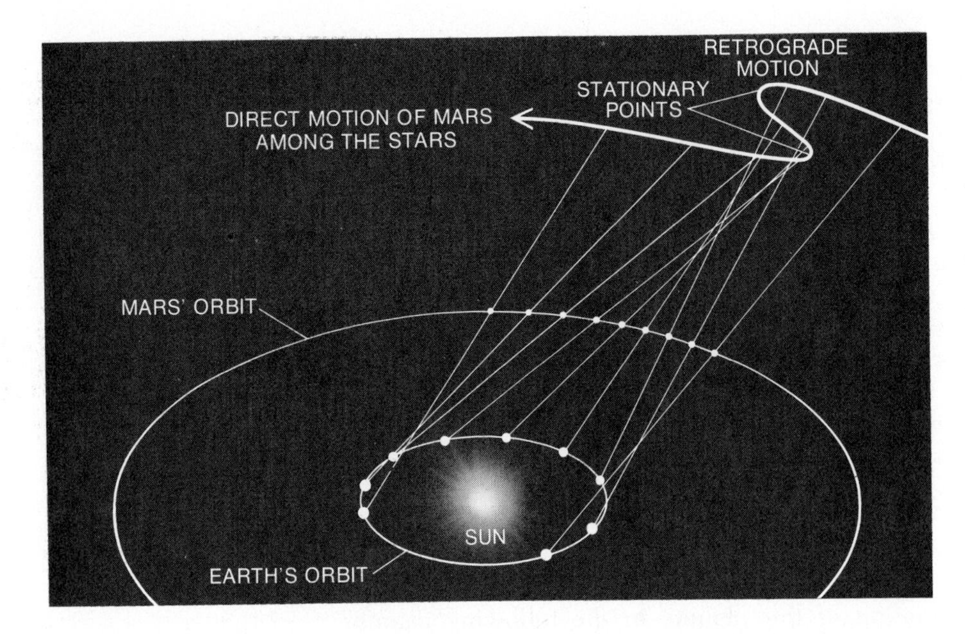

Figure 10–16 The Copernican theory explains retrograde motion as an effect of projection. The drawing shows the explanation of retrograde motion for Mars or for the other *superior planets,* that is, planets whose orbits lie outside that of the Earth. Similar drawings can explain retrograde motion for *inferior planets,* that is, planets whose orbits lie inside that of the Earth (namely, Mercury and Venus).

Mars line to the stars (which are essentially infinitely far away compared to the planets) moves slightly against the stellar background. As the Earth in its orbit comes closest to Mars, and then passes it, the projection of the line joining the two planets can actually seem to go backwards since the Earth is going at a greater speed than Mars. Then, as the Earth continues around its orbit, Mars appears to go forward again. A similar explanation can be demonstrated for inner as well as outer planets.

The Italian scientist Galileo Galilei began to believe in the Copernican system in the 1590's, and later provided important observational confirmation of the theory. In 1610, simultaneously with the first settlements in the American colonies, Galileo was the first to use a telescope for astronomical observation. His discovery of the moons of Jupiter (Fig. 10–17) showed that not everything in the solar system orbited the Earth, and his discovery of sunspots showed that the Sun had "blemishes" on it. This observation was a key disagreement with the religious and philosophical Aristotelian view that the Sun was perfect. Galileo's discovery that Venus went through an entire set of phases (Fig. 10–18), like the Moon, could not be explained with the Ptolemaic Earth-centered system. If Venus traveled in an epicycle located between the Earth and the Sun, Venus should always appear as a crescent since it is observed never to stray far in the sky from the Sun. This rationale was contrary to Galileo's observations. But while the Roman

Figure 10–18

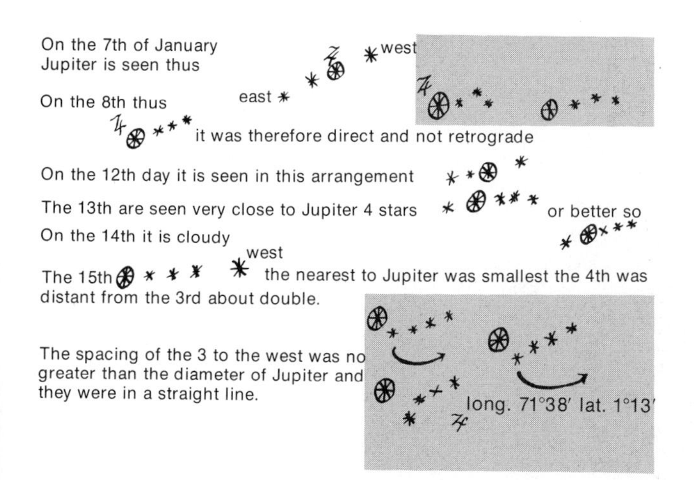

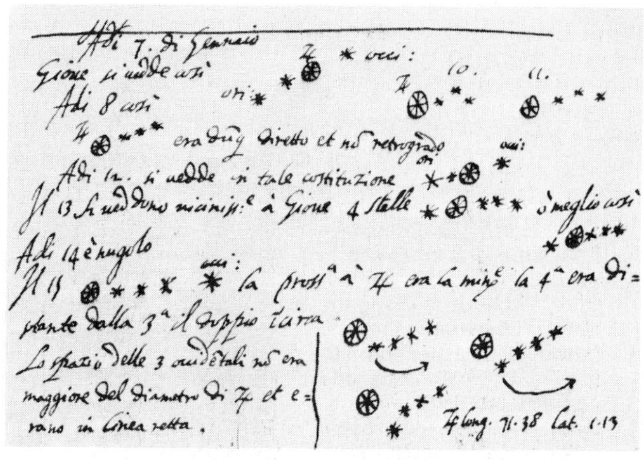

Figure 10–17 A translation (*left*) of Galileo's original notes (*right*) summarizing his first observations of Jupiter's moons in January 1610. The shaded areas were probably added later. It had not yet occurred to Galileo that the objects were moons in revolution around Jupiter.

Catholic Church was not concerned with anyone's making models to explain observations, they were concerned with assertions that those models represent physical truth. Galileo was forced, in his old age, to recant his belief in the Copernican theory.

In the last part of the 16th century, not long after Copernicus's death, Tycho Brahe had made a series of observations of Mars and other planets at his observatory, named Uraniborg, on an island off the mainland of Denmark (Fig. 10–19). These positional observations were considerably more precise than any observations that had been made up to that time, even though the telescope had not yet been invented. In 1597, Tycho lost his financial support in Denmark, and moved to Prague. Johannes Kepler came there as a young assistant to work with him. At Tycho's death, in 1601, Kepler was left to analyze all the observations that Tycho and his assistants had made.

Tycho was known all over Europe as an astronomer, and had advanced his own, Earth-centered, cosmology (Fig. 10–20). In it, though the Sun revolved around the Earth, the other planets revolved around the Sun. This cosmology is a compromise that allows various advantages of the Copernican system while retaining what was then the philosophical advantage of having the Earth remain immobile and at or near the center of the universe. Tycho's cosmology was shortly forgotten, and a large part of Tycho's current reputation results from his having provided the observational base for Kepler's research.

Tycho's observational data showed that the tables then in use did not adequately predict planetary positions. Laboring for many years, without benefit of the modern computers that could now solve the problem in minutes, Kepler was finally able to make sense out of the observations of

Figure 10–18 The phases of Venus. Note that Venus is a crescent only when it is in a part of its orbit that is relatively close to the Earth, and so it looks larger at those times.

Figure 10–19 Tycho's observatory at Uraniborg, on the island of Hveen in Denmark. In the illustration, Tycho is showing the mural quadrant that he used to measure the altitudes at which stars and planets crossed the meridian. The pre-telescopic observations made here enabled Kepler to discover his laws. The drawing dates from 1598.

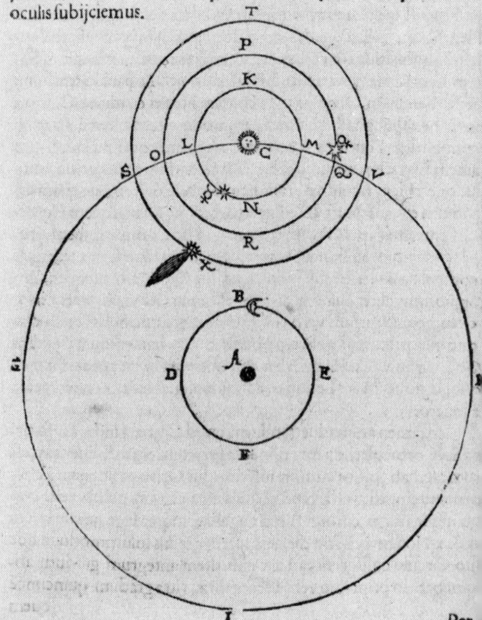
DE COMETA ANNI 1577. 181

motui coniunctus fuisset, in infima Orbis sui parte & Terris proxima constitutus assumatur, atq; hinc per consequentiā Signorum, aliter quàm in Venere & Mercurio vsuuenit, versus eiusdē Orbis Apogæum perrexisse, centro huius reuolutionis Solis simplici motui perpetuò concurrente, admittatur. Quæ omnia vt rectiùs percipiantur, nunc orbiū huc aliquid facientiū oportunā dispositionem oculis subijciemus.

Per.

Figure 10–20 Tycho's theory, which had the Moon traveling around the Earth but the other planets traveling around the Sun. The Comet of 1577 is also illustrated in this plate from Tycho's book about that comet, published in 1588.

Mars that Tycho had made and clear up the discrepancy. In 1609 and 1618 Kepler presented three laws of motion that, to this day, we consider to be the basis of our understanding of the solar system.

Box 10.2 Kepler's Laws of Planetary Motion

(1) The planets orbit the Sun in ellipses, with the Sun at one focus.
(2) The line joining the Sun and a planet sweeps through equal areas in equal times.
(3) The period of revolution of a planet, and the semi-major axis of its orbit (half the longest dimension of the ellipse), are related to each other such that the square of the period is proportional to the cube of the semi-major axis of the orbit.

The major axis *of an ellipse is a line bounded by the ellipse and drawn through the two foci, or the length of that line; the* semimajor axis *is half that length. The* minor axis *is the perpendicular line bisecting the major axis, or the length of that part of the line included within the ellipse. In the special case when the major and minor axes are the same length, the ellipse is a circle.*

Until Kepler worked out these laws, even the heliocentric calculations assumed that the planets followed "perfect" orbits, namely, circles. The realization that the orbits were in fact ellipses (Fig. 10–21), as stated in Kepler's first law, greatly improved the accuracy of the calculations.

The second law, also known as *the law of equal areas*, governs the speeds with which the planets travel in their orbits. When a planet is at its greatest distance from the Sun in its elliptical orbit, it must travel relatively slowly because the weaker gravity at the greater distance from the Sun accelerates the planet less. The line joining it with the Sun sweeps out a

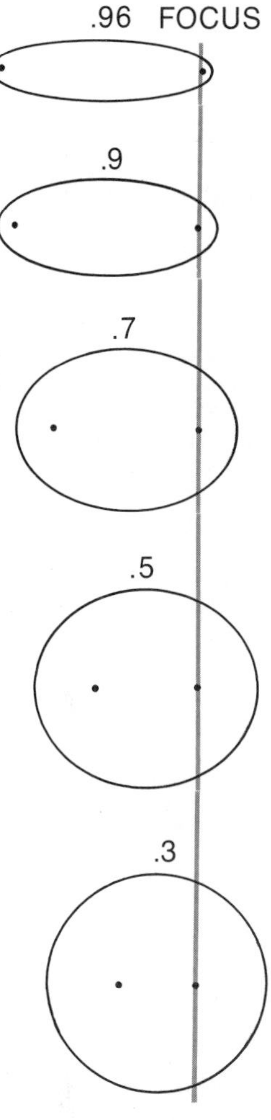

Figure 10–21 A series of ellipses of the same major axis but different eccentricities. The *foci* are marked; these are the two points inside with the property that the sum of the distances from any point on the circumference to the foci is constant. As the eccentricity—distance between the foci divided by the major (longer) axis—approaches 1, the ellipse approaches a straight line. As the eccentricity approaches zero, the foci come closer and closer together. A circle is an ellipse of zero eccentricity.

long skinny sector. This long skinny sector has the same area as the short fat sector formed when the planet is closer to the Sun, when the planet travels faster in its orbit (Fig. 10–22).

To understand the third law (it is ordinarily referred to as *Kepler's third law* and is demonstrated in Figure 10–23), let us compare the orbit of, say, Jupiter with that of the Earth. We can choose to work in units that are convenient for the Earth. We call the average distance from the Sun to the Earth *1 Astronomical Unit.* Similarly, the unit of time that the Earth takes to revolve around the Sun is defined as *one year.* Using these values, if we know from observation that Jupiter's period of revolution around the sun is 11.86 years, we can use Kepler's third law. If P is the period of revolution of a planet and R is the radius of its orbit,

$$\frac{\text{P (Jupiter)}^2}{\text{P (Earth)}^2} = \frac{\text{R (Jupiter)}^3}{\text{R (Earth)}^3}.$$

Substituting,

$$\frac{(11.86)^2}{1} = \frac{\text{R (Jupiter)}^3}{1}.$$

Now $(11.86)^2$ can be rounded off to 12^2, which is approximately 150. (Astronomers estimate calculations all the time, and it is important for students to realize that they need not do a calculation very exactly when only an approximate answer is required. At some other times, an exact answer may be required.) So the radius of Jupiter's orbit around the Sun is approximately the cube root of 150, or approximately 5 A.U. The actual measured value is 5.2 A.U.—the small difference was almost entirely introduced in the rough estimation process. So it checks.

Note that although we have calculated in a system that uses a "year" as a unit of time, it is a unit that has a special physical meaning only for us on Earth. Saturn takes 30 (Earth) years to orbit the Sun, and if we were to live

A sector is the area bounded by two straight lines from a focus of an ellipse (or, for a circle, two radii) and the part of the ellipse joining their outer ends.

TABLE 10–1 THE NUMBER OF A PLANET'S DAYS IN ONE OF ITS YEARS

Planet	Length of Sidereal Day in Earth Days	Length of Year in Earth Years	Number of the Planet's Sidereal Days in the Planet's Year
Mercury	59	.241	1.5
Venus	−243*	.615	.9
Earth	1	1	365
Mars	1.03	1.88	667
Jupiter	.41	11.86	10,564
Saturn	.43	29	24,600
Uranus	−.51*	84	60,000
Neptune	.66	165	114,000
Pluto	6.39	248	14,000

*Retrograde rotation.

Note that the intuitive idea that we have of the relation of a day and a year would be different if we lived on a different planet. If we made a table like this for solar days on a planet instead of for sidereal days, then Mercury, for example, whose solar day (the time from one noon to the next) is 176 Earth days long (as we shall discuss in Section 12.1), would have .5 solar Mercury day per Mercury year.

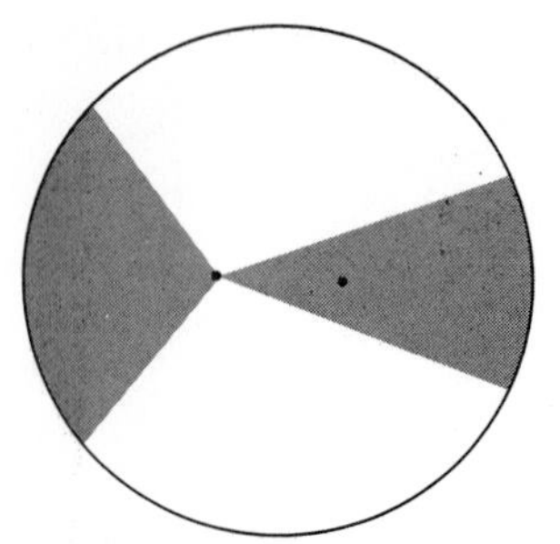

Figure 10–22 Kepler's second law states that the two shaded sectors, which represent the areas covered by a line drawn from a focus of the ellipse to an orbiting planet in a given length of time, are equal in area. The Sun is at this focus; nothing is at the other focus.

Box 10.3 Kepler's Third Law and Planetary Masses

We have consistently seen how astronomers make rough calculations to test if physical processes under consideration could conceivably be valid. Astronomy has also had a long tradition of exceedingly accurate calculations. Pushing accuracy to yet one more decimal place sometimes leads to important results.

For example, Kepler's third law—the period of a planet squared is proportional to its distance from the Sun cubed ($P^2 = \text{constant} \times R^3$)—holds to a reasonably high degree of accuracy and seemed completely accurate when Kepler did his work. But now we have more accurate observations. If we consider each of the planets in turn and if a term involving the sum of the masses of the Sun and the planet under consideration is included in the equation—$P^2 = \frac{\text{constant}}{m_{Sun} + m_{planet}} \times R^3$—the agreement with observation is improved in the fourth decimal place for the planet Jupiter and to a lesser extent for Saturn. The masses of the other planets are too small to have an effect even the size of the tiny effect for the massive planets.

Isaac Newton derived the equation in its general form:

$$P^2 = \frac{4\pi^2}{G(m_1 + m_2)} R^3$$

for a body with mass m_1 revolving in an elliptical orbit around a body with mass m_2. The constant G is called the *constant of universal gravitation* because to the best of our knowledge it is constant throughout the Universe. The equation can be used to derive the masses of celestial bodies. Not only the Sun and its family of planets but also a planet and its moons and the components of multiple star systems can be "weighed" with this equation.

Figure 10–23 Kepler's third law relates the period of an orbiting body to the size of its orbit. The outer planets orbit at much slower velocities than the inner planets and also have a longer path to follow in order to complete one orbit. The distances the planets travel in their orbits in 1 year is shown here.

on Saturn we would call that longer length of time by the name "year." If we were then to visit Earth, we would say that the Earth "year" was very brief—only 1/30th of what we Saturnians would call a "year" (Table 10–1).

Kepler's third law applies not only to planets and the Sun, but also to all pairs of bodies in which one body containing relatively little mass is orbiting another, more massive, body. Planets and their moons are such

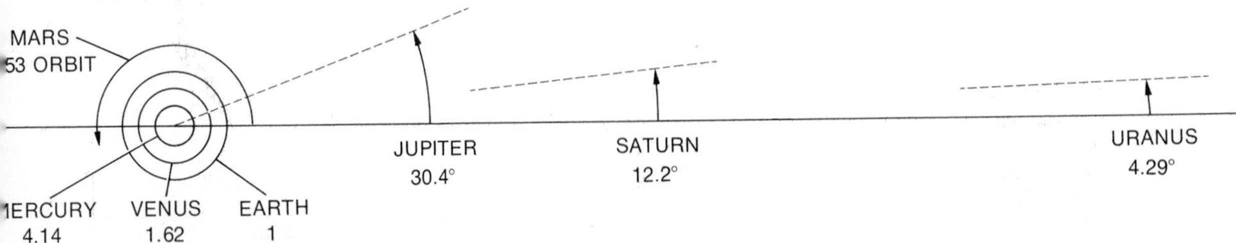

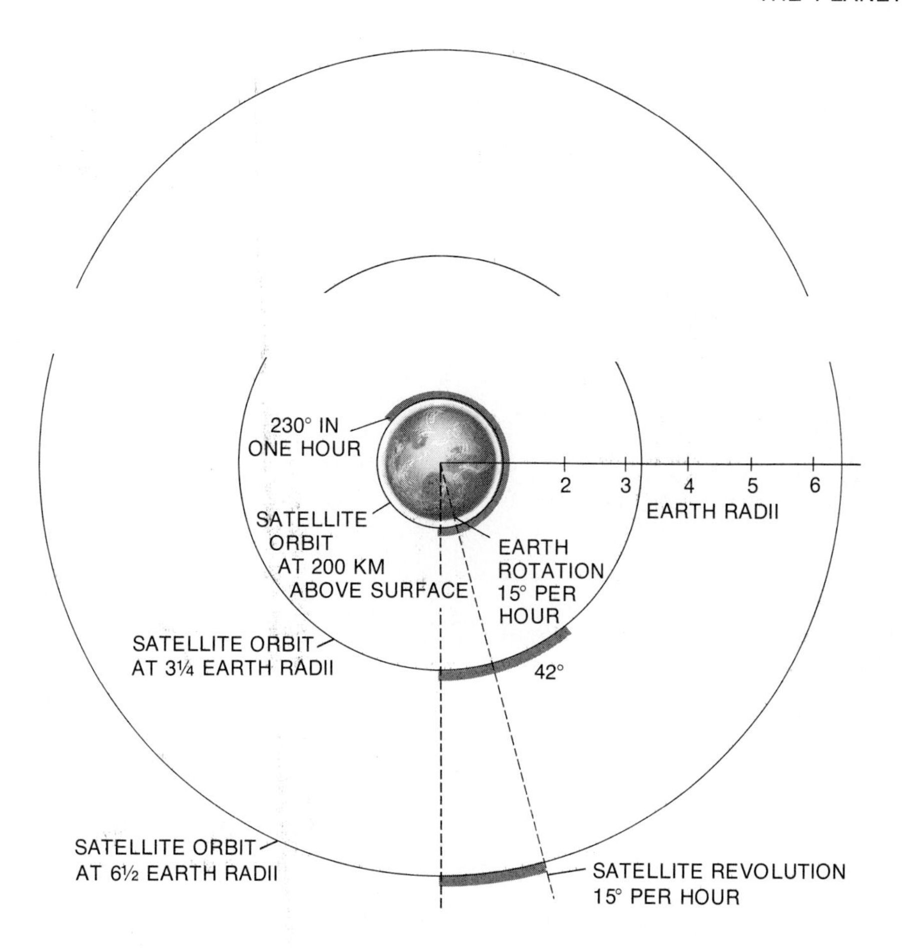

Figure 10–24 Most orbiting satellites are only 200 km or so above the Earth's surface, and orbit the Earth in about 90 minutes. From Kepler's third law we can see that the velocity in orbit of a satellite decreases as the satellite gets higher and higher. At about 6½ Earth radii, the satellite orbits at the same velocity as that at which the Earth's surface rotates underneath. Thus the satellite is in *synchronous rotation,* and always remains over the same location on Earth. This property makes such synchronous satellites useful for relaying communications.

cases. The form of the equation is the same; only the constant—which depends on the mass of the central body—is different.

You may be familiar with the fact, for example, that most artificial Earth satellites orbit in about 90 minutes; this is the period that corresponds to an orbit at approximately the radius of the Earth plus two hundred kilometers or so, about 6600 kilometers. But we now also have synchronous satellites that always stay over the same place on the Earth so that they can beam radio or television pictures, or take photographs of surface changes over time. We want these satellites to orbit at the same speed that the Earth below them is rotating, so that they always stay over the same point on the surface. Their period must be twenty-four hours. Thus by Kepler's third law, the satellites must be in much higher orbits for their periods to be that much longer than those of the other satellites. The numbers work out that they must be about 40,000 kilometers from the Earth's center, equivalent to approximately 7 Earth radii (Fig. 10–24).

It was not until the time of Isaac Newton, many years after these three laws were laboriously calculated and discovered by Kepler, that the laws

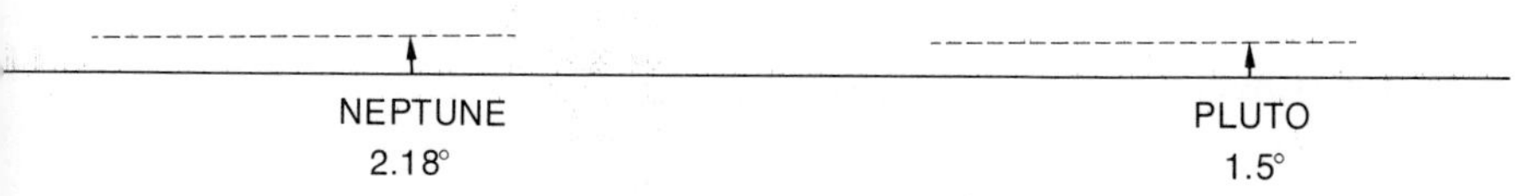

Figure 10–23

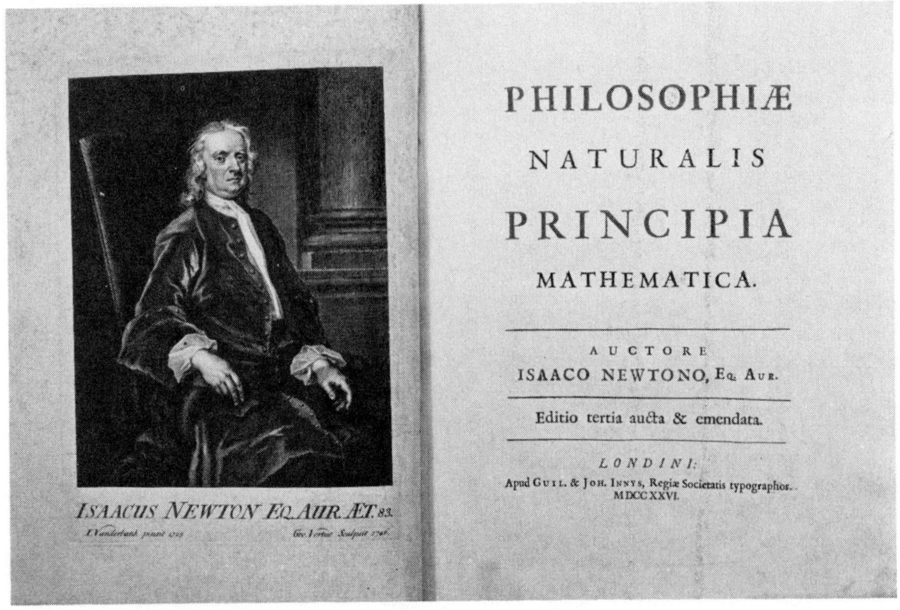

Figure 10–25 The title page and frontispiece of the third edition of Isaac Newton's *Principia Mathematica,* published in 1726.

were derived mathematically from basic principles. Newton invented and used the calculus to do this in his *Principia*, published in 1687 (Fig. 10–25). He used the law of gravitation that he had discovered, for Newton—whether or not you believe that an apple fell on his head—was the first to realize the universality of gravity. He formulated the law that the force of gravity between two bodies varies directly with the product of the masses of the two bodies and inversely with the square of the distance between them: force of gravity $\propto m_1m_2/d^2$.

$\propto$ means "is proportional to"; it means that the left hand side is equal to a constant times the right hand side. The constant, 6.67×10^{-8} when masses are in grams and distances are in centimeters, is called G *or the* universal gravitational constant. *We can write, $F = Gm_1m_2/d^2$. This is known as the* law of universal gravitation; *it is a universal law in that it applies all over the Universe rather than being limited to local applicability.*

10.3 THE REVOLUTION AND ROTATION OF THE PLANETS

The orbits of all the planets lie in approximately the same plane. They thus take up only a disk whose center is at the Sun, rather than a full sphere. Little is known of the parts of the solar system away from this disk, although the comets may originate in a spherical cloud around the Sun.

The plane of the Earth's orbit around the Sun is called the *ecliptic plane* (Fig. 10–26). Of course, since we are on the Earth rather than outside the solar system looking in, we rather see the Sun appear to move across the sky with respect to the stars. (Remember that on a given day of 24 hours, the Sun moves at almost the same rate as the stars as it rises, moves across the sky, and sets; one has to plot the position of the Sun with respect to the stars from day to day to see that it is in fact moving among them.) The path that the Sun takes among the stars, as seen from the Earth, is called *the ecliptic* (see also Section 3.11).

The motion of the planets around the Sun in their orbits is called revolution. *The spinning of a planet is called* rotation. *We must be careful and precise when using these words (Fig. 10–27).*

The fact that the planets all have orbits close to the ecliptic plane means that they are all in the part of our sky close to the ecliptic. Since the ecliptic is divided into 12 *signs of the zodiac* (see Fig. 3–43), each associated with a distinct constellation, the planets always fall in one of these 12 *houses*, and never stray into other constellations.

The *inclinations* of the orbits of the other planets with respect to the ecliptic are small, with the exception of Pluto (Fig. 10–28). Of the other

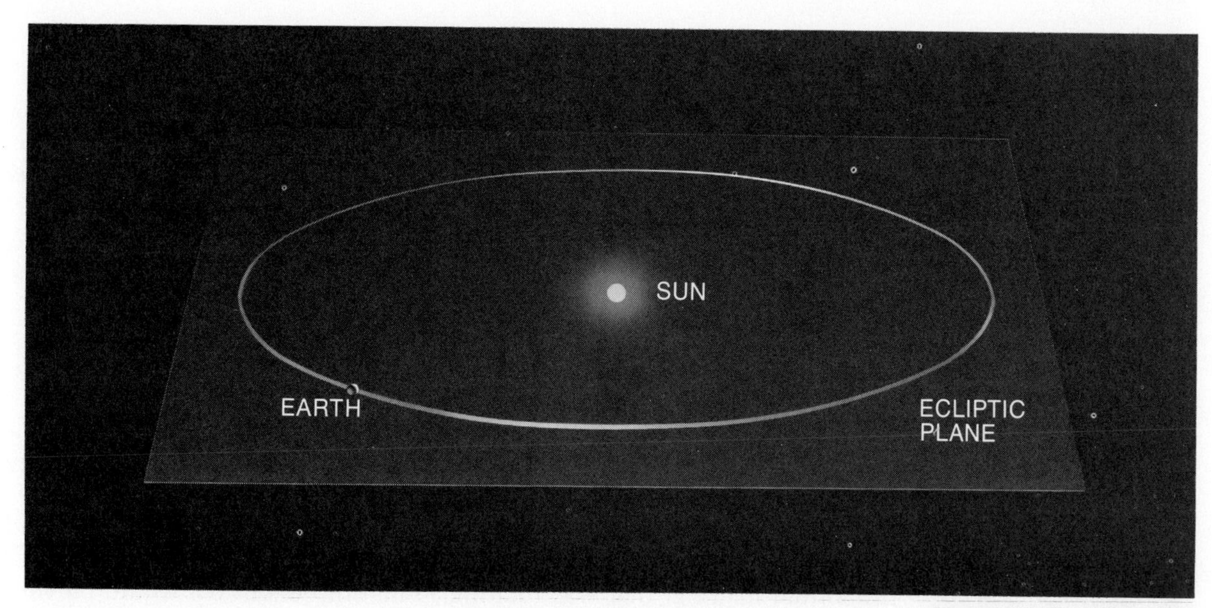

Figure 10–26 The ecliptic plane is the plane of the Earth's orbit around the Sun.

The inclination *of the orbit of a planet is the angle that the plane in which its orbit lies makes with the plane in which the Earth's orbit lies.*

planets, Mercury has an inclination of 7°; the remainder, except for Pluto, have inclinations of less than 4°. Pluto's much larger inclination of 17° is discrepant and is just one of the pieces of evidence suggesting that Pluto may not have formed in the same circumstances as the other planets in the solar system.

The fact that the planets all orbit the Sun in essentially the same plane is one of the most important facts that we know about the solar system. Its explanation is at the base of most cosmogonical models, and central to that explanation is a property that astronomers and physicists use in analyzing spinning or revolving objects: *angular momentum.* The amount of angular momentum of a small body revolving around a large central body is (distance from the center) × (velocity) × (mass). The importance of angular momentum lies in the fact that it is conserved, that is, the total angular momentum of the system (the sum of the angular momenta of the different parts of the system) doesn't change even though the distribution of angular momentum among the parts may change. The total angular momentum of the solar system should thus be the same as it was in the past, unless some mechanism is carrying angular momentum away.

The most familiar example of angular momentum is an ice skater. She may start herself spinning by exerting force on the ice with her skates, thus giving herself a certain angular momentum. When she wants to spin faster, she draws her arms in closer to her (Fig. 10–29). This changes the distribution of her mass so that it is effectively closer to her center (the axis around

Figure 10–27 It is all too easy to "misspeak" when discussing rotation, which refers to a body turning on its axis, and revolution, which refers to a body orbiting around another body.

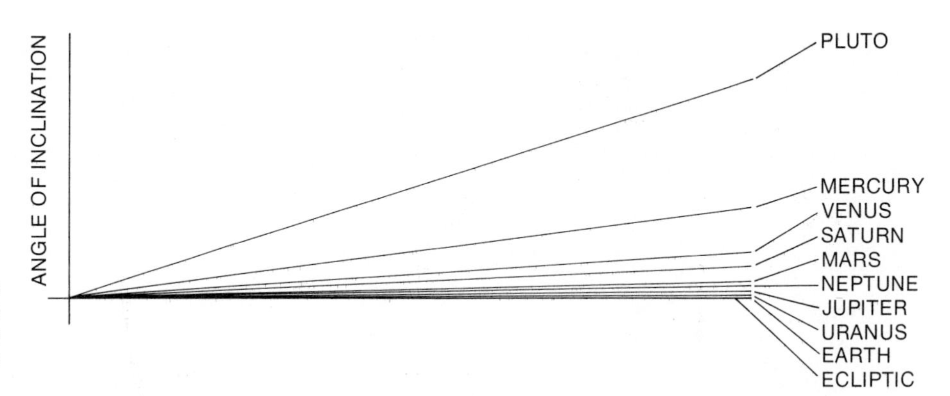

Figure 10–28 The orbits of the planets, with the exception of Pluto, have only small inclinations to the ecliptic plane.

Figure 10–29 An ice skater draws her arms in, in order to redistribute her mass. Because her angular momentum is conserved, she begins to spin faster in order to compensate for the new mass distribution. Dorothy Hamill is shown here during her gold medal performance in the free style event at the Innsbruck Olympics in 1976.

which she is spinning), and to compensate for this she starts to rotate more quickly so that her angular momentum remains the same.

Thus from the fact that every planet is revolving in the same plane and in the same direction, we deduce that the solar system was formed of primordial material that was rotating in that direction. We would be very surprised to find a planet revolving around the Sun in a different direction—and we do not—or, to a lesser extent, to find a planet rotating in a direction opposite to that of its fellows. There are, however, two examples of such backwards rotation, which must each be carefully considered. Such rotation in the opposite sense is called *retrograde rotation.* Do not confuse it with *retrograde motion,* which has to do with the apparent motion of the position of a planet in the sky as seen from the Earth.

10.4 BODE'S "LAW"

For over two hundred years, a numerical relation called "Bode's law" has been known to give the approximate distances of the planets from the Sun, though it has never been theoretically understood.

Bode's law says to write down first 0, then 3, and then keep doubling the previous number: 6, 12, 24, etc. Now add 4 to each number to get the series 4, 7, 10, 16, 28, and so on. Then divide each number by 10. The results: .4, .7, 1.0, etc., give the approximate radii of the planetary orbits in astronomical units (Table 10–2).

The relation was first published anonymously by Titius of Wittenberg in 1766 in a translation of a book by a Swiss scientist, but Bode's name became attached to the relation a half-dozen years later when Bode, director of the Berlin Observatory, popularized it. Sometimes we now call it the Titius-Bode law. The observational tests of Bode's law are connected with the outer planets and asteroids, which will be discussed in Chapters 17 and 18. When Uranus was discovered by William Herschel in England in 1781, its orbital distance fit Bode's law and the discovery spurred theoretical work to derive the "law." Bode's law also predicted a planet between Mars and Jupiter. The asteroids, some of which are at this distance, as predicted by Bode's law, began to be discovered twenty years after Uranus. The fact that there were many asteroids (minor planets), instead of one major planet, was disturbing to many. To have a perspective in time, let us realize that this active discussion of Bode's law was going on in Europe at the time of the American Revolution.

But Neptune and Pluto, which were discovered later, do not fit Bode's

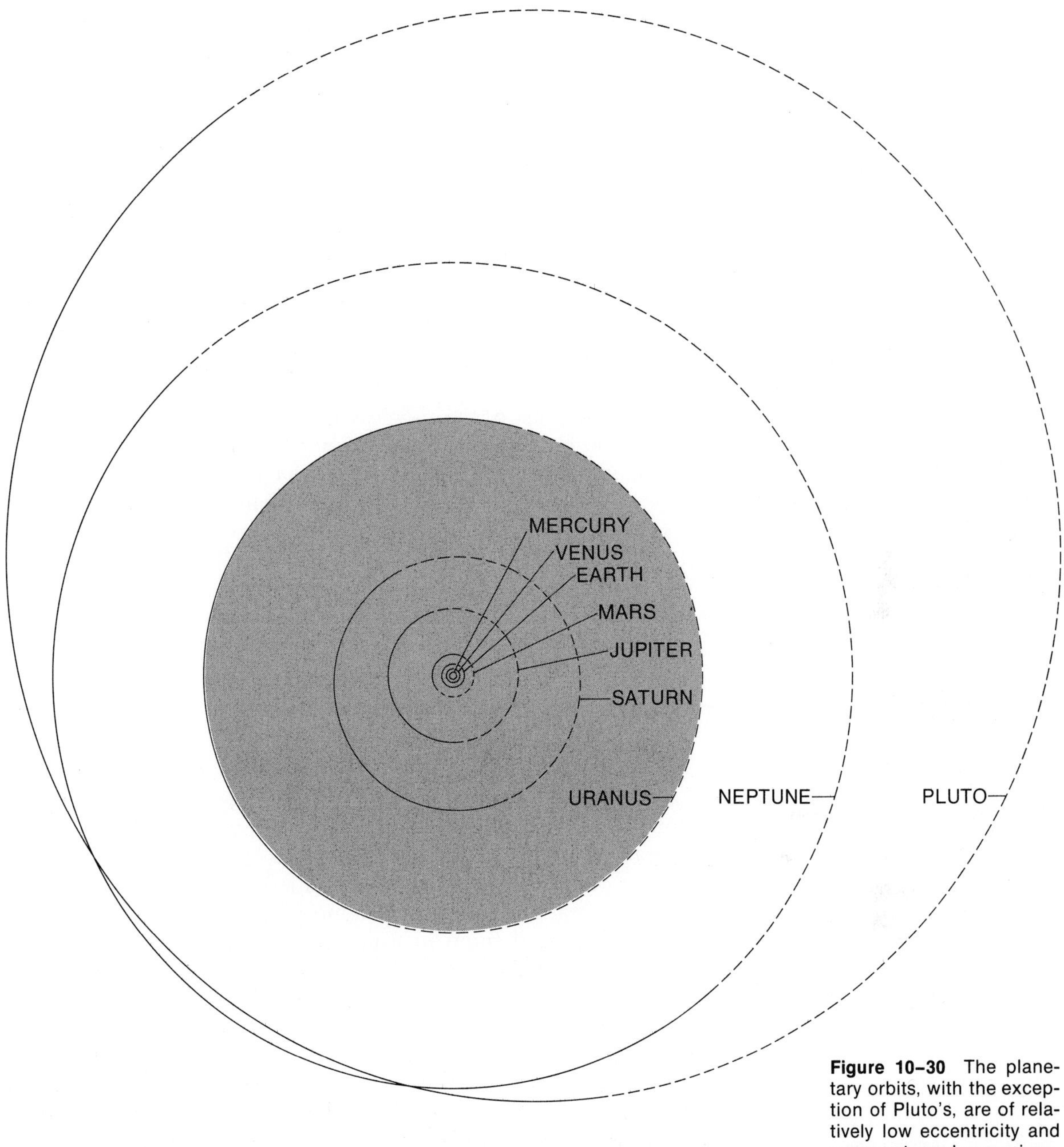

Figure 10–30 The planetary orbits, with the exception of Pluto's, are of relatively low eccentricity and appear at regular spacings. The portion of the orbits that are below the ecliptic plane are shown with dotted lines.

TABLE 10–2 BODE'S LAW

Planet	Distance by Bode's Law (A.U.)	Actual Mean Distance (A.U.)
Mercury	0.4	0.39
Venus	0.7	0.72
Earth	1.0	1.00
Mars	1.6	1.52
Asteroids	2.8	–
Jupiter	5.2	5.20
Saturn	10.0	9.54
Uranus	19.6	19.2
Neptune	38.8	30.6
Pluto	77.2	39.4

1 A.U., the average distance from the Earth to the Sun, equals 150,000,000 km.

law, and no mathematical derivation of Bode's law has ever been found. (Some numerical calculations with large computers do show planetary spacings that follow different rules.) Many astronomers now think that Bode's law is merely a numerical coincidence, devoid of deeper meaning. Statisticians point out that since it had to explain only 6, and then only 8 planets, and since there were no limitations on the form that a law could take, one had enough freedom merely to try out ways of combining a few additions, subtractions, multiplications, cosines, and so on, in some sequence that "predicts" this small set of data. Of course, each different "law" will have a different prediction for the next number in the series – in this case the radius of the orbit of an as yet unknown planet. And even though Bode's law seemed to work for Uranus, and possibly for the asteroids, it later began to fail since it is not satisfied by the orbits of Neptune or Pluto.

It is no surprise that the planetary orbits have spaces between them such that we don't find two planets in orbits very close to each other. Most matter that would exist between the planets would be swept up by the gravitational forces of the planets as they go around the Sun.

For the moment, many astronomers think of Bode's law as merely a historical curiosity or, at best, a way of memorizing planetary distances (should one want to do so). That is not to say that the interactions of gravity in the solar system over billions of years may not have brought the planets to a distribution of orbits for which a simple formula can be found, but rather that we do not know whether there is any significance in the particular combination of small integers enunciated by Bode.

10.5 THEORIES OF COSMOGONY

The solar system exhibits many regularities, and theories of its formation must account for them. The orbits of the planets are almost, but not quite, circular, and all lie in essentially the same plane. All the planets revolve around the Sun in the same direction, which is the same direction in which the Sun rotates. Moreover, almost all the planets and planetary satellites rotate in that same direction. Some planets have families of satellites that revolve around them in a manner similar to the way that the planets revolve around the Sun. And cosmogonical theories must explain the spacing of the planetary orbits and the distribution of planetary sizes and compositions.

René Descartes, the French philosopher, was one of the first to consider the origin of the solar system in what we would call a scientific manner. In his theory, proposed in 1644, circular eddies of all sizes called vortices were formed in a primordial gas at the beginning of the solar system and eventually settled down to become the various celestial bodies.

After Newton proved that Descartes's vortex theory was invalid, it was 60 years before the next major developments. The Comte de Buffon suggested in France in 1745 that the planets were formed by material ejected from the Sun when what he called a "comet" hit it. (At that time, the composition of comets was unknown, and it was thought that comets were objects as massive as the Sun itself.) Later versions of Buffon's theory, called *catastrophe theories*, followed a similar line of reasoning although they spoke explicitly of a collision with another star. The requirement of a collision could be loosened by postulating that the material for the planets was drawn out of the Sun by the gravitational attraction of a passing star. This latter possibility is called a *tidal theory*.

But catastrophe and tidal theories are currently out of fashion, for they predict that only very few planetary systems would exist, since calculations have been made that show that only very few stellar collisions or near-collisions would have taken place in the lifetime of the galaxy. There is some observational evidence that many stars have planets, which would require a more common method of formation. There is new evidence contradictory to some of these observational results, though, as we shall see later on (in Section 19.2). Moreover, theoretical calculations show that gas drawn out of a star in a collision or by a tidal force would not condense into planets, but would rather disperse.

The theories of cosmogony that astronomers now tend to accept stem from another 18th century idea. Immanuel Kant, the noted German philosopher, suggested in 1755 that the Sun and the planets were formed in the same way as each other. In 1796, the Marquis Pierre Simon de Laplace, the French mathematician, independently advanced a similar kind of theory to Kant's when he postulated that the Sun and the planets all formed from a spinning cloud of gas called a *nebula.* Laplace called this the *nebular hypothesis,* using the word "hypothesis" because he had no proof that it was correct. The spinning gas supposedly threw off rings that eventually condensed to become the planets. But though these beginnings of modern cosmogony were laid down in the time of Benjamin Franklin and George Washington, the theory still has not been completely understood or quantified, for not all the stages that the primordial gas would have had to follow are understood.

Note that "cosmogony" is pronounced with a hard "g."

Further, some of the details of Laplace's theory were later thought for a time to be impossible. For example, it was calculated in the last century, by the great British physicist James Clerk Maxwell, that the planets would not have been able to condense out of the rings of gas, and also that they could not have been set to rotate as fast as they do. Furthermore, because of the conservation of angular momentum, if the gas from which the Sun itself condensed were rotating sufficiently fast to throw off rings of gas in the first place, why is it rotating so slowly now?

So for a time, the nebular hypothesis was not accepted. But the current theories of cosmogony again follow Kant and Laplace (Fig. 10–31). In these *nebular theories* the Sun and the planets condensed out of what is called a *primeval solar nebula.* Some five billion years ago, billions of years after the galaxies began to form, smaller clouds of gas and dust began to contract out of interstellar space. Similar interstellar clouds of gas and dust can now be detected at many locations in our galaxy. (In Section 6.1 we discussed the formation of stars out of such interstellar matter; in this section we shall concentrate on planetary formation.)

Because of random fluctuations in the gas and dust from which it formed, the primeval solar nebula probably would have had a small net spin from the beginning. As it contracted, it would have begun to spin faster because of the conservation of angular momentum (the same reason that the ice skater spins faster). Gravity would have contracted the spinning nebula into a disk, for in the directions perpendicular to the plane in which the nebula is rotating there was no force to oppose gravity's pull.

Perhaps rings of material were left behind by the solar nebula as it contracted towards its center. Perhaps there were additional agglomerations beside the *protosun,* the part of the solar nebula that collapsed to become the Sun itself. These additional agglomerations, which may have been

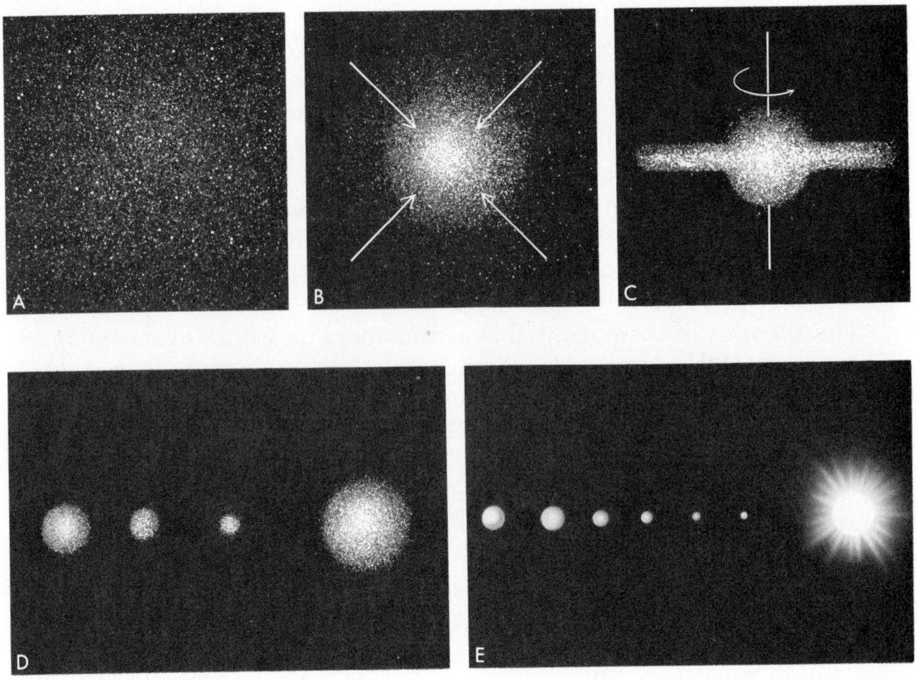

Figure 10–31 The leading model for the formation of the solar system has the protosolar nebula condensing and, between stages C and D, contracting to form the protosun and a large number of small bodies called planetesimals. The planetesimals clumped together to form protoplanets (D), which in turn contracted to become planets. Some of the planetesimals may have become moons or asteroids (which are discussed in Section 18.3).

formed from interstellar dust, would have grown larger and larger. Millimeter-size particles would have clumped into larger bodies and then larger still; the size of the clumps would have increased until the bodies were kilometers and then thousands of kilometers across. The intermediate bodies, hundreds of kilometers across, are called *planetesimals.* (The idea that planetesimals exist and are an intermediate stage in the formation of the planets was borrowed from collision theories.) Laplace's old idea that the protosun had thrown off rings that condensed to become planets, an idea that had been rejected by Maxwell, was not necessary to provide planets after all.

The planetesimals combined under the force of gravity to form *protoplanets.* These protoplanets may have been larger than the planets that resulted from them because they had not yet contracted, though gravity would ultimately cause them to do so. Some of the larger planetesimals may themselves have become moons.

As the Sun condensed, the energy it gained from its contraction went to heat it until its center reached the temperature at which nuclear fusion reactions began taking place. The planets, on the other hand, were simply not massive enough to heat up sufficiently to have nuclear reactions start.

Several modifications of this basic theory have been worked out to accommodate particular observational facts that are known about the solar system. For example, one has to explain the fact that the inner planets are small, rocky, and dense, while the next group of planets out are large and are made of light elements. Since the protosun would have been made primarily out of hydrogen and helium, with just traces of the heavier elements formed in earlier cyclings of the material through stars and supernovae, one simply has to provide a mechanism for ridding the inner regions of the solar system of this lighter material. One possible way is to say that the Sun flared fiercely and/or often in its younger days (it was then a T Tauri star, as described in Section 6.1), and that the lighter elements were blown out of the inner part of the solar system. But proto-Jupiter and the other outer protoplanets, which were much farther away from the Sun, would have retained thick atmospheres of hydrogen and helium because of their high gravity.

A major modern method of calculation considers how the temperature decreases with distance from the center of the solar nebula, ranging from almost 2000 K close to the protosun, down to only about 20 K at the distance of Pluto. Various elements are able to condense, solidify, at different locations because of these differing temperatures. For example, in the positions occupied by the terrestrial planets, it was too hot for icy substances to form, though such ices are present in the giant planets. Such calculations are perhaps the dominant consideration in much current cosmogonic research.

Until recently it was assumed that the abundances of the elements relative to each other were constant throughout the primeval solar nebula. But new studies, particularly analyses of the composition of meteorites, are leading many astronomers to believe that the abundances could have been different at different places in the primeval solar nebula. These variations might have resulted from the presence of dust grains, and a small amount (up to 2 per cent) of the heavier elements could even have been formed in a supernova explosion that might have taken place nearby shortly before the solar nebula began to condense.

Many of the difficulties in theories of cosmogony have to do with angular momentum. No one understands why so much of the angular momentum of the solar system should be in the orbital motion of the planets, and so little in the rotation of the Sun itself. That is one reason why some astronomers were pleased when the solar oblateness experiment we have described (see Section 5.11) reached the conclusion that the solar interior is rotating very rapidly, and thus has a higher angular momentum than had been thought. But the latest experiments do not agree with this result about the solar oblateness. Most current cosmogonical theories invoke some mechanism to transfer angular momentum from the Sun toward the outer parts of the solar system. Now that we know that the solar wind flows outward from the Sun, for example, the solar wind is a likely culprit. Perhaps the strong flares that supposedly rid the inner protoplanets of their light gases were accompanied by a strong solar wind that carried away most of the original solar angular momentum.

The cosmogonical ideas we have discussed are all speculative, for the nature of the formation of the solar system is not definitively understood, nor are the details known. Professional astronomers join students of astronomy in wishing that the theory could be laid out neatly and conclusively for all to see, but we are not yet able to do that.

There are still more possibilities that bedevil cosmogonical research. Some of the planets and moons may not have been formed in their present configurations. Perhaps the Earth captured another protoplanet, which became the Moon. Perhaps Pluto was ejected from an orbit around Neptune into an orbit as an independent planet. Perhaps gravitational encounters put some of the planets and moons into retrograde rotation.

It is, of course, always best to test a scientific theory by seeing if it explains facts that were not known when the theory was advanced. With the recent exploration of the planets from space, we are finding tremendous amounts of basic information and checking to see if this information was predicted by any of the theories. For example, as we shall see, space observations can give us direct information on the composition of not only the surfaces but also the interiors of some of the planets. We can look for more similarities among planets or among satellites. In the early days of the space program, when the Moon landing program was just getting under way, it may have seemed to some people that the Moon would soon give us the answers to our questions about the origin of the solar system. The observations that have been made have been very valuable for the interpretation of cosmogonical theories, but they have turned out to have opened as many questions as they have answered. Some theories have been ruled out, but all the new data are not directly explained by any theory. As we proceed, now, to discuss the planets individually and in detail, we must always be aware when a particular set of observations that we discuss is of special significance in the understanding of our origins.

SUMMARY AND OUTLINE OF CHAPTER 10

Aims: To discuss the scale of the solar system, the history of our understanding of the heliocentric theory, Kepler's laws, and the origin of the solar system

The scale of the solar system and the planets themselves
The phases of the moon and lunar eclipses (Section 10.1)
The apparent motions of the planets; retrograde motion (Section 10.2)
The development of our knowledge of the solar system (Section 10.2)
- Earth-centered system: Aristotle and Ptolemy
  - Epicycles and deferents
- Heliocentric system: Copernicus
  - *De Revolutionibus*, 1543
  - Retrograde motion explained as a projection effect
- First telescopic observations: Galileo
- Improved positional observations and deductions from them: Tycho and Kepler

Mathematical and physical deductions
- Kepler's three laws and how to use them
  - (1) Planets orbit the Sun in ellipses
  - (2) Law of equal areas
  - (3) Period^2 proportional to distance^3
    - (Period is a time interval; think of Times Square, New York)
- Newton and the universality of gravity

The ecliptic; inclinations of the orbits of the planets (Section 10.3)
- Pluto has largest inclination by far
- Other planets have only slight inclinations

Conservation of angular momentum (Section 10.3)
- All planets revolve in same direction
- Most planets rotate in the same sense

Bode's "law" (Section 10.4)
- 0, 3, 6, 12, and so on
- Add 4 and divide by 10
- Not known if the "law" has basic significance

Theories of cosmogony (Section 10.5)
- Catastrophe theories
- Nebular theories
  - Kant and Laplace
  - Newer versions: protosun, planetesimals, and protoplanets

QUESTIONS

1. What are the features that distinguish the terrestrial from the giant planets? What features does Pluto have in common with either group?

2. Suppose that you live on the Moon. Sketch the phases of the Earth that you would observe for various times during the Earth's month.

3. If you lived on the Moon, would the motion of the planets appear any different than from Earth?

4. If you lived on the Moon, how would the position of the Earth change in your sky over time?

5. If you lived on the Moon, what would you observe during an eclipse of the Moon. How would an eclipse of the Sun by the Earth differ from an eclipse of the Sun by the Moon that we observe from Earth?

6. If you lived on Saturn, which planets would appear to go through phases? Which planets would exhibit retrograde motion?

7. Discuss the following statement: "With the addition of epicycles, the geocentric theory of the solar system could be made to agree with observations. Since it was around first, and therefore better known, it should have been kept."

8. Do Kepler's laws permit circular orbits?

9. At what point in its orbit is the Earth moving fastest?

10. Use Kepler's 3rd law and the fact that Mercury's orbit has a semimajor axis of 0.4 A.U. to deduce the period of Mercury. Show your work.

11. (a) Use the data in Appendix 5 to show for which planets the square of the period does not equal the cube of the semimajor axis to the accuracy given.
    (b) Use the formula in Box 10.3 to show that including the effects of planetary masses removes the discrepancy.

12. The occupants of Planet X note that they are 1 greel from their sun (a greel is the Planet X unit of length), and they orbit with a period of 1 fleel. They observe Planet Y which orbits in 8 fleels. How far is Planet Y from their sun (in greels)?

13. Would an observer on the Sun see the Earth moving along the ecliptic? Would the apparent path of the Earth across the sky be east to west, or west to east? Explain.

14. Explain how conservation of angular momentum applies to a diver doing somersaults or twists. What can divers do to make sure they are vertical when they hit the water?

15. If two planets are of the same mass but different distances from the Sun, which will have the higher angular momentum around the Sun? (Hint: Use Kepler's third law.)

16. If the planets condensed out of the same primeval nebula as the Sun, why didn't they become stars?

17. Which planets are likely to have their original atmospheres? Explain.

11

The Moon

The Earth's nearest celestial neighbor—the Moon—is only 380,000 km (238,000 miles) away from us on the average, close enough that it appears sufficiently large and bright to dominate our nighttime sky. The Moon's stark beauty has called attention to it since the beginning of history, and studies of the Moon's position and motion led to the earliest consideration of the solar system, to the prediction of tides, and to the establishment of the calendar.

11.1 THE APPEARANCE OF THE MOON

The fact that the Moon's surface has different kinds of areas on it is obvious to the naked eye. Even a small telescope—Galileo's, for example—reveals a surface pockmarked with craters. The *highlands* are heavily cratered; other areas, called *mare* (pronounced mar′-ā; plural, *maria,* pronounced mar′ē-a), are relatively smooth, and indeed the name comes from the Latin word for sea (Fig. 11–1). But there are no ships sailing on the lunar seas and no water in them; the Moon is a dry, airless, barren place. The Moon's mass is only 1/81 that of the Earth, and the gravity at its surface is only 1/6 that of the Earth. Any atmosphere and any water that may once have been present would long since have escaped into space. The Moon is 3476 km (2160 miles) in diameter, about one-fourth the diameter of the Earth. No other moon in the solar system is such a large fraction of the size of its planet.

When the Moon is full, it is especially bright in the nighttime sky—bright enough, in fact, to cast shadows or even to read by. But full moon is a bad time to try to observe surface structure, for we are looking along the same line that is being traveled by the sunlight, and any shadows we see are short. When the Moon is in a crescent phase or even a half-moon, however, the part of the lunar surface where the shadows are long faces us. At those times, the lunar features stand out in bold relief (Fig. 11–2).

Shadows are longest near the *terminator,* the line separating day from night. Since the Moon goes around the Earth and returns to the same position with respect to the Earth and Sun once every 29 1/2 days (this is the

---

Scientist-astronaut Harrison Schmitt collecting small rocks and rock chips with a lunar rake during the Apollo 17 mission to the Taurus-Littrow region of the Moon.

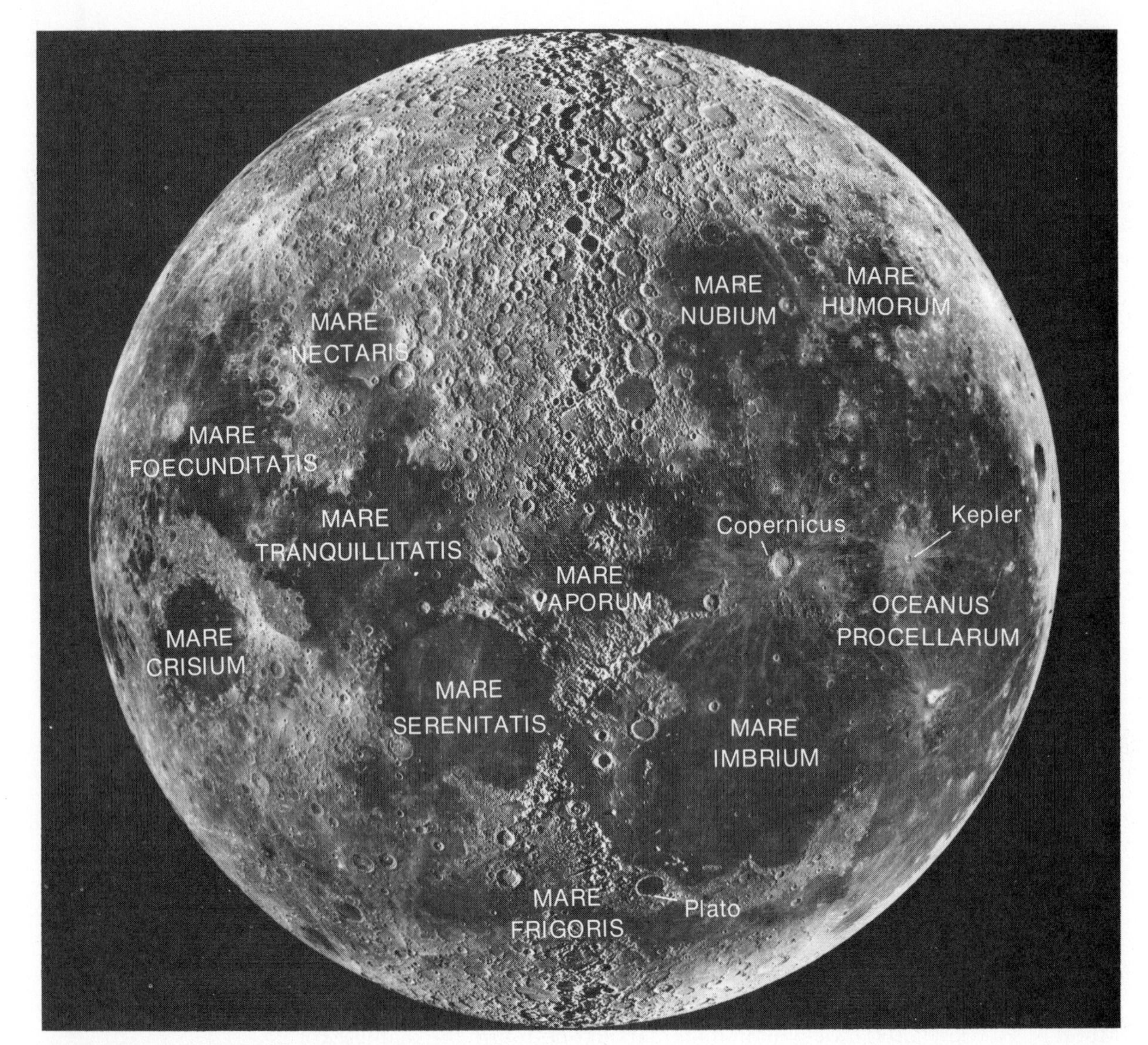

Figure 11–1 In order to show the whole Moon, but still show the detail that does not show up well at full moon, the Lick Observatory has put together this composite of first and third quarters. Note the dark maria and the lighter, heavily cratered highlands. Two young craters, Copernicus and Kepler, can be seen to have rays of light (that is, darker than the background) material emanating from them.

Box 11.1 Sidereal and Synodic Periods

Sidereal: with respect to the stars. For example, the Moon revolves around the Earth in about 27 1/3 days with respect to the stars.

Synodic: with respect to some other body, usually the Earth. For example, by the time that the Moon has made one sidereal rotation around the Earth, the Earth has moved about 1/13 of the way around the Sun (about 27°). Thus the Moon must travel a little farther to catch up (Fig. 11–3). The synodic period of the Moon, the synodic month, is about 29 1/2 days.

Another example concerns revolution rather than rotation. Consider Mars, which has a sidereal period of revolution around the Sun of 687 days. However, only every 70 days does the Earth, which is traveling in orbit faster than Mars, catch up with Mars. (This is equivalent to stating that it is 780 days from opposition to opposition.) The synodic period of Mars is thus 780 days.

Figure 11–2 The crescent phases of the Moon, 3 and 5 days, respectively, after new moon.

synodic revolution period of the Moon), the terminator moves around the Moon in that time. The phases repeat with this period, which is called the *synodic month.* Most locations on the Moon are thus in about 15 days of sunlight, during which time they become very hot, and then 15 days of darkness, during which time their temperature drops to as low as 160 K.

The Moon rotates on its axis at the same rate as it revolves around the Earth, always keeping the same face in our direction. The Earth's gravity has locked the Moon in this pattern, interacting with a bulge in the distribution of the lunar mass to prevent the Moon from rotating freely.

As a result of this interlock, we always see the same side of the Moon from our vantage point on Earth. Actually, we can see a bit more than half. For one thing, the Moon's speed in its orbit varies because of the orbit's eccentricity and Kepler's second law, but its rate of spin is constant, so sometimes we can see around one side of the Moon or the other. Also, its

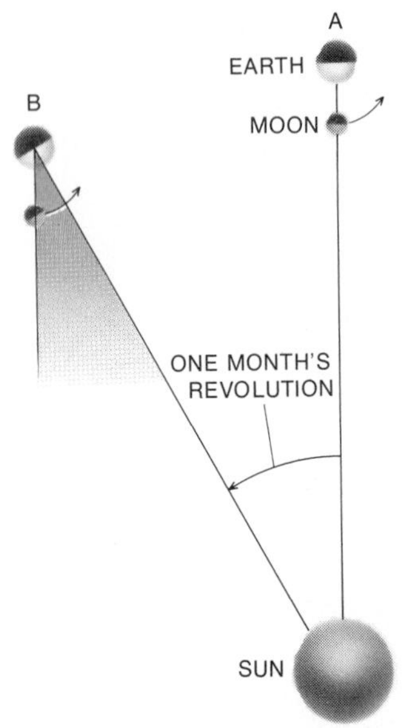

Figure 11–3 After the Moon has completed one rotation on its axis with respect to the stars, which it does in 27⅓ days, it has moved from A to B. A given feature has still not swung far enough around to again be in the same position with respect to the Sun because the Earth has revolved one month's worth around the Sun. It takes about an extra two days for the Moon to complete its rotation with respect to the Earth, which gives a synodic rotation period of about 29½ days. The extra angle it must cover is shaded. In the extra two days, both Earth and Moon have continued to move around the Sun (toward the left in the diagram).

equator is inclined by about 5° to the plane of its orbit, so we can see over the north pole some of the time and, when the Moon has moved around to the other side of the Earth two weeks later, see over the south pole. These motions are called *librations* (Fig. 11–4); as a result we can see 59 per cent of the lunar surface. Another contribution to libration comes from the fact that because of the Earth's diameter we look from different vantage points in the course of each day. An actual small variation of the Moon's period of rotation is a fourth, minor contribution to libration.

In some sense, before the period of exploration by the Apollo program we knew more about almost any star than we did about the Moon. As a solid body, the Moon reflects the solar spectrum rather than emitting one of its own, so we were hard pressed to determine the composition or the physical properties of its surface. On the other hand, stars are much easier to understand because of the spectra of their gases. Before we landed on the Moon we could not even know how firm the surface was. There were those who thought it might be powdery, that a spacecraft would sink and would never be heard from again. Others thought that the lunar surface was hard and solid, as is much of the surface of the Earth. This question was resolved by the first unmanned landing on the Moon (see Section 11.3a).

Another debate that raged was the origin of the craters. One side thought that they were formed by the impact of *meteorites* (Section 18.2). Others thought they were volcanic in nature and were the result of internal lunar activity. This has been finally settled (see Section 11.3b) in favor of impacts, but we also see numerous signs of volcanic activity.

11.2 LUNAR EXPLORATION

The space age began on October 4, 1957, when the U.S.S.R. launched its first Sputnik (the Russian word for *satellite*) into orbit around the Earth. The shock galvanized the American space program into action and within months American spacecraft were also in Earth orbit. We have seen in our

Figure 11–4 Libration allows us to see part way around the far side of the Moon. From the Earth, we can see 59 per cent at one time or another. Note, for example, the position of Mare Crisium on the lower left of both photographs.

discussions of stars, and will continue to see in our discussions of the Milky Way Galaxy and of other galaxies, how the ability to observe from space has benefited astronomy.

In the case of the Moon and planets, the ability to travel into space did not merely free us from the obscuration of the Earth's atmosphere but also allowed us to explore the planets directly. The Moon, as the closest celestial body, was obviously the place to begin.

In 1959, the Soviet Union sent its Luna 3 spacecraft around the Moon; it radioed back the first murky photographs of the Moon's far side. Now that we have high-resolution maps of most of the lunar surface, it is easy to forget how big an advance that was.

In 1961, President John F. Kennedy announced that it would be a national goal of the United States to put a man on the Moon, and bring him safely back to Earth, by 1970. This grandiose goal led to the largest coordinated program in the history of the world, and had an outstandingly successful conclusion.

The American lunar program, under the direction of the National Aeronautics and Space Administration (NASA), proceeded in gentle stages. On one hand, the ability to carry out manned space flight was developed with single-astronaut sub-orbital and orbital capsules, called Project Mercury, and then with two-astronaut orbital spacecraft, called Project Gemini. Simultaneously, on the other hand, a series of unmanned spacecraft were sent to the Moon, starting with Ranger. It was only the soft landing of the Soviet Luna and American Surveyor spacecraft on the lunar surface in 1966 and the photographs they sent back that settled the argument over the strength of the lunar surface; the Surveyor perched on the surface without sinking in more than a few centimeters. A series of American Lunar Orbiters and Soviet Zonds radioed back photographs of much of the lunar surface.

The manned and unmanned trains of development came together with Apollo 8, which circled the Moon on Christmas Eve, 1968, and returned to Earth. The next year, Apollo 11 brought humans to land on the Moon for the first time. It went into orbit around the Moon after a three-day journey from Earth, and a small spacecraft called the Lunar Module (LM) separated from the larger Command Module. On July 20, 1969—a date that from the long-range standard of history may be the most significant of the last millennium—Neil Armstrong and Buzz Aldrin left Michael Collins orbiting in the Command Module and landed on the Moon (Fig. 11–5). In the preceding days there had been much discussion of what Armstrong's historic first words should be, and millions listened as he said "One small step for a man, one giant leap for mankind."

Figure 11–5 Neil Armstrong, the first person to set foot on the Moon, took this photograph of his fellow astronaut Buzz Aldrin climbing down from the Lunar Module of Apollo 11 on July 20, 1969. The site is called Tranquility Base, as it is in the Sea of Tranquility (Mare Tranquillitatus).

The Lunar Module carried a variety of experiments, including devices to test the soil, a camera to secure close-up stereo photos of the lunar soil, a sheet of aluminum with which to capture particles from the solar wind, and a seismometer. Unfortunately, the mission seemed to get confused in the popular mind with just one of the experiments—the collection of rocks. These rocks and dust, returned to Earth for detailed analysis, were indeed important, but represented only a fraction of the mission.

The first three missions, Apollos 11, 12, and 14 (Apollo 13 suffered an explosion en route to the Moon, and barely returned its crew safely to Earth), carried out a variety of experiments but were mainly directed to establishing the conditions under which humans could work on the lunar surface and to getting all the hardware working. The next three missions, Apollos 15, 16, and 17, were devoted to science to a greater extent. They

A B

Figure 11–6 *(A)* Eugene Cernan riding on the Lunar Rover during the Apollo 17 mission. The mountain in the background is the east end of the South Massif. *(B)* Drawing by Alan Dunn; © 1971 *The New Yorker Magazine, Inc.*

benefited by their longer stay on the lunar surface—three days each instead of the three hours of Apollo 11—and by the addition of the Lunar Rover (Fig. 11–6 and Color Plate 26), a sort of car with which they could greatly increase the distances they could cover. Also, they went to locations on the lunar surface that were less flat than the locations of the first lunar landings, which had been chosen for safety. For example, the Apollo 17 site in the lunar highlands near Taurus-Littrow seemed more scientifically interesting than the Apollo 11 site in the Sea of Tranquility, the flattest site that could be found.

Originally, there were supposed to be three more Apollo missions, numbers 18, 19, and 20. Astronomers were really looking to these later missions to carry out the bulk of the lunar exploration, for the techniques would then be well in hand, the capabilities of the spacecraft and associated equipment would be increased, and the stay on the surface would be relatively long. Unfortunately, because of funding limitations placed on NASA, something had to give, and scientists found that these last three lunar missions were cut. The era of manned lunar exploration ended almost as soon as it had begun. Only after tremendous struggle on the part of the scientific community was one lone scientist-astronaut (who is shown in the photograph opening this chapter), as distinguished from the test pilot-astronauts who had manned all the earlier missions, assigned to Apollo 17 (whose launch is shown in Color Plate 28).

At present, the United States has no definite plans of any kind for sending manned or even unmanned spacecraft to the Moon, though there is a project being discussed to send an unmanned vehicle into a polar orbit around the Moon in order to map regions at high lunar latitudes.

The Soviet Union, which has sent three unmanned spacecraft to land on the lunar surface, collect lunar soil, and return it to Earth, seems to be continuing lunar work. Its first two round-trip spacecraft each collected a few grams of lunar soil. Luna 24, which went to the Moon in August 1976, drilled to a depth of 1/2 meter (2 feet) below the lunar surface and brought this long, thin cylinder of material back to Earth. In 1970 and 1973, two

TABLE 11–1 MISSIONS TO THE LUNAR SURFACE AND BACK TO EARTH

Apollo 11	U.S.A.	1969	manned
Apollo 12	U.S.A.	1969	manned
Luna 16	U.S.S.R.	1970	unmanned
Apollo 14	U.S.A.	1971	manned
Apollo 15	U.S.A.	1971	manned
Luna 20	U.S.S.R.	1972	unmanned
Apollo 16	U.S.A.	1972	manned
Apollo 17	U.S.A.	1972	manned
Luna 24	U.S.S.R.	1976	unmanned

other Soviet spacecraft had carried remote-controlled rovers, Lunokhods 1 and 2, that traveled over 10 or so kilometers of the lunar surface over a period of many months each.

11.3 THE RESULTS FROM APOLLO

The kilometers of film exposed by the astronauts, the 384 kilograms (846 pounds) of rock brought back to Earth, the rolls and rolls of magnetic tape recording the results from lunar seismographs, and other data, all studied by hundreds and hundreds of scientists from countries all over the Earth, have led to new views of several basic questions. They have raised many new questions about the Moon and the solar system as a whole. Let us consider (a) the composition and chronology of the lunar surface, (b) the origin of the lunar craters, (c) the structure of the lunar interior and whether it is hot or cold, (d) how and where the Moon was formed, and (e) other matters.

11.3a The Composition and Chronology of the Lunar Surface

The types of rocks that were encountered on the Moon are types that are familiar to terrestrial geologists. In the maria, the rocks are mainly *basalts,* a type of rock that results from the cooling of molten material like lava. In the highlands, the rocks are mainly *breccias,* which are mixtures of fragments of several different types of rock that have been compacted and welded together. The fragments in a breccia, of course, date from a time longer ago than the time at which the breccia was welded together.

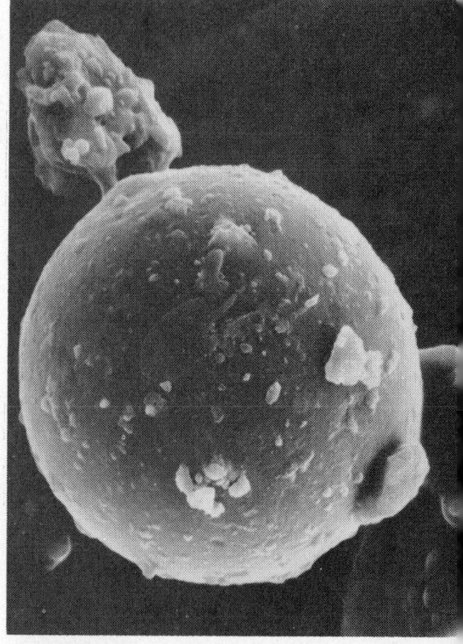

Figure 11–7 An enlargement of a glassy spherule from the lunar dust collected by the Apollo 11 mission. The shape and composition of the spherule indicate that it was created when a meteorite crashed into the Moon, melting lunar material and splashing it long distances. The glassy bead is enlarged 3200 times in this photograph taken with a scanning electron microscope.

The astronauts also collected some *lunar soils,* bits of dust plus larger fragments from the Moon's surface. Some of these proved, on examination through the microscope, to contain a type of structure that is not common on Earth. Small glassy globules (Fig. 11–7) are mixed in. In some of the soils (Color Plate 27) glass coatings and glass globules were present; the bubbly globules in these cases undoubtedly resulted from the melting of rock during its ejection from the site of a meteorite impact and the subsequent cooling of the molten material. In some cases, where most of the soil consisted of glass beads, a volcanic origin is probable.

Almost all of these rocks and soils have lower proportions of elements with low melting points (*volatile* elements) than does the Earth. On the other hand, there are relatively high proportions of elements with

The fragments that make up lunar breccias are of a type of rock called anorthosites. *Anorthosites are rare on Earth, though the Adirondack Mountains are made of them. Anorthosites, like basalts, also result from the cooling of molten material, though under different conditions; anorthosites have taken longer to cool.*

Elements 57 to 71 (Appendix 4) are known as rare earths.

high melting points (*refractory* elements) like calcium, aluminum, and titanium compared to terrestrial abundances. Titanium, which is only a minor constituent of Earth rocks, amounted to 10 per cent of some lunar samples. Of elements that are even rarer on Earth, the abundances of uranium, thorium, and the rare-earth elements are also increased. On the other hand, analyses of basalts from the maria, rocks that may have been formed in the lunar interior instead of on the surface, are not so very rich in such refractory elements as aluminum and calcium. For some time it was generally thought that uranium, thorium, and the group of elements known as rare earths were increased in some places by more than an order of magnitude (i.e., by more than a factor of 10). But estimates of the amount of uranium in the interior, based on the amount of heat measured by the astronauts to be flowing out through the layers just under the surface, were revised after restudy of the data and after new data became available to better interpret the original lunar results. It now seems that the total uranium concentration of the Moon may be the same or only twice as much as that of the Earth. So the Moon and the Earth now seem to be more similar than had been thought for the last few years, though some differences are real. Since the comparison of abundances on the Earth and the Moon has changed drastically over the last few years, it seems wise not to be too confident that the accepted abundance values won't change again.

Something that seems fairly definite is that none of the lunar rocks contain any trace of water bound inside their minerals. This ends all hope that water existed on the Moon at any time in the past, and so seems to eliminate the possibility that life evolved there.

One way of dating the structures on the surface of a moon or planet is to observe the number of craters on them. If we assume that the events that cause craters—whether they are impacts of meteors or the eruptions of volcanoes—continue over a long period of time and that there is no strong erosion, surely those locations with the greatest number of craters must be the oldest. Relatively smooth areas—like the maria—must have been covered over with volcanic material at some relatively recent time (which is still billions of years ago). Even from the Earth we can count the larger craters. And when one crater is superimposed on another (Fig. 11–8), we can be certain that the superimposed crater is the younger one.

Figure 11–8 This view from the orbiting Apollo 15 Command Module shows a smaller crater, Krieger B, superimposed on a larger crater, Krieger. Obviously, the smaller crater is younger than the larger one. Several *rills* (clefts along the lunar surface that can be hundreds of kilometers in length) and *ridges* are also visible. Sometimes the areas in the centers of craters, the *floors,* are smooth but sometimes craters have *central peaks.*

Crater counts can be carried out more completely from near the Moon, since then smaller craters can be seen. Among the important tasks of the Apollo missions was to actually sample lunar rock and soil. The time elapsed since the rock and soil were formed can be measured in laboratories on Earth from detailed studies of the relative abundances of radioactive and non-radioactive isotopes.

A few craters on the Moon, notably Copernicus (Figs. 11–1 and 11–9), have thrown out quite obvious rays of lighter matter. Since these rays extend over other craters, they are younger than these other craters. The few rayed craters may be very young indeed—perhaps only a few hundred million years. The rays darken with time, so rays that may have once existed near other craters are now indistinguishable from the rest of the surface.

All along it had been hoped that the astronauts would find a "genesis rock," a rock from the time of the origin of the solar system some 4.6 billion years ago (see Section 10.4). But although at one point the astronauts thought they had found such a rock, analysis showed that it was some hundreds of millions of years younger than they originally thought.

The oldest rocks that were found at the locations sampled on the Moon were formed 4.42 billion years ago. The youngest rocks were formed 3.1 billion years ago.

There are overall differences between the ages of highland and maria rocks. The highland rocks were formed between about 3.9 and 4.4 billion years ago, and the maria between 3.1 and 3.8 billion years ago. One highland rock even showed signs of having been melted and re-formed twice—mainly at 3.93 billion years but with some crystals that were 4.4 billion years old included. Several highland rocks are exactly the same age, all 4.42 billion years old. So 4.42 billion years ago may therefore have been the origin of the lunar surface material, that is, the time when it last cooled.

Highland rocks from several different locations on the Moon all underwent fundamental changes 3.95 billion years ago, so something catastrophic—such as the huge impact that caused the basin of Mare Imbrium (Fig. 11–1)—may have occurred at that time.

One can use information like the above to build a model of what happened on the Moon. For example, we know that the lunar surface either could have been melted by the heat of impact of meteorites or could have been formed by lava from below the surface, but it would have taken at least half a billion years for the internal heat to generate lava. So any features that were formed before the origin of the Moon plus half a billion years would have had to have been heated by meteoritic impact or been molten because of the heat that resulted when the Moon was formed.

Meteorites hit the Moon with such high velocities that huge amounts of energy are released at the impact. The effect is that of an explosion, as though it had been TNT or an H-bomb exploding.

All the observations can be explained on the basis of the following general picture. The Moon formed 4.6 billion years ago. We do not know if it was then hot or cold. We know that the top 100 km or so of the surface was molten after about 200 million years. This could have been caused by the original heat or by an intense bombardment of meteorites (or debris from the period of formation of the Moon and the Earth) that melted the surface entirely. The presence of radioactive aluminum (Al^{26}) also provided heat. Then the surface cooled definitively. From 4.2 to 3.9 billion years ago, meteorite bombardment caused most of the craters we see today. About 3.8 billion years ago, the interior of the Moon heated up sufficiently (from radioactive elements inside) that vulcanism began; lava flowed on the lunar surface and filled the largest basins that resulted from the earlier bombardment, thus forming the maria. By 3.1 billion years ago, the era of vulcanism was over, and the Moon has been geologically pretty quiet since then (Fig. 11–10).

Figure 11–9 The crater Copernicus, seen in this ground-based photograph, has rays of light material emanating from it. This light material was thrown out radially when the meteorite that formed the crater impacted.

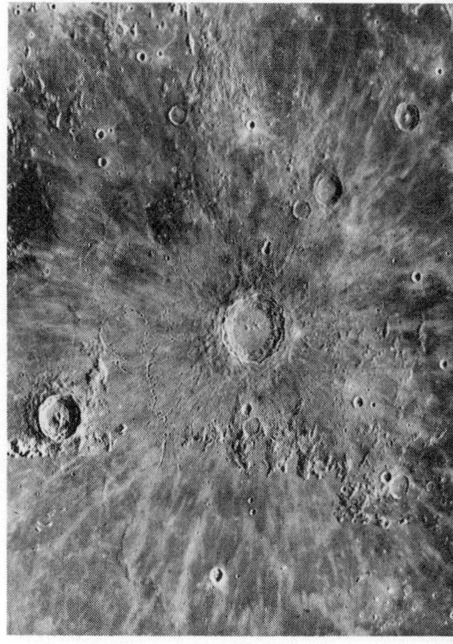

Up to this time, the Earth and the Moon shared similar histories. But active lunar history stops about 3 billion years ago, while the Earth continued to be geologically active. Further, because the Earth's interior continued to send gas into the atmosphere and because the Earth's higher gravity retained that atmosphere, the Earth developed conditions in which life evolved. The Moon, because it is smaller than the Earth, presumably lost its heat more quickly and also generated a thicker crust.

Almost all the rocks on the Earth are younger than 3 billion years of age; erosion and the remolding of the continents as they move slowly over the Earth's surface, according to the theory of *plate tectonics* (known colloquially as *continental drift*), have taken their toll. The oldest single rock ever discovered on Earth has an age of 3.7 billion years. So we must look to extraterrestrial bodies—the Moon or meteorites—that have not suffered the effects of plate tectonics or erosion by such processes as occur in the presence of water or an atmosphere. Only in locations that did not suffer the effects of major erosive processes can we study the first billion years or

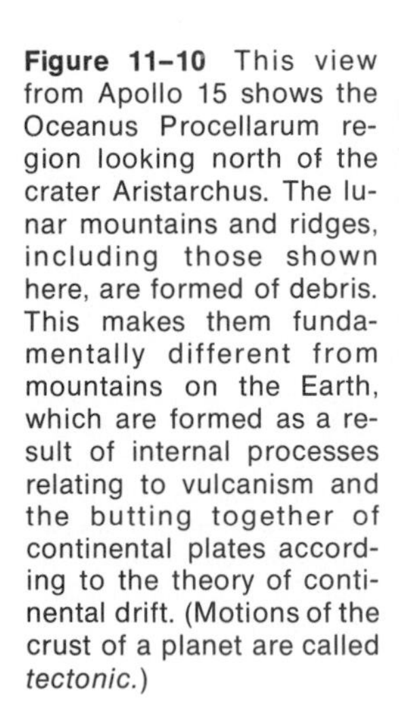

Figure 11–10 This view from Apollo 15 shows the Oceanus Procellarum region looking north of the crater Aristarchus. The lunar mountains and ridges, including those shown here, are formed of debris. This makes them fundamentally different from mountains on the Earth, which are formed as a result of internal processes relating to vulcanism and the butting together of continental plates according to the theory of continental drift. (Motions of the crust of a planet are called *tectonic.*)

more of the solar system. On the Moon, bombardment by micrometeorites —thought to be the major cause of erosion—removes only 1 millimeter of the surface in a million years, a rate much slower than erosive processes on Earth. Still, in the 3 billion years since the most recent lava flows, the top few meters of the lunar surface have been churned up.

11.3b The Origin of the Craters

The debate over whether the craters were formed by meteoritic impact or by volcanic action began in pre-Apollo times. The results from Apollo indicate that most craters resulted from meteoritic impact, even though some very small fraction may represent the effects of vulcanism. It is almost as though the wrong question was being asked all along. The question should not have been, "Are the craters volcanic?" but rather, "Is there evidence of vulcanism?" The answer to the latter question is "yes." Only a very few craters result from vulcanism, but there are many other signs of volcanic activity, including the lava flows that filled the maria. In any case, over 99 per cent of the lunar craters that we see with our telescopes were caused by the impacts of meteorites.

The photographs of the far side of the Moon have shown us that the near and far hemispheres are quite different in overall appearance. The maria that are so conspicuous on the near side are almost absent from the far side, which is cratered all over (Fig. 11–11). Since the many ways that the Earth affects the Moon extend even to the distribution of mass in the outer lunar layers, the asymmetry in the distribution of maria may arise inside the Moon itself. Indeed, once any asymmetry is set up, then the Earth's gravity would lock one side toward us. It is interesting to note that not only the Moon but also the Earth, Mars, and probably Mercury have asymmetric hemispheres. The cause of this is unknown.

11.3c The Lunar Interior

Before the Moon landings, it was widely thought that the Moon was a simple body, with the same composition throughout. But we now know it to be a *differentiated* body like the planets, with a metallic *core* at its center, a silica-rich *mantle* making up the rest of the interior, and a *crust* at the surface of lighter material. The crust is perhaps 65 kilometers thick on the near side and twice as thick on the far side; this asymmetry may explain the different appearances of the near and far sides.

Differentiated *bodies have layers of materials in different states and of different compositions, as opposed to bodies that are homogeneous throughout.*

Figure 11–11 The far side of the Moon looks very different from the near side in that there are few maria (compare with Fig. 11–1). This photograph was taken from Apollo 16, and shows some of the near-side maria at the edge.

The interior of a moon or planet cannot be observed directly, though there are many indirect methods of deduction that can be brought to bear. One of the best methods has scientists using the seismograph, the device employed on Earth to detect earthquakes. The speeds at which different types of waves are transmitted through the lunar interior tell us of the condition of the interior.

There are working seismometers left by Apollo astronauts at four widely spaced locations on the Moon. The seismometers enable us to locate the origins of the thousands of weak moonquakes that occur each year —almost all of which are magnitudes 1 to 2 on the Richter scale used on Earth. Perhaps three moonquakes per year reach Richter 4, a strength which on Earth could be felt but which would not cause damage. On one occasion (July 21, 1972) a meteorite hit the far side of the Moon and generated seismic waves strong enough that we would have expected them to travel through to the near side, where the seismometers are all located. From the fact that one type of seismic wave—so-called *shear waves*—did not travel through the core while the other types did, most researchers deduce that the core is molten or at least plastic in consistency.

Moonquakes of one type all arise 800 to 1100 kilometers under the surface (halfway to the center of the Moon and a distance ten times deeper than most terrestrial earthquakes). This fact can be used to interpret conditions in the lunar interior. If too much of the interior of the Moon were molten, the source of moonquakes probably could not have remained suspended 800–1100 km below the surface. However, if the crust of the Moon were sufficiently thick, then the locations of moonquakes can be accounted for even with a molten core. The moonquakes at each location are triggered twice a month by tidal forces, which result from gravitational effects caused by the variation in the Earth-Moon distance.

Another observation that can be brought to bear here is the fact that the Moon has no general magnetic field. If the interior of the Moon were molten, as is the Earth's interior, it might be expected that a more intense magnetic field than is detected would be set up. The weak magnetic field we do detect could be left over from a core that could have been molten long ago but has since cooled. We shall see in subsequent chapters that the unexpected detection of a magnetic field on Mercury has shown us that we do not understand as well as we had thought how planetary magnetic fields are formed.

Tracking the orbits of the Apollo Command Modules and other satellites that orbited the Moon also tells us about the lunar interior. If the Moon was a perfect, uniform sphere of mass, the spacecraft orbits would have been perfect ellipses. Any deviation of the orbit from an ellipse can be interpreted as an effect of an asymmetric distribution of lunar mass.

It might be noted that although the finding of mascons on the Moon was unexpected, there are two dozen locations on Earth where terrestrial "mascons" have even larger gravitational effects than does the largest lunar mascon.

One of the major surprises of the lunar missions was the discovery under most maria of *mascons*, regions of **mass con**centrations under the surface that lead to anomalies in the gravitational field, that is, deviations of the gravitational field from a sphere (Fig. 11–12). The mascons may be lava that is denser than the surrounding matter. The existence of the mascons is evidence that the whole lunar interior is not molten, for if it were, then these mascons could not remain near the surface. However, the mascons could be supported by a crust of sufficient thickness.

In sum, we have contradictory evidence as to whether the lunar interior is or was hot or cold. Still, as a result of the meteorite detected with the seismic experiment, most scientists believe that the Moon's core is molten. If such a small body as the Moon formed with a hot interior, then

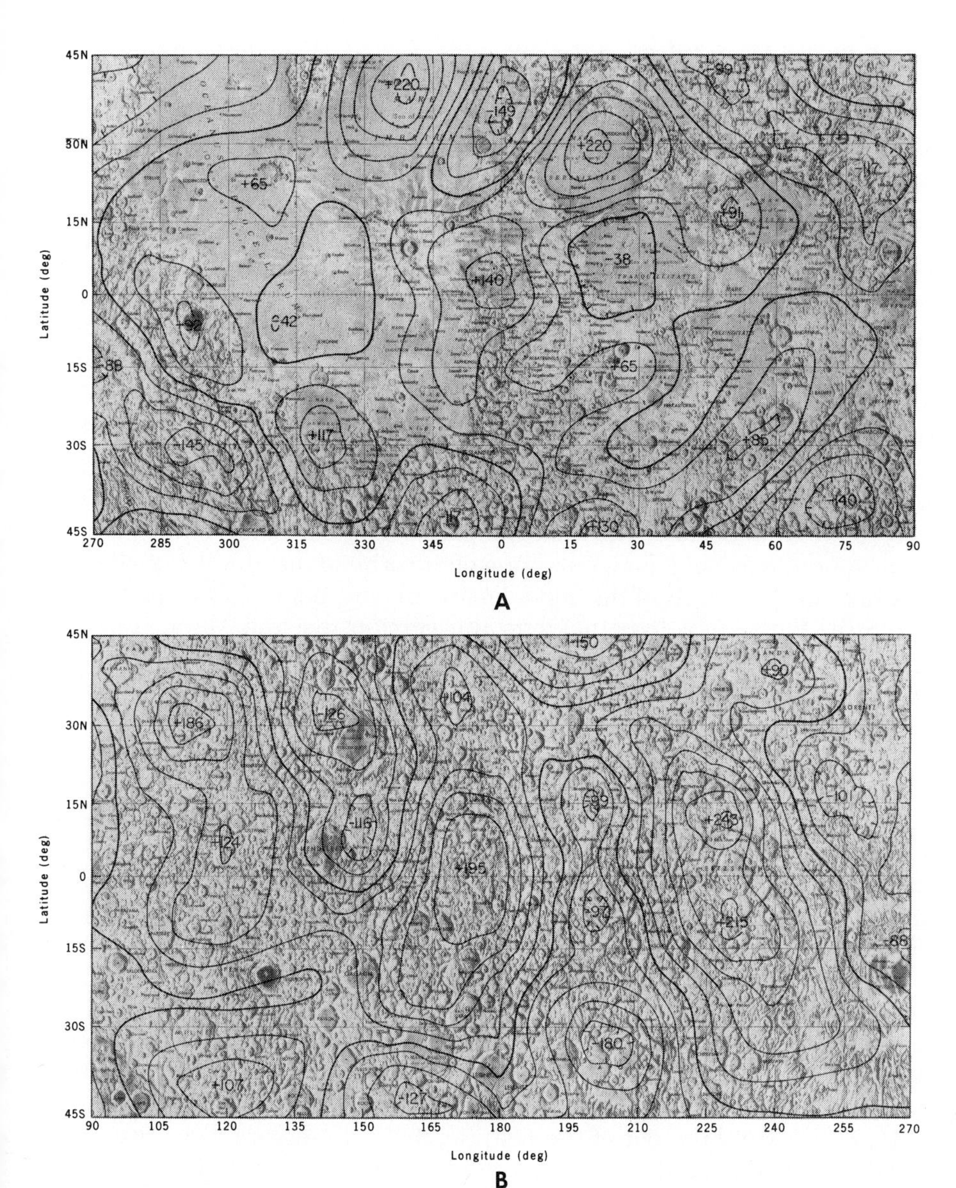

Figure 11–12 The contour lines superimposed on a lunar map show the gravitational field, derived from analysis of the orbits of various lunar orbiting spacecraft. The areas of greatest gravitational field lie over the mascons. Fig. 11–12*A* shows the near side and Fig. 11–12*B* the far side of the Moon.

we know that the planetesimals (Section 10.4) would have had to coalesce very rapidly into planets, since the heat did not have time to be radiated away. The study of the Moon gives us insight into the formation of the whole solar system.

11.3d The Origin of the Moon

Among the models that have been considered in recent years for the origin of the Moon are the following:

(1) *Condensation*: the Moon was formed near to and simultaneously with the Earth in the solar system;

(2) *Fission*: the Moon was separated from the material that formed the Earth; and

(3) *Capture*: the Moon was formed far from the Earth in another part of the solar system, and was later captured by the Earth's gravity.

Studies of the chemical composition of the lunar surface, and compari-

Figure 11–13 A crystal of armalcolite (**Arm**strong-**Al**drin-**Col**lins, the crew of Apollo 11), examined under a polarization microscope. This mineral has been found only on the Moon.

son with the composition of the terrestrial surface, have been important in narrowing down the possibilities. The mean lunar density of 3.3 grams/cm^3 is close to the average density of the Earth's mantle, which had led to some belief in the fission hypothesis. One version had the Moon drawn out of the Earth by a tidal force caused by a passing star. (An early version of this had the hole left behind becoming the Pacific Ocean, but the subsequent development of theories of continental drift made this theory obsolete.) Detailed examination of the lunar rocks and soils indicates that the abundances of elements are sufficiently different to rule out the possibility that the Moon formed directly from the Earth.

Some minerals (Fig. 11–13) that do not exist on Earth have been discovered on the Moon, though this results from different conditions of formation rather than from abundance differences.

Another type of fission model, in which the Earth was rotating very rapidly and threw off the matter that condensed into the Moon, was once very popular. The idea was that the excess angular momentum from the spin went into the angular momentum of the orbit of the Moon, but attempts to work out the details of this model in detail have not been successful.

Still, the fission hypothesis in the version that the Moon separated from the Earth very early on cannot be excluded. The length of the interval in which this could have happened, however—given the extreme ages of lunar soil and some of the rocks—was only a short one.

The third model, the capture model, is also currently thought by most astronomers to be unlikely. The conditions necessary for the Earth to capture such a massive body by gravity seem too stringent for this to have occurred. However, the evidence against the model is not conclusive—one can't apply statistical methods to one example—and several possible ways have been suggested in which the Moon could have been captured by the Earth.

The discovery that the Moon has a higher concentration of refractory minerals (those with high melting points) led to a need to provide some different conditions for the protomoon and the protoearth. The analysis led to a variation of the condensation theory in which the lunar orbit was originally sharply tilted with respect to the ecliptic. When the Moon was at the parts of its orbit high out of the ecliptic, the gas pressure was lower. This would have allowed the refractory minerals to condense better.

Another, more probable version of the condensation theory points out that the planetesimals would have become a thick cloud and gravity would have caused heavier elements to have become relatively more concentrated toward the center of the cloud. This inner region would have become the Earth and the outer region, containing lower concentrations of the heavy elements, would have become the Moon.

Although there are chemical differences between the Moon and the Earth, they are not so overwhelming as to exclude the condensation possibility, and most astronomers currently believe in some version of this model. The idea that the Moon has an iron core adds backing to this theory. In sum, it seems most probable that the Earth and the Moon formed near each other as a double planet, probably by accretion of planetesimals. The condensation model thus closely connects the question of the origin of the Moon to the larger question of the origin of the solar system, which we also think involved the accretion of planetesimals.

It had been hoped that landing on the Moon would enable us to clear up the problem of the lunar origin. But though the lunar programs have led to modifications and updating of the models described above, none of these models has been entirely ruled out. Nobody can say definitely which, if any, of the three is correct.

11.3e Other Lunar Results

Some of the American and Soviet landers have carried sets of retroreflectors, small cubes that very efficiently reflect light back in exactly the direction from which it came (Fig. 11–14). This enables scientists to fire powerful lasers at the Moon and then detect the return of the laser pulses. Accurate timing of the round trip gives the Moon's distance from Earth to an accuracy within centimeters, a manyfold improvement over previous accuracy. Such measurements lead to a better understanding of the gravitational workings of the Earth-Moon system and of the Earth's irregularities of rotation (which is important for the accurate keeping of time). The retroreflector measurements even give such accurate positions on Earth that we will soon be able to measure continental drift directly (Fig. 11–15); such measurements may one day help us with earthquake prediction.

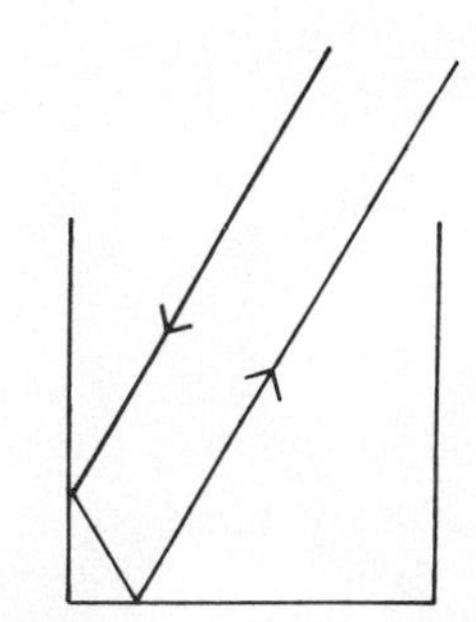

Figure 11–14 A cube, coated on the inside with reflecting material, has the property of reflecting an incoming light ray back in exactly the same direction from which it came, no matter what that direction was. The phenomenon is shown at top in two dimensions.

The discovery of many thousands of new features on the Moon—small craters on the near side and features of all sizes on the far side—led to a need for names, which are the responsibility of the International Astronomical Union. At each of its last triennial meetings, new sets of names for the newly discovered features have been assigned, mainly those of deceased scientists. The astronauts (the American title) and cosmonauts (the Soviet title) who perished in space exploration have also been so honored.

11.4 THE VALUE OF MANNED LUNAR RESEARCH

The issue of manned versus unmanned space research has long been controversial among astronomers, who tend to favor unmanned research because its substantially lower cost would allow more work to be carried out. The success of the Apollo missions plus the ability of the Skylab astronauts to rescue a damaged spacecraft and operate complex equipment requiring on-the-spot judgment has tempered this opinion for many.

Further, it must be recalled that the decision to start the program of manned lunar exploration was largely the result of political factors, and the sending of a person to the Moon was a necessary part of the project. While it is true that much of the cost of the project was related to the cost of keeping humans alive and safe, it is also true that this part of the cost was justified on this ground and the equivalent amount of money would not have been put into straight astronomical research.

We have also succeeded through the Apollo program (and in the succeeding Skylab program, which used rockets and spacecraft developed and built for Apollo) in showing that humans can function in space and carry out useful work over long periods of time. Weightlessness and other effects of spaceflight (such as calcium depletion in bones) do not seem to be major barriers to space exploration. Perhaps this knowledge of our human capabilities will be the longest-lasting benefit of the Apollo and Skylab programs.

One day we may be able to use the far side of the Moon as a site for observatories. Because the Moon has no atmosphere, it is unsurpassed for optical, ultraviolet, or infrared observations. Furthermore, the bulk of the

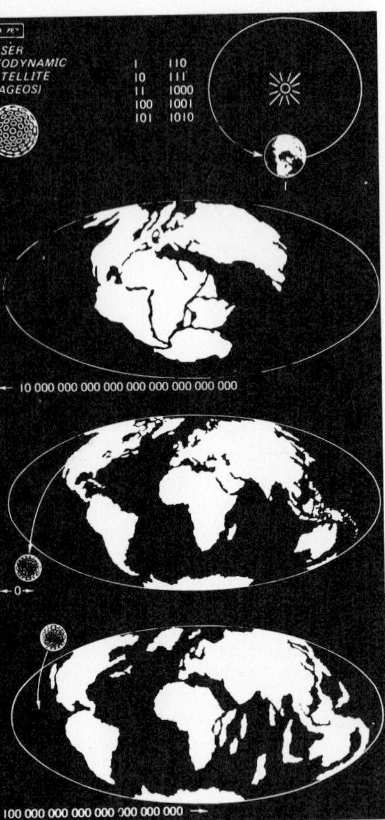

Figure 11–15 A NASA satellite bearing 426 retroreflectors was launched into a circular Earth orbit in 1976. It is being used to provide information on the Earth's rotation and crustal movement. The satellite, called LAGEOS (Laser Geodynamic Satellite), is expected to survive in orbit for 8 million years. In case the satellite is discovered millions of years hence, it bears a series of 3 views (bottom) of the continents in their locations 200 million years in the past, at present, and 200 million years in the future, according to the theory of continental drift. In the future, we expect California to separate from the rest of the United States, Australia to be linked to Asia, and the Italian "boot" to disappear.

Moon blocks radio waves from Earth, so the lunar far side would be a quiet, interference-free site for a radio observatory.

For those of us who remember the shock of Sputnik, and followed each small step farther and farther off the Earth's surface and into space, it is hard to believe that the era of manned lunar exploration not only has begun but also has already ended. At present we have no plans to send more people to the Moon. Perhaps by the turn of the 21st century, manned lunar exploration will resume. Twenty or thirty years from now, we may each be able to visit the Moon as researchers or even as tourists.

SUMMARY AND OUTLINE OF CHAPTER 11

Aims: To see how direct exploration of the Moon has increased our knowledge manyfold, but how such fundamental questions as how the Moon was formed remain unanswered

Lunar features (Section 11.1)
- Fixed: mare, highlands, mountains
- Dependent on sun angle: terminator

Revolution and rotation (Box 11.1)
- Sidereal: with respect to the stars
- Synodic: with respect to another body
- Libration allows us to see 59% of the surface from the Earth

Lunar exploration (Section 11.2)
- Unmanned series: Ranger, Surveyor, Lunar Orbiter, Luna, Zond
- Manned series in Earth orbit: Mercury, Gemini
- Manned and unmanned lines of development merge in Apollo
- Six Apollo landings: 1969 to 1972
- No more manned landings planned
- Three Soviet unmanned spacecraft brought samples back to Earth
- Two Soviet Lunokhods roved many kilometers over the lunar surface

Composition of the lunar surface (Section 11.3a)
- Mare basalts: from cooling of lava
- Highland breccias: broken up and re-formed rock
- Soils
- Abundances of refractory elements had been thought to be much higher than terrestrial abundances, but now thought to be similar

Chronology (Section 11.3a)
- Dating by crater counting
- Dating by ratios of radioactive and non-radioactive isotopes
- Oldest soils: 4.6 billion years; oldest rocks: 4.2 billion years, almost back to origin of the solar system
- Highland rocks (3.9–4.4 billion years) older than maria (3.1–3.8 billion years)
- Probable model: Moon formed 4.6 b.y. At 4.4 b.y., the surface, which had been molten, cooled. Meteorite bombardment from 4.2 to 3.9 b.y. made most of the craters. Interior heated up from radioactive elements; the resulting vulcanism from 3.8 to 3.1 b.y. caused lava flows that formed the maria. Then lunar activity stopped.

Craters (Section 11.3b)
- Almost all formed in meteoritic impact
- Signs of vulcanism also present on surface, though only very few craters are volcanic
- Far side has no maria, quite different in appearance from near side

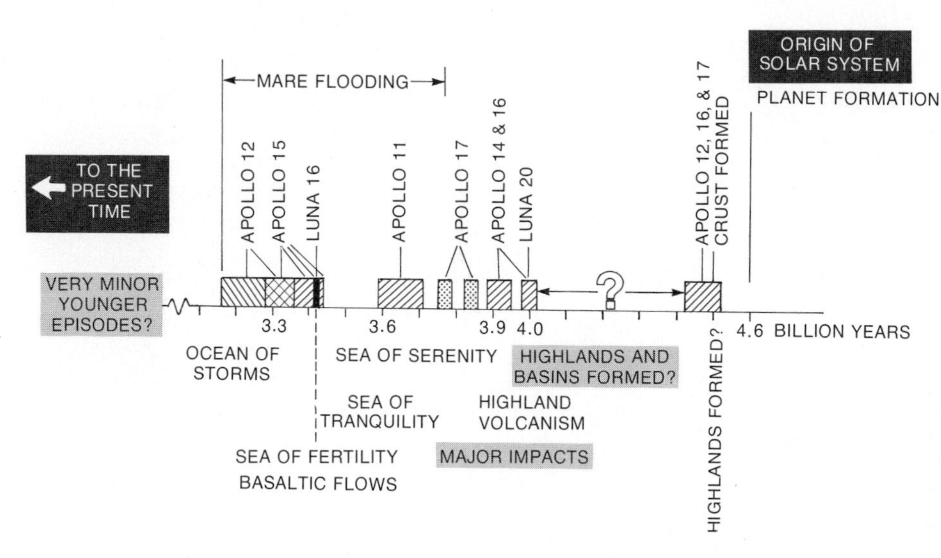

Figure 11–16 The chronology of the lunar surface, based on work carried out at the Lunatic Asylum, as the Caltech laboratory of Gerald Wasserburg is called.

Interior (Section 11.3c)
- Moon is differentiated into a core, a mantle, and a crust
- Seismographs left by Apollo missions reveal that interior is molten
- Lack of lunar magnetic field, on the other hand, is an indication that interior is not molten, though seismic evidence seems definitive
- Weak moonquakes occur regularly
- Mascons discovered from their gravitational effects

Origin (Section 11.3d)
- Condensation, fission, and capture theories
- All still viable, with most astronomers tending to believe in a condensation theory

Retroreflectors for laser beams (Section 11.3e)
- Can measure Earth-Moon distance and continental drift

QUESTIONS

1. Compare the lengths of sidereal and synodic months. Which is longer? Why?
2. If the mass of the Moon is 1/81 that of the Earth, why is the gravity at the Moon's surface as great as 1/6 that at the Earth's surface?
3. To what location on Earth does the terminator on the Moon correspond?
4. Why is the heat flow rate related to the uranium content of the Moon?
5. What does cratering tell you about the age of the surface of the Moon?
6. What method do we use to date individual rocks? Could we have used this method without actually having people or spacecraft visit the Moon?
7. Why is it not surprising that the rocks in the lunar highlands are older than those in the maria?
8. Why are we more likely to learn about the early history of the Earth by studying the rocks from the Moon than those on the Earth?
9. What do the mascons tell us about the interior of the Moon?
10. Choose one of the proposed theories to describe the origin of the Moon and discuss the evidence pro and con.

TOPIC FOR DISCUSSION

Discuss the scientific, political, and financial arguments for resuming manned exploration of the Moon.

☿ 12

Mercury

Mercury is the innermost planet, and until very recently has been one of the least understood. Except for distant Pluto, its orbit around the Sun is the most elliptical (the difference between the maximum and minimum distances of Mercury from the Sun is as much as 40 per cent of the average distance, compared with less than 4 per cent for the Earth). Its average distance from the Sun is 58 million kilometers (36 million miles), which is 4/10 of the Earth's average distance. In order to use convenient numbers, we have defined the Earth's average distance from the Sun to be 1 Astronomical Unit. Thus Mercury is 0.4 A.U. from the Sun.

In order to know when Mercury will be favorably located in the sky for you to observe it, you should watch the sky descriptions published in monthly magazines like Sky & Telescope *or* Astronomy *and in some local newspapers.*

Since we on the Earth are outside Mercury's orbit looking in at it, Mercury always appears close to the Sun in the sky (Fig. 12–1). At times it rises just before sunrise, and at times it sets just after sunset, but it is never up when the sky is really dark. The maximum angle from the Sun at which we can see it is 28°, which means that the Sun always rises or sets within about two hours of Mercury's rising or setting. Of course, the difference in

Since a full rotation of 360° takes 24 hours, it takes 28°/360° or about 30/360 = 1/12 of 24 hours, which equals 2 hours, for Mercury to set after sunset in the most favorable circumstances possible.

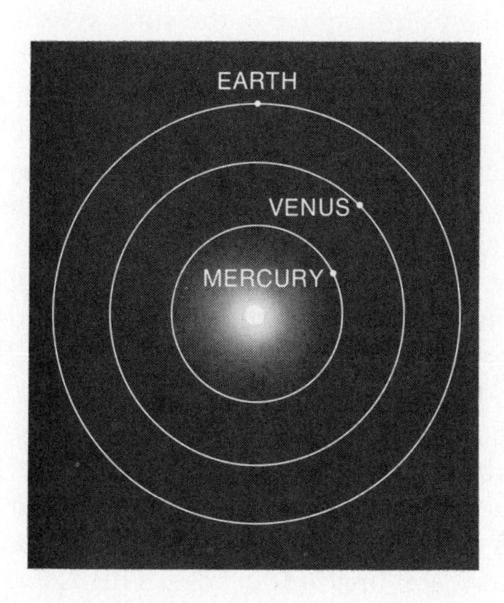

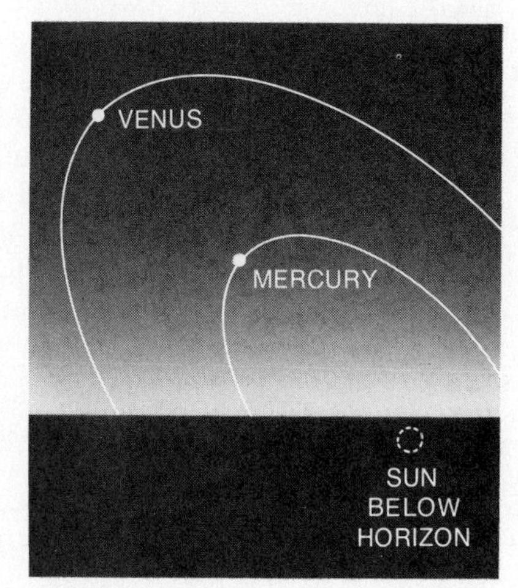

Figure 12–1 Since Mercury's orbit is inside that of the Earth, Mercury is never seen against a really dark sky. A view from the Earth appears at right, showing Mercury and Venus at their greatest respective distances from the Sun. The open view of their orbits is an exaggeration; we actually see the orbits nearly edge on.

Mercury, photographed at a distance of 200,000 km from Mariner 10, revealed a cratered surface. The symbol for Mercury appears at the top of this page.

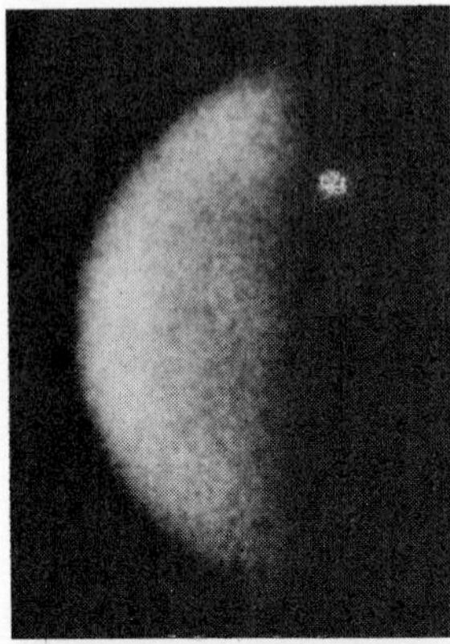

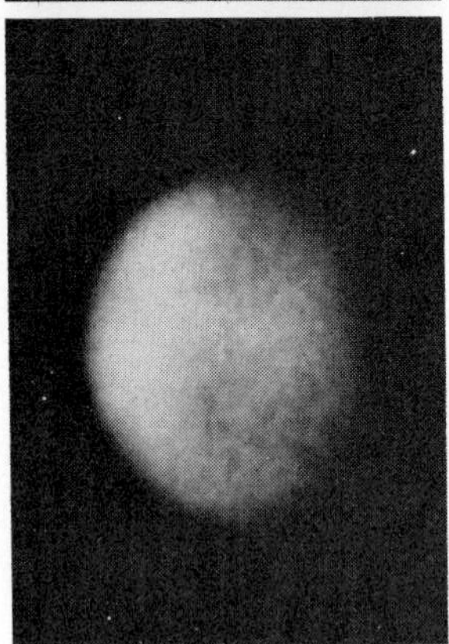

Figure 12–2 From the Earth, we cannot see much surface detail on Mercury. These views, among the best ever taken on Earth, were made by the New Mexico State University Observatory. Mercury was only 7.1 (*top*) and 5.1 (*bottom*) seconds of arc across.

time is usually even less than this maximum. Because of Mercury's closeness to the Sun in the sky, astronomers have never gotten a really good view of Mercury from the Earth, even with the largest telescopes. Many people have never seen it at all. (Copernicus's deathbed regret was that he had never seen Mercury.) The best photographs taken from the Earth show Mercury as only a fuzzy ball with faint, indistinct markings (Fig. 12–2). These photographs could be taken only when Mercury was low in the sky near sunrise and sunset. Consequently the light from Mercury had to pass obliquely through the Earth's atmosphere, making a long path through turbulent air. Hence the photographs are blurred.

12.1 THE ROTATION OF MERCURY

Astronomers, from studies of drawings and photographs, did as well as they could to describe Mercury's surface. A few features could barely be distinguished, and the astronomers watched to see how long those features took to rotate around the planet. From these observations they decided that Mercury rotated in the same amount of time that it took to revolve around the Sun. It seemed reasonable that there could be a bulge in the distribution of mass of Mercury. The side that was bulging would be attracted to the Sun by gravity, locking the rotation to the revolution, just as the Moon is locked to the Earth. This is called *synchronous rotation.* An 88 day period of rotation for Mercury, matching its known 88 day period of revolution around the Sun, thus appeared in all the reference works and textbooks. This led to the fascinating conclusion that Mercury could be both the hottest planet and the coldest planet in the solar system. The patch of Mercury's surface that is closest to the Sun would always be that particular patch closest to the Sun since it would never rotate around. Because the solar radiation is over 6 times more intense there than it is at the Earth, this would be the hottest point in the solar system aside from the Sun itself. But the point on the opposite side of Mercury, the bit of surface farthest from the Sun, would never receive sunlight and, in the absence of an atmosphere to conduct and convect heat, would be the coldest point. (All other planets in the solar system rotate with periods different from their periods of revolution, thus allowing each part of their surfaces to be warmed by the sunlight at regular intervals.)

Synchronous rotation implies that the periods of rotation and revolution are equal, and so the less massive body would always keep the same face toward the more massive body.

In the 1960's, a new method of studying some of the planets—radar—advanced to the point where it could be used to study Mercury. Generally, radio telescopes just sit and listen; that is, they receive signals from the planets and outer space. The intensity of the signal received in this passive manner indicated that the dark side of Mercury was hotter than had been thought when it had been assumed that one half of Mercury's surface was never struck by light from the Sun.

Radio signals can also be used in an active, radar mode, sending out signals and receiving their echoes. This is the same method that is routinely used to track airplanes. (*Radar* is an acronym for "**ra**dio **d**etection **a**nd **r**anging.") When radio radiation is sent out of a radio telescope, it travels across space in a widening cone. Sharp bursts of very intense radiation are sent out from a large radio telescope on Earth (the radiation travels, of course, at the speed of light), a small fraction hits the planet being observed and is bounced back toward Earth, and an even smaller fraction of the original burst is received by the same radio telescope on Earth, minutes later. Since Mercury is a sphere, the point nearest Earth reflects the signal back a fraction of a second before points farther away (Fig. 12–3). Thus by following the change of the returned signal, weak as it is, over time, we can tell something about specific regions on the surface, as the radar signals spread in concentric circles around the planet starting at the point facing the Earth. (Remember that since light travels at 300,000 kilometers per second, and Mercury is only 5000 kilometers in diameter, there is only about one hundredth of a second between the time the radiation striking the nearest point reaches us and the time the radiation striking the edge we can see, which is halfway around the planet, reaches us. Therefore, the signal returned must be recorded on tape very carefully to allow the variations of the signal received during that crucial one-hundredth of a second to be accurately analyzed later on.) Through complex computer studies of the time delays on the returned signals, astronomers can make maps of Mercury's surface.

There is even one more important item that can be told from this radar technique. Since Mercury is rotating, one side of the planet is always receding relative to the other. (Mercury's orbital motion may give an overall approach or recession, which can be ignored when making a radar map.) As a result of the rotation, the signal from one side has a Doppler shift in its frequency when returned that is different from the Doppler shift of the signal from the other side (Fig. 12–4). The information from both the time analysis and the Doppler analysis of the radar signal can be put together to deduce the height and roughness of each point on the visible surface of Mercury. Actually, there is an ambiguity in that two separate points on the planet's disk satisfy both criteria—the time analysis and the Doppler analysis—at each specific time, but this apparent problem can be overcome. (Since this ambiguity does not affect the determination of the period of rotation, we shall defer further discussion until Chapter 13, when we shall discuss radar mapping.)

The results of the radar experiment were a surprise: scientists had been wrong about Mercury's rotation. It actually rotates in 59 days. This is about ⅔ of the period of its revolution, so the planet rotates three times for each two times it revolves around the Sun. Although the inertia of its rotation—its tendency to continue rotating—is too strong to be overcome by the gravitational grip of the Sun, the Sun's steadying pull is strongest every 1½ rotations. At those times, Mercury is in its perihelion position, so the gravitational bulge on Mercury is as near to the Sun as it can be. Mercury's spin was probably once much faster, and was slowed down by the fact that the Sun's gravity attracted the bulge more than it attracted the rest of the planet. Later studies of the rotation than those mentioned above have narrowed down the uncertainty in Mercury's rotation period and tend to confirm this idea because the revolution:rotation ratio is so precisely 2:3 that it can't be accidental.

If Mercury were in a circular orbit, it would be in synchronous rotation. But because its orbit is elliptical, its bulge is locked toward the Sun mainly when it is at perihelion.

The theory that as Mercury moves in its orbit a bulge in the planet is interlocked to the Sun, creating a type of *spin-orbit coupling,* can still be applied. This is a type of *gravitational interlock.* In this case, the planet has a little too much spin to be stopped in a 1 to 1 relationship of rotation and revolution.

Note that because of Mercury's eccentric orbit, it is not as easy for Mercury to have its periods of rotation and revolution equal each other in the way that the Moon's do. For the rotation and revolution periods to be equal, the rate at which the planet spins and the rate at which it moves in its orbit must be equal. Mercury's

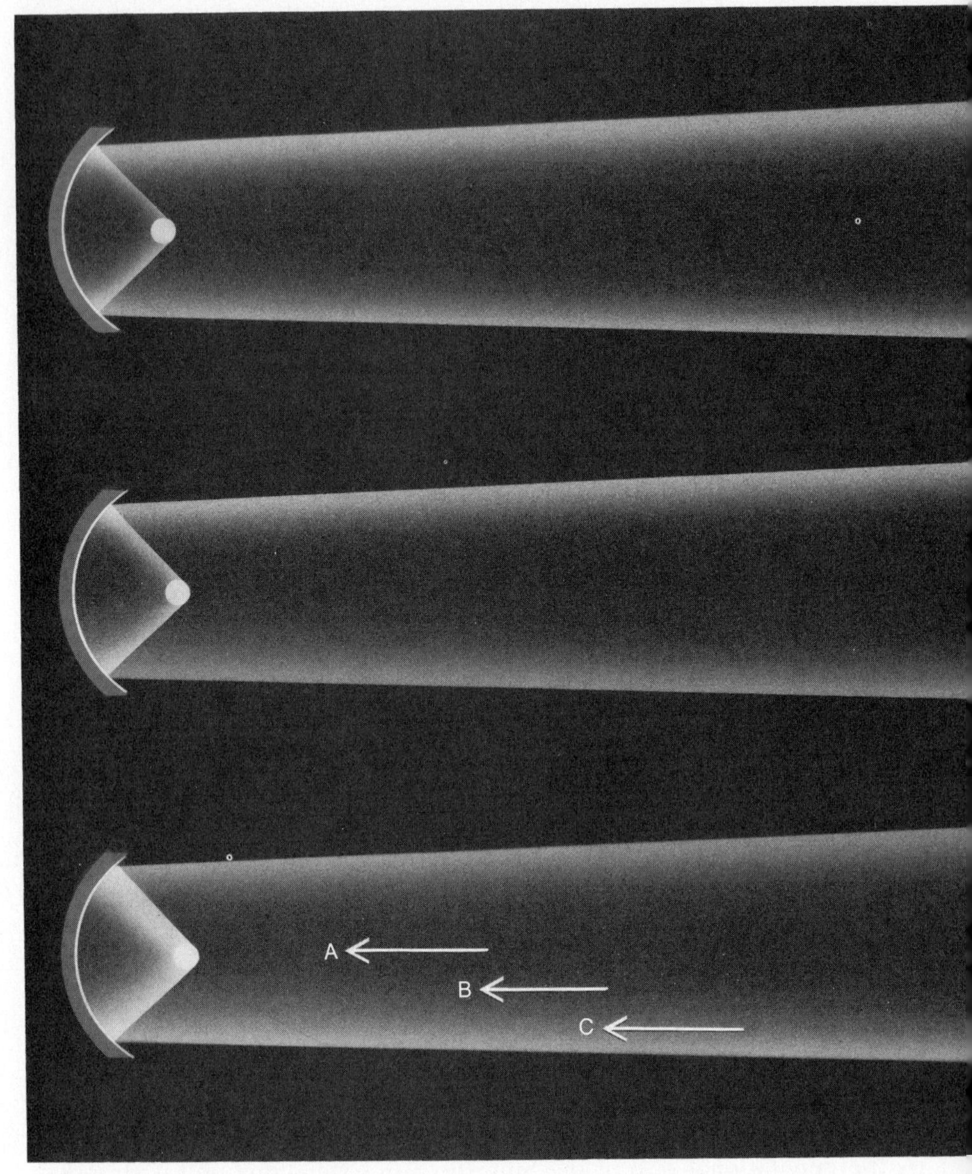

Figure 12–3 A pulse of radio waves sent from the telescope on Earth, shown at left, is reflected from Mercury, shown on the right. The radar pulse first hits the point on Mercury nearest to the Earth (A), and then spreads as time goes on in concentric circles (B and C). The reflected pulse is thus spread out over a longer time interval than was the transmitted pulse. By observing the time of arrival of each part of the reflected pulse back at the Earth, we can tell from which concentric circle that part of the returned pulse was reflected.

Figure 12–4 Everything the same distance from the axis of rotation of a planet, i.e., the surface of the cylinder, rotates at the same velocity. Our radar signal from Earth travels parallel to the plane shown at right. Any signal reflected on the part of Mercury that intersects the plane has the same Doppler shift. We see only the side of the circle shown that is facing us; from our vantage point it looks like a straight line. Thus by observing the signal at a given frequency, which has undergone a certain Doppler shift, we define a line across the surface of the planet. This line intersects the circle shown in Fig. 12–3 at two points.

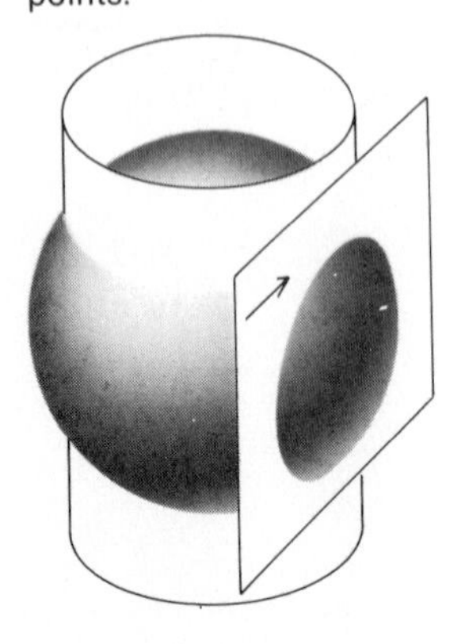

eccentricity is 0.2, and according to Kepler's second law the planet moves in its orbit at greatly varying speeds. Its rate of spin, on the other hand, is constant.

Because rotation is always measured from the time any given star is directly overhead of a position on Mercury until it is again at the same position in the sky (a sidereal day), Mercury's rotation and revolution combine to give a value for the rotation of Mercury relative to the Sun (that is, a Mercurian solar day) that is neither the 59 day *sidereal rotation period* nor the 88 day period of revolution (Fig. 12–5). Actually, if we lived on Mercury we would measure each day and each night to be 88 Earth days long; we would alternately be fried and frozen for 88 Earth days at a time. For Mercury, the *solar rotation period* (not to be confused with the rotation period of the sun) is thus 176 days long.

Because of Mercury's slow rotation, the subsolar point is not always at the same place on the surface and so is not eternally heated. Still, we

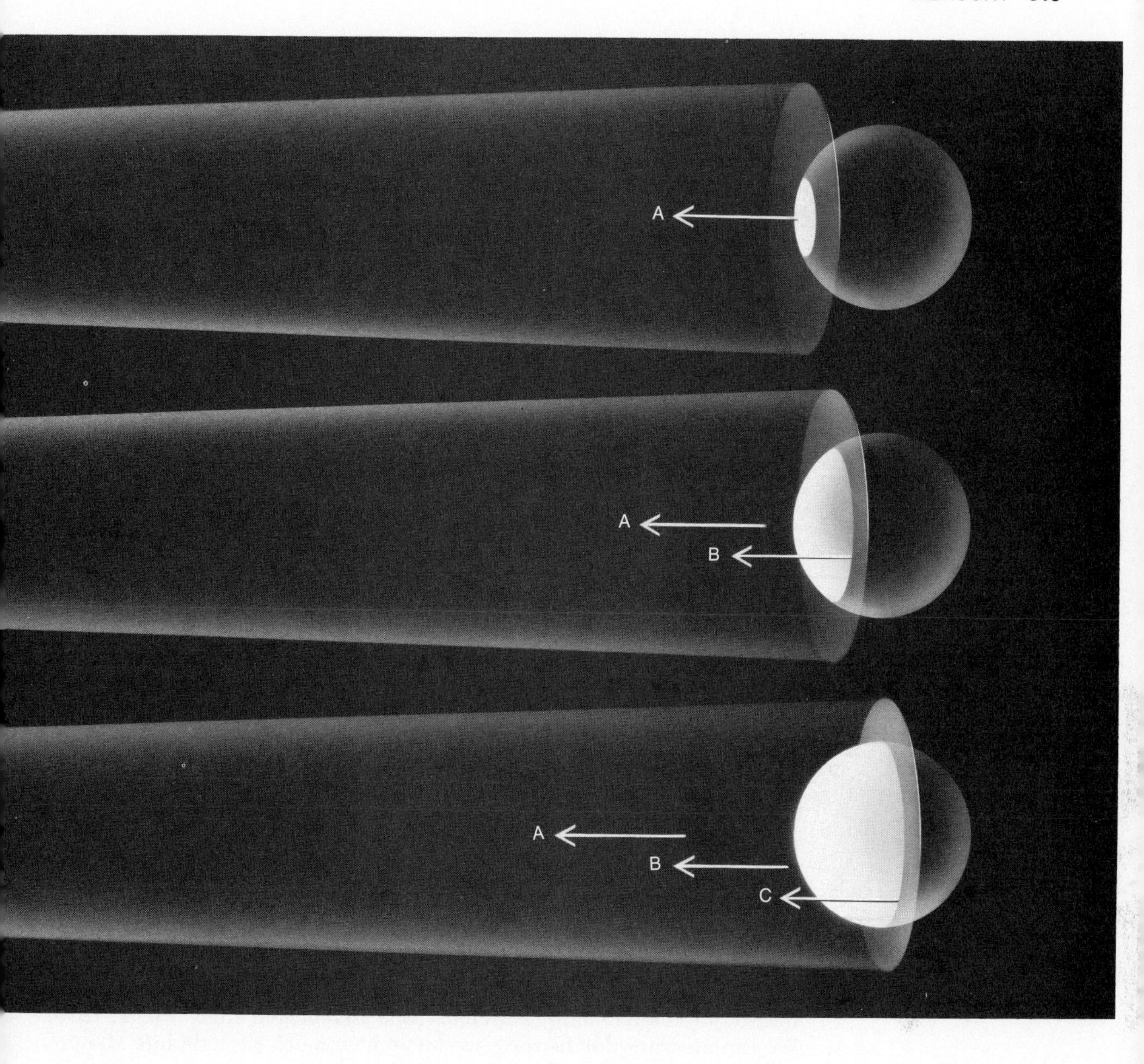

measure a temperature of about 700 K at the subsolar point. Since Mercury has no atmosphere to conduct heat, and since the surface is not a good heat conductor, the temperature drops with increasing distance from this point.

The subsolar point *is the location on a planet where the Sun is directly overhead; this is the point closest to the Sun at any given instant.*

As we can see from Kepler's second law, Mercury travels around the Sun at different speeds at different times in its eccentric orbit. This effect, coupled with its slow rotation on its axis, could lead to an interesting effect if we could stand on Mercury's surface. From some locations we would see the Sun rise for an Earth day or two, then retreat below the horizon from which it had just come, and then rise again.

No harm was done by the scientists' misconception of Mercury's rotational period for all those years, but the story teaches all of us a lesson: we should not be too sure of so-called facts, even when they are stated in all the textbooks. Don't you believe everything you read here, either.

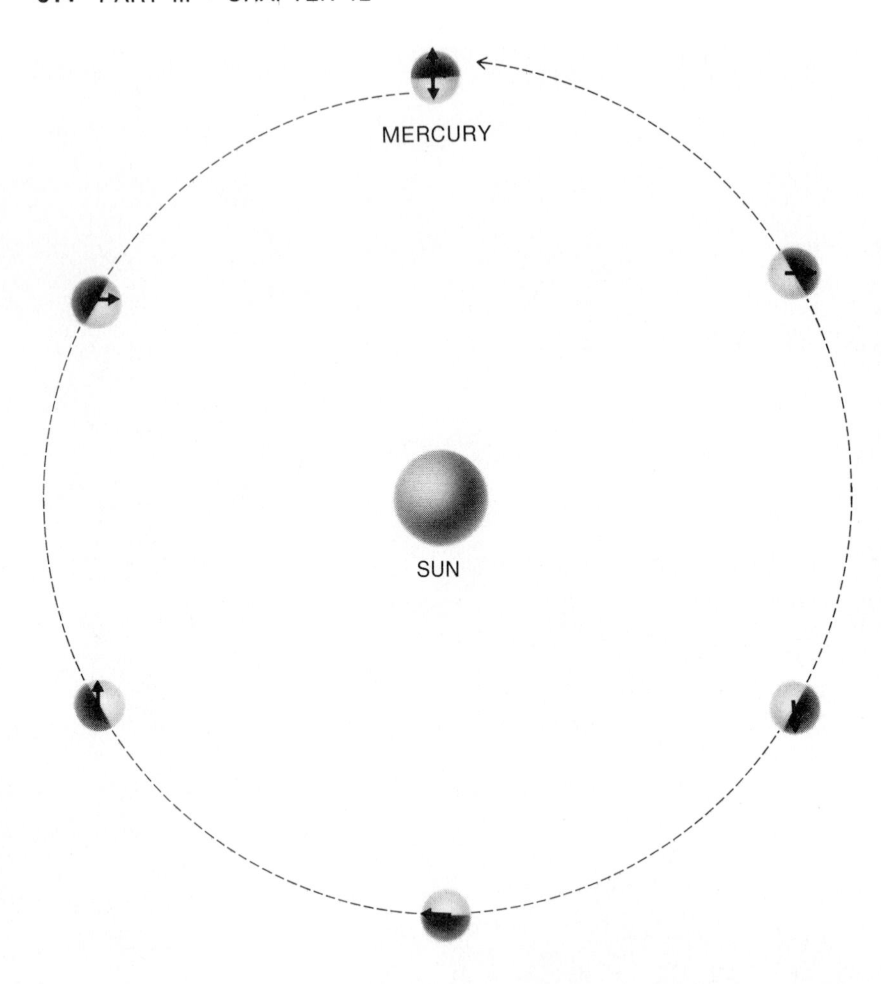

Figure 12–5 Follow the arrow that starts facing down toward the Sun in the image of Mercury at the top of the figure, as Mercury revolves along the dotted line. Mercury, and thus the arrow, rotate once with respect to the stars in 59 days, when Mercury has moved only 2/3 of the way around the Sun. (Our view is as though we were watching from a distant star.) Note that after one full revolution of Mercury around the Sun, the arrow is facing away from the Sun. It takes another full revolution, a second 88 days, for the arrow to be again facing the Sun. Thus the rotation period with respect to the Sun is twice 88, or 176, days.

12.2 OTHER KNOWLEDGE FROM GROUND-BASED OBSERVATIONS

Even though the details of the surface of Mercury can't be studied very well from the Earth, there are other properties of the planet that can be better studied. For example, we can measure Mercury's *albedo*, the fraction of sunlight hitting Mercury that is reflected from it (Fig. 12–6). We can measure this because we know how much sunlight hits Mercury (we know how bright the Sun is and how far away Mercury is from it). Then we can easily calculate at any given time how much light Mercury reflects, from both (1) how bright Mercury looks to us and (2) its distance from the Earth. Once we have a measure of the albedo, we can compare it

Figure 12–6 *Albedo* is the fraction of radiation reflected. A surface of low albedo looks dark.

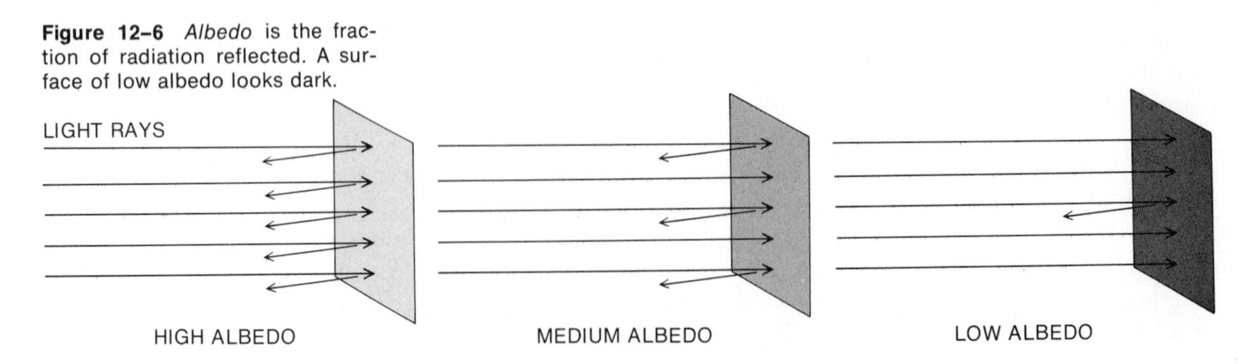

with the albedo of materials on the Earth and on the Moon and thus learn something of what the surface of Mercury is like.

Let us consider some examples of albedo. An ideal mirror reflects all the light that hits it; its albedo is thus 100 per cent (the very best real mirrors have albedos of as much as 96 per cent). A black cloth reflects essentially none of the light; its albedo is almost 0 per cent. Mercury's overall albedo is only about 6 per cent. Its surface, therefore, must be made of a dark—that is, poorly reflecting—material. The albedo of the Moon is similar. In fact, Mercury (or the Moon) appears bright to us only because it is contrasted against a relatively dark sky; if it were silhouetted against a bedsheet, it would look relatively dark, as if it had been washed in Brand X instead of Tide.

The albedo (from the Latin for whiteness) is the ratio of light reflected from a body to light received by it.

Mercury's mass can also be measured without getting up close to it, because the amount of gravity that a body has depends on its mass. In other words, by observing the gravitational effect of Mercury on the orbits of other bodies in the solar system, we can in effect "weigh" Mercury. (Actually, mass and weight are not the same, for the mass of a body is an intrinsic property of that body, while the weight of a body depends not only upon its mass but also both on where it is located with respect to another body and on the mass of the other body. Thus you would weigh more than you weigh on Earth if you were to stand on Jupiter and less if you were to stand on the Moon (Table 12–1), but your mass would not change. In any case, the difference need not concern us here. We are really finding not Mercury's weight but rather Mercury's mass.)

Venus is the planet closest to Mercury and is thus the planet that is most affected by its gravity, but the asteroid Icarus passed relatively close to Mercury in 1968 and gave us a more accurate measure of Mercury's mass. Mercury has a mass of 5½ per cent that of the Earth. Since we can calculate its size simply by measuring its angular size in the sky and its distance from Earth (it is only slightly bigger than the Moon), we can divide its mass by its volume (the volume of a sphere is $\frac{4}{3}\pi r^3$) to get its density. The density turns out to be 5.5 grams/cubic centimeter, about the same density as that of Venus and the Earth. Thus Mercury's core, like the Earth's, must be heavy; perhaps it too is made of iron. Since the surface gets hot enough to melt even lead and tin, only materials that have very high evaporation points can remain on the surface. We can surmise that the surface must be made of rocks with a high content of such materials as iron, magnesium, and silicon.

TABLE 12–1 WEIGHT ON THE SUN AND PLANETS

If on Earth you weigh you would weigh on	50 kg	(110 lbs)	65 kg	(143 lbs)	80 kg	(176 lbs)
Mercury	19 kg	(42 lbs)	25 kg	(54 lbs)	30 kg	(67 lbs)
Venus	46 kg	(100 lbs)	59 kg	(130 lbs)	73 kg	(160 lbs)
Earth	50 kg	(110 lbs)	65 kg	(143 lbs)	80 kg	(176 lbs)
The Moon	8 kg	(18 lbs)	11 kg	(24 lbs)	13 kg	(29 lbs)
Mars	19 kg	(42 lbs)	25 kg	(54 lbs)	30 kg	(67 lbs)
Jupiter	127 kg	(278 lbs)	164 kg	(362 lbs)	202 kg	(445 lbs)
Saturn	53 kg	(117 lbs)	69 kg	(152 lbs)	85 kg	(187 lbs)
Uranus	46 kg	(101 lbs)	60 kg	(132 lbs)	73 kg	(161 lbs)
Neptune	60 kg	(132 lbs)	78 kg	(172 lbs)	95 kg	(210 lbs)
Pluto	29 kg	(63 lbs)	37 kg	(82 lbs)	46 kg	(100 lbs)
The Sun	1400 kg	(3090 lbs)	1820 kg	(4010 lbs)	2240 kg	(4940 lbs)

12.3 MARINER 10

So astronomers, in typical fashion, had deduced a lot from limited data. But in 1974, we learned much more about Mercury in a brief time. We flew right by.

The tenth in the series of Mariner spacecraft launched by the United States went to Mercury. First it passed by Venus, as will be described in the next chapter, and then had its orbit changed by Venus's gravity to direct it to Mercury (Fig. 12–7). Tracking the orbits improved our measurements of the gravity of these planets and thus of their masses. Further, the 475-kilogram (1042-pound) spacecraft had a variety of instruments on board, in-

A

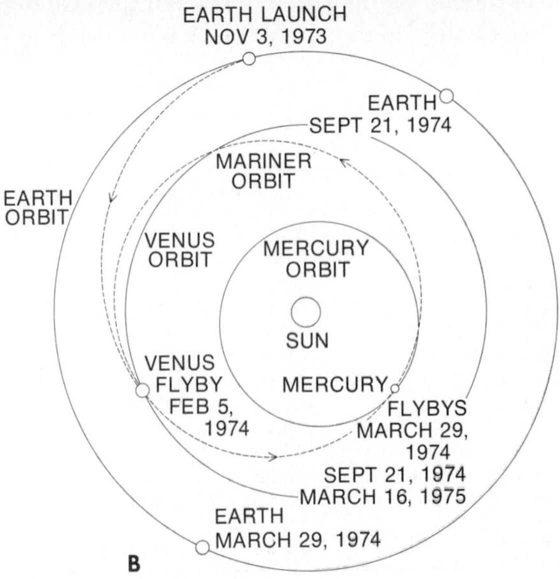

B

Figure 12–7 *(A)* An artist's sketch of the Mariner 10 spacecraft and its trajectory from the Earth, past Venus, to Mercury. *(B)* Mariner 10 flew by Mercury on three separate occasions. The position of the Earth is shown at the time of launch, and at the time of two of the flybys.

cluding devices to measure the magnetic fields in space and near the two planets, and devices to measure the infrared emission of the planets and thus their temperatures. Two of these instruments—a pair of television cameras—provided not only the greatest popular interest but also many important data.

Mariner 10 flew by Mercury the first time (yes, it went back again) from March 29th to April 3rd, 1974, and took 1800 photographs that were transmitted to Earth. It came as close as 750 km from Mercury's surface.

The most striking overall impression is that Mercury is heavily cratered (see the photograph that opened this chapter). At first glance, it looks like the Moon! But there are several basic differences between the features on the surface of Mercury and those on the lunar surface. We can compare how the mass and location in the solar system of these two bodies affected the evolution of their surfaces.

Mercury's craters seem flatter than those on the Moon, and have thinner rims (Figs. 12–8 and 12–9). This is largely an effect of Mercury's higher gravity. The craters may have been eroded by any of a number of methods, including the impacts of meteorites or micrometeorites (large or small bits of interplanetary rock); the solar wind (the expansion of the particles in the solar corona into interplanetary space); or, perhaps, something that occurred during a much earlier period when Mercury may have had an atmosphere, or internal activity, or been flooded by lava.

Most of the craters themselves seem to have been formed by impacts of meteorites. In many areas they appear superimposed on relatively smooth plains. There is a class of smaller, brighter craters that are sometimes in turn superimposed on the larger craters and thus must have been made afterwards (Figs. 12–10 and 12–11). The rate at which objects hit and formed craters is probably about the same for all the terrestrial planets. Judging from the rate of cratering on the lunar maria, it seems unlikely that

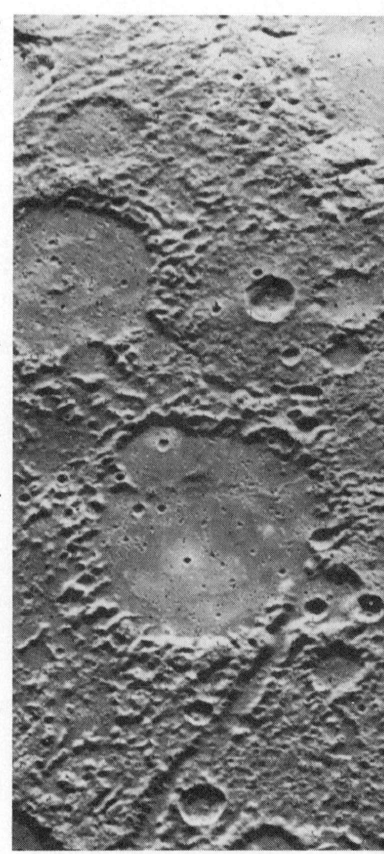

Figure 12–8 This 150 × 300 km area of Mercury, photographed from a distance of 35,000 km as Mariner 10 approached the planet for the first time, shows a heavily cratered surface with many low hills. The valley at bottom is 7 km wide and over 100 km long. The large flat-floored crater is about 80 km in diameter.

Figure 12–9 This is one of the highest resolution pictures obtained on the first pass of Mariner 10; craters as small as 150 meters across can be seen. An area 50 × 40 km is shown in this photograph, which was taken from a distance of only 5900 km.

Figure 12–10 Figure 12–11

Figure 12–10 A fresh new crater, about 12 km across, in the center of an older crater basin.

Figure 12–11 Another fresh crater on Mercury, about 120 km across. Because Mercury's gravitational field is higher than that of the Moon, material ejected by an impact on Mercury does not travel as far across the surface.

there were enough chunks of rock floating around in space near Mercury to have caused the many larger craters in recent times. ("Recently," when speaking of planetary or geologic time scales, means "only" millions of years rather than the billions since the formation of the solar system.) Thus the surface of Mercury that we see is probably very little changed from the way it was long ago (Fig. 12–12), though whether there was an episode of melting about 4 billion years ago, followed by recratering (as there was on the Moon), is controversial. In any case, there must have been a higher rate of crater formation (from planetesimals?) in the first billion years of the solar system.

Some of the craters have light rays emanating from them (Fig. 12–13), just as some of the craters on the Moon do. The ray material has an albedo two or three times as high as the 6 per cent average albedo of the surface, and represents relatively recent crater formation (that is, within the last hundred million years). The ray material must have been tossed out in the impact that formed the crater. Many of the smaller craters were undoubtedly also formed by such secondary impacts. Larger craters on Mercury are much more closely surrounded by secondary craters than are large craters on the Moon, which makes sense because of Mercury's higher surface gravity.

There are some maria and craters flooded with some material—presumably lava. This observation also indicates that there may well have been a period of internal heating and vulcanism—although it couldn't have been too extensive or too recent, or else the many craters that now exist would have been obliterated. But lava could also have been formed by the heat generated in impacts of large meteorites, so the presence of lava does not prove the presence of vulcanism.

Figure 12–12 The Caloris Basin on Mercury, a ring basin 1300 km in diameter that is bounded by mountains that rise as high as 2 km. This feature is similar to Mare Imbrium on the Moon in both size and appearance.

Figure 12–13 *(A)* Many rayed craters can be seen on this photograph of Mercury, taken six hours after Mariner 10's closest approach on its first pass. The north pole is at the top and the equator extends from left to right about 2/3 of the way down from the top. (*B*) A field of light (that is, not dark) rays radiating from a crater off to the top left, photographed on the second pass of Mariner 10. The crater at top is 100 km in diameter.

A

B

Figure 12–14 Figure 12–15

Figure 12–14 This photomosaic of the heavily-cratered south polar region of Mercury was obtained on the second pass of Mariner 10. The south pole is just out of the field of view at the bottom, and north is toward the top. Some of the numerous scarps that can be seen cross and distort large craters; they are hundreds of kilometers long. A ray system associated with a fresh crater can also be seen.

Figure 12–15 A scarp more than 300 km long extends from top to bottom in this second-pass picture.

One interesting kind of feature that is visible on Mercury is a line of cliffs hundreds of miles long; on Mercury as on Earth such lines of cliffs are called *scarps*. The scarps are particularly apparent in the region of Mercury's south pole (Figs. 12–14 to 12–16). Unlike fault lines we know on the Earth, such as the San Andreas Fault in California, there are no signs of geologic tensions like rifts or fissures nearby. Further, sometimes craters are superimposed upon the scarps, indicating that the scarps were formed before these craters. The more that Mercury has been photographed and the photographs analyzed, the more we realize that these cliffs or scarps are global in scale, and not just isolated occurrences.

It seems that these scarps may actually be wrinkles in the crust of the planet. Mercury's core, judging by the fact that its density is about the same as the Earth's, is probably iron and takes up much of the central volume, perhaps 50 per cent of the volume or 70 per cent of the mass. At one time, perhaps the core was molten, and shrank by 1 or 2 km as it cooled. The crust would have settled down with it, making the scarps in the quantity that is in fact observed. Alternatively, it seems possible that Mercury formerly rotated more rapidly. As the rotation slowed down, the shape of the planet would have become less oblate. The crust might have cracked and wrinkled as it tried to match the new shape of the interior.

One part of the Mercurian landscape seems particularly different from the rest (Fig. 12–17). It seems to be grooved, with relatively smooth areas between the grooves. It is called the "weird terrain." Just a couple of areas of this type have been found on the Moon, and no others are known on Mercury. The weird terrain is 180° around Mercury from the Caloris Basin, the site of a major meteorite impact. Shock waves from that impact may have been focused halfway around the planet.

Figure 12–16 This first-pass view of Mercury's northern limb shows a prominent scarp extending from the limb near the middle of the photograph. The photograph shows an area 580 km from side to side.

Box 12.1 Naming the Features of Mercury

The mapping of the surface of Mercury leads to a need for names. The scarps are being named for historical ships of discovery and exploration, such as Endeavour (Captain Cook's ship), Santa Maria (Columbus's ship), and Vostok (a Soviet spacecraft). Plains are being given the name of Mercury in many different languages, such as Tir, Odin, and Suisei. Craters are being named for non-scientific authors, composers, and artists, in order to complement the lunar naming system, which honors scientists.

The Mariner 10 mission was a navigational coup not only because it used the gravity of Venus to get the spacecraft to Mercury, but also because scientists and engineers were able to find an orbit around the Sun that brought the spacecraft back to Mercury several times over. Every six months the Mariner 10 and Mercury returned to the same place at the same time. As long as the gas jets for adjusting and positioning Mariner functioned, it was able to make additional measurements and to send back additional pictures in order to increase the photographic coverage. On its second visit, for example, in September 1974, Mariner 10 was able to study

Figure 12–17 This mosaic of pictures from the first pass of Mariner 10 shows a terrain unique to Mercury—hills and ridges cut across many of the craters and the inter-crater areas.

By "high resolution," we mean that smaller details are visible than would be visible under "low resolution."

the south pole and the region around it for the first time. This pass was devoted to photographic studies. The spacecraft came within 48,000 kilometers of Mercury, farther away than the 750 kilometer minimum of the first pass, but the data were still very valuable. On its third visit, in March 1975, it had the closest encounter ever—only 300 km above the surface (Fig. 12–18). Thus it was able to photograph part of the surface with a high resolution of only 50 meters. Then the spacecraft ran out of gas for the small jets that control its pointing, so even though it still passes close to Mercury every few months, it can no longer record data or send them back to Earth.

The cameras were not the only instruments on board Mariner. The infrared radiometer, for example, gave data that indicate that the surface of Mercury is covered with fine dust, as is the surface of the Moon, to a depth of at least several centimeters. Astronauts sent to Mercury, whenever they go, will leave footprints behind them.

Figure 12–18 *(A)* A high resolution view of the fractured and ridged plains of the Caloris Basin, photographed from a distance of 19,000 km on Mariner 10's third pass.

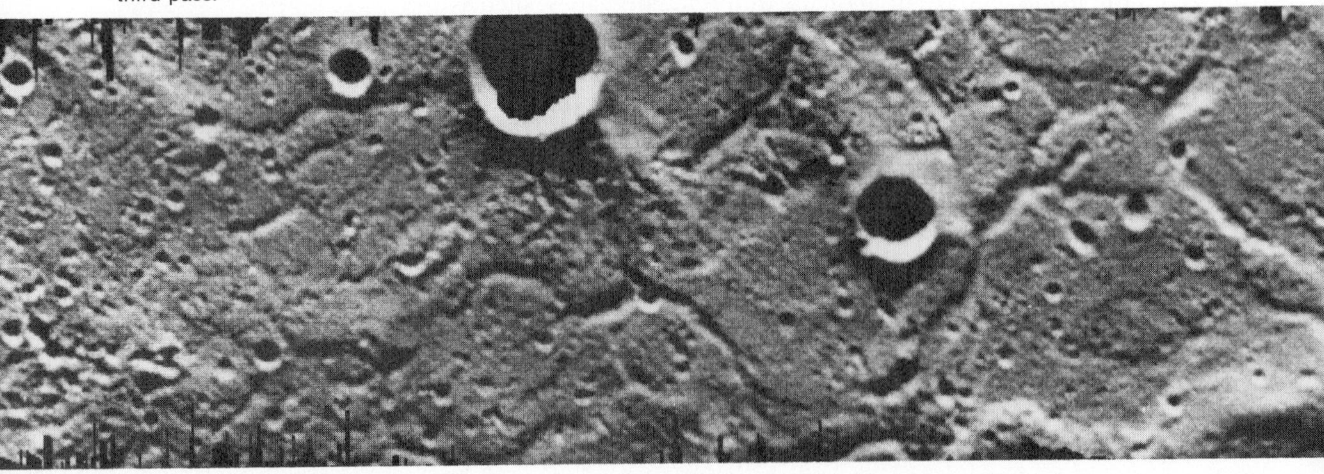

A

The Mariner 10 mission gave more accurate measurements of the temperature changes across Mercury than had been determined from the Earth. In a few hundred kilometers at the terminator, the line between Mercury's day and night, the temperature falls from about 775 K (500°C) to about 425 K (150°C), and then drops even lower farther across into the dark side of the planet.

The ability to study Mercury from so close up has led to the discovery that Mercury even has an atmosphere, although an all-but-negligible one. It is only a few billionths as dense as the Earth's, so slight that even someone standing on Mercury would need special instruments to detect it. Traces of helium, oxygen, carbon, argon, nitrogen, and xenon have been detected with a spectrometer that operated in the ultraviolet. The presence of helium is a particular surprise, because helium is a light element and would be expected to escape from Mercury's weak gravity within a few hours. So there must be a constant source to replace it. Either it comes from the Sun in the solar wind, or, more likely, it comes from radioactive decay of uranium and thorium on Mercury.

Mariner 10 detected lots of electrons near Mercury. Perhaps they are trapped in some sort of belt by the magnetic field, similar to the Van Allen belts around the Earth. But perhaps they are bound to Mercury for shorter times than electrons trapped by the Earth's magnetic field.

The Van Allen belts *are belts of charged particles surrounding the Earth. They were perhaps the first major discovery of orbiting Earth satellites, the earliest of which was Sputnik, launched on October 4, 1957. American satellites soon followed; James Van Allen of the University of Iowa analyzed data from them and discovered that the satellites regularly passed through belts of particles around the Earth.*

One more surprise—perhaps the biggest of the mission—was that a magnetic field was detected in space near Mercury. It was discovered on Mariner 10's first pass, and then confirmed on the third pass. The field is weak; extrapolated down to the surface it is about 1 per cent of the Earth's. It had been thought that magnetic fields were generated by the rapid rotation of molten iron cores in planets, but Mercury doesn't rotate fast enough to cause this effect. Perhaps Mercury rotated faster once upon a time, and the magnetic field generated has been frozen into Mercury since then. If, as some scientists think, Mercury was never hot enough for iron to separate out from other material, and the presence of an iron core is inferred from the existence of the magnetic field, then Mercury might have been formed in layers with the iron core formed first. If this is true we could also conclude that the Earth probably formed in layers, which would resolve (at least temporarily) one of the basic questions we have had about our own planet. Other scientists are not convinced by this line of reasoning. There is no agreement on the reason for Mercury's magnetic field.

No moons of Mercury have ever been detected, and Mariner 10 did an

Figure 12–18 *(B)* This third-pass photograph, taken from 67,000 km away, includes a multi-impact feature of three different-sized craters nested within one another. The smallest of the three is about 15 km across.

especially careful job of searching. The amount of light that any moon reflects depends on both its size and its albedo; we now know that there is no moon any bigger than 5 kilometers across with the same low albedo as Mercury itself. A higher albedo would correspond to a smaller upper limit on size.

Spacecraft encounters like that of Mariner 10 have revolutionized our knowledge of Mercury. The kinds of information that we can bring to bear on the basic questions of the formation of the solar system and of the evolution of the planets are much more varied now than they were just a short time ago.

SUMMARY AND OUTLINE OF CHAPTER 12

Aims: To discuss the difficulties in studying Mercury from the Earth, and what space observations have told us

Difficulty in observing from the Earth
Never far from the Sun in the sky
Radar astronomy (Section 12.1)
 The rotation period of Mercury is 2/3 its orbital period
 How radar can map the surface of a planet
 Time-delay effect gives concentric circles
 Doppler effect gives lines
 Circles and lines intersect at two points
 Sidereal and solar rotation periods and their relationship
Albedo (Section 12.2)
 Low albedo means a poor reflector
 Mercury has a low albedo, only 6 per cent
Mass and density (Section 12.2)
Mariner 10 observations (Section 12.3)
 Photographic results
  Types of objects
   Craters
   Maria
   Scarps
   "Weird terrain"
  Mechanisms
   Impact
  Vulcanism
   Shrinkage of the crust
 Results from infrared observations
  Dust on the surface
  Temperature measurements
 Results from other types of observations
  A small atmosphere
   Where does the helium come from?
  A big surprise: a magnetic field
   What this tells us about the histories of Mercury and of the Earth

QUESTIONS

1. Assume that on a given day, Mercury sets after the Sun. Draw a diagram, or a few diagrams, to show that the height of Mercury above the horizon depends on the angle that the Sun's path in the sky makes with the horizon. Discuss how this depends on the latitude or longitude of the observer.

2. If Mercury did always keep the same side towards the Sun, does that mean that the night side would always face the same stars? Draw a diagram to illustrate your answer.

3. Explain why a day on Mercury is 176 Earth days long.

4. Estimate the angular size of Mercury as viewed from Earth (at its closest approach). In a radar ranging experiment, the radar pulse is spread out over a cone whose point (technically, apex) has an angle of one arc minute. What fraction of this radiation strikes Mercury?

5. If you increased the albedo of Mercury, would its temperature increase or decrease? Explain.

6. List those properties of Mercury that could best be measured by spacecraft observations.

7. Think of how you would design a system that transmits still photographs taken by a spacecraft back to Earth and then produces the picture. The system can be simpler than your TV since still pictures are involved.

8. How would you distinguish an old crater from a new one?

9. What evidence is there for erosion on Mercury? Does this mean there must have been water on the surface?

10. List three major findings of Mariner 10.

13 Venus

Venus and the Earth are sister planets; their sizes, masses and densities are about the same. But they are as different from each other as the wicked sisters were from Cinderella. The Earth is lush, has oceans and rainstorms of water, an atmosphere containing oxygen, and creatures swimming in the sea, flying in the air, and walking on the ground. On the other hand, Venus is a hot, foreboding planet with temperatures constantly over 750 K, a planet on which life seems unlikely to develop. Why is Venus like that? How did these harsh conditions come about? Can it happen to us here on Earth?

Venus orbits the Sun at a distance of 0.7 A.U. Although it comes closer to us than any other planet—it can approach as close as 45 million kilometers (30 million miles)—we still do not know much about it because it is always shrouded in heavy clouds (Fig. 13–1). Observers in the past saw faint hints of structure in the clouds, which seemed to indicate that these clouds might circle the planet in about 4 days, revolving in the opposite sense from Venus's orbital revolution, but the clouds never parted sufficiently to allow us to see the surface.

Figure 13–1 A crescent Venus, observed with the 5-m Hale telescope in blue light. We see only a layer of clouds.

13.1 THE ROTATION OF VENUS

In 1961, radar astronomy penetrated the clouds and provided an accurate rotation period for the surface of Venus. Venus, because of its relative proximity to Earth, is an easier target for radar than Mercury. Venus rotates in 243 days with respect to the stars in the direction opposite from the other planets; this backward motion is called retrograde rotation, to distinguish it from forward (direct) rotation. Venus revolves around the Sun in 225 Earth days. These periods combine, in a way similar to that in which Mercury's sidereal day and year combine (see Section 12.1), so that a solar day on Venus would correspond to 127 Earth days; that is, the planet's rotation would bring the Sun back to the same position in the sky every 127 days.

A sidereal day on Venus is 243 Earth days long.

A mosaic of photographs of Venus, made in ultraviolet light from the Mariner 10 spacecraft in 1974. The symbol for Venus appears at the top of this page.

The notion that Venus is in retrograde rotation seems very strange to astronomers, since the other known planets revolve around the Sun in the same direction, and almost all the planets and satellites also rotate in that same direction. Because of the conservation of angular momentum (which was described in Section 10.3) and since the original material from which the planets coalesced was undoubtedly rotating, we expect that its original angular momentum should now be divided up among the Sun and the planets. We thus expect all the planets to revolve and rotate in the same sense.

It is thus a problem to explain why Venus rotates "the wrong way." (The rotation of Uranus, as we shall see in Section 17.2, is also a problem.) Nobody knows the answer. One possibility depends on the idea that when Venus was in the process of forming, the planetesimals formed clumps of different sizes. Perhaps the second largest clump struck the largest clump at such an angle as to cause the result to rotate backwards. Scientists do not like *ad hoc* ("for this special purpose") explanations like this one, because they are constructed to explain specific situations and do not permit generalization. Nevertheless, that's all we can do in this case.

The slow rotation of the solid surface contrasts with the rapid rotation of the clouds. The clouds rotate in the same sense as the surface of Venus but much more rapidly, once every 4 days. The tops of the clouds have been studied from the Earth, and the composition of the clouds—a mixture of droplets of water and of sulfuric acid—was deduced. Sulfuric acid takes up water very efficiently, so there is little water vapor above the clouds. It took especially careful work from high-altitude sites on the Earth, and from balloons, to detect the presence of the small amount of Cytherean water vapor above the clouds. This observation is difficult, because the spectral lines of water vapor from Venus were masked by the spectral water vapor lines that arose from the Earth's own atmosphere.

Astronomers often use the adjective Cytherean to describe Venus. Cythera was the island home of Aphrodite, the Greek goddess who corresponded to the Roman goddess Venus.

13.2 THE SURFACE OF VENUS

Since we are speaking in approximate values, numbers are rounded off to the nearest multiple of 5.

The surface of Venus can be detected from Earth with radio telescopes. Radio waves emitted by the surface penetrate the clouds to give us information about the surface of this planet. Just as we can determine the surface temperature of a star from studying the distribution of its radiation over optical wavelengths and by the strength of the emission, so can we also determine the temperature of the Cytherean surface by studying the radio emission (the optical radiation doesn't pass through the clouds). The surface is very hot, about 750 K (475°C).

In addition to directly measuring the temperature on Venus, we can calculate theoretically approximately what it should be if the Cytherean atmosphere did not impede the radiation. We know that the amount of radiation that Venus receives from the Sun must be in balance with the amount of radiation that Venus gives off; we say that the situation is in equilibrium (Fig. 13–2). Since the amount of radiation that an object gives off depends on the temperature of that object, we can calculate the amount of radiation that Venus would give off if it were at various temperatures and if it reacted to radiation in the same ways that the Earth does. Matching these output values with the input, we find a theoretical value for the temperature that Venus should have. This value—less than 375 K (100°C)—is much lower than the measured values. So the high temperatures derived

The carbon dioxide, plus water vapor and sulfuric acid particles, makes the atmosphere of Venus opaque to infrared radiation and leads to a large greenhouse effect. The Earth's atmosphere is not so opaque in the infrared, and its greenhouse effect is much smaller, about 35 K.

from radio measurements indicate that energy does not leave Venus as readily as it leaves the Earth.

What accounts for this difference? What is happening on Venus is similar to the process that is generally—though incorrectly—thought to occur in greenhouses here on the Earth. The process is thus called the *greenhouse effect* (Fig. 13–3). Sunlight passes through the Cytherean atmosphere in the form of radiation in the visible part of the spectrum. The sunlight is absorbed by, and so heats up, the surface of Venus. At the temperatures that result, the radiation given off is mostly in the infrared. But the carbon dioxide in Venus's atmosphere is opaque to infrared radiation, and the energy is trapped. Thus Venus heats up above the temperature it would have if the atmosphere were transparent; the surface radiates more and more energy, and eventually an equilibrium is reached at which the energy that gets out through the atmosphere balances the energy that is coming in. (Greenhouses on Earth don't work quite this way. The closed glass of greenhouses on Earth prevent convection and the mixing in of cold outside air. The trapping of energy by the "greenhouse effect" is a less important process in an actual greenhouse. Try not to be bothered by the fact that the greenhouse in your backyard is not heated by the "greenhouse effect.") Understanding such processes involving the transfer of energy is but one of the practical results from the study of astronomy.

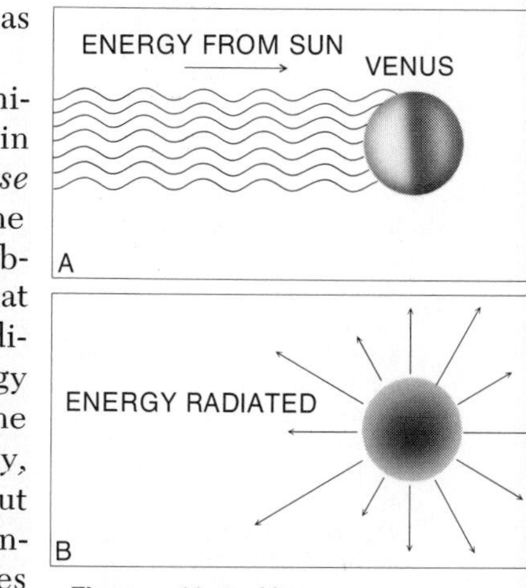

Figure 13–2 Venus receives energy from only one direction, and heats up and radiates energy (mostly in the infrared) in all directions. From balancing the energy input and output, astronomers can calculate what Venus's temperature would be if Venus had no atmosphere. This type of calculation is typical of those often made by astronomers.

Some additional information we have about the surface of Venus comes from radar observations. Maps of the surface have painstakingly been built up at several of the huge radio telescopes capable of radar work. The 305-meter dish at Arecibo in Puerto Rico (shown in Fig. 19–6), the 36-meter "Haystack" dish in Massachusetts, and the 64-meter dish at Goldstone, California, are equipped for radar. Using these large dishes together with smaller radio telescopes a few kilometers away as "interferometers" (which are discussed in Section 23.6) allows us to have the high resolution that a single telescope that same few kilometers across would have. With such high resolution the ambiguity of two areas on the surface mentioned in Section 12.1 can be resolved. Signals from the two points that represent the

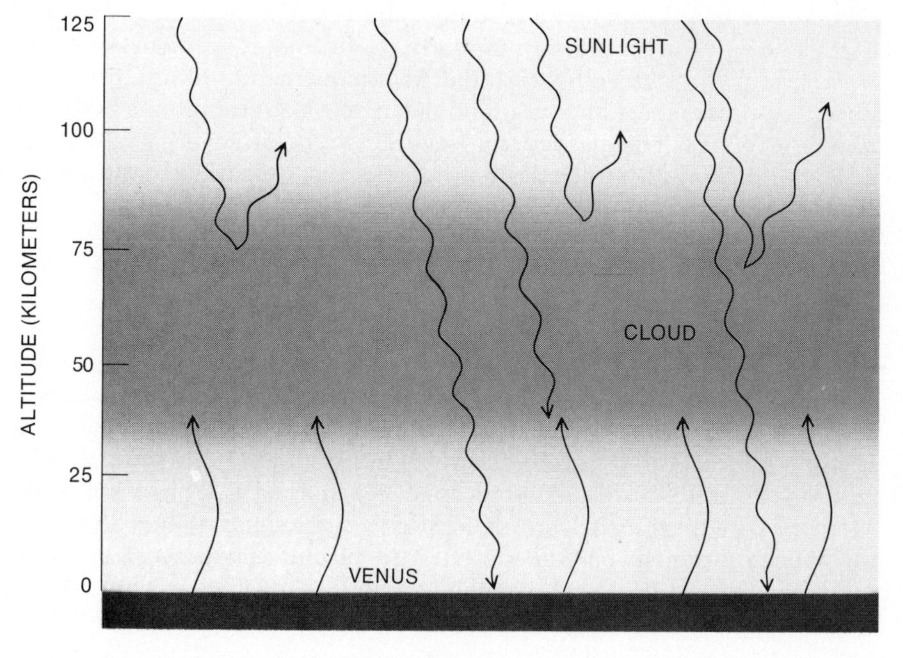

Figure 13–3 Sunlight can penetrate Venus's clouds, so the surface is illuminated with radiation in the visible part of the spectrum. Venus's own radiation is mostly in the infrared, and the presence of carbon dioxide in the atmosphere blocks the transmission of most of this radiation. This is called the greenhouse effect. Water vapor and sulfuric acid particles block the rest of the infrared. The Earth's atmosphere is not as opaque as is Venus's in the infrared, so the Earth's greenhouse effect is much smaller, about 35 K of heating.

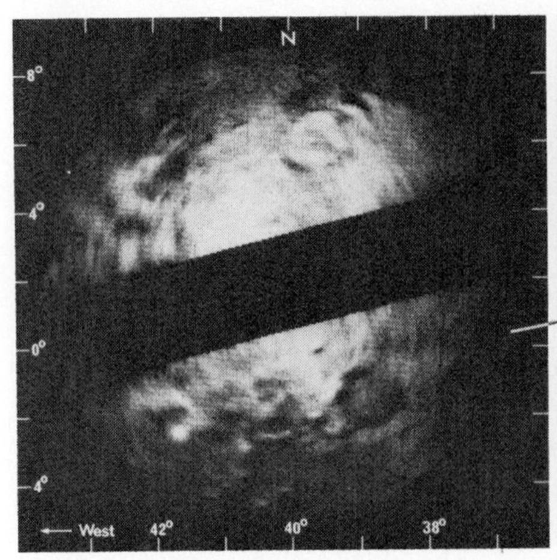

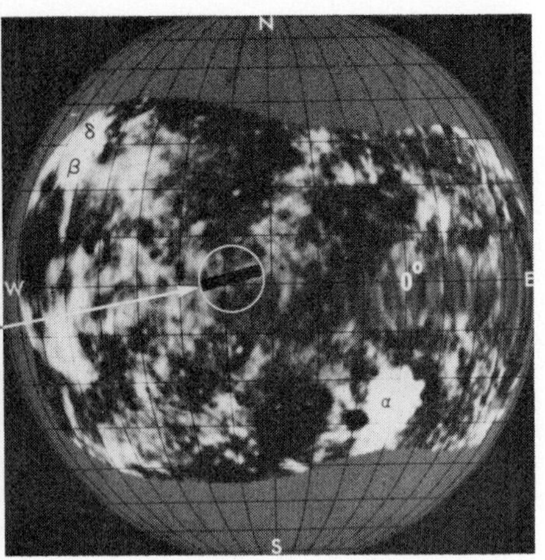

Figure 13–4 Radar maps of Venus. The area circled in the large-scale map at right is enlarged at left. An area that could not be well observed is shown as black. Radar results only show the radar reflectivity of the surface, but the results can be interpreted as showing shallow craters as big as 160 km across.

intersections of the concentric circles from time-delay measurements and the straight line from Doppler measurements (as discussed in Section 12.1) can be distinguished by this method. Radio interferometry, as we shall see, is also of great value in studying objects far beyond the solar system.

The radar maps of Venus (Fig. 13–4) show large-scale surface features, some very rough and others relatively smooth, just as surfaces on the Moon. Many craters have been found, all relatively shallow, but ranging up to 1000 km across. Most have probably been formed by meteoric impact.

From Venus's size and from the fact that its mean density is similar to that of the Earth, we conclude that its interior is also probably similar to that of the Earth. This means that we might expect to find volcanoes and mountains on Venus, and that venusquakes probably occur too. A huge peak whose base is over 150 km in radius and which has a depression 40 km in radius at its summit has been detected by radar, and may be a giant volcano. A cluster of some 20 smaller peaks resembles the configuration of clusters of volcanoes on Earth. At present, we cannot tell whether any of the volcanoes are currently active. A long trough at Venus's equator, 1500 km in length, seems to resemble the Rift Valley in East Africa, the Earth's largest canyon. It may similarly result from movements of Venus's crust, just as the Rift Valley resulted from movements of the Earth's crust. Recent improvements have enabled radars to reach resolutions better than 20 km on Venus. This has led, for example, to the discovery of a large area that is probably a lava flow, which tells us that volcanic conditions have existed in the interior.

Figure 13–5 Transits of Venus.

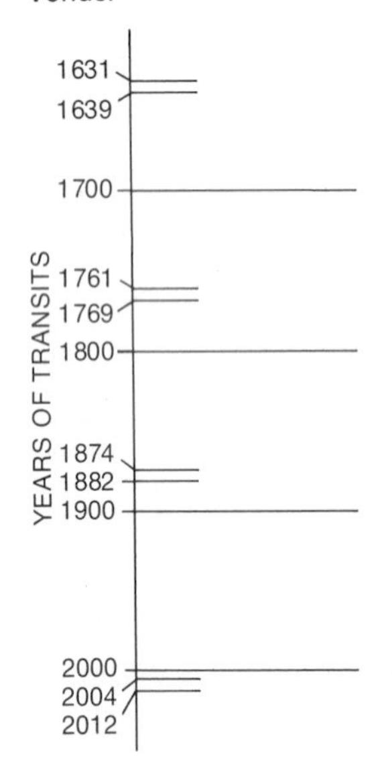

13.3 TRANSITS OF VENUS

Every so often Venus passes in front of the surface of the Sun, as seen from Earth. These transits of Venus are rare events. They occur in pairs, the second following the first by 8 years. The pairs are spaced at very long intervals. Transits occurred in 1631 and 1639, in 1761 and 1769, and then in 1874 and 1882 (Fig. 13–5). Since Venus appears just as a small black dot, as does Mercury when it transits across the Sun every few years (every 8 years on the average), transits of Venus are not spectacular events to watch in the same way that eclipses are, even though they are much rarer. One sees only a small black dot traveling across the face of the Sun. Of course, only these

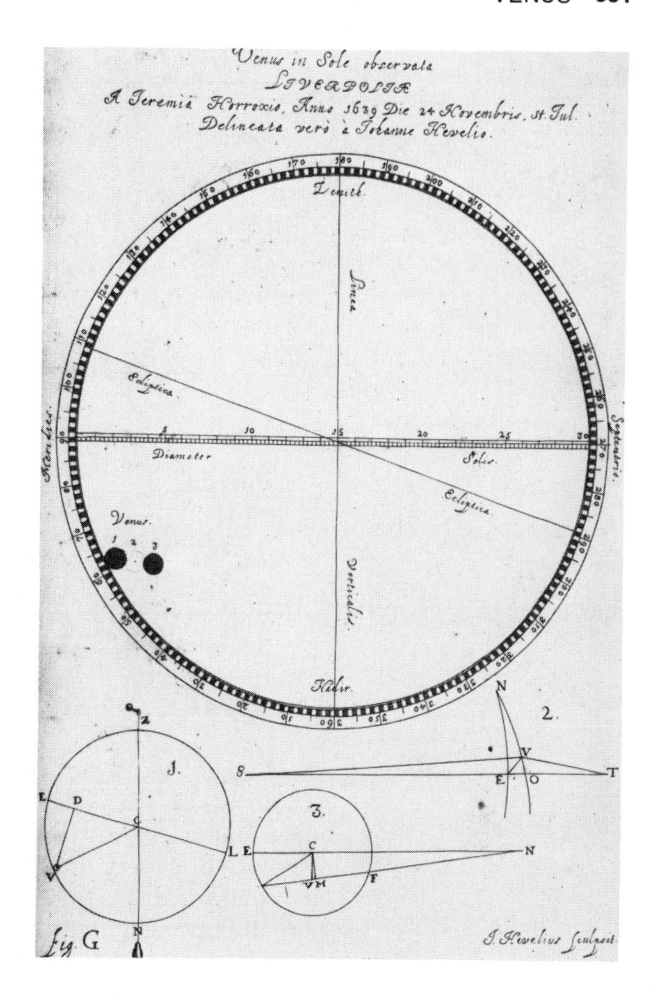

Figure 13–6 A drawing of the 1639 transit of Venus.

two inner planets can have transits, as seen from the Earth; the other planets never pass between the Earth and the Sun.

In times past (Fig. 13–6), astronomers used transits of Mercury and Venus to get information about the size of the solar system (by triangulation). Venus's atmosphere, in addition, was discovered at the transit of 1761 when it appeared silhouetted against the Sun. International cooperation in scientific pursuits was fostered by the many expeditions to observe that transit. In hope of further success, Captain Cook sailed to Tahiti from England on one of many expeditions that observed the transit of 1769. Captain Cook discovered Hawaii and a few other places as bonuses along the way; it is not often that the side benefits of astronomical research are so apparent. Traveling off in a jet plane to see a solar eclipse anywhere in the world is certainly less arduous than was this lengthy expedition by ship.

In recent years, we have been able to send spacecraft, such as the earlier Mariners, to fly behind Venus so that the radio signals they sent back passed through the Cytherean atmosphere en route to the Earth. We use these *occultations* to determine some of the properties of the Cytherean atmosphere, and of course the very act of sending a spacecraft and then following its trajectory gives us accurate planetary positions and masses. Because of the results from space exploration, the transits of the future will be of less scientific importance. But nonetheless, in 2004 and 2012 when the next pair of transits comes, I hope to be able to observe them.

An occultation occurs when one celestial object passes in front of, occults, another.

TABLE 13–1 UNMANNED PROBES LAUNCHED TO VENUS

Spacecraft	Launched by	Arrival Date	Comments
Venera 1	USSR	1961	Failed en route
Mariner 2	USA	1962	Flyby
Venera 2	USSR	1966	Flyby
Venera 3	USSR	1966	Crash-landed
Venera 4	USSR	1967	Probed atmosphere
Mariner 5	USA	1967	Flyby
Venera 5 and 6	USSR	1969	Probed atmosphere
Venera 7	USSR	1970	23 minutes of operation on surface
Venera 8	USSR	1972	50 minutes of operation on surface
Mariner 10	USA	1974	Flyby
Venera 9	USSR	1975	53 minutes of operation on surface; photograph
Venera 10	USSR	1975	65 minutes of operation on surface; photograph

13.4 SPACE OBSERVATIONS

Venus was an early target of both American and Soviet space missions (Table 13–1). The American craft were in the Mariner series, a heavier and more elaborately instrumented series than the Pioneers. The Soviet series was called Venera, the plural form, because each launch was of a double craft that separated into two when it reached the vicinity of Venus, with one part attempting a landing, and the other traveling farther into space.

The Soviet Union has invested great effort in the exploration of Venus. Each Venera lander radioed information concerning temperatures and pressures back to Earth as it descended through the atmosphere of Venus. The spacecraft were designed to withstand the tremendous pressures expected at the base of Venus's atmosphere. Unfortunately, the earlier Soviet spacecraft didn't make it to the surface. Venera 7 to 10 were designed to withstand the higher pressures that we now know are typical of Venus: 100 times that of Earth.

In 1970, the Venera 7 spacecraft radioed 23 minutes of data back from the surface of Venus. It confirmed the high temperatures of 750 K and high pressures. The Cytherean atmosphere (Fig. 13–7) is composed of 90 or more per cent carbon dioxide (CO_2), less than 5 per cent nitrogen (N_2), less than 1 per cent oxygen (O_2), and 0.01 per cent water. The pressure is 100 times higher than the pressure of Earth's atmosphere, because of the high amount of carbon dioxide in the atmosphere of Venus (Fig. 13–8); carbon dioxide makes up less than 1 per cent of the terrestrial atmosphere. Carbon dioxide on Earth dissolved in sea water and eventually formed our terrestrial rocks in the form of carbonates. If this carbon dioxide were to be released from the Earth's rocks, along with other carbon dioxide trapped in sea water, our atmosphere would become as dense and have as high a pressure as that of Venus. Venus, of course, has no oceans in which the carbon dioxide can dissolve.

Figure 13–7 The composition of Venus's atmosphere.

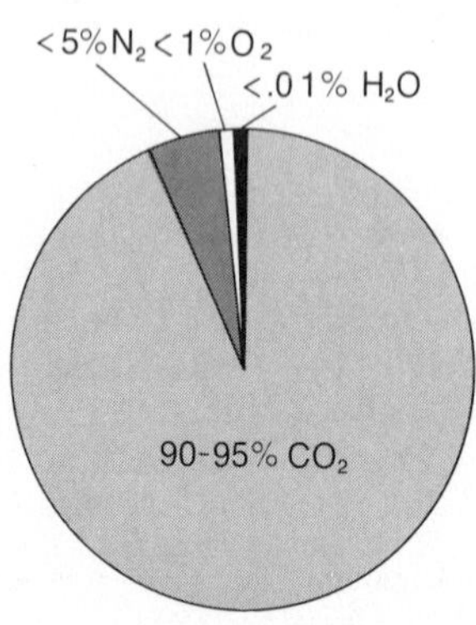

The spacecraft observations have improved our understanding of the clouds that surround Venus, high above its surface. Terrestrial clouds rarely go above 10 km but the denser Cytherean clouds may not start until 50 km

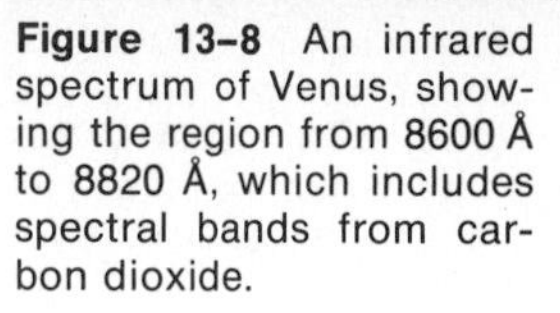

Figure 13-8 An infrared spectrum of Venus, showing the region from 8600 Å to 8820 Å, which includes spectral bands from carbon dioxide.

above the surface and then thinner clouds extend upward for an additional 30 km or so. These clouds may be composed of liquid droplets of sulfuric acid.

Subsequently, in 1972, the lander from Venera 8 survived on the surface of Venus for 50 minutes and confirmed the findings of Venera 7.

The United States spacecraft Mariner 10 (Fig. 13–9) took thousands of photographs of Venus in 1974 as it passed by en route to Mercury. It went sufficiently close and its imaging optics were good enough that it was able to study the structure in the clouds with resolution of up to 200 meters, about an eighth of a mile (Fig. 13–10). The structure shows only when viewed in ultraviolet light (see the figure opening this chapter). We could observe much finer details from space than we can see from the Earth (Fig. 13–11).

The clouds, when viewed in the ultraviolet, appear as long, delicate streaks, as do cirrus clouds viewed in visible or ultraviolet light on Earth. In the Cytherean tropics the clouds can also show a mottling, which suggests that convection, the boiling phenomenon, is going on. A big "eye" of convection (Fig. 13–13) is visible just downwind from the point at which the Sun is overhead and, therefore, at which its heating actions are at the maximum. The situation of the "eye" downwind from the subsolar point is almost a textbook example of what atmospheric scientists would expect, for hot air rises and starts the convection. Strong winds blow the clouds at these upper levels around the planet at 300 km/hr, as rapidly as the jet stream blows on Earth; the surface is rotating only very slowly. The raging winds and clouds observed on Venus stay at high levels above the surface, without much interaction with atmospheric regions below.

Studies of Venus like these have great practical value. The better we understand the interaction of solar heating, planetary rotation, and chemical composition in setting up an atmospheric circulation, the better we will understand our Earth's atmosphere. We then may be better able to predict the weather and discover jet routes that would aid air travel, for example. The potential saving of large amounts of money from this knowledge is enormous: it would be many times the investment we have made in planetary exploration.

Figure 13-9 The Mariner 10 spacecraft.

Figure 13-10 Several layers of haze at the limb of Venus, photographed by Mariner 10.

Figure 13–11 A series of photographs taken with the 4-m reflector at the Lick Observatory, showing cloud markings of Venus in the part of the ultraviolet that can be observed from the ground.

In 1975 Soviet scientists succeeded in landing a spacecraft on Venus and having it survive long enough to send back photographs. Venera 9 and Venera 10 traveled over four months from Earth to reach Venus, and arrived in October 1975. Each sent down a lander (Fig. 13–12). The single photograph that the Venera 9 lander took in the 53 minutes before it succumbed to the tremendous temperature and pressure showed a clear image of sharp-edged, angular rocks (Fig. 13–14). This came as a surprise to some scientists, who had thought that erosion would be rapid in the dense Cytherean atmosphere, and that the rocks should therefore have become smooth or disintegrated into sand.

Three days later, the Venera 10 lander reached the surface and transmitted data for 65 minutes, including a photograph (Fig. 13–15). The landing site was 2000 kilometers distant on the surface of Venus from the Venera 9 site. The rocks in the Venera 10 site were not sharp; they resembled huge pancakes, and between them were sections of cooled lava or debris of weathered rock.

Figure 13–12 A model of the Soviet Venera 9 lander.

Thus we have two photographs of the surface of Venus: mankind's first photographs from the surface of another planet. The one from Venera 9 looked typical of a "young mountainscape," with the sharp rocks possibly ejected from volcanoes. On the other hand, the one from Venera 10 "showed as a landscape typical of old mountain formations," in the words of Soviet scientists. They measured the surface wind velocity to be only 1 to 4 km/hr, although, since the atmosphere is much denser than that of Earth, more material would hit a rock in a given time on Venus than on Earth at the same wind velocity. The low wind speed we see makes it seem likely that erosion is caused not by sand blasting but by melting, temperature changes, chemical changes (which can be very efficient at the high temperature of Venus), and other mechanisms.

The Venera landers also made measurements of the soil, determining that its chemical composition and density correspond to that of basalt, in

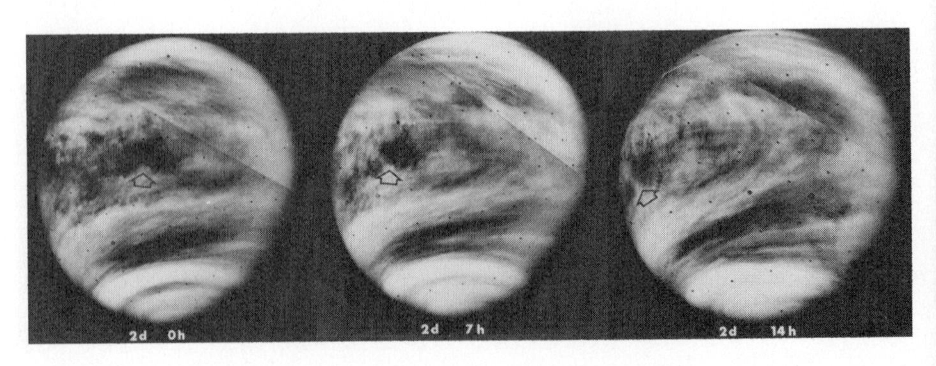

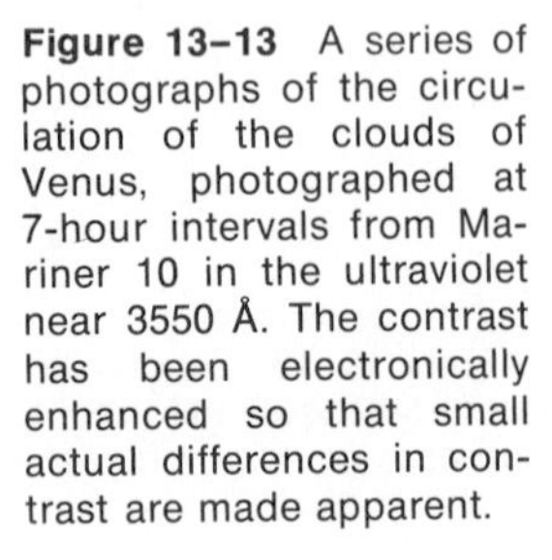

Figure 13–13 A series of photographs of the circulation of the clouds of Venus, photographed at 7-hour intervals from Mariner 10 in the ultraviolet near 3550 Å. The contrast has been electronically enhanced so that small actual differences in contrast are made apparent.

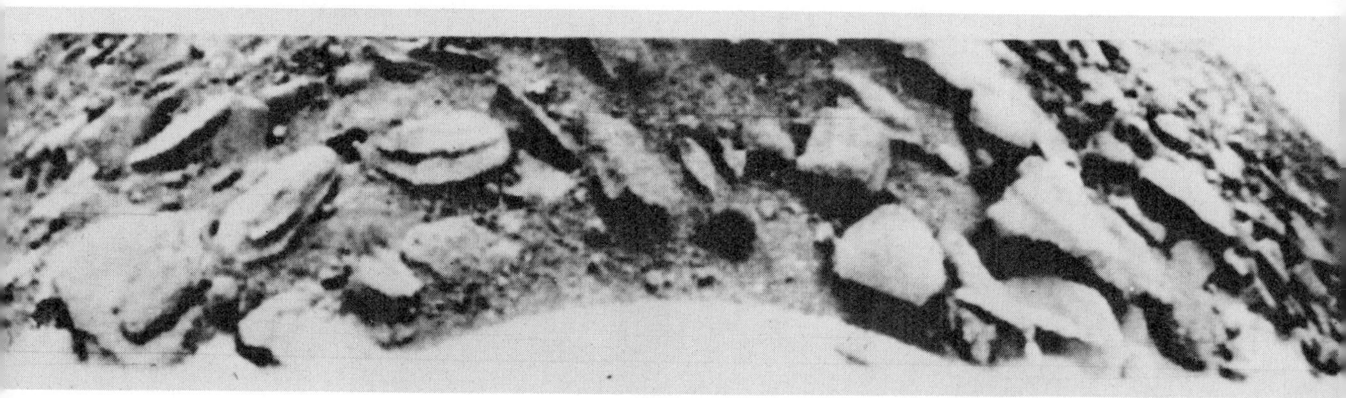

Figure 13–14 The surface of Venus photographed from Venera 9. The photograph is reproduced in a flat strip, but was actually a scan from upper left, down toward the center, and up toward the right. Think of looking to your left at the horizon, then down at your feet, and then up to the horizon at the right. That is why the horizon is visible as tilted lines at upper left and at upper right.

common with the Earth, the Moon, and Mars. They measured temperatures of 760 K and 740 K at the two sites, and a pressure over 90 times that of the Earth's atmosphere. The other halves of the Venera spacecraft (Fig. 13–16) remained in orbit around Venus, and undertook mapping missions to study the clouds.

The atmosphere of Venus is so dense that it should play tricks on our eyes, if we could see long distances through it. It has been predicted that if we were to stand on the surface and look around us, the horizon would seem to stretch above us in all directions because of extreme refraction. The horizon would seem to be elevated rather than straight ahead. It would be like living at the bottom of a giant concave mirror. When refraction causes such optical effects on Earth, we call them mirages. But mirages are transient, and these strange perspectives on Venus would be of a permanent nature.

Actually the Cytherean atmosphere is so dense that we couldn't see very far through it. Only after seeing the Venera 9 and 10 photographs could we know that there was enough light for photography and that no distortions are detectable in close-up photography. The intensity of light on the surface is as much as reaches the Earth under an overcast sky. This was a surprise, for it is a higher value than had been expected. Another surprise was that the horizon, when photographed, was straight. This was in contradiction to the prediction above. We do not yet know whether the spacecraft's horizon was in fact so close to it that the refractive properties would not have led to a large effect, or whether the prediction of extreme refraction is incorrect.

Figure 13–15 The surface of Venus photographed from Venera 10.

Figure 13–16 The Soviet Venera 9 spacecraft being assembled prior to launching.

13.5 OF WATER AND LIFE

Why is it that there is so much water on Earth, in our atmosphere, and in our oceans, and so little liquid water on the surface of Venus? Is it there, and just trapped inside the rocks, or is it trapped under the surface? What features that have made Earth a hospitable planet for life as we know it have made Venus so unattractive? Can we make sure we are not changing the Earth's atmosphere so that we will wind up with an atmosphere like that of Venus? One speculative model of the planet Venus does have large reservoirs of water under a surface layer that traps the water.

On the Earth, the origin of life changed the future course of our planet. The atmosphere of Earth in its primitive state may well have been of carbon dioxide, methane, ammonia, and other gases we now find distasteful or even fatal, but life forms produced much of the oxygen that we now have. One form of an oxygen molecule, ozone, consists of three oxygen atoms bound together (O_3). In the presence of ozone, water (H_2O) breaks down into oxygen and hydrogen. The hydrogen is so light that it has long since escaped from Earth. That's fine with us, as the oxygen is left behind for us to breathe. Perhaps if life had arisen on Venus at the same time that it arose on Earth, it would have modified the Cytherean atmosphere as it did the terrestrial one.

A catalyst *is a substance that causes or influences a chemical change without itself being affected by it.*

The water on the Earth has acted as a catalyst to spur the fixation of carbon dioxide into certain types of rocks. On Venus the carbon dioxide is still in the atmosphere—in fact, makes up 95 per cent of the atmosphere. The carbon dioxide causes the greenhouse effect to occur, thus keeping the surface of Venus very hot. Apparently water and plant life have saved the Earth from suffering the same fate as Venus.

The ozone plays another vital role on Earth: it keeps out ultraviolet radiation from the Sun. The part of the ultraviolet that has wavelengths slightly longer than 3000 Å, where the Earth's atmosphere becomes transparent, causes suntanning at the beach. None of the wavelengths shorter than 3000 Å pass through the ozone. If they did, they would kill us. Most scientists think that amounts of ultraviolet radiation even slightly increased over the amount we already receive, whether longer or shorter than 3000 Å, would increase the incidence of skin cancer.

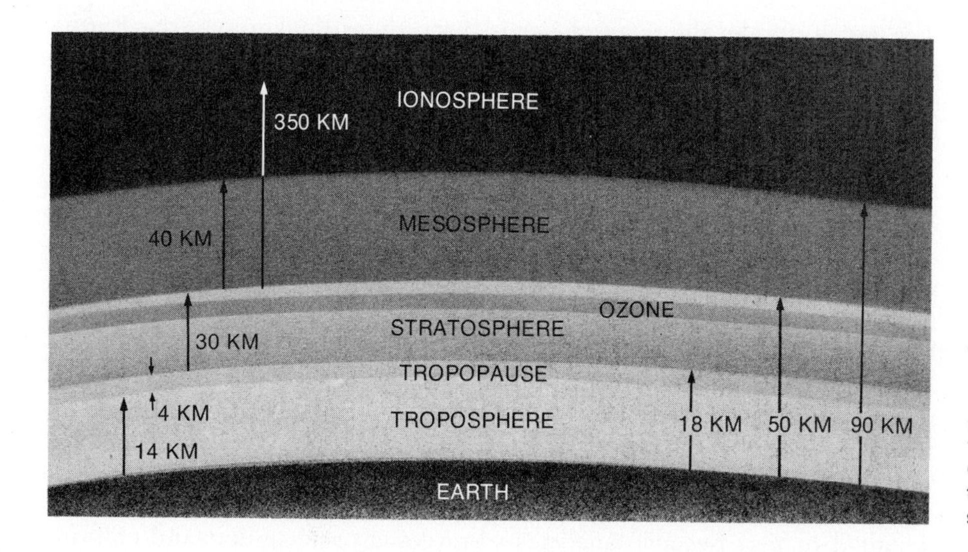

Figure 13–17 The layers of the Earth's atmosphere showing the location of the part of the stratosphere where ozone is formed. Once formed, the ozone is transported to lower stratospheric layers.

Scientists have recently been concerned about the way that man's technology may threaten the ozone layer of our Earth. For example, airplanes exhaust nitrous oxides, and if a plane should fly in the stratosphere the nitrous oxides could stay around for a long time, migrate to the upper stratosphere, and destroy some of the ozone, which would allow more ultraviolet radiation to reach the Earth's surface. Since supersonic transports fly at stratospheric heights, far higher than ordinary jets, there is concern that a fleet of SST's in commercial service, putting out nitrous oxides for half-a-dozen hours a day, could seriously deplete the ozone. This possibility was one of the major reasons why the American SST program was terminated. But in the 1980's there could be fleets of commercial Soviet SST's and Anglo-French supersonic Concordes flying in the stratosphere, in addition to the military flights. The concern persists, even though our own SST program is over.

An official report of the National Research Council, released in 1976, says that the fluorocarbon problem is a definite hazard to the Earth's ozone layer. Plans to restrict the use of fluorocarbon propellants are under way.

Another worry that surfaced more recently is the possibility that the gases that are used as refrigerants in refrigerators and air conditioners and as propellants in some aerosol cans are accumulating in the stratosphere and depleting the ozone. These gases were originally chosen for aerosol use because they were thought to be inert, that is, non-interacting. However, it now seems that they are not as inert as they originally had been thought to be. Once they are released into the atmosphere, the propellant gases (such as some types of Freon, a commonly used trade name for chloro-fluorocarbons including $CFCl_3$ and CF_2Cl_2), eventually break down, and the individual chlorine atoms can transform the ozone into other molecules. The chlorine returns to its atomic state after breaking down the ozone (chemically, it acts as a catalyst), and so each chlorine atom can wreak havoc on the ozone for years. We may already have put enough gases into the atmosphere to make a permanent change in the ozone layer. The Copernicus Orbiting Astronomical Observatory is being used to measure the chlorine content of the upper atmosphere in order to improve our assessment of the situation, radio astronomers are measuring molecules in the atmosphere, and other astronomical methods are also being applied. Gradually we experiment and observe to verify that the above mechanism actually takes place.

These problems must be balanced against other risks we take every

The use of Freons in aerosol cans seems non-essential and will be limited. Essential uses, such as refrigeration, will still be permitted.

day. For example, the ozone layer is three times thicker over the Earth's poles than it is over the equator. Thus by moving to a location where the ozone layer is thinner (from New York to Florida, for example), one is protected by less ozone. But the slightly increased risk of skin cancer that may occur by moving south would be offset by numerous other possible complications, including diminished risks of accidents in snowstorms in the northeast and earthquakes in California. So it is difficult to make truly rational decisions based on all the evidence. Still, we certainly wouldn't want our atmosphere to be transformed to one similar to that of Venus.

One lesson to learn from all these considerations is that we can be glad that the basic scientific research is being done, because it exposes the possible side effects of ordinary things we otherwise take for granted: fast airplanes and cans of deodorant. It's easier to say "go out and design an airplane that can go 2000 miles per hour" than to be sure you have thought of all the possible side effects that the decision might have on everything else on Earth. The dangers from refrigerants and aerosol propellants were understood in part because analyses were going on to understand chlorine compounds in Venus's atmosphere. Studies of the Cytherean atmosphere make us appreciate the fundamental benefits of our own atmosphere. With the new technology continuing to be developed, we must make certain that basic scientific research goes forward on all fronts so that we will be prepared in the future for whatever problems might arise. Research limited to specific problems won't be enough.

13.6 FUTURE EXPLORATION

NASA is planning an elaborate mission to Venus in 1978 that will send four probes through the atmosphere of Venus. The probes will be launched from a Pioneer spacecraft and will radio back signals via a second Pioneer that will orbit Venus. The probes will send back data as they pass through the Cytherean atmosphere but are not being designed to survive the impact of the surface. The orbiter will contain a radar for high resolution mapping of the surface.

Soviet plans for space exploration are ordinarily not disclosed in advance, and the fact of a mission is disclosed only when the rocket is already en route to a planet. But on the basis of their past record, we can assume that the Soviets will also continue their launches of spacecraft to Venus.

SUMMARY AND OUTLINE OF CHAPTER 13

Aims: To study Venus's clouds and surface, and to understand why and how Venus became inhospitable to life as we know it

The rotation period (Section 13.1)
The surface of Venus (Section 13.2)
  The high temperature: 750 K
  The greenhouse effect
  Surface maps from radar observations
Transits: international efforts of long ago (Section 13.3)
Space observations (Sections 13.4 and 13.6)
  U.S. Mariner 10 observations of clouds and atmosphere

Soviet Venera 9 and 10 landers: temperature measurements and photographs of the surface
Why Venus and the Earth evolved differently (Section 13.5)

QUESTIONS

1. Make a table displaying the major similarities and differences between the Earth and Venus.
2. Two main factors determine how bright Venus appears from the Earth: how much of the lighted side faces the Earth and how far Venus is from Earth. What are the relative positions of the Earth, Sun, and Venus when Venus appears brightest? You may wish to draw diagrams to demonstrate this.
3. Why do we think that there have been significant external effects on the rotation of Venus?
4. If observers from another planet tried to gauge the rotation of the Earth by watching the clouds, what would they find?
5. Suppose a planet had an atmosphere that was opaque in the visible but transparent in the infrared. Describe how the effect of this type of atmosphere on the planet's temperature differs from the greenhouse effect.
6. Why do radar observations of Venus provide more data about the surface structure than a Mariner flyby?
7. If one removed all the CO_2 from the atmosphere of Venus, the pressure of the remaining constituents would be how many times the pressure of the Earth's atmosphere?
8. The Earth's magnetic field protects us from the solar wind. What does this tell you about the charge (or lack of it) of the solar wind particles?
9. Some scientists have argued that if the Earth had been slightly closer to the Sun, it would have turned out like Venus; outline the logic behind this conclusion.
10. Outline what you think would happen if we sent an expedition to Venus that resulted in the loss of almost all of the CO_2 from the Cytherean atmosphere.

TOPIC FOR DISCUSSION

1. Evaluate the consequences of continued use of chloro-fluorocarbons in aerosol cans. Distinguish essential from non-essential uses of chloro-fluorocarbons.

14

The Earth

Since it doesn't seem fair to know so much more about the Earth than the other planets simply because we live on it, let us consider our own planet as though it were being described by an inhabitant of Mars.

REPORT OF THE MARTIAN ACADEMY OF SCIENCE

For centuries we have known Earth as an interesting object in our sky; sometimes it is the morning star and sometimes it is the evening star. Since it is an inner planet for us, Earth is always within a few hours of rising or setting, so we can never observe it high overhead in a dark sky and through the minimum amount of our atmosphere.

The major problem we have in observing Earth is not our atmosphere but its own. Often major portions of Earth are covered with white clouds that prevent us from seeing the surface.

When we can see through the clouds, we see that Earth is mostly covered with a blue-greenish dark substance of much lower albedo than the clouds (as shown on the photograph on the facing page and on Color Plate 31). A smaller fraction of the surface is covered with lighter-colored material, and this lighter material changes in color somewhat with the seasons. As springtime comes to each hemisphere of Earth, the lighter material becomes greenish.

There is much less of the life-giving carbon dioxide on Earth than there is in our own atmosphere. But oxygen, a gas that is deadly to us, does exist by itself in Earth's atmosphere, both as O_2 and O_3. Earth has polar caps, one in the north and another in the south, and their sizes change with the season. Perhaps they are tremendous reservoirs of carbon dioxide in the form of ice (frozen CO_2), as are our polar caps. Some Martian scientists, but a minority, think that the polar caps on Earth might partially be made of frozen dihydrogen oxide (H_2O), which we call "wet ice," and are planning to take better spectra to study this possibility. They claim that their infrared measurements of Earth's temperature indicate that the planet is too warm for ordinary carbon dioxide ice.

Earth is accompanied by a remarkable moon, called Selene. Selene is not too much smaller than the planet Mercury, and ranks in size with the giant moons of Jupiter and Saturn. This is unusual because Earth is a martian planet rather than a giant planet, and planets ordinarily have moons

The Earth, photographed from Apollo 17. The symbol for the Earth appears at the top of this page.

Figure 14–1 A negative print of Hawaii, showing the shield volcano Mauna Kea. Measured from the ocean floor, it is the highest mountain on Earth. (Everest rises from a high plain.) The largest canyon-like feature on Earth is the Great Rift Valley in Africa.

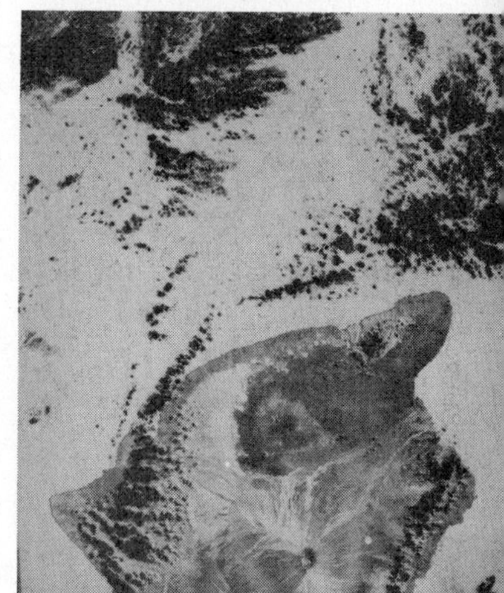

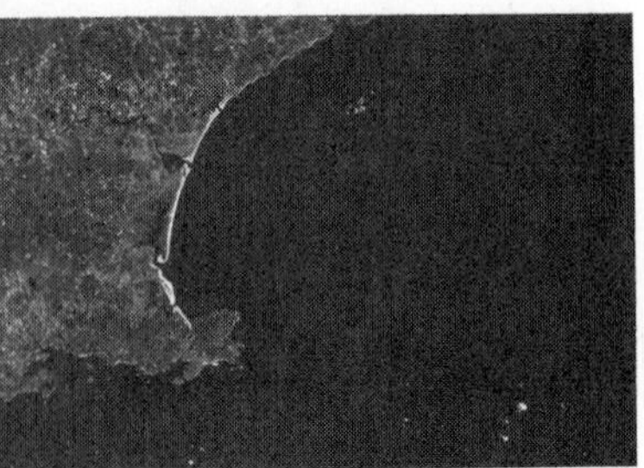

Figure 14–2 The Martian Terra 5 presumably landed in the ocean, shown here off the coast of Massachusetts and New Hampshire.

Figure 14–3 At first, the Martians sighted the Great Wall of China. This view over Chicago and Lake Michigan shows more distinct signs of civilization.

that are only about one-thousandth their sizes. Selene, on the contrary, is approximately one-fourth the diameter of its primary, and as much as 1/81 of the primary's mass. The seasonal changes that we have detected on Earth have not been seen on Selene.

For the last few years, we have sent a series of rockets to Earth. The Terra 1 and Terra 2 did not succeed in traveling the long distance to Earth, but Terra 3 flew by at a distance of 8 billion centirams (remember that 1 ram is the length of the left antenna of Queen Schrip, who reigned from the year 15,363 to 16,437) and succeeded in getting a series of photographs from close up. They showed a planet that is mostly covered with maria, which correspond to the darker areas. Radar reflectivities indicate that they may be covered with dihydrogen oxide (H_2O). There are mountain areas and very few craters. The largest peaks (Fig. 14–1) are smaller than our own volcanoes, and the largest canyon, in a raised land mass that extends from slightly above the equator far into the southern hemisphere, is about the size of our own Great Canyon. Since Earth is a larger planet than ours, these canyons and mountains are smaller than ours with respect to the size of the planet.

Terra 4 went into orbit around Earth, and took a series of photographs over its ten phobon lifetime (one phobon, of course, is the length of time it takes for Phobos to orbit the Mars, and corresponds to about 1/100 of an Earth selenth). Four relatively smooth areas were chosen as prime landing sites.

Terra 5 attempted a crash landing on Earth last year, and succeeded in slowing its velocity as it passed through Earth's considerable atmosphere. But contact with it was lost a few seconds after landing; perhaps it was covered over by whatever material makes up the greenish areas (Fig. 14–2). Those of our scientists who say that this material is dihydrogen oxide claim that this mysterious disappearance supports their theory.

We have not been able to establish the presence of any intelligent life on the planet. Indeed, the presence of life would seem to depend on establishing that for Earth, as on the Mars, the seasonal blowing of dust that takes place provides shelter from solar radiation for individual Earthians, assuming that they, as we, cannot stand exposure to sunlight. We look for signs of Earthian work. Some of our intelligence analysts think that they have detected a long serpentine streak traversing one of the continents and some signs of a checkerboard pattern on a large scale, which could indicate the presence of agriculture. But these detections are marginal, and must be checked further. More recent observations seem to show traces of cities (Fig. 14–3).

A somewhat later look in the Martian archives might turn up the following ideas, as reported by Paul A. Weiss of the Rockefeller Institute.

A summary of our report is that we have discovered life on Earth! It is the discovery of the millennium for us Martians. From a height of 5 million rams, we could see streaks of light moving like waves across the landscape, often in two channels right next to each other and flowing in opposite directions. When we came closer, we saw that each luminous knot had an independent existence and had two white lights in front and two red lights behind. They were phototactic, attracted by a flickering light source. But rather than rush into the source, they stopped just short of it and formed a crystalline array. Then they remained immobile, perhaps sleeping. Their luminous activity diminished (Fig. 14–4).

Figure 14–4 A drive-in movie must look strange to a visiting alien.

Inside these Earthians, we eventually discovered even smaller objects, probably parasites (Fig. 14–5). They are lodged, for the most part, in the interior of Earthians. They never stray far off from the Earthians, even when they are disgorged, so they must be dependent on the Earthians for sustenance. The parasites are more numerous in larger hosts (Fig. 14–6); since host size reflects host age, the accessory bodies obviously multiply inside the hosts as the latter grow.

We can show that the Earthians are alive, for they metabolize by taking in sustenance sucked in through tubes inserted in their rears (Fig. 14–7). They give off wastes mostly as gas and smoke. Some areas of Earth are positively fouled with the waste products. We also noted signs of grooming, with a wiping motion, but only in front (Fig. 14–8). Astonishingly, this was started and stopped in synchrony by all the members of the population, which may imply that the Earthians have brains.

Figure 14–5 If cars are considered to be the dominant life form on Earth, then the people would often be considered to live inside them.

OUTLINE AND SUMMARY OF CHAPTER 14

Aims: To understand the Earth on a planetary scale, and to realize the limitations of our studies of other planets from seeing how someone on another planet would interpret observations of the Earth

Terrain and atmosphere
Signs of life difficult to detect from afar

Figure 14–6 Larger vehicles have more people inside them.

QUESTIONS

1. (a) Imagine that you are observing the Earth from an altitude of 100 kilometers. Estimate the angular size of a house, a football field, and an average size city. What features do you think would point to the existence of intelligent life? (b) What about from 1000 km up?

2. Consult an atlas and compare the size of the Grand Canyon in Arizona with the Rift Valley in Africa. How do they compare in size with the giant canyon on Mars, which is about the diameter of the United States?

3. Plan a set of experiments or observations that you, as a Martian scientist, would have an unmanned spacecraft carry out on Earth. What data would your spacecraft radio back if it landed in a corn field? In the Sahara? In the Antarctic? In Times Square?

Figure 14–7 Cars have many of the characteristics that we consider to imply that something is alive.

Figure 14–8 A car grooming itself, or at least grooming its windshield.

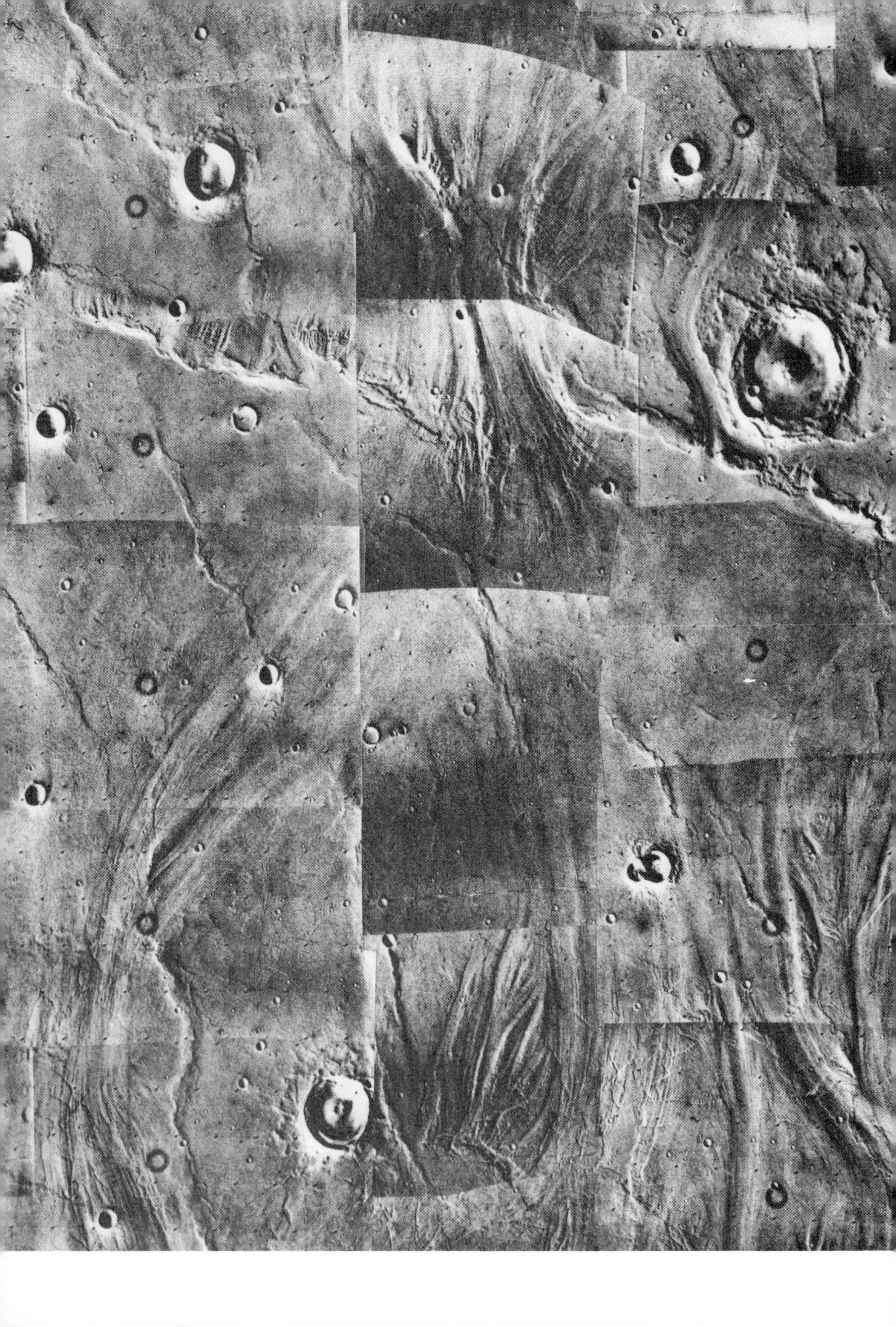

♂ 15

Mars

Mars has long been the planet of greatest interest to scientists and nonscientists alike. Its interesting appearance as a reddish object in the night sky and some of the past scientific studies that have been carried out have made Mars the prime object of speculation as to whether or not extraterrestrial life exists there.

The study of life in the universe outside of the Earth, now grown into a respectable scientific discipline, is known as *exobiology* and will be discussed in Chapter 19.

In 1877, the Italian astronomer Giovanni Schiaparelli published the results of a long series of visual telescopic observations he had made of Mars. He reported that he had seen *canali* on the surface. When this Italian word for "channels" was improperly translated into "canals," which seemed to connote that they were dug by intelligent life, public interest in Mars increased.

In this country, Percival Lowell grew very interested in the problem and in 1894 established an observatory in Flagstaff, Arizona, to study Mars. Over the next decades, there were endless debates over just what had been seen.

We now know that the channels or canals Schiaparelli and other observers reported are not present on Mars—the positions of the *canali* do not even always overlap the spots and markings that are actually on the Martian surface (Fig. 15–1). But hope of finding life in the solar system springs eternal, and the latest studies have indicated the presence of considerable quantities of liquid water in Mars's past, a fact that leads many astronomers to hope that life could have formed during those periods.

Figure 15–1 A drawing of Mars and a photograph, both made at the opposition of 1926. We now realize that the "canals" seen in the past do not usually correspond to real surface features.

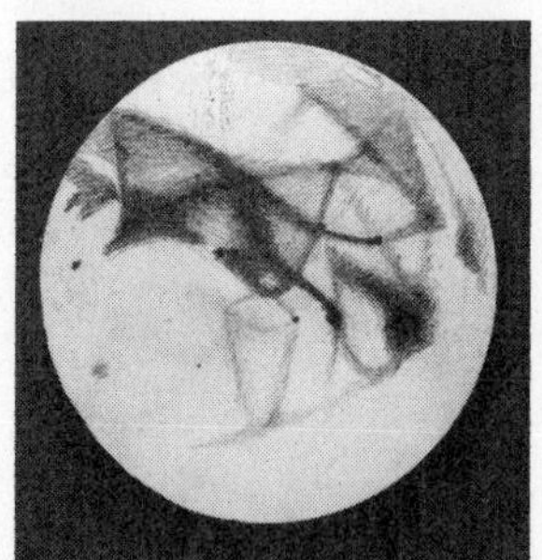

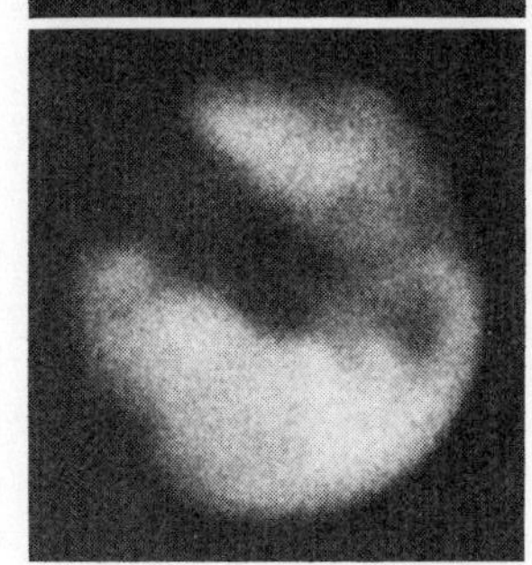

15.1 CHARACTERISTICS OF MARS

Mars is a small planet, 6800 km across, which is only about half the diameter of Earth or Venus, although one-and-a-half times that of Mercury.

A mosaic of a 250 by 200 km area of Mars taken from the Viking 1 orbiter. Lava flows are broken by faults that form ridges and are peppered with meteorite impact craters. Sinuous river channels cross the area. The symbol for Mars appears at the top of this page.

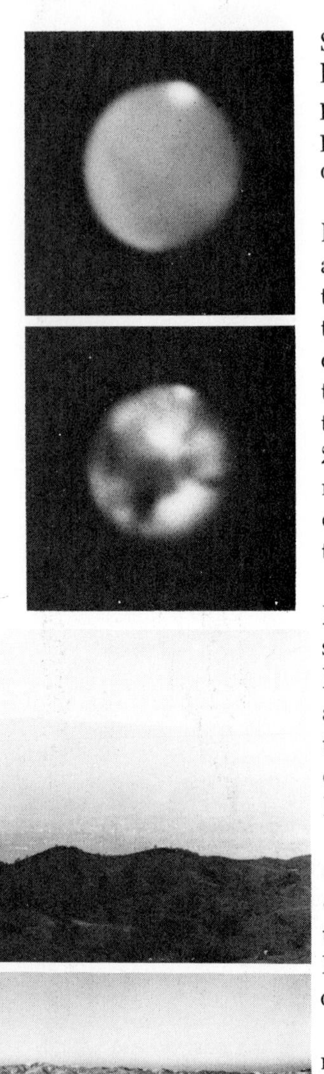

Figure 15–2 Photographs of Mars and of San Jose, California, taken from the Lick Observatory on Mount Hamilton in violet (top of each pair) and in infrared (bottom of each pair) light. The infrared better penetrates both the Earth's atmosphere and Mars's atmosphere.

Since Mars is much farther from the Sun than Mercury, it receives much less solar heat and so has been able to retain an atmosphere. The atmosphere is thin—at the surface its pressure is only 1 per cent of the surface pressure of Earth's atmosphere—but it might be sufficient for certain kinds of life.

From the Earth, we have relatively little trouble in seeing through the Martian atmosphere to inspect the planet's surface (Fig. 15–2), except when a Martian dust storm is raging. At other times, we are limited mainly by the turbulence in our own thicker atmosphere, which causes unsteadiness in the images and limits our resolution to 60 km at best (in other words, we cannot see detail smaller than 60 km even under the best observing conditions). We can, however, follow features on the surface of Mars and measure the rotation period quite accurately. It turns out to be 24 hours 37 minutes 22.6 seconds from one Martian noontime to the next, so a day on Mars lasts nearly the same length of time as does our own—it is just a few minutes a day longer. With an albedo of 16 per cent, Mars is darker than the Earth, though lighter than the Moon.

Unlike the orbits of Mercury or Venus, the orbit of Mars is outside the Earth's, so it is much easier to observe in the night sky (Fig. 15–3). We can simply observe it in the direction opposite from that of the Sun. We are then looking at the lighted face. When three celestial bodies are in a line, the alignment is called a *syzygy*, and the particular case when the planet is on the opposite side of the Earth from the Sun is called an *opposition*. These oppositions occur at intervals of 780 days on the average, equivalent to 26 Earth months.

Mars's orbit, about 1.5 A.U. in radius, has an eccentricity of 9 per cent, so at some of its oppositions it is closer to the Earth than at others. Obviously these more favorable oppositions are better for observing, because at these times the disk of Mars appears greater in angular size. It varies from 14 to 25 seconds of arc at opposition, from about 1/130th to 1/60th the size of the full moon.

Mars revolves around the Sun in 687 Earth days, equivalent to 23 Earth months. The axis of its rotation is tipped at a 25° angle from the plane of its orbit, nearly the same as the Earth's 23 1/2° tilt. Because the tilt causes the seasons, we know that Mars goes through a year with four seasons just as the Earth does, except that the Martian year is 23 Earth months long.

From the Earth, we have long watched the effect of the seasons on Mars. In the Martian winter, in a given hemisphere, there is a polar cap. As the Martian spring comes to the northern hemisphere, for example, the north polar cap shrinks and material at more temperate zones darkens. The surface of Mars is always mainly reddish, with darker gray areas that appear blue-green for physiological reasons of color contrast (see Color Plate 24). In the spring, the darker regions spread. Half a Martian year later, the same thing happens in the southern hemisphere.

One possible explanation that previously seemed obvious for these changes is biological; Martian vegetation could be blooming or spreading in the spring. But there are other explanations, too. The theory that presently seems most reasonable is that each year at the end of northern-hemisphere springtime, a global dust storm starts, covering the entire surface with light dust. Then the winds get as high as hundreds of kilometers per hour and blow fine, light-colored dust off the slopes. This exposes the dark areas underneath. Gradually over the next Martian year, dust is also stripped away in certain places, such as near craters and other obstacles.

Figure 15–3 When Mars is at opposition, it is high in our nighttime sky.

The synodic period of Mars is 780 Earth days. Mars will be getting farther and farther away from Earth in its oppositions of January 1978 and February 1980, and then the oppositions will become better and better until in September 1988 Mars will be only 59 million km away from Earth.

The sidereal period of Mars is 687 Earth days.

Finally, a global dust storm starts up again to renew the cycle. If the dust were of certain kinds of materials, like limonite ($2Fe_2O \bullet 3H_2O$), the reddish color would be explained.

Mars has 1/10 the mass of the Earth, and from the mass and radius one can easily calculate that it has an average density of slightly less than 4 grams/cm^3, not much higher than that of the Moon. This is substantially less than the density of 5½ grams/cm^3 shared by the three innermost terrestrial planets, and indicates that Mars's overall composition must be fundamentally different from that of these other planets. Mars probably has a small iron-nickel core, or perhaps a core of iron sulfide. The core may or may not be molten. In any case, Mars probably has a smaller core and a thicker crust than the Earth.

15.2 SPACE OBSERVATIONS OF MARS

Mars has been the target of a series of spacecraft launched by both the United States and the Soviet Union. The earliest of these spacecraft were flybys, which attempted to photograph and otherwise study the Martian surface and atmosphere in the few hours they were in good position as they passed by (Table 15–1).

TABLE 15–1 UNMANNED PROBES LAUNCHED TO MARS

Spacecraft	Launched by	Arrival Date	Comment
Mariner 4	USA	July 1965	Flyby
Mariner 6	USA	July 1969	Flyby
Mariner 7	USA	August 1969	Flyby
Mariner 9	USA	November 1971	Orbiter
Mars 2	USSR	November 1971	Orbiter; lander lost
Mars 3	USSR	December 1971	Orbiter and lander
Mars 4	USSR	February 1974	Flew by
Mars 5	USSR	February 1974	Orbiter
Mars 6	USSR	March 1974	Orbiter; lander failed
Mars 7	USSR	March 1974	Orbiter; lander failed
Viking 1	USA	June 1976	Orbiter and lander
Viking 2	USA	August 1976	Orbiter and lander

Mariner 4, for example, sent back close-up photographs of one line of passage across the surface. In those few photographs, craters were visible, so the astronomers said, "Aha, Mars is covered with craters like the Moon." It was a little like the tale of the blind men and the elephant: one man touches a leg and finds it is like a tree, leading him to conclude that an elephant is like a tree; another, finding the tail like a rope, reaches a different conclusion. However, in 1969, Mariners 6 and 7 radioed back information that showed that not all the areas of the surface of Mars are covered with craters.

In 1971, the United States sent out a spacecraft not just to fly by Mars for only a few hours, but actually to orbit the planet and send back data for a year or more. This spacecraft, Mariner 9, went into orbit around Mars in November 1971, after a 5-month voyage from Earth.

But when the spacecraft successfully reached Mars, our first reaction was as much disappointment as it was elation. While Mariner 9 was en route, a tremendous dust storm had come up, almost completely obscuring

Figure 15–4 The observations made by Mariner 9 permitted high resolution maps to be drawn by the U. S. Geological Survey.

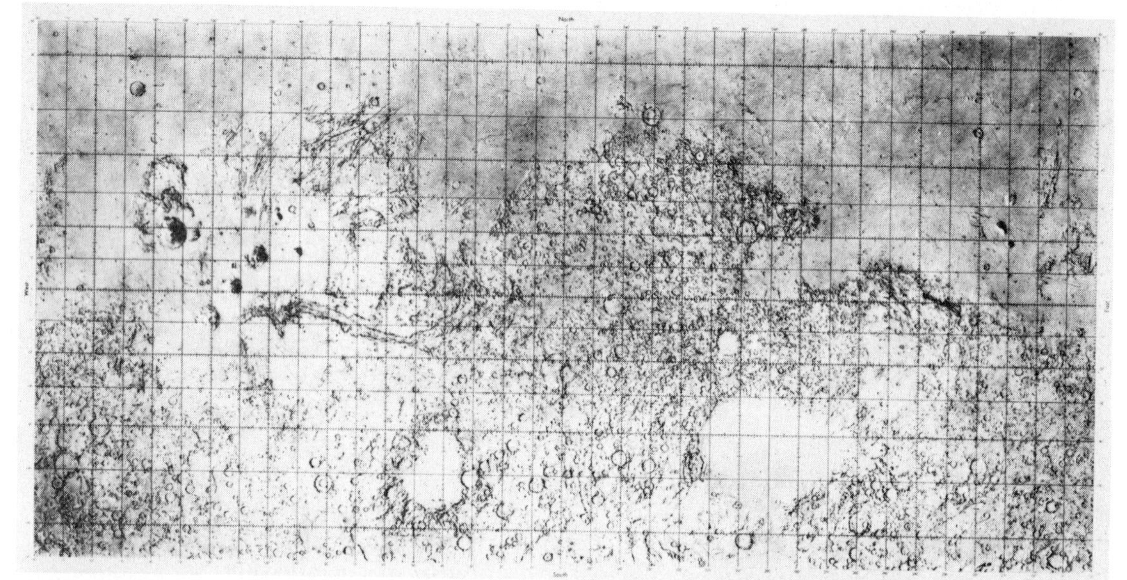

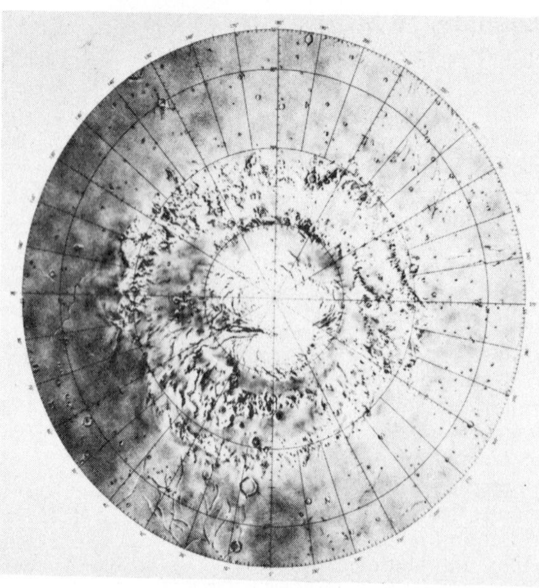

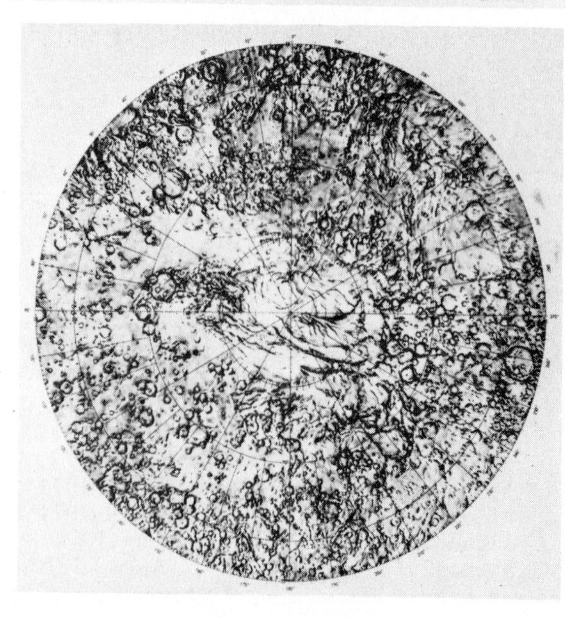

the entire surface of the planet. Only the south polar cap and 4 dark spots were visible. We had hoped for detailed photographs of the craters, but were at first rewarded with just murky photographs. Of course, astronomers knew that the storm would end eventually, but even though the spacecraft was designed to work for at least a year, one never knows when some vital component might fail. As the craft's cameras waited out the storm below, photographs were taken of Mars's moons, and many nonphotographic observations were made.

The storm began to settle after a few weeks, with the polar caps best visible through the thinning dust. Finally, three months after the spacecraft arrived, the surface of Mars was completely visible, and Mariner 9 could proceed with its mapping mission. It mapped the entire surface at a resolution of 1 km (Fig. 15–4), and photographed selected areas at a resolution 10 times better than this limit. The spacecraft was to last another year, and so more than completed its assigned tasks—in fact, we even got the scientific bonus of being able to see how the dust acted at the end of the storm. These observations gave us information both about the winds aloft and about the heights and shapes of the surface features. And studies of the temperature changes on Mars that resulted from the dust storm are helping us assess the possible results of some kinds of atmospheric pollution on Earth's temperatures. During the dust storm, the temperature of the Martian atmosphere increased while the surface temperature decreased. This may have resulted from the change in the albedo of Mars caused by the dust in the atmosphere. Also, absorption of sunlight by the suspended dust particles helped to heat the atmosphere.

Now that we have the data taken after the dust storm had ended, we can classify four major "geological" areas on Mars: volcanic regions, canyon areas, expanses of craters, and terraced areas near the poles.

"Geo-" is from the Greek for Earth. Strictly speaking, "geology" applies only to the Earth, but the term is now more widely used.

Even through the storm, there had been an indication of four dark spots on the Martian surface. When the dust cleared, these proved to be volcanoes. A chief surprise of the Mariner 9 mission was the discovery of extensive areas of vulcanism on Mars. The largest of the volcanoes, which corresponds in position to the surface marking long known as Nix Olympica, "the snow of Olympus," is named Olympus Mons, "Mount Olympus." It is a huge volcano—600 kilometers across at its base and about 25 kilometers high (Fig. 15–5). The tallest volcano on Earth is Mauna Kea, in the Hawaiian islands, if we measure its height from its base deep below the ocean. Mauna Kea is only 9 kilometers high, taller than Everest. And remember, it is on the Earth, a much bigger planet than Mars. Olympus Mons is crowned with a crater 65 kilometers wide; Manhattan Island could be easily dropped inside the crater.

Mauna Kea is shown in Fig. 14–1.

Near Olympus Mons are three smaller volcanoes in a straight line. The volcanoes are important pieces of evidence in figuring out the history of Mars. The volcanic activity on Mars is located in several discrete areas. One Martian expert, Bruce Murray of the Jet Propulsion Laboratory, believes that the volcanic activity and internal heating on Mars have simply not yet spread to the rest of the planet. If this is so, then the inside of Mars is heating up now, has been colder in the past, and will be hotter in the future. Perhaps, over the next thousands or hundreds of thousands or millions of years, it will be more volcanically active. Shifting continental plates, such as those on the Earth, may develop. Mars may not have reached that stage yet. Other astronomers do not agree and point to evidence that

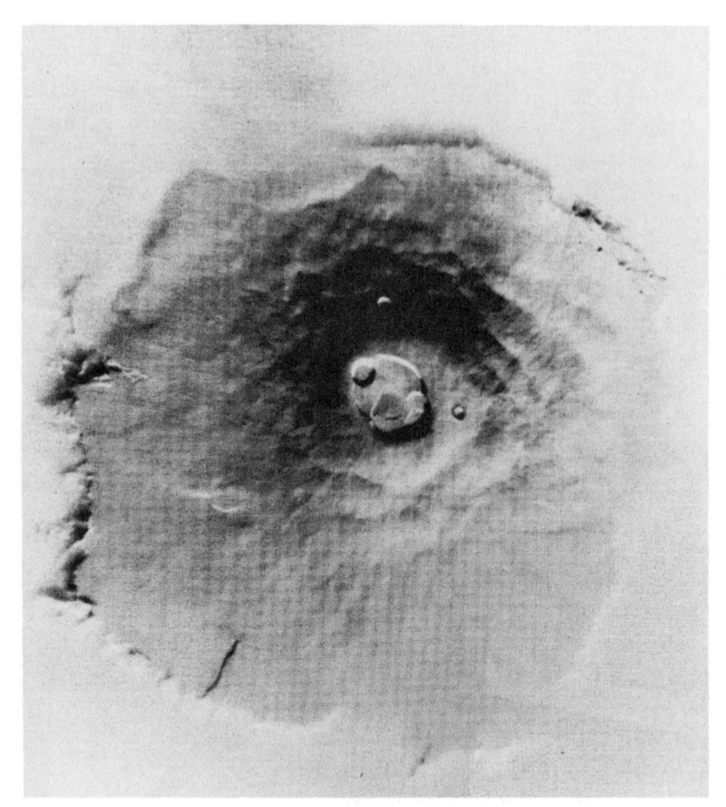

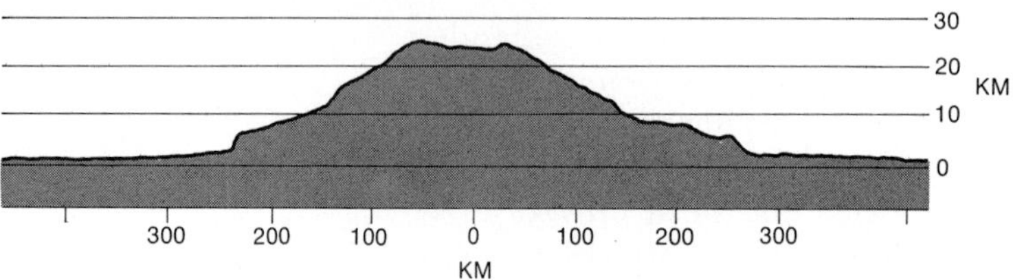

Figure 15–5 Olympus Mons. Its height profile, measured from a radar on Mariner 9, is also shown.

Mars may have been internally active for much longer than Murray thinks. Most of the northern hemisphere shows signs of vulcanism—either volcanoes or plains flooded with lava. Perhaps the explanation of the preference for vulcanism in this hemisphere is that the surface is at a lower altitude than the surface of the rest of the planet.

The volcanoes on Mars would release oxygen and hydrogen into the atmosphere in the form of water and carbon dioxide. In Murray's model, Mars's present atmosphere was ejected during the recent formation of volcanoes such as Olympus Mons. If plants should evolve to photosynthesize oxygen, the atmosphere might become much more Earthlike in some thousands of years. Of course, at this point it seems quite unlikely that we shall leave Mars and its atmosphere untampered with for all this time. Already suggestions have been made to transform the atmosphere with plants or microorganisms, or to melt the polar caps by distributing a black substance on them. Let us hope that we don't tamper too quickly.

Some areas of Mars are covered with extensive sand dunes (Fig. 15–6). The dunes are sometimes aligned in the same direction as each other, showing that the wind at such locations must always be from the same direction. We can also observe the wind direction by studying surface streaks on the leeward sides of craters.

Besides the volcanoes, another surprise on Mars was the discovery of

Figure 15–6 Sand dunes on Mars, showing complex effects of Martian winds.

systems of canyons. One tremendous canyon (Fig. 15–7)—about 5000 kilometers long—is as big as the United States and comparable in size to the Rift Valley in Africa, the longest geological fault on Earth. Here again the size of the canyon with respect to Mars is proportionately larger than any geologic formations on the Earth.

Perhaps the most amazing discovery on Mars was the presence of sinuous channels. These are on a smaller scale than the *canali* that Schiaparelli had seen, and are entirely different phenomena. Some of the channels show tributaries (Fig. 15–8), and the bottoms of some of the channels show the characteristic features that stream beds on Earth have. Even though water cannot exist on the surface of Mars under today's conditions, it is difficult to think of other ways to explain them satisfactorily than to say that the channels were cut by running water in the past.

This is particularly interesting because biologists feel that water is necessary for the formation and evolution of life. The presence of water on

Figure 15–7 The tremendous canyon, now named Valles Marineris—Mariner Valley—is about 5000 km long, nearly the diameter of the United States. The canyon is 120 km wide and 6 km deep.

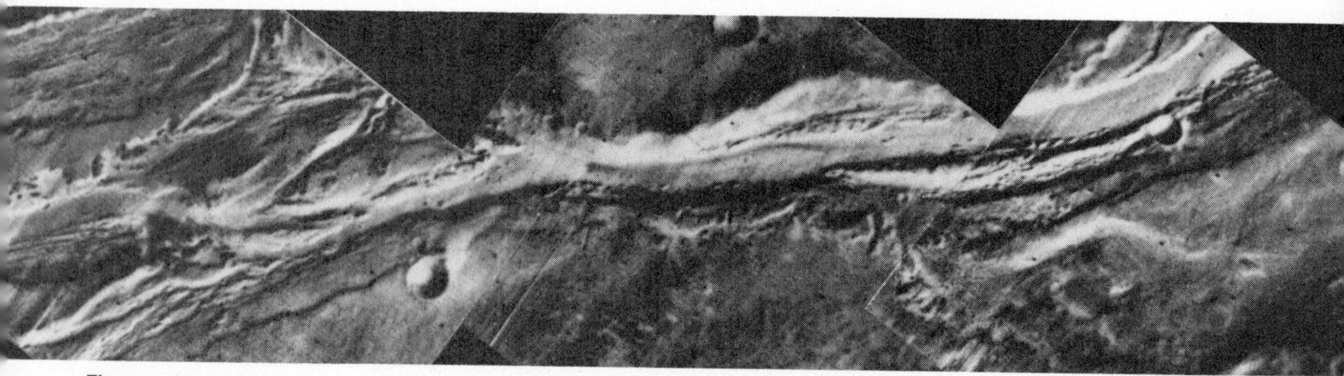

Figure 15–8 Mariner 9 photographed stream beds that showed many signs of the past presence of running water.

Mars, therefore, even in the past, may indicate that life could have formed and may even have survived. If we could discover life on another planet that arose independently of life on Earth, comparison of the life forms would tell us what is important to cells and to life, and what things may be of peripheral importance. For us on the Earth, the implications of these discoveries for advances in medicine, not to mention theology, are obvious.

If there had been water on Mars in the past, and in quantities great enough to cut the river beds, where has it all gone? One possible solution is that it has been bound into the rocks by an interaction with the carbon dioxide that is known to make up the largest part of the Martian atmosphere. Or the water could have been absorbed by the surface soil.

Some of the water is bound in the polar caps (Fig. 15–9). At the time of Mariner 9, we thought that the polar caps were mostly frozen carbon dioxide—"dry ice"; the presence of water was controversial. In the winter these caps extend down to latitude 50°.

If the water trapped in the polar caps or under the surface of Mars was released at some time in the past, as liquid water on the surface or as water vapor in the atmosphere, it may have been sufficient to make Mars verdant. Perhaps the polar caps melted, releasing quantities of water because dust

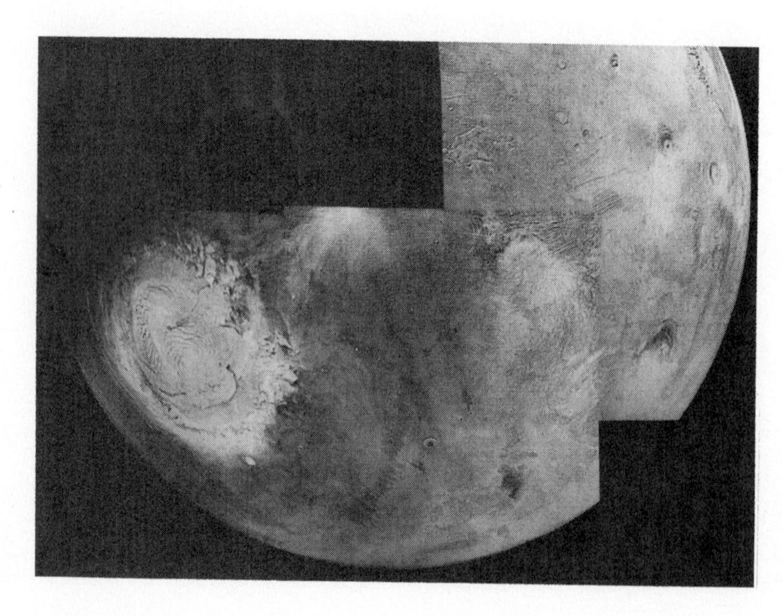

Figure 15–9 This mosaic of Mariner 9 views showed the north polar cap, which is shrinking as a result of Martian springtime. The volcanoes and the west end of the great canyon can also be seen.

that settled on the ice caused more sunlight to be absorbed or a widespread episode of vulcanism caused a general heating. Climatic change could also have resulted from a change in the orientation of Mars's axis of rotation, from a change in the solar luminosity, or from a variation in atmospheric composition (which would affect the greenhouse effect).

Infrared detectors had been placed aboard the various spacecraft to study temperatures on Mars because the Planck curves (Section 2.4) that correspond to planetary temperatures peak in the infrared, as had been known from lower-resolution Earth-based observations. The data show that the temperature can rise as high as 300 K (70° F) at the equator at noon, but usually temperatures are below freezing, even at local noon. The spectral intensity distribution curves for Mars also have a second peak, in the yellow, from reflected sunlight. They are thus the superposition of two Planck curves, one peaking in the infrared and one in the visible.

The composition of the atmosphere is measured by looking at the spectrum. This can even be done from the Earth, but problems arise because the spectrum of the Earth's own atmosphere interferes, particularly at the wavelengths of the water vapor lines.

We found that the Martian atmosphere is composed of 90 per cent carbon dioxide (CO_2) with small amounts of carbon monoxide (CO), oxygen (O, O_2, and O_3), water (H_2O), and hydrogen (H). The surface pressure is 1 per cent of that which exists on the Earth's surface, but the pressure decreases with altitude more slowly than it does on Earth.

We measure the density of the atmosphere by sending a spacecraft around behind Mars and monitoring the changes of its radio signal as the spacecraft is occulted by the atmosphere of the planet. This has been done with all the Mariner craft, including flybys and Mariner 9 as it orbited. The occultation tells the rate at which the density drops. From the composition and the way the density varies with height, we derive the variations of pressure and temperature.

Another method of studying the atmosphere is to measure it by actually flying through it. The two Soviet missions Mars 2 and 3 were designed to do just that. Each mission had two connected craft; each went into orbit around Mars and ejected a lander. The lander from Mars 2, ejected on November 27, 1971, was never heard from again. Mars 3, reaching the planet a week later, also ejected a lander into the Martian atmosphere. It took some hours to make its entry. Its parachute unfurled (the atmosphere is just dense enough to allow a parachute to work), a braking rocket fired, and another rocket was supposed to shoot the parachute off to the side so that it would land separately from the spacecraft. The lander reached the surface, turned on its cameras and began to beam information. It sent signals for 20 seconds, and then stopped. Just what happened is not known. The lander could have been damaged on impact, even though it was braced to withstand a 200 mph crash. It could have been buried in the surface dust within a few seconds. Perhaps it was buffeted by the high winds and damaged or turned over. Or perhaps its own parachute landed on top of it, even though it was designed to shoot out to the side.

The orbiting parent ships remained in orbit around Mars, and together with the orbiting Mariner 9 continued to send back vital information about the temperatures on Mars, the constituents of the atmosphere, and other data.

Scientists have been concerned for many years about the effects of our spacecraft landing on other planets. This discussion has particular immediacy for Mars. Perhaps a spacecraft would bring along micro-organisms that would contaminate the planet's atmosphere. When later spacecraft come to explore, they might find life forms that had come from the contamination introduced earlier rather than from life indigenous to the planet.

Different sides are taken in the controversy over how the possibility of contamination should affect our future plans. Some scientists feel that we should not send landers until we are ready for more detailed exploration. Others believe that the risk of contamination is low and say we are ready right now to land on other planets. As a compromise, certain sterilization procedures have been worked out for spacecraft to Mars and the other planets. The spacecraft are baked under high temperatures for some time.

Lest we grow complacent and feel that we have but to launch rockets to the planets to get the information we want, we have the sad story of Mars 4, 5, 6, and 7. These four Soviet spacecraft were launched within a three week period in 1973. Hopes were high that at least one would successfully land, and that much data would be radioed back.

In February 1974, the braking rockets of Mars 4 did not work properly, and the spacecraft went past Mars, while Mars 5 successfully went into orbit and photographed the surface for 10 days. The next month, Mars 7 launched a probe probably meant to be a lander (the Soviets never give advance notice), but it left the vicinity of Mars and went into solar orbit. A few days later, Mars 6 descended, the braking rocket fired, the descent engine fired, and the parachutes opened as scheduled, but Tass reported that "radio contact with the descent module broke off when it was almost on the surface."

Although none of the four spacecraft completed their missions, Mars 6 did radio back interesting information, including reports of the varying amounts of water vapor over different parts of the surface, and of the strength of Mars's weak magnetic field, which turned out to be only 7 to 10 times greater than the magnetic field of ordinary interplanetary space.

Mariner 9, the hero of this section, made 698 orbits of Mars, and then exhausted the gas in the little jet thrusters it used for alignment. It will remain in orbit around Mars for another 50 or 100 years, but is no longer in communication with Earth.

Figure 15–10 Part of the first panoramic view of the surface of Mars, taken on July 20, 1976, by NASA's Viking 1 lander. The site is in Chryse Planitia (the Plain of Chryse).

15.3 VIKING!

In the summer of 1976, two U.S. spacecraft named Viking reached Mars after flights of about 10 months. Each spacecraft contained two parts: an orbiter and a lander. The orbiter served two roles; it not only used its cameras and other instruments to map and analyze the surface but also served as a relay station for the radio signals from the lander to Earth. The lander's role was surely no less significant: its major task was to sample the surface and decide whether there was life on Mars!

Viking 1 reached Mars in June of 1976 after a flight that led to spectacular large-scale views of the Martian surface (see the cover of this book and

Figure 15–11 This mosaic shows a part of the Chryse region near the landing site of Viking 1. The tear-drop shapes on one side of the craters (from left to right) Gold, Bok, and Lod, show the presence of erosion from large volumes of water in the past. Nothing like this is seen on the Moon.

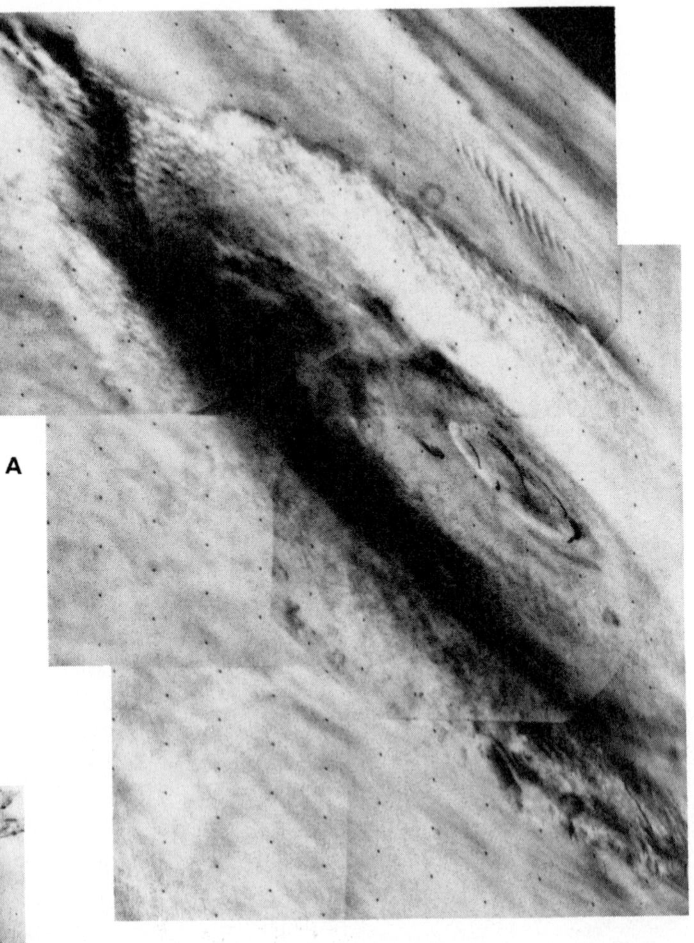

Figure 15–12 (*A*) Olympus Mons photographed by the Viking 1 orbiter. Clouds extend up most of its 25-km height. The volcanic crater, 80 km across, extends into the stratosphere and was cloud free. Mars's limb shows several haze layers.

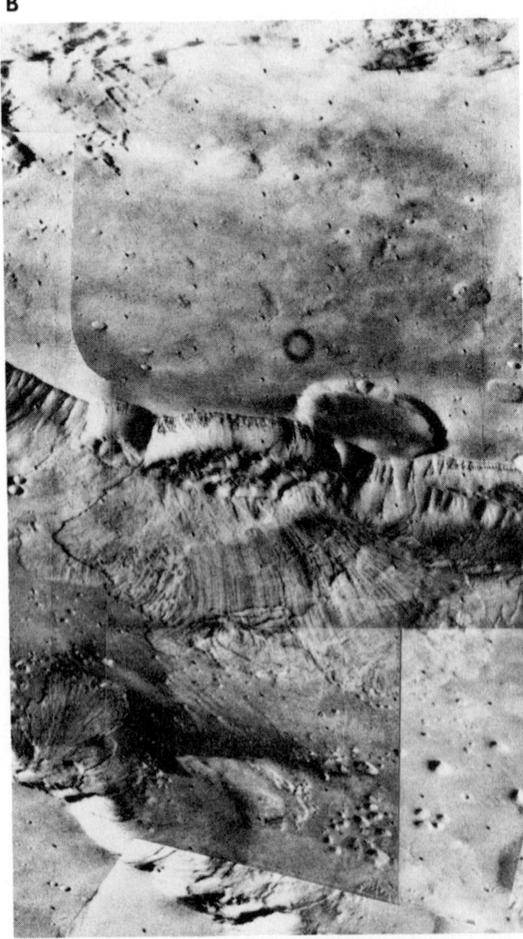

Figure 15–12 (*B*) A Viking 1 view of Valles Marineris. The walls appear as though they had collapsed at intervals to form huge landslides that flow down and across the canyon floor. On the canyon's far wall, one landslide seems to have ridden over a previous one. This photograph shows a 70 km × 150 km area.

Figure 15–13 This oblique view from the Viking 1 orbiter shows Argyre, the smooth plain at left center. The Martian atmosphere was unusually clear when this photograph was taken, and craters can be seen nearly to the horizon. The brightness of the horizon results mainly from a thin haze. Detached layers of haze can be seen to extend from 25 to 40 km above the horizon, and may be crystals of carbon dioxide.

Color Plate 35). Viking 1 orbited Mars for a month while it radioed back pictures to scientists on Earth who were trying to determine a relatively safe spot to land.

The orbiter part of the mission alone was extremely successful in that it sent back pictures not only confirming but also greatly extending the earlier observations. The effects of flows of surface material that involved the past presence of water on or in the Martian surface was also revealed (Fig. 15–11). The sinuous river channels (p. 344) showed clearly. Observations were obtained at higher resolution than that provided by Mariner 9 of the giant volcanoes (Fig. 15–12A and Color Plate 34) and of the huge 5000-km-long valley (Fig. 15–12B and the cover). Even the atmosphere could be observed (Fig. 15–13).

These detailed views of Mars allow us to better interpret the similarities and differences that this planet of extremes – such as huge canyons and gigantic volcanoes – has with respect to the Earth. For example, Mars has exceedingly large, gently sloping volcanoes but no signs of the long mountain ranges or the equivalent of the deep mid-ocean ridges that on Earth tell us that plate tectonics has been and is taking place. All the volcanoes on Mars are of a type called "shield volcanoes" that have gently sloping sides, similar to Kilimanjaro and Mauna Kea (Fig. 14–1) on Earth. The slopes are

A

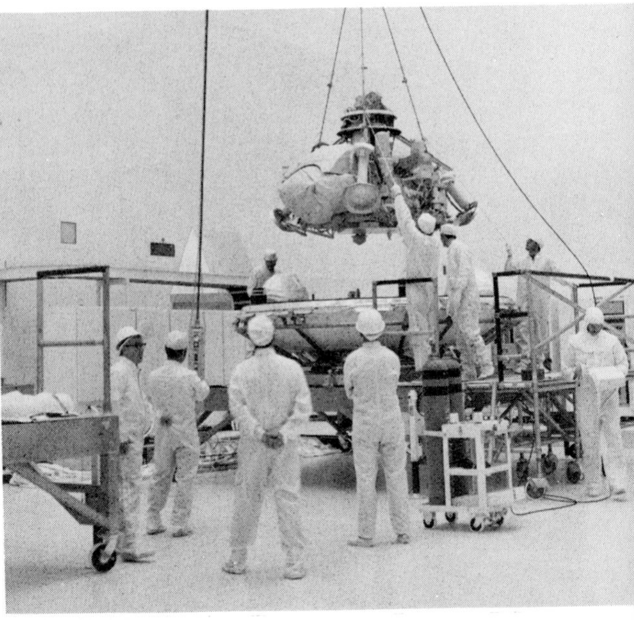

B

Figure 15–14 (*A*) The Viking 1 lander before launch. (*B*) The Viking 2 lander being inspected at Cape Kennedy before launch.

gradual because the lava spread rapidly. On Earth, we also have steep-sided volcanoes, which occur where the continental plates are overlapping, as in the Aleutian islands or for Mount Fujiyama. No Martian volcanoes are of this type.

Perhaps the volcanic features on Mars are so huge because continental drift is absent there. If molten rock flowing upward causes volcanoes to form, as is thought to be the cause, on Mars the features just get bigger and bigger rather than move away from the underlying source.

The first choices for a landing site, made long in advance based on Mariner 9 pictures, had to be rejected when Viking's pictures revealed dangerous boulders or craters that could overturn the spacecraft. But finally a site was selected, and on July 20, 1976, exactly seven years after the first manned landing on the Moon, Viking 1's lander (Fig. 15–14) descended safely onto a plain called Chryse.

The views showed rocks of several kinds (Color Plate 37), apparently covered with reddish material that is probably limonite, an iron oxide compound. Sand dunes were also visible (Fig. 15–15). The sky on Mars turns out to be pink, from reddish dust suspended in the air as a result of one of Mars's frequent dust storms.

A series of experiments aboard the lander was designed to search for signs of life. A long arm was deployed (Color Plate 36) and a shovel at its end dug up a bit of the Martian surface (Fig. 15–16). The soil was dumped into three experiments that searched for such signs of life as respiration and

Figure 15–15 This 100° view of the Martian surface was taken from the Viking 1 lander, looking northeast at left and southeast at right. It shows a dune field with features similar to many seen in the deserts of Earth. From the shape of the peaks, it seems that the dunes move from upper left to lower right. The large boulder at the left is about 8 meters from the lander and is 1 × 3 meters in size. The boom that supports Viking's weather station cuts through the center of the picture. ("Chance of precipitation," the local newscaster would say, "0 per cent.")

metabolism. The results were astonishing. The biological experiments sent back signals that were similar to those that would be caused on Earth by biological rather than by mere chemical processes. Still, scientists searched hard for non-biological explanations. It is possible that some strange chemical process mimics life in these experiments.

One important experiment was much more negative in evaluating the chance that there is life on Mars. It analyzed the soil and looked for traces of organic compounds. On Earth, many organic compounds left over from dead forms of life remain in the soil; the life forms themselves are only a tiny fraction of the organic material. On Mars, who knows? Perhaps life forms evolved that efficiently used up their predecessors. Still, the absence of organic material from the Martian soil is a strong argument against the presence of life on Mars.

Anyway, we mustn't always ask for a yes or no answer. Even if the life signs detected by Viking come from chemical rather than biological processes, we have still learned of some fascinating new chemistry going on. When life arose on Earth, it probably took up chemical processes that had been previously in existence. Similarly, if life arose on Mars in the past or would normally arise there in the future (assuming our visiting there doesn't contaminate Mars and ruin the chances for the beginning of indigenous life), we might expect the life forms to use the chemical processes that already existed. So even if we haven't detected life itself, we may well have learned important things about its origin.

Figure 15–16 The first trench, 7 cm wide × 5 cm deep × 15 cm long, excavated by Viking 1 to collect Martian soil for the life-detection experiments. From the fact that the walls have not slumped, we can conclude that the soil is as cohesive as wet sand.

The Viking 2 lander descended on September 3, 1976, on a site that was thought to be much more favorable to possible life forms than was the Viking 1 site. Again, Viking 2 found a variety of types of rocks (Fig. 15–17 and Color Plate 38) and a pink sky. The atmosphere at this second site, Utopia Planitia (Fig. 15–18), contains three or four times more water vapor than has been observed near Chryse. As of this writing, Viking 2's experiments were sending back data similar to Viking 1's, with positive results for the biological experiments and negative ones from the chemical ones.

Viking 2 had the additional scientific value that its seismograph could measure marsquakes. The seismograph on Viking 1 had been the only instrument on that craft that did not work on the Martian surface. The time-delay information that would have been afforded by two working seismographs has been lost, but still the seismic information from Viking 2 should be exciting.

The presence of two orbiters circling Mars allowed a division of duties. One continued to relay data from the two Viking landers to Earth, while the orbit of the other was changed so that it passed over Mars's north pole. The main polar cap, made of dry ice, had retreated as a result of summertime and only a residual ice cap was left. From the amount of water vapor detected with its spectrograph, and from the fact that the temperature was too high for this residual ice cap to be frozen carbon dioxide, the conclusion has been reached that it is made of water ice. This important

Figure 15–17 A panoramic view covering 90 per cent of the scene from the Viking 2 lander. The rocks range in size up to several meters. Many have pitted surfaces, resembling lava, while others appear smooth. Some of the rocks appear to have grooves that may have been cut by the impact of windborne sand and dust grains. The dome at right is the cover of the lander's radioactive power generator. A radio antenna is at the extreme right.

Figure 15–18 The first photograph of the surface of Mars taken from Viking 2 on September 3, 1976. A wide variety of rocks are seen to lie on a surface of fine-grained material. One of the lander's footpads can be seen at the lower right.

result indicates the probable presence of lots of water on the Martian surface in the past—in accordance with the existence of the stream beds—and seems to raise the probability that life has existed or does exist on Mars.

What's the next step? There are no definite plans, but the Viking missions have been so spectacularly successful that it seems reasonable that further exploration of Mars will be given priority. Carl Sagan, a Cornell astronomer who not only has worked with the Viking observations but also has become the foremost spokesman to the public about the project and its results, points out the frustration at not being able to see over the horizon. He asks for a roving vehicle that could go over many kilometers of the Martian surface. Enough spare parts exist from Viking that this could be done at a cost that is reasonable for such complex space experiments. So perhaps we can hope for a Martian rover in the 1980's. Manned space flights to Mars would be much more expensive, and at present seem much farther off.

An unmanned mission to bring soil back to Earth could be launched in the late 1980's if we start planning now.

The more we learn about it, the more interesting Mars becomes.

15.4 SATELLITES OF MARS

Mars has two moons. In a sense, these are the first moons we have met in our study of the solar system, for Mercury and Venus have no moons at all, and the Earth's Moon is so large relative to its parent that we may consider the Earth and Moon as a double planet system. The moons of Mars, Phobos (from the Greek for Fear) and Deimos (from the Greek for Panic), are mere chunks of rock, only 27 and 15 kilometers across, respectively. (In Greek mythology, Phobos and Deimos were companions of Ares, the equivalent of the Roman war god, Mars.)

Phobos and Deimos are very minor satellites. They revolve very rapidly, Phobos only 6000 and Deimos only 20,000 km above the surface. Phobos completes an orbit of Mars in only 7 hours and 40 minutes, more rapidly than a Martian day, which is, as we have said, about the same length as an Earth day. Deimos orbits in about 30 hours, slower than Mars's rotation period, just as our Moon revolves more slowly than the Earth's rotation period. An observer on the surface of Mars would see Phobos moving conspicuously backwards from the east-to-west direction that Deimos, the other planets, and the stars would move across the sky.

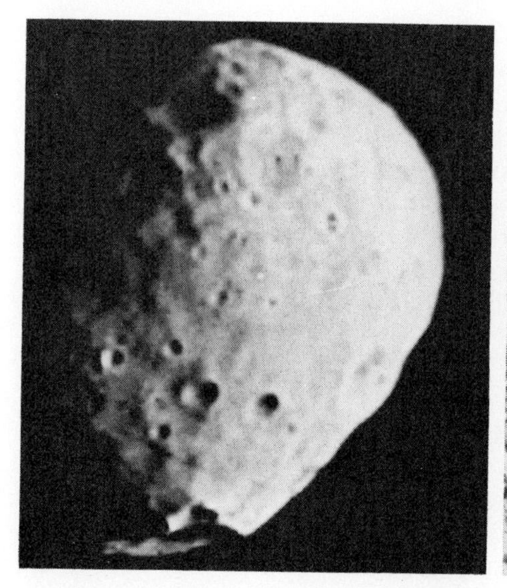

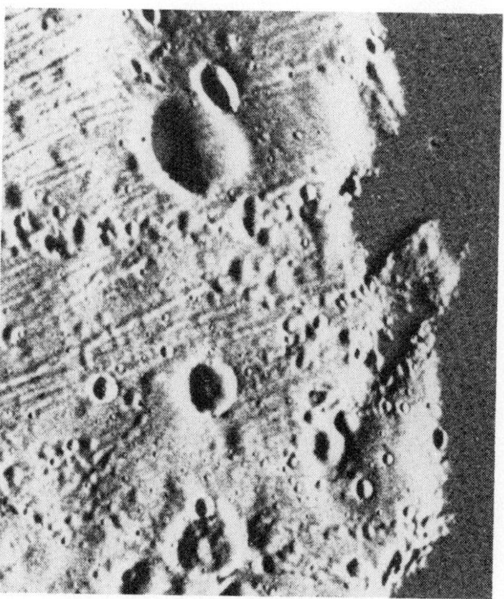

Figure 15–19 (*A*) A Viking 1 photograph of Phobos, showing a heavily cratered side that was not viewed by Mariner 9. The largest two craters on Phobos have been named after Asaph Hall and after Angelina Stickney, his wife, who encouraged the search. (*B*) A close-up view of Phobos from Viking Orbiter 2. Phobos' north pole is near upper left. It is heavily cratered as was expected but, surprisingly, shows striations and chains of small craters. Similar chains of small craters on Earth's Moon, on Mars, and on Mercury were formed by secondary cratering from a larger impact.

It is amusing to note that in 1727 Jonathan Swift invented two moons of Mars for *Gulliver's Travels,* long before the moons were discovered. This is widely considered to be a lucky guess, but it was a reasonable guess at that time, for numerological reasons. It may even have dated back to Kepler, who had reasoned in 1610 that the planets interior to the Earth, Mercury and Venus, had no known moons, the Earth had one, while Jupiter, the next planet out from Mars, had 4 known moons. Two is an intermediate value. Both Swift's moons and the real ones are very small and revolve very quickly. The two non-fictional moons were discovered in 1877 by Asaph Hall of the U.S. Naval Observatory in Washington, D.C.

Mariner 9 and the Vikings made closeup photographs and studies of Phobos and Deimos (Figs. 15–19 and 15–20). Each turns out to be not at all like our Moon, which is a spherical, planet-like body. Phobos and Deimos are just cratered chunks of rock. Chunks may have been broken off on impact with meteoroids.

Phobos, which is only about 27 km in its longest dimension and 19 km in its shortest, has a crater on it that is 8 kilometers across, a large fraction of Phobos's circumference. Deimos is even smaller than Phobos, only around 15 km in its longest dimension and 11 km in its shortest. Now that we know

Figure 15–20 This photograph of Deimos was obtained by Viking Orbiter 1 from a range of 3300 km. About half of the side facing the camera is illuminated, and that lighted portion measures about 12 × 8 km. While Mariner 9 pictures taken nearly five years ago from greater distances showed only a few large craters on Deimos, the improved resolution here reveals a heavily-cratered and presumably very old surface. At least a dozen craters are prominent—the two largest measuring 1.3 km and 1 km. Craters as small as 100 meters pockmark the surface.

their sizes, we can calculate their albedos; Phobos and Deimos turn out to be about as dark as the darkest maria on the Moon. From more detailed but similar comparisons of the variation of the albedos over the spectrum, we deduce that they may be made of the same kind of basaltic rock that the Moon is made of, though other possibilities also exist.

We can conclude that the moons are fairly old because they have been around long enough for the surface to be extensively cratered. The rotation of both moons is gravitationally linked to Mars with the same side always facing the planet, just as our Moon is linked to Earth.

SUMMARY AND OUTLINE OF CHAPTER 15

Aims: To study Mars, and to assess the chances of finding life there

- Observations from the Earth (Section 15.1)
  - Rotation period
  - Surface markings
  - Dust storms
  - Seasonal changes
- Space observations
  - Many space probes sent from the U.S. and U.S.S.R.
  - Mariner 9 orbiter mapped all of Mars (Section 15.2)
  - Volcanic regions indicating geological activity
  - Canyons larger than Earth counterparts
  - Craters and terraced areas near the poles
  - Signs of water evident
  - Underlying layer of polar caps
  - Sinuous channels, some with tributaries
  - Channel bottoms show markings similar to terrestrial stream beds
  - Viking (Section 15.3)
  - Orbiters and landers
  - Closeup photographs of surface rocks
  - Analysis of soil
    - Both positive and negative results for tests for indigenous life
  - Seismograph to study marsquakes
  - Residual polar cap is made of water ice
- Satellites of Mars (Section 15.4)
  - Phobos and Deimos: small, irregular chunks of rock
  - Close-up photographs from Mariner 9 and Viking

QUESTIONS

1. Outline the features of Mars that make scientists think that it is a good place to search for life.

2. If Mars were closer to the Sun, would you expect its atmosphere to be more or less dense than it is now? Explain.

3. Would you expect the difference between a solar and a sidereal day on Mars to be more, less, or the same as on Earth? Explain.

4. From the relative masses and radii (see Appendix 5), verify that the density of Mars is about 85 per cent that of Earth.

5. Approximately how much more solar radiation strikes a square meter of the Earth than a square meter of Mars?

6. As viewed from the Earth, the Sun has an angular diameter of half a degree. How large does it appear from Mars?

7. Compare the tallest volcanoes on Earth and Mars relative to the diameters of the planets.

8. Describe the importance of radioactive material in the evolution of a planet. How does this apply to Mars?

9. What evidence is there that there is, or has been, water on Mars?

10. Compare the fractional abundances of various major gases in the atmospheres of Mars and the Earth.

11. How does the absolute abundance of carbon dioxide in the atmosphere of Mars, that is, the total amount of carbon dioxide, compare with the absolute abundance of carbon dioxide in the Earth's atmosphere.

12. List the various techniques for determining the composition of the atmosphere of Mars.

13. List the evidence from Viking for and against the existence of life on Mars.

14. (a) Aside from the biology experiments, list three types of observations made from the Viking landers. (b) What are two types of observations made from the Viking orbiters?

15. Why aren't Phobos and Deimos regular, round objects like the Earth's Moon?

TOPICS FOR DISCUSSION

1. Discuss the problems of sterilizing spacecraft, and whether we should risk contaminating Mars by landing spacecraft.

2. Discuss the additional problems of contamination that manned exploration would bring. Analyze whether, in this context, we should proceed with manned exploration in the next decade or two.

16

Jupiter

The planets beyond the asteroid belt—Jupiter, Saturn, Uranus, and Neptune—are very different from the four "terrestrial" planets. These *giant planets,* or *Jovian planets,* not only are much bigger and more massive, but are also less dense. (Since density is mass divided by volume, they are less dense because there is a proportionally larger increase in volume over the volumes of the terrestrial planets than there is an increase in mass.) This suggests that the internal structure of these giant planets is entirely different from that of the four terrestrial planets. Jupiter and Saturn in turn are more similar to each other than they are to Uranus and Neptune. The ninth planet, Pluto (to be discussed in Section 17.3), represents still another different type, and may not even be an "original" planet but rather an escaped moon of one of the giant planets.

Jupiter, also called Jove, was the chief Roman deity.

16.1 FUNDAMENTAL PROPERTIES

The largest planet, Jupiter, dominates the Sun's planetary system. Jupiter is 5 A.U. from the Sun and revolves once every 12 years. It alone contains two-thirds of the mass in the solar system outside of the Sun, 318 times as much mass as the Earth. Jupiter has at least fourteen moons of its own and is a miniature planetary system in itself. It is often seen as a bright object in our night sky, and observations with even a small telescope reveal bands of clouds across its surface and show four of its moons. Jupiter's albedo is very high: 51 per cent. Jupiter is so large that even at its great distance from us it can subtend an angle of up to 47 seconds of arc.

Jupiter's mass can be readily measured by following the orbits of its satellites and using Kepler's third law. Although it has 318 times the mass of Earth, it has only 1/1000 the mass of the Sun. (This is too little mass for nuclear fusion to have begun.)

The north pole of Jupiter, photographed from the Pioneer 11 spacecraft. The Great Red Spot is also visible. The symbol for Jupiter appears at the top of the page.

Figure 16–1 Jupiter, photographed from the Earth, shows belts and zones of different shades and colors. By convention (that is, the set of terms we arbitrarily choose to use) the bright horizontal bands are called *zones* and the dark horizontal bands are called *belts.*

Jupiter is more than 11 times greater in diameter than the Earth, so its volume is $(11+)^3$ or 1400 times greater than that of our planet. Thus its density is calculated to be 1.3 grams/cm³, not much greater than the 1 gram/cm³ density of water. This tells us that any core of heavy elements (such as iron) that Jupiter may have does not make up as substantial a fraction of Jupiter's mass as the cores of the inner planets make up of their planets' masses. Jupiter, rather, is mainly composed of the lighter elements hydrogen and helium, sufficiently compressed to reach this density.

The bands on the surface of Jupiter appear in subtle shades of orange, brown, gray, yellow, cream, and light blue, and are beautiful to see (Fig. 16–1). They are in constant turmoil; the shapes and distribution of bands continually change in a matter of days. For example, in 1971, one area of the equatorial belt seemed to erupt. A brightening of clouds occurred, and spread to cover a large part of the equatorial belt. This kind of eruption had been seen previously in 1919, 1928, and 1943.

Jupiter rotates very rapidly, once every 10 hours. Undoubtedly, this rapid spin rate is a major reason for the colorful bands, which are clouds spread out parallel to the equator. Interestingly, the clouds do not rotate as would the surface of a solid body: the clouds at the equator rotate slightly faster than the clouds nearer the poles, completing their rotation about five minutes earlier each time around. Thus every fifty days, the clouds at the equator have made an additional rotation. (The Sun also shows differential rotation, as described in Section 5.6.) Not only the uppermost layers of Jupiter's clouds but also much deeper levels probably rotate at differing velocities. Jupiter isn't solid; it probably has no crustal surface at all. Its gas just gets denser and denser, eventually liquefying, at deeper and deeper levels.

Figure 16–2 Jupiter's differential rotation, which causes the points that were aligned at time T_1 to be separated at time T_2, spreads out the clouds into bands.

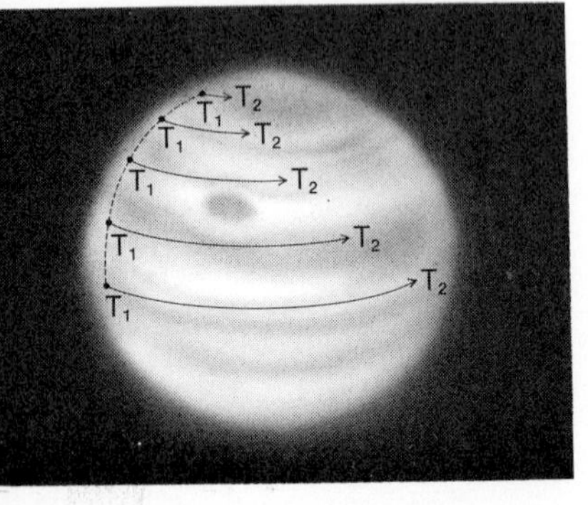

The rapid rotation makes the planet bulge at the equator. The distance from pole to pole is substantially less—7 per cent—than the equatorial diameter; we say that Jupiter is *oblate* (Fig. 16–3). (If the polar diameter were longer rather than shorter, the planet's shape resembling that of a cigar, it would be called *prolate.*) The axis of rotation is only 3° from the axis of Jupiter's orbital revolution around the Sun, so that, unlike the Earth and Mars, Jupiter has no seasons.

The most prominent feature of the visible cloud surface of Jupiter is a large reddish oval known as the *Great Red Spot* (see Fig. 16–1 and Color Plate 38). It is about 14,000 km × 30,000 km, many times larger than the Earth, and drifts about slowly with respect to the clouds as the planet rotates. The Great Red Spot is a relatively stable feature, for it has been visible for at least 150 years, and may be the same as a feature that was reported by Cassini in France and possibly Hooke in England 300 years ago. Sometimes it is relatively prominent, and at other times the color may even disappear for a few years. The Great Red Spot also changes in shape.

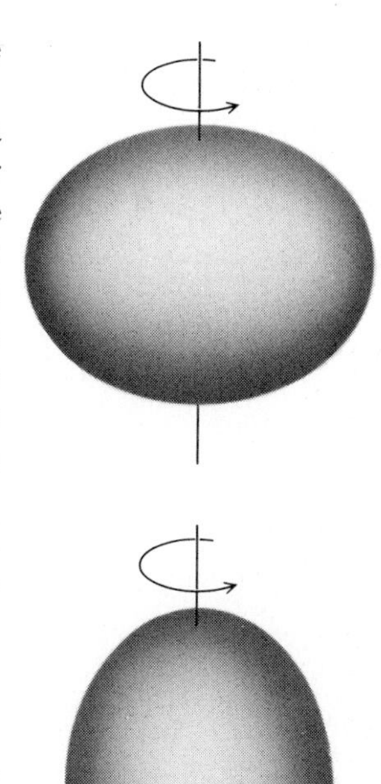

Figure 16–3 The top ellipsoid is *oblate;* the bottom ellipsoid is *prolate.*

There has been endless speculation as to the cause of the Great Red Spot, and recent investigations from space have narrowed down the possibilities considerably, as we shall see.

From the study of spectra of Jupiter taken from the Earth, we have long known that the atmosphere contains some ammonia (NH_3, a molecule with one nitrogen atom bound to three hydrogen atoms) and methane (CH_4, one carbon atom bound to four hydrogen atoms). The presence of these chemical compounds with their high hydrogen content and of molecular hydrogen as well, together with evidence of the low density, makes it clear that Jupiter is composed mainly of hydrogen. Its chemical composition is closer to that of the Sun and other stars than it is to that of the Earth. Jupiter's atmosphere consists of almost 85 per cent hydrogen molecules and 15 per cent helium atoms, with a remaining one per cent consisting of methane, ammonia, and several other molecules (Fig. 16–4). The composition of Jupiter's atmosphere is of particular interest because scientists believe the hydrogen and helium to be present in relative abundance comparable to the relative abundance that existed as a result of the big bang that began the Universe and of later element-forming processes in stars, which are discussed in Chapter 25 on cosmology.

Jovian is the adjectival form of Jupiter.

One can also study the Jovian atmosphere by measuring the characteristics of the light that we get from Jupiter, and comparing this measurement with laboratory measurements of the characteristics of light reflected off the surfaces of different kinds of dust particles and of ices. It seems reasonable to assume that many of the cloud colors are caused by reflection of light off dust and ices in suspension in the atmosphere.

Twenty years ago, in 1955, intense bursts of radio radiation were discovered to be coming from Jupiter. Though it was quite a surprise to detect any radio signals at all, several kinds of signals were detected. Some signals seemed to come from a particular location above Jupiter. Some sharp bursts seemed to come from the atmosphere, similar to the radiation we get from lightning discharges here on Earth. It has been discovered that the bursts of radio emission are correlated with passages of Jupiter's innermost Galilean moon, Io, above the regions from which the bursts are coming. At shorter radio wavelengths, Jupiter emits continuous radiation.

Figure 16–4 The composition of Jupiter.

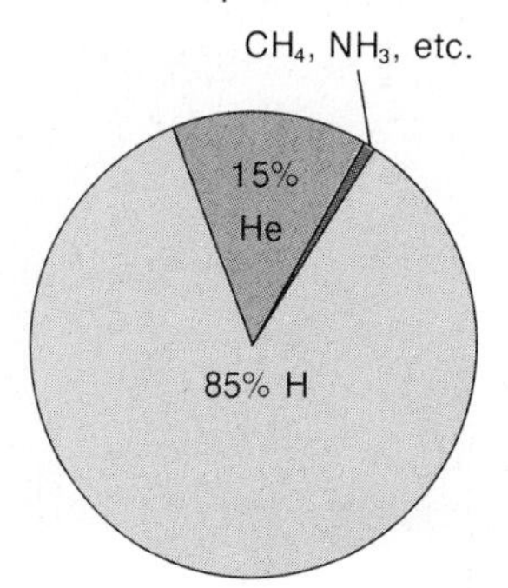

The fact that Jupiter emits radio waves indicated that Jupiter, even more so than the Earth, has a strong magnetic field and a strong *radiation belt* (actually, a belt filled with magnetic field in which particles are trapped, a large-scale version of the Van Allen belts of Earth), because we can account for the presence of the radio radiation only with mechanisms that involve such magnetic fields and belts. Some of the radio emission comes from interactions of high-energy particles passing through space with Jupiter's magnetic field.

16.2 JUPITER'S MOONS

Jupiter has fourteen satellites. Four of the innermost satellites are very nicely named: Io, Europa, Ganymede, and Callisto. At 5216 km in diameter, Ganymede is one of the biggest satellites in the solar system (Saturn's Titan is larger), and has a strong gravitational field.

These four moons are called the *Galilean satellites* (Fig. 16–5). They were discovered by Galileo in 1610 when he first looked at Jupiter with his telescope, and they have played a very important role in the history of astronomy. In particular, it was the fact that these particular satellites were noticed to be going around another planet, like a solar system in miniature, that supported Copernicus's heliocentric model of our solar system. Not everything revolved around the Earth! And in the 17th century the moons were used to measure the speed of light. In 1675, Olaus Roemer in Denmark noticed that the moons reached their predicted positions later than expected when Jupiter was on the far side of the sun and earlier when it was on the near side. He attributed this difference to a finite speed of light and derived a value for the speed of light close to the currently accepted value.

The largest moon, Ganymede, has been thought to have a thin atmosphere since it occulted a faint star one night in 1972. The occultation of this eighth-magnitude star was observed at observatories in Indonesia and India. The star's light dimmed gradually, rather than abruptly as it would have if it was merely being hidden behind Ganymede's solid surface, though the data were not of high enough quality to allow us to be certain. Io, too, has been found to have a thin atmosphere.

It was not until 1892 that the fifth moon of Jupiter, Amalthea, was discovered, and by 1951, twelve moons were known. These other satellites have been, not very inventively, identified by number instead of by name. In 1976, the International Astronomical Union certified the new names Himalia, Elara, Pasiphae, Sinope, Lysithea, Carme, and Ananke for moons VI through XII.*

*These are all classical allusions. Pasiphae, for example, was the wife of Minos and the mother of the Minotaur. One of her daughters was Phaedra. All the moons except Amalthea are named after lovers of Jupiter; Amalthea was Jupiter's nurse.

Figure 16–5 (*A*) Jupiter, with the four Galilean satellites. These satellites were named Io, Europa, Ganymede, and Callisto by Simon Marius, a German astronomer who independently discovered them. (*B*) Galileo's sketches of the changes in the positions of the satellites.

A

B

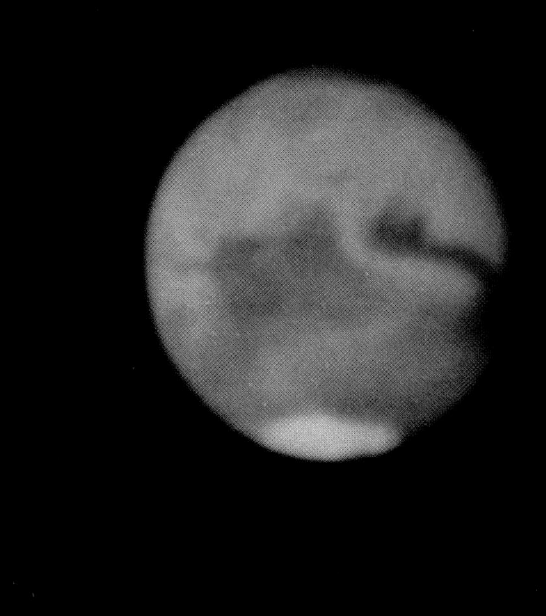

Color Plate 31 (left): Earth photographed from Apollo 11. Africa and the Middle East are clearly visible. (NASA photo)

Color Plate 32 (right): Mars photographed from Earth as part of the International Planetary Patrol.

Color Plate 33: An aurora borealis photographed in the Goldstream Valley, Alaska. (Gustav Lamprecht photo)

Color Plate 34: Mars, photographed from Viking 1 on June 17, 1976 when the spacecraft was at a distance of 560,000 km from the planet. Olympus Mons, the large volcano, is toward the top of the picture. The Tharsis Mountains, a row of 3 other volcanoes, are also visible. To the left of these volcanoes, the irregular white area may be surface frost or ground fog. The large impact basin, Argyre, is the circular feature at the bottom of the disk. (NASA photo)

Color Plate 35 (below): Viking 1's sampler scoop is in the foreground of this view from the Martian surface. Angular rocks of various types can be seen. Large blocks one to two meters across can be seen on the horizon, which is about 100 meters from the spacecraft. The horizon may be the rim of a crater. (NASA photo)

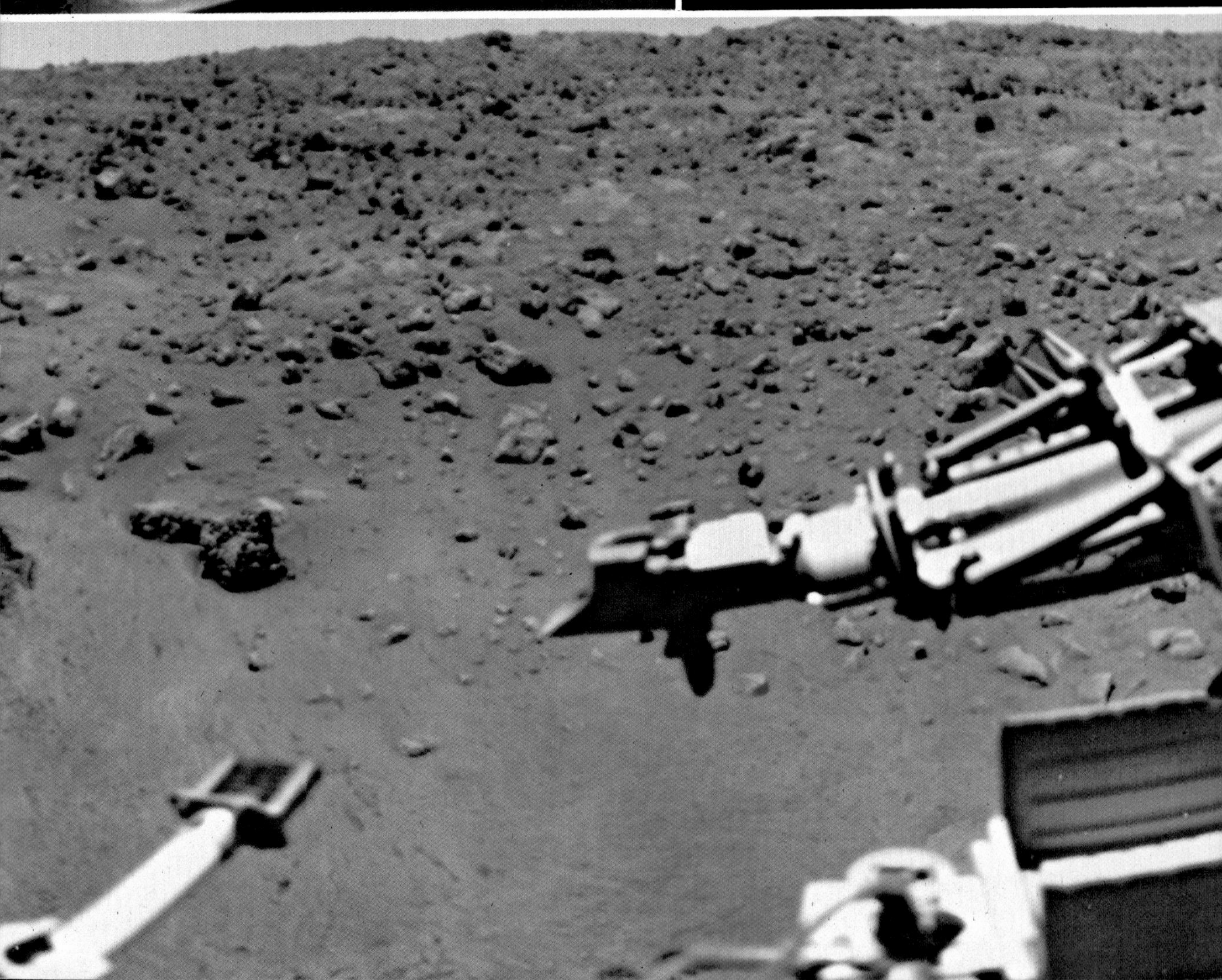

Color Plate 36: Mars's Chryse Planitia as seen by Viking 1. The surface material is probably limonite (hydrated ferric oxide). Reddish dust suspended in the atmosphere makes the sky pink. (NASA photo)

Color Plate 37: Mars's Utopia Planitia seen by Viking 2. Many of the rocks are porous and sponge-like, similar to some of Earth's volcanic rocks. (NASA photo)

Color Plate 38: Jupiter as observed by Pioneer 10. Bands, zones, and the Great Red Spot show clearly. (NASA photo)

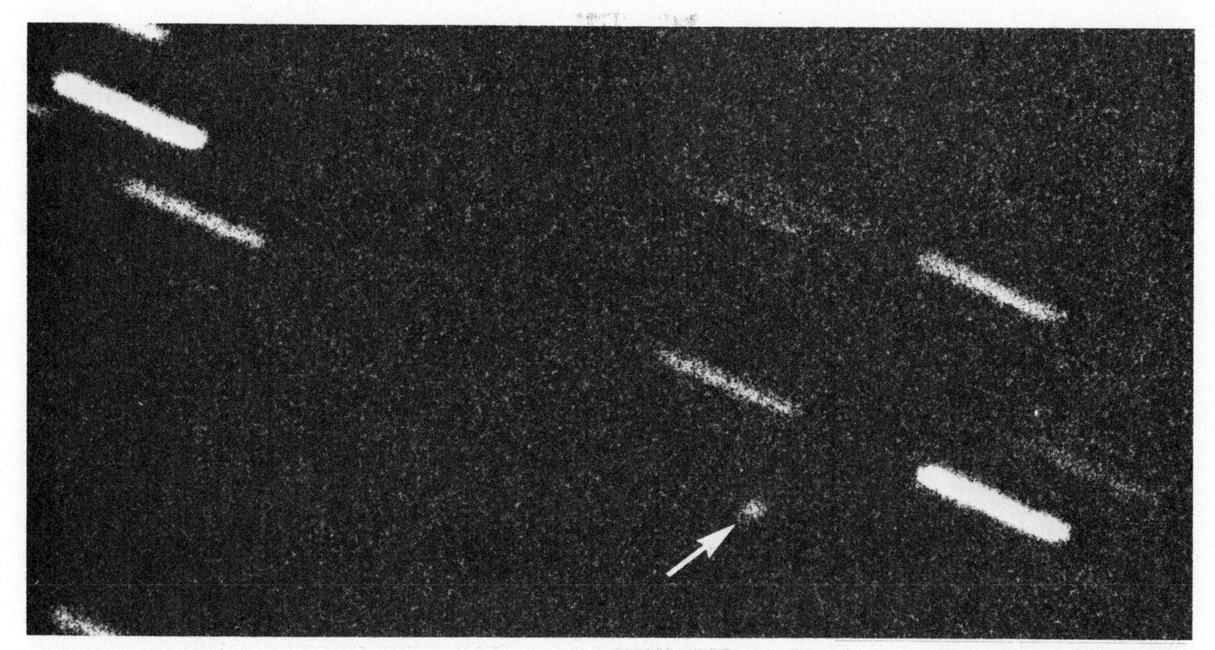

Figure 16–6 The photograph on which the thirteenth moon of Jupiter, shown with an arrow, was discovered. The moon is named Leda. The telescope was set to track across the sky along with Jupiter; thus stars show as trails. The trails here correspond to anonymous—unnamed and unnumbered—stars between about 14th and 20th magnitudes.

These twelve moons of Jupiter fall into three groups. The first group includes the innermost five satellites (Amalthea and the Galilean satellites), which are also the largest bodies. Their orbits are all less than 2 million kilometers in radius. The orbits of the three moons in the second group are all about 12 million kilometers in radius. The orbits of the four outermost moons, which are in the third group, are 21 to 24 million kilometers in radius. These four outermost moons revolve in the direction opposite from that of Jupiter's rotation (that is, in the retrograde direction), while the other eight moons revolve prograde. Moons in the outer two groups may be bodies captured by Jupiter's gravity after their formation, for they have high inclinations and/or eccentricities.

Recently, in 1974, Charles Kowal was studying the outermost Jovian moons with the Schmidt telescope at Palomar. He was trying to detect moons all the way down to about 22nd magnitude, a level of intensity fainter than had been previously studied. Kowal tracked the telescope on Jupiter; thus stars appear as trails on the photograph, since Jupiter is moving with respect to the positions of the stars. Any moon would appear as a stationary dot, since it would be moving at approximately the same speed as Jupiter. Sure enough, besides the known moons there was an extra dot: the thirteenth Jovian satellite (Fig. 16–6). In 1975, Kowal found a fourteenth.

The thirteenth moon, which Kowal named Leda after a mistress of Jupiter, is a very dim object and is only about 21st magnitude. On the basis of its magnitude it is estimated that the moon is only 3 to 5 km in diameter. Its orbit was calculated on the basis of 10 different observations made both at Palomar and at the Steward Observatory on Kitt Peak. It joins 3 of the previously known moons in the middle group of orbits. The fourteenth moon is similarly small and even fainter. Its orbit must be determined before it is named, as a new I.A.U. convention requires that all new discoveries of satellites of Jupiter be given names ending in "a" if they revolve prograde and in "e" if they revolve retrograde. Kowal expects that there is an even chance that he will discover a new faint moon at every opposition of Jupiter for the next few years. These oppositions, of course, occur at intervals of approximately one year. (Jupiter moves only 1/12 of the way around the Sun in a year, while the Earth makes a full revolution.) If a faint moon, brighter than the 22nd magnitude limit, is at or near its maximum separation from Jupiter as seen from the Earth, Kowal should find it. There may actually be a very large number of chunks of rock and dust grains in orbit around Jupiter, so the ultimate question of just how many moons there are may be a matter of semantics.

16.3 SPACECRAFT OBSERVATIONS

We have seen that each of the planets we have discussed so far has recently been visited by spacecraft that have greatly advanced our knowledge about them. Jupiter is no exception. Two spacecraft, Pioneer 10 and Pioneer 11, gave us our first close-up views of the colossal planet in 1973 and 1974. The Pioneers did not go into orbit around Jupiter; in fact they are traveling fast enough to escape the solar system. By 1980, Pioneer 10 will be as far away from the Sun as Uranus is; beyond that, radio signals from the satellite will be too weak for us to detect. But until that time, the spacecraft will beam back valuable information on interplanetary conditions. In about 80,000 years the spacecraft will be about one parsec (about three light years) away from the solar system. A plaque that gives some information about the Pioneer's origin and about life on Earth was affixed to the spacecraft in case some interstellar traveler from another solar system might pick it up (Fig. 16–7).

By studying the Pioneers' measurements of Jupiter's reflected ultraviolet light, scientists have elaborated on earlier measurements made from the Earth of the composition of the Jovian atmosphere.

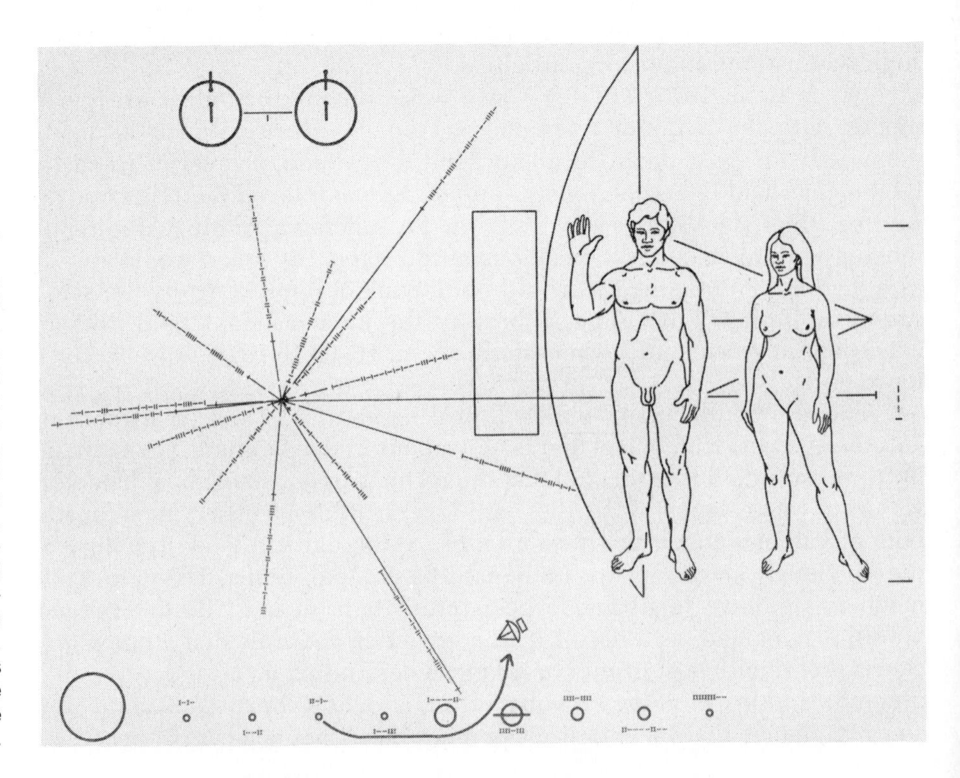

Figure 16–7 The plaques borne by Pioneers 10 and 11. A man and a woman are shown standing in front of an outline of the spacecraft, for scale. The spin-flip of hydrogen, described in Section 21.3, is shown at top left, also for scale, because even travelers from another solar system would know that the wavelength of the transition is 21 cm. The sun and planets, including distinctively ringed Saturn, and the spacecraft's trajectory from Earth, are shown at bottom. The directions and periods of several pulsars are shown at left. Numbers are given in the binary system.

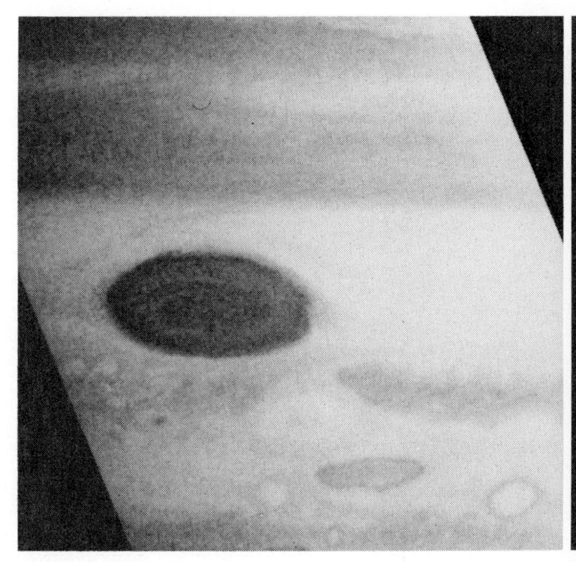

Figure 16–8 Pioneer 11 (*left*) and 10 (*right*) views of the Great Red Spot.

The experiment on Pioneer 10 and 11 that is of the most popular interest is the one in which close-up pictures of Jupiter were beamed back to Earth. The resolution of the best images is at least five times better than the best resolution we can get in photographs taken from the surface of the Earth. (Though the spacecraft came much closer to Jupiter than the telescopes on Earth do—Pioneer 10 came within 130,000 km—the telescopes on board were necessarily much smaller than telescopes we can use on the Earth.) The images were telemetered back to Earth in the form of raster scans. (When an image is scanned from side to side and top to bottom, like a TV picture, the result is called a *raster pattern.*) Detailed computer work on them was able to enhance to a considerable degree the contrast of lighter areas to darker areas.

Images were taken through two colored filters, blue and red. Full color pictures could be plotted by a computer at the University of Arizona, where computer enhancement of the images was accomplished, by adding an appropriate amount of a third color to make the results comparable to Earth-based photographs. The results are truly spectacular (see Color Plate 38).

The Great Red Spot shows very clearly in many of the images (Fig. 16–8). The spot is a gaseous island about 35,000 km across, whose top is about 8 km above the rest of the atmosphere. It is now generally believed that the Spot is the vortex of a violent, long-lasting cyclonic storm.

On Jupiter heat energy flows into the storm from below it, maintaining its energy supply; this does not occur to as great an extent with storms on Earth, though terrestrial hurricanes may share this mechanism. The competing theory that the Red Spot was a column connected to a large solid core was ruled out by our new ideas of the extent of liquid in the core. Many of the alternative theories explaining the Red Spot suggested that the underlying layers differed in density. These explanations have been ruled out by such Pioneer observations as the failure of a gravity-sensing experiment to detect any variations in the density of the Jovian atmosphere or its interior.

The circulation of gas around the Spot is clearly visible on the photographs, and we can even see that the Spot is streamlined in a manner that allows other matter to flow around it. A time-lapse series of photographs seems to show that the Spot itself is in counterclockwise rotation, spun by

Figure 16–9 Many other spots and ovals are visible on this Pioneer 11 photograph of the surface of Jupiter, besides the Great Red Spot.

the differential motion of the cloud bands above and below it. The atmospheric winds can be as fast as 500 km/hour. Other storms as large as the diameter of the Earth are visible in other areas of the photographs (Fig. 16–9), but the Great Red Spot is by far the largest.

There have been several suggestions as to the causes of the color of the Spot. Some of the possibilities are predicated on the idea that the Spot is a giant storm in which material from lower atmospheric layers may be rapidly brought up, changing color in the process because solar radiation acts on it or because it interacts with the water vapor in Jupiter's atmosphere. The water vapor was discovered not from Pioneer, but rather from infrared observations made in 1974 from a jet aircraft flying above most of the Earth's own water vapor. Two hydrocarbons, ethane and acetylene, were also discovered in Jupiter's atmosphere in 1974, as was phosphine (PH_3), one phosphorus plus 3 hydrogen atoms. Since the acetylene is not a stable compound, it must be produced constantly in order to have the abundance we observe today. It may be produced by ultraviolet light acting on methane or by lightning in Jupiter's atmospheric storms. An alternative explanation for the color of the Great Red Spot is provided by the phosphorus, which would condense at the relatively great altitude of the top of the Spot, which is elevated above the top of the cloud deck.

Data from the Pioneers increased our understanding not only of the atmosphere but also of the interior of Jupiter (Fig. 16–10). It appears that most of the interior is in liquid form. The central temperature may be between 13,000 and 35,000 K, with a pressure of 100 million times the pressure of the Earth's atmosphere measured at our sea level. Because of this high pressure, Jupiter might even have an interior composed of ultracompressed hydrogen surrounding a rocky core consisting of 20 Earth masses of iron and silicates. The heavy elements should have sunk to the center, but we have no direct evidence that they are there. The liquid hydrogen is probably in a state called "metallic" because it would conduct heat and electricity; these properties are basic to our normal definition of "metal." (The unusual use of the word "metal" by stellar astronomers to mean all elements heavier than helium is not relevant here.) This metallic region is probably where Jupiter's magnetic field is generated by dynamo action. The metallic region makes up 75 per cent of Jupiter's mass.

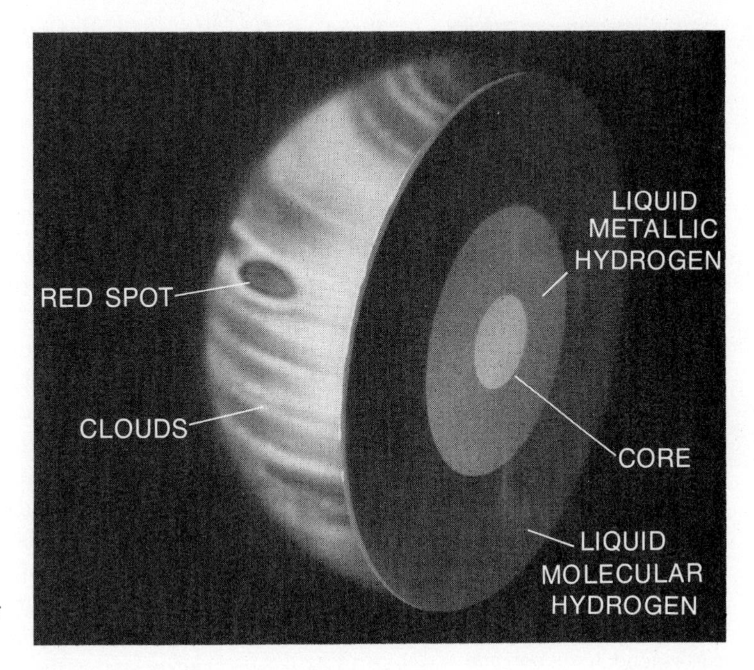

Figure 16–10 The current model of the interior of Jupiter.

Jupiter radiates twice as much heat as it receives from the Sun. No other planet comes even close to doing this. This fact is the key to understanding Jupiter's complex and beautiful cloud circulation pattern. There must be some internal energy source—perhaps the energy remaining from Jupiter's collapse from a primordial gas cloud 20 million km across to a protoplanet 700,000 km across, 5 times the present size of Jupiter. This catastrophic phase of collapse started, it has been suggested, when the temperature grew sufficiently high to break up hydrogen atoms. The rapid phase may have taken only 3 months to occur, following the 70,000 years it had previously taken to shrink from a more diffuse cloud. Jupiter is undoubtedly still contracting. Jupiter lacks the mass necessary, however, to have heated up enough to have become a star. If Jupiter had been about 75 times more massive, nuclear reactions would have been able to start. Then it would have become a star, and we would have been in a double star system (some of which are described in Section 4.1).

The heat emanating from the interior of Jupiter produces huge convection currents. The bright zones are rising currents of gas driven by this convection. The belts are falling gas; the tops of these dark belts are somewhat lower (about 20 km) than the tops of the zones and are about 10 K cooler. More evidence of the thermal activity is seen in a prominent plume that was photographed by the Pioneers (Fig. 16–11). Gas rises thermally at the nucleus of the plume, and is swept back about 30,000 km by the differential rotation of Jupiter. The white ovals seen at higher latitudes (Fig. 16–9) are more evidence of thermal convection. The cloud bands themselves are cyclonic patterns, pulled out to surround the planet by Jupiter's rotation.

Earth-based infrared observations measure temperatures only 100 K to 200 K in the uppermost atmosphere far above Jupiter's clouds. Yet Pioneer's data at other infrared wavelengths (Fig. 16–12) reveal that at a pressure of 1/2 that of the Earth at sea level, the temperature of supposedly frigid Jupiter reaches a boiling 400 K.

Figure 16–11 Belts and zones photographed from Pioneer 11. The dark *belts* are about 20 km lower than the light *zones.* The plume of gas that extends from upper center to upper left photographed by Pioneer 10 indicates a rising current of gas.

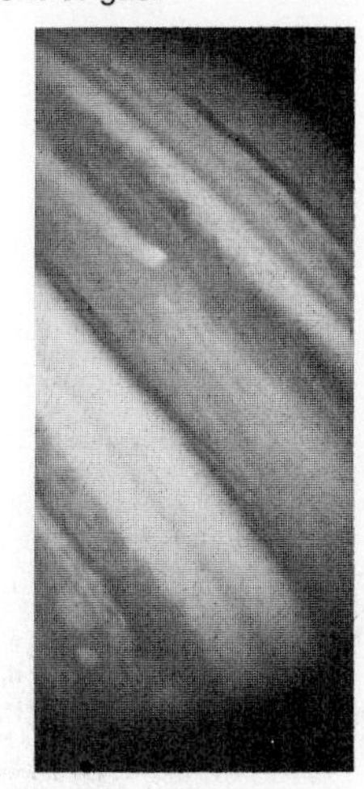

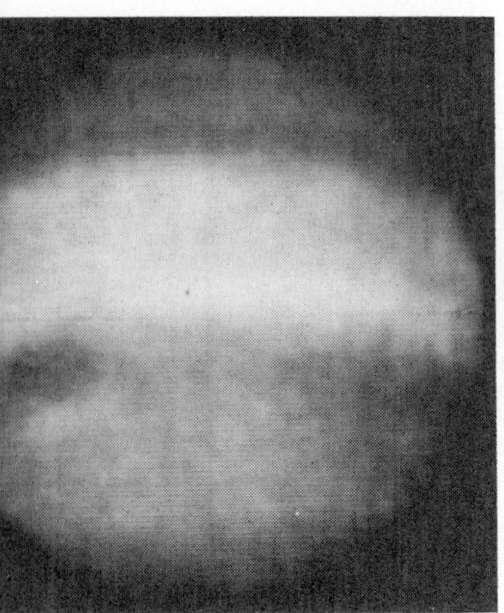

Figure 16–12 Observations at an infrared wavelength of 20 microns, made from Pioneer 11, show the temperature distribution at and below the cloud tops of Jupiter. Lighter areas correspond to hotter and darker areas to cooler regions.

Pioneer 10 took spectacular pictures, but took them only in the plane of Jupiter's rotation, so that the photographs were taken from the same point of view that we have from Earth. But Pioneer 11 took a different approach to Jupiter. It came in over the polar region. It took this path for two reasons. First, when Pioneer 10 swept by Jupiter in late 1973, its instruments narrowly escaped crippling damage during its passage through the intense Jovian radiation belts. Pioneer 11 came three times closer to Jupiter, passing 40,000 km (0.6 Jupiter radii) above the cloud tops. Kepler's second law shows that Pioneer 11 was traveling much faster than Pioneer 10, and so spent less time in the radiation zone. It survived. Another reason it took this path was scientific. We can never see the polar regions from Earth, as Jupiter's axis is nearly perpendicular to our orbit. So scientists were anxious to get their first look at Jupiter's poles (Fig. 16–13). The results were noteworthy; at high latitudes the circulation pattern of cloud bands that we see at the equator is destroyed and the bands break up into eddies. With information like this, we can now study "comparative atmospheres": those of the Earth, Venus, Mars, and Jupiter.

As a result of the Pioneer missions, we now know that Jupiter possesses a tremendous magnetic field, more intense than many scientists had expected (Fig. 16–14). (The existence of this magnetic field was known because of Jupiter's radio emission.) The magnetic field of Jupiter interacts with the solar wind as far as 7,000,000 km from the planet and forms a shock wave, as does the bow of a ship plowing through the ocean. At the height of Jupiter's clouds the magnetic field is 10 times that of the Earth, which itself has a strong field. (We extrapolate this value to the cloud levels from meas-

Figure 16–13 The polar views from Pioneer 11 show that the cloud bands break up into small, turbulent mottles.

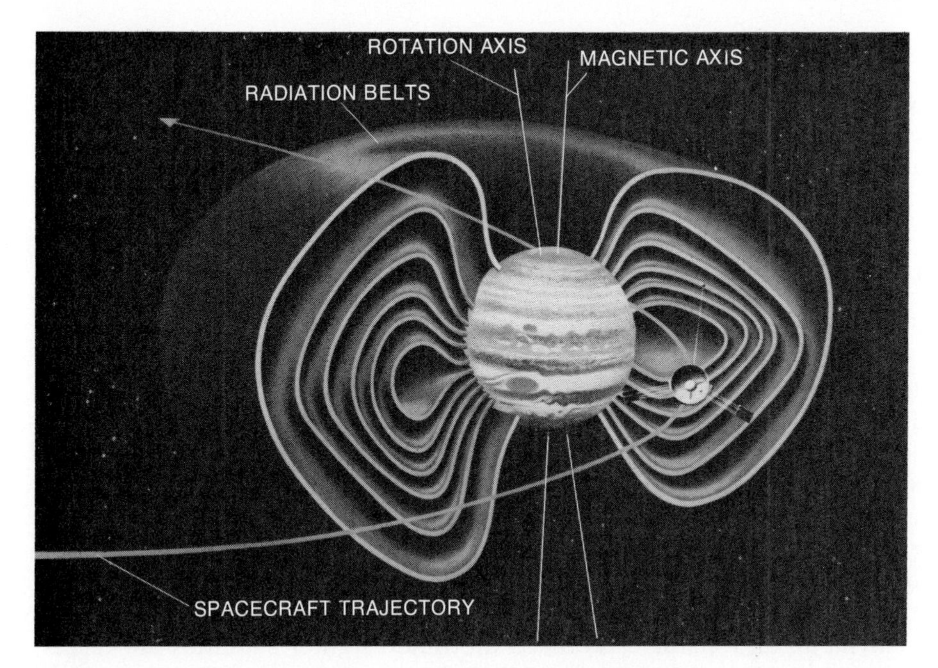

Figure 16–14 Jupiter's magnetosphere, the region of space occupied by the planet's magnetic field. It rotates at hundreds of thousands of miles per hour along with the planet (like a big wheel with Jupiter as the hub). The inner magnetosphere is shaped like a doughnut with the planet in the hole. The highly unstable outer magnetosphere is shaped as though the outer part of the doughnut had been squashed. The outer magnetosphere is "spongy." It pulses in the solar wind like a huge jellyfish, and often shrinks to one-third of its largest size.

urements made at the distance of the spacecraft.) Pioneer 11 observed a 10-hour fluctuation in the magnetic field. It discovered that this fluctuation occurs because the magnetic axis is tilted 10 degrees from the axis of rotation and the center of the field is not exactly at the center of the planet. The magnetic field thus resembles a wobbly disk. The field is toroidal, shaped like a doughnut. Also, Jupiter's magnetic poles are reversed compared with the Earth's: our north pole is on the same side of the plane of the planetary orbits as is Jupiter's south pole. Geiger counters aboard the Pioneers studied the high energy electrons and protons trapped in the inner magnetic belts, which are super-versions of the Earth's Van Allen belts.

At the time that Pioneer 10 reached the distance of Saturn's orbit in 1976, it passed into a zone where the solar wind was absent. This is undoubtedly the effect of a long magnetic tail that Jupiter has. The tail may be a billion kilometers long and results from the stretching of some of Jupiter's magnetic field lines by the solar wind. These field lines prevent the solar wind from entering the tail region, which extends over 400 million miles from Jupiter away from the Sun, 100 times longer than the Earth's magnetic tail. Saturn enters Jupiter's magnetic tail once every 20 years. When it next does so, in April 1981, scientists will try to observe the effects that it might have on Saturn's radiation belts.

All these discoveries, and more, were made with spacecraft that were subject to the most severe environmental conditions and design limitations. Because they had to fight the Sun's gravity in order to travel outward from the Earth to Jupiter, the limits on their weights set by our launching capabilities were very tight. The Pioneers weigh only 270 kg, including only 30 kg of scientific instruments (Fig. 16–15). Because they must function so far from the Sun, they carry nuclear power generators instead of the solar cells that we can use on spacecraft sent to the terrestrial planets.

Figure 16–15 An artist's conception of the Pioneer spacecraft.

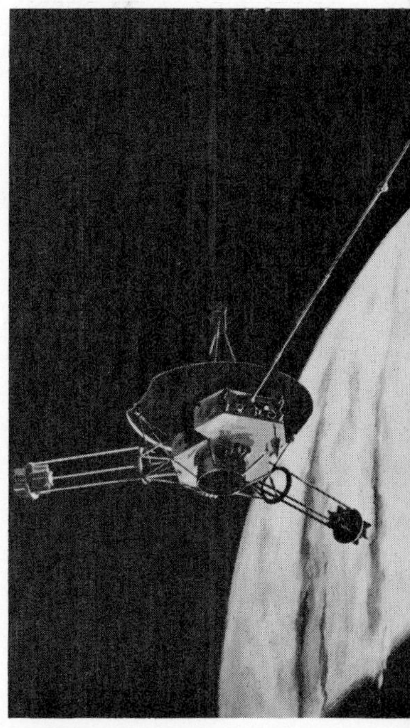

The spacecraft had to be especially reliable because of the long time duration of their journeys. And they are on their own to a large extent during the most crucial moments of their Jupiter flyby because they are then so far away that radio commands take about 45 minutes to travel from the spacecraft to Earth. Thus even if something went wrong, it would take at least 45 minutes for this information to reach the flight controllers at the Jet Propulsion Laboratory in Pasadena, and an additional 45 minutes for any new instructions to travel back to the spacecraft, at least an hour and a half in all.

Even detecting the signal from the Pioneers is a good trick. The amount of energy collected from a Pioneer by JPL's 63-meter radio telescope is so small that it would have to be collected for 20 billion years in order to light a Christmas tree bulb for just one second!

16.4 RECENT OBSERVATIONS OF JUPITER'S MOONS

One of the side benefits of sending a probe to Jupiter is that merely tracking the rocket's trajectory tells us about the gravitational influences on it. Thus new improved values have been derived for the mass of the Galilean satellites, revising the old values by as much as 22 per cent. Compared with our own Moon, Io has 1.22 times its mass, Europa 0.67 times, Ganymede 2.02 times, and Callisto 1.44 times. Callisto is the same size as the planet Mercury, and Ganymede is even larger. The density of these Galilean satellites drops with increasing distance from Jupiter, from 3 to 3½ times that of water for Io and Europa, to 1½ to 2 times that of water for Ganymede and Callisto. Perhaps the outer, lighter satellites are largely water ice, which cannot be formed on the innermost satellites because the high temperature that Jupiter would have had soon after its formation would have made temperatures at their distances too great. The resulting water vapor would have escaped from them into space, and these innermost moons would now be mainly rock.

All these Galilean satellites have been photographed and otherwise studied from the Pioneers. The four satellites all have surface temperatures of only 130 K to 155 K. From Pioneer 10, Ganymede showed a bright polar feature that may be a polar cap, plus other features that are probably craters and maria (Fig. 16–16). We are limited by the resolution of the photograph, about 400 km on Ganymede. Results from Earth-based radars are also consistent with Ganymede having a rocky surface perhaps covered with ice or ice rubble. From Pioneer 11, a feature that is probably a polar cap was photographed on Callisto as well. Ganymede and Callisto have densities too low, however, to allow them to be solid rock. Ice and water may be mixed in.

The most interesting moon may be Io, which is in any case thought to be the trigger of Jupiter's radio bursts. Measurements of Io's ionosphere were made by following the radio signal from Pioneer 10 as it was occulted by Io.

Io's albedo is very high. An ice cover would explain this, but no absorption lines typical of ice have been detected. The sodium D lines have been detected, on the other hand, so some astronomers think that Io may be covered with salt, probably a more complex type of salt than we use at our dinner tables. We think that the salts may have been deposited after being concentrated in water; both water and salt would have come from the interior. At present, we cannot definitely say what is covering Io. Europa, also closer to Jupiter than Ganymede and Callisto, has a similarly high albedo.

Figure 16–16 Pioneer 10 observations of Ganymede. The three marked round spots resemble the appearance of maria.

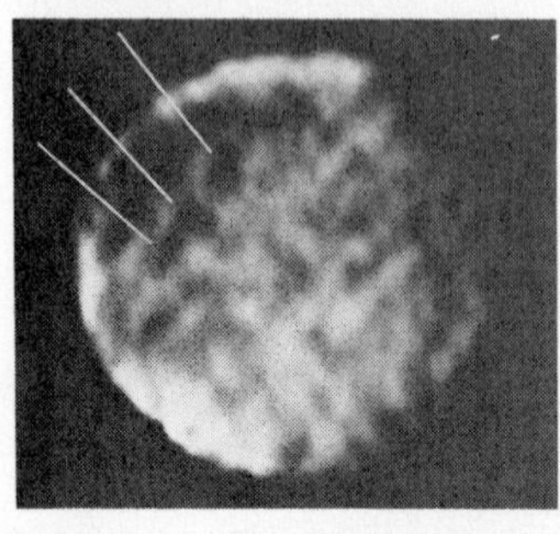

From the Earth, it has been unexpectedly discovered that a ring of sodium fills much of Io's orbit. Protons from Jupiter's strong radiation belts may be knocking sodium atoms off the surface of Io. A potassium ring has also been found. Ultraviolet space measurements of the Lyman lines from Pioneer showed that a band of hydrogen stretches a third of the way around Io's orbit. The distribution of the hydrogen is somewhat different from the distribution of the sodium and potassium. Perhaps the protons from the radiation belts picked up electrons from Io to become neutral hydrogen atoms.

16.5 FUTURE OBSERVATIONS OF JUPITER

The Pioneers are lighter spacecraft than the Mariners and spin around one axis, while the Mariners are completely stabilized. The next step that NASA plans is to send two Mariner craft to Jupiter and Saturn, with their launches scheduled for 1977 and arrivals at Jupiter in 1979. Jupiter's large mass makes it a handy source of energy to use in order to send probes to more distant planets. Just as a sling, like David's, transfers energy to a stone, some of Jupiter's energy can be transferred to a spacecraft through a gravitational interaction. We will get more observations of Jupiter as a bonus in much of our exploration of the outer reaches of the solar system because of the use of this *gravity assist* method. Already, Pioneer 11 is making use of this method to get to Saturn.

"And David put his hand in his bag, and took thence a stone, and slung it, and smote the Philistine [Goliath] in his forehead."
First Samuel *17:49*

Since Jupiter's temperature varies from thousands of degrees in its interior to below freezing at the top of its atmosphere, there may even be regions where the temperatures are suitable for the formation of life. Some scientists have suggested that a form of life may have developed that could live in the clouds, and be carried around in benign temperature regions by atmospheric currents. The measurements of Jupiter's atmosphere at least give us hope that the atmosphere would slow up a capsule we might eject into it, and so in some future year we may get firsthand evidence. There is hope of launching an orbiter and atmospheric probe in 1982.

We might also want to study the moons as possible places where life might have arisen, or at least use them as staging platforms for the exploration of Jupiter. But only Callisto of the Galilean satellites lies outside the region of the most intense radiation belts and so is the closest moon to Jupiter on which a person could hope to stand in the foreseeable future.

SUMMARY AND OUTLINE OF CHAPTER 16

Aims: To study Jupiter, the dominant planet of the solar system, and to study an example of a class of planets very different from the inner, terrestrial planets

Fundamental properties (Section 16.1)
- Highest mass and largest diameter of any planet
- 14 moons, the most of any planet
- High albedo: 51 per cent
- Low density: 1.3 grams/cm^3
- Composition primarily of hydrogen and helium; some ammonia and methane present
- Rapid rotation, causing oblateness of disk
- Colored bands on surface; Great Red Spot
- Intense bursts of radio radiation
- Strong magnetic field and radiation belt

- Jupiter's moons (Section 16.2)
  - At least 14 satellites
  - Galilean satellites are largest
  - Three groups of satellites
    - 5 innermost moons: orbits less than 2 million km in radius
    - 4 next moons: orbits about 12 million km in radius
    - 4 outermost moons: orbits 21 to 24 million km in radius
    - Orbit of 14th moon unknown
  - Moons still being discovered
- Spacecraft observations (Section 16.3)
  - Pioneer 10 and 11 flybys in 1973 and 1974
  - Many photographs taken of cloud structure at higher resolution
  - Great Red Spot believed to be giant cyclonic storm
  - No solid surface detected: gaseous atmosphere and liquid interior
  - Jupiter radiates 2 to 3 times the heat it receives from the Sun (confirmed previous observations)
  - Colored bands the result of huge convection currents and the rapid rotation
  - Temperatures of 400 to 450 K measured in middle layers of atmosphere
  - Pioneer 11 photographed the Jovian polar regions, a task impossible from Earth
  - Tremendous magnetic field detected
- Recent observations of Jupiter's moons (Section 16.4)
  - Galilean satellites from 2/3 to twice the size of the Moon
  - Innermost satellites are probably rocky, perhaps covered with ice
  - Outer moons may be a mixture of ice and rock
  - Polar caps on Ganymede and Callisto
  - Io has an atmosphere and an ionosphere, and is the triggering mechanism for Jupiter's radio bursts. Rings of hydrogen, sodium, and potassium follow in its orbit.
- Future observations of Jupiter (Section 16.5)
  - Plans to send two Mariner spacecraft in 1977
  - Possibility of life in certain levels of Jovian atmosphere

QUESTIONS

1. Why does Jupiter appear brighter than Mars despite its greater distance from the Earth?

2. Assume that the Jovians put a dye in their atmosphere such that there is a green line running from the north pole to the south pole. Sketch the appearance of this line a few days later.

3. Even though Jupiter's atmosphere is very active, the Great Red Spot has persisted for a long time. How is this possible?

4. How do we know that Jupiter has a magnetic field?

5. In Roemer's measurement of the speed of light, compare how "late" the moons of Jupiter appeared to arrive in their predicted positions when Jupiter was at its farthest point from the Earth compared to when Jupiter was closest to the Earth.

6. It has been said that Jupiter is more like a star than a planet. What facts support this statement?

7. What advantages did Pioneers 10 and 11 have for photography over the 5-m telescope on Earth? How much clearer were the Pioneer photos?

8. Aside from photography, what are three other types of observations made from the Pioneers to Jupiter?

9. How do the temperatures of the larger moons of Jupiter compare with that of Jupiter? How do you explain the differences?

10. Which of Jupiter's moons might prove to be interesting targets for future manned exploration? Why?

TOPICS FOR DISCUSSION

1. Although many of the most recent results about Jupiter came from Pioneers 10 and 11, prior ground-based studies had told us many things. Discuss the status of our pre-1973 knowledge of Jupiter, and specify both some things about which space research did not add appreciably to our knowledge and some things about which space research led to a major revision of our knowledge.

2. If we could somehow set up a space station suspended in Jupiter's clouds, the astronauts would find a huge gravitational force on them. What are some of the effects this would have? (The novel *Slapstick* by Kurt Vonnegut (Delacorte Press, 1976) deals with some of the problems that would arise if gravity were different in strength from what we are used to.)

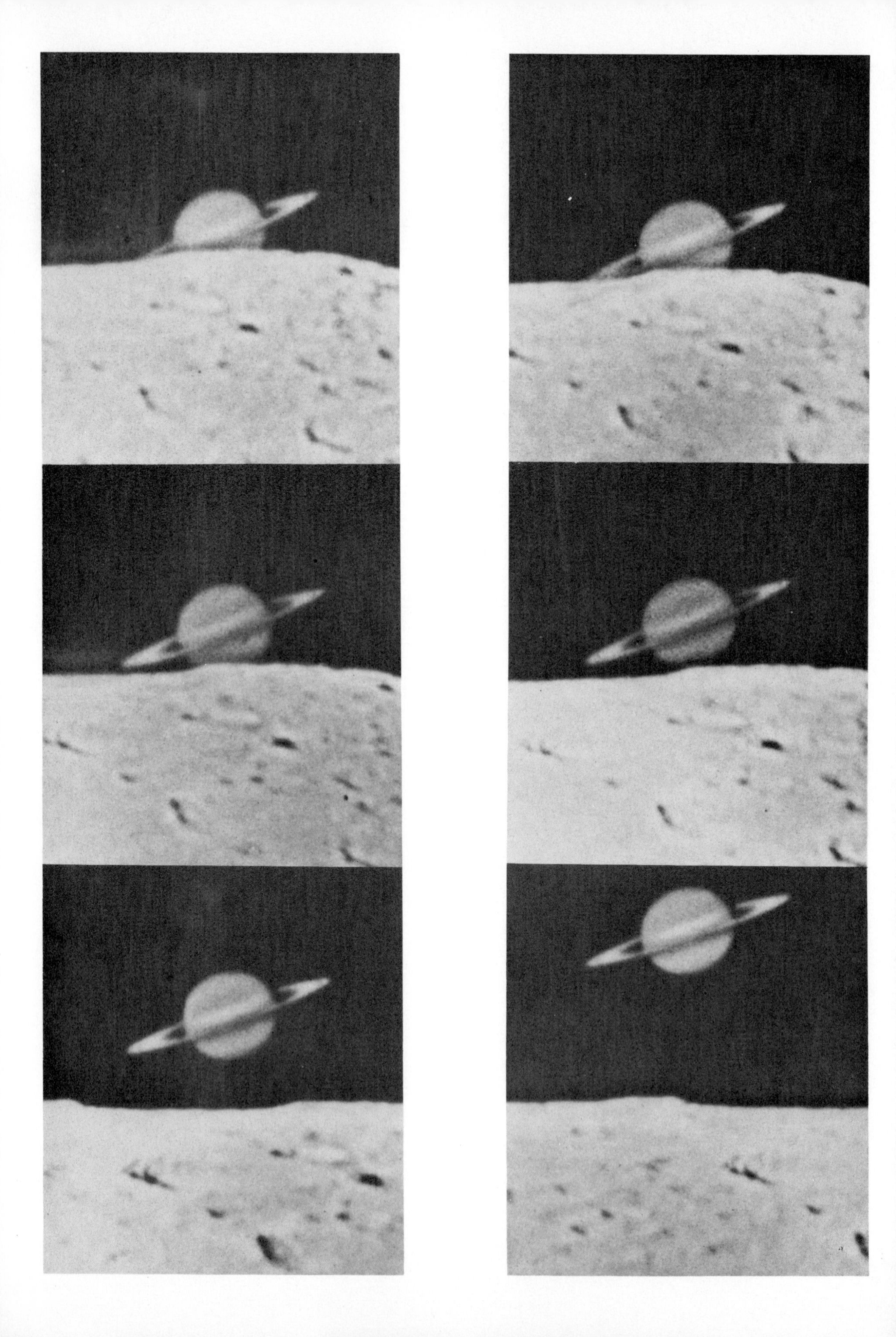

17

The Outer Planets — Saturn, Uranus, Neptune, and Pluto

17.1 SATURN

Saturn is the most beautiful object in our solar system, and possibly even the most beautiful object we can see in the sky. The glory of its system of rings makes it stand out even in small telescopes.

Figure 17–1 Saturn's density is lower than that of water.

Saturn, like Jupiter, Uranus, and Neptune, is a giant planet. Saturn is 9.5 A.U. from the Sun, and so, as we can deduce from Kepler's third law or by simply observing, it has a lengthy year, equivalent to 30 Earth years.

The giant planets are characterized by low densities. Saturn has the lowest density of any planet in our solar system: only 0.7 gm/cm^3, which is 70 per cent the density of water. Thus, if we could find a big enough bathtub, Saturn, like Ivory Soap, would float (Fig. 17–1). It could have a core of heavy elements making up 20 per cent of its interior.

Saturn's mass is 95 times that of Earth, and its diameter is 9 times greater. (It is from these facts that we derive the lower density: the volume of Saturn is greater by a factor of $9^3 = 81 \times 9 = 729$, a much greater factor than the mass. Thus the density is lower than the Earth's by $95/729 = 1/8$, and 1/8 of 5.5 is about 0.7.)

Saturn is shown in Color Plate 39.

The diameter just mentioned, 9 times greater than Earth's, is that of the planet itself excluding the rings. The rings extend far out in the equatorial plane of Saturn, and range from about 70,000 to over 135,000 km from the planet's center (Fig. 17–2). Saturn's rings are inclined to the planet's orbit by 27°. Over a thirty-year period, we sometimes see them from a vantage point of 27° above their northern side, sometimes from 27° below their southern side, and at intermediate angles at intermediate times. When seen edge on, they are all but invisible (Fig. 17–3).

Saturn, emerging from an occultation by the Moon. The symbols for Saturn, Uranus, Neptune, and Pluto appear from left to right, respectively, at the top of this page.

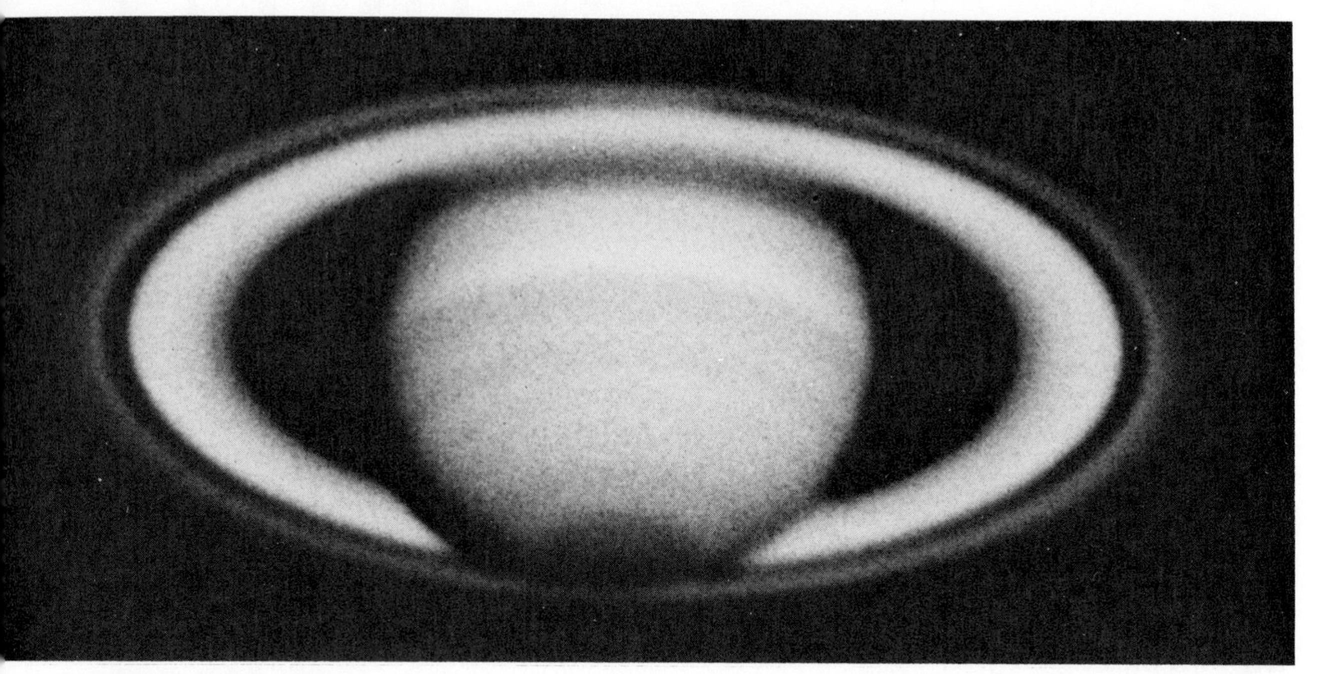

Figure 17–2 The rings of Saturn, photographed on January 19, 1973. (New Mexico State University Observatory photograph)

The rings of Saturn are either the bits of a moon that were torn apart by Saturn's gravity or chunks that failed to accrete into a moon at the time when the planet and the other moons were forming. These bits or chunks formed individual rocks and spread out in concentric rings around Saturn. There is a sphere for each planet, called *Roche's limit*, or *the Roche limit*, inside of which blobs of gas are separated from each other; an agglomeration of blobs cannot form and be held together by the blobs' own gravity. The forces that tend to tear the blobs apart from each other are called *tidal forces;* they arise because some blobs are closer to the planet than others and are thus subject to higher gravity. (The differing gravitational forces tend to pull the blobs apart; similarly, differential gravitational forces from the Moon and the Sun cause the ocean tides on Earth.) The radius of the Roche limit varies with the amount of mass in the parent body. The Sun also has a Roche limit, but all the planets lie outside it. All the moons of the various planets lie outside the respective Roche limits, but Saturn's rings lie inside Saturn's Roche limit, so it is not surprising that the material in the rings is not collected into a single orbiting body.

Artificial satellites that we send up to orbit around the Earth are constructed of sufficiently rigid materials that they do not break up even though they are within the Earth's Roche limit.

Each rock in the ring independently follows Kepler's laws, so the ones closer to Saturn orbit more quickly than the ones farther out.

There are several concentric rings around Saturn. The brightest ring is separated from a fainter outer ring by what is called *Cassini's division* (Fig. 17–2). Two fainter rings are inside the brightest ring. The gaps in the

   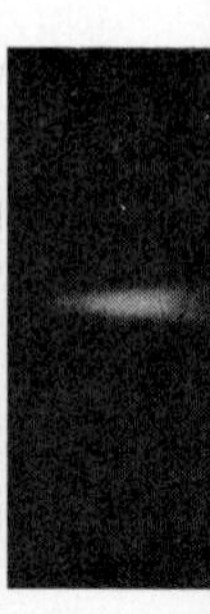

Box 17.1 Cassini's Division

In 1610, Galileo had used his new invention, the telescope, to discover that Saturn was not round; it seemed to have "ears." The Dutch astronomer Christian Huyghens published an anagram in 1656 explaining that Saturn has a ring; Huyghens's more conventional publication of his result in a book followed three years later. Cassini, who worked in Paris some years later, observed the largest gap in the ring in 1675, and this gap is named after him (the Cassini division). The idea that the ring is composed of many bodies orbiting independently was proposed by James Clerk Maxwell in England in 1856. The existence of the rings of Saturn had an important role in the development of the nebular hypothesis by Kant and Laplace.

ring structure probably result from gravitational effects from Saturn's satellites, and finer structure surely exists in the rings than the few divisions that are visible from the Earth.

We can see that the rings are not solid objects, for stars occulted by the rings can be seen shining dimly through. Even though the rings are 275,000 km across, they are very thin from top to bottom: not more than 10 km and perhaps less. Some astronomers think they may be only 10 to 100 meters thick from top to bottom, a good trick for something hundreds of thousands of km across! Radar waves were first bounced off the rings in 1973. Subsequently, in 1975, observers transmitted a powerful radar beam from the 305-m (1000-ft) radio telescope at Arecibo, Puerto Rico, and received the signal, after it had traveled 2 hours and 14 minutes at the speed of light, with the 64-m (210-ft) JPL antenna at Goldstone, California. The results of the radar experiments show that the particles in the ring are probably rough chunks of ice or rock at least a few centimeters and possibly a meter across. Earlier ideas that the rings were made of dust and smaller pieces of ice were ruled out by these observations, though there might be ice on a rocky base.

Like Jupiter, Saturn rotates very quickly on its axis, also in about 10 hours; thus it is oblate by as much as 10 per cent. Saturn has delicately colored bands of clouds. Both methane (CH_4) and molecular hydrogen (H_2) have been detected spectrographically in its atmosphere. Ammonia, which we might also expect to find, solidifies at higher temperatures than methane and hydrogen and so may have fallen out of the atmosphere as "snow." This would explain why ammonia cannot be detected in the atmosphere.

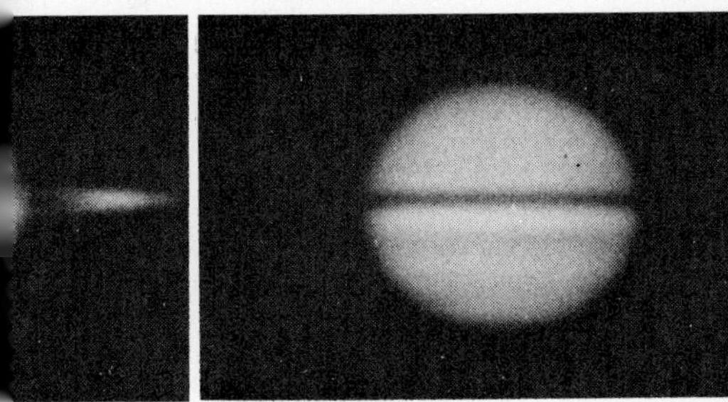

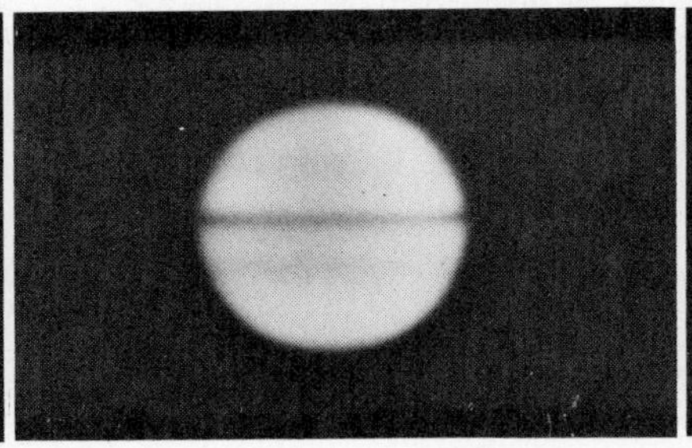

Figure 17–3 The rings of Saturn photographed at various times. (Lowell Observatory photographs)

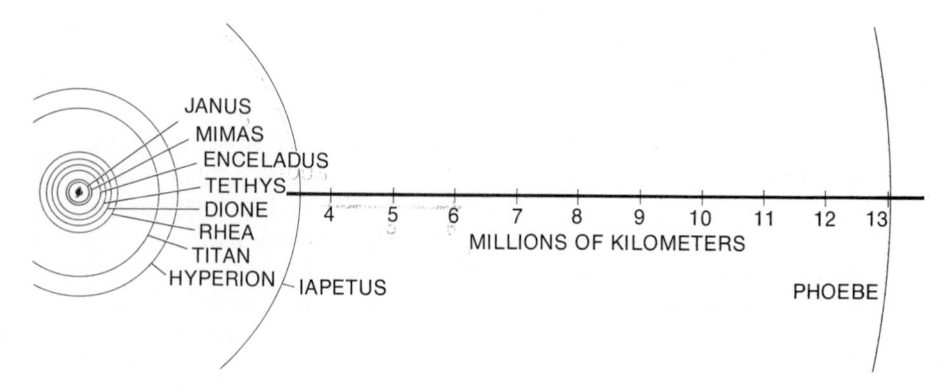

Figure 17–4 Saturn's moons.

In 1975, infrared studies disclosed the presence of ethane (C_2H_6) in Saturn's atmosphere, which tend to support theories that predict the production of hydrocarbons from methane in the upper atmospheres of the outer planets. Although we know very little about what is under Saturn's clouds, we calculate that Jupiter and Saturn should turn liquid under the tremendous pressures present at great depths, but there is no evidence that there is a solid surface at any level. Consequently, if you fell into Jupiter or Saturn, you would descend into denser and denser slush.

Saturn gives off radio signals, as does Jupiter, so we presume that it also has a magnetic field. The data indicate that Saturn's magnetic field may be one-tenth that of Jupiter. Just as for Jupiter, Saturn seems to have a source of internal heating.

Saturn has ten moons (Fig. 17–4), all of them named—which is much more poetic than having a number, as was the situation until recently with Jupiter's moons VI through XII. All but the innermost moon of Saturn have beautiful names taken from Greek mythology. They are named after the Titans, the children and grandchildren of Gaea, goddess of the Earth, who had been fertilized by drops of Uranus's blood. The 9 outermost moons are called Mimas, Enceladus, Tethys, Dione (mother of Aphrodite), Rhea, Titan, Hyperion (the father of Helios), Iapetus, and Phoebe (Fig. 17–5).

The tenth moon, the innermost, was discovered in 1966 when Saturn's rings were seen edge-on, and thus did not interfere with the view of the space nearest the planet (Fig. 17–6). The new moon was named Janus, after the Roman deity who presided over beginnings and endings. (Janus is the figure commonly represented with two faces oriented in opposite directions.) All the moons except Phoebe and Iapetus are in orbits inclined no more than 1° or 2° to the plane of Saturn's equator. Iapetus's orbit is inclined 15°. Phoebe's orbit is inclined 150° and is very eccentric, so Phoebe may well be an asteroid captured at some time in the past.

Figure 17–5 Saturn and seven of its ten moons. We can see (1) Mimas, (2) Enceladus, (3) Tethys, (4) Dione, (5) Rhea, (6) Titan, and (8) Iapetus. Only (7) Hyperion, (9) Phoebe, and the innermost moon, (0) Janus, cannot be seen in this photograph. A faint star is visible at top.

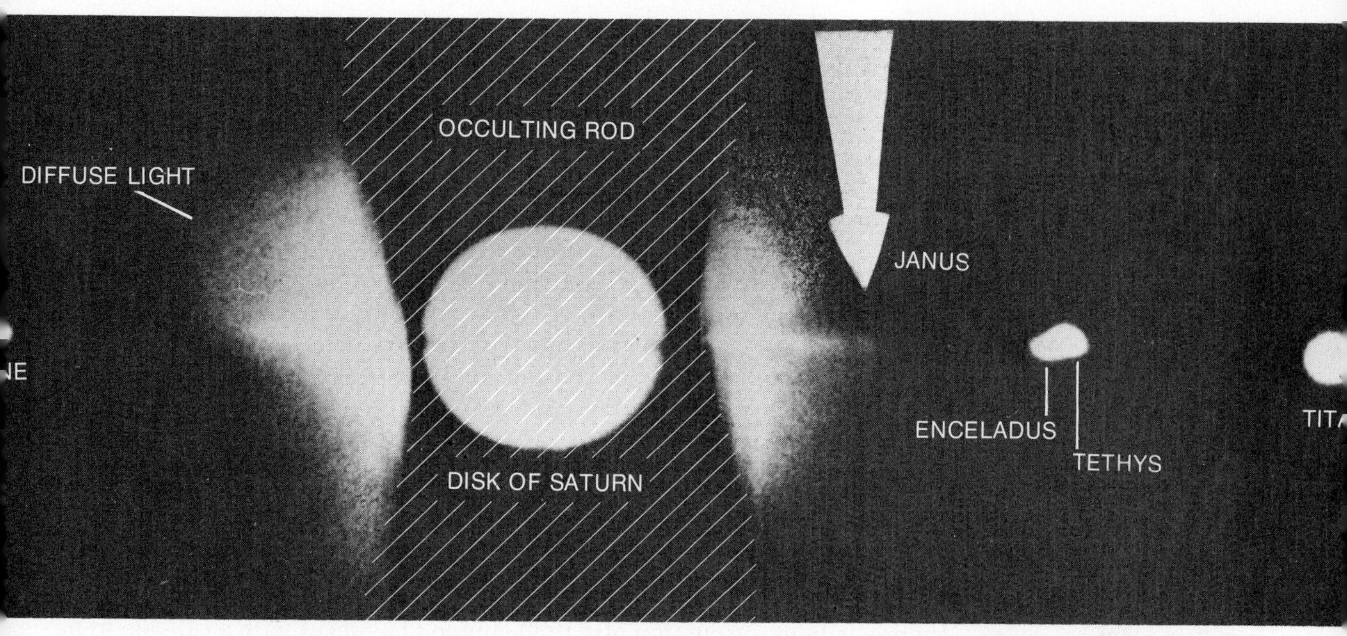

Figure 17–6 The photograph on which Janus was discovered, taken on October 15, 1966. A filter was placed over the center of the photograph to diminish the brightness of Saturn's disk; some of the overexposure from the disk is visible to the left and right of the filter. The rings, which were nearly edgewise, appear as a narrow bright bar. The arrow shows Janus. Four other satellites can also be seen; Dione is at the left, overlapping images of Enceladus and Tethys are on the right, and Titan is at the far right.

All the moons, except Titan, range from about 130 km to 1600 km across. Planet-size Titan, however, is a different kind of body. An atmosphere has been detected on Titan, and methane has been found in it. Because of this atmosphere, a greenhouse effect may have warmed the surface of Titan, and so Titan has become one of the more interesting places in the solar system on which we can search for life—some astronomers think that there may even be more chance of finding life there than on Mars. Recently, the strength of radio radiation at very short radio wavelengths has been measured, and it indicated that Titan's surface temperature is about 175 K. This is about 75 K warmer than we would expect from the calculations that astronomers have made based on balancing the input of solar radiation to Titan against the amount it would radiate if it had no atmosphere. From the observational fact that the measured temperature is so high (though still about −100°C, cold for humans), we can deduce the existence of a greenhouse effect.

Carl Sagan, a Cornell astronomer, conceived of a complex scenario for what could be happening on Titan. He suggests that the thick clouds could contain some organic compounds floating in an atmosphere of hydrogen, methane, and ammonia. Where did this atmosphere come from? Well, volcanoes would do it. Then the ultraviolet solar radiation could have turned the gases into complex organic compounds, and even into amino acids. This last step has proved possible in laboratories on Earth, so perhaps it happened—or is happening—on Titan.

The rotation of Saturn's rings can be measured directly from Earth with a spectrograph, for one side of the rings has a different Doppler shift than the other side (Fig. 17–7). Indeed, Jupiter and Saturn subtend large enough angles in the sky that this velocity effect can even be measured for the cloud layers from side to side of the planet. The clouds of Saturn rotate in about 10 hours at the equator, and about 10 per cent more slowly at high latitudes.

A similar spectrographic method tells us about the rotation of spiral galaxies.

Pioneer 11, the second of the two craft that passed by Jupiter, has been targeted to fly to Saturn, and will arrive there in September 1979

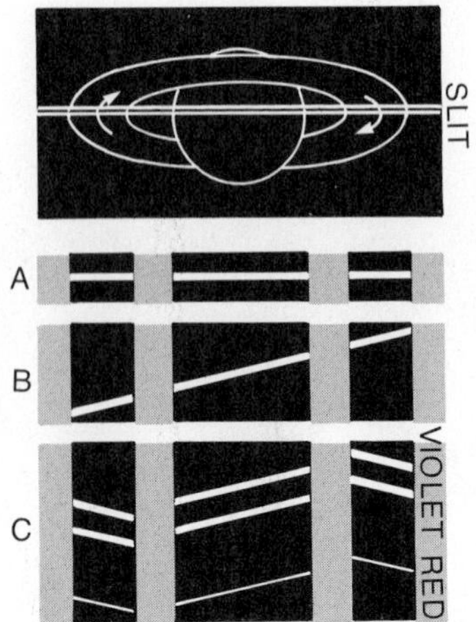

Figure 17–7 If a slit is laid across Saturn, as shown at top, then the spectrum appears as shown at C on the bottom. If Saturn were not rotating at all, then no Doppler shift would be seen, as in A. If Saturn and its rings rotated as a solid body, then the spectrum would appear as in B. The fact that the spectrum looks like C indicates that Saturn itself rotates as a solid body but that each rock in the rings has an orbit that obeys Kepler's laws.

Figure 17–8 The trajectories of Pioneers 10 and 11 are shown with dotted lines and the orbits of the Earth, Saturn, and Jupiter are shown with solid lines. The positions of the planets and the two spacecraft are shown for each January 1st.

(Fig. 17–8). Jupiter's gravity sped up the spacecraft in a manner opposite to the way that Venus's gravity slowed down Mariner 10 en route to Mercury. Because of this gravity assist, the spacecraft is able to go all the way to Saturn even though we only expended enough power at launch to send it to Jupiter. However, its trip to Saturn is a lengthy one because Saturn, during the period of this voyage, is on the side of the solar system opposite to Jupiter. The closeups of the rings should be spectacular. Pioneer 11 will also concentrate on observing Titan, but will not be targeted too close to Titan for fear that it could be pulled in by Titan's gravity, crash, and contaminate the surface, thus preventing us from eventually investigating an independent origin of life there. The nuclear batteries of Pioneer 11 should be expended not long after the visit to Saturn, so the opportunity to use yet another gravity assist to send it to Uranus might not give additional useful data.

Current U.S. plans for further study of Saturn are for the two Mariner-type spacecraft that will be launched in 1977 to reach Jupiter in 1979, and Saturn in 1980 and 1981. (The continuing progress of Jupiter and Saturn around their orbits makes the trip between them, at that time, shorter than it is for Pioneer 11.) Beyond that no definite plans have been made, but there has been discussion of a 1980 launch of a rocket with a capsule to be ejected into Saturn's atmosphere, and a 1985 launch of a pair of spacecraft that would orbit Saturn.

17.2 URANUS AND NEPTUNE

There are two other giant planets beyond Saturn: Uranus (U'ranus) and Neptune. Both Uranus and Neptune are large planets, about 25,000 km across and about 15 times more massive than the Earth. The densities of Uranus and Neptune are low (1.2 and 1.7 g/cm³, respectively). Their al-

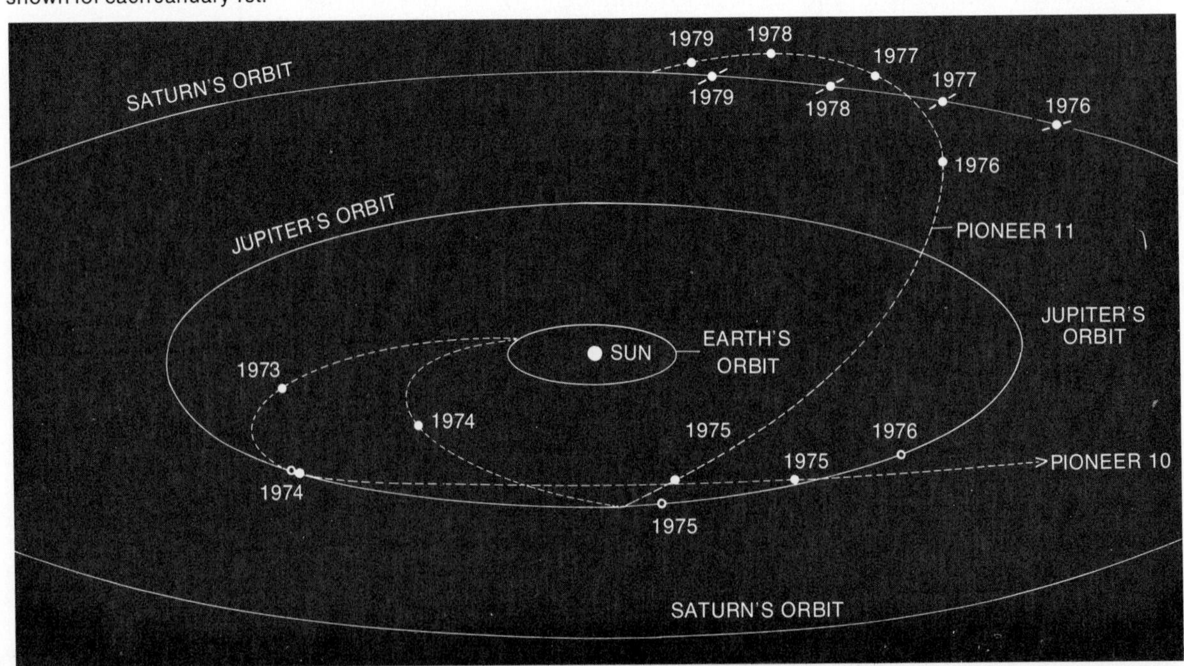

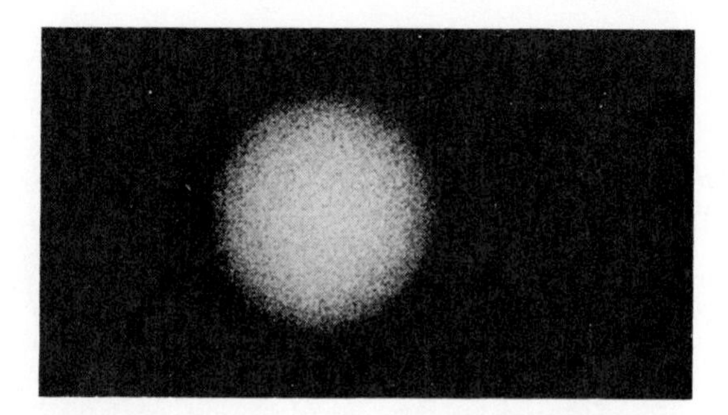

Figure 17–9 Uranus, photographed from the ground in blue light, shows no surface markings.

bedoes are between 60 per cent and 70 per cent in the visible part of the spectrum, which indicates that they are covered with clouds.

Uranus was discovered to be a planet by William Herschel in England in 1781, but it had been plotted on sky maps for about a hundred years prior to that with the thought that it was just another star. It revolves around the Sun in 84 years, at an average distance of more than 19 A.U. from the Sun. Even at the most favorable possible place in its orbit, which has an eccentricity of almost 5 per cent, it is never closer than 17 A.U. (2.5 billion km) to the Earth. Thus Uranus never gets any larger in the sky than 3.6 seconds of arc, and studying its surface structure from the surface of the Earth is very difficult (Fig. 17–9).

Uranus, in Greek mythology, was the personification of Heaven and ruler of the world, the son and husband of Gaea, the Earth. Neptune, in Roman mythology, was the god of the sea, and the planet Neptune's astronomical symbol reflects that origin.

In 1972, the balloon Stratoscope II carried a 36-inch telescope up to an altitude of 24 km, above most of the Earth's atmosphere. Photographs were taken of Uranus, as well as of other objects. The 17 photographs of Uranus had resolutions of 1/6 second of arc, and give us our most accurate measurements of Uranus's diameter, but even with this resolution no detail on the surface can be seen. Thus we know that Uranus does not have belts of clouds as do Jupiter and Saturn. Molecules in its atmosphere may simply be scattering the incoming sunlight; this would also account for the high albedo. The Stratoscope observations suggest that Uranus is surrounded by a thick layer of methane clouds, with a semi-transparent atmosphere of molecular hydrogen (H_2) above the methane. Both methane and molecular hydrogen have been observed with ground-based spectrographs. Uranus and Neptune both have greenish casts when observed from the Earth, which led in times past to many diffuse objects in the sky (some of which also appear green but which are not planets at all) being named "planetary nebulae."

Even though the photography does not reveal structure on Uranus, there is strong evidence from visual observations that there may be structure after all. Direct observation with the eye at the telescope is certainly out of favor at present, but nonetheless the eye and brain can take advantage of especially fine moments of observing, when the Earth's atmosphere is particularly still, and can use color differences to differentiate one region from the next. Over the years, many visual observers have reported seeing structure on the surface of Uranus, and even though they do not agree about the exact form of that structure they may well be correct that structure does exist. Perhaps the fact that the Stratoscope observations were made through a filter that allowed only one color to pass through prevented those balloon observations from detecting structure on the surface of Uranus. On Venus, after all, the detailed cloud structure that is detectable in ultraviolet light is undetectable in visible light.

The other planets rotate such that their axes of rotation are roughly

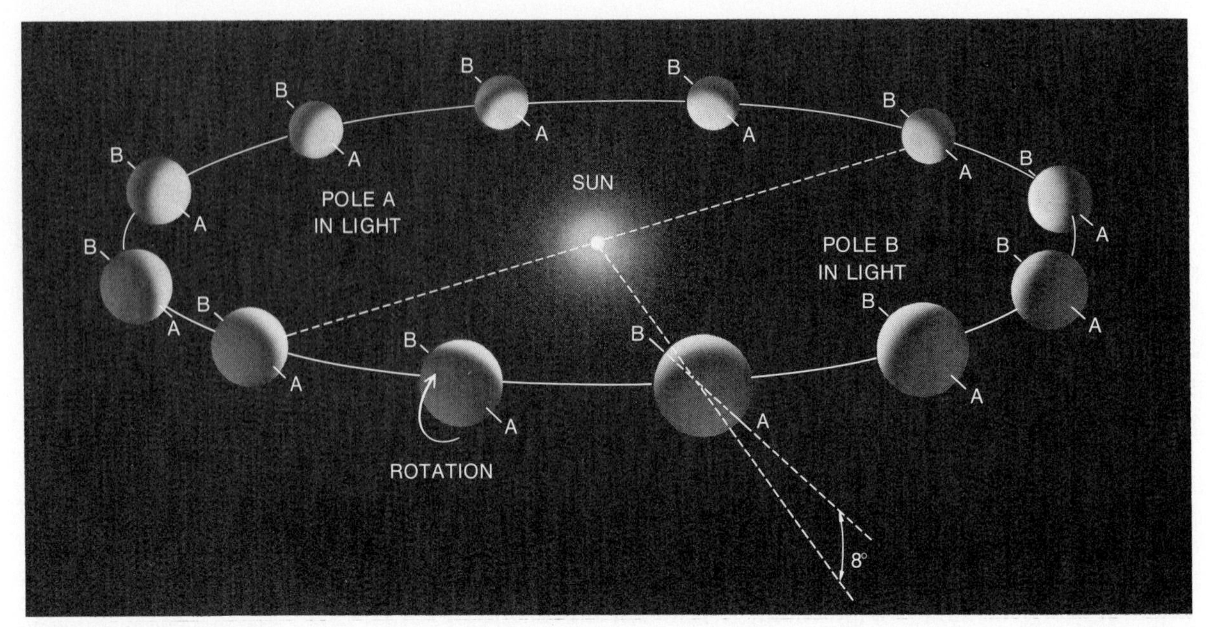

Figure 17–10 Its axis of rotation lies in the plane of Uranus's orbit.

parallel to their axes of revolution around the Sun; Uranus is different, for its axis of rotation is roughly perpendicular to the other planetary axes. Uranus's axis of rotation lies in the plane of its orbit (Fig. 17–10). Sometimes, one of Uranus's poles faces the Earth, 21 years later its equator crosses our field of view, and then another 21 years later we face the other pole. Thus there are strange seasonal effects on Uranus, and the differences in the heating pattern from year to year may be the reason that there are no cloud bands. When we understand just how the heating affects the clouds, we will be closer to understanding our own Earth's weather systems. Of course, even though there is an effect because the angle at which sunlight hits different parts of the surface of Uranus varies with the season, we must recall that Uranus is very distant from the Sun, so the intensity of sunlight is never very great. Uranus is always very cold, perhaps 90 K (–185° C).

Figure 17–11 The moons of Uranus.

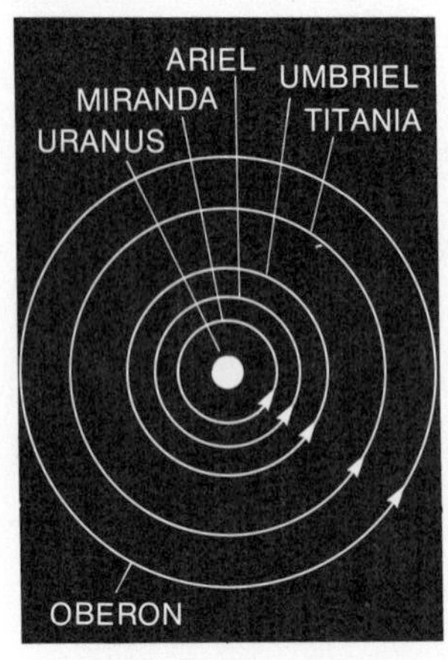

It is said that Uranus rotates in the retrograde direction, and so it does—but barely. Mainly, from our point of view, we see it rotating sometimes as do the hands of a clock and sometimes from top to bottom, while the other planets always appear to rotate from side to side. In 1975 it was discovered from studies of how the Doppler effect broadens a sharp spectral absorption line that Uranus rotates in 12.3 hours.* The correction from the previously accepted value was over 10 per cent, which indicates how inaccurate our knowledge is of the other members of our solar system.

Uranus has five moons (Fig. 17–11), each with a beautiful name. From the innermost to the outermost, they are Miranda, Ariel, Umbriel, Titania, and Oberon. The innermost, Miranda, was discovered fairly recently, in 1948, and is the smallest. The other four moons range from 400 to 1000 km across. Very little is known about them (Fig. 17–12).

At present it is planned for the second of the Mariners to be launched in 1977 to pass not only Jupiter and Saturn, but also Uranus, arriving there

*The light from the entire planet enters the spectrograph simultaneously, and includes Doppler-shifted contributions from each side of the planet in addition to the bulk of the light from nearer the center of the disk.

Figure 17–12 Photographs of Uranus and its moons.

in 1985 or 1986. The definite decision need not be made until we see in 1980 how well the first of the Mariners has succeeded in studying Saturn. The option will remain until then to target the second Mariner for another close-up of Titan instead of sending it on toward Uranus.

Some scientists had been excited about plans for a "Grand Tour" of the planets. Spacecraft would have taken advantage of rare planetary alignments and gravity assists to approach all the otherwise unvisited planets without excessive fuel requirements. But these plans, and other plans to explore Uranus, were dropped by NASA because of budgetary restraints.

Neptune is even farther away than Uranus, 30 A.U. compared to about 20 A.U. (actually 19 A.U.). Its orbital period, by Kepler's third law, is thus $\sqrt{(30/19)^3}$ as long, 165 years. Its discovery was a triumph of the modern era of Newtonian astronomy. Neptune had not been known until mathematicians analyzed the deviations from an elliptical orbit shown by Uranus. These deviations are small, but were detectable, and could have been caused by gravitational interaction with another, as yet undetected, planet.

The denouement is one of astronomy's most famous anecdotes. John C. Adams in England (Fig. 17–13) predicted positions for the new planet in 1845, but the astronomy professor at Cambridge did not bother to try to observe this prediction of a recent college graduate. The Astronomer Royal's butler put off Adams from interrupting the dinner of the A.R. (as he is referred to), so Adams never saw the A.R. in person. Though Adams left a copy of his calculations, when the A.R. requested further information, partly to test Adams's abilities, Adams did not take the request seriously and did not respond at first. The very "proper" Astronomer Royal took offense and did not choose to have further dealings with Adams. The story then continues in France, where a year later Urbain Leverrier was independently working on predicting the position of the undetected planet.

When the Astronomer Royal saw in the scientific journals that Leverrier's work was progressing well (Adams's work had not been made public), for nationalistic reasons he began to be more responsive to Adams's calculations. But the search for the new planet, though begun in Cambridge, was carried out half-heartedly. Neither did French observers take up the search. By this time it was 1846. Leverrier sent his predictions to an acquaintance at Berlin, who enthusiastically began observing and discovered Neptune within hours.

Years of acrimonious debate followed over who (and which country) should receive the credit for the discovery. We now tend to credit both

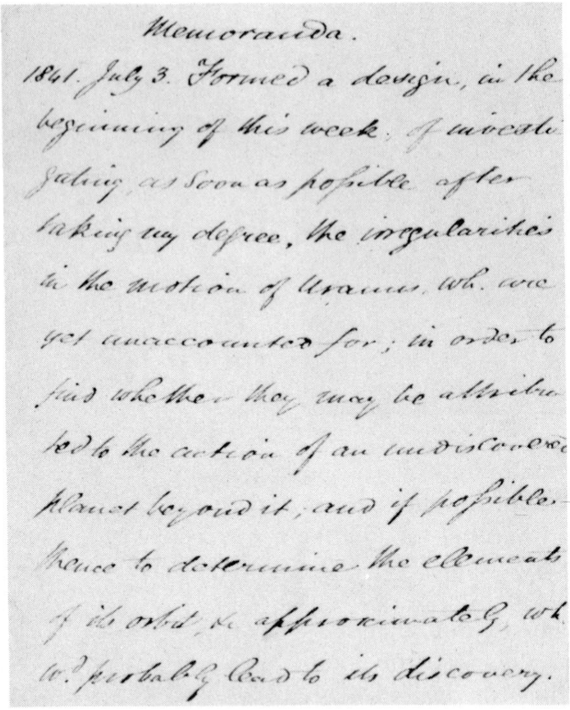

Memoranda.
1841. July 3. Formed a design, in the beginning of this week, of investigating, as soon as possible after taking my degree, the irregularities in the motion of Uranus, wh. are yet unaccounted for; in order to find whether they may be attributed to the action of an undiscovered planet beyond it, and if possible thence to determine the elements of its orbit, &c. approximately, wh. wd. probably lead to its discovery.

Figure 17–13 The memorandum that John Couch Adams wrote to mark his determination to investigate the possible existence of a planet outside the orbit of Uranus.

Adams and Leverrier (and mock that particular Astronomer Royal and Cambridge professor).

We now have the benefit of hindsight, and can examine the calculations of Adams and Leverrier more carefully. One assumption was necessary for them to make their calculations—they had to guess a radius for the orbit of Neptune in order to calculate its expected direction in the sky. They used the value from Bode's law (Section 10.4). This value was 39 A.U., substantially larger than the value we have since measured for Neptune. The error happened not to be of importance for the configuration of the planets at the time they were working, but at other times their calculations would not have given the right position. So there certainly was an element of luck in their successful prediction.

Neptune's orbit is so large that it takes a very long time to travel around the Sun, and it has not yet made a full orbit since it was discovered. It never appears larger than 2 seconds of arc across in our sky, so is always very difficult to study. Even measuring its diameter accurately is hard. In 1968, Neptune occulted (passed in front of) an eighth magnitude star, a star about as bright as Neptune itself. This occultation was visible only from Japan, Australia, and New Zealand. From the fact that the star is known only by its catalogue number, BD −17° 4388 (pronounced BD minus seventeen degrees, four, three, eight, eight), you can tell that it was a humdrum and unspectacular star. However, when it was occulted by Neptune it became very important. From observations of the rate at which the star dimmed, astronomers could deduce information about Neptune's upper atmosphere and eventually the atmosphere's temperature and pressure structure. Further, from the length of time that the star was hidden by the disk of Neptune, they deduced that Neptune's diameter is 2.3 seconds of arc, equivalent to 49,200 km, accurate to ±2300 km. The astronomers could calculate the diameter from the occultation because they knew how fast

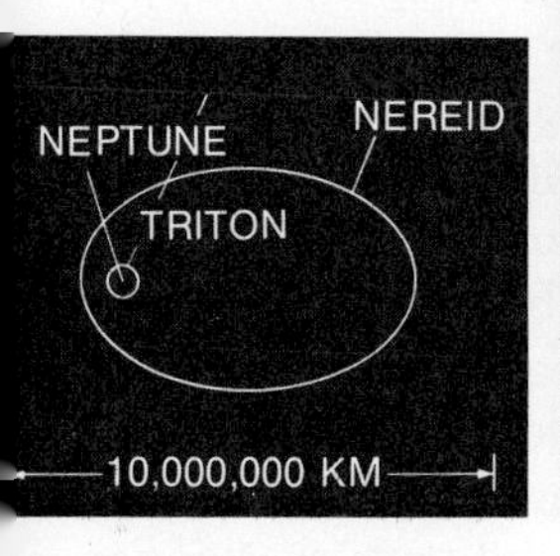

Figure 17–14 Neptune and its satellites. Triton is close to Neptune, and Nereid is at upper right.

Neptune moves across the sky with respect to the fixed stars. The longer the star was hidden from view, the greater the diameter of Neptune must be.

The diameter had been known to be about this value, similar to Uranus's diameter, but it was interesting to get a more accurate value. Accurate values are important for calculating the planet's density, thus leading to deductions about its composition. (Since the volume of a sphere is proportional to the cube of its radius, and the density is the mass divided by the volume, any inaccuracy in the radius leads to a much larger inaccuracy in the density.)

The value adopted by the International Astronomical Union in 1976 (see Appendix 5) gives heavy weight to the occultation value but also considers other results.

Neptune, like Uranus, appears greenish in a telescope. Only hydrogen and methane have thus far been detected with spectrographs. Neptune has two moons (Fig. 17–14). Triton (a sea god, son of Poseidon) is large, perhaps 6000 km across. Its orbit is 350,000 km in diameter. A second moon, Nereid (a sea nymph), is small. Nereid is perhaps only 500 kilometers across, and is in a very eccentric orbit with an average radius of 5.5 million kilometers. The orbits of Triton and Nereid are inclined 160° (i.e., 20° and retrograde) and 28°, respectively, with respect to the ecliptic plane. Nereid never gets brighter than 20th magnitude, which is near the limit of our observational capabilities even with the largest telescopes.

Our hope for getting better information about Neptune and its moons is that a spacecraft that might be launched to Uranus in the mid-1980's could continue on to Neptune, but such a mission is not currently scheduled.

17.3 PLUTO

Pluto, the outermost known planet, is a deviant. It has the most eccentric orbit: an eccentricity of .25 compared with Mercury's .21 and a maxi-

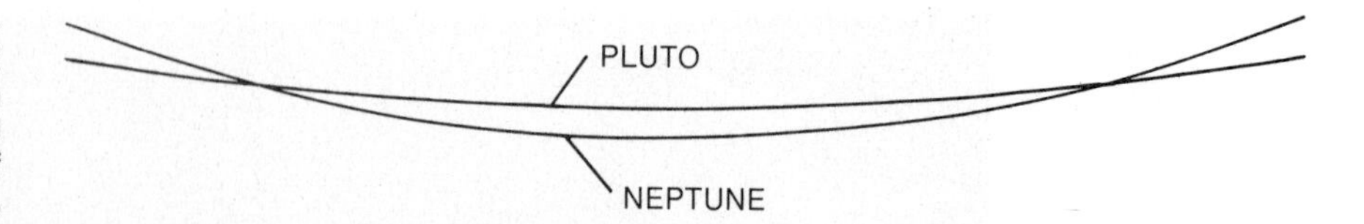

Figure 17–15 Pluto's orbit is so eccentric that it sometimes goes inside the orbit of Neptune. Pluto will enter the mean orbit of Neptune on November 28, 1978, and exit from within the mean orbit of Neptune on May 12, 2000.

mum of .09 for the other planets. Its orbit has the greatest inclination with respect to the ecliptic plane, in which the other planets revolve. Pluto's orbit is inclined by 17°, while Mercury's is inclined by only 7° and the other planets are all inclined by 4° or less. In a drawing of planetary orbits (see Fig. 10–29) Pluto sticks out.

Pluto will reach perihelion, its closest possible distance from the Sun, in 1989. Its orbit is so eccentric that for many years at the end of this century, up to 1998, the planet will be on the part of its orbit that is inside the orbit of Neptune (Fig. 17–15). Thus in a sense, Pluto will be the eighth planet for a while (though, of course, the semi-major axis of Pluto's orbit is greater than that of Neptune).

Even at perihelion, no features can be seen on Pluto. It appears only as a dot in the sky, and its observed diameter seems to be about 0.2 second of arc, right at the limit of our seeing (astronomical seeing) capability. It does fluctuate in brightness periodically, and this has been interpreted to be the result of a rotation of the planet bringing areas of differing albedos toward us. Thus the rotation can be accurately determined to be 6 days 9 hours 16 minutes 54 seconds, with an uncertainty of only 26 seconds.

Figure 17–16 Small sections of the plates from which Tombaugh discovered Pluto. On February 18, 1930, Tombaugh noticed that one dot among many had moved between January 23, 1930 *(top)* and January 29, 1930 *(bottom)*.

Even such basics as the mass and diameter of Pluto are very difficult to determine. The discovery of Pluto was a result of a long search for an additional planet which, together with Neptune, was causing perturbations in the orbit of Uranus. The best known searchers in the first three decades of this century were Percival Lowell in Arizona and W. H. Pickering of Harvard, but they were not successful in locating the unknown planet near its predicted position, which was fairly uncertain. Finally, in 1930, Clyde Tombaugh, after a year of diligent study of photographic plates at the Lowell Observatory, found the dot of light that is Pluto. From its slow motion with respect to the stars from night to night, it was identified as a new planet (Fig. 17–16 and Fig. 17–17).

But the predictions that led to the discovery also predicted that the planet would be 6.6 times the mass of the Earth. If Pluto's diameter is as small as it seemed to be from direct measurement of the size of the disk—much smaller than the giant planets just inside it—its density would be impossibly high, hundreds of grams/cm^3. This density would be many times greater than that of any other object in the solar system. To resolve this difficulty, we must be as certain as we can be that we are using correct values of mass and radius.

It is very difficult to deduce the mass of Pluto because the procedure requires measuring Pluto's effect on Uranus, a more massive body. Moreover, Pluto has made less than one revolution around the Sun since its discovery. As recently as 1968, Pluto was thought to have a mass 91 per cent that of the Earth. The latest studies of the orbit of Uranus indicate that Pluto may have a mass only 11 per cent that of the Earth, but these observations are very uncertain. On the basis of current data, we are not able to reliably determine Pluto's mass; the value given may be off by a factor of two or more.

The best method for determining the radius of Pluto, as it is for Nep-

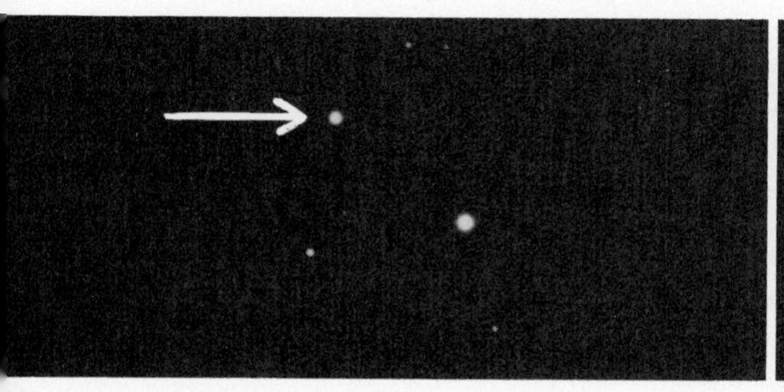

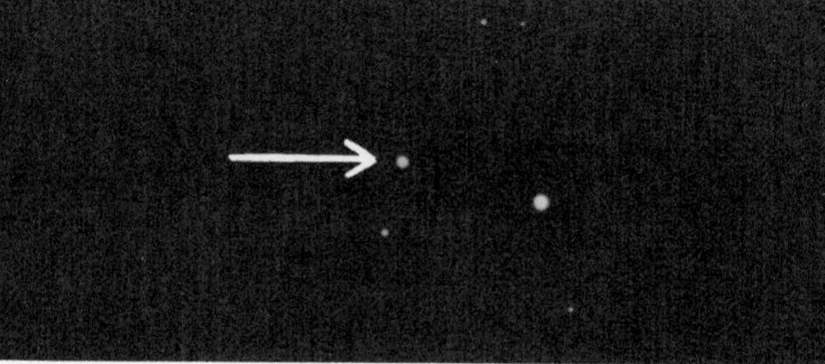

Figure 17–17 Pluto's motion can be seen in these photographs taken on successive nights.

tune, is to observe a stellar occultation. It was predicted that Pluto would pass near a 15th magnitude star in 1965, and so the passage was observed very closely by several observatories to see if the star would be occulted. We knew the orbital path of Pluto in the sky very accurately, but since a planet's gravity acts as though it were concentrated at the center of the planet, we only knew the path that the **center** of Pluto would take. According to the prediction, the center of Pluto would pass within one second of arc from this star, which appears, of course, as only a point of light. If Pluto's radius subtended a large enough angle at this distance from the Earth, at some moment Pluto's surface would hide the star from view. But the star was never hidden from view, and from this fact astronomers knew that Pluto appeared smaller in the sky than the minimum angular separation (Fig. 17–18). Since we know the distance to Pluto, simple trigonometry gives a limit to the radius of Pluto. This observation showed that Pluto had to be smaller than 6800 kilometers across, which confirmed that Pluto was closer in size to the terrestrial than to the Jovian planets.

Pluto could be substantially smaller than this. In 1976, infrared spectral studies carried out at Kitt Peak showed the presence of methane ice on Pluto's surface. Since ice has a high albedo, the planet would not have to be very large to reflect the amount of light that we measure. The smaller the true value of Pluto's diameter, the larger the density we derive.

Thus the diameter of Pluto is uncertain, and the mass of Pluto is even more uncertain. We cannot, therefore, make a good estimate of the density.

We can calculate the effect that Pluto would have on the orbits of Uranus and Neptune by using the latest value of mass instead of the value that had been deduced in the original discovery predictions. The calculations show that Pluto's mass is far too small to cause the perturbations in Uranus's orbit that originally led to Pluto's discovery. These earlier measurements were apparently inaccurate. It is now thought by many astronomers, therefore, that the prediscovery prediction was wrong and that the

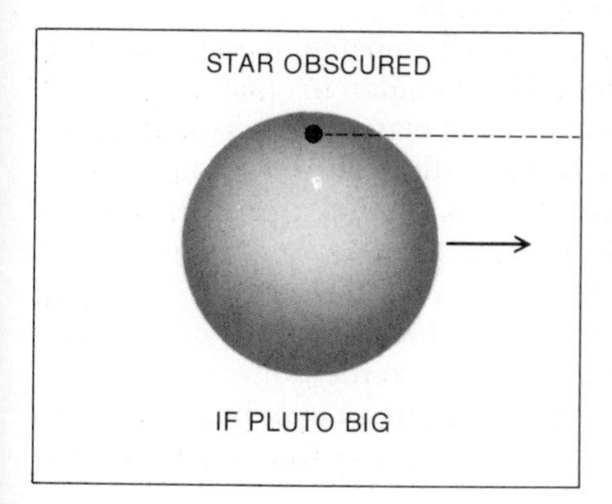

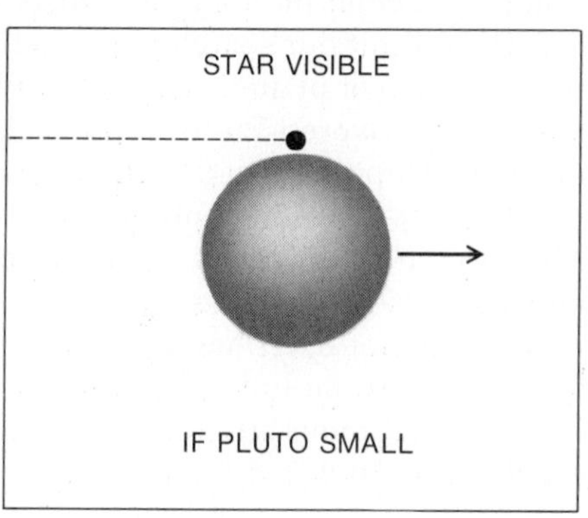

Figure 17–18 The fact that a 15th magnitude star was not occulted by Pluto gives us a limit for how large the diameter of Pluto could be.

discovery of Pluto was purely the reward of hard work in conducting a thorough search in a zone of the sky near the ecliptic.

If we were standing on Pluto, the Sun would be over a thousand times fainter than it is to us on Earth, and we would need a telescope to see its disk. Pluto is so far away from the Sun, almost 40 A.U., and therefore so cold, about 43 K, that any of the common gases except neon would be frozen out of an atmosphere. But Pluto is probably not massive enough to retain much of an atmosphere anyway. Perhaps it has a thin atmosphere of neon.

Pluto is so strange a planet—very small, and with an eccentric and inclined orbit—that it might not even be a true planet in the sense of being formed in a planetary orbit along with all the others. It might rather be an escaped moon of Neptune. But it has been found that at present the orbits of Neptune and Pluto seem to be affected by their mutual gravities in such a way that their relative orbital positions repeat in a cycle every 20,000 years; also, the two planets can never come closer to each other than 18 A.U. This makes it less likely that Pluto was a satellite of Neptune, although we do not know what might have happened in the distant past.

Pluto is so far away that, the Grand Tour having been cancelled, there are no current plans to send spacecraft there. So it will be a long time before we know very much more than the very little that we now know.

17.4 OTHER PLANETS?

Every few years one sees newspaper headlines reporting the possible discovery of "Planet X," a tenth planet. So far, none of these reports has turned out to be true.

If a tenth planet had a sufficiently small mass, or was sufficiently far away, then we could not rule out its existence on the basis of current observations of planetary orbits. We could discover it only by a very lucky accident of happening to look in the right place for it. Of course, if an object we might discover is much smaller than Pluto we might not even want to call it a full-fledged planet.

One recent prediction of "Planet X" was based on the study of orbits of comets, and a mass and orbit for a perturbing planet were deduced. But it was soon realized that a planet of this mass and with an orbit at the inferred inclination would have long since disrupted the orbits of the outer planets that we do see. And besides, a search made with the 1.2-meter (48-inch) Schmidt telescope at Palomar didn't turn up the object. It would seem that we must treat all predictions of "Planet X" with great skepticism until and unless an actual photographic discovery is made.

If we limit our search to the ecliptic, then Tombaugh has set a limit on the presence of planets exterior to Pluto. He extended his planetary search after he discovered Pluto, and his observational material rules out the presence of a Neptune-sized body in the ecliptic within 270 A.U. from the Sun.

But what about a tenth planet very close to the Sun, inside the orbit of Mercury? For hundreds of years people have searched for such a planet, and its "discovery" has been reported on several occasions. This nonexistent planet even has a name: Vulcan, after the Roman god of fire, for it would be extremely hot because of its closeness to the Sun.

One ingenious attempt to explain why Vulcan has never been seen is to say that it is always directly on the other side of the Sun from the Earth. But if it is at an orbital distance less than Mercury's, Kepler's laws show that it could not remain directly opposite the Earth. And if Vulcan is of appreciable size, it would have revealed itself by gravitationally perturbing Mercury's orbit. As we have seen (in Section 5.11), it took Einstein's general theory of relativity to account for the details of Mercury's orbit, but the orbit is now satisfactorily explained.

The best time to search for an inner planet is during a total solar eclipse, when one can see stars in the sky near the Sun. No new planet has ever been seen during eclipses, although several experiments have been carried out to look for one. Such experiments have only turned up an occasional new comet. Based on these observations, we know that any additional body in orbit around the Sun could not be bigger than about ten kilometers across, far from real planetary size.

For the present, it seems fair to conclude that the Sun has nine, and only nine, planets.

SUMMARY AND OUTLINE OF CHAPTER 17

Aims: To study the three other giant planets: Saturn, Uranus, and Neptune; to understand Saturn's rings; and to see how little we know about Pluto

Saturn (Section 17.1)
- Giant planet with system of rings
- Lowest density of planets: 0.7 g/cm^3
- Ring system
  - 27° inclination to orbit
  - Orbiting chunks of rock and ice
  - Inside Roche's limit for Saturn
  - Cassini's division and other divisions caused by gravitational effects of moons
  - Very thin from top to bottom
  - Radar waves bounced off rings indicate that they are made of rocks of certain sizes and roughness
- Rapid rotation period, causing oblateness
- Methane, ethane, and molecular hydrogen detected in atmosphere
- No evidence of solid surface
- Magnetic field present
- Internal heating
- 10 moons, including Titan, largest moon in solar system
- Atmosphere detected on Titan giving rise to speculation as to the chance of finding life there
- Pioneer 11 en route to 1979 flyby

Uranus (Section 17.2)
- Large planet with low density and high albedo
- Discovered by Herschel in 1781
- Atmosphere of methane and molecular hydrogen
- No surface structure can be seen
- Axis of rotation is in the plane of its orbit and retrograde
- Short rotation period
- 5 known moons

Neptune (Section 17.2)
- Large planet with low density and high albedo
- Discovery by Adams and Leverrier part science and part luck
- Atmosphere of methane and molecular hydrogen
- Diameter measured more accurately as the result of a stellar occultation
- 2 known moons

Pluto (Section 17.3)
- Most eccentric orbit in solar system
- Greatest inclination to ecliptic
- Rotation accurately determined due to differing albedo
- Discovered in 1930 after diligent search along the ecliptic
- Radius, mass (and thus density) very difficult to determine
- Near occultation of star indicates that diameter is less than 6800 km
- Possibly similar to Mars in radius, mass, and density
- May be an escaped moon of Neptune

Search for additional planets (Section 17.4)
- Search by Tombaugh failed to turn up any new planets within 270 A.U. from Sun
- Existence of planet inside Mercury's orbit ruled out

QUESTIONS

1. What are the similarities between Jupiter and Saturn? What are the differences?
2. What is the angular size of the Sun as viewed from Saturn? How many times smaller is this than the angular size of the Sun we see from the Earth?
3. When Jupiter and Saturn are closest to each other, what is the angular size of Jupiter as viewed from Saturn? How does this compare with the angular size of Jupiter as viewed from the Earth?
4. What conditions in the early moments of the solar system would have led to other planets in addition to Saturn having rings?
5. Why are the moons of the giant planets more appealing for exploration than the planets themselves?
6. What is strange about the direction of rotation of Uranus, and how might that affect Uranus's weather?
7. Why do we know so little about Uranus and Neptune compared with Jupiter and Saturn?
8. Explain how the occultation of a star can help us learn the diameter of a planet.
9. The inaccuracies in measuring the diameter of Neptune mean that we know only that Neptune's diameter is probably between 46,900 km and 51,500 km. That is, it is known to ±5 per cent (±2300/49,200). To what percentage of accuracy is the density known?
10. What fraction of its orbit has Neptune traversed since it was discovered?
11. What fraction of its orbit has Pluto traversed since it was discovered?
12. What evidence suggests that Pluto is not a "normal" planet?
13. The position of Pluto in the sky was accurately known, so why were astronomers unsure whether it would occult a particular star in 1965?
14. Summarize the evidence that suggests that Pluto is not a giant planet.
15. Why might we expect to find the atmosphere of Pluto composed primarily of neon?

REVIEW QUESTIONS ON THE PLANETS

1. List the planets in order of
   (a) increasing size,
   (b) increasing density,
   (c) increasing distance from the Sun.

2. Which planets probably have significant internal energy sources?

3. Mars and Venus are the planets potentially "most like" the Earth, but there are substantial differences. To what degree do you think the various differences can be ascribed to their different distances from the Sun, and to what extent must other explanations be invoked?

4. Which two planets come closest together?

5. If you were an astronomer and could set up an observatory on any planet, which would you choose? To what degree would your decision be influenced by the part of the spectrum in which you are interested?

6. Of all the moons in the solar system, the one whose existence is hardest to understand is the Earth's. Why?

7. Which planets have radiation belts like the Earth's? What property of the planet determines whether or not it will have radiation belts?

8. Classify the planets according to those whose atmosphere is thicker or thinner than that of the Earth.

9. Which of the three photographs in Set A shows (a) the Moon, (b) Mars, and (c) Mercury? Which of the three close-up views in Set B shows (a) the Moon, (b) Mars, and (c) Mercury? How did you know in each case?

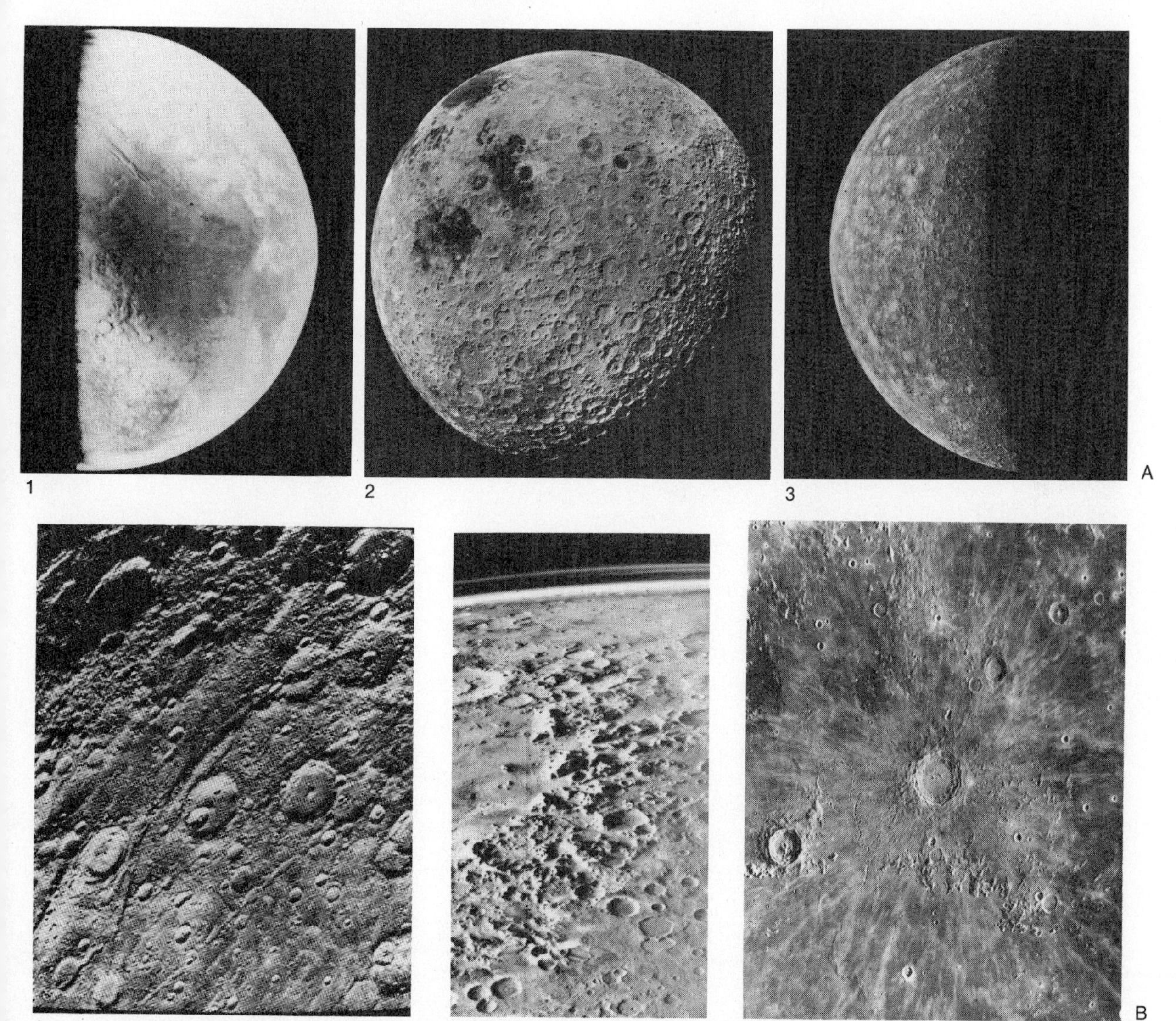

18

Comets, Meteoroids, and Asteroids

Besides the planets and their moons, there are many other objects in the family of the Sun. The most spectacular, as seen from Earth, are some of the comets. Bright comets have been noted throughout recorded history, instilling in observers great awe of the heavens.

It has been realized since the time of Tycho Brahe, who studied the comet of 1577 (see Fig. 10–20), that comets are phenomena of the solar system rather than of the Earth's atmosphere. From the fact that the comet did not show a parallax when observed from different locations on Earth, Tycho deduced that the comet was at least three times farther away from the Earth than the Moon.

Asteroids and meteoroids are other residents of our solar system. We shall see how, along with the comets, they may prove to be storehouses of information about the solar system's origin.

Comets have long been seen as omens.
"When beggars die, there are no comets seen;
The heavens themselves blaze forth the death of princes."
Shakespeare,
Julius Caesar

18.1 COMETS

Every few years, a bright comet fills our sky with its tremendous tail (Fig. 18–1). From a small, bright area called the *head,* a *tail* may extend gracefully over one-sixth (30°) or more of the sky. Although the tail may give an impression of motion, because it extends out to only one side, the comet does not move visibly across the sky as we watch. With binoculars or a telescope, however, an observer can accurately note the position of the head with respect to nearby stars and detect that the comet is moving at a

The "long hair" that is the tail led to the name comet, *which comes from the Latin for "long-haired star,"* aster kometes.

Comet West, a bright comet that was visible to pre-dawn observers in the northern hemisphere in 1976.

Figure 18–1 Comet Ikeya-Seki over Los Angeles in 1965, observed from one of the solar towers at the Mount Wilson Observatory. Photographs like this are taken with ordinary 35-mm cameras; this one was a 32-sec exposure on Tri-X film at f/1.6.

slightly different rate from the stars as comet and stars rise and set together. A photograph of even a few seconds duration will show the relative motion. By the next night, the comet may have moved 2°, four times the diameter of the Moon. (Both its right ascension and its declination change.)

Within days or weeks this bright comet will have faded below naked-eye brightness, though it can be followed for additional weeks with binoculars and then for additional months with telescopes.

The tail of a comet is generally directed away from the Sun. Thus, if we see the comet setting in the western sky after sunset, its tail would extend easterly, directed upward toward the zenith and away from the Sun. Conversely, if we see the comet rising in the eastern sky before dawn, its tail would extend westerly toward the zenith.

Most comets are much fainter than the one described above. Half a dozen or so new comets are discovered every year, and most become known only to astronomers. An additional few are "rediscovered"—that is, from the orbits derived from past occurrences it can be predicted when and approximately where in the sky a comet will again become visible. Up to the present time, over 600 comets have been discovered, 13 in 1975 alone. The returns of four already known comets were also detected in that year.

Box 18.1 Discovering a Comet

A comet is generally named after its discoverer, the first person to see it (or the first two or three people if they independently find it without too much time separating the first observation). Comets are also assigned letters and numbers. First the letters are assigned in order of discovery in a given year. Then, a year or two later when all the comets that passed near the Sun are likely to be known, Roman numerals are assigned in order of their perihelion passage. The two comets, for example, discovered by Lubos Kohoutek in March 1973, were both named Comet Kohoutek. The one that eventually went close to the Sun was first known as 1973f. It was assigned the number 1973XII at the end of 1974, a year after it passed perihelion.

Many discoverers of comets are amateur astronomers, including some amateurs who examine the sky each night in hope of finding a comet. To do this, one must know the sky very well, so that one can tell if a faint, fuzzy object is a new comet or a well-known nebula. It is for that reason that Messier made his famous eighteenth century list of nebulae (described in Chapter 23).

If you find a comet, telegraph the International Astronomical Union Central Bureau for Astronomical Telegrams, at the Smithsonian Astrophysical Observatory in Cambridge, Massachusetts (telegraph address—RAPID SATELLITE CAMBMASS). If it is during the daytime, you may alternatively telephone (617) 864-5758, though this is a less desirable method. In any case, you should identify the direction of the comet's motion, its position, and its brightness. Don't forget to identify yourself with your name, address, and telephone number. If you are the first (or maybe even the second or third) to find the comet, it will be named after you.

The record for discovering comets belongs to the former caretaker of the Marseilles Observatory, Jean Louis Pons, who discovered 37 of them between 1801 and 1827. Several Japanese amateur astronomers have found many comets: Minoru Honda has found a dozen. After work, he spends many hours each night scanning the sky. But the comet may become sufficiently bright for discovery only while it is daylight or cloudy in Japan. So there is hope for less dedicated observers.

Many comets, particularly the ones discovered by professional astronomers, are discovered not by eye but only by examination of photographs taken with telescopes. The photographs are usually those taken for other purposes, and so the discovery of a comet is serendipity—a fortuitous extra discovery—at work.

18.1a The Composition and Origin of Comets

At the center of the head of a comet is the *nucleus*, which is at most a few kilometers or so across. It is composed of chunks of matter. The most widely accepted theory of the composition of comets, advanced in 1950 by Fred L. Whipple of the Harvard and Smithsonian Observatories, is that the

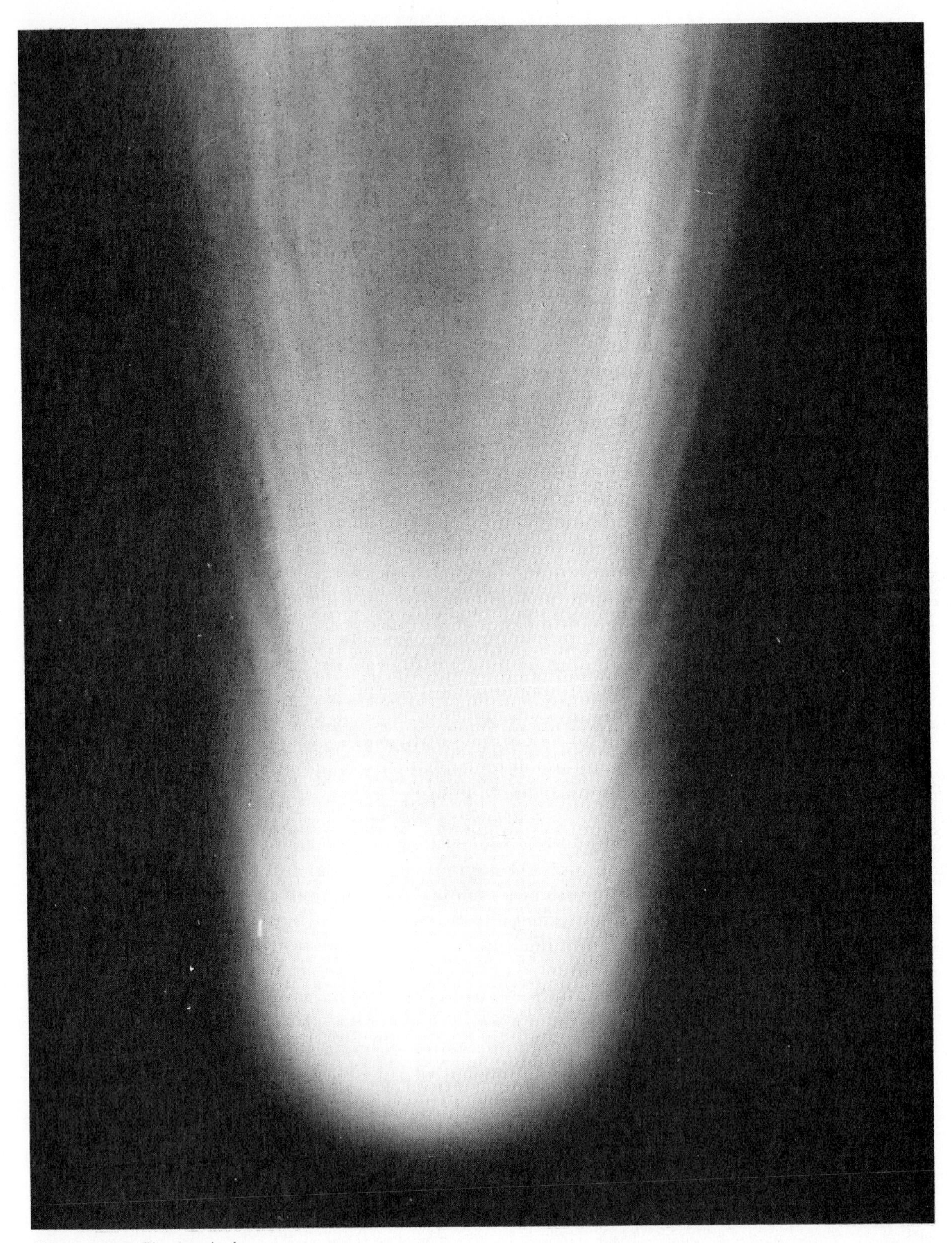

Figure 18–2 The head of Halley's Comet in 1910.

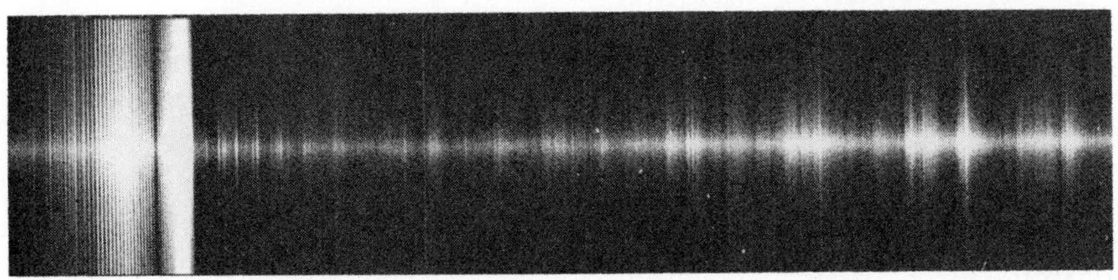

Figure 18–3 Part of the spectrum of Comet Tago-Sato-Kosaka, photographed in 1969 at the European Southern Observatory in Chile. Molecules give not just single lines but rather bands (groups) of lines; one of the bands of CN is visible at the left. Bands of CH and C_3 are seen in the rest of the photograph.

nucleus is like a *dirty snowball.* The nucleus may be ices of such molecules as water (H_2O), carbon dioxide (CO_2), ammonia (NH_3), and methane (CH_4), with dust mixed in.

This model explains many observed features of comets, including the fact that the orbits of comets do not appear to accurately follow the laws of gravity. When sunlight evaporates the ices, the molecules are expelled from the nucleus. This action generates an equal and opposite reaction, the same force that runs jet planes. Since the comet nucleus is rotating, the force is not always directly away from the Sun, even though the evaporation is triggered on the sunny side. In this way, comets show the effects of nongravitational forces in addition to the effect of the force of the solar gravity.

The nucleus itself is so small that it is impossible to observe directly from Earth. It is surrounded by the *coma* (pronounced cō′ ma), which may grow to be as large as 100,000 km or so across. The coma shines partly because its gas and dust are reflecting sunlight toward us and partly because gases liberated from the nucleus are excited enough that they radiate. (Since they are excited by solar ultraviolet radiation and radiate in the visible, this is an example of fluorescent processes, which are described in Box 20.1.) The spectrum of a comet head shows sets of lines from simple molecules (Fig. 18–3).

The nucleus *and* coma *together are the* head *of a comet (Fig. 18–2).*

AUGUST 24, 1957

AUGUST 22, 1957

Figure 18–4 In Comet Mrkos, the straight *ion tail,* extending toward the top, and the *dust tail,* gently curving toward the right, were clearly distinguished.

The tail can extend as much as 1 A.U. (150,000,000 km), and so comets can become the largest objects in the solar system. But the amount of matter in the tail is very small—the tail is a much better vacuum than we can make in laboratories on Earth.

Many comets actually have two tails (Fig. 18–4). Both extend generally in the direction opposite to that of the Sun, but are different in appearance. The *dust tail* is caused by dust particles that had been impurities in the ices of the nucleus, released when the ice was vaporized. The dust particles are left behind in the comet's orbit; they are blown slightly away from the Sun by the pressure caused by photons of solar light hitting the particles (this is known as "radiation pressure"). As a result of the comet's continued orbital motion, the dust tail usually curves smoothly behind the comet. The *gas tail* (also called the *ion tail*) is composed of ions (such as CO^+, N_2^+, CO_2^+, and CH^+) blown out more or less straight behind the comet by the solar wind (which was described in Section 5.7). As puffs of ionized gas are blown out and as the solar wind varies, the ion tail takes on a structured appearance. Each puff of matter can be seen. The magnetic fields carry only the ionized matter along with them; the neutral atoms are left behind in the coma.

With the second Orbiting Astronomical Observatory, we became able to observe in comets the Lyman alpha line of hydrogen, which is in the ultraviolet. In this way, it was discovered in 1970 that a huge hydrogen cloud (Fig. 18–5) surrounds the head. This hydrogen cloud may be a million miles in diameter! It probably results from the break up of water molecules by ultraviolet light from the Sun.

A comet—head and tail together—contains less than a billionth of the mass of the Earth. It has been said that comets are as close as something can come to being nothing.

It is now generally accepted that there are hundreds of millions of incipient comets surrounding the solar system in a sphere perhaps 50,000 A.U. (almost 1 light year) in radius. This sphere is known as the *Oort comet*

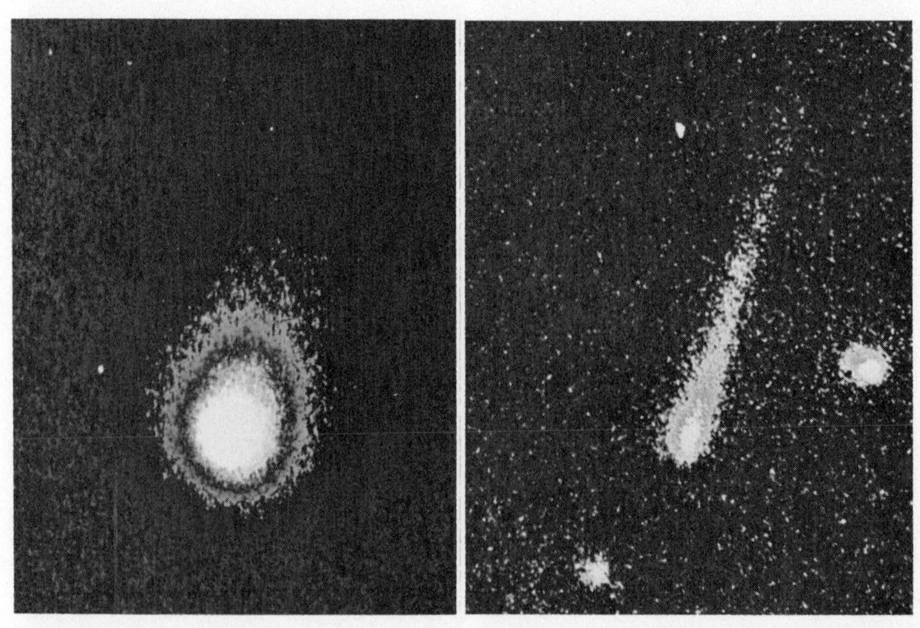

Figure 18–5 *(A)* These contours of intensity result from a photograph of Comet Kohoutek taken by the Skylab astronauts outside the Earth's atmosphere. The photograph was taken in the Lyman alpha line of hydrogen at 1216 Å in the ultraviolet, and shows the hydrogen halo. This halo was about 1° across, about 2,500,000 km in diameter, at the time of this observation. (*B*) These contours result from a photograph that was taken immediately after Fig. 18–5 *A*, but through a filter that did not pass the Lyman alpha line. Thus the hydrogen halo does not show, though the tail does. The tail was about 2° across, about 5,000,000 km in diameter. Hot stars from the background constellation Sagittarius also show.

cloud after Jan H. Oort, the Dutch astronomer who advanced the theory in 1950. The total mass of matter in the cloud is only 10 to 100 times the mass of the Earth. Occasionally one of the incipient comets leaves the Oort cloud, perhaps because gravity of a nearby star has tugged it out of place, and the comet approaches the Sun. Its orbit is a long ellipse (that is, an ellipse of high eccentricity). If the comet passes near the Jovian planets, its orbit is altered by the gravity of these massive objects. Because the Oort cloud is spherical, comets are not limited to the plane of the ecliptic and come in randomly from all angles.

As the comet gets closer to the Sun, the solar radiation that reaches it begins to vaporize the molecules in the nucleus. We have not generally been able to detect the molecules in the nucleus directly—the *parent molecules*—though we would clearly love to do so, but we have mostly detected in the head the simpler molecules into which the parent molecules break down—the *daughter molecules*.

Examples of daughter molecules are H, OH, O, CN, C_2, C_3, CO^+, NH, NH_2, CH, and N_2^+. We think the parents may have been H_2O, NH_3, CH_4, C_2, N_2, and CO_2, and perhaps C_2H_2, C_2N_2, and C_3H_4 (to beget C_3). The spectra change as the comet changes its distance from the Sun. Certain of the lines are seen only in the tail.

The tail begins to form, and as more and more of the nucleus is vaporized, the tail grows longer and longer. Still, even though the tail can be millions of kilometers long, it is so tenuous that only 1/500 of the mass of the nucleus may be lost. Thus a comet may last for many passages around the Sun.

The comet is brightest and its tail is generally longest at about the time it passes perihelion (the closest point in its orbit to the Sun). However, because of the angle at which we view the tail from the Earth, it may not appear the longest at this stage.

Following perihelion, as the comet recedes from the Sun, its tail fades; the head and nucleus receive less solar energy and fade as well. The comet may be lost until its next return, which could be as short an interval as 3.3 years (as it is for Encke's Comet) or as long as 80,000 years (as it is for Comet Kohoutek) or more. Periodic comets lose their mass reappearance after reappearance, and the comet eventually disappears. We shall see in Section 18.2 that some of the meteoroids are left in its orbit.

Because new comets come from the places in the solar system that are farthest from the Sun and thus coldest, they probably contain matter that is unchanged since the formation of the solar system 4.6 billion years ago. So the study of the constituents of comets is important for understanding the early stages of solar system formation.

Figure 18–6 Edmund Halley.

18.1b Halley's Comet

In 1705, the English astronomer Edmund Halley (Fig. 18–6) applied a new method developed by his friend Isaac Newton to determine the orbits of comets from observations of the positions of the comets in the sky. He reported that the orbits of the bright comets that had appeared in 1531, 1607, and 1682 were about the same. Because of this, and because the intervals between appearances were approximately equal, Halley suggested that we were observing a single comet orbiting the Sun, and predicted that it would again return in 1758. The reappearance of this bright comet on Christmas Night of that year, 16 years after Halley's death, was the proof of Halley's hypothesis (and Newton's method); it has since been known as Halley's Comet. It seems probable that the bright comets reported every 74 to 79 years since 240 B.C. (and possibly even before then) were earlier appearances. The fact that Halley's Comet has been observed at least 29

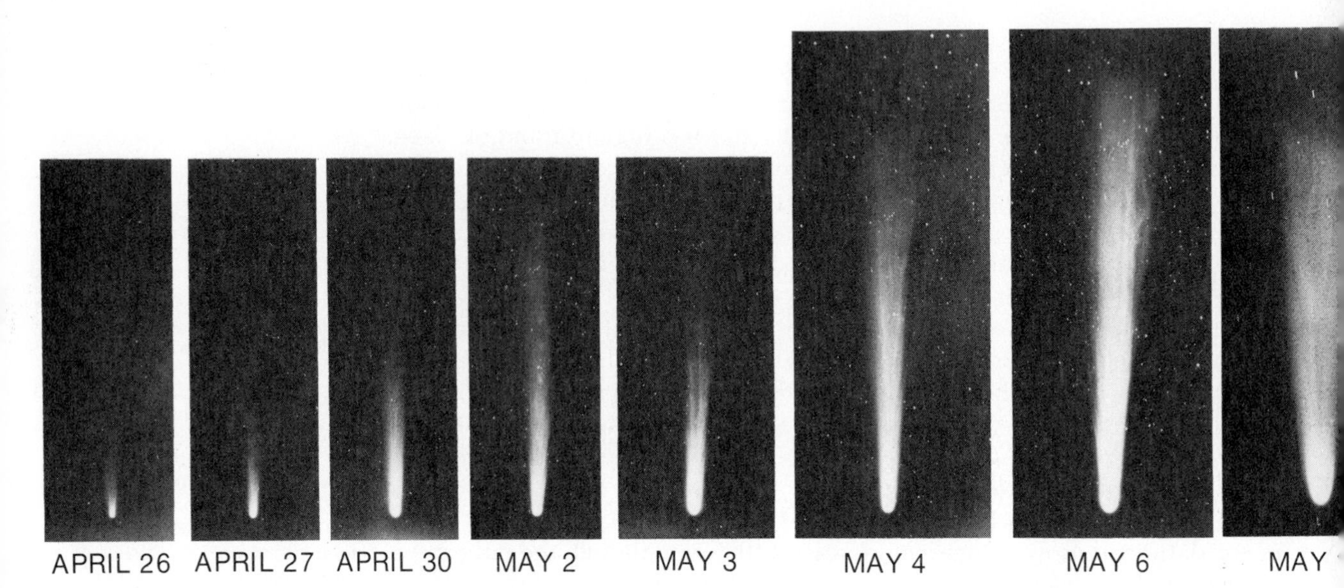

times endorses the calculations that have been made that show that less than one per cent of a cometary nucleus's mass is lost at each perihelion passage.

Mark Twain was born during the appearance of Halley's Comet in 1835, and often said that he came in with the comet and would go out with it. And so he did, dying during the 1910 return of Halley's Comet.

Halley's Comet went especially close to the Earth during its 1910 return (Fig. 18–7), and the Earth actually passed through its tail. Many people had been frightened that the tail would somehow damage the Earth or its atmosphere, but the tail had no noticeable effect. It was known to most scientists even then that the gas and dust in the tail were too tenuous to harm our environment.

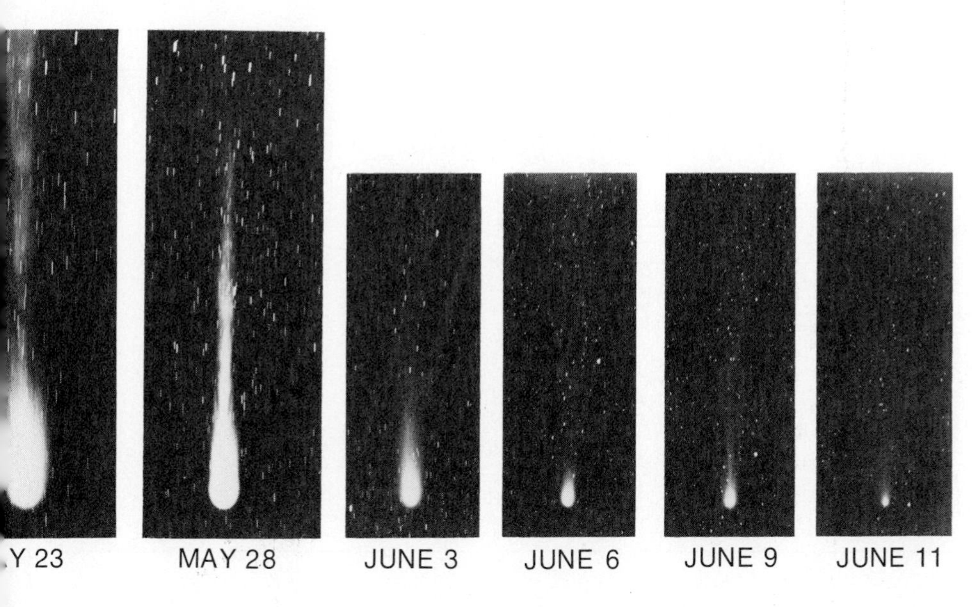

Figure 18–7 Halley's comet in 1910.

Halley's Comet reappears every 74 to 79 years; Jupiter and Saturn perturb its orbit enough to cause the variation. Since we can count on its reappearance in 1985 in time for its 1986 perihelion (Fig. 18–8), we can make plans long in advance to observe it. The possibility of launching a spacecraft to fly near it or even through its tail is being discussed, though no definite plans yet exist.

The 1986 reappearance will not be as spectacular as its 1910 passage, however, for the Earth will be in a part of its orbit such that the comet is not seen broadside. Thus the tail will appear foreshortened, and is not expected

Figure 18–8 *(A)* The orbit of a comet.

Illustration continued on following page.

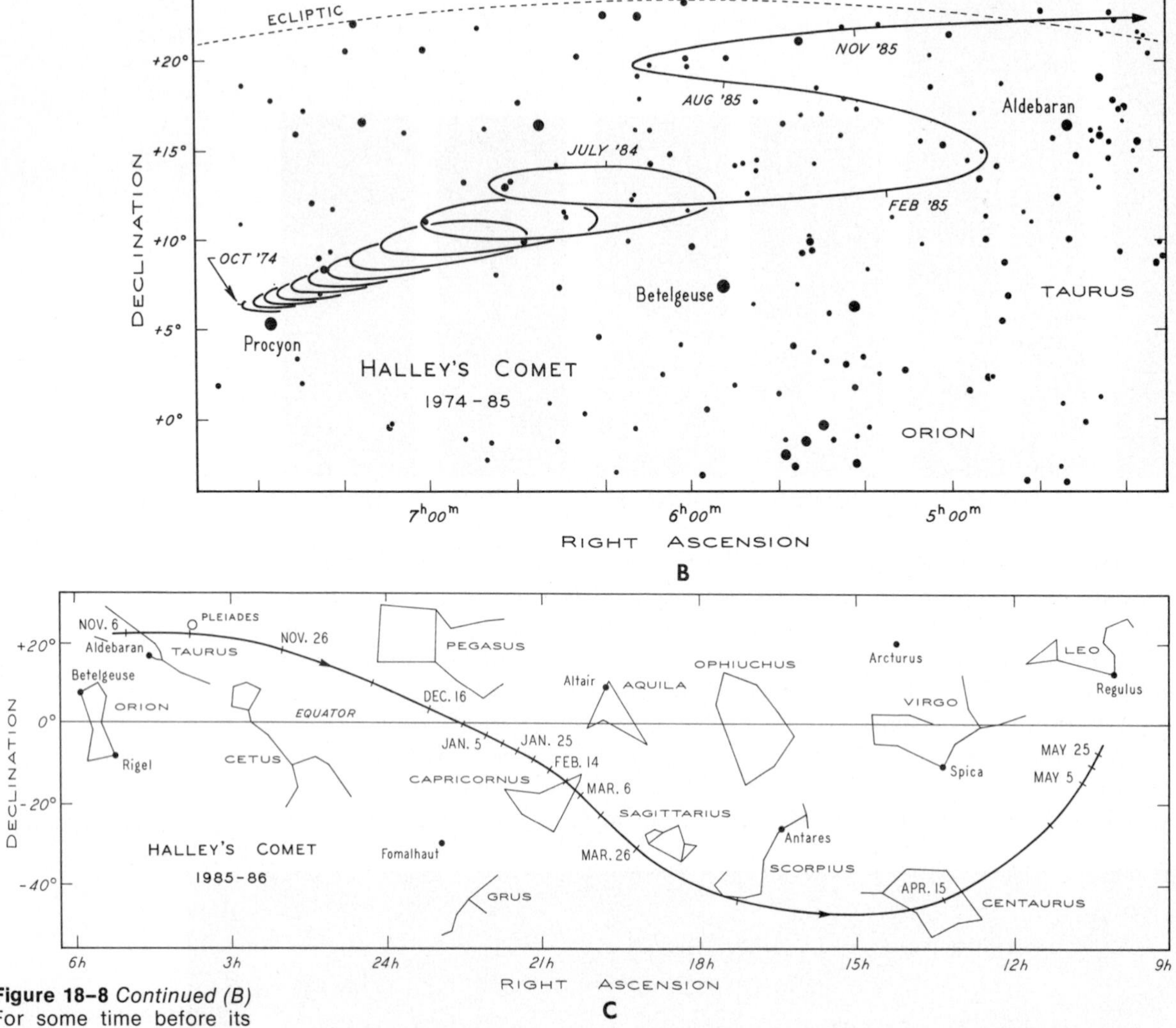

Figure 18–8 *Continued (B)* For some time before its perihelion, Halley's Comet will follow an orbit that seems to circle in the sky as a result of the Earth's revolution around the Sun. *(C)* When Halley's Comet draws close to us, its apparent motion across the sky will grow more rapid. Observers can use these charts to search for Halley's Comet in 1985, when we expect to be able to see it coming.

to appear as spectacular as, for example, 1975n, Comet West (which is shown in the photograph opening this chapter and is discussed in the following section).

18.1c Comet Kohoutek and Comet West

In March of 1973, Lubos Kohoutek was studying faint asteroids on photos he had taken at the Hamburg Observatory in Germany where he works. On two of these plates he discovered faint comets—about 16th magnitude—which were named for him when he sent word of the discoveries to the IAU Central Bureau for Astronomical Telegrams.

Further analysis of his plates of the second comet, including a further set of plates he took later in the month and a two-month-old plate that he found also showed the comet, revealed that he had made a most unusual discovery: the comet would not reach perihelion for about nine months and would pass very close to the Sun—within only 21 million km (.14 A.U.)—thus becoming very bright. Because of these predictions the second comet, 1973f, became known the world over as *Comet Kohoutek.*

Never before had a potentially bright comet been discovered this far from the Sun—at the distance of Jupiter's orbit—or this long before peri-

Figure 18–9 Sketches of Comet Kohoutek and its sunward spike by the astronauts aboard the third Skylab mission.

helion. Thus it was impossible to make an accurate prediction of the amount of brightening, but it seemed that it might brighten from 16th to −5th magnitude, a factor of 250,000,000. On the basis of this prediction, scientists readied all kinds of experiments that required months to prepare, and reserved time on large optical and radio telescopes for the period of anticipated brightness. The long advance notice even allowed the schedule of the third crew of astronauts due to visit Skylab to be delayed so that they would be aloft during Comet Kohoutek's perihelion passage. Anyone who has ever worked with a space experiment will especially realize how major an event it is to change such a schedule.

After a few months, the Earth in its orbit swung around to the side of the Sun opposite to Kohoutek, and we could temporarily no longer follow the comet's development. When the comet eventually re-emerged from the Sun's glare it had not brightened quite as much as had been expected, and so the astronomers revised toward fainter levels their predictions of the comet's eventual maximum brightness. Somehow these revised and updated predictions never became as widely circulated as the original hopes that the comet would be brighter than Venus.

Comet Kohoutek was expected to be at its brightest within a few days of its perihelion passage. During that time it was too close to the Sun to be seen by anyone except the astronauts in Skylab, who photographed (see Color Plate 21) and sketched the comet. The shape and size of the comet and the tail appeared to change rapidly, and these Skylab data on this phase of the development of the comet are unique. The comet was actually almost as bright as Venus at this time, but it was too close to the Sun for anybody except the astronauts to see it.

A day after perihelion, the astronauts discovered a brilliant spike—an anti-tail—pointing toward the Sun, the opposite direction from that in which comet tails point (Fig. 18–9). This rare phenomenon had previously been seen to such an extent in only one other modern comet, and to any extent in only a dozen comets. It had been thought that sunward spikes were caused only by effects of perspective, and that such spikes were actually directed away from the Sun. But analysis of the Kohoutek observations showed that the spike contained meteoroid particles as large as 1 mm across, far larger than the dust in a comet tail. These particles are controlled by different mechanisms from those that control the particles in the tail, and the spike was probably actually pointing in the general direction of the Sun.

A week after perihelion the comet drew far enough away in the sky from the Sun that its tail could be seen by observers on Earth. Unfortunately, it was just barely bright enough to be seen with the naked eye, and was not the spectacular object to which many had been looking forward. In particular, it could not be seen by observers in cities, where light pollution

Figure 18–10 A photograph of Comet Kohoutek with the 1.2-meter Schmidt telescope on Palomar Mountain. Although the comet was too faint to be seen well with the naked eye, it was a fine photographic object and was bright enough to enable unique scientific observations to be made.

made the sky brighter than the comet's tail. Nevertheless, the tail stretched gracefully over many degrees of sky (Fig. 18–10). Observers with optical telescopes and radio telescopes on the ground were able to get good spectra and discover molecules that had never before been seen in comets. Radio astronomers reported the detection of two molecules—methyl cyanide (CH_3CN) and hydrogen cyanide (HCN)—that had previously been detected only in interstellar space (as will be discussed in Section 21.6). If these reports could be fully accepted, they would indicate that comets were formed far from the Sun, because otherwise these molecules would have been torn apart. In particular, the observations would tend to favor the model that the comets formed at interstellar distances, even though it seems that they have always been bound to the solar system.

Why did Kohoutek appear so bright so early on? It has been hypothesized that Kohoutek may have been a "virgin comet," a comet making its first passage around the Sun. Thus, an especially great amount of material was vaporizing when the comet was first discovered, making it appear brighter than average for that distance from the Sun and from the Earth. Once that layer of material escaped, lower layers were more ordinary. We can think of the Sun vaporizing layer after layer of the material of a comet's nucleus, similar to the way in which we can peel off layer after layer from an onion. The idea that the comet was shedding dust was borne out by infrared observations; the fact that a subsequent comet's infrared brightness dropped by a factor of a hundred within a few weeks lends support to the layered-onion theory.

Comet Kohoutek is now receding into space, and will go out as far as 4000 A.U. While it is far from the Sun, we can see from Kepler's second law that it will move very slowly. It won't be back in our vicinity for another 80,000 years.

Astronomers often say that comets are unpredictable. Comet Kohoutek, which was a popular bust and an astronomer's delight, bore out that idea.

Bright comets ordinarily appear with much less notice than Kohoutek had. In late 1975, Richard West, of the European Southern Observatory, discovered on a plate taken in Chile a comet that within a couple of months became an object very easy to see with the naked eye. Before perihelion, Comet West was visible only to observers in the southern hemisphere, but after perihelion it was visible in the predawn sky to northern hemisphere observers (see Color Plate 40). For a few days its tail extended over 30°. Bits of dust in the nucleus were ejected at an irregular rate, which led to the delicate structure in the tail that is visible in the photograph opening this chapter. Indeed, the nucleus itself broke into at least four pieces. On the basis of the orbit, it might not be back for a million years; it might even have been set on a course of ejection from the solar system, in which case it will never return.

18.2 METEOROIDS

There are many small chunks of matter in interplanetary space, ranging up to tens of meters across. When these chunks are in space, they are called *meteoroids*. When one hits the Earth's atmosphere, friction slows it down and heats it up—usually at a height of about 100 km—until all or most of it is vaporized. Such events result in streaks of light in the sky, which we call *meteors*. (Meteors are popularly known as *shooting stars*.) The brightest meteors can reach magnitude −15 or even −20, which is brighter than the full moon. We call such bright objects *fireballs* (Fig. 18–11). We can sometimes even hear the sounds of their passage and of their breaking up into smaller bits. When a fragment of a meteoroid survives its passage through the Earth's atmosphere, the remnant that we find on Earth is called a *meteorite*. We now also refer to objects that hit the surfaces of the Moon or of other planets as meteorites.

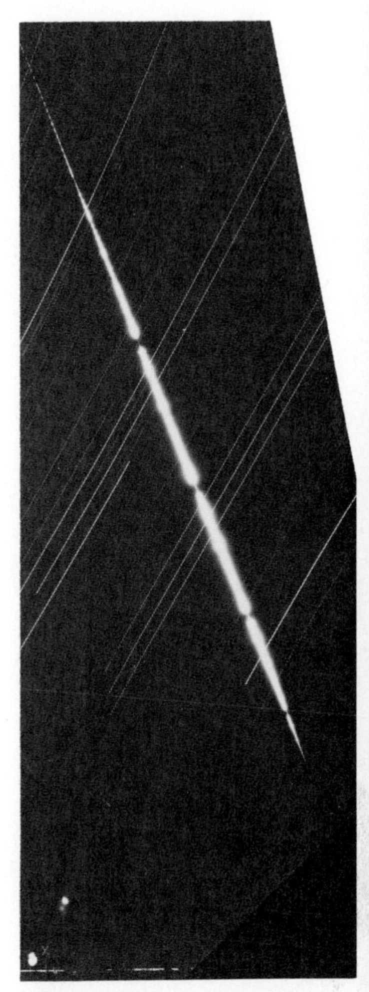

Figure 18–11 A fireball observed in 1970 by the "Prairie Network" of the Smithsonian Astrophysical Observatory, a network of wide-field cameras spaced around the mid-west of the United States in hope of pinpointing a meteor's path and thus permitting the meteorite to be found. A shutter in the camera rotates so that breaks occur in the image of the meteor trail at intervals of 1/20 sec. This permits timing of the motion. The fireball pictured here was photographed by two of the cameras of the Network, which led to the discovery of a meteorite at Lost City, Oklahoma. The trail persisted for about 8 seconds, though the short breaks in the trail from which this could normally be measured do not show because of film overexposure except at the top.

We have seen in Chapter 11 that meteorites have formed almost all of the craters on the Moon and that tiny meteorites less than a millimeter across, called *micrometeorites*, are the major cause of erosion on the Moon. Micrometeorites also hit the Earth's upper atmosphere all the time, and remnants of the material they deposit can be collected for analysis from balloons or airplanes. The micrometeorites, which may have been only the size of a grain of sand when they first hit the Earth's atmosphere, are often sufficiently slowed down before they are vaporized that they can reach the ground. Dust resulting from micrometeorites can be sampled by collecting bits of ice from the Arctic or the Antarctic, or from mountain tops.

There are several major kinds of meteorites. Most of the meteorites that are found have a very heavy content—about 90 per cent—of iron; the rest is nickel. These *iron meteorites* (or, for short, *irons*) are thus very dense—that is, they weigh quite a lot for a given volume.

Most meteorites that hit the Earth are stony in nature and are often referred to simply as *stones*. Most of these stony meteorites are of a type called *chondrites*, because they contain rounded particles called "chondrules." Because stony meteorites resemble ordinary rocks, they are not usually discovered unless their fall is observed. That explains why we more usually discover irons. But when a fall is observed, most meteorites recovered are stones. The stony meteorites have a high content of silicates; only about 10 per cent of their mass is iron and nickel.

The largest meteorite crater on Earth may be a depression over 400 km across deep under the Antarctic ice pack. This is comparable with the size of lunar craters. Another very large crater, in Hudson's Bay in Canada, is filled with water. Most meteorite craters on Earth are either disguised in such ways or have eroded away. A large crater that is obviously meteoritic in origin is the Barringer Crater in Arizona (Fig. 18–12). It is the result of

TABLE 18–1 METEORITES

Composition	Seen Falling	Finds
Irons	6%	66%
Stony-irons	2%	8%
Stones	92%	26%

Figure 18–12 The Baringer meteor crater in Arizona.

what was perhaps the most recent large meteor to hit the Earth, for it was formed only 25,000 years ago.

Tektites, *small, rounded glassy objects that are found at several locations in the Earth's southern hemisphere, may have splashed out from this impact. The origin of tektites, including whether they came from the Earth or from the Moon, has long been controversial, though a terrestrial origin seems most probable.*

Every few years a meteorite is discovered on Earth immediately after its fall (Fig. 18–13). The chance of a meteorite's landing on someone's house is very small, but it has happened! Often the positions of fireballs in the sky are tracked in the hope of finding fresh meteorite falls. The newly discovered meteorites are rushed to laboratories for chemical analysis of their constituents before the Earth's atmosphere or human handlers can contaminate them.

One *carbonaceous chondrite* (a rare kind of stony meteorite with a high carbon content) was found to contain simple amino acids. The object was the Murchison meteorite, which fell near Murchison, Victoria, Australia in 1969. The formation of such complicated building blocks of life in cold,

Figure 18–13 A meteor crater 2 meters wide and 5.5 meters deep, one of a hundred from the fall in Kirin Province of China on March 8, 1976. A fireball was seen before the meteorites landed. The largest meteorite is the largest stony meteorite ever recovered; it weighs 1770 kg (3894 lbs).

isolated places like meteoroids is one of several indications that the precursors of life develop naturally. We shall be developing more of this evidence in the following chapter.

The question was raised whether these amino acids were truly extraterrestrial or had rather entered the samples on Earth. Analysis showed that the amino acids contained equal quantities of types that cause the polarization of light to be rotated in right-handed and left-handed senses; as a result, no rotation of the plane of polarization is seen. Since all amino acids in living beings on Earth cause polarized light to be rotated in the left-handed sense, this indicates that the amino acids in the Murchison meteorite are extraterrestrial.

As we can measure the ages of the meteorites through study of the ratios of radioactive and non-radioactive isotopes contained in them, we know that the meteoroids were formed at times up to 4.6 billion years ago, the time of the beginning of the solar system. Furthermore, the chondrites show no sign of ever having melted, and may date from the formation itself; some may be the planetesimals. The analysis of the abundances of the elements in meteorites thus tells us about the solar nebula from which the solar system formed. In fact, up to the time of the first landing on the Moon, meteorites were the only extraterrestrial material we could get our hands on.

The largest carbonaceous chondrite available for scientific study is the Allende meteorite, which fell in Mexico in 1969. Sufficient quantities have been available for analysis that our knowledge of the abundances in meteorites has been greatly improved. For example, we can demonstrate that enough radioactive aluminum (Al^{26}) existed long ago to provide enough heat to melt those meteorites that show signs of melting.

Most of our knowledge of the abundances of the elements in the solar system has come either from analysis of solar radiation or from analysis of meteorites. Until recently it was thought that all the analyses were consistent with uniform abundances of each element throughout the solar nebula. But evidence is growing, including in particular the high quality data from the Allende meteorite, that abundances varied from place to place in the solar nebula. The abundances of common elements like oxygen may vary by 5 per cent; the abundances of rare earths can vary by greater amounts.

Figure 18–14 A meteor crossing the field of view while the Palomar Schmidt was taking a 15-minute exposure of Comet Kobayashi-Berger-Milon on August 11, 1975. The cluster of galaxies in Ursa Major and Canes Venatici is also in the field; M106 is the most prominent galaxy visible. This photograph shows objects that are at three entirely different scales of distance. The meteor is in the Earth's atmosphere; the comet is in interplanetary space; and the cluster of galaxies is extremely far away.

TABLE 18–2 METEOR SHOWERS*

Name	Date of Maximum	Duration Above 25% of Maximum	Approximate Limits	Number per Hour at Maximum
Quadrantids	Jan 4	1 day	Jan 1–6	110
Lyrids	April 22	2 days	April 19–24	12
Eta Aquarids	May 5	3 days	May 1–8	20
Delta Aquarids	July 27–28	–	July 15–Aug 15	35
Perseids	Aug 12	5 days	July 25–Aug 18	68
Orionids	Oct 21	2 days	Oct 16–26	30
Taurids	Nov 8	–	Oct 20–Nov 30	12
Leonids	Nov 17	–	Nov 15–19	10
Geminids	Dec 14	3 days	Dec 7–15	58

*The number of sporadic meteors per hour is 7 under perfect conditions. The visibility of showers depends mostly on how bright the Moon is on the date of the shower, which depends on its phase. Meteors are best seen with the naked eye; using a telescope or binoculars merely restricts your field of view.

Space is full of meteoroids of all sizes, with the smallest particles being the most abundant. Most of the small particles, less than about 1 mm across, come from comets. Most of the large particles, more than about 1 cm across, come from collisions of asteroids in the asteroid belt. It is generally these larger meteoroids that become meteorites. But some of the carbonaceous chondrites may be cometary debris.

Meteors often occur in *showers*, that is, times when the rate at which meteors are seen is far above average. On any clear night a naked-eye observer may see a few *sporadic* meteors an hour, that is, meteors that are not part of a shower (Fig. 18–14). (Just try going out to a field in the country and watching the sky for an hour.) During a shower many meteors may be visible to the naked eye each minute. Meteor showers occur at the same time each year (Table 18.2), and probably represent the passage of the Earth through the orbits of defunct comets. The meteoroids in the orbit of a former comet may be the products of the decay of the comet. Meteorites usually do not result from showers, so presumably the meteoroids that cause a shower are very small.

Dust in interplanetary space can be seen, from locations with exceptionally clear skies, as the zodiacal light, *the reflection of sunlight off the dust grains. The F-corona seen at solar eclipses is also interplanetary dust.*

The rate at which meteors are seen usually increases after midnight on the night of a shower, because that side of the Earth is then facing and plowing through the oncoming interplanetary debris. If the Moon is gibbous or full, then the sky is too bright to see the shower well. The meteors in a shower are seen in all parts of the sky, but their trajectories all seem to emanate from a single point in the sky called the *radiant*. A shower is usually named after the constellation that contains its radiant. The existence of a radiant is just an optical illusion because perspective makes a set of parallel paths approaching you appear to emanate from a point.

Figure 18–15 Asteroids leave a trail on a photographic plate when the telescope is tracking at a sidereal rate, that is, with the stars.

The seismic network on the Moon occasionally—about every six months—detects diffuse groups of meteoroids hitting the Moon. Analysis of one such encounter indicates that the cloud of meteoroids may have been 15,000,000 km (0.1 A.U.) in diameter and contained in its many objects a total mass equivalent to a single object 300 meters in diameter. The individual objects in these storms seem much more massive than the meteoroids that cause the meteor showers that we observe on Earth; we do not observe meteorite storms on Earth corresponding to these lunar meteorite storms because the objects arrive only one at a time at intervals of a few days and for the most part do not penetrate our atmosphere.

18.3 ASTEROIDS

The nine known planets were not the only bodies to result from the agglomeration of planetesimals 4.6 billion years ago. Thousands of *minor planets,* called *asteroids,* also resulted. They are detected by their small motions in the sky relative to the stars (Fig. 18–15).

Most of the asteroids are found in elliptical orbits whose average size is between the sizes of the orbits of Mars and Jupiter. The zone where most asteroids are found is called the *asteroid belt.* Indeed, it was not a surprise when the first asteroid was discovered, on January 1, 1801, the first day of the nineteenth century, because Bode's law (described in Section 10.4) predicted the existence of a new planet in approximately that orbit.

This first asteroid was discovered by a Sicilian monk, Giuseppe Piazzi, and was named Ceres after the patron saint of Sicily. Though the discovery of Ceres was a welcome surprise, even more surprising was the subsequent discovery in the next few years of three more asteroids. Bode's law hadn't called for **them!** The new asteroids were named Pallas, Juno, and Vesta, also after goddesses, which began the generally observed tradition of assigning female names to asteroids.

The next asteroids weren't discovered until 1845, and over 2000 are now named and numbered. They are assigned numbers when their orbits become well determined, leading to such names as 1 Ceres, 16 Psyche, and 433 Eros. Only half a dozen asteroids discovered are known to be larger than 300 km across, and over 200 more are larger than 100 km across. Per-

1 Ceres has a mass of 5.9×10^{-10} that of the Sun, 2 Pallas has 1.1×10^{-10} the mass of the Sun, and 4 Vesta has 1.2×10^{-10} the mass of the Sun.

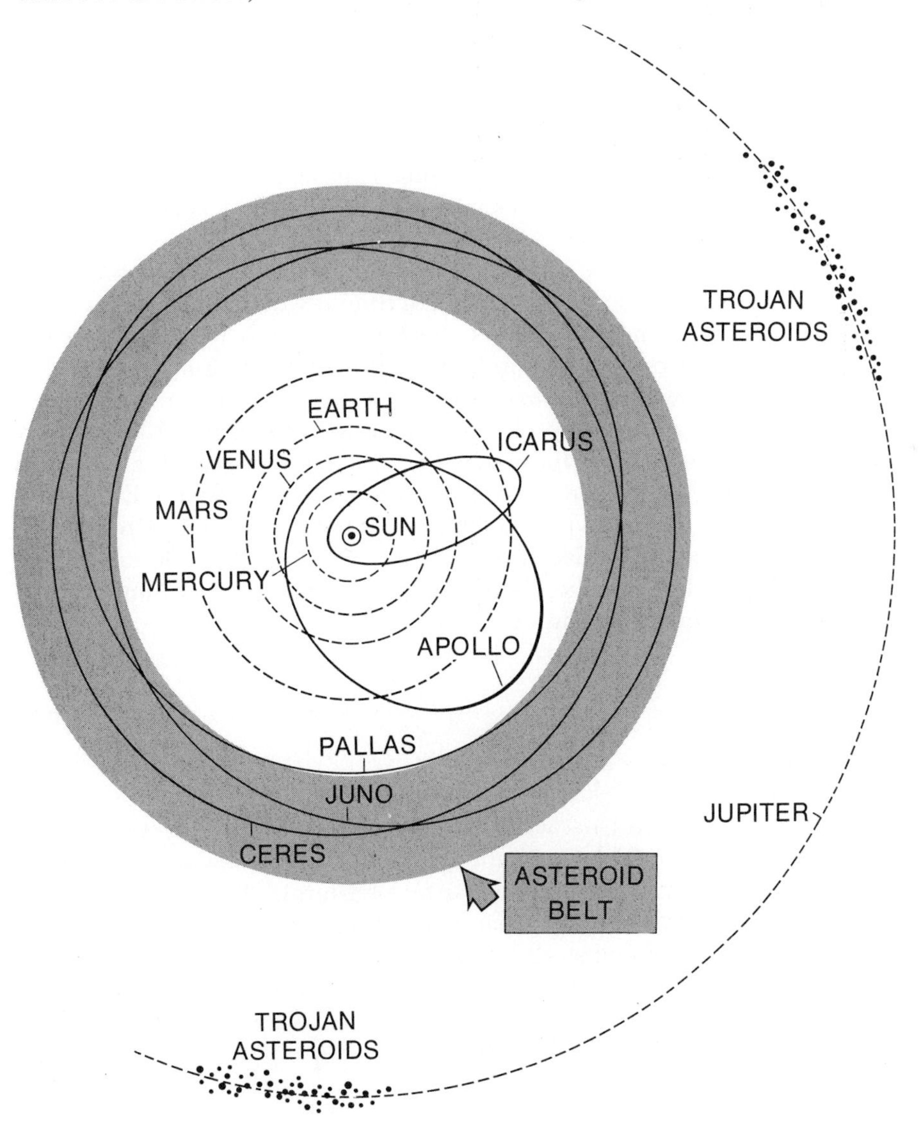

Figure 18–16 Some asteroid orbits. The asteroid belt, in which most asteroids lie, is shaded.

Figure 18–17 The sizes of the larger asteroids and their relative albedos.

haps 100,000 asteroids could be detected with Earth-based telescopes if we wanted to work on it.

Though most asteroids are located between the orbits of Mars and Jupiter, from 2.2 to 3.2 A.U. from the Sun (Fig. 18–16), there is plenty of empty space among them. Asteroids rarely come within a million kilometers of each other. Still, asteroids may collide with each other occasionally, forming meteoroids.

Pioneers 10 and 11, en route to Jupiter and beyond, traveled through the asteroid belt for many months and showed that the amount of dust among the asteroids is not increased over the amount of interplanetary dust in the vicinity of the Earth. Particles the size of dust grains are only about three times more plentiful in the asteroid belt, and smaller particles are less plentiful. So the asteroid belt will not be a hazard for space travel to the outer parts of the solar system.

One type of asteroid, known as the *Apollo asteroids* (after the prototype, Apollo), has orbits whose perihelia (the plural of perihelion) are inside the orbit of the Earth. We know of about two dozen asteroids of this type, of which there may be 1000 examples in all. One asteroid even goes inside the orbit of Mercury, and has been named Icarus, after the inventor in Greek mythology who flew too near the Sun. (Asteroids outside the asteroid belt sometimes have names other than feminine.) On June 14, 1968, Icarus came within 6 million km of the Earth.

One Apollo asteroid discovered in 1976, called 1976AA pending assignment of a name and number, has an orbit that is almost identical with that of the Earth. It revolves around the Sun every 347 days and it ranges from 117 million km to 170 million km from the Sun. This 3 km chunk of rock may be a planetesimal. It is a logical place to think about sending spacecraft for its next close approach in 1995.

There is a fair chance that this asteroid will hit the Earth sometime during the next 25 million years, at which time it would gouge out a crater about 30 km across and send lethal shock waves out over a much greater area. Most Apollo asteroids will probably collide with the Earth eventually; luckily, there are only a few dozen of them greater than 1 km in diameter.

A somewhat larger group of asteroids crosses the orbit of Mars. Only three dozen are known, but there may be 20,000 of these greater than 1 km across. These may be the source of the meteoroids that strike the Earth. The asteroid 433 Eros passed within 23 million km of the Earth in 1975. It was found from optical and radar observations to be $13 \times 15 \times 36$ km across. Eros probably looks a bit like Phobos (Fig. 15–20), a moon of Mars. Eros is rotating every 5 hours 16 minutes.

Box 18.2 The Size of Eros

On the night when Eros was at its closest to the Earth, it occulted the 3.6th magnitude star κ Geminorum (kappa of the Twins), the first well observed occultation of a star by an asteroid. Because the star is so much farther from the solar system than Eros is from the Earth, Eros casts a shadow of starlight that hardly tapers. Thus the area on the Earth from which the star was hidden is the projection of the asteroid, and is almost exactly the same size as the asteroid.

The predictions of where the shadow would fall were revised

only four hours before the occultation was to take place, and groups of observers rushed to take up positions along a line that ran from Amherst, Massachusetts, southwest to Waterbury, Connecticut. Three of the groups saw the star abruptly disappear for 2.6 to 3.4 seconds. Analysis of the observations directly gives the dimension of Eros at the orientation it then had; this cross section is 22 km across. More such occultations of stars by asteroids could be observed if effort were put into calculations and observations of this type.

Another unusual type is the Trojan asteroids, which always remain in orbit around the Sun oscillating about the point 60° ahead of or behind Jupiter. At these locations, known as the Lagrangian points, the gravitational pulls of the Sun and of Jupiter balance out; we have seen in Section 5.10 how it has been proposed that a space station in the Earth's Lagrangian points could collect solar energy and transmit it to Earth.

The first Trojan asteroid to be discovered was named Achilles. Later it was decided to name the asteroids in the Lagrangian point ahead of Jupiter after Greek heroes and those in the point trailing Jupiter after Trojan heroes; but 617 Patroclus, a Greek, is in the Trojan camp. Similarly, there is one Trojan, 624 Hector, in the Greek camp.

When the solar system was forming, proto-Jupiter probably became so massive that it disrupted a large region of space around it. The planetesimals in that region may have been sent into eccentric orbits, and may have collided at high velocity. In this way, they broke each other apart and sent the fragments off in orbits of high eccentricity and inclined to the ecliptic plane. Small asteroids may only be 1 km or less across, but the largest asteroids known (Fig. 18–17) are of the size of some of the moons of the planets. All the asteroids together contain less mass than does the Moon.

Many of the meteorites that we find on Earth may be chips of asteroids. If this is the case, then we already have some kinds of asteroid material on hand to analyze. But other types of asteroids, such as the Trojan asteroids, seem to be located in only one part of the solar system, and we probably don't have bits of them.

Since 1970, new methods have enabled us to better measure the sizes and albedos of asteroids. These methods are based on the amount of infrared radiation emitted by asteroids at wavelengths of 10 or 20 microns and on the polarization of visible light (which seems to change more as the asteroid's position changes for darker asteroids than for lighter ones). Some asteroids are as dark as coal, with albedos of only 3 per cent. Indeed, the surface may be covered with carbon or carbon compounds. Spectra of such asteroids show only a continuum, with no bands typical of various minerals. Other asteroids have albedos of 50 per cent.

Such considerations, and also asteroid spectra, lead us to the conclusion that asteroids are made of different materials from each other, and represent the chemical compositions of different regions of space. The asteroids at the inner edge of the asteroid belt are mostly stony in nature, while the ones at the outer edge are darker (as a result of being more carbonaceous). Most of the small asteroids that pass near the Earth—Icarus, Eros, Toros, Geographos, and Alinda—are among the stony group. Three of the largest asteroids—Ceres, Hygiea, and Davida—are among the carbonaceous group. A third group may be mostly composed of iron and nickel. The differences may be a direct result of the variation of the solar nebula with distance from the protosun. Differences in chemical composition also show that we must discard the old theory that the asteroids represent the breakup of a planet that once existed between Mars and Jupiter.

Asteroids have also had different types of histories since their formation. 4 Vesta, for example, seems to be made of basalt, which would have resulted from lava. Thus some asteroids are differentiated, that is, layered. However, we do not know how Vesta could have been heated to the 1000 K necessary to cause the lava; it is too small for radioactive decay inside it to have provided the heat.

The asteroids and their meteoritic offspring may provide the data to show us the details of how the solar system was formed.

SUMMARY AND OUTLINE OF CHAPTER 18

Aims: To study the non-planetary members of our solar system, and to see how the conditions they have been under may provide us with our best information about the origin of the solar system

Comets (Section 18.1)
- The head (the nucleus + the coma together); the tail
- Dust tail is sunlight reflecting off particles blown back by radiation pressure; ion tail (gas tail) is sunlight re-emitted by ions blown back by the solar wind
- Dirty snowball theory: nucleus composed of rock covered with ices
- Coma composed of gases vaporized from nucleus; spectrum shows molecules
- Ultraviolet space observations revealed huge hydrogen cloud around coma
- Origin of comets: Oort comet cloud; comets detached from cloud by gravity of passing star
- Comet orbits are ellipses
- Halley's Comet followed for last 29 returns, but 1986 return will not be as spectacular as some past passages
- Comet Kohoutek did not become the popular display that had been hoped for, but nonetheless gave unique data because of its relatively high brightness and because it was discovered long before it passed perihelion, giving time for planning
- Comet West became very bright with more typical notice of 2 months

Meteoroids (Section 18.2)
- *Meteoroids*: in space; *meteors*: in the air; *meteorites*: on the ground
- Most meteorites that hit the Earth are chondrites, which resemble stones and are therefore often undetected. Iron meteorites are more often found.
- Meteors may be chips off asteroids
- Meteorites provide us with much of our information about abundances in the solar system. We could use them to map out inhomogeneities in the solar nebula if we only knew what parts of the solar nebula the different meteorites came from.
- Some meteors arrive in showers, others are sporadic

Asteroids, also called minor planets (Section 18.3)
- Most in asteroid belt, 2.2 to 3.2 solar radii, between orbits of Mars and Jupiter
- Apollo asteroids come within orbit of the Earth
- Some asteroids, such as Eros, don't quite cross the orbit of the Earth but nonetheless come very close to us

Trojan asteroids are in the Lagrangian points 60° before and after Jupiter in its orbit

Asteroids have different compositions, different albedos, and different histories

QUESTIONS

1. In what part of its orbit does a comet travel head first?

2. Why is the Messier catalogue important for comet hunters?

3. Would you expect comets to follow the ecliptic? Explain.

4. How far is the Oort comet cloud from the Sun, relative to the distance from the Sun to Pluto?

5. The energy that we see as light from a comet comes from what source or sources?

6. Many comets are brighter after they pass close to the Sun than they are on their approach. Why should that be?

7. Which part of a comet has the most mass?

8. Explain why Comet West, seen in the photograph opening this chapter, showed delicate structure in its tail. Are we observing mainly the dust tail or the ion tail?

9. Why was the large cloud of hydrogen that surrounds the head of a comet not detected until 1970?

10. Suggest which of the daughter molecules listed in Section 18.1a might result from which of the parent molecules.

11. Why do most meteorites not reach the surface of the Earth?

12. Why do some meteor showers last only a day while others can last several weeks?

13. Why are meteorites important in our study of the solar system?

14. Does the asteroid belt fit Bode's law? How might this be interpreted in terms of the history of the solar system?

15. What does the occultation of κ Gem by Eros tell us about the possibility that Eros has an atmosphere?

19

Life in the Universe

We have discussed the nine planets and the dozens of moons in the solar system, and have found most of them to be places that seem very hostile to terrestrial life forms. Yet some locations besides the Earth—Mars with its signs of ancient running water and perhaps Titan—have characteristics that allow us to convince ourselves that life may have existed there in the past or might even be present now.

Since life may well preferentially arise on planets, this chapter is included in Part III of this book, which discusses the solar system.

Our first real attempt to search for life on another planet has led to fascinating results. The Viking landers, described in Section 15.3, have carried out a series of biological and chemical experiments with Martian soil, and some of these experiments have provided data that cannot be readily interpreted in terms of chemical processes that are common on the Earth. Whether the proper interpretation is that there is biological activity—life—on the surface of Mars, or that chemical processes can provide the complete explanation must await further experimentation on Earth and probably further exploration and experimentation on Mars. Later in this chapter, we shall discuss the criteria that we use to decide whether or not to accept such radical new conclusions as the existence of extraterrestrial life.

Since it seems reasonable that life, as we know it, anywhere in the Universe would be on planetary bodies, let us first discuss the chances of life arising elsewhere in our solar system. Then we will consider the chances that life has arisen in some more distant parts of the galaxy in which we live or elsewhere in the Universe.

Recently the study of life in other locations in the Universe, a study now called *exobiology,* has become a respectable science. The billion dollars spent to send the Viking probes to Mars equipped to search for signs of life is a sure sign that the possibilities of finding life are taken very seriously indeed.

From *The Day the Earth Stood Still.*

19.1 THE ORIGIN OF LIFE

It would be very helpful if we could state a clear, concise definition of life, but unfortunately that is not possible. Biologists state several criteria that are ordinarily satisfied by life forms, reproduction, for example. Still, there exist forms on the fringes of life—viruses, for example, which need a host organism in order to reproduce—and scientists cannot always agree whether some of these things are "alive" or not.

In writing science fiction, authors sometimes conceive of beings that show such signs of life as the capability for intelligent thought, even though the beings may share few of the criteria that we ordinarily recognize. In Fred Hoyle's novel *The Black Cloud,* for example, an interstellar cloud of gas and dust is as alive as—and smarter than—you or I. But we can make no concrete deductions if we allow such wild possibilities, and exobiologists prefer to limit the definition of life to forms that are more like "life as we know it."

This rationale implies, for example, that extraterrestrial life is based on complicated chains of molecules that involve carbon atoms. Life on Earth is governed by deoxyribonucleic acid (DNA) and ribonucleic acid (RNA), two carbon-containing molecules that control the mechanisms of heredity. Chemically, carbon is able to form "bonds" with several other atoms simultaneously, which makes these long carbon-bearing chains possible. In fact, we speak of compounds that contain carbon atoms as *organic.*

Carbon is one of the more abundant elements on a cosmic scale (see Table 5–1), although its abundance is far below those of hydrogen or helium. Another reasonably abundant element that can make several chemical bonds simultaneously is silicon, and some scientists have theorized that silicon-based life may exist somewhere. Serious consideration of life forms seems limited to life based on chains of carbon or silicon atoms.

If we accept this limitation, then we ask how hard it is to build up the long organic chains. To the surprise of many, an experiment was performed twenty-five years ago that showed that the construction of organic molecules was much easier than had been supposed.

At the University of Chicago, at the suggestion of Harold Urey, Stanley Miller put several simple molecules in a glass jar. Water vapor (H_2O), methane (CH_4), and ammonia (NH_3) were included along with hydrogen gas. Miller exposed the mixture to electric sparks, simulating the lightning that may exist in the early stages of the formation of a planetary atmosphere. After a few days, he found that long chains of atoms had formed in the jar, and that these organic molecules were even complex enough to include simple amino acids, the building blocks of life.

It was later found that exposing a similar mixture to ultraviolet light

Figure 19–1 It is all in the point of view. © 1974 United Features Syndicate, Inc.

had the same effect of causing the formation of organic molecules. It seems reasonable that the ultraviolet radiation reaching the ground in the earlier stages of the Earth's atmosphere was much stronger that that striking us today, because the ozone layer may not yet have formed. (In fact, strong ultraviolet radiation has reached the surface of Mars, and continues to reach it even now.) Thus it seems reasonable that ultraviolet radiation and lightning may have led to the formation of amino acids from simpler molecules in the primitive atmosphere of the Earth.

It seems clear, however, that mere amino acids or even DNA molecules are not life itself. A jar containing a mixture of all the atoms that are in a human being is not the same as the human being in person. This is the vital gap in the chain; astronomers certainly are not qualified to say what supplies the "spark" of life.

Still, it seems reasonable to many astronomers that since it is not very difficult to form complex molecules, life may well have arisen not only on the Earth but also in other locations in our solar system. Even if life is not found in our solar system, there are so many other stars in space that it would seem that some of them could have planets around them, and it would seem that life could have arisen on some of the planets independently of the origin of life on the Earth. The interesting nature of the results of the biological experiments on Viking add support to this view.

19.2 OTHER SOLAR SYSTEMS?

We have seen how difficult it was to detect even the outermost planets in our own solar system. At the four light year distance of Proxima Centauri, the nearest star to the Sun, any planets would be too faint for us to observe directly.

But even though we cannot hope to see the light reflected by even a giant planet, there is a faint hope of detecting such a planet by observing its gravitational effect on the star itself. One such method is the same that has been used to discover white dwarfs (see Section 7.4). We study the proper motion of a star—its apparent motion across the sky with respect to the other stars (as discussed in Section 2.16)—and see if the star follows a straight line or appears to wobble slightly.

This is a much more difficult observation to make when searching for planets than it is to make when searching for white dwarfs, simply because planets have much less mass. Even Jupiter, the largest planet in our solar system, has only one-thousandth the mass of the Sun, while a white dwarf has nearly a solar mass.

The most likely stars to observe in a search for planets are the stars with the largest proper motion, for these are likely to be the closest to us and thus any wobble would cover the greatest angular extent we could hope for. The star with the largest proper motion, Barnard's star, is only 1.8 parsecs (6 light years) away from the Sun. It is the fourth nearest star to the Sun; only the members of the Alpha Centauri triple system are nearer.

The criteria for a telescope used to study proper motion are different from those of a giant telescope used to study distant objects. To study proper motions, one wants a field of view large enough to include not only the object being studied but also other stars for comparison, a scale on the photographic plate such that proper motions correspond to measurable displacements, and a very stable telescope system. In particular, one

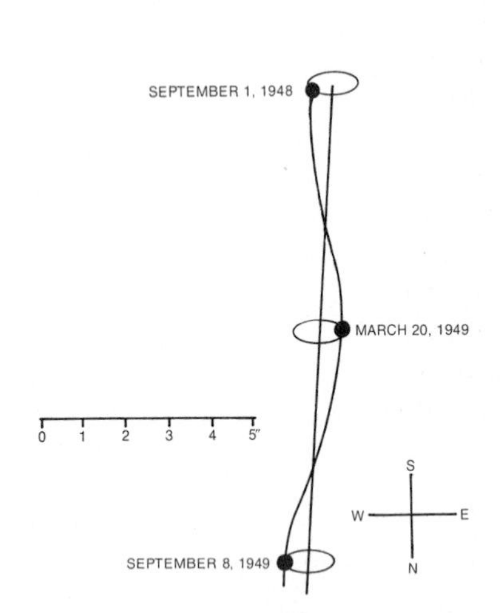

Figure 19–2 The photograph is a composite of three photos of Barnard's star, taken at the Sproul Observatory at intervals of approximately 6 months. The stars at the top and right have small proper motions and small parallaxes, and appear in the same position on each of the three negatives. Barnard's star, on the other hand, shows its proper motion of over 10 arc sec per year. It also shows a lateral displacement caused by parallax; the result of the Earth's orbit around the Sun.

wants a system that is not changed over a period of many years, so that photographs taken now can be directly compared with photographs taken decades ago. The part of astronomy that involves the measurement of the positions of objects in the sky is called *astrometry*. Until recently, refractors rather than reflectors have most often been used for astrometric work.

The Sproul Observatory in Swarthmore, Pennsylvania, is a major astrometric center, and Peter van de Kamp has studied the many thousands of stellar positions measured there over the years. Van de Kamp considered observations of Barnard's star taken from 1937 on, and reported in 1962 that he had discovered a wobble in its proper motion (Fig. 19–2). He interpreted this wobble to indicate the presence of a giant planet, larger even than Jupiter, in a very elliptical orbit. After further observations and analysis, he reported that the data were even better explained if there were two giant planets present; then the orbits could be more nearly circular. Van de Kamp also found evidence, though less confidently, for a planet around another of the nearest stars, ϵ Eridani.

If we accept the nebular theory for the formation of planets, then it comes as no surprise that many stars have planets. Still, we like to see direct observational backing for our theories. Since the nearest three systems are multiple—our Sun has planets, the next nearest star is a triple system and the next star out seems to have planets—we might reasonably extrapolate that planets are a common rather than an uncommon phenomenon. If this is true, there certainly would be billions of planets in our galaxy alone, and billions more in other galaxies.

Within the last few years, however, George Gatewood of the Allegheny Observatory in Pittsburgh, and Heinrich Eichhorn of the University of South Florida, have reported that their study of other long-term observations of Barnard's star did not show the wobble in its proper motion. A restudy of the Sproul astrometric plates by van de Kamp shows that for a period of a few years the measured values were different from the trend of preceding and following years; the difference may have started when the telescope's lens was remounted in 1949. A restudy of the data with this in mind significantly reduced the amount of wobble that may be present.

At present, the matter is unresolved. Gatewood now finds that his own data contain the suggestion of a deviation from straight-line proper motion for Barnard's star, but not conclusively so. He admits the possibility that the current data could imply the presence of planets, though this current indication of a wobble does not seem sufficiently convincing that he would have reported it in the absence of the historical context. Van de Kamp's new analysis of his own data still indicates a wobble that can best be interpreted in terms of planets orbiting Barnard's star, but this is a different wobble from the one he had reported previously.

Astrometry has been undergoing a revolutionary change in methods recently, with the advent of new machines that automatically measure positions on photograph plates. Photocells can be used to "center" an automatic measuring engine on the center of a star image, and the position can be recorded directly by a computer, or at least on punched cards. These new procedures greatly increase not only the accuracy but also the speed of studying the astrometric plates that are taken.

For the moment, the data seem to indicate the presence of two planets around Barnard's star, one with the approximate mass of Jupiter and the other with the approximate mass of Saturn. Any Earth-sized planets would not have enough gravity for us to be able to detect them. At present at least some astronomers feel less certain about these results than they did a few years ago. The extrapolation based on proper motion studies that almost all stars have solar systems is less certain than it had been.

But a new spectrographic survey carried out at Kitt Peak seems to endorse the idea that many stars have planets. One third of the 123 solar-type stars under study showed variations over time in their radial velocities, even though the invisible companions have masses that are too small to

Figure 19–3 A spectroscopic study of the radial velocities of solar-type stars, carried out at the Kitt Peak National Observatory, shown here, has indicated that many have companions. One of the 0.9-m telescopes, barely visible at the center, was used in the study.

allow them to be stars. Some of these may be planets; just how many, it is difficult to say.

19.3 THE STATISTICAL CHANCES FOR EXTRATERRESTRIAL INTELLIGENT LIFE

Instead of phrasing one all-or-nothing question about life in the Universe, we can break down the problem into a chain of questions. This procedure was developed by Frank Drake, a Cornell astronomer, and extended, among others, by Carl Sagan of Cornell and Joseph Shklovskii, a Soviet astronomer.

First we ask what the probability is that stars at the centers of solar systems are suitable to allow intelligent life to evolve. For example, an O star would probably stay on the main sequence for too short a time to allow the evolution of intelligent life. We have seen in the discussion of the stellar luminosity function (Section 4.2) how most stars have spectral types F, G, K, and M; these might be suitable sites for planets on which life can evolve, though M stars, like Barnard's star, might be too cool. Some scientists rule out multiple stars because the temperature at planetary distances from multiple stars might vary rapidly with time as the stars and planets orbited. As a result, the planetary orbits in multiple star systems may be unsatisfactory because it would be difficult for life to evolve on the planets. Even disregarding such systems could still leave a large number of candidate stars. And satisfactory orbits around a multiple system might exist after all.

Second, we ask what are the chances of a suitable star having planets. The import of the previous section is that the chances are probably pretty high, but that we are less certain of this conclusion than we had thought.

Third, we need planets with suitable conditions for the origin of life. A planet like Jupiter might be ruled out, for example, because of its lack of a solid surface and because of its high surface gravity, though alternatively one could consider a liquid region, if it were at a suitable temperature, to be as advantageous as were the oceans on Earth to the development of life here.

Fourth, one has to consider the fraction of systems on which life begins. This is the biggest uncertainty, for if this fraction is zero (with the Earth being a unique exception), then we can get nowhere with this entire line of reasoning. Still, the discovery that amino acids can be formed in laboratory simulations of primitive atmospheres, and the discovery of complex molecules containing half a dozen or more atoms even in interstellar space (see Section 21.3) indicate to many astronomers that it is not as difficult as had been thought to form complicated molecules. Amino acids, much less complicated than DNA but also basic to life as we know it, have even been found in meteorites. Many astronomers choose to think that the fraction of planetary systems on which life begins may be high.

At this point we have already reached an interesting result, for if one tries to assess (or, really, guess) fractions for each of the above steps in the chain of reasoning, and multiplies the fractions together, one finds that there may be billions of planets in this galaxy on which life may have evolved, and that the nearest one may be within dozens of light years of the Sun.

Box 19.1 The Probability of Life in the Universe

The discussion we have been carrying out follows the lines of an equation written out to estimate the number of civilizations in our galaxy that would be able to contact each other. In 1961, Frank Drake of Cornell wrote the equation:

$$N = R_{*} f_p n_e f_l f_i f_c L,$$

where R_{*} is the rate at which stars form in our galaxy, f_p is the fraction of these stars that have planets, n_e is the number of planets per solar system that is suitable for life to survive (for example, an Earth-like atmosphere), f_l is the fraction of these planets on which life actually arises, f_i is the fraction of these life forms that develop intelligence, f_c is the fraction of the intelligent species that choose to communicate with other civilizations and develop adequate technology, and L is the lifetime of such a civilization. The largest uncertainty is f_l, which could be essentially zero or could be close to 1. The result is also very sensitive to the value one chooses for L—is it 1 century or a billion years? Do civilizations destroy themselves? Depending on the alternatives one chooses, one can find that there are dozens of communicating civilizations within 100 light years or that the Earth is unique in having one. Most people choose their values such that the result comes out closer to the former than to the latter situation.

If we want to have meaningful conversations with the aliens, however, we must have a situation where not just life but intelligent life has evolved. We cannot converse with algae or paramecia. Furthermore, the life must have developed a technological civilization capable of interstellar communication. These considerations reduce the probabilities somewhat, but it has still been calculated that there are likely to be technologically advanced civilizations of intelligent life within a few hundred light years of the Sun.

What are the chances of our visiting or being visited by representatives of these civilizations? As we have seen in Chapter 16, Pioneers 10 and 11 are even now carrying plaques out of the solar system in case an alien interstellar traveler should happen to encounter these spacecraft. Still, it seems unlikely that humans can travel the great distances to the stars, unless some day we develop spaceships to carry whole families and cities on indefinitely long voyages into space. We must recall, however, that the Sun has another 4 or 5 billion years to go on the main sequence, and recorded history on Earth has existed for only about 5000 years, one millionth of the age of the solar system. There is plenty of time in the future for interstellar space travel to develop. Gerard O'Neill of Princeton has recently led a series of investigations that seem to show that we could start to think of setting up cities in space (Fig. 19–4). (A major purpose of these cities would be to collect solar energy and beam it to Earth.)

We can hope even now to carry out interstellar communication by

Figure 19-4 An artist's conception of the lunar station from which materials are lifted to the Lagrangian points, where Gerard O'Neill has suggested that we build space stations. (From *Science Year, The World Book Science Annual.* © 1975 Field Enterprises Educational Corporation.)

means of radio signals. Much thought has been given to what signals to send out, and to what is the best frequency to use, as will be discussed in the following section. What is the chance of hearing another civilization or having our own messages heard? Note that we have known the basic principles of radio for only a hundred years, and that powerful radio and television transmitters have existed for less than 50 years. Even now we are sending out signals into space on the normal broadcast channels. A wave bearing the voice of Caruso is expanding into space, and at present is 50 light years from Earth. And once a week a new episode of *Rhoda* is carried into the depths of the Universe. The Sun would appear to be a variable radio source as seen from most directions in space, because the Earth's signals would come from the vicinity of the Sun. A periodicity of 24 hours would be detectable since the signal would peak each day as specific concentrations of transmitters rotated to the side of the Earth facing our listener.

But how long does the phase last when the radio radiation of technological civilizations escapes into space? Already signals that once were broadcast through the atmosphere are being put underground in cables and so no longer are radiated into space. If civilizations on other planets follow our pattern, then perhaps the period over which a planet is noisy in the radio spectrum lasts only the few decades between the discovery of radio and the development of cable TV. If so, the chances of detecting such unintentionally broadcast signals would be reduced considerably.

Moreover, one comes to the important question of the lifetime of the technological civilization itself. We now have the capability of destroying our civilization either dramatically in a flurry of hydrogen bombs or more slowly by, for example, altering our climate or increasing the level of atmospheric pollution. It is a sobering question to ask whether the lifetime of a technological civilization is measured in decades, or whether all the problems that we have, political and otherwise, can be overcome, leaving our civilization to last for millions or billions of years.

19.4 THE SEARCH FOR LIFE

For planets in our solar system, we can search for life by direct exploration, but how would you go about trying to detect radio signals from creatures located outside our solar system? It would be too overwhelming a task to listen for signals at all frequencies in all directions at all times. One must make some reasonable guesses.

There are a few frequencies in the radio spectrum that seem especially fundamental. Neutral hydrogen in space, for example, has a basic spectral line at 1420 MHz, a frequency over ten times higher than stations at the high end of the normal FM band. (This hydrogen radiation is of great importance in studying our galaxy, and will be discussed in Chapter 21.) We might conclude that creatures on a far-off planet would decide that we would be most likely to listen near this frequency because it is so fundamental. On this theory, they would broadcast at about 1420 MHz.

On the other hand, perhaps the great abundance of radiation from hydrogen itself would clog this frequency, and one or more of the other frequencies that correspond to strong natural radiation should be preferred. Frequencies that correspond to radiation or absorption by water vapor (H_2O) in space ("the water hole") or by the hydroxyl radical (OH) have been considered. If we ever detect radiation at any frequency, I am sure that we will immediately start to wonder why we did not realize that this

Figure 19–5 An artist's conception of the Project Cyclops array.

Figure 19–6 An artist's conception of some of the dishes in the Project Cyclops array.

frequency was the obvious choice. If only we now had hindsight on this matter.

In 1960, Frank Drake used a telescope at the National Radio Astronomy Observatory to listen for signals from two of the nearest stars. He was searching for any abnormal kind of signal, a sharp burst of energy, for example, such as pulsars emit. He was able to devote to this investigation, known as Project Ozma (after the queen of the land of Oz in L. Frank Baum's stories) only 200 hours of observing time, distributed over a few months.

The observations are now being extended in a more methodical search called Ozma II. Since 1973, Ben Zuckerman of the University of Maryland and Patrick Palmer of the University of Chicago have been systematically monitoring the 500 nearest stars of supposedly suitable spectral type. They use computers to search the data they record for any sign of special signals. Needless to say, nothing significant has turned up yet; the world will know it rapidly if anything ever does.

Figure 19–7 The Arecibo telescope in Puerto Rico.

A grandiose proposal has been made to build a huge array of over 1000 radio telescopes, each 100 meters (100 yards) in diameter (Fig. 19–5), covering an area about 10 kilometers on a side (Fig. 19–6). The total area collecting radiation would be almost that of a single telescope with a 10-kilometer diameter! The array would be fantastically sensitive. This plan, called Project Cyclops, would cost 5 billion dollars. Its construction would depend on a decision that the search for interstellar life in this way was an important national priority, a decision that seems unlikely to be made in the foreseeable future. Of course, valuable astronomical research could be carried out with the Cyclops array when it was not being used for interstellar monitoring, but nobody is about to spend 5 billion dollars on straight astronomy either. Still, the amount is not out of line with the cost of space research.

In 1974 the giant radio telescope at Arecibo, Puerto Rico (Fig. 19–7), was upgraded, in that the precision and quality of the mesh of the bowl-like surface of the dish was improved. The telescope is 305 meters across, the largest telescope on Earth. At the rededication ceremony, a powerful signal was sent out into space at the hydrogen frequency, bearing a message from the people on Earth (Fig. 19–8). This is the only major signal that we have purposely sent out as our contribution to the possible interstellar dialogue.

The signal from Arecibo was directed at the globular cluster M13 in the constellation Hercules (see page 90) on the theory that the presence of 300,000 closely packed stars in that location would increase the chances of our signal being received by a civilization on one of them. But the travel time of the message (at the speed of light) is 24,000 years to M13, so we certainly could not expect to have an answer before twice 24,000, or 48,000 years, have passed. If anybody (any**thing**) is observing our Sun when the signal arrives, the radio brightness of the Sun will increase by 10 million times for a 3-minute period.

Not only astronomers in the United States but also astronomers in the Soviet Union are interested in "communication with extraterrestrial intelligence," which is becoming known by the acronym CETI, and in the simpler and less expensive "search for extraterrestrial intelligence," SETI. The first phase of the Soviet program will last until 1985, and includes scanning a wide range of frequencies in all directions in the sky for suitable strange signals. In the second phase, which is scheduled to last from 1980 to 1990, radio telescopes in space would continue the search, perhaps from a position behind the Moon where the Moon's bulk prevents interference by radio signals from the Earth. In late 1976, NASA announced a SETI program, a major part of which will be the construction of a special computer to analyze the incoming radio waves for signals from far-off civilizations.

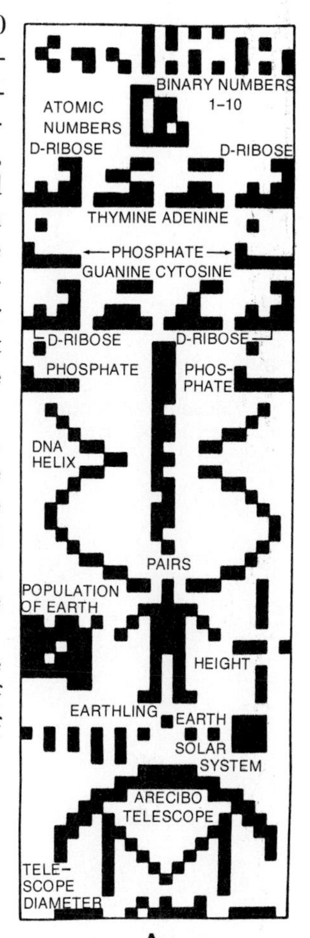

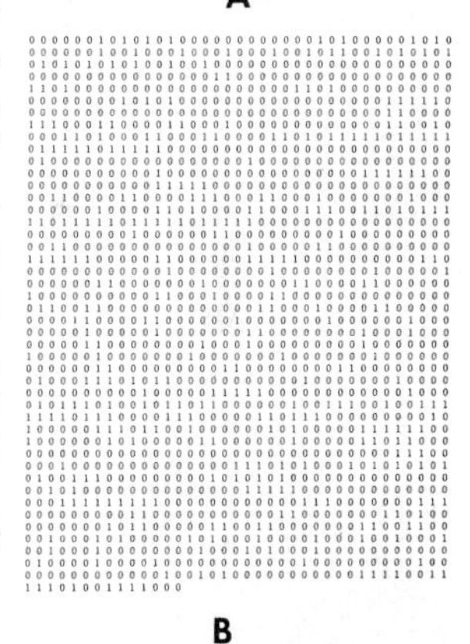

Figure 19–8 *(A)* The message sent to M13, with a translation into English added. *(B)* The message was sent as a string of 1679 consecutive characters, in 73 groups of 23 characters each. There were two kinds of characters, each represented by a frequency; the two kinds of characters are reproduced here as 0's and 1's.

19.5 UFO'S AND OCCAM'S RAZOR

But why, you may ask, if most astronomers accept the probability that life exists elsewhere in the Universe, do they not accept the idea that unidentified flying objects (UFO's) represent visitations from these other civilizations (Fig. 19–9)? The answer to this question leads us not only to explore the nature of UFO's but also to consider the nature of knowledge and truth. The discussion that follows is a personal view, but one that is shared by many scientists.

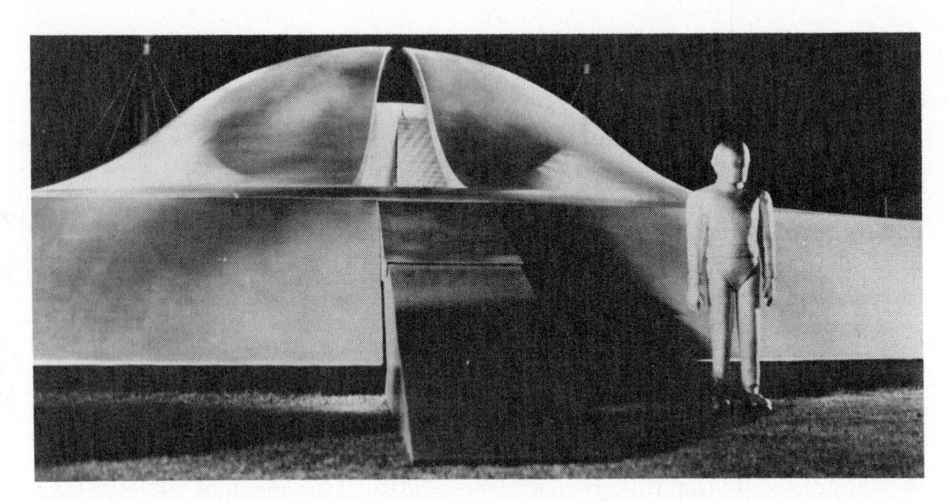

Figure 19-9 Gort, from *The Day the Earth Stood Still.* (Museum of Modern Art/Twentieth-Century Fox, Inc.)

First of all, most of the sightings of unidentified flying objects that are reported can be explained in terms of natural phenomena. Astronomers are experts on strange effects that the Earth's atmosphere can display, and many UFO's can be explained by such effects. For example, I know that every time Venus shines brightly on the horizon, in my capacity as local astronomer I will get a lot of telephone calls asking me about the UFO. A planet or star low on the horizon can seem to flash red and green because of atmospheric refraction, a phenomenon not generally realized by non-scientists. Atmospheric effects can affect radar waves as well as visible light.

Often, eyewitness accounts are untrustworthy. On my campus, the civil air patrol and the police have picked me up to take me to the telescope in case I could verify a report that a bright star had moved back and forth across the sky at tremendous speed. The star in question proved to be listed in star catalogues and had thus been in its present location in the sky for years. I am afraid that this demonstration, which showed the star was not newly in its location as claimed, did not convince the person who reported the motion, although it certainly convinced me. Every so often, groups of students release balloons carrying candles into the air, and it is interesting to read the reports of sightings that are made to the press and to the police. How a candle aloft can be seen as a spaceship with windows and aliens sitting inside is beyond me. But such reports are routinely made.

Figure 19-10 (*A*) A UFO fabrication: an aluminum plate topped by a cottage cheese container on which black dots have been placed to simulate port-holes. The landing gear is constructed from hemi-spheres of ping-pong balls. (*B*) A UFO fabrication.

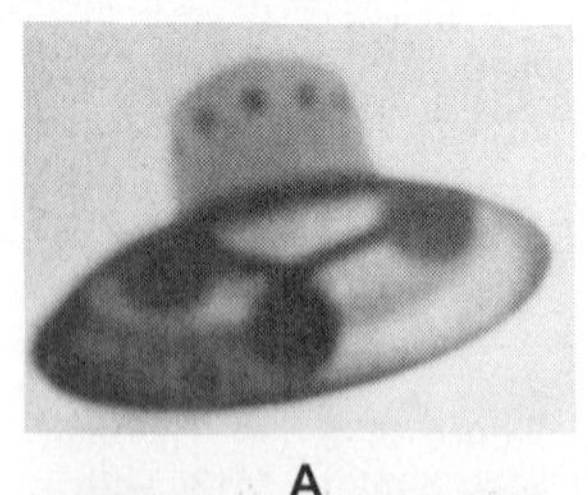

A

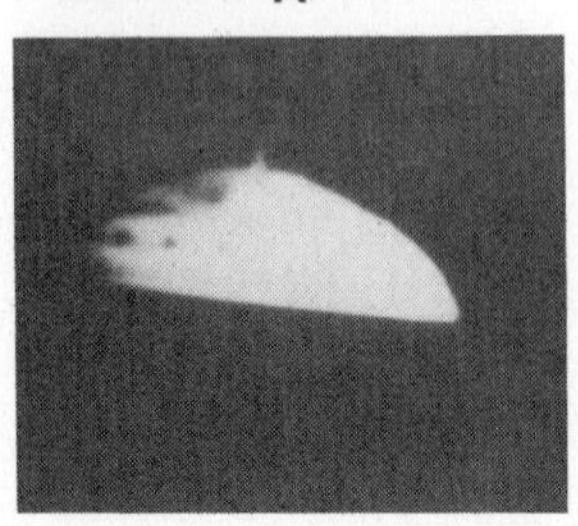

B

Sometimes other natural phenomena—flocks of birds, for example—are reported as UFO's. One should not accept explanations that UFO's are flying saucers from other planets before more mundane explanations—including exaggeration and fraud—are exhausted.

The topic is of tremendous public interest. As the result of public pressure, a committee was appointed under governmental auspices a few years ago to investigate UFO's. A distinguished physicist, the late E. U. Condon, headed the group, which worked from 1966 to 1969. They asked the leading flying saucer groups to send them the information about the cases that seemed to prove most convincingly that UFO's really have an extraterrestrial identification. After exhaustive analysis, the Condon group was able to account for these cases on the basis of more commonplace effects. Only when too little information is available—as when one has only a report that "something bright was seen in the northern sky for ten minutes and it moved around"—do the UFO reports become impossible to explain without invoking flying saucers.

Further, it has been shown that for many of the effects that have been reported, the UFO's would have been defying well-established laws of physics if their reported motions actually took place. Where are the sonic booms, for example, from rapidly moving UFO's? Scientists treat challenges to well-established laws of physics very seriously, as much of our science and technology is based on these laws. We can ordinarily explain new discoveries in terms of these known laws. If the laws of physics are seriously challenged by UFO's, then we would have to start questioning other applications we make of these laws, such as the safety of skyscrapers, or the lift of airplanes. The reports of UFO's are not sufficiently well documented to force us to assume that the laws of physics are inadequate to provide the explanation. Only when a new theory explains facts that are not understandable on the basis of the old theory—like the Copernican explanation of retrograde motion (Section 10.2) or Einstein's explanation of the excess advance of the perihelion of Mercury (Section 5.11)—do we accept the new theory.

Figure 19–11 The two UFO's are really lights shining through the hole in a phonograph record viewed obliquely. The background was added later.

It is particularly interesting that almost every professional astronomer feels that UFO's can so obviously and completely be explained by natural phenomena that they are not worthy of more of our time. Astronomers would be among the most excited people in the world at the discovery of life coming from elsewhere in the Universe. The discovery of a flying saucer would surely be the key astronomical discovery of the millennium, yet scientists feel that they have no hope of finding one. Only a handful of scientists even think that the odds of finding a flying saucer are high enough to make it worthwhile to investigate, even if a few unexplained reports remain. Astronomers just know too much about the properties of the atmosphere to give credence to most of the UFO reports or to think that further investigation is a worthwhile investment of time or money that could be put to better pursuits.

It is mainly non-scientists who feel that UFO's are worth investigating. And it should be pointed out that there is a lot of money involved in reporting sightings of UFO's—since the topic is so popular there are books and articles to be written, and interviews to be given, all of which have a financial return. One must be skeptical of the motives of those who push reports of UFO sightings or want their organizations to continue UFO investigations.

Some individuals may ask why we reject the identification of UFO's with flying saucers, when that explanation is "just as good an explanation as any other." Let us go on to discover what we mean by "truth" and how that applies to the above question.

At every instant, whatever is happening can be explained in a variety of ways. When we flip a light switch, for example, we assume that the switch closes an electric circuit in the wall and allows the electricity to flow. But it is certainly possible, although not very likely, that the switch activates a relay that turns on a radio that broadcasts a message to an alien on Mars. The Martian then sends back a telepathic message to the electricity to flow, and the light goes on. The latter explanation sounds unlikely to the point that we don't seriously consider it. We would even call the former explanation *true,* without qualification. Even the fact that the signal sent in 1976 to cut the ribbon to open the National Air and Space Museum went to the Viking spacecraft orbiting Mars and back before activating a robot arm a few feet away from the signal's starting point, doesn't lead us to doubt what happens when we flip a wall switch.

We regard as *true* the simplest explanation that satisfies all the data we

Occam's Razor, sometimes called the Principle of Simplicity, *is a razor in the sense that it is a cutting edge that allows a distinction to be made among theories.*

have about any given thing. This principle is known as *Occam's Razor;* it is named after a fourteenth-century British philosopher. Without it, we would always be subject to such complicated doubts that we would accept nothing as known.

Science is based on Occam's Razor, though we don't usually bother to think about it. For example, we accept the notions of Copernicus and Kepler that the planets orbit around the Sun largely because they provide a relatively simple explanation for all the planetary motions. It is possible to conceive of ways to explain the solar system with everything orbiting around the Earth, or around even Pluto, but the equations would be much more complicated. We say without hesitation that the Earth revolves around the Sun, or that it is a fact that the Earth revolves around the Sun, or that the fact that the Earth revolves around the Sun is true, even though we cannot absolutely, completely, and forevermore rule out all other descriptions.

Sometimes we call something true that might be more accurately described as a *theory.* The scientific method is based on hypotheses and theories. A *hypothesis* is an explanation that is advanced to explain certain facts. When it can be shown that the hypothesis actually explains most or all of the facts known, then we call it a *theory.* We usually test a theory by seeing whether it can predict things that were not previously observed, and then by trying to confirm whether the predictions are valid. We might call the theory, if it passes these tests, a *well-established theory.* Others might prefer to reserve the word *theory* for this stage and call the previous stages *hypotheses* or *conjectures.* The dividing line between hypotheses and theories is not always clear.

An example of a theory is the Newtonian theory of gravitation, which succeeded for many years in explaining almost all the planetary motions. Only a discrepancy of 43 seconds of arc per century in the advance of the perihelion of Mercury, as was described in Section 5.11, remained unexplained. Now, since approximately 1919, we accept Einstein's general theory of relativity as a better explanation of gravitation, and say that Newtonian theory is only approximately valid in regions of space that are limited in size and in which the force of gravitation present is not very large. Is Newton's theory "true"? Yes, in most regions of space. Is Einstein's theory "true"? We may say so, although we may also think that one day a new theory will come along that is more general than Einstein's in the same way that Einstein's is more general than Newton's.

How does this view of truth tie in with our discussion of UFO's? Every moment we must make decisions about what to do, based on what we think will happen next. We wouldn't even take a step if we thought that a yawning chasm would open up before us in the midst of our living rooms. Scientists have assessed the probability of UFO's being flying saucers from other worlds, and have decided that the probability is so low that the possibility

Figure 19–12 Cartoon by Charles Schulz. © 1975 by United Features Syndicate, Inc.

is not even worth considering. We have better things to do with our time and with our national resources. We have so many other, simpler explanations of the phenomena that are reported as UFO's that we apply Occam's Razor and decide to call the identification of UFO's with extraterrestrial visitation *false*. UFO's may be unidentified, but they are probably not flying, nor for the most part are they objects.

19.6 OF ASTRONOMY AND ASTROLOGY

Neither astrology nor UFO's are at all connected with astronomy, except in a historical context, so neither really deserves a place in a text on contemporary astronomy. But since so many people associate astrology with astronomy, and since astrologers claim to be using astronomical objects to make their predictions, let us use our astronomical knowledge to assess astrology's validity.

Most professional astronomers would privately agree to sentiments something like those expressed in this section, or at least reach the same conclusion. Many astronomers don't even think that astrology is worth discussing, or at best, that discussions have no effect on those who "believe" and so are a waste of time. We will not take this latter position here.

Astrology is an attempt to predict or explain our actions and personalities on the basis of the positions of the stars and planets now and at the instants of our births. Astrology has been around for a long time, but it has never been shown to work. Believers may cite incidents that reinforce their faith in astrology, but no successful scientific tests have ever been carried out. If something happens to you that you had expected because of an astrological prediction, you would more certainly notice that this event occurred than you would notice the thousands of other unpredicted things that happened to you that day. And likewise with hindsight. We do enough things, have sufficiently varied thoughts, and interact with enough people that if we make many predictions in the morning, some of them are likely to be at least partially fulfilled during the day. We simply forget that the rest ever existed. Besides, we always have the easy out of being able to say, "well, we just didn't have a good astrologer."

Studies have shown that superstition actively constricts the progress of science and technology in various countries around the world and is therefore not merely an innocent force. It is not just that some people harmlessly believe in astrology. Their lack of understanding of scientific structure may actually impede the proper scientific training of people needed to work on the problems of our age, including the problems of pollution and the energy crisis. A recent paper even shows how widespread superstitious beliefs can impede smallpox-prevention programs. Thus many scientists are not content to ignore astrology, but actively oppose its dissemination. Further, if large numbers of citizens do not understand the scientific method and the difference between science and pseudoscience, how can they intelligently vote on or respond to scientific questions that have societal implications?

Gloria: Oh Ma, you don't believe in astrology, do you?
Edith: No, but it's fun to know what's going to happen to you, even if it don't.
from All in the Family

A major reason why scientists in general and astronomers in particular don't believe in astrology is that they cannot conceive of a way in which it would work. The human brain is so complex that it seems most improbable that any celestial alignment can affect people, including newborns, in an overall way. The celestial forces that are known cannot be sufficient to set

Figure 19–13 The signs of the zodiac are shown on this map dating from 1700, when it was drawn by Frederik de Wit. (Courtesy of the Royal Library of Copenhagen)

personalities nor influence day to day events. And even if people reply that they do not think astrology is true but merely find it interesting, many scientists feel that so many strange and exciting explainable things are going on in the Universe that we wonder why anybody should waste time with far-fetched astrological concerns.

After all, we have already discussed such fascinating things as neutrinos from the Sun, pulsars, and black holes. Later in this book we will consider complex molecules that have spontaneously formed in interstellar space, and try to decide whether the Universe will expand forever. We have sent a rocket into interstellar space bearing a portrait of humans, and have beamed a radio message toward a group of stars 24,000 light years away. These topics and actions are part of modern astronomy, what contemporary, often conservative, scientists are doing and thinking about the Universe. How prosaic and fruitless it thus seems to spend time pondering celestial alignments and wondering whether they can affect individuals. In fact, even the alignments that most astrologers use are not accurately calculated, for the precession of the Earth's pole has changed the stars that are overhead at a given time of year from what they were millennia ago when astrological tables that are often still in current use were computed.

Moreover, astrology just doesn't work. Bernie I. Silverman, a psychologist at Michigan State University, tested specific values: Do Libras and Aquarians rank "Equality" highly? Do Sagittarians especially value "Honesty"? Do Virgos, Geminis, and Capricorns (Fig. 19–13) treasure the value "Intellectual"? Several astrology books agreed that these and other similar examples are values typical of those signs. Although believers often criticize objections on the ground that these group horoscopes are not as valuable or accurate as individualized charts, surely some general assumptions and rules hold in common.

The subjects, 1600 psychology graduate students, did not know in advance just what was being tested. They gave their birthdates, and the questioners determined their astrological signs. The results: no special correlation was apparent for any of the signs with the values they were supposed to hold. This was confirmed by statistical tests. As many subjects, regardless of their astrological signs, when asked to what extent they shared the qualities of each given sign, ranked themselves above average as below.

Furthermore, Silverman tested whether individuals from particular pairs of signs were especially compatible or incompatible. To do this, he sampled marriage and divorce rates in Michigan in 1967 and 1968. He compared these rates with the predictions of two astrologers as to compatible or incompatible combinations. His sample was large—2978 marriages and 478 divorces—so even slight effects should show up in his statistical treatment. And none did—those born under "compatible" signs neither married more frequently nor divorced less frequently than would be expected by chance, nor did those born under "incompatible" signs marry less frequently or divorce more frequently. Silverman concluded that "the position of the Sun in the zodiac at birth does not affect later personality," and in general that "astrology is invalid." Thus astrological predictions were shown by his studies to be no better than chance.

Similar statistical studies of signs "ruled" by Mars, the planet named for the god of war, and military reenlistment rates were recently carried out by James T. Bennett and James R. Barth, economists at George Washington University. They also found no empirical support for astrology.

But if astrology is so meaningless, why does it still have so many adherents? Well, it could be the bandwagon effect, and Silverman had an ingenious test that endorsed this idea. He took twelve personality descriptions from astrology books, one for each astrological sign, and displayed them to two groups of individuals. The first group, composed of 51 subjects, was told to which astrological signs the descriptions pertained, and was asked to write their own signs on the covers of the questionnaires. More than half the members of this group thought that the descriptions listed under their own signs were, for each individual, among the four best descriptions of themselves out of the twelve choices. It would seem, if one considered only this phase of the experiment, that astrology was working.

Yet when the second group was given the twelve descriptions without mention of astrology, being told that they came from a book entitled "Twelve Ways of Life," their choices were random. Only 30 per cent chose their own sign's description as being in the group most closely describing them. So the idea that astrology can predict personality types seems to be the result of self-delusion. When people know what they are expected to be like, they tend to identify themselves with the description. But that doesn't mean that they actually satisfy the description that astrology predicts for them better than any other description.

A historian of science has suggested that the bandwagon effect by itself is insufficient to explain belief in astrology, because astrological beliefs have persisted for so long. Astrology, following this line of reasoning, joins other pseudosciences in satisfying intuitional needs of individuals who do not clearly understand what science is, on what it is based, and what it has to offer us.

From an astronomer's view, astrology is meaningless, unnecessary, and impossible to explain if we accept the broad set of physical laws we have conceived over the years to explain what happens on the Earth and in the sky. Astrology snips at the roots of all pure science. Moreover, astrology patently doesn't work. If people want to believe in it on an *a priori* basis, as a religion, or have a personal astrologer act as a psychologist, let them not try to cloak their beliefs in scientific astronomical gloss. The only reason people may believe that they have seen astrology work is that it is a self-fulfilling means of prophecy, conceived of long ago in times when we knew less about the exciting things that are going on in the Universe. Let's all learn from the stars, but let's learn the truth.

SUMMARY AND OUTLINE OF CHAPTER 19

Aims: To discuss the possibility of intelligent life existing elsewhere in the Universe besides the Earth, and to assess our chances of communicating with such life; to give an astronomer's view of UFO's and of astrology; to give scientific criteria for "truth," and definitions of "hypotheses" and "theories"

The origin of life (Section 19.1)
- Earth life based on complex chains of carbon-bearing (organic) molecules
- Carbon, as well as silicon, forms such chains easily
- Organic molecules are easily formed under laboratory conditions that simulate primitive atmospheres

Evidence for other solar systems (Section 19.2)
- Search for wobbles in proper motions of stars
- Evidence for a planet around Barnard's star once was positive, but is currently not as certain as had been thought previously

Statistical chances for extraterrestrial intelligent life (Section 19.3)
- Problem broken down into stages
- Major uncertainties include what the chance is that life will form given the component parts, and what the lifetime of a technological civilization is likely to be

The search for life (Section 19.4)
- Experiments on Viking spacecraft to Mars
- Projects Ozma and Ozma II
- Signal beamed toward globular cluster M13 from Arecibo radio telescope
- Soviet and American CETI and SETI plans

UFO's and Occam's Razor (Section 19.5)
- UFO's explainable as natural phenomena

- Application of Occam's Razor—the Principle of Simplicity—rules out the existence of UFO's
- Definitions of truth, hypothesis, and theory

Astronomy and astrology (Section 19.6)

- No scientific basis known for astrology
- Belief in such pseudoscience impedes the advance of science and technology
- Statistical arguments show that astrology doesn't work

QUESTIONS

1. What is the significance of the Miller-Urey experiments?

2. Assume that a star has a planet with a mass one tenth that of the star. Sketch the path of the star and the planet in the sky.

3. Does a star with a large proper motion necessarily have planets? Explain.

4. Discuss the evidence that Barnard's star has planets around it.

5. If one tenth of all stars are of suitable type for life to develop in the case that planets exist around them, and 1 per cent of all stars have planets, and 10 per cent of all planetary systems have a planet at a suitable distance from the star for a comfortable environment, what fraction of stars have a planet with conditions suitable for life? How many such stars would there be in our galaxy?

6. List the means by which we might detect extraterrestrial civilizations.

7. Describe how Occam's razor might have been applied during the debate over the geocentric and heliocentric pictures of the solar system.

8. Give a non-scientific example of a hypothesis, and suggest how it might be tested. At what stage would the hypothesis become a theory?

TOPICS FOR DISCUSSION

1. Comment on the possibility of scientists in another solar system observing the Sun, suspecting it to be a likely star to have populated planets, and beaming a signal in our direction.

2. Discuss the extent of the astrological beliefs of the members of your class. If any students believe in astrology, do their beliefs affect their daily lives?

Part IV

The Milky Way Galaxy

The Milky Way from Sagittarius to Cassiopeia.

(left) The Lagoon Nebula, M8, in Sagittarius.

The Milky Way Galaxy . . .

On the clearest nights, when we are far from city lights, we can see a hazy band of light stretched across the sky. This is the *Milky Way*—the aggregation of dust, gas, and stars that make up the galaxy in which the sun is located.

Don't be confused by the terminology: the Milky Way itself is the band of light that we can see from the earth, and the Milky Way Galaxy is the whole galaxy in which we live. Like other galaxies, our Milky Way Galaxy is composed of a hundred billion stars plus many different types of gas, dust, planets, etc. The Milky Way is that part of the Milky Way Galaxy that we can see with the naked eye in our nighttime sky.

The Milky Way appears very irregular in form when we see it stretched across the sky—there are spurs of luminous material that stick out in one direction or another, and there are dark lanes or patches in which nothing can be seen. This is just the manifestation of the splotchy distribution of the dust, stars, and gas.

The Origin of the Milky Way by Tin
circa 1578. (The National Gallery, L

Here on earth, we are inside our galaxy together with all of the matter we see as the Milky Way. Because of our position, we see a lot of matter when we look in the plane of our galaxy. On the other hand, when we look "upwards" or "downwards" out of this plane, our view is not obscured by matter, and we can see past the confines of our galaxy. We are able to see distant galaxies only by looking in parts of our sky that are away from the Milky Way, that is, by looking out of the plane of the galaxy.

As seen from the earth, the center of our galaxy is located in the direction of the constellation Sagittarius. This constellation never rises high in the sky seen from the United States, but in the southern hemisphere it can appear directly overhead. Since the denser clouds of stars, gas, and dust are in the direction of the galactic center, the Milky Way appears more spectacular in the sky when observed from points south of the equator.

The gas in our galaxy is more or less transparent to visible light, but the small solid particles that we call "dust" are opaque. So the distance we can see through our galaxy depends mainly on the amount of dust that is present. This is not surprising: we can see great distances through our gaseous air on earth, but let a small amount of particulate material be introduced in the form of smoke or dust thrown up from a road and we find that we can no longer see very far. Similarly, the dust between the stars in our galaxy dims the starlight by scattering it in different directions besides absorbing some of it. This property of the dust is called *extinction.* The extinction of visible light in the plane of our galaxy is about 1 magnitude per thousand parsecs (1 magnitude per kiloparsec)—that is, a star a thousand parsecs away would appear about one magnitude fainter than it would if there were no interstellar dust, a star two thousand parsecs away would appear two magnitudes dimmer, and so on. Of course, the amount of extinction is really irregular, in that it varies from this average value as one looks in different directions.

The presence of so much dust in the plane of the Milky Way Galaxy actually prevents us from seeing very far toward its center—we can see only 1/10 of the way toward the galactic center itself. This explains the dark lanes across the Milky Way, which are just areas of dust, obscuring any emitting gas or stars. The net effect is that we can see just about the same distance in any direction we look in the plane of the Milky Way. It follows that these direct optical observations by themselves might fool us into thinking that we are in the center of the galaxy, or indeed in the center of the universe. Using this reasoning, scientists at the turn of the century thought that the earth was near the center of the galaxy.

It was not until the 1920's that the American astronomer Harlow Shapley realized that we were not in the center of the galaxy. He was studying the distribution of globular clusters and noticed that they were all in the same general area of the sky as seen from the earth. They mostly appear above or below the galactic plane and thus are not obscured by the dust. When he plotted their distances and directions (using the methods described in Part II), he saw that they formed a spherical halo around a point many thousands of light years away from us (see Fig. 20–3). Shapley's touch of genius was to realize that this point must be the center of the galaxy.

In recent years astronomers have also been able to use wavelengths other than optical ones, especially radio wavelengths, to study the Milky Way Galaxy. We shall first discuss the types of objects that we find in the Milky Way Galaxy, and then go on to describe how radio astronomy and other methods of observation have given us new information about our galaxy.

20

Components of the Milky Way

20.1 CLUSTERS OF STARS

In Part II, we described stars, which are important constituents of any galaxy. In this chapter, we describe other material in our galaxy and also briefly review the contribution of stars and their debris. We also describe the overall structure of the Milky Way Galaxy and how, from our location inside it, we detect this structure.

There are two main types of clusters of stars: galactic clusters and globular clusters. The properties of stars in these types of clusters and the division of stars into Stellar Populations were discussed in Section 4.5, and we shall only review the basic information about clusters here.

Galactic clusters are asymmetric groups of stars spread through a limited volume of space, mostly in the galactic plane. Many galactic clusters have about 1000 members, but some have as few as ten or as many as two thousand members. To an observer, galactic clusters appear as an increased number of stars in a limited area. A *rich* cluster (Fig. 20–1*A*) has many more stars than a *poor* cluster (Fig. 20–1*B*). Sometimes gas and dust can be seen associated with the stars in a young galactic cluster. Galactic clusters are also called *open clusters.* Do not confuse "galactic clusters," which are clusters of stars in our galaxy (the individual objects are stars), with "clusters of galaxies," which are, as their name shows, clusters of separate galaxies (the individual objects are galaxies).

The Pleiades is the best known example of a galactic cluster. Known as the Seven Sisters, this cluster contains six stars that are readily visible to the naked eye (see Section 4.5). However, even a small telescope shows dozens of stars in the Pleiades, and larger telescopes show hundreds of fainter stars. Long exposure photographs show patches of dust surrounding some of the stars and reflecting light from these stars.

Studies of their Hertzsprung-Russell diagrams (which were discussed in Section 4.5) show that galactic clusters are relatively young compared to globular clusters—the gas and dust observed in galactic clusters is presumably that from which the stars were formed. The abundances of the "metals" (everything other than hydrogen and helium) are relatively high in the stars. (By relatively high we mean that the total abundance of

The Milky Way in Sagittarius.

Figure 20–1 *(A)* A rich galactic cluster, M37, in the constellation Auriga, the Charioteer. It contains about 150 stars brighter than magnitude 12.5 and perhaps over 500 stars in all. *(B)* A poor galactic cluster, NGC 7510.

A

B

"metals" in stars in the galactic clusters is a higher fraction of a per cent than is the total abundance of these trace elements in stars in the globular clusters. The percentages, in both cases, are very small.) The higher percentage for galactic clusters indicates that these young stars were formed out of second generation gas, that is, recycled material. The material has already gone through a whole stage of stellar evolution and has been enriched in the heavy elements via fusion processes in an earlier generation of stars or in supernova explosions.

Globular clusters (Fig. 20–2 and Color Plate 4) are older clusters and are composed of tens or hundreds of thousands of stars tightly bound together in a spherical shape. Several globular clusters can readily be seen through small telescopes. Studies of the globular clusters led to the discovery that the sun was not in the center of the Milky Way Galaxy (Fig. 20–3).

The globular clusters contain older stars than do the galactic clusters. Again, we know this from studies of their color-magnitude diagrams. In globular clusters, the stars have relatively low abundances of the heavy elements, and there is no interstellar gas and dust; the stars were formed before the gas was enriched with heavy elements and the gas and dust has been used up.

About one-third of the globular clusters contain RR Lyrae stars (discussed in Section 4.4c), which are also known as cluster variables. The RR Lyrae stars are used for measuring the distances to these clusters.

Figure 20–2 Omega Centauri, a globular cluster prominent in the sky seen from the southern hemisphere.

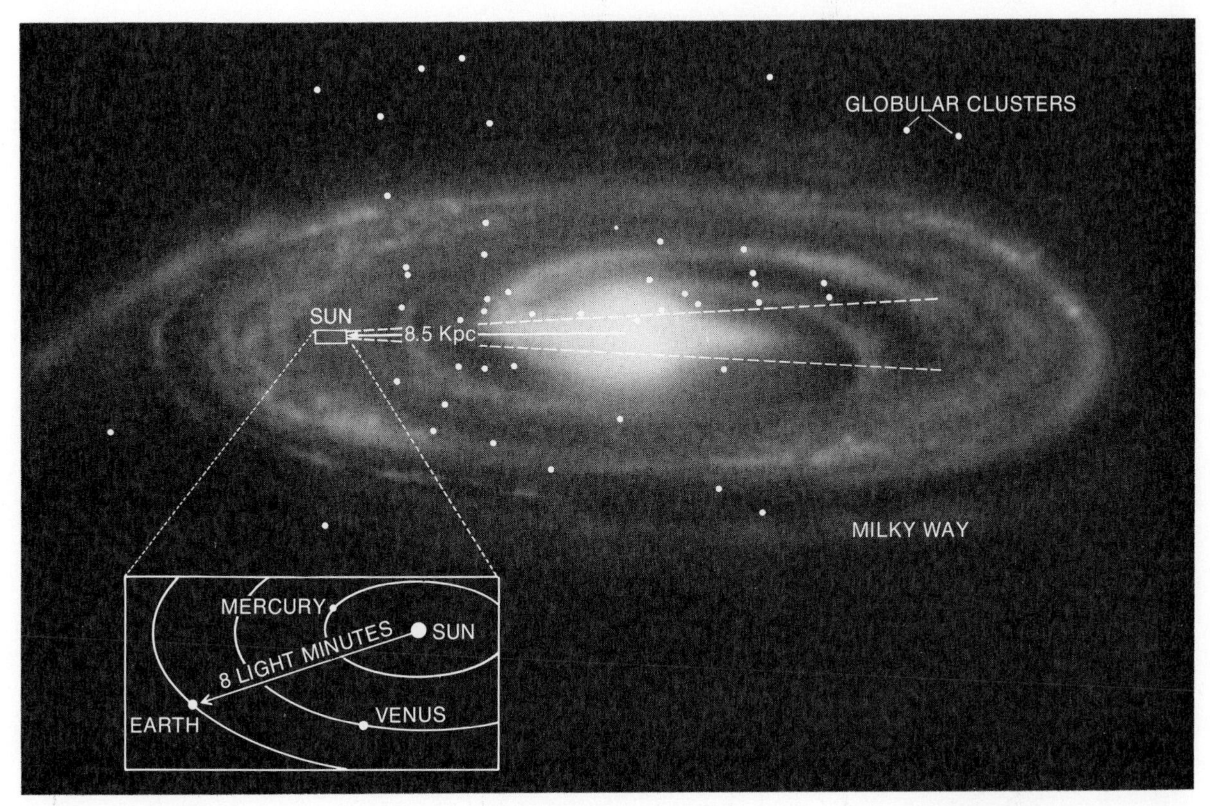

Figure 20–3 An artist's conception of the Milky Way Galaxy, with the actual positions of the globular clusters for which distances are known. Absorption by interstellar matter blocks our view of most objects between the dotted lines. Many of the clusters seem concentrated near the center of our galaxy and the rest are far from the plane of the galaxy. From the fact that most of the clusters appear in less than half of our sky, Shapley deduced that the galactic center is in the direction indicated.

20.2 NEBULAE

A *nebula* (pl: *nebulae,* not "nebulas") is a cloud of gas and dust that we see in visible light. Sometimes we see the gas actually glowing, sometimes we see the dust appear as a dark silhouette against other glowing material, and sometimes we see the dust reflecting light from other objects toward us. These are called *emission nebulae, absorption* (or *dark*) *nebulae,* and *reflection nebulae,* respectively.

"Nebula" is Latin for fog or mist.

Another class of objects was once known as "spiral nebulae," since they looked like glowing gas with "arms" spiraling away from their centers. (We still speak of the Great Nebula in Andromeda.) However, these spiral nebulae are now known to be galaxies in their own right (historically, the debate was whether they were part of our own galaxy or "island universes"), and we shall discuss them in Chapter 23. The use of the term "nebula" for these particular objects is of historical interest only.

The North America Nebula (Fig. 20–4), gas and dust that has a shape in the sky similar to the shape of North America on the earth's surface, is a good example of both an emission and an absorption nebula. The red emission comes from glowing gas spread across the sky. This gas is not completely opaque to visible light; of the stars that are visible in the nebula, some are between the sun and nebula but others are behind the nebula.

The boundary of the nebula as we see it, however, is not just the point where the glowing gas stops. There is actually dark absorbing dust that forms some of the boundaries. For the North America Nebula, there is dust absorbing in the area that corresponds to the Gulf of Mexico. We can tell

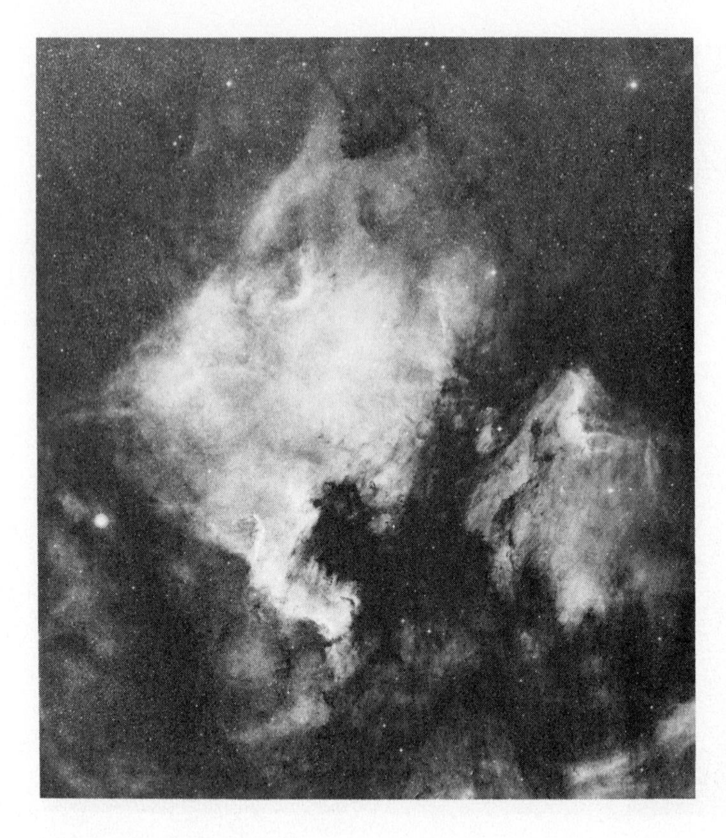

Figure 20–4 The North America Nebula, NGC 7000, in Cygnus.

this because the number of stars that we see in that direction is very small. We think that the average number of stars we could see if there were no dust is fairly constant in a given area of the sky. Thus when we see fewer stars in a part of that area, we conclude that there must be dust absorbing the radiation from stars behind it.

The Horsehead Nebula (Figs. 3–27 and 20–5 and Color Plate 41) is another example of emission and absorption nebulae. A bit of absorbing dust intrudes onto emitting gas, outlining the shape of a horse's head for us. We can see that the horsehead is a continuation of a dark area in which very few stars are visible. A reflection nebula is also visible nearby.

The Great Nebula in Orion (Color Plate 12) is an emission nebula. It can readily be observed even with small telescopes in the winter sky, but only with long photographic exposures or large telescopes can we study the structure in the nebula in detail. In the center of the nebula are four closely grouped bright stars called the Trapezium, which provide the energy to make the nebula glow. The Orion Nebula is an exciting place where we think stars are being born this very minute.

The clouds of dust surrounding some of the stars in the Pleiades (Fig. 20–6 and Color Plate 5) are examples of reflection nebulae—they merely reflect the starlight toward us without emitting much radiation of their own. Many other reflection nebulae are known (Fig. 20–7).

The nebulae are particularly beautiful objects because of their interesting configurations, and because the different processes of emission and absorption cause color effects that, although they are too faint to be visible to the naked eye, can be captured on photographs. Other nebulae are shown in Color Plates 42, 43, 44, and 45.

The nebulae represent regions of our galaxy in which the density of gas and dust is slightly higher than in contiguous regions. But even though the nebulae are comparatively denser, they are still considerably less dense than is the gas in our air. In fact, even the nebulae with the highest densities are still many orders of magnitude less dense than the best vacuums that can be made in laboratories on earth, so that the nebulae provide scientists with a way of studying the basic properties of gases.

The emission nebulae glow because the gas is heated by one of several mechanisms. Bright spectral lines are emitted at certain wavelengths by a *fluorescent* process. The hydrogen that makes up the overwhelming proportion of the gas is ionized—that is, the hydrogen is separated into protons and electrons. Neutral hydrogen is known as H I (the "first state," with all its electrons), and ionized hydrogen as H II (the "second state," with one electron missing), so these ionized nebular regions are known as *H II regions* (Fig. 20–8).

Fluorescence is discussed in Box 20.1 on p. 452.

Figure 20–5 The Horsehead Nebula, IC 434, in Orion.

Figure 20–6 Reflection nebulae in the Pleiades, M45.

We may consider interstellar space as filled with hydrogen at an average density of about 1 atom per cubic centimeter, although individual regions may have densities departing greatly from this average. Regions in which the atoms of hydrogen are predominantly neutral are called *H I regions*. Where the density of an H I region is high enough, pairs of hydrogen atoms combine to form molecules (H_2). The densest part of the molecular cloud associated with the Orion Nebula might have 10^6 or more hydrogen molecules per cubic centimeter. On the other hand, wherever a hot star provides enough energy to ionize hydrogen, an H II region results.

Figure 20–7 A reflection nebula, NGC 7129, in Cepheus.

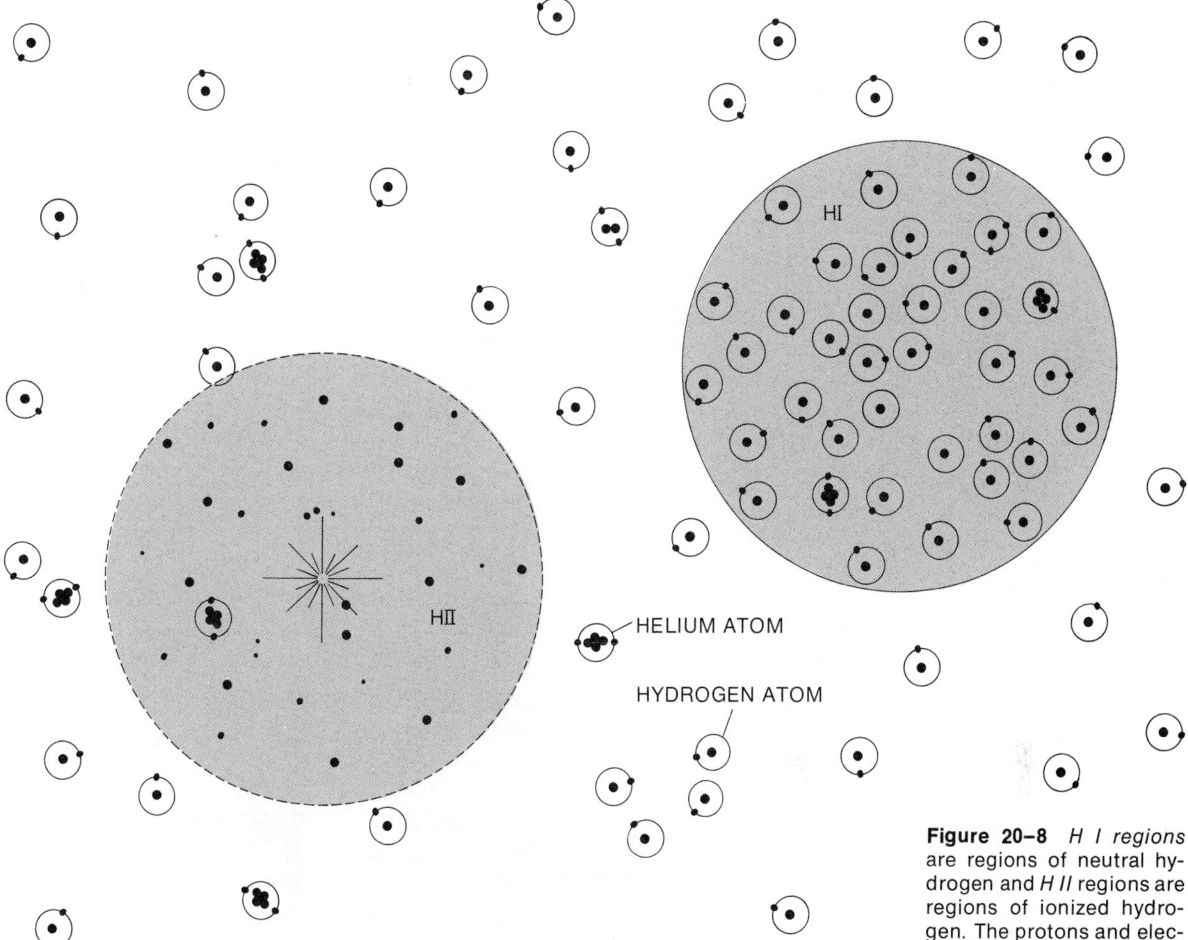

Figure 20–8 *H I regions* are regions of neutral hydrogen and *H II* regions are regions of ionized hydrogen. The protons and electrons that result from the ionization of hydrogen by a hot star, and the neutral hydrogen atoms, are shown schematically. The outlines of the regions are shown for illustrative purposes only; in space, of course, the regions are not outlined. The densities in H I and H II regions are factors of 10 and more times, ranging up to thousands of times, higher than average interstellar densities. Note that heavier elements, such as helium, are mixed in with interstellar hydrogen in their normal abundances and require different amounts of energy to be ionized than does hydrogen.

20.3 PLANETARY NEBULAE AND SUPERNOVA REMNANTS

Some of the most beautiful shapes in the sky are simply composed of gas thrown off by certain types of stars in late stages of their evolution. These are known as *planetary nebulae* (Fig. 20–9 and Color Plates 6, 7, and 8) since they appeared as greenish disks when viewed in the small telescopes of 19th century observers, similar in appearance to Uranus and Neptune. Actually, the planetary nebulae—sometimes just called "planetaries" for short—are not related to planets at all. About 1000 are now known. Much of their radiation comes from a process called fluorescence, which gives them their greenish color.

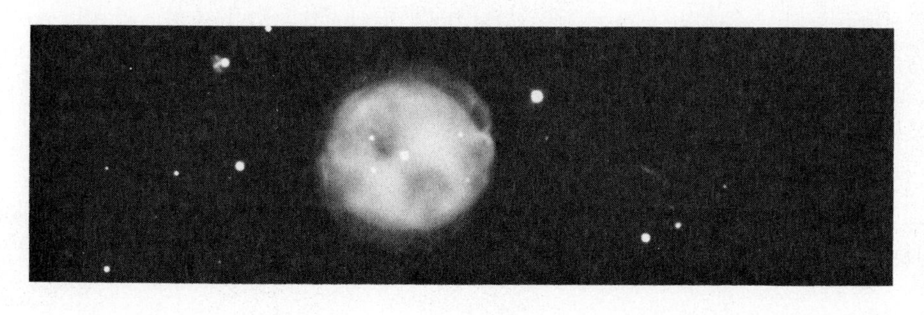

Figure 20–9 The Owl Nebula, M97, a planetary nebula in Ursae Majoris.

Orthographic note: remember that "fluorescent" is spelled with a "uo," opposite to the "ou" that appears in the word "flour."

Box 20.1 Fluorescence

A *fluorescent* process is one in which photons of a high energy (which correspond to short wavelengths) are transformed, through interactions with atoms, to photons of lower energy (which correspond to longer wavelengths). In particular, very hot stars give off many photons at wavelengths shorter than 912 Å, which is the wavelength that is the longest at which photons can ionize hydrogen. These short-wavelength photons ionize the hydrogen near the hot stars, which makes H II regions. We cannot observe these high energy photons directly, for even those that get out of the H II region without giving up their energy to a hydrogen atom cannot penetrate the earth's atmosphere. But we can observe their effect through the following fluorescent process.

The electrons continually recombine with the hydrogen ions, and the resultant neutral atoms are usually on their higher energy levels. They then jump down through a series of middle energy levels until they reach the ground state. This means that emission lines in the Balmer series appear in the visible (and other hydrogen emission lines appear in other parts of the spectrum). The atoms are quickly ionized again, and the fluorescent process takes place all over again.

In addition, a specific ultraviolet transition excites doubly ionized oxygen ions (O III), and emission lines of O III that happen to be in the green part of the spectrum also show. This process in oxygen is thus also fluorescent, for it begins with ultraviolet photons and winds up with visible photons. Singly ionized nitrogen (N II), singly ionized oxygen (O II), and doubly ionized neon (Ne III) are also observed. Until this process was understood, the lines had been of unknown origin and were spoken of as the "nebulium lines."

Figure 20–10 The Ring Nebula, M57, a planetary nebula in Lyra.

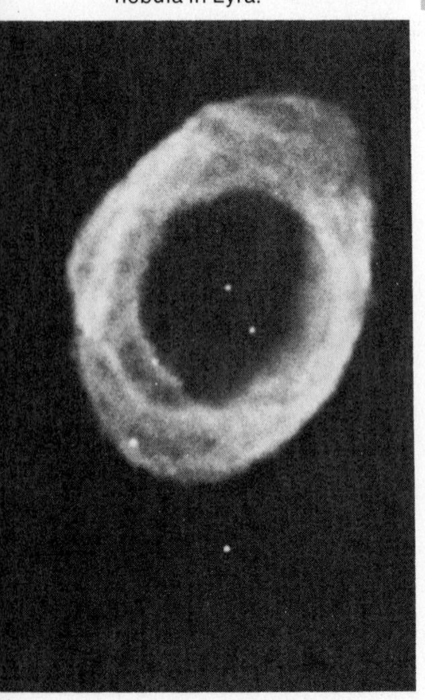

The Ring Nebula (Fig. 20–10 and Color Plate 6) in the constellation Lyra is the best known planetary nebula, for it can be seen even with small telescopes. The colors can be seen on long-exposure photographs but not with the eye directly, even through a telescope.

The gaseous shells can actually be seen to expand over a 10 or 20 year time scale, and we can extrapolate back to the time when the shells were blown off, assuming that their motion away from their central stars has not slowed down or accelerated very much. Such calculations indicate that planetary nebulae can be only 20,000 to 50,000 years old—not very old on an astronomical time scale. (Planetary nebulae are discussed more fully in Section 7.2.)

More catastrophic stellar explosions also throw matter into interstellar space. Supernovae remnants (Fig. 20–11 and Color Plates 10 and 11) are described in Section 8.2.

Figure 20–11 The Veil Nebula, NGC 6960, a supernova remnant.

20.4 THE INTERSTELLAR MEDIUM

Not only are distant stars in the plane of our galaxy obscured from our vision, but also the amount of obscuration of stars that are still close enough to be visible varies with the wavelength at which we observe. The blue light is *scattered* by dust in space more efficiently than the redder light is; that is, for a given distance through the dust more of the blue light has been bounced around in every direction. Thus less of the blue light comes through to us than the red light, and the stars look redder—we say that the stars are *reddened*. This reddening is thus a consequence of the scattering properties of the dust. It has nothing to do with "red shifts," since in the present case the spectral lines are not shifted in wavelength.

Over the years many studies have been made of the *interstellar reddening*, and it has been discovered that in the visible, the amount of reddening varies approximately inversely with the wavelength; that is, the amount of reddening is proportional to one divided by the wavelength. For example, red light of 7000 Å wavelength is dimmed about half as much as ultraviolet light of 3500 Å.

The amount of scattering together with the amount of actual absorption of visible radiation is known as the *extinction*. The amount of extinction has been found empirically (that is, on the basis of observation rather than theoretically) to be proportional to the amount of reddening. So by measuring the colors of stars, which gives the reddening, we can derive the extinction. In the blue part of the spectrum, the total extinction is about 25 magnitudes between the center of the galaxy and the sun. Thus even a tremendously bright object located there would be dimmed too much to be seen from the earth.

A similar scattering by the electrons in air molecules in the terrestrial atmosphere makes the sky blue. Further, when we look a long way through the air, most of the blue is scattered out before it reaches us and relatively more red light reaches us. This accounts for the reddish color of sunsets. Electrons scatter visible wavelengths of light much more efficiently than dust does. Electrons scatter light based on the fourth power of the wavelength. Thus ultraviolet light at 3500 Å is scattered $2^4 = 16$ times more efficiently than light at 7000 Å in the infrared.

Traditionally, studies of reddening and extinction have been used to find the distances to stars. Let us consider two stars of identical spectral type (and thus having the same distribution of energy with wavelength), one out of the galactic plane and one in the plane. The first, which is out of the plane, is not reddened because there isn't much interstellar material between it and us. The light from the second star, located in the galactic plane, is reddened. From the relative amount of reddening and the amount of extinction, we can determine how much interstellar material the light has passed through and can, in principle, calculate the approximate distance to the second star.

Unfortunately, the reddening is splotchy in nature, and we cannot be certain that a volume of gas of unusually high or unusually low density is not included in our line of sight. The edges of dark nebulae are good examples of how the amount of reddening of two stars almost but not quite in the same direction could be very different. Also, reddening can be caused by shells of gas surrounding the star in addition to arising in interstellar space. This method of determining distances must be used with caution or its use limited to statistical studies.

We have just been discussing how to observe the dust by studying how it affects starlight that passes through it. It is difficult to observe the dust directly, but the dust is heated a bit by radiation, thus causing the dust to radiate. The dust never gets very hot, so its radiation peaks in the infrared. The radiation from dust scattered among the stars is too faint to be detected, but the radiation coming from clouds of dust surrounding stars has been observed. We shall discuss some of the special techniques of infrared astronomy in Section 22.1.

Similarly, as the interstellar gas is "invisible" in the visible part of the spectrum (except at the wavelengths of certain weak spectral lines), special techniques are needed to observe the gas in addition to observing the dust. Radio astronomy is the most widely used technique for studying the gas, and we shall discuss it at length in the next chapter.

The gas and dust between the stars is known as the *interstellar medium.* Note that the nebulae discussed in Section 20.2 are part of the interstellar medium.

20.5 THE SPIRAL STRUCTURE OF THE GALAXY

20.5a Bright Tracers of the Spiral Structure

When we look out past the boundaries of the Milky Way Galaxy, usually by observing in directions above or below the plane of the Milky Way, we can see a number of galaxies external to our own system. Some of these appear as though bright arms spiral off a central core. We shall discuss these *spiral galaxies* (Color Plates 46, 47, 49, and 50) in Chapter 23; in this section we shall discuss some of the evidence that our own Milky Way Galaxy also has spiral structure.

It is always difficult to tell the shape of a system from a position inside it. Think, for example, of being somewhere inside a maze made of tall hedges. We might be able to see through some of the foliage and be reasonably certain that there were layers and layers of hedges surrounding us, but we would find it difficult to trace out the pattern. If we could fly overhead in a helicopter, though, the pattern would become very easy to see.

Similarly, we have difficulty tracing out the spiral pattern in our own galaxy, even though the pattern would presumably be apparent to someone located outside our own galaxy. Still, by meticulously noting the distances and directions to objects of various types, we can tell something about the Milky Way's spiral structure.

One type of object that can be studied for this purpose is a galactic cluster, which is known to always be in the spiral arms. We can find the distance to a galactic cluster from its color-magnitude diagram (Section 4.5a). We have now determined the distances to over 200 of these clusters.

In addition, it is known from studies of other galaxies that H II regions are also preferentially located in spiral arms. In studying the locations of the H II regions, we are really studying the locations of the O stars and the hotter B stars, since it is ultraviolet radiation from these hot stars that provides the energy for the H II regions to glow. These O and B stars, which we know must be relatively young because their lifetimes are short, must tend to be formed in the spiral arms; in the next section we shall discuss a theory that explains why this should be the case.

We can find the distances to H II regions, in some cases, by studying the intensities of their hydrogen spectral lines (Hβ, for example), and deducing how distant the H II regions must be in order for these lines to have their observed intensities. Only a few dozen distances have been calculated by this method, however.

When the galactic clusters and H II regions of known distance are plotted, they appear to trace out three spiral arms (Fig. 20–12). The spacings between the arms and the widths of the arms appear consistent with spacings and arm widths that we can observe in other galaxies.

These observations are carried out in the visible part of the spectrum.

Figure 20–12 The positions of young galactic clusters and H II regions in our own galaxy are projected on a photograph of the spiral galaxy NGC 1232, which has the same linear diameter as our own galaxy.

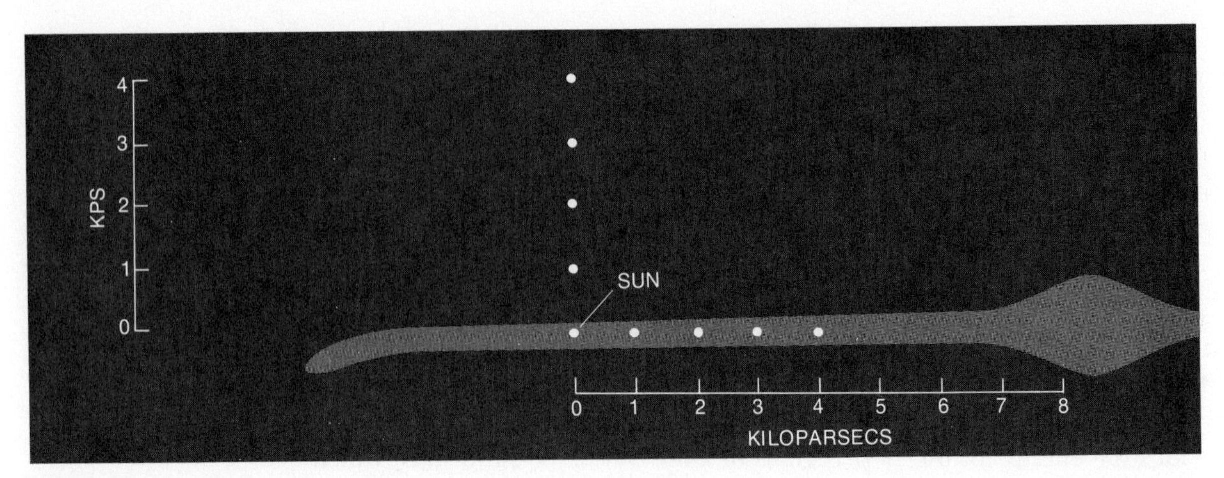

Figure 20–13 The sun is 8500 parsecs from the center of our galaxy.

Note that even when we are observing the O and B stars, which are very bright and therefore can be seen at distances that are relatively great compared to other stars, the optical map shows only regions of our galaxy close to the sun. The interstellar extinction prevents us from studying parts of the spiral arms farther away from the sun. Another very valuable method of studying the spiral structure in our own galaxy involves a spectral line of hydrogen in the radio part of the spectrum, and will be discussed in Section 21.4. Radio waves penetrate the interstellar dust, and we are no longer limited to studying the local spiral arms.

The current picture that we have of our own galaxy (Fig. 20–13) is that the inner 5000 parsecs or so, called the *nucleus*, represents a bulge and does not show spiral structure. Outside the nucleus, the next 10 kiloparsecs or so contain the spiral arms; the sun is about in the middle of this region, which is called the *disk*. The disk is only about 600 parsecs thick, 2 per cent of its width, and contains stars of Population I (Section 4.5). It is slightly bent, like a warped record, perhaps by interaction with the Magellanic Clouds. The Population II stars, including the globular clusters, form a *halo* around the disk.

20.5b Differential Rotation

The latest calculations indicate that the sun is approximately 8½ kiloparsecs from the center of our galaxy. (An older estimate of 10 kiloparsecs is often still used for calculations. It is much more convenient to calculate with a round number like 10 instead of 8½, and one can easily make the correction necessary in order to adopt other values for the distance of the sun from the galactic center.)

The local standard of rest, defined in Section 2.16, is defined by the average velocity of the stars in the vicinity of the sun. We consider the local standard of rest in order to eliminate the particular effect of the peculiar velocity of the sun.

From spectroscopic observations of the Doppler shifts of globular clusters, which do not participate in the galactic rotation, or of distant galaxies, we can tell that the sun is revolving around the center of our galaxy at a speed of approximately 250 kilometers per second. The velocity of revolution of our local standard of rest is such that it is heading toward the stars in the constellation Cygnus, although of course the stars there are revolving too, so we never get any closer to them than we are now. In our sky, Cygnus is at an angle of about 90 degrees away from the direction toward the center of our galaxy, which is in Sagittarius. So it seems as though we are revolving around the center in a more-or-less circular orbit.

At this velocity, it would take the sun about 250 million years to travel once around the center; this period is called the *galactic year.* But not all stars revolve around the galactic center in the same period of time. The central part of the galaxy rotates like a solid body, with the angular velocity remaining constant as you go out from the center. Since the angular velocity (in degrees/sec) of this central part is constant, the linear velocity (in km/sec), called the *tangential velocity,* increases with distance from the center because a given star or bit of gas has a longer path to travel. Beyond the central part, the stars that are farther out have longer galactic years than stars closer in, following Kepler's third law (which is described in Section 10.1).

If this system of *differential rotation,* with differing rotation speeds at different distances from the center, has persisted since the origin of the galaxy, we may wonder why there are still just a few spiral arms in our galaxy and in the other galaxies we observe. The sun could have made fifty revolutions during the lifetime of the galaxy, but points closer to the center would have made many more revolutions. Thus the question arises: why haven't the arms wound up very tightly?

Figure 20–14 Each part of the figure includes the same set of ellipses; the only difference is the relative alignment of their axes. Consider that the axes are rotating slowly and at different rates. The compression of their orbits takes a spiral form, even though no actual spiral exists. The spiral structure of a galaxy may arise from an analogous effect.

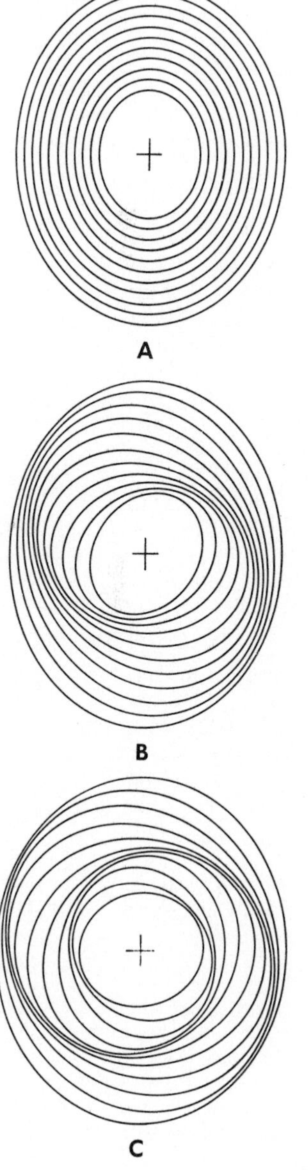

20.5c Density-wave Theory

The leading current solution to this conundrum is a theory first suggested by the Swedish astronomer B. Lindblad and elaborated mathematically by the American astronomers C. C. Lin and Frank Shu. They say, in effect, that the spiral arms we now see are not the same spiral arms that were previously visible. In their model, the spiral arm pattern is caused by a spiral *density wave,* a wave that moves through the stars and gas in the galaxy. This density wave is a wave of compression, not of matter being transported. It rotates more slowly than the actual material, and causes the density of material to build up as it passes by. A shock wave (defined in Section 5.3) is also built up, heating the gas.

We can think of the analogy of a crew of workers painting the white line down the center of a busy highway. A bottleneck occurs at the location of the painters, and an observer in an airplane would see an increase in the number of cars at that place. As the sign painters continued slowly down the road at, say, 8 km/hr (5 mph), the airborne observers would seem to see the place of increased density move down the road at that slow speed. But they would not be seeing the individual cars, which could still speed down the highway at 80 km/hr (50 mph), slow down briefly as they cross the region of the bottleneck, and then resume their high speed. Similarly, we might only be viewing some galactic bottleneck at the spiral arms. The gas is compressed there, both from the increased density itself and, to a greater extent, from the shock wave formed by other gas piling into this slowed-up gas from behind. The compression causes heating, and also leads to the formation of protostars that collapse to become stars. Thus the spiral density wave leads to star formation, though the stars themselves in their revolution around a galactic center would move across the arms.

It may help to think of the analogy of a sound wave, which is also a compression wave. The sound wave travels through the air at a rapid rate, but the air itself doesn't move at that rate in the direction of the sound wave. The mass of air has no net velocity at all.

Similarly, the density wave can pass through the galactic matter even though the galactic matter does not share in its motion. As the density wave passes a given point, stars begin to form there because of the increased density. The distribution of new stars can take a spiral form (Fig. 20–14). The new stars heat the interstellar gas so that it becomes visible. In fact, we do see young, hot stars and glowing gas outlining the spiral arms, which is one check of this prediction of the density-wave theory.

Some other astronomers do not accept the density-wave theory. Some of them think that the spiral arms in galaxies may be gas thrown off by explosions inside the galaxies in times past. Others think that in at least some cases the spiral arms are pulled out of the cores during collisions or near passages of two galaxies in space. It seems unlikely, however, for enough of these gravitational interactions of galaxies to have taken place to account for the large number of spiral galaxies that we observe. We shall discuss the effects of gravitational interactions in Section 23.2c.

SUMMARY AND OUTLINE OF CHAPTER 20

Aims: To understand the Milky Way as the visible part of our galaxy and to discuss the various aspects of the matter that exists between the stars in our galaxy; to understand spiral structure

- Clusters of stars (Section 20.1; see also Chapter 4 for a fuller discussion)
  - Galactic clusters
  - Globular clusters
- Nebulae (Section 20.2)
  - Emission
  - Absorption (dark)
  - Reflection
  - "Spiral"
  - H I and H II regions
- Planetary nebulae (Section 20.3)
- The interstellar medium (Section 20.4)
  - Reddening and extinction used to determine stellar distances
  - Blue sky caused by scattering of blue light by atmosphere
  - Dust and gas perform the same scattering effect in space
- Spiral structure of the galaxy (Section 20.5)
  - Bright tracers of the spiral structure (Section 20.5a)
    - Difficult to determine shape of our galaxy because of obscuring dust and gas
    - H II regions used to determine spiral structure
    - Our galaxy has a nucleus, disk, and halo
  - Differential rotation (Section 20.5b)
    - Sun approximately 8 1/2 kpc from center of galaxy
    - 1 galactic year equals about 250 million years for the sun
    - Central part of galaxy rotates as a solid body, i.e., angular velocity remains constant as one travels outward
    - Outer part, including the spiral arms, is in differential rotation

Density-wave theory (Section 20.5c)
Leading current solution to why spiral arms have not wound up very tightly
We see the effect of a wave of compression; the distribution of mass itself is not spiral in structure

QUESTIONS

1. Why do we think that our galaxy is a spiral?
2. How would the Milky Way appear if the sun were close to the edge of the galaxy?
3. Contrast the properties of galactic clusters and globular clusters.
4. If the older stars are located in the "galactic halo," what does this tell you about the evolution of the galaxy?
5. How is the metal content of a star related to its age? Why?
6. Compare: (a) absorption (dark) nebulae, (b) reflection nebulae, and (c) emission nebulae.
7. If you see a blue star surrounded by a blue nebula, would you expect this nebula to be emission or reflection? Explain.
8. If you see a blue star surrounded by a red nebula, is this nebula emission or reflection? Explain.
9. Which spectral type of stars would you expect to best excite H II regions?
10. Does a fluorescent light have emission lines or absorption lines in its spectrum?
11. First roughly graph the spectrum—wavelength vs. intensity—for an A star. (a) Sketch what the spectrum would look like if the radiation were reddened. (b) Sketch what the spectrum would look like if the radiation were redshifted. (c) Could the radiation be both reddened and redshifted? Explain.
12. Where in the sky, relative to the sun, do you expect the sky to be "bluest"?
13. If you didn't know about the interstellar medium, what is the observational evidence to point to its existence?
14. (a) What types of objects trace out spiral arms? (b) Why?
15. What is another system, besides the Milky Way Galaxy, that exhibits differential rotation?

21

The Interstellar Medium

From antiquity until 1930, observations in a tiny fraction of the electromagnetic spectrum, the visible part, were the only way that observational astronomers could study the universe. Most of the images that we have in our minds of objects in our galaxy (e.g., nebulae) are based on optical studies, since most of us depend on our eyes to discover what is around us.

Optical astronomy has shown us how to sort the objects in the universe into stars and galaxies. In our own galaxy we can tell the nebulae apart from the stars, distinguish the spectral types of the stars, and, from another point of view, recognize different types of clusters. We have used the Doppler effect and the period-luminosity relation to measure the distances to stars and clusters of stars. Astronomy has made good use of the limited amount of data available to it, and most of the descriptions in this text up to this point have been based on optical studies.

If they had to make a choice, however, between observing only optical radiation or only everything else, many astronomers might choose "everything else." The rest of the electromagnetic spectrum carries more information in it than do the few thousand angstroms that we call visible light. In this chapter we will first discuss the basic techniques of radio astronomy, and then go on to see how radio astronomy joins with other observing methods to investigate interstellar space.

21.1 RADIO ASTRONOMY

Many people still think of astronomers as scientists who peer through big telescopes with glass lenses atop high mountains. The discovery of radio astronomy in the 1930's started the major change in the methods used by working astronomers. Now, with radio astronomy flourishing and many other types of observing also possible, only a minority of present day observational astronomers fit the traditional image.

One cannot always pinpoint the discovery of a whole field of research as decisively as that of radio astronomy. In 1931, Karl Jansky of the Bell Telephone Laboratories in New Jersey was experimenting with a radio

---

The 100-meter radio telescope at Effelsburg, near Bonn, Germany.

Figure 21–1 Karl Jansky with the rotating antenna with which he discovered radio astronomy.

antenna to track down all the sources of noise that might limit the performance of short-wave radiotelephone systems (Figs. 21–1 and 21–2). After a time, he noticed that a certain static appeared at approximately the same time every day. Then he made the key observation: the static was actually appearing four minutes earlier each day. This was the link between the static that he was observing and the rest of the universe. The stars and galaxies rise four minutes earlier every day, with respect to solar time. The four-minute difference arises from the fact that sidereal time and solar time differ, as described in Section 3.11. Jansky's static kept sidereal time, and thus was coming from outside our solar system! Jansky was actually receiving radiation from the center of our galaxy.

Figure 21–2 Karl Jansky lecturing about his discovery of a source of radio signals moving across the sky according to sidereal time.

Radio astronomy, which was to become so important, didn't grow right away. In fact it was an amateur astronomer, Grote Reber, who alone advanced radio astronomy in the late 1930's and early 1940's. Reber built a radio telescope in his back yard in Illinois. (Both Reber's and a reconstruction of Jansky's early radio telescopes are now enshrined at the National Radio Astronomy Observatory in Green Bank, West Virginia.) Professional astronomers were not quick to join in the activity at that time.

Radio astronomy received its impetus during World War II. Something was jamming the British radars, and that "something" turned out to be the sun! Of course, this discovery was top secret and remained so until the end of the war. But in the meantime a lot was learned about solar radio astronomy. Even more important, new receivers had been developed for wartime radar work and they could be adapted for use in radio astronomy. After the war, radio astronomy flourished.

The principles of radio astronomy are exactly the same as those of optical astronomy: radio waves and light waves are exactly alike – only the wavelengths differ. Both are simply forms of electromagnetic radiation. But different technologies are necessary to detect the signals: light waves can blacken silver grains on a photographic plate or cause a response in the human eye, while radio waves have such long wavelengths that they pass undetected right through cameras or humans. Radio waves do cause electrical changes in antennas, and these faint electrical signals can be detected

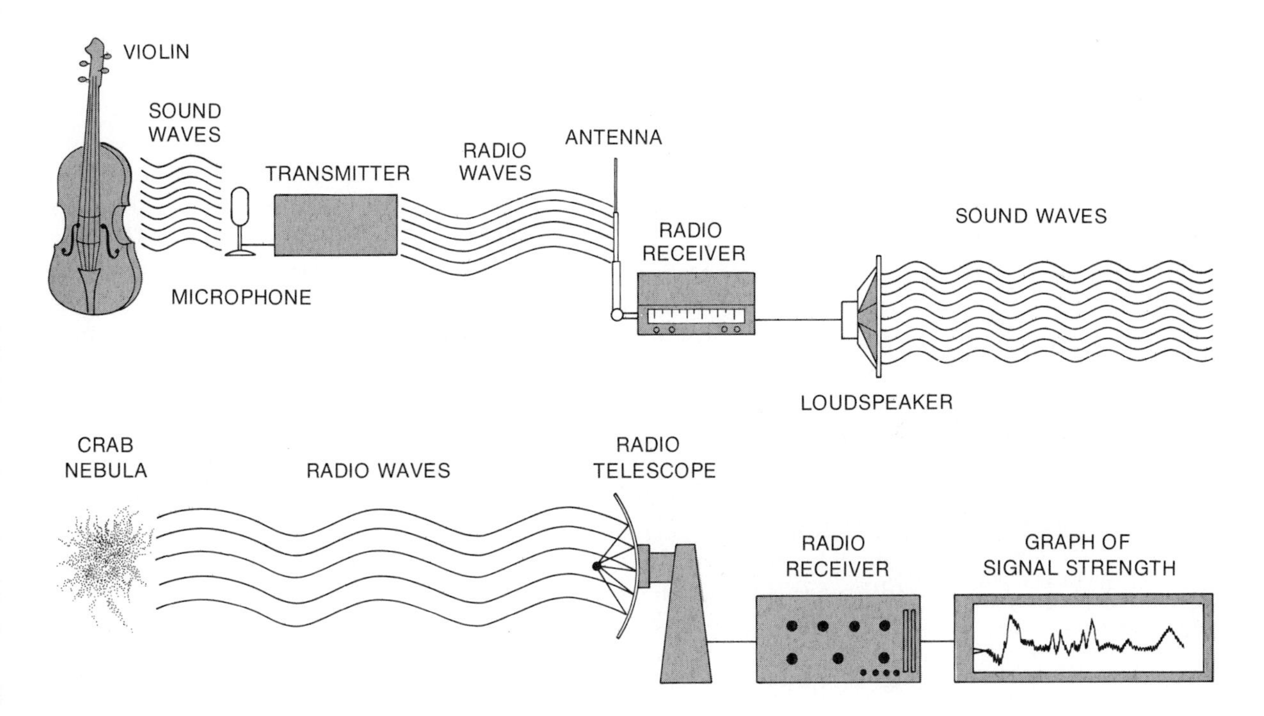

Figure 21–3 A radio telescope system is similar to a home hi-fi system in that radio waves are electronically amplified. We tend to use the output of a hi-fi, though, to drive a loudspeaker, while astronomers usually record the strength of the signal received by a radio telescope in other ways. In radio astronomy, a reflector is usually used to concentrate the weak radio signals and focus them onto the actual antenna that converts the radio waves into an electric signal.

with instruments that we call radio receivers. We ordinarily choose to have the radio signals of the Boston Symphony's concerts changed into sound waves by having the electrical signals put out by the receiver drive a loudspeaker, which converts them into sound waves. The basic operation of a radio astronomy receiver (Fig. 21–3) is not very different from that of your home radio; the parts just cost a lot more since they are made with more precision and are not mass produced. In order to record the intensity of the radio signal that causes an electrical signal to be generated, radio astronomers may simply graph the intensity of the electrical signal rather than use a loudspeaker to generate a sound signal from the radio waves.

Radio waves are "received" by antennas not unlike television aerials. If we want to collect and focus radio waves just as we collect and focus light waves, we must find a means to concentrate the radio waves at a point at which we can place an aerial. Lenses to focus radio waves are impractically heavy, so refracting radio telescopes are not used. However, radio waves will bounce off metal surfaces. Thus we can make reflecting radio telescopes that work on the same principle as the 5-meter optical telescope on Palomar Mountain. The mirror, called a "dish" in common radio astronomy parlance, is usually made of metal rather than the glass used for optical telescopes. The dish needn't look shiny to our eyes, as long as it looks shiny to incoming radio waves. To appear shiny, the surface of a radio telescope has to be smooth to within a small fraction of the wavelength of radiation, one-twentieth or less. (Optical telescopes need not have quite one-twentieth-wavelength accuracy since they are limited by "seeing" rather than intrinsic roughness.) Since radio waves can be, say, 10 cm or a meter long, a radio astronomy dish only has to be smooth to 1/2 cm or 5 cm, in those two cases. It is obviously much easier to make a big metal structure to such an accuracy than to an accuracy of a fraction of a micron (1 micron = 1 micrometer = 10^{-6} meter), and thus we can make huge radio telescopes that cover acres and acres (like the Arecibo telescope shown in

Figure 21–4 Even though the holes in this radio telescope may look large to the eye, they are much smaller than the wavelength of radio radiation being observed. Thus the telescope is smooth and shiny to the incoming radio radiation.

100 meters/10 centimeters per wave = 1000 wavelengths.

Fig. 19–6). In fact, a dish need not have a solid surface – an open mesh will do, as long as the size of the holes is substantially smaller than the wavelengths at which we are observing (Fig. 21–4). The radio waves then "see" the dish as solid.

We want dishes as big as possible for two reasons: first, a larger surface area means that the telescope will be that much more sensitive. The second reason is also simple but takes a few more sentences to explain. The basic point is that measurements the size of a telescope are most meaningful when they are in units of the wavelength of the radiation the telescope is being used to observe (Fig. 21–5). Thus an optical telescope one meter across used to observe optical light, which is about one two-millionth of a meter in wavelength, is two million wavelengths across. A radio telescope, even one 100 meters across (the size of the currently largest fully steerable dish, at the Max Planck Institute for Radio Astronomy in Bonn, Germany), is only 1000 wavelengths across when used to observe radio waves 10 centimeters in wavelength. The larger the telescope, measured with respect to the wavelength of radiation observed, the narrower is the width of the beam of radiation that the telescope receives. (The reflecting radio telescope dish receives radiation only from within a narrow cone. This cone is called the *beam;* see Fig. 21–6.) Radio astronomers normally work with broad beams and cannot resolve any spatial details smaller than the size of their beams. The 100-meter telescope at 10 cm has a beam that is four minutes of arc across; at longer wavelengths or with smaller telescopes one cannot even resolve an area the size of the moon. Obviously, this is looking at the sky with less than the optimum resolution, given that large optical telescopes ordinarily resolve 1 arc sec. However, under certain circumstances, one can now use the technique known as interferometry to

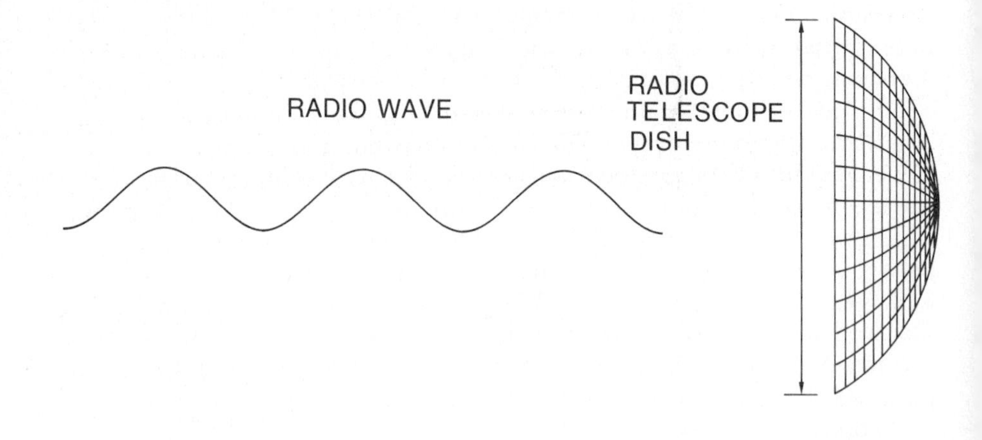

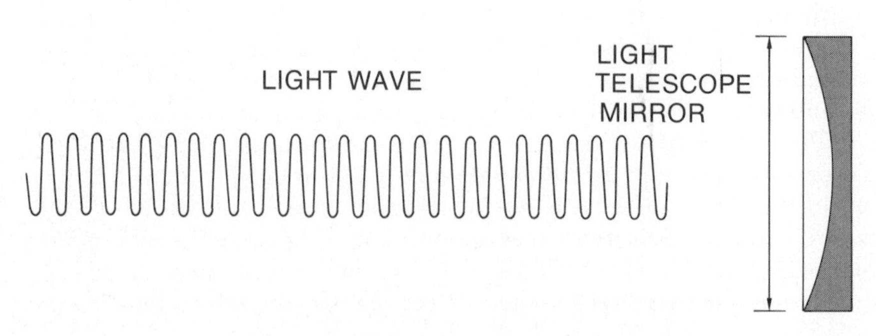

Figure 21–5 It is more meaningful to measure the diameter of telescope mirrors in terms of the wavelength of radiation that is being observed than it is to measure it in terms of units like centimeters that have no particularly relevant significance.

Color Plate 39: Saturn, on March 11, 1974, showing bands on its disk and Cassini's division in its rings. (Lunar and Planetary Laboratory, University of Arizona photo)

Color Plate 40: Comet West, a bright comet visible in 1976. The solar telescope of the Kitt Peak National Observatory and the lights of Tucson, Arizona are visible in the foreground. (Photo by Saul G. Levy)

Color Plate 41: The Horsehead Nebula, NGC 2024, in Orion, is an absorbing region superimposed on one emitting red radiation characteristic of hydrogen. (Hale Observatories photo with the 1.2-m Schmidt camera)

Color Plate 42: The Trifid Nebula, M20, in Sagittarius is glowing gas divided into three visible parts by absorbing lanes. The blue nebula at the top is unconnected to the Trifid. (Hale Observatories photo with the 5-m telescope)

Color Plate 43 (top): M16, the Eagle Nebula in the constellation Serpens. Hydrogen radiation makes it appear red. The bright stars at the upper right are hot and young, and are part of a galactic cluster. The small, dark regions may be protostars. (Hale Observatories photo with the 5-m telescope)

Color Plate 44 (bottom): M17, the Omega Nebula in Sagittarius. (Hale Observatories photo with the 1.2-m Schmidt camera) improve the resolution. This technique will be discussed later on, in Section 23.5b.

Resolution is discussed in Section 3.1, where it is stated that resolution is proportional to wavelength and inversely proportional to telescope diameter.

21.2 CONTINUUM RADIO ASTRONOMY

All radio astronomy research in the early days was of the continuum (Fig. 21–7). (In radio astronomy, as in optical astronomy, that means that we consider the average intensity of radiation at a given frequency without regard for variations in intensity over small frequency ranges, that is, we ignore any spectral lines.) When radio "continuum spectra" were measured, the continuum levels were measured at widely separated frequencies: for example, at 5, 10, 20, and 50 cm. These continuum values, and the spectra derived from them, provided information about the mechanism that causes the continuum emission.

Since radio waves cannot be focused into an image on a photographic plate, there is an important difference in the way radio and optical observations must be made. At a radio telescope, the observer simply measures the total amount of radiation that is coming from the direction in which the telescope is pointing. To make a radio map with a single dish, the pointing of the telescope must be physically changed; the telescope may be scanned from point to point across the image.

In one sense, an analogue in optical astronomy to continuum radio astronomy is not photography, but rather *photoelectric* recording of data. There too one can only measure the total intensity of light that is coming into the telescope from the direction in which the telescope is pointed; we do not get a "picture" without rastering. New methods in both optical and radio astronomy are finally allowing astronomers to make images without using photographic plates.

To take a radio continuum spectrum of an object, one must keep the telescope pointing at the object as it passes across the sky, and measure the intensity of radiation at different frequencies one after the other. To make these frequency changes one is sometimes required to physically climb up to the focal point of the telescope where the antenna (called the "feed") is located, and change the piece of receiving equipment that is installed there. In order to take a radio spectrum that covers a particularly large range in frequency, one may even have to use several different telescopes, each of which works efficiently at a particular part of the frequency range.

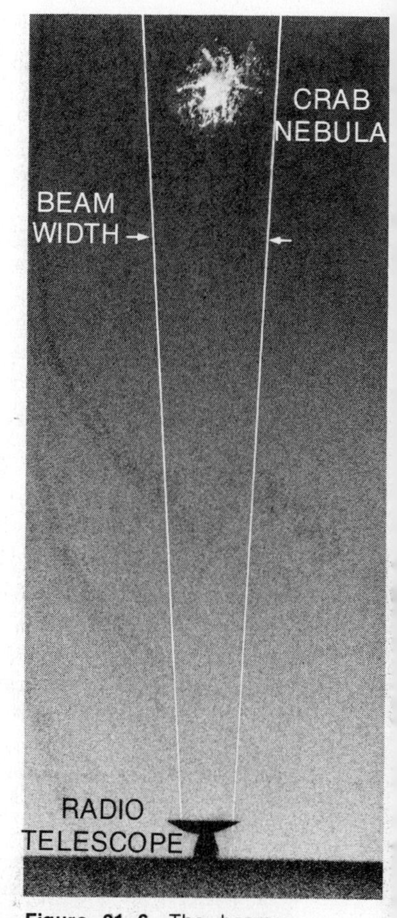

Figure 21–6 The beams of radio telescopes are ordinarily relatively large compared with radio sources. Thus single-dish radio telescopes cannot resolve—detect detail on—radio sources to the same extent that optical telescopes resolve optical objects.

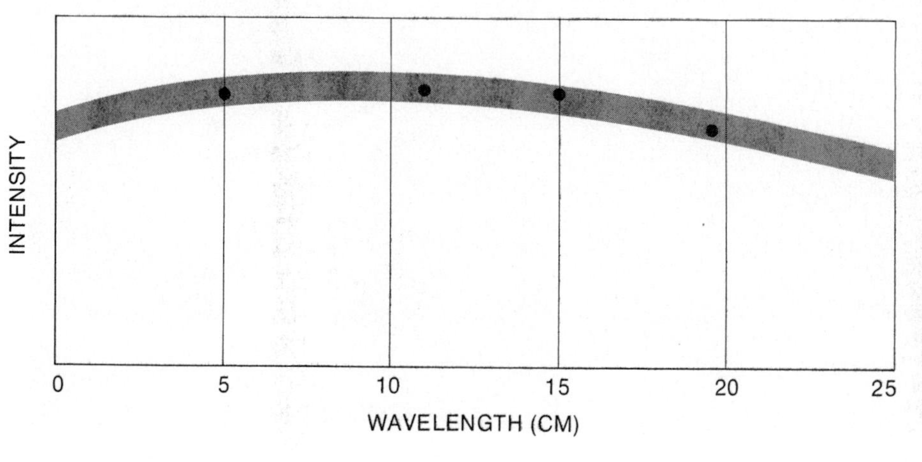

Figure 21–7 When we speak of a *radio spectrum*, we often mean the way the intensity of radiation changes over a wide range of wavelengths. In the graph shown, the intensity was measured only at four widely separated wavelengths, and the shaded curve was fit to the observed points.

As radio astronomical observations got under way, it was immediately apparent that the brightest objects in the radio sky, that is, the objects that give off the most intense radio waves, are not identical with the brightest objects in the optical sky. The radio objects were named with letters and with the names of the constellations in which they are located. Thus Taurus A is the brightest radio object in the constellation Taurus; we now know it to be the Crab Nebula. Sagittarius A is the center of our galaxy; Sagittarius B is another radio source nearby, whose emission is caused by clouds of gas near the galactic center.

Many of the radio objects have been discovered to lie outside our galaxy, and we shall discuss them in subsequent chapters. Quasars (Chapter 24), for example, are an important group of extragalactic objects that radiate strongly in the radio region of the spectrum.

Extragalactic means outside our galaxy.

Continuum radio radiation can be generated by any of several processes. One of the most important is *synchrotron emission*, the process that produces the radiation from Taurus A (Fig. 21–8). Taurus A is a supernova remnant, and lines of magnetic field extend throughout the visible Crab Nebula and beyond. Electrons, which are electrically charged, tend to spiral around magnetic lines of force. Electrons of high energy spiral very rapidly, at speeds close to the speed of light. We say that they move at "relativistic speeds," since the theory of relativity must be used for calculations when the electrons are going that fast. Under these conditions, the electrons radiate very efficiently. (This is the same process that generates the light seen in electron synchrotrons in laboratories on earth, hence the name synchrotron radiation.) The suggestion that the synchrotron mechanism causes radiation from various astronomical sources was first made by several Soviet theoreticians about 1950. Synchrotron radiation is highly polarized, and the discovery a few years later that the optical radiation from the Crab Nebula is highly polarized (Fig. 21–9) was an important confirmation of this suggestion. The radio radiation from the Crab and many other sources is also highly polarized.

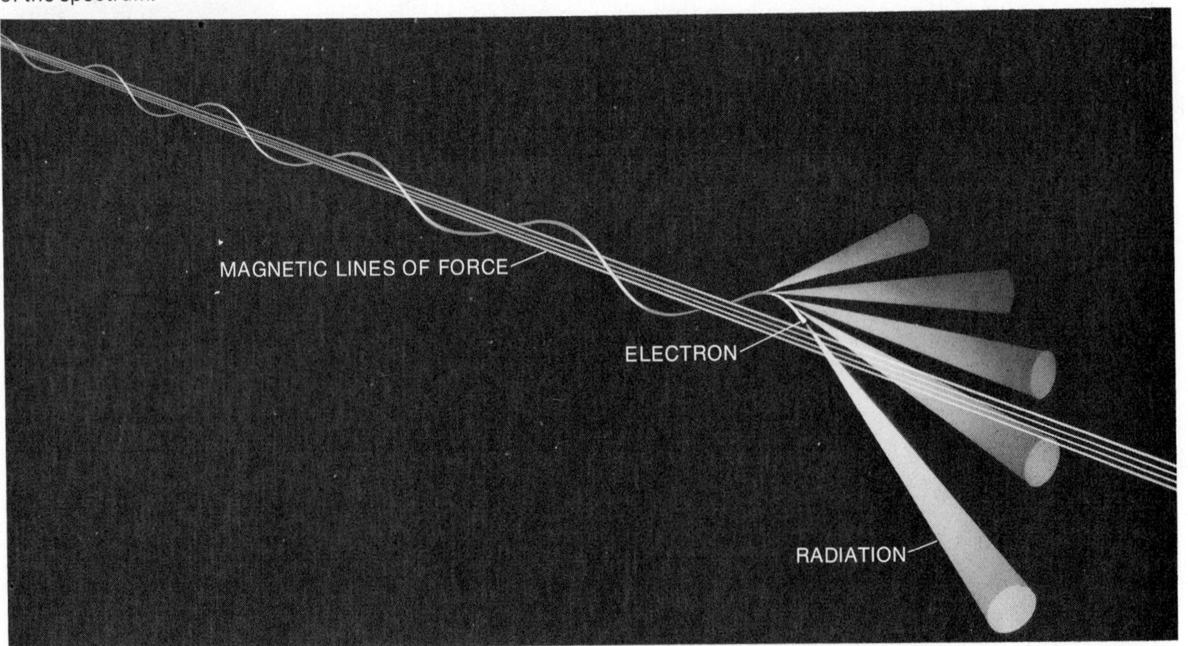

Figure 21–8 Electrons spiralling around magnetic lines of force at velocities near the speed of light (we say "at relativistic velocities") emit radiation in a narrow cone. This radiation, which is continuous and highly polarized, is called synchrotron radiation. Synchrotron radiation has been observed in both optical and radio regions of the spectrum.

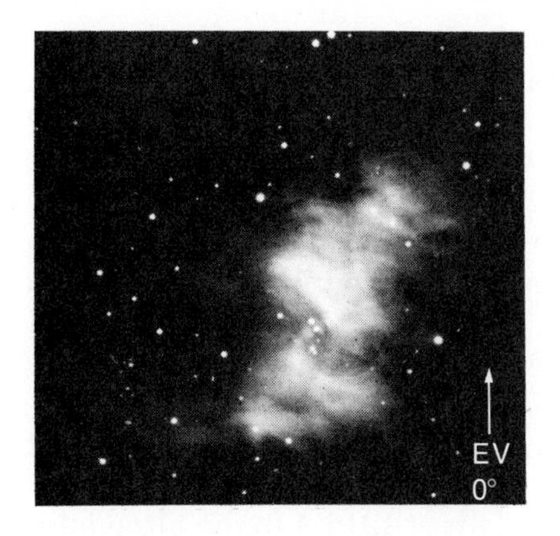

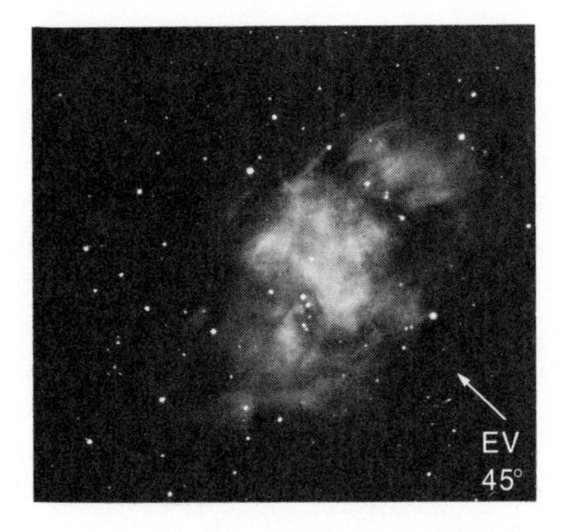

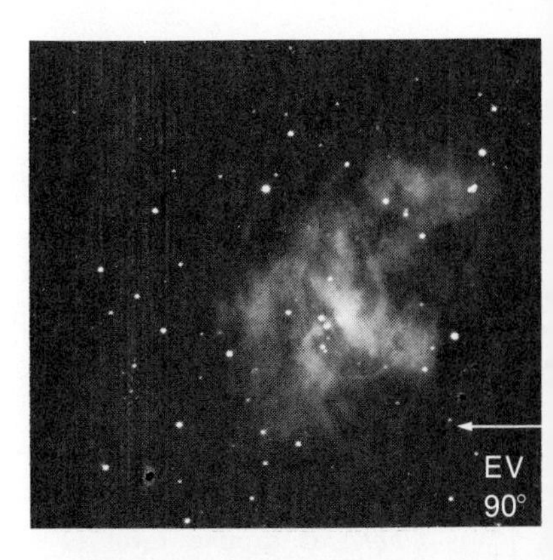

Figure 21–9 Photographs of the Crab Nebula that show visible light polarized at different angles are very different from each other. This shows that the light from the Crab is highly polarized, which implies that it is caused by the synchrotron mechanism. A non-polarized source would appear the same when viewed at any angle of polarization. EV is the "electric vector"; its direction is shown with an arrow.

The intensity of synchrotron radiation is related not to the temperature of the astronomical body that is emitting the radiation, but rather to the strength of its magnetic field and to the number and energy distribution of the electrons captured in that field. Since the temperature of the object cannot be derived from knowledge of the intensity of the radiation, we call this radiation *non-thermal* radiation. Synchrotron radiation is but one example of such non-thermal processes. The synchrotron process can work so efficiently that a relatively cool astronomical body can give off a tremendous amount of such radiation, at a given frequency, perhaps so much that it would have to be heated to a few million degrees before it would radiate as much *thermal* radiation at that frequency (by thermal radiation we mean radiation that is consistent with Planck's law).

21.3 THE RADIO SPECTRAL LINE FROM INTERSTELLAR HYDROGEN

In about 1950, though radio astronomers were very busy with continuum work, there was still a hope that a radio spectral line might be discovered. This discovery would be important for many reasons, but especially because it would allow Doppler shift measurements to be made.

What is a radio spectral line? Just as an optical spectral line corresponds to a wavelength (or frequency) in the optical spectrum that is more (for an emission line) or less (for an absorption line) intense than neighboring wavelengths or frequencies, a radio spectral line corresponds to a frequency (or wavelength) at which the radio noise is slightly more, or slightly less, intense. If a radio station broadcasted just a hum, it would appear as an emission line on our home radios; terrestrial transmissions are normally "modulated" (the voice or music is carried as a modulation on a steady "carrier" signal).

The most likely candidate for a radio spectral line that might be discovered was a line from the lowest energy levels of interstellar hydrogen atoms. This line was predicted to be at a frequency of 1420 MHz, equivalent to a wavelength of 21 cm. Since hydrogen is by far the most abundant element in the universe, it seems reasonable that it should produce a strong spectral line. Furthermore, since most of the interstellar hydrogen has not been heated by stars or by any other strong mechanism, it is most likely that this hydrogen is in its state of lowest possible energy.

Let us return to the formation of the hydrogen spectrum, which we discussed in Section 2.7. We have seen how the Balmer series of hydrogen is visible in the optical spectrum of the sun and other stars. This series of lines comes from transitions of electrons that cause the hydrogen atom to change between its second lowest principal energy state and other, higher states. Thus the transition from level 3 to level 2 is called Balmer alpha. Actually, since for many years the Balmer series was the only one that could be observed, it is usually called hydrogen alpha, or Hα (H alpha). The transition from level 4 to level 2 is Hβ (H beta), from level 5 to level 2 is Hγ (H gamma), and so on. (Astronomers all know the Greek alphabet or at least the first few letters.)

One speaks interchangeably of ground level *or* ground state.

The hydrogen lines that result from transitions to or from the lowest principal energy state (the *ground state*) to the higher states are called the Lyman series. Ly α (Lyman alpha), which falls at 1216 Å in the ultraviolet, does not come through the earth's atmosphere, so Ly α from celestial sources must be observed from satellites. In particular, the Copernicus satellite has made valuable studies of the Lyman series of hydrogen.

As we discussed, spectral lines come not from the energy levels themselves but from changes between one energy level and another. The difference in energy represented by the change is transferred into a bundle of electromagnetic energy called a *photon.* According to the quantum theory, this photon has a dual nature—it sometimes acts like a particle, but it also sometimes acts like a wave. Inasmuch as it acts as a wave, it can be thought of as having a wavelength. For most astronomical purposes it is sufficient to think of radiation as waves. But when we try to understand the pressure that can be exerted by radiation, which is important in forming comet tails, for example, then it is easier to think of photons acting as particles.

The energy that corresponds to a transition of an electron in an atom from one energy state to another, and thus the energy emitted or absorbed in the overall process of an atom changing from one energy state to another, appears as a photon of one and only one wavelength. The wavelength of this radiation is determined, as we have seen in Section 2.7, by the formula: E = constant ÷ wavelength. The constant is Planck's constant, h, times the speed of light, c.

Box 21.1 Conversion from Frequency to Wavelength

Although optical astronomers usually use wavelength units, radio astronomers usually use *frequency* units. If a wave travels at a constant speed—in this case at the speed of light—fewer peaks will cross a given point in a given time when the wavelength is longer. Thus the frequency at which peaks in the wave pass a given point is decreased. The converse is also true. The simple equation that links wavelength and frequency is wavelength × frequency = c. Wavelength is usually denoted with the Greek letter λ (lambda) and frequency is usually denoted with the Greek letter ν (nu), so the equation is written $\lambda\nu = c$ (Fig. 21–10). The energy E is simply expressed as hν, $E = h\nu$, where h is Planck's constant.

Students often find this dual terminology—sometimes in wavelengths and sometimes in frequencies—confusing. So do astronomers. After doing research in a given field, one actually thinks in the appropriate units; radio astronomers think in frequency units. So at

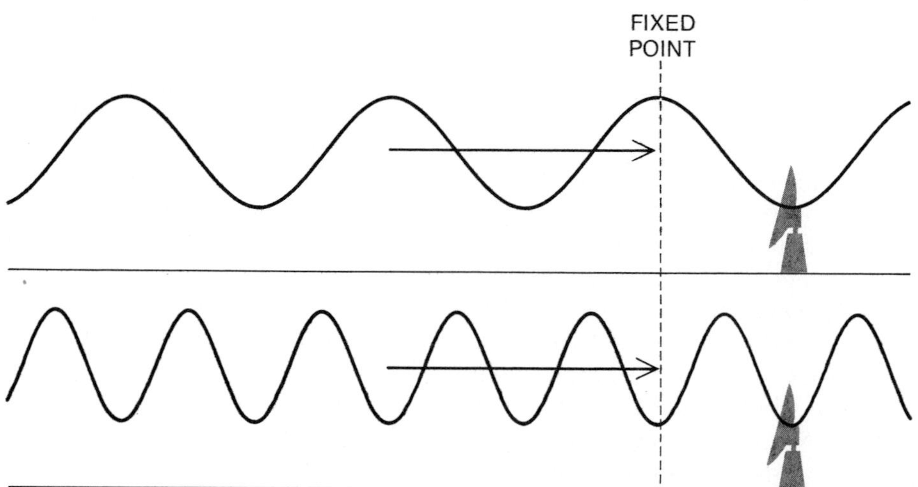

Figure 21–10 Since all electromagnetic radiation travels at the same speed (the speed of light) in a vacuum, fewer waves of longer wavelength *(top)* pass an observer in a given time interval than do waves of a shorter wavelength *(bottom)*. If the wavelength is half as long, twice as many waves pass, i.e., the frequency is twice as high. The wavelength times the frequency is constant, with the constant being the speed of light: $\lambda\nu = c$.

meetings at which both optical and radio astronomers are present, there is considerable translating of units back and forth.

The one basic number that must be remembered is the speed of light: 3×10^{10} cm/sec. From that, frequency-wavelength conversions can be derived easily. For example, if a radio wave has a wavelength of 1 cm, its frequency is 3×10^{10} cm/sec $\div$ 1 cm $= 3 \times 10^{10}$ hertz (hertz is the name for cycles per second, and is abbreviated Hz). Since M is the symbol for mega (10^6), and G is the symbol for giga (10^9), the frequency of this line would be 3×10^{10} Hz $=$ 30×10^9 Hz $=$ 30 GHz (see Appendix 2).

Let us now return to consideration of the hydrogen line at 1420 MHz $=$ 1420×10^6 Hz. To find its wavelength, we divide the frequency into the speed of light, 3×10^{10} cm/sec $\div$ 1.42×10^9 Hz, and get 21 cm.

This line at 21 cm comes not from a transition to the ground state from one of the higher states, not even from level 2 to level 1 as does Lyman alpha, but rather from a transition between the two sublevels into which the ground state of hydrogen is divided. Thus it is called a *hyperfine structure* line, since the line arises from transitions between two "hyperfine" structure levels in the lowest principal energy level of the hydrogen atom (Fig. 21–11).

An atom of hydrogen, you will recall, simply consists of an electron orbiting around a proton. (Actually, in quantum mechanics, there is an electron cloud with a certain set of probabilities that the electron is at given places at a given time, but for this astronomical purpose it is good enough to think in terms of the classical picture.) Both the electron and the proton have the property of spin; each one has angular momentum (Section 10.3) as if it were spinning on its axis. When we describe a spinning object, we describe the direction of the spin by what is called the *right-hand rule.* This rule can be explained as follows: if you wrap your right hand around the spinning object with your fingers pointing in the direction of the rotation, the direc-

Figure 21–11 21-cm radiation results from an energy difference between two sub-levels, called *hyperfine levels,* in the lowest principal energy state of hydrogen. The energy difference is much smaller than the energy difference that leads to Lyman α. So (because $E = h\nu = hc/\lambda$) the wavelength is much longer.

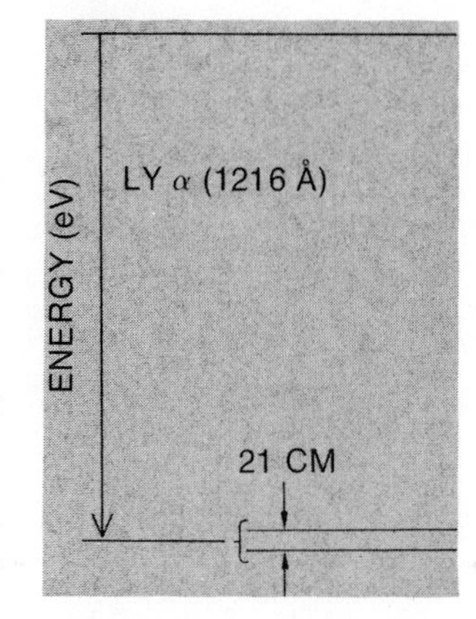

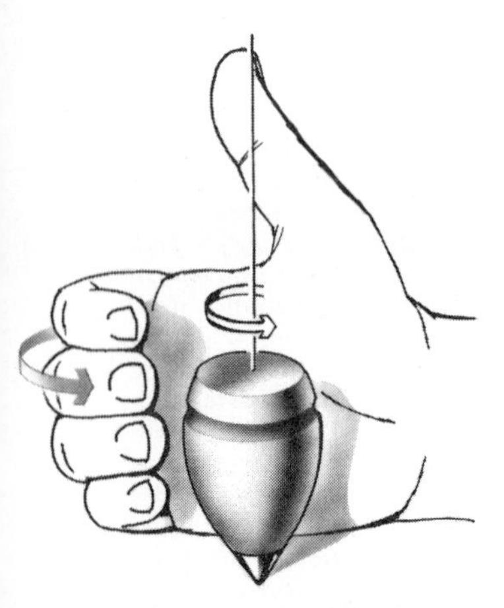

Figure 21–12 We characterize the *direction of spin,* according to a convention called the *right-hand rule,* as the direction in which the thumb is pointing when the fingers are wrapped around in the direction in which an object is spinning.

tion that your thumb, held out from your hand, is pointing is defined as the direction of spin (Fig. 21–12). Thus if you spin a top on a table so that a dot on its side passes across your field of view from left to right, if you wrap your right hand about it you will see that your thumb is pointing up. We say that the "spin of the top is in the upward direction."

With this terminology, the spin of the electron can be either in the same direction as the spin of the proton or in the opposite direction. Intermediate orientations are prohibited according to the rules of quantum mechanics. If the spins are in opposite directions, the energy state of the atom is very slightly lower than the energy state occurring if the spins are in the same direction. The two energy states are called the lower and upper hyperfine states. The energy difference between the two states is equal to $h\nu = h \times 1420 \times 10^6$ Hz, which is the same as saying that a transition from the upper to the lower state gives rise to a 1420 MHz photon.

If an atom is sitting alone in space in the upper of these two energy states, with its electron and proton spins aligned in the same direction, it has a certain small probability of having the spinning electron spontaneously flip over to the lower energy state and emit a photon. We thus call this a *spin-flip* transition (Fig. 21–13). The photon corresponds to radiation at a wavelength of 21 cm (Fig. 21–14). If we were to watch any particular group of hydrogen atoms, we would find that it would take 11 million years before half of the electrons had undergone spin-flips; we say that the *half-life* is 11 million years for this transition. But even though this *transition probability* is so very low, there are so many hydrogen atoms in space that enough 21-cm radiation is given off to be detected.

We have described how an *emission line* can arise at 21 cm. But what happens when continuous radiation passes through neutral hydrogen gas? In this case, some of the electrons in atoms in the lower state will absorb a 21-cm photon and flip over, putting the atom into the higher state. Then the radiation that emerges from the gas will have a deficiency of such photons and will show the 21-cm line in *absorption* (Fig. 21–13).

In 1944, H. C. van de Hulst, a Dutch astronomer, predicted that the 21-cm emission would be strong enough to be observable as soon as the proper equipment was developed. It took seven more years for the instrumental capability to be built up. In 1951, scientists both in Holland at Leiden and in the United States at Harvard were building equipment to detect the 21-cm radiation. An unfortunate fire set back the Dutch effort many months. In the meantime, the work at Harvard went on. Electronic equipment was built in the physics lab to observe in the direction of the galactic center through a small antenna stuck out a window (into which passing undergraduates occasionally lobbed snowballs). Finally, the Harvard team, consisting of a graduate student named Harold Ewen and his advisor, Edward M. Purcell, went "on the air" and succeeded in observing the 21-cm line in emission (Fig. 21–15). Soon the Dutch group and then a group in Australia confirmed the detection of the 21-cm line. Spectral-line radio astronomy had been born.

21-cm hydrogen radiation has proven to be a very important tool for studying our galaxy because it passes unimpeded through the dust that prevents optical observations very far into the plane of the galaxy. Using 21-cm observations, astronomers can study the distribution of gas in the spiral arms. We can detect this radiation from gas located anywhere in our galaxy, even on the far side, whereas light waves penetrate the dust clouds in the galactic plane only about 10 per cent of the way to the galactic center.

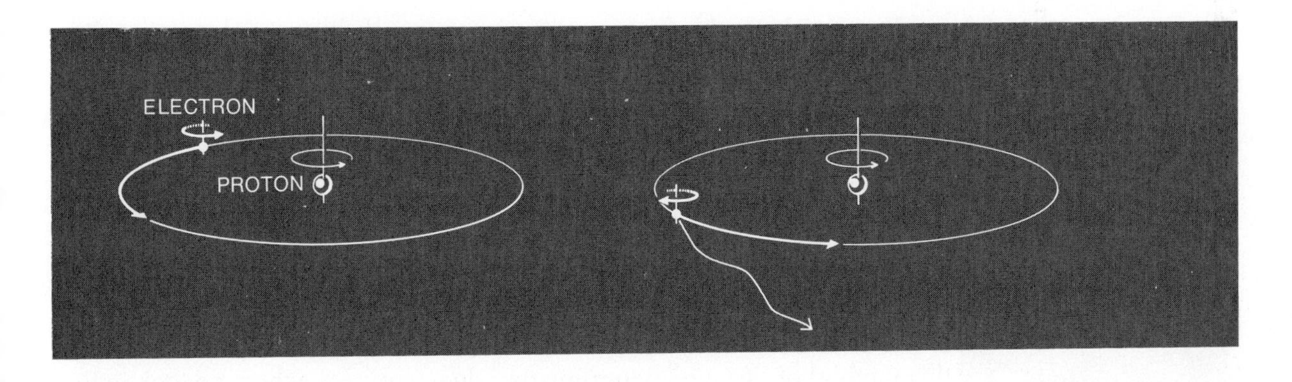

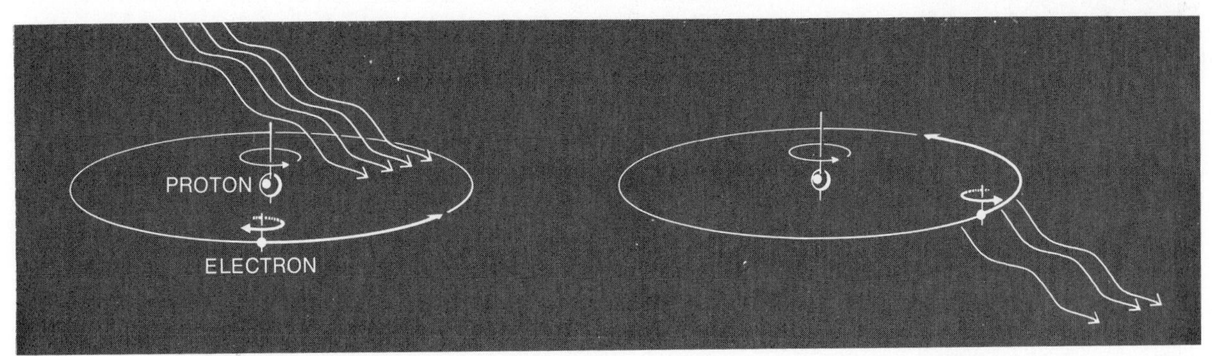

Figure 21–13 When the electron in a hydrogen atom flips over so that it is spinning in the opposite direction from the spin of the proton (*top*), an emission line at a wavelength of 21 cm results. When an electron takes energy from a passing beam of radiation to flip from spinning in the opposite direction from the proton to spinning in the same direction (*bottom*), then a 21-cm line in absorption results.

But here again we come to the question that bedevils much of astronomy: how do we measure the distances? Given that we detect the 21-cm radiation from a gas cloud (since it is neutral hydrogen, it is technically called an H I region), how do we know how far away the cloud is from us?

The answer can be found by using a model of rotation for the galaxy. As we have already learned, the outer regions of galaxies rotate differentially, that is, the gas nearer the center rotates faster than the gas farther away from the center. It is as though the gas were revolving around a central gravitational mass in Keplerian orbits, with the period increasing as you go farther from the center. Note that because we can consider ourselves either as being in the galaxy and revolving around its center or as part of the galaxy and rotating around its center, there is a confusion between the words "rotation" and "revolution" here that cannot be avoided.

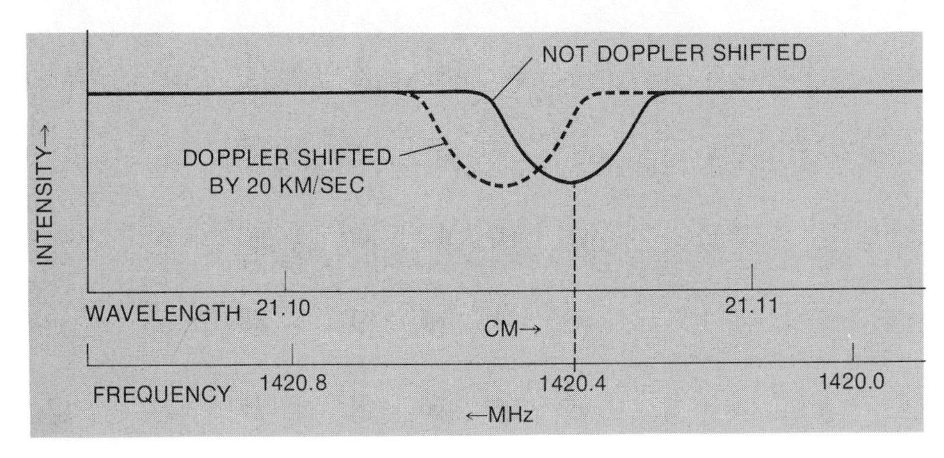

Figure 21–14 The position of the hydrogen line, shown here in absorption, can be given in either frequency or wavelength units. These numbers are commonly rounded to 21 cm or 1420 MHz in conversation. Any Doppler shift, of course, shifts the absorption in wavelength and frequency.

Figure 21–15 Harold Ewen with the horn radio telescope he and Edward Purcell used to discover the 21-cm line.

Figure 21–16 shows a simplified version of differential rotation. Because of the differential rotation, the distance between us and point A is decreasing. Therefore, from our vantage point at the sun, point A has a net velocity toward us. Thus its 21-cm line is Doppler shifted toward shorter wavelengths. If we were talking about light this shift would be in the blue direction; even though we are discussing radio waves we say "blueshifted" anyway. If we look from our vantage point at gas cloud C, we see a redshifted 21-cm line, because its higher speed of rotation is carrying C away from us. But if we look straight toward the center, clouds B_1 and B_2 are both passing across our line of sight in a path parallel to that of our own orbit. They have no net velocity toward or away from us. Thus this method of distance determination does not work when we look in the direction of the center, nor indeed in the direction of the anti-center.

Once we measure the redshift, we can deduce the net velocity. In the inner part of the galaxy, we can deduce the law of differential rotation with distance from the center since the highest velocity cloud along each line of sight must be the cloud along the line of sight that is closest to the center. We can get the distance of that cloud from the center by simple geometry. Once we work out the law of differential rotation, then we can tell how far away each cloud is from the center of the galaxy by figuring out where along our line of sight in a given direction the cloud would have the proper velocity to match the observations. Clouds farther from the galactic center take longer to revolve than do clouds closer to the center, in a manner similar to Kepler's third law (Fig. 21–17).* By observing in dif-

*We cannot use Kepler's third law in as simple a form as it is used for our solar system, because in the case of the solar system we have orbiting planets of masses that are negligible with respect to a central mass, the sun. In the case of our galaxy, the mass inside the orbit of an outer cloud is greater than the mass inside the orbit of an inner cloud. (The mass internal to an orbit acts as though it were concentrated at a point in the center. The mass external to an orbit has no effect, since the gravitational pull in one direction has exactly balanced out the gravitational pull in the opposite direction. Also, these considerations are exactly true only for spherical distributions of mass, which is not the case for our galaxy.)

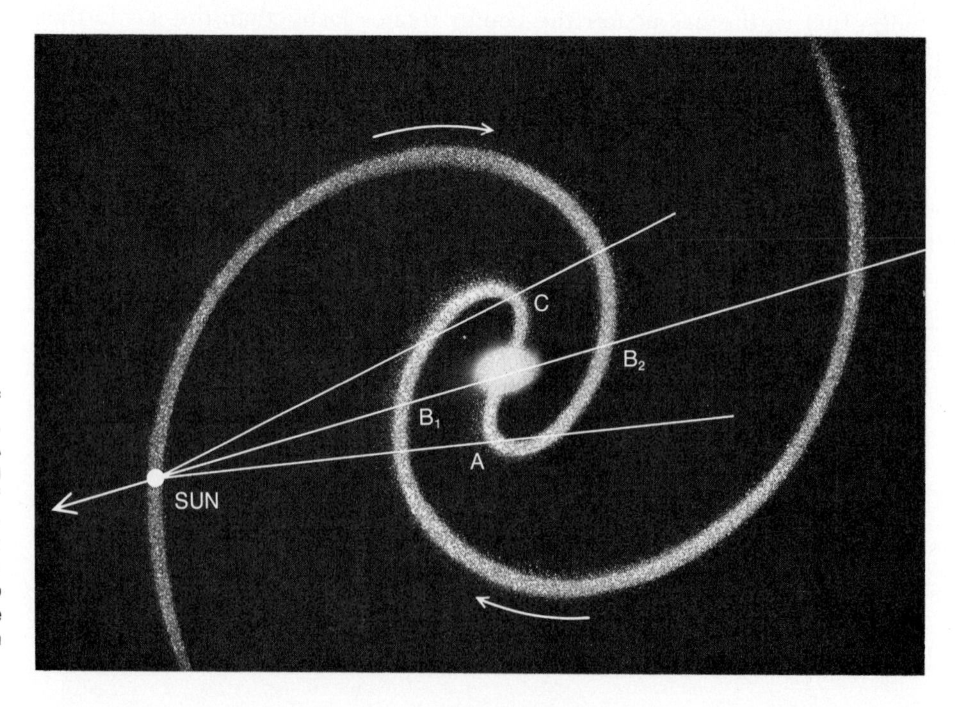

Figure 21–16 Because of the differential rotation, the cloud of gas at point A appears to be approaching the sun, and the cloud of gas at point C appears to be receding. Objects at points B_1 or B_2 have no net velocity with respect to the sun, and therefore show no Doppler shifts in their spectra.

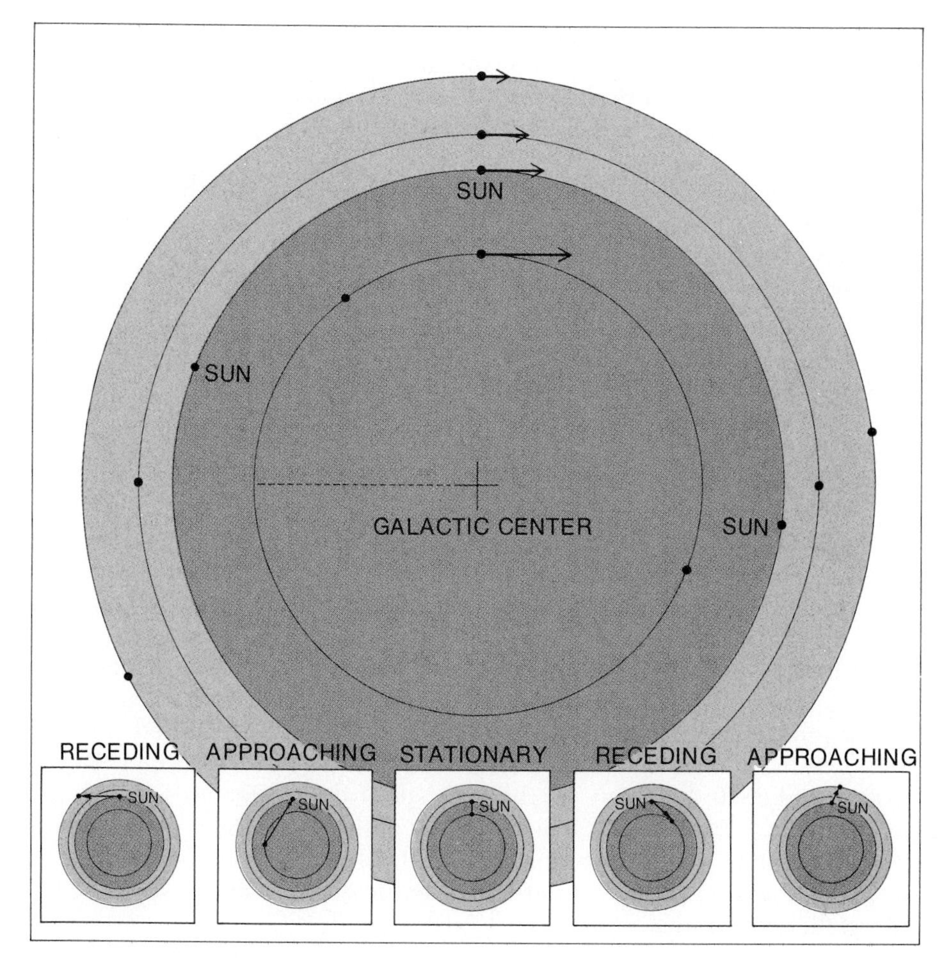

Figure 21–17 Because of differential rotation in our galaxy, stars or gas clouds that were lined up at one time (upper dots on top part of diagram) are spread out by the time they have gone about 1/4 of the way *(right)* or about 3/4 of the way *(left)* around the galaxy. The inserts show the net velocity that stars or gas clouds at different distances from the galactic center would have with respect to the sun. Objects interior to the sun's orbit around the galactic center (darker shaded area) are orbiting faster than does the sun, and objects farther out than the sun's orbit (lighter shaded area) are orbiting more slowly than does the sun. Thus measurements of the Doppler shift of 21-cm radiation can be used to construct a map of the hydrogen clouds, which in turn map out spiral structure.

ferent directions, we can build up a picture of the spiral arms. (The stars, which cannot be observed at the radio wavelength of 21 cm, also rotate differentially, as was mentioned in Section 20.5b.)

Unfortunately, gas clouds have not only a velocity of revolution around the center of the galaxy but also random velocities to and fro (defined in Section 2.16 as peculiar velocities). When we look at the center of the galaxy, we see that there are some clouds of gas moving outward with an average velocity of 50 km/sec. These motions, both systematic and random, place a fundamental uncertainty on the conclusions from this method. But this is the best that we can do.

Note that each line of sight in which we look crosses a circle of gas at a given distance from the center at two points. Thus an ambiguity arises, because a cloud at either of these two points would have the same Doppler velocity with respect to the sun. We have to resort to methods other than 21-cm studies to resolve this ambiguity. We know, for example, the average size of H I regions, and so can often tell whether the emitting cloud is at the farther or nearer point by its angular size in the sky. Alternatively, we can sometimes tell by noting whether the 21-cm radiation seems relatively weak or strong.

Of course, the whole sky cannot be seen from either the northern hemisphere or the southern hemisphere of the earth. Therefore maps must be made from observatories in both hemispheres and then correlated. The major workers in this field were originally in Holland and in Australia.

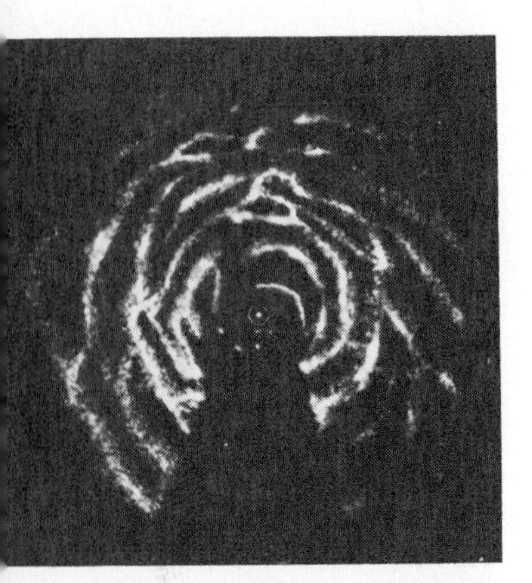

A

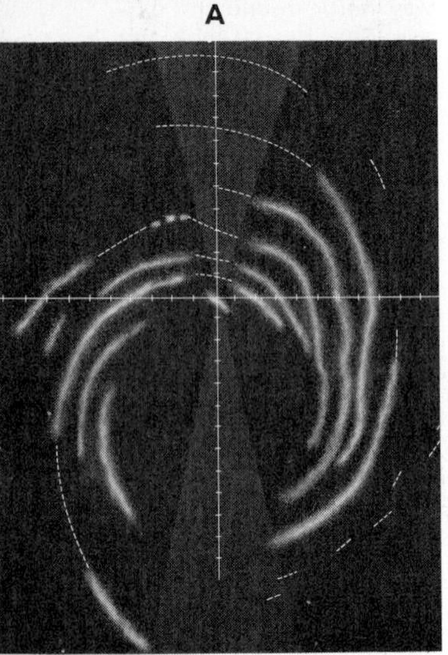

B

Figure 21–18 Two artists' impressions of the structure of our galaxy based on 21-cm data. *A* was compiled almost 20 years ago by Gart Westerhout of the University of Maryland, and *B* was recently compiled by Gerrit Verschuur of the University of Colorado. Differences between the two maps are probably not real and give an idea of the accuracy of the method. Because hydrogen clouds located in the directions either toward or away from the galactic center have no radial velocity with respect to us, we cannot find their distances.

There are considerable regions of overlap that are visible from both hemispheres. Unfortunately, the maps that were made do not correspond completely in the overlap regions, and this raises questions in many people's minds about the validity of the results. This is a topic of current research.

Also, the 21-cm maps show many narrow arms (Fig. 21–18) but no clear pattern of a few broad spiral arms like those we see in other galaxies. Is our galaxy really a spiral at all?

Even with all these questions and uncertainties waiting to be resolved, it is clear that the 21-cm radiation is at the base of our mapping efforts for our own galaxy, and has allowed us to make major advances in our understanding of the Milky Way Galaxy.

21.4 INTERSTELLAR OPTICAL MOLECULAR LINES

Over the last few decades, several optical spectral lines were discovered that originated from interstellar space rather than from the stars themselves. These lines can be observed through ordinary optical telescopes, and until 1963, they were the only interstellar lines known besides the line at 21 cm (Fig. 21–19). They come from atoms, molecules or radicals (other multi-atomic units) in space, including CN, CH, and CH^+. Besides these lines, there are an additional 3 dozen unidentified very diffuse (i.e., fuzzy) features in the optical spectrum, all of which lie between 4400 Å and 6800 Å. It may be that this set of diffuse features, which has been primarily studied by George Herbig of the Lick Observatory, results from absorption by interstellar dust particles. These particles may be only 300 Å across, about 1/10th the size of the dust particles that cause the interstellar reddening.

21.5 MOLECULAR HYDROGEN

While individual atoms of hydrogen, whether neutral or ionized, have been extensively observed in the interstellar gas, hydrogen molecules (H_2) have been observed only recently. Even though astronomers have long thought that molecular hydrogen could be a major constituent of the interstellar medium, it was simply not possible to observe hydrogen in molecular form. This is because at the low temperature of interstellar space, only the lowest energy levels of molecular hydrogen are excited, and no lines linking these levels fall in the optical or radio regions of the spectrum. We had to wait to observe from space in order to observe lines from H_2 because these lines occur in the far ultraviolet. Our atmosphere prevents these lines from reaching us on earth.

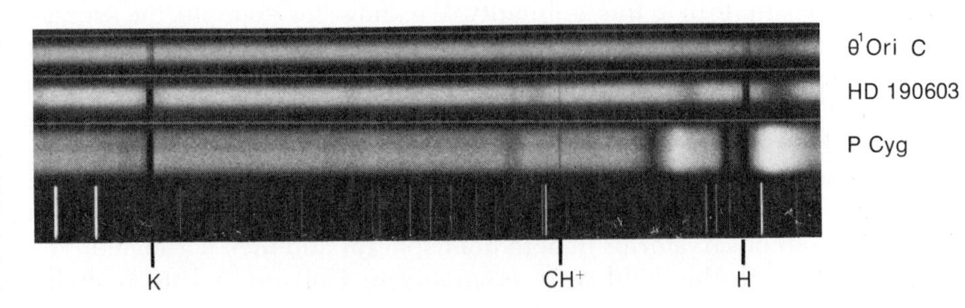

Figure 21–19 Absorption lines in the visible part of the spectrum caused by ionized calcium (H and K lines) and the molecular ion CH^+ in clouds of interstellar gas lying in the direction of the 3 stars listed.

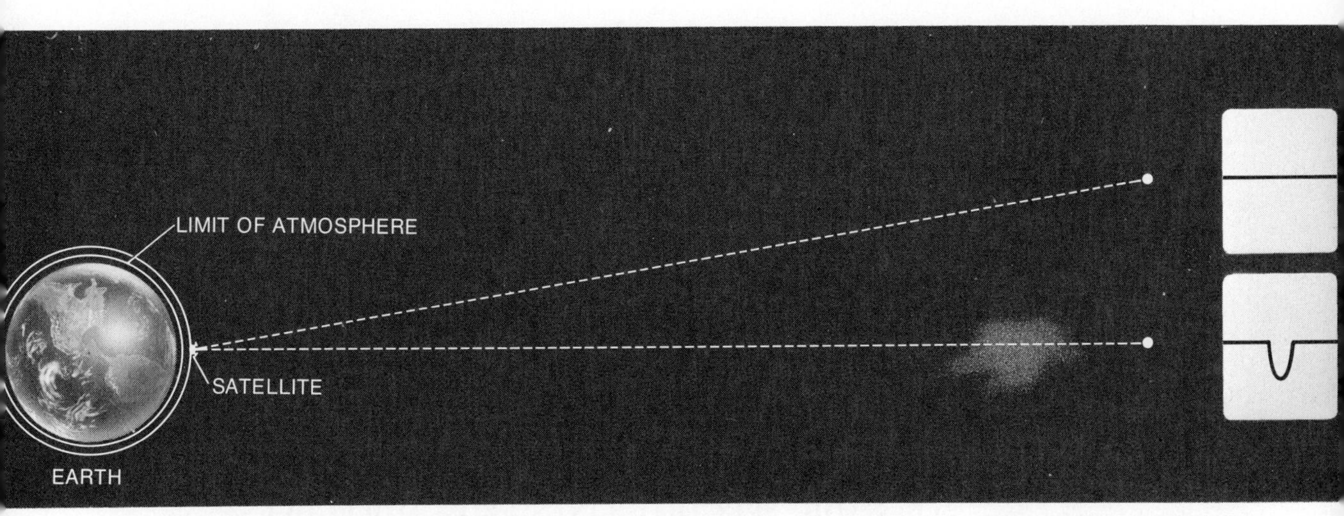

Figure 21–20 From above the atmosphere of the earth, the Copernicus satellite looks toward a star that provides a more-or-less continuous spectrum. (A rapidly-rotating B star is usually chosen, because B stars have few lines and those lines are washed out by Doppler shifts resulting from the velocity of rotation.) When a cloud of gas is in the path, then absorption or emission from molecules or atoms in the cloud results.

In 1970, George Carruthers of the Naval Research Laboratory in Washington succeeded in using a rocket to observe interstellar ultraviolet absorption lines from H_2. However, the rocket flight lasted only a few minutes, so only one star could be studied.

In 1972, a 90-cm (36-inch) telescope was carried into orbit aboard NASA's third Orbiting Astronomical Observatory, named "Copernicus" in honor of that astronomer's 500th birthday (which occurred a year later). The telescope, operated by a group of scientists from the Princeton University Observatory, was largely devoted to observing interstellar material. The observers could point the telescope at a star (directing it by remote control from the ground, of course) and look for absorption lines caused by gas in interstellar space as the light from the star passed through the gas en route to us. Since it is easiest to pick up interstellar absorption lines if the star itself has no lines of its own, the scientists observed in the direction of B stars, which have few lines and are very bright (Fig. 21–20). (O stars would do as well, but there are many fewer of them.) The scientists preferred B stars that were rapidly rotating, so that even the few lines they do have would be smeared out by the Doppler effect. Further, because of their brightness, B stars are visible at greater distances than are stars of most spectral types.

Since the hydrogen molecule is easily torn apart by ultraviolet radiation, the observers did not find much molecular hydrogen in most directions. But whenever they looked in the directions of highly reddened stars, they found a very high fraction of hydrogen in molecular form: more than 50 per cent. Presumably, in these regions the dust shields the hydrogen from being torn apart by ultraviolet radiation. There are also theoretical grounds for believing that the molecular hydrogen is formed on dust grains (other molecules are probably formed on dust grains too, though this is less certain), so it seems reasonable that the high fraction of H_2 is found in the regions where there are more dust grains.

The Copernicus satellite also made observations of deuterium, a rare isotope of hydrogen (described in Section 6.3). In the same shielded areas in which they observed molecular hydrogen, the astronomers also found a small fraction (10^{-6}) of the H_2 in the HD form, that is, a molecule consisting of one atom of the ordinary isotope of hydrogen tied to one atom of deuterium (${}_1H^2$).

As we shall see later on, the amount of deuterium in interstellar space gives us information about the origin of the universe, so the observation of HD caused some excitement in the astronomical community. But the results proved difficult to interpret, because D and H can combine at a different rate from the rate at which H combines with other H's. Thus the amount of D relative to the amount of H in the HD form could be different from the total amount of D relative to the total amount of H. Thus the relative abundances of HD and H_2 do not necessarily indicate the overall ratio of D to H.

Later on, however, the Copernicus satellite was able to detect deuterium in the form of individual atoms by studying the Lyman lines from hydrogen and deuterium, which also lie in the ultraviolet. We will analyze the importance of these observations in the chapter on cosmology, in Section 25.7.

21.6 RADIO SPECTRAL LINES FROM MOLECULES

Even though optical studies of the interstellar medium have been going on for decades, recently much more information has come from ultraviolet observations made from space and radio studies made from the earth. For several years after the 1951 discovery of 21-cm radiation, spectral-line radio astronomy continued with just the one spectral line. Astronomers tried to find others. One prime candidate was OH, the hydroxyl radical, which should be relatively abundant, as molecules go, because it is a combination of the most abundant element, hydrogen, with one of the most abundant of the remaining elements, oxygen. OH has four lines close together at about 18 cm in wavelength, and the relative intensities expected for the four lines have been calculated.

It wasn't until 1963 that other radio spectral lines were discovered, and four new lines were indeed found at 18 cm. But the intensity ratios were all wrong to be OH, according to the predicted values, and for a time we spoke of the "mysterium" lines. The idea of discovering a new element was not unprecedented—after all, unknown lines at the solar eclipse in 1868 were assigned to an unknown element, "helium," because they occurred only (as far as was known at that time) on the sun. Of course, the periodic table of the elements has been filled in during the last 100 years, so a new element wasn't really expected. Mysterium turned out to be OH after all, but with strange processes affecting the excitation of the energy levels of OH and thus amplifying certain of the lines at the expense of others, which are weakened. This in itself is very interesting. The process is that of masering, which will be described below. *Masers* (maser is an acronym for microwave amplification by stimulated emission of radiation) and lasers, their analogue using light instead of radio waves, are of great practical use on the earth. For example, masers are used as sensitive amplifiers. Masers were "invented" on earth not long before they were found in space. If astronomical research had been a little more advanced, a multi-billion dollar industry that now exists, and which may provide the key to solving our energy problems through laser-induced fusion, might have been set up earlier. This is another lesson on the value of basic research.

For maser action to take place, the numbers of molecules or atoms in the various possible energy states must be different from the numbers that would normally be there. By "normally" we mean that under most conditions we can readily calculate the numbers of molecules or atoms in the various possible energy states according to standard methods that depend mainly on the temperature of the gas. Using these standard methods, we can predict the average number in each level; we call this number "the *population* of each level."

In particular, for maser action to take place in a molecule, we must build up a concentration of molecules at a given energy level higher than the ground (lowest) level. The population of this "excited" level can be millions of times greater than the population would be under ordinary circumstances. Under certain conditions, the excited molecules can be triggered to all jump down together to a lower energy state, generating an intense emission line.

To build up this concentration, there must be some method of *pumping* enough energy into the system to excite the molecules to the given excited level, and there are various mechanisms that can come into play in particular situations for particular molecules. For example, in some cases, ultraviolet radiation from a nearby star can pump molecules from the ground state to a given excited state, and then they can jump down together to another excited state through maser action. If infrared instead of ultraviolet radiation does the pumping, we say that we have an "infrared pump" to excite the maser. In any case, by allowing for pumping to specific levels we can see why the ratios of intensities of the lines can be different from the intensities that would be expected in the absence of pumping. We may have to consider different kinds of pumping mechanisms for OH in different regions.

So the four "mysterium" lines turned out to be OH undergoing maser action. OH lines have now been detected not only in absorption against background emission, but also in emission. They are quite widespread and are used to tell us the physical conditions in the clouds from which the OH radiation comes.

The abundance of OH can be calculated from the measurements of OH radiation. It was very much less than that of isolated hydrogen or oxygen atoms, which seems reasonable since in the virtual vacuum of interstellar space there is little chance for oxygen and hydrogen atoms to interact. Still, radio astronomers were disappointed with the low OH densities: only one OH molecule for every billion H atoms. It seemed quite unlikely that the quantities of any molecules composed of three or more atoms would be great enough to be detected. The chance of three atoms getting together in the same place, it seemed, would be very small.

In 1968, however, Charles Townes, Al Cheung, and David Rank used Berkeley's Hat Creek Observatory to observe at the radio frequencies in the centimeter range that were predicted to be the frequencies of water (H_2O) and ammonia (NH_3). The lines of these molecules proved surprisingly strong, and were easily detected. In fact, we had had the capability for some years to make equipment sensitive enough to detect them, but nobody had tried to make the observations.

At about this time, a group of radio astronomers used the 43-meter telescope (Fig. 21–21) at the National Radio Astronomy Observatory in Green Bank, West Virginia, to search for interstellar formaldehyde (H_2CO) at a wavelength of 6 cm. They succeeded in detecting the formaldehyde molecule, the first molecule that contained two "heavy" atoms, that is, two atoms that were of an element other than hydrogen.

By this time it was apparent that the earlier notion that it would be difficult to form molecules in space was wrong. There has been much re-

A

Figure 21–21 *(A)* The 43-m (140-ft) telescope at Green Bank.

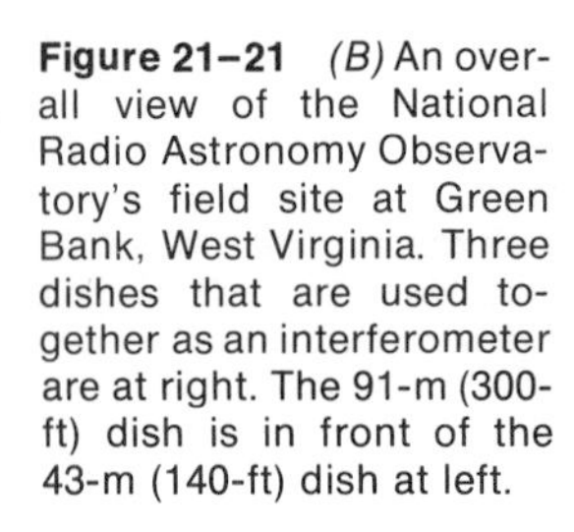

B

Figure 21–21 *(B)* An overall view of the National Radio Astronomy Observatory's field site at Green Bank, West Virginia. Three dishes that are used together as an interferometer are at right. The 91-m (300-ft) dish is in front of the 43-m (140-ft) dish at left.

search on this topic, but the mechanism by which molecules are formed has not yet been satisfactorily determined. For some molecules, including molecular hydrogen, it seems that the presence of dust grains is necessary. In this scenario, one atom would hit a dust grain and stick to it (Fig. 21–22). It may be thousands of years before a second atom hits the same dust grain, and even longer before still more atoms hit. But these atoms may stick to the dust grain rather than bouncing off, which gives them time to join together. Complex reactions may take place on the surface of the dust grain. Then, somehow, the molecule must get off the dust grain back into space as a gas. Perhaps incident ultraviolet radiation or the energy released in the formation of the molecule allows the molecule to escape from the grain surface.

It is not yet even definitively determined that the molecules in fact form on dust grains; there is a strong body of opinion that holds that the molecules are somehow formed in the interstellar gas. Recent work on some of the simpler molecules indicates that some can be formed on the grain while others cannot, and similarly, some molecules can be formed in the gas while others cannot. None of the current theories succeeds in satisfactorily explaining many of the measured abundances of the various molecules that we have observed in space. It is likely that the final result will be a mixture of processes.

After the discoveries of water, ammonia, and formaldehyde in interstellar space, further discoveries came one after another. Many different radio astronomers looked up the wavelengths of likely spectral lines from

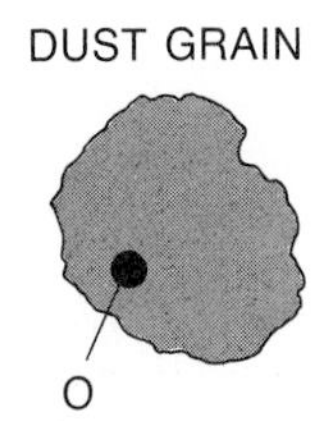

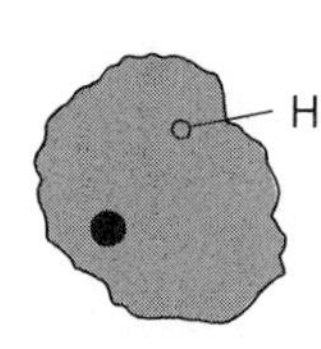

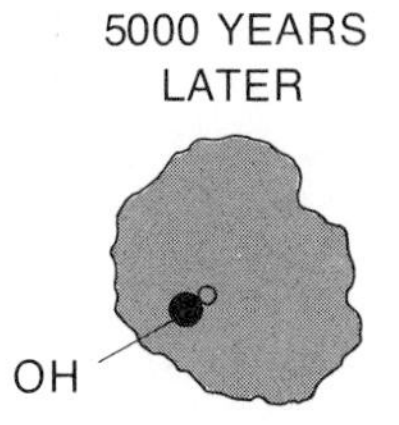

Figure 21–22 One major model of the formation of interstellar molecules invokes the presence of a dust grain at intermediate stages. The mechanism by which the molecule is freed from the dust grain is not known.

molecules containing abundant elements and were able to observe the radiation. Sometimes the lines were in absorption and sometimes in emission. The list of molecules discovered gradually expanded from three-atom molecules like ammonia and water, and four-atom molecules like formaldehyde, to even more complex molecules, like formic acid (HCOOH) with five atoms, methyl cyanide (CH_3CN) with six, and cyanodiacetylene (HC_5N) with seven. The discovery of methyl alcohol, CH_3OH, in 1970, was greeted jocularly, and cases of liquor were bet on the discovery of ethyl alcohol, CH_3CH_2OH, the drinking kind. In 1974, the ethyl alcohol molecule was discovered too. (There are 10^{28} fifths, calculated at 200 proof, in the molecular cloud in which it was observed.)

At first most discoveries were made in the centimeter range of the spectrum, and attempts to work at the longer wavelengths closer to a meter proved fruitless. Then the technology at the shorter wavelengths—from 2 to 10 millimeters—improved, largely because of work done in developing sensitive instrumentation at the Bell Telephone Labs, which was also where Jansky worked.

Recently, new devices that make infrared spectra have been used to discover molecules in solar-system size shells around sources that radiate continuous spectra in the infrared. Acetylene, C_2H_2, was discovered in this way.

Many more molecules were found in these short *millimeter radio wavelengths.* With the new technology, the leadership in discovering molecules passed from the 43-meter (140-foot) radio telescope of NRAO at Green Bank to an 11-meter (36-foot) radio telescope (Fig. 21–23) that NRAO operates at Kitt Peak in Arizona. The 11-meter telescope is able to observe

Figure 21–23 The 11-m (36-ft) telescope of the National Radio Astronomy Observatory, seen here through a fish-eye lens, is on Kitt Peak in Arizona (see Fig. 3–19). It is used for observations at millimeter wavelengths, and has been used to discover many interstellar molecules.

TABLE 21–1 INTERSTELLAR MOLECULES

Name of Molecule	Chemical Symbol	Year of Discovery	Part of Spectrum	First Wavelength Observed	Telescope Used for Discovery	
methylidyne	CH	1937	visible	4300 Å	2.5-m	Mt. Wilson
cyanogen radical	CN	1940	visible	3875 Å	2.5-m	Mt. Wilson
methylidyne ion	CH^+	1941	visible	4232 Å	2.5-m	Mt. Wilson
hydroxyl radical	OH	1963	radio	18 cm	26-m	Lincoln Lab
ammonia	NH_3	1968	radio	1.3 cm	6-m	Hat Creek
water	H_2O	1968	radio	1.4 cm	6-m	Hat Creek
formaldehyde	H_2CO	1969	radio	6.2 cm	43-m	NRAO/Green Bank
carbon monoxide	CO	1970	radio	2.6 mm	11-m	NRAO/Kitt Peak
hydrogen cyanide	HCN	1970	radio	3.4 mm	11-m	NRAO/Kitt Peak
cyanoacetylene	HC_3N	1970	radio	3.3 cm	43-m	NRAO/Green Bank
hydrogen	H_2	1970	ultraviolet	1013-1108 Å		NRL rocket
methyl (wood) alcohol	CH_3OH	1970	radio	36 cm	43-m	NRAO/Green Bank
formic acid	$HCOOH$	1970	radio	18 cm	43-m	NRAO/Green Bank
"X-ogen"	HCO^+	1970	radio	3.4 mm	11-m	NRAO/Kitt Peak
formamide	$HCONH_2$	1971	radio	6.5 cm	43-m	NRAO/Green Bank
carbon monosulfide	CS	1971	radio	2.0 mm	11-m	NRAO/Kitt Peak
silicon monoxide	SiO	1971	radio	2.3 mm	11-m	NRAO/Kitt Peak
carbonyl sulfide	OCS	1971	radio	2.7 mm	11-m	NRAO/Kitt Peak
methyl cyanide, acetonitrile	CH_3CN	1971	radio	2.7 mm	11-m	NRAO/Kitt Peak
isocyanic acid	$HNCO$	1971	radio	3.4 mm	11-m	NRAO/Kitt Peak
methylacetylene	CH_3C_2H	1971	radio	3.5 mm	11-m	NRAO/Kitt Peak
acetaldehyde	CH_3CHO	1971	radio	28 cm	43-m	NRAO/Green Bank
thioformaldehyde	H_2CS	1971	radio	9.5 cm	64-m	Parkes
hydrogen isocyanide	HNC	1971	radio	3.3 mm	11-m	NRAO/Kitt Peak
hydrogen sulfide	H_2S	1972	radio	1.8 mm	11-m	NRAO/Kitt Peak
methanimine	H_2CNH	1972	radio	5.7 cm	64-m	Parkes
sulfur monoxide	SO	1973	radio	3.0 mm	11-m	NRAO/Kitt Peak
(no name)	N_2H^+	1974	radio	3.2 mm	11-m	NRAO/Kitt Peak
ethynyl radical	C_2H	1974	radio	3.4 mm	11-m	NRAO/Kitt Peak
methylamine	CH_3NH_2	1974	radio	3.5, 4.1 mm	11-m, 6-m	NRAO and Tokyo
dimethyl ether	$(CH_3)_2O$	1974	radio	9.6 mm	11-m	NRAO/Kitt Peak
ethyl alcohol, ethanol	CH_3CH_2OH	1974	radio	2.9-3.5 mm	11-m	NRAO/Kitt Peak
sulfur dioxide	SO_2	1975	radio	3.6 mm	11-m	NRAO/Kitt Peak
silicon sulfide	SiS	1975	radio	2.8, 3.3 mm	11-m	NRAO/Kitt Peak
acrylonitrile, vinyl cyanide	H_2CCHCN	1975	radio	22 cm	64-m	Parkes
methyl formate	$HCOOCH_3$	1975	radio	18 cm	64-m	Parkes
nitrogen sulfide radical	NS	1975	radio	2.6 mm	5-m	Texas
cyanamide	NH_2CN	1975	radio	3.7 mm	11-m	NRAO
cyanodiacetylene	HC_5N	1976	radio	3.0 cm	46-m	Algonquin
formyl radical	HCO	1976	radio	3.5 mm	11-m	NRAO/Kitt Peak
acetylene	C_2H_2	1976	infrared	2.4 μ	4-m	KPNO

Notes: Only the first wavelength or wavelengths observed are listed. Discoveries of forms including isotopes not included. Mt. Wilson is one of the Hale Observatories. Hat Creek is the site of the University of California's radio observatory. NRAO has most of its telescopes at Green Bank, West Virginia, its millimeter-wave telescope on Kitt Peak in Arizona, and the ULA in New Mexico. The Naval Research Laboratory (NRL) is in Washington, D.C. The Australian National Radio Astronomy Observatory is at Parkes, N.S.W. The millimeter telescope at Fort Davis, Texas, is operated by the University of Texas. The Herzberg Institute's radio telescope is at Algonquin Park, Canada. The Kitt Peak National Observatory (KPNO) is in Arizona.

at millimeter wavelengths. This radio telescope, run by the national radio observatory, is located on the same mountain that is used for the telescopes of the national optical observatory. The 43-meter telescope cannot observe at millimeter wavelengths.

Note that the 43-meter telescope operating at a wavelength of 10 cm is 430 wavelengths across, whereas the 11-meter telescope operating at 2 mm is 5500 wavelengths across. Thus the 11-meter telescope is effectively larger than the 43-meter telescope, relative to the wavelengths at which observations are being made. This has the practical effect of providing higher angular resolution. There is currently emphasis on millimeter radio astronomy, and new millimeter telescopes are under construction or have just been completed. The University of Massachusetts at Amherst opened a 14-meter (45-foot) telescope in 1976.

Also, there have been further advances in the technology of designing and building sensitive radio receivers. Maser action in the receiver itself

Figure 21–24 The contours show the molecular cloud associated with the Orion Nebula. The molecular cloud is actually on the far side of the glowing gas of the nebula, but the radio waves from the molecules penetrate the nebula and are observed with radio telescopes like the NRAO 11-m (36-ft) dish on Kitt Peak. The contours of radio emission correspond roughly to regions of different density. The densest part (the smallest region) also includes a region of infrared radiation called the Kleinmann-Low Nebula and a strong point infrared emitter called Becklin-Neugebauer object (which will be discussed in Section 22.1). Weaker emission from the molecules actually extends beyond the last contour and even far beyond the range of this photograph. The bright region at the lower part of the photo, NGC 1977, appears to the naked eye as the northern star in the sword of Orion but actually consists of a number of stars and a small H II region. The shape of the outermost contour shown here, indicates that this H II region may be expanding against the molecular cloud.

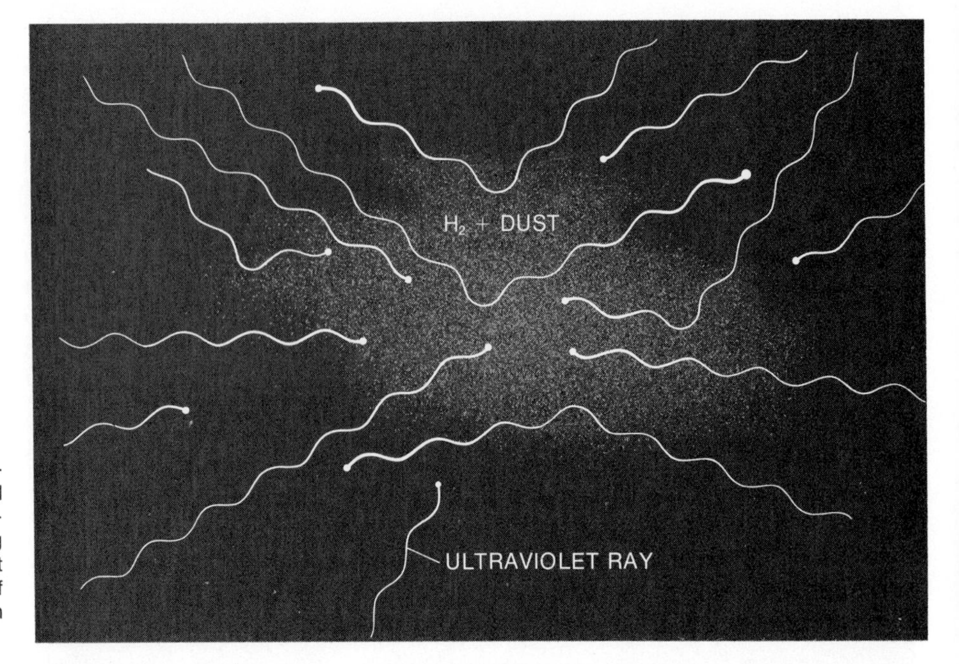

Figure 21–25 The presence of a dense dust cloud shields interstellar molecules like H_2 from being broken apart by ultraviolet radiation (the photons of which have relatively high energies).

has been used to amplify very faint signals observed at centimeter wavelengths. Some receivers are cooled by surrounding them with liquid nitrogen or liquid helium—the technology of ultracooling is called *cryogenics*. The advances in receiver technology continue to be linked to research on more mundane communications problems.

There is much more to spectral-line radio astronomy than simply discovering new spectral lines. For one thing, studying the lines provides information about physical conditions in the gas clouds that emit the lines. These clouds are usually so dense that hydrogen atoms have combined into hydrogen molecules and very little 21-cm radiation is emitted. Some of the lines, for instance those of carbon monoxide (CO), have been detected in many directions in space, and maps have been made of the distribution of these molecules. Sensitive radio receivers can do this, even though the densities of these molecules in the densest clouds may be only one in every cubic meter.

Most spectral lines seem to come only from a very limited number of places in the sky; CO is the major exception. Many lines have been detected only on a particular cloud of gas located in the constellation Orion, not very far from the main Orion Nebula. This *Orion molecular cloud* (Fig. 21–24) is itself buried deep in nebulosity, and it is presumed that there is so much dust present that radiation is prevented from breaking the molecules apart (Fig. 21–25). This allows the molecules to accumulate in number. Probably this cloud is also a site of star formation—new stars are being born there. The properties of the molecular cloud can be deduced by comparing the radiation from various molecules and by studying the radiation from each molecule individually. The density, 1000 particles per cubic centimeter in the outer limits at which the cloud is visible to us, increases toward the center, and it may actually be as dense as 10^6 particles/cm^3 at its center. This is still billions of times less dense than is our earth's atmosphere, though it is substantially denser than the average interstellar density of 0.1 to 1 particle per cm^3.

An additional contribution from observations of interstellar molecules has come from studies of the relative abundances of different isotopes. For example, not only can formaldehyde (H_2CO) be studied in its usual form, which contains the most common isotope of carbon, C^{12}, but also it can be studied in the form $H_2C^{13}O$. Knowledge of the relative abundances of the different isotopes is very important in developing theories of the formation of the elements. The study of these abundances had been the subject of much previous work with other methods. Who would have thought ten years ago that these isotopic abundances would best be measured with this—then unknown—method?

H_2CO normally consists of 2 atoms of hydrogen, 1 atom of C^{12}, and 1 atom of O. It can, more rarely, have another isotope of carbon, such as C^{13}, instead of C^{12}. It could also contain other hydrogen or oxygen isotopes.

It even turns out that basic parameters that describe the structure of some of these molecules may best be measured through astronomical spectral lines. Some molecules don't have spectral lines that are amenable to the measurement processes that can be used on laboratory samples on earth, and others can't even be formed on earth.

Not the least important conclusion that is reached by these spectral-line studies is that complex molecules can be readily formed by natural processes. Some of the molecules that have been discovered in space are "organic" in nature, in that they contain carbon, and we are approaching the level of complexity at which we can find simple amino acids, the building blocks of life. And we know (Section 19.1), even though we have not found amino acids themselves, that amino acids can be readily formed in a laboratory. All we need do is pass electricity through mixtures containing some of the molecules that have been discovered in space, or illuminate such a mixture with ultraviolet light. So it seems that it is much easier than we had originally thought to form the molecules from which life is made. These considerations add support to the beliefs of the many astronomers who feel that life has probably arisen spontaneously at many different locations in the universe.

These spectral-line molecular investigations have brought a new type of immigrant into the astronomy profession—organic chemists. Sometimes these scientists have used their expertise in the lab to measure the frequencies of the lines, something that must be known in order to observe them in radio sources. Sometimes they use their knowledge and training to help interpret the formation of these complex molecules. In any case, this is an example of how astronomy can incorporate another field of science. It also leaves traditionally trained astronomers with a whole new set of names with which they are otherwise unfamiliar—carbonyl sulfide, for example—names which neither astronomers nor students of astronomy need remember unless they intend to enter this specific field of study.

21.7 RADIO RECOMBINATION LINES

In addition to the molecular lines, a host of other radio spectral lines have been observed recently. They are called *recombination lines* because they result after an electron recombines with an ion in an H II region. Although the hydrogen is mostly ionized in an H II region, there is continual recombining and reionization. Actually, the observed radiation is emitted not in the recombination itself, but rather from changes in the energy states of the recombined atoms. The H II regions are studied photographically from their Hα recombination emission in the optical part of the spectrum (Fig. 21–26).

Figure 21–26 The recombination lines that we observe in the radio part of the spectrum are transitions among very high energy states of hydrogen (and other light atoms). H 109α, for example, is at a frequency of 5 GHz, which is equivalent to a wavelength of 6 cm. Lyman α, on the other hand, is a transition between the lowest two levels.

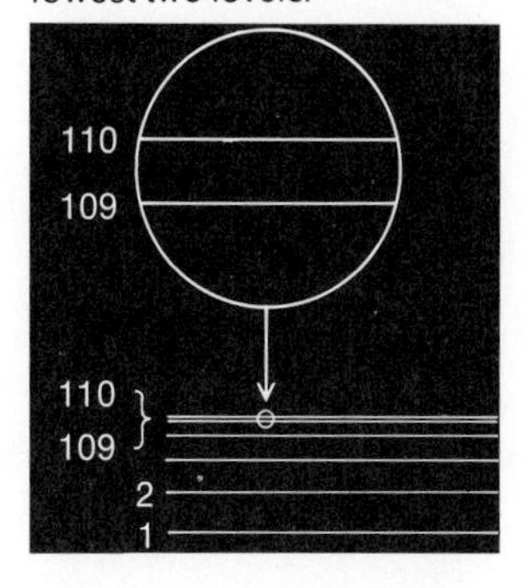

Recombination lines in the radio part of the spectrum can be detected for hydrogen and also for helium and for carbon. There are hundreds of such lines, distributed all up and down the radio spectrum. In 1959, the Soviet astronomer N. Kardashev first suggested that these lines could be detected. Their subsequent study has taught us a lot about the conditions in H II regions.

Let us consider, for the moment, the hydrogen recombination lines. We have seen how Lyman alpha results from a transition from the second energy level to the first level; let us call it 1α (1 alpha), the 1 from the fact that the lowest level was the lowest, or number 1, level in hydrogen and α from the fact that the difference in energy levels was 1. (We are now again considering principal energy levels and not the hyperfine structure in which the 21-cm line arises.) If we generalize this notation, always allowing the number to indicate the lower of the two principal energy levels and the Greek letter to represent the jump in level number, then Lyman beta is now 1β (1 beta), and Balmer alpha is 2α.

With this notation, typical recombination lines that astronomers observe are, for example, 109α, or 148β. When Kardashev first suggested that these high level recombination lines could be detected, the idea seemed ridiculous to many astronomers. They were accustomed to dealing with lower atomic number lines, and they assumed that these higher level lines would be so broad and faint that they would merge and be too weak to be detectable. But soon the radio recombination lines were definitely found, and they have been studied from all levels ranging from about 35α, at millimeter wavelengths, to about 250α, at about 70 cm.

When free electrons recombine with the ions, they can find themselves in excited states. They return to the ground state gradually, in steps of one or more level numbers at a time, thus emitting various "recombination lines."

Close to the hydrogen recombination lines in the spectrum are weaker recombination lines from helium and carbon. They are slightly shifted in frequency because of an effect of the different masses of the hydrogen, helium, and carbon atoms. The elements of atomic masses between helium and carbon (lithium, beryllium, and boron) have much lower abundances than that of carbon, and the heavier elements are also less abundant. No recombination lines other than those of hydrogen, helium, carbon, and perhaps silicon have been detected thus far.

The recombination lines arise primarily in H II regions, and from their intensities astronomers deduce temperatures and densities in the regions where the gas that is emitting the lines is located. Thus the recombination lines give us an important new way to investigate the physical conditions, that is, the electron temperature and density, in gas clouds in interstellar space.

In addition to the intensities of the lines, their Doppler shifts can be studied. This gives us information about the motions of the gas clouds in the galaxy, and can be used in a manner similar to using the Doppler velocities of 21-cm radiation to assess distances of the H II regions under study.

21.8 AT A RADIO OBSERVATORY

What is it like to go observing at a radio telescope? First, you decide just what you want to observe, and why. You have probably been working in the field before, and your reasons might tie in with other investigations under way. Then you need to know the frequencies of the spectral lines; they may be available in books or tables, or they may have to be measured specially for you in chemical laboratories. Perhaps it was a newly available set of radio frequencies received in the mail from a colleague that made you decide to observe a particular molecule.

Then you decide at which telescope you want to observe; let us say it is the 11-meter (36-foot) antenna of NRAO at Kitt Peak. You would send in a written proposal to the NRAO headquarters, where it would be read and evaluated. If the proposal is approved, it would be placed in a queue waiting for observing time. You might be scheduled to observe for a three-day period to begin six months to a year after you have submitted your proposal.

At the same time you might be applying for support to carry out the research, usually to the National Science Foundation. Your proposal would possibly contain requests for some support for yourself, perhaps for a summer, and support for a student or students to work on the project with you. It might also contain requests for funds for computer time at your home institution, some travel support, and funds to support the eventual publication of the research. The requests would always include an expense called "overhead." This would be approximately 50 per cent of the other costs, or perhaps just of the salary items, as your project's fair share of the costs that your home institution is incurring to have you there while the research is going on—the equivalent of office rent, secretarial help, telephone bills, and the like. An overhead rate of 50 per cent means that you must apply for $15,000 to have $10,000 available for actual research expenses.

If you will be observing at one of the NRAO telescopes, your travel support does not have to be included in the proposal to the NSF, because one or two members of each team granted observing time is automatically funded by NRAO for one trip to the telescope. This is because NRAO is

Figure 21–27 The 11-meter telescope.

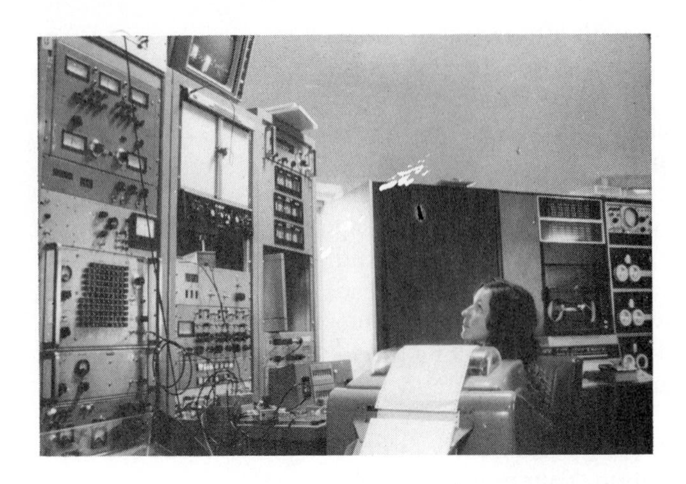

Figure 21–28 The taking of spectral line data is often controlled by a scientist or observing assistant at a keyboard linked to a computer. A screen on which the spectra can be continually seen is at top left. The view here is inside the control room of the 67-m telescope at Parkes, Australia.

a national observatory, and its policy is that no astronomer should have important advantages because of mere geographical proximity to the telescope.

When your observing time comes, you pack your bags and go to the airport. It doesn't really matter whether you are at a big university or at a small college, or even on the staff of the national observatory itself—you still have to pack your bags and go off to the telescope for a few days. With a multimillion dollar telescope, it is important that it be located at a good observing site and that the observing time be used as efficiently as possible. The cost of a few additional plane trips is small compared to the other costs. You are not charged directly for telescope time—that is covered in the overall budget of the observatory itself.

The National Radio Astronomy Observatory (NRAO) has its headquarters at Charlottesville, Virginia, and telescopes at Green Bank in West Virginia, Kitt Peak in Arizona, and Socorro in New Mexico. (The Kitt Peak National Observatory (KPNO), the national **optical** observatory, also has its telescopes at Kitt Peak, an hour's drive from Tucson.) Both optical and radio national observatories are funded entirely by the National Science Foundation. The funding level is very tightly controlled, and in recent years has been dropping in real terms. Current NSF astronomy budgets are about $10 million for support of individual grantees through their institutions. A report of the President of the American Astronomical Society suggested that a doubling of the support could easily be put to very good use, but there is no sign that support will be increased in the near future.

Figure 21–29 Radio astronomy depends on electronics as much as optical astronomy has depended on optics. Shown here is the inside of the autocorrelator—a device used for observing radio spectral lines—at the 64-m telescope at Parkes. A failure of any of the thousands of electronic parts shown can mean a breakdown of the whole procedure of collecting data.

When you arrive in Tucson, you might stay a day or two in town to meet with the other astronomers there, or you might go straight to the mountain. NRAO has lodgings on the mountain top, distinct from the more elaborate quarters that the optical observatory has there, though now KPNO feeds the visiting radio astronomers in the same dining hall as their own staff.

At the telescope, you will meet the other members of your team if they have come from different places; sometimes your whole team will have traveled with you. If this is your first time at the telescope, care will have been taken to see that an astronomer experienced at observing is present to help you get started.

The astronomers sit at a computer console monitoring the data as they come in from the telescope (Fig. 21–28). The electronics that are used to treat the signal incoming from the feed (Fig. 21–29) are as important a part of the system as the dish itself. A trained observing assistant actually runs the mechanical aspects of the telescope. These observing assistants are regular employees of the observatory and work in shifts. The astronomers arrange their own time schedules so that they can observe around the clock—one doesn't want to waste any observing time. This is unlike optical astronomers who can work all night and sleep for part of the daytime. (Optical astronomers, though, often have other tasks to perform during part of the day, such as developing photographic plates.)

Unlike the optical sky, which is blue in the daytime, the radio sky background remains dark even when the sun is up. As long as they don't point their telescopes to within a few degrees of the sun, radio astronomers can observe anywhere in the sky at any hour of the day or night.

You give the observing assistant the coordinates of the point in the sky that you want to observe, and the telescope is pointed for you. The electronic systems at the 11-meter are particularly advanced, and a computer can display the incoming spectral data on a video screen for you to see. You can even use the computer to manipulate the data a bit, perhaps adding together the results from different five minute chunks of "integration time," that is, exposures, that you have made. The data are stored on magnetic computer tape and can be graphed on paper very quickly.

Depending on just what you are observing, your results may or may not be immediately apparent to you on the computer screen. Some spectral lines are so intense in certain sources that you can see the emission line on the screen in just a few minutes. Some lines are so faint that you may have to integrate for many hours, or even days or months, to reach an acceptable level of sensitivity.

When you have finished observing one source, or it has set below the horizon, you ask the operator to point the telescope to another source, and off you go again.

At the end of your observing run, you take the data back to your home institution to complete the analysis. You are expected to publish the results as soon as possible in one of the standard scientific journals and to report at professional meetings.

SUMMARY AND OUTLINE OF CHAPTER 21

Aims: To study the interstellar medium, and to understand the techniques (especially radio astronomy and ultraviolet astronomy) used to study it

Radio astronomy (Section 21.1)
- Discovered by Jansky in 1931
- Radio waves reflected by a metal dish to a focus where an antenna collects the signal; the signal is then amplified and analyzed with electronic equipment
- Larger dishes have larger collecting areas and higher resolutions

Continuum radio astronomy (Section 21.2)
- Spectra are measured over a broad frequency range
- Radiation generated by synchrotron emission, electrons spiraling rapidly in a magnetic field, found in many non-thermal sources

Radio spectral line from interstellar hydrogen (Section 21.3)
- 21-cm spectral line from neutral hydrogen was discovered in 1951
- Line occurs at 21 cm through a spin-flip transition, corresponding to a change between energy levels of the hyperfine structure of hydrogen
- Both emission and absorption have been detected
- 21-cm radiation used to map the galaxy

- Distances measured using differential rotation and the Doppler effect
- Interstellar optical molecular lines (Section 21.4)
  - CN, CH, CH^+
  - Diffuse lines
- Molecular hydrogen (Section 21.5)
  - Interstellar ultraviolet absorption lines from H_2 with abundances up to 50 per cent observed from above the earth's atmosphere
  - Copernicus satellite detected not only normal H_2 but also deuterium, which has cosmological importance
- Radio spectral lines from molecules (Section 21.6)
  - "Mysterium" lines discovered at 18 cm—later identified as OH affected by masering process
  - Masers first developed artificially on earth, but later discovered to exist in space
  - Maser action occurs after atoms are excited by energy pumping mechanism so that many are in the same excited state
  - Three dozen molecules have been discovered in space, including some as massive as amino acids
  - Many molecules found at short millimeter radio wavelengths
  - Molecules associated with dark clouds, such as the Orion molecular cloud, where the molecules are shielded from being torn apart by ultraviolet radiation
  - Distribution of CO is widespread
- Radio recombination lines (Section 21.7)
  - Result after an electron recombines with an ion in an H II region
  - Recombination lines have been discovered from H, He, C, and perhaps Si
  - Analysis tells us physical conditions (e.g., temperature and density)
  - Doppler shifts of the lines can be used to estimate distances to the H II regions

QUESTIONS

1. Estimate the wavelengths for which a window screen would be a good reflector.
2. List two relative advantages and disadvantages of radio astronomy and optical astronomy.
3. The angular resolution of the 100-meter telescope at a wavelength of 21 cm is approximately one arc minute. How large a telescope would you need to get the same angular resolution at a wavelength of 2 mm?
4. How does the procedure of making a radio map of a region differ from taking a photograph?
5. Why is emission from cool regions of space most likely to be detectable at radio wavelengths?
6. What is the frequency of 1 mm waves?
7. What is the wavelength of 108 MHz waves, which are those you tune in at the right end of your FM radio dial.
8. What determines whether the 21-cm line will be observed in emission or absorption?
9. If our galaxy rotated like a rigid body, would we be able to use the 21-cm line to determine distances to H I regions? Explain.
10. Why are dust grains important for the formation of interstellar molecules?

11. (a) How many interstellar molecules have 2 atoms, 3 atoms, etc?

    (b) How many interstellar molecules contain one heavy (i.e., non-H) atom, 2 heavy atoms, etc?

    (c) What fraction of the known interstellar molecules are organic?

    (d) How many interstellar molecules are chemicals of which you have heard previously?

12. Why can't we observe 21-cm radiation from interstellar molecular clouds?

13. What type of information can interstellar molecules provide?

14. Would you expect to receive more photons from H 107α or H 108α? Explain.

15. What information can we learn from recombination lines?

TOPICS FOR DISCUSSION

1. What does the discovery of fairly complex molecules in space imply to you about the existence of extraterrestrial life?

2. Consider the observing schedules of astronomers who study different parts of the spectrum – for example, optical stellar astronomers work in the nighttime and optical solar astronomers work in the daytime. What are the observing schedules of x-ray astronomers, astronomers using balloons to study the infrared, solar astronomers using space telescopes, and so on? How many different kinds of specialties can you think of?

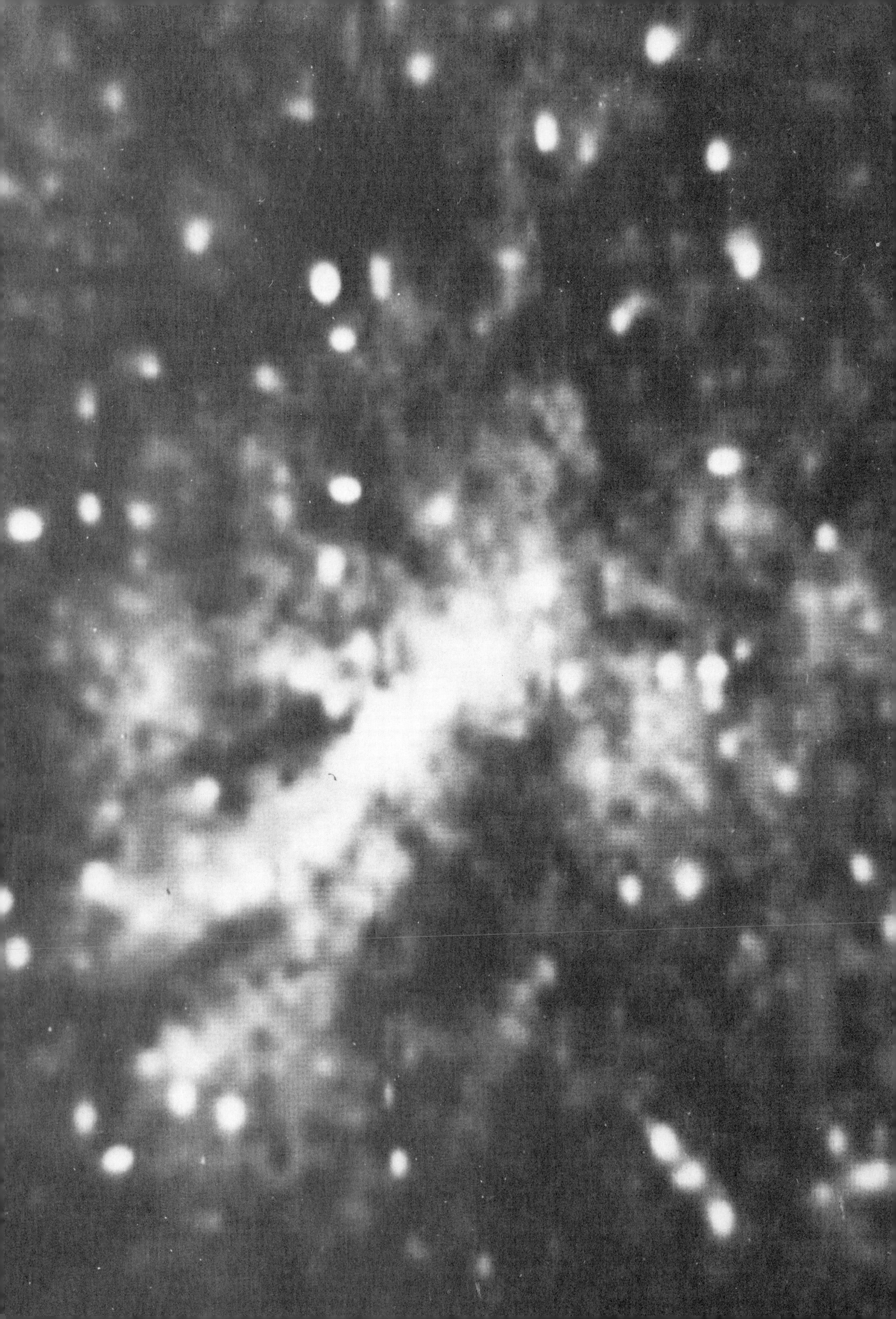

22

Infrared, X-Ray, and Gamma-Ray Studies of Our Galaxy

Optical and radio astronomical methods have become standard for studying our galaxy. In recent years, studies in other parts of the electromagnetic spectrum have become increasingly important.

22.1 INFRARED OBSERVATIONS

For many years the sky has been intensively studied at optical wavelengths up to about 8500 Å (0.85 micron) and to a lesser extent up to 1.1 microns, which is in the near infrared and beyond which film is not sensitive. The sky has also been studied for decades at radio wavelengths down to one or two centimeters (10,000 or 20,000 microns). But until the last few years, the sky has been studied very little at the wavelengths in between, which are the millimeter and the infrared wavelengths.

One reason that the infrared (about 1 micron to about 1 mm) was not studied is the fact that the earth's atmosphere doesn't cooperate. Most infrared radiation that hits the top of the earth's atmosphere doesn't get through; rather, it is absorbed by water vapor, carbon dioxide, or ozone. The earth's atmosphere is transparent in only a few limited regions of this broad range of the infrared spectrum. These regions of transparency are called *windows* (Fig. 5–77). One window is at about 1.65 microns, another is from 2.0 to 2.4 microns (known as the 2.2 micron window), and others lie at about 3.6 microns, 4.8 microns, from 8 to 14 microns, and from 17 to 22 microns. The atmosphere is then opaque for a long wavelength stretch from about 22 up to 350 microns; there are then windows near 0.4, 1.2, 2.0, and 3.3 mm. (Fig. 2–8).

One can observe better in the infrared from locations where there is little water vapor overhead. A prime location, for example, is Mauna Kea in Hawaii, which is an especially dry site at its 4145-meter (13,600-foot) peak (that is, it is characterized by very low absorption by water vapor) even though it is surrounded by ocean. Alternatively, one can try to get above as much of the earth's atmosphere as possible by sending telescopes aloft in balloons (Fig. 22–1), or by flying as high as possible in airplanes.

A high resolution map of the center of the Milky Way Galaxy at a wavelength of 2.2 microns. The resolution is about 1.2 arc min, and the width of the entire image is about 1.1°.

Figure 22–1 A balloon being launched to make infrared studies.

Another difficulty in studying the infrared involves the detectors—it is hard to develop sufficiently sensitive instruments with which to study the infrared radiation. Remember (from Section 2.7) that $E = hc/\lambda$; since the infrared wavelengths are longer than the visible wavelengths, the energy that each infrared photon has is relatively low compared to that of a visible photon. Thus many types of detectors are simply not sensitive to infrared radiation. The longer the infrared wavelength that you want to study, the more difficult it is to find a suitable detector.

Yet another difficulty comes from the fact that ordinary heat radiation is in the infrared. The earth's atmosphere radiates conspicuously in the infrared, so the radiation coming into a telescope includes both infrared radiation from the atmosphere and from the source being observed. The telescope itself contributes its own infrared radiation. To limit the telescopic contribution of radiation, the equipment is usually bathed in liquid nitrogen, or even in liquid helium, which is colder and more expensive.

The necessity to work with very cold liquids, that is, to use cryogenic technology, certainly makes infrared astronomy technologically difficult. When some astronomers tried to do an experiment to study the solar corona at 500 microns during the 1966 eclipse, they brought their equipment to the middle of a plain in the Bolivian Andes. They not only had to get all the equipment there, but they also had to get a supply of liquid nitrogen to the site. The liquid nitrogen was liable to boil away at any instant and was hazardous to transport; this created major logistic problems. Another astronomer had to use liquid helium in the desert in northern Kenya at the 1973 eclipse in order to work at 1 millimeter, which is a wavelength on the boundary between infrared and radio radiation. It is hard enough to work with liquid nitrogen or liquid helium in the laboratory or at an established observatory, much less at a field site or in a high flying plane or in a balloon. Infrared astronomers doing field work must contend with these additional problems.

Robert Leighton and Gerry Neugebauer of Caltech mapped the sky in the 2.2 micron window during the mid 1960's. To carry out this lengthy sky survey inexpensively, Leighton elected to build his own large telescope mirror. Since infrared wavelengths are longer than optical wavelengths, the mirror did not have to be ground to the same accuracy as a telescope mirror for visible light, though it did have to be more accurate than the metal surface of a radio telescope.

Leighton realized that he could make a sufficiently accurate mirror out of epoxy. First, he machined out of metal a very deep dish, 1.5 meters (60 inches) across. He coated it with epoxy and for three days he spun it around an axis perpendicular to its center while the epoxy hardened—the surface of a spinning bowl of liquid is a parabola. Finally, he had a good enough telescope for the infrared. Such a telescope, which has a surface sufficiently accurate in shape to collect light and direct it toward a focus but not good enough to form high quality images, is known as a *light bucket.*

Leighton and his colleagues surveyed the whole part of the sky that can be observed from the top of Mount Wilson. In this survey, they were able to detect some 20,000 infrared sources! Remember that there are only about 6000 stars that can be seen with the naked eye.

Most of the infrared sources that they discovered do not coincide in space with known optical sources. The map of the radio sky doesn't look like the map of the optical sky, and the map of the infrared sky doesn't resemble either of the others. Most of these infrared-emitting objects, however, unlike the radio objects, are in our galaxy (Fig. 22–2). Following

Figure 22–2 Hubble's Variable Nebula, which is associated with the variable star and infrared source R Monocerotis. This was the first official photograph taken with the 5-m Hale telescope.

the survey program, some of the more interesting infrared objects have been individually studied much more carefully.

For example, an infrared source in the constellation Leo varies in brightness by a couple of magnitudes from year to year. No spectral lines can be observed in this source, which limits the information we can hope to measure. If this infrared object is a thermal source—that is, as we saw in Section 21.2, its radiation simply results from the fact that it is hot and the radiation follows Planck's law—this variable object is at a temperature of 650 K. Theoreticians have calculated what the equilibrium temperature would be on the assumption that a dust cloud surrounding a certain kind of star would radiate away just as much energy in the infrared as it received at all wavelengths from the underlying star. This temperature is also calculated to be about 650 K. Thus this infrared radiation, and presumably the infrared radiation from many other infrared sources, comes from warm dust shells surrounding stars; the dust shells are warm enough to radiate even though they are much cooler than the stars they surround. Such objects are sometimes known as *cocoon stars*.

The brightest infrared source at a wavelength of 20 microns is associated with Eta Carinae (Color Plate 45). From observations in the infrared, it has been deduced that Eta Carinae is surrounded by a cloud of dust that is about 0.1 ly across and is at a temperature of 250 K.

One of the strangest infrared objects is known as the Becklin-Neugebauer object (Fig. 22–3), named after Eric Becklin and Gerry Neugebauer, the Caltech astronomers who discovered it. Even though this star is very bright in the infrared, it is completely invisible at optical wavelengths. It is in the region of the Orion molecular cloud behind the Orion Nebula. If the Becklin-Neugebauer object is a bright star at the distance of the Orion Nebula and is about to enter the main-sequence part of its lifetime, we can calculate how much its light would have to be reddened to appear to us the way it does. It could be reddened by dust either between it and us in interstellar space, or else in a shell or shells around it. In either case, the extinction in the visible between the Becklin-Neugebauer object and the earth is at least 50 magnitudes (a factor of 10^{20}), a tremendous amount. Alternatively to the idea of reddening by dust, the Becklin-Neugebauer object could just be a very cool star, i.e., a star in formation that is not yet heated enough to start nuclear burning. Either possibility is very exciting.

Perhaps the most interesting infrared source that we know is a source of very small diameter located in the direction of the center of our galaxy. (It is the photograph opening this chapter.) Since the amount of scattering by dust varies with wavelength, we can see farther through interstellar space in the infrared than we can in the visible. In this case, it seems that in the infrared we can see 8.5 kiloparsecs to the center of our galaxy. The

Figure 22–3 The Becklin-Neugebauer object in the Orion Nebula cannot be seen on photographs taken in visible light, but is located behind the region marked with a circle.

brightest infrared source is located at the position of the radio source Sagittarius A (Fig. 22–4).

This source subtends an angle of 1 arc min, and so is about 3 parsecs (about 10 light years) across. This makes it a very small source for the prodigious amount of energy it emits: as much energy as if there were 80 million suns radiating. We don't know just what it is in the center of our galaxy that causes this radiation, but this small source does seem to give off as much as 0.1 per cent of the total radiation from our galaxy. It is interesting to know that the center of our galaxy is qualitatively different from the outer parts: the center is a different kind of place from the arms, and its greater density of stars isn't the only difference. Other galaxies have even more prominent infrared sources in their nuclei.

There are several models to account for the infrared radiation from the galactic nucleus. One model is that there are bright sources of ultraviolet or visible radiation there, such as a dense group of stars. These sources would be surrounded by shells of dust that heat up to about 50 K. One needs the equivalent of about 10 solar masses of dust, not an inconceivable amount, to make this model work. Similar processes, on a smaller scale, are known for shells surrounding hot stars.

Another model, much more speculative, says that both matter and anti-matter exist in the center of our galaxy. They would annihilate each other, as matter and anti-matter do, and a tremendous amount of energy would be liberated in this annihilation. The mechanical reaction from the annihilation could tend to separate the matter and anti-matter along their common boundary, preventing the situation from "running away," that is, continuously escalating at an increasing rate. This model certainly cannot be

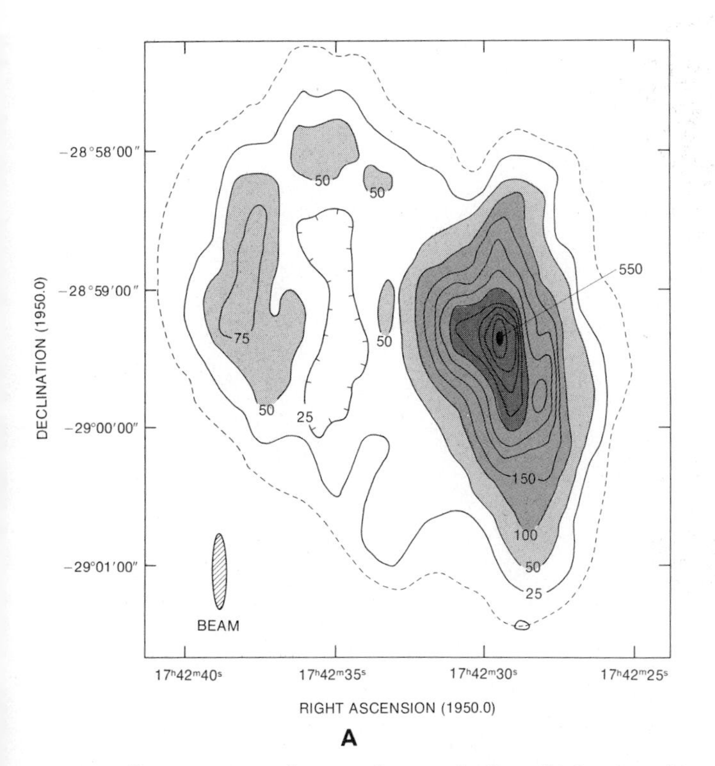

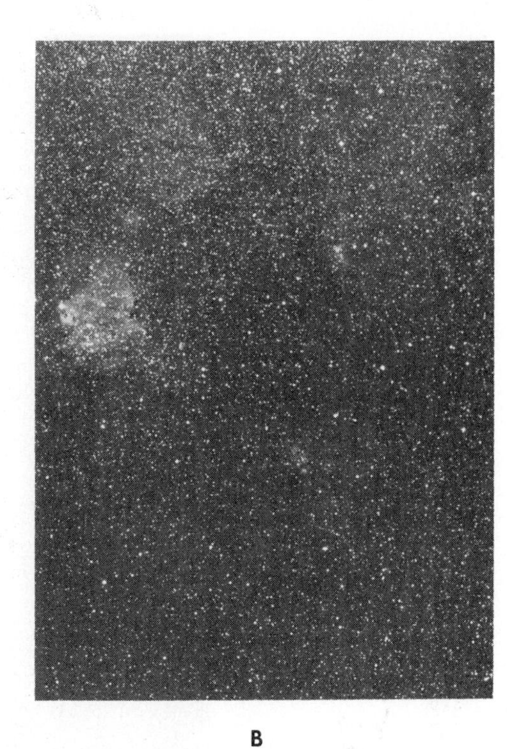

Figure 22–4 *(A)* A radio map of the galactic center, made at a wavelength of 6 cm with the 100-m Bonn telescope. The intense source at right is called Sgr A. *(B)* A photograph in visible light taken in the direction of the center of our galaxy, showing exactly the same region observed in the infrared that is displayed as the opening photograph to this chapter. In visible light, interstellar dust completely hides the galactic center, and we are seeing only relatively nearby stars and gas.

accepted at present, for we do not definitely know of any large scale existence of anti-matter at all, but neither can it be ruled out.

In any case, infrared observations together with radio observations are showing that the center of our galaxy is a very interesting place.

22.2 HIGH-ENERGY ASTRONOMY

Thus far we have discussed how to study our galaxy not only at wavelengths in the visible region of the spectrum but also at longer wavelengths. We also study our galaxy, as of the last few years, at wavelengths shorter than those of visible radiation.

Radiation immediately below the violet part of the visible spectrum, from about 4000 Å down to 100 Å, is called *ultraviolet.* The earth's atmosphere prevents radiation shorter than 3000 Å from penetrating down to the surface, so observations of shorter wavelengths must be made from rockets, balloons, or satellites above the atmosphere. We continue to call the region of the spectrum down to about 100 Å by the name "ultraviolet," but unlike the case for the "near ultraviolet" (3000 Å to 4000 Å), we must go above the atmosphere to study the shorter ultraviolet radiation. Sometimes this shorter radiation is called *XUV* or *EUV*, both of which are short for "**E**xtreme **U**ltraviolet." XUV can also refer to the boundary region between **x**-rays and **u**ltraviolet radiation.

Chemists studying atoms join astronomers in using angstroms as a measuring unit, although purists now claim that we should only use SI units, which allow only powers of 1000 or 1/1000 meters. In this use of the metric system 1 angstrom = 0.1 nanometer. Astronomers won't give up the use of angstroms without a struggle.

Below about 100 Å, and ranging down to about 1 Å, we speak of the radiation as *x-rays.* This name survives from the discovery of x-rays in 1895 by Wilhelm Roentgen. The 1 Å length we use as a lower limit is about the diameter of a small atom.

Below about 1 angstrom, we speak of *γ-rays* (gamma rays). Again, this is just electromagnetic radiation of extremely short wavelength, and is subject to the same laws as longer wavelength radiation. The dividing lines between uv and x-rays, or between x-rays and gamma rays, are not sharp ones.

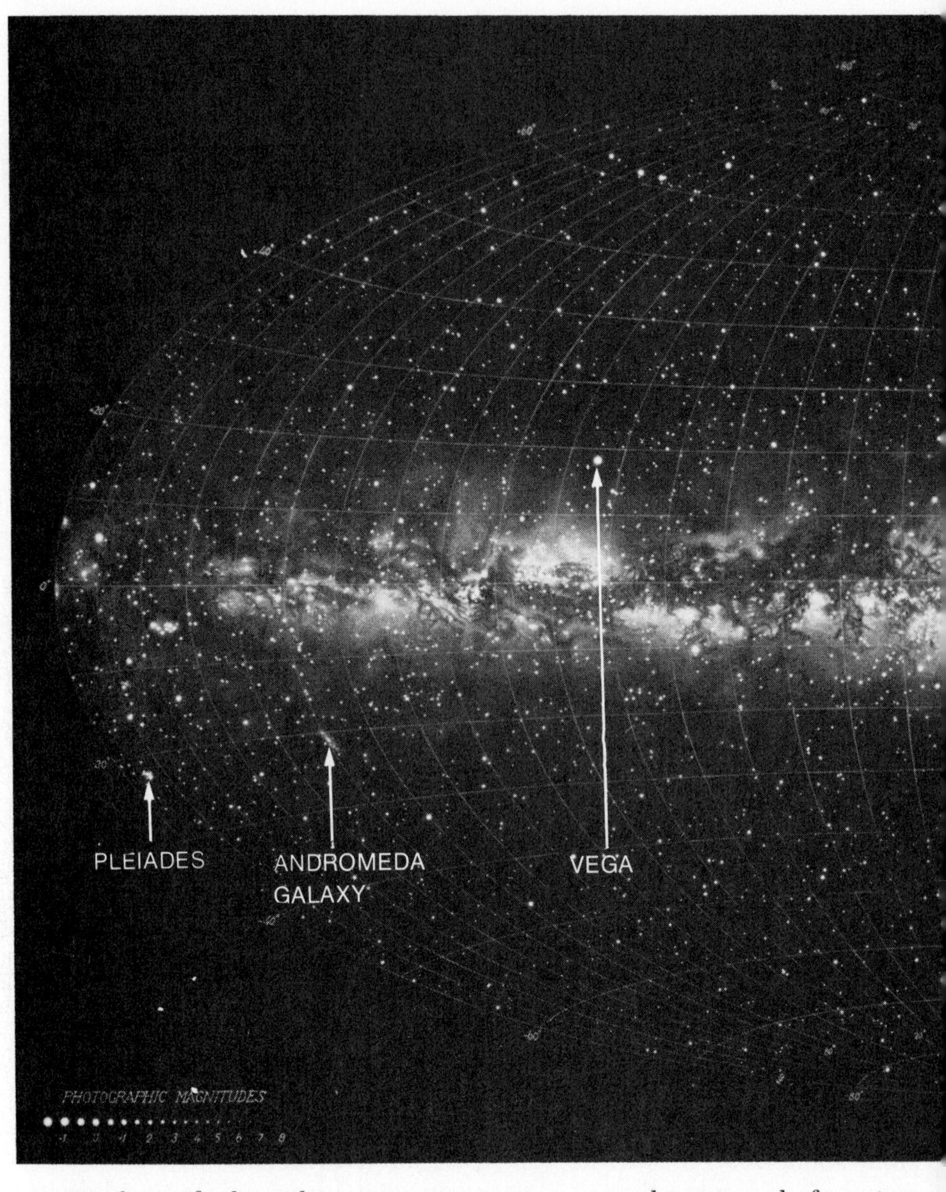

Just as radio astronomers prefer to work in frequency instead of wavelength units, gamma-ray astronomers also use $E = h\nu$ but work in energy units. (See Section 2.7 for a definition of electron volts.) So the technical papers talk of gamma rays in millions of electron volts (abbreviated MeV; M means millions to contrast with m from milli-, which means thousandths), but we can ignore this technical point for the purposes of our general discussion. It may be useful to note that a photon with a wavelength of 1 Å has an energy of about 12,000 eV = 12 keV = 0.012 MeV.

We have dealt with x-ray astronomy in several contexts before, including discussions of the sun and of black holes. Most of the x-ray sources that we detect are located throughout our galaxy (Fig. 22–5). The most thorough map of the x-ray sky (Fig. 22–6) that we have was made with a NASA satellite in its Small Astronomy Satellite series (SAS-1). We have already mentioned (in Section 9.5 on black holes) some of the results provided by this satellite, which is named Uhuru. Since 1970, the Uhuru satellite (Fig. 22–7) has observed hundreds of x-ray sources. Its most recent successor satellite, SAS-3, was launched in 1975. Other satellites to study x-rays have been launched by the Soviet Union, by Britain from a launch pad in Kenya, and by the Netherlands from the United States. (Some of the observations of x-ray bursts from globular clusters detected by these satellites have been discussed in Section 4.5c.)

Instead of film, Uhuru used counters filled with argon. When x-rays hit the counters, electrons were given off in numbers proportional to the energy of the incoming x-rays. Detecting and counting the electrons thus gave the energy of the x-rays.

Hundreds of x-ray sources have now been mapped, and many of them

Figure 22–5 A drawing of the Milky Way, made under the supervision of Knut Lundmark at the Lund Observatory in Sweden. 7000 stars plus the Milky Way are shown in this panorama, which is in coordinates such that the Milky Way falls along the equator.

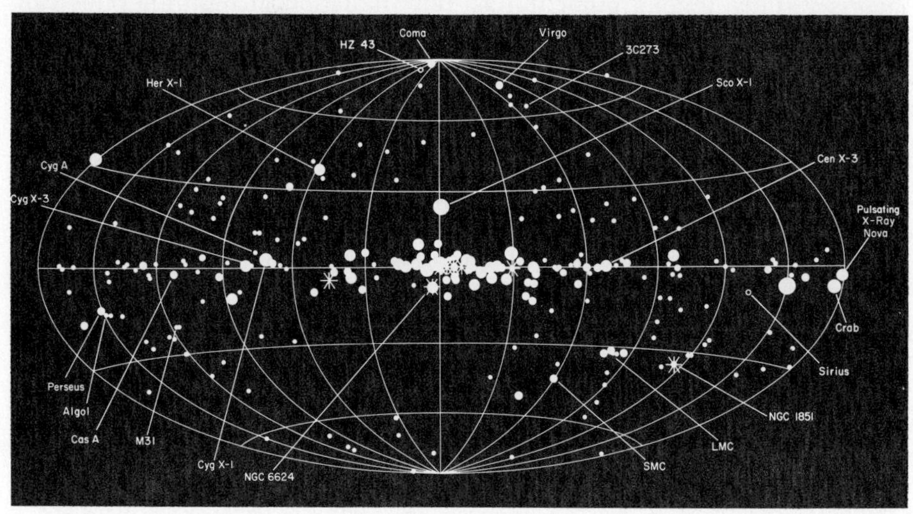

Figure 22–6 An x-ray map of the Milky Way showing the objects observed by the Uhuru spacecraft as of mid-1976. X-ray "burstars" (see Section 4.5c) are shown with asterisks. X-ray sources observed only by other satellites are shown as open circles; these sources are detectable only at relatively long x-ray wavelengths.

Some two dozen burstars are now known; only a handful are in globular clusters. None are in binary systems, though many x-ray sources that do not give off bursts of x-rays are in binaries. The x-ray binary systems seem to involve neutron stars; some scientists think that the burstars involve black holes, though this is very hard to prove.

have been studied over periods of time. Some of the sources vary with time (Fig. 22–8). The center of our galaxy is a strong and variable x-ray source; we have seen that it is also an infrared and a radio source. This again indicates that violent processes must be going on there in order to generate such powerful radiation. Although many of the x-ray sources mapped by Uhuru are in our galaxy, others are extragalactic and are discussed in Section 23.5 and 24.1. The nuclei of some other galaxies are much more active than that of the Milky Way Galaxy.

γ-ray astronomy is even harder to carry out than x-ray astronomy, and is only now getting going. Although an occasional strange object, like the Crab Pulsar, can be detected at every wavelength of the spectrum, including the x-ray and γ-ray regions, astronomers were hard put to find optical counterparts of other γ-ray sources. γ-ray telescopes are aboard various satellites, including the later Orbiting Solar Observatories (the γ-ray telescopes were not on the part of the OSO's that always points at the sun but rather were on the second part that rotates and scans the sky) and the second Small Astronomy Satellite, SAS-2, which was launched in 1972. Many γ-ray observations have also been made from Vela satellites, U.S. satellites whose prime purpose is to detect nuclear explosions in space.

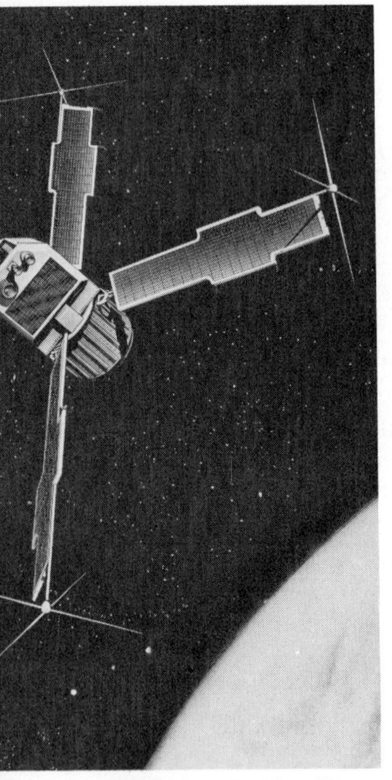

Figure 22–7 The Uhuru satellite.

One interesting γ-ray observation is the discovery of a general background of γ-rays concentrated along the plane of our galaxy. This diffuse background is not concentrated in particular objects, not even in the direction of the galactic center. Presumably, these γ-rays are caused by the interaction of *cosmic rays*, nuclei of atoms that travel around space at tremendous velocities, with interstellar matter. The origin of the primary cosmic rays, which have long been studied at the top of the earth's atmosphere from balloons or rockets or from the earth's surface through detection of secondary cosmic rays, is a major question. We have discussed (in Section 8.3) their probable relation to supernovae. Even though the cosmic rays are already formed with high energies at such sites, many of them are brought to even higher energies by processes taking place throughout the galaxy.

Studies of electromagnetic radiation like x-rays and γ-rays and of rapidly-moving cosmic ray particles are part of the new field of *high-energy astrophysics.* ($E = h\nu$, and ν is very large for radiation in this part of the spectrum, so we speak of high-energy radiation.) High-energy astrophysics is an especially active field for theorists, as well as observers. The theorists study, among other things, how radiation of high energies can be generated in interactions between matter and other matter or between matter and radiation.

High-energy astrophysics has been given a considerable priority in studies of the kinds of research to be supported in the next few years. In particular, a panel of the National Academy of Sciences chaired by Jesse Greenstein of Caltech submitted a major report in 1972 that evaluated the importance of different fields of future astronomical research and endorsed an especially intensive effort in high-energy astrophysics.

Circular No. 2957

CENTRAL BUREAU FOR ASTRONOMICAL TELEGRAMS
INTERNATIONAL ASTRONOMICAL UNION

POSTAL ADDRESS: CENTRAL BUREAU FOR ASTRONOMICAL TELEGRAMS,
SMITHSONIAN ASTROPHYSICAL OBSERVATORY, CAMBRIDGE, MASS. 02138, USA
CABLE ADDRESS: SATELLITES, NEWYORK - WESTERN UNION: RAPID SATELLITE CAMBMASS

MX1803-24

G. Jernigan, Massachusetts Institute of Technology, reports that a new transient x-ray source was observed by SAS 3 during May 13-19 at $\alpha = 18^h03^m47^s$, $\delta = -24°36'.3$ (equinox 1950.0; error radius 1'.5). The maximum intensity was equal to that of the Crab.

Figure 22–8 X-ray astronomy is currently so exciting that new sources as bright as any in the sky appear without notice.

NASA had also given this area a high priority and had a very elaborate and expensive satellite called HEAO (High-Energy Astronomy Observatory) in preparation for launch in the late 1970's. But financial limitations led them to cut this program very severely and very suddenly—its very expense ($250 million for a single spacecraft) made it an obvious place to economize. Protests from the astronomical community were prompt and fierce. At present, NASA plans to launch many of the experiments that would have been in the big HEAO in other, smaller, HEAO satellites during the next few years (see Section 3.9). HEAO-A, scheduled for launch in 1977, will largely take spectra, while HEAO-B, with imaging capabilities, is scheduled for launch in 1979. Discoveries are expected to concern not only the center of our galaxy but also neutron stars, black holes, and clusters of distant galaxies.

SUMMARY AND OUTLINE OF CHAPTER 22

Aims: To study infrared, x-ray, and γ-ray observations of our galaxy, which are among the newest methods of research, and to see that cosmic rays permeate the galaxy

Infrared observations (Section 22.1)
- Windows in the spectrum
- Technological difficulties
- Special infrared telescope–a light bucket–produced an IR map
- Cocoon stars, B-N object
- Radiation from the galactic nucleus indicates a lot of energy is generated in a small volume there

High-energy astronomy (Section 22.2)
- The Uhuru satellite and an x-ray map
- Radiation from the galactic nucleus indicates violent activity there
- γ-ray background radiation the result of cosmic ray interaction with matter
- Growing importance of high-energy astrophysics
- Plans for HEAO satellites

QUESTIONS

1. Why is it possible to do some infrared observations from mountain observatories while all x-ray observations must be made from space?
2. Illustrate, by sketching the appropriate Planck radiation curves, why cooling an infrared telescope from room temperature to 4 K greatly reduces the background noise.
3. Does mapping a region of the sky in the infrared more closely resemble optical or radio techniques? Explain.
4. From what you know about interstellar reddening, explain why you can see some stars in the infrared that you can't see in the visible.
5. What is the property of x-rays that makes them suitable for looking into the body?
6. Why will x-rays expose a photographic plate while infrared radiation will not?
7. What is the advantage of using proportional (argon) counters, such as those used on the Uhuru satellite, to detect x-rays?
8. What types of objects give off strong infrared radiation?
9. What types of objects give off x-rays?
10. Discuss how observations from space have added to our knowledge of our galaxy.

Part V

Galaxies and Beyond

M104, the Sombrero Galaxy in Virgo.

(*left*) The cluster of galaxies in the constellation Hercules.

Galaxies and Beyond . . .

Though the individual stars that we see are all part of the Milky Way Galaxy, discussed in the preceding three chapters, we cannot be so categorical about the locations of the conglomerations of gas and stars that can be seen through telescopes. Once they were all called "nebulae," but we now restrict the meaning of this word to gas and dust in our own galaxy. Some of the objects that were originally classed as nebulae turned out to be huge collections of gas, dust, and stars located far from our Milky Way Galaxy and of a scale comparable to that of our galaxy. These objects are galaxies in their own right, and are both fundamental units of the universe and the stepping stones that we use to extend our knowledge to tremendous distances.

The large reflector, with a mirror 6 feet (1.8 meters) across, built by the Earl of Rosse in Ireland in 1845. Problems with maintaining

In the 1770's, a French astronomer named Charles Messier was interested in discovering comets. To do so, he had to be able to recognize whenever a new fuzzy object appeared in the sky. He thus compiled a list of about 100 diffuse objects that could always be seen. To this day, these objects are commonly known by their *Messier numbers.* Messier's list contains the majority of the most beautiful objects in the sky, including nebulae, star clusters, and galaxies. The list, compiled to search for comets, turns out to have this much more general importance.

Soon after, William Herschel, in England, compiled a list of 1000 nebulae and clusters, which he expanded in subsequent years to include 2500 objects. Herschel's son John continued the work, incorporating observations made in the southern hemisphere. In 1864, he published the *General Catalogue of Nebulae.* In 1888, J. L. E. Dreyer published a still more extensive catalogue, *A New General Catalogue of Nebulae and Clusters of Stars,* the *NGC,* and later published two supplementary *Index Catalogues, IC's.* The 100-odd non-stellar objects that have Messier numbers are known by them, or else by their numbers in Dreyer's catalogue. Thus the Great Nebula in Andromeda = M31 = NGC 224. The Crab Nebula = M1 = NGC 1952.

When larger telescopes were turned to the Messier objects in later years, especially by Lord Rosse in England in about 1850, some of the objects showed traces of spiral structure, like pinwheels. They were called "spiral nebulae." But where were they located? Were they close by or relatively far away?

an accurate shape for the mirror, which was made of metal, led to the telescope's abandonment.

When such telescopes as the 0.9-meter Crossley reflector at Lick in 1898, and later the 1.5-meter and 2.5-meter reflectors on Mount Wilson, began to photograph the "spiral nebulae," they revealed many more of them. The shapes and motions of these "nebulae" were carefully studied. Some scientists thought that they were merely in our own galaxy, while others thought that they were very far away, "island universes" in their own right, so far away that the individual stars appeared blurred together. (The name "island universes" originated with the philosopher Immanuel Kant in 1755.)

The debate raged until 1925, when observations made at the Mount Wilson Observatory by Edwin Hubble proved that there were indeed other galaxies in the universe besides our own. In fact, we think of galaxies and clusters of galaxies as fundamental units in the universe. The galaxies are among the most distant objects we can study; quasars are objects on a galactic scale that are, for the most part, even farther away.

Galaxies and quasars can be studied in most parts of the spectrum. Radio astronomy, in particular, has long proved a fruitful method of study. The study across the spectrum of galaxies and quasars provides tests of physical laws at the extremes of their applications and links us to cosmological consideration of the universe on the largest scale.

23

Galaxies

Figure 23–1 Harlow Shapley.

On April 26, 1920, Harlow Shapley (Fig. 23–1) and Heber D. Curtis were brought together at the National Academy of Sciences in Washington to discuss the scale of our own galaxy and the nature of the "spiral nebulae," matters on which they had become known in the preceding years as the major protagonists. Shapley, then on the staff of the Mount Wilson Observatory, was soon to go to the Harvard College Observatory; Curtis was then on the staff of the Lick Observatory and was soon to go to the Allegheny Observatory in Pittsburgh and later on to the University of Michigan Observatory. The *Shapley-Curtis debate* is an interesting example of the scientific process at work, though it did not settle the question of the nature of the "spiral nebulae."

The arguments used by Shapley and Curtis were several in number, and involved many of the concepts that we have dealt with in earlier chapters. Shapley and Curtis mainly debated our own galaxy's size. We have seen in the Introduction to Part IV how Shapley's earlier research had led him to realize that the Milky Way was ten times larger than had been thought, though Curtis did not accept this result. The nature of the "spiral nebulae" was treated in the final paragraphs of the published versions of their statements. Curtis's conclusion that the "spiral nebulae" were external to our galaxy was based in large part on the notion that our galaxy is much smaller than we now know it to be. Curtis had several reasons for the view that the "spiral nebulae" were actually far-off galaxies. These included his analysis of what were then called "novae" that in 1917 were discovered to be going off from time to time in the "spiral nebula" in Andromeda. He reasoned that if the "spiral nebulae" were external to our galaxy, then the absolute magnitudes of the novae would be consistent with the absolute magnitude of novae in our own galaxy. Shapley, on the other hand, argued that the "spiral nebulae" were close by because proper motion for points, probably stars, in several of them had been detected by his colleague Adriaan van Maanen. This proper motion was presumably

An edge-on view of a galaxy of type Sb, NGC 4565, in the constellation Coma Berenices.

All novae in our galaxy reach the same maximum absolute magnitude.

caused by the rotation of these "nebulae." Also, Shapley had earlier noted that one of the "novae"—S Andromedae—that had erupted in the Andromeda "spiral nebula" in 1885, would have had to be much brighter than the other novae observed in that "nebula." This made Shapley feel that the evidence from "novae" was not internally consistent, that is, some pieces of evidence were inconsistent with other pieces.

In fact, both astronomers were using incomplete or fallacious evidence. Nobody at that time knew about interstellar absorption (page 443). This dimmed distant stars and thus made them seem farther away, which led to a general underestimate of the distance scale. Van Maanen's observations of proper motions in "spiral nebulae" were incorrect; later observations were to show that these objects do not show proper motions. Some of Shapley's feelings were based on incomplete knowledge, for S Andromedae was actually a supernova rather than a nova. (At that time one spoke of "Tycho's nova," which we now call Tycho's supernova.) The distinction between novae and supernovae was not realized until the work of the Swedish astronomer Knut Lundmark, published in 1920. Even Lundmark's research left unanswered questions, and Curtis did not agree that the "novae" fell into two such well-defined classes.

Even as late as 1929, Hubble wrote that the fact that the visual magnitude at maximum of S And was about 8.0 "places it at once in that mysterious class of exceptional novae which attain luminosities that are respectable fractions of the total luminosities of the systems in which they appear."

Curtis was correct that the "spiral nebulae" were comparable objects to our own galaxy, but for the wrong reasons. Shapley, on the other hand, came to the wrong conclusion but followed a proper line of argument that was unfortunately based on incorrect and inadequate data.

The question of the distance to the "spiral nebulae" was settled only in 1924, when Edwin Hubble (who was shown in Fig. 3–25), who had used the Mount Wilson telescopes to observe Cepheid variables in 3 of the "spiral nebulae," presented his definitive conclusion (along the lines we described in Section 4.4b) that the "spiral nebulae" were outside our own galaxy and of dimensions not overwhelmingly different from it, and even this distance scale has since been increased by the discovery of two types of Cepheids. Since Hubble's work there has been no doubt that the spiral forms we see in the sky are galaxies like our own. For the rest of the book we shall strictly use the term *spiral galaxies;* the incorrect, historical term, "spiral nebula," often hangs on in certain contexts, chiefly when we discuss the "Great Nebula in Andromeda," which is actually a spiral galaxy.

Figure 23–2 A galaxy of Hubble type E0, NGC 4486, in the constellation Virgo. Globular clusters can be seen in the outer regions. This is actually a peculiar elliptical galaxy, and is a strong source of radio radiation.

23.1 TYPES OF GALAXIES

Hubble went on to use the Mount Wilson telescopes to study the different types of galaxies. Actually, spiral galaxies are in the minority; there are many galaxies that have elliptical shapes and others that are irregular or abnormal in appearance. In 1925, Hubble set up a system of classification of galaxies that we shall discuss below; we normally describe a galaxy by its *Hubble type.*

23.1a Elliptical Galaxies

Most of the galaxies of which we know are elliptical in shape (Fig. 23–2). The largest of these *elliptical galaxies* may contain 10^{13} solar masses and may be 10^5 parsecs across (approximately the diameter of our own galaxy); these *giant ellipticals* are rare. Much more common are *dwarf*

Figure 23–3 M31, the Great Galaxy in Andromeda, a type Sb spiral, with its accompanying elliptical galaxies NGC 205 (*top left*) and M32 (*bottom right*).

ellipticals, which may contain "only" a few million solar masses and be only 2000 parsecs across.

Elliptical galaxies range from nearly circular in shape, which Hubble called *type E0,* to very elongated, which Hubble called *type E7,* with galaxies of various amounts of apparent oblateness (a measure of the difference between the longest and shortest diameters) in between. The spiral Andromeda Galaxy, M31, shown in Figure 23–3 and in Color Plate 46, is accompanied by two elliptical companions. It is obvious on the photographs that the companions are much smaller than M31 itself.

23.1b Spiral Galaxies

Although spiral galaxies, with arms unwinding smoothly from the central regions, are a minority of all the galaxies in the universe, they form a majority in certain particular groups of galaxies. Also, since they are brighter than the more abundant small ellipticals, we tend to see the spirals as dominant in a given volume of space, while really the fainter ellipticals may make up the majority.

Sometimes the arms are tightly wound around the nucleus; Hubble called this type *Sa*, the *S* standing for "spiral" (Fig. 23–4). Categories of spirals with arms less and less tightly wound (that is, looser and looser) are called *types Sb* (Color Plates 49 and 50) and *Sc* (Color Plate 47). The nuclear bulge as seen from edge-on is less and less prominent as we go from Sa to Sc galaxies, as can be seen from the difference between the figures opening this chapter and this part of the book. It has been found spectroscopically that galaxies rotate in the sense that the arms trail.

Spiral galaxies can be 10,000 to 30,000 parsecs across. They contain

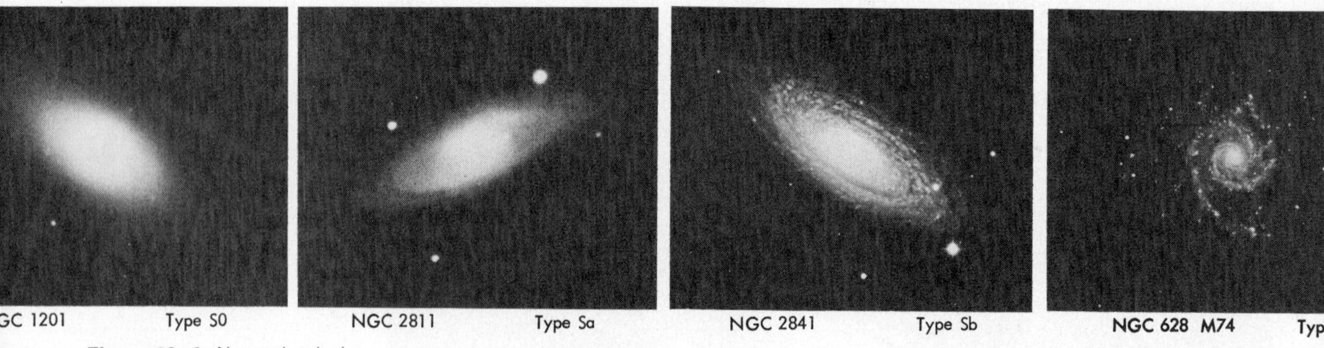

Figure 23–4 Normal spiral galaxies.

10^9 to over 10^{11} solar masses; since most stars are of less than 1 solar mass this means that spirals contain over 10^9 to over 10^{11} stars – we think our own galaxy has 10^{11} (100 billion).

In about one-third of the spirals, the arms unwind not from the nucleus but rather from a straight *bar* of stars, gas, and dust that extends to both sides of the nucleus (Fig. 23–5). These are again classified in the Hubble scheme from *a* to *c* in order of increasing openness of the arms, but with a *B* for "barred" inserted: *SBa, SBb*, and *SBc.* There is actually a complete range in the size of the bar from not visible to dominant in the appearance of a galaxy, so non-barred ("normal") spirals and barred spirals may not really be distinct types from each other.

23.1c Irregular Galaxies

A few per cent of the galaxies show no regularity, neither spiral nor elliptical. The Magellanic Clouds, for example, are irregular galaxies; they were shown in Figure 4–18 and enlargements appear in Figure 23–6. Sometimes traces of regularity – perhaps a bar – can be seen. Irregular galaxies are classified as *Irr.*

Irregular galaxies that can be resolved into nebulae, stars, and clusters are called *type Irr I.* Other galaxies appear amorphous and cannot be resolved into nebulae, stars, and clusters. They are called *type Irr II.*

23.1d Peculiar Galaxies

In some cases, as in M82 (Fig. 23–7), it appears at first as though an explosion has taken place in what might have been a regular galaxy. But the

Figure 23–5 Barred spiral galaxies.

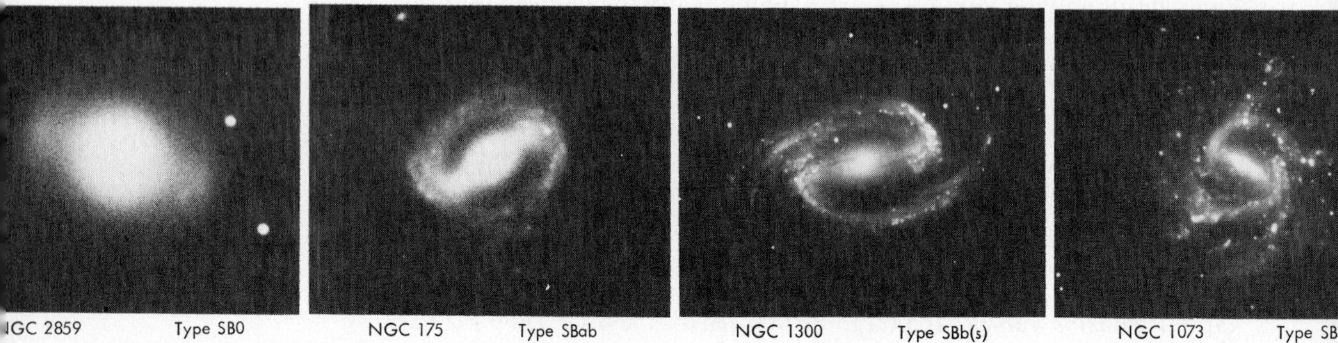

Figure 23–6 The Large Magellanic Cloud (*left*) and the Small Magellanic Cloud (*right*), photographed with the new British 1.2-m Schmidt telescope at Siding Spring, Australia, as part of the current project to extend the National Geographic Society–Palomar Observatory Sky Survey to the southern hemisphere. The Royal Observatory, Edinburgh, Scotland, operates the telescope as a national facility, and the exposed plates from Siding Spring are sent to Edinburgh for analysis.

form of the Hα radiation that has been detected from the filaments does not fit with the model of a hot explosive blast. We must rather be seeing light from the galaxy's nucleus scattered toward us by dust in the filaments. In this new model, the energy required to expel the dust from the nucleus is much less than the amount of energy that would have been necessary to cause the supposed explosion.

For other peculiar galaxies, such as those shown in Figures 23–8 and 23–9, we have little idea of what might have gone on. Theoretical studies may provide explanations for other examples (Fig. 23–10). Peculiar galaxies are classified as the corresponding Hubble type followed by (*pec*): for example, *Sa (pec)*.

Figure 23–7 M82, a most unusual galaxy that is a powerful source of radio radiation. It was once thought to be exploding, but now gentler processes are thought to cause its form and non-thermal radiation.

Figure 23–8

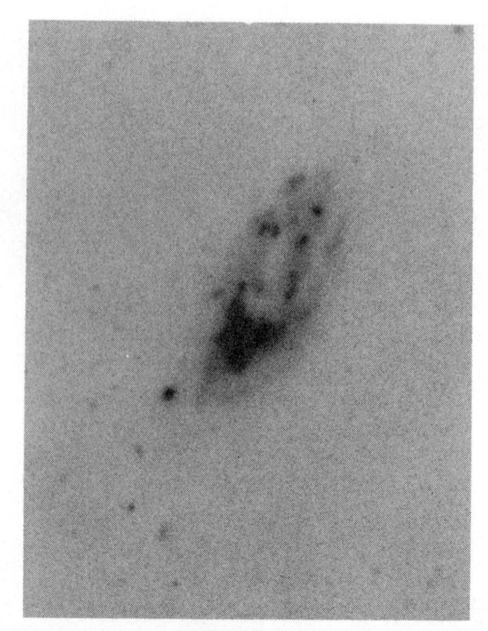

Figure 23–9

Figure 23–10

Figure 23–8 NGC 2685, a peculiar barred spiral nebula, type S0 (pec), in Ursa Major. It seems to be wrapped with helical filaments around a second axis of symmetry; the reasons for this are not understood.

Figure 23–9 A peculiar galaxy, IC 3862, photographed with the 5-m telescope by Halton Arp for his *Atlas of Peculiar Galaxies.* Stars or knots of gas are resolved. This is a negative print.

Figure 23–10 A ring galaxy; the ring has about the same diameter as does our own Milky Way Galaxy. The galaxy is called ESO-034 IG 11. This ring galaxy can be explained as the result of the collision of the large dot seen inside the ring with another galaxy, which became the ring.

23.1e The Hubble Classification

Hubble drew out his scheme of classification in a *tuning-fork diagram* (Fig. 23–11). The transition from ellipticals to spirals is represented by *type S0,* which resembles spirals in having the shape of a disk but does not have spiral arms.

It has since been shown, from optical observation and from studies of the 21-cm hydrogen line, that the amount of gas between the stars in galaxies depends on the type of galaxy. Elliptical galaxies have essentially no gas or dust between their stars, while spiral galaxies have interstellar

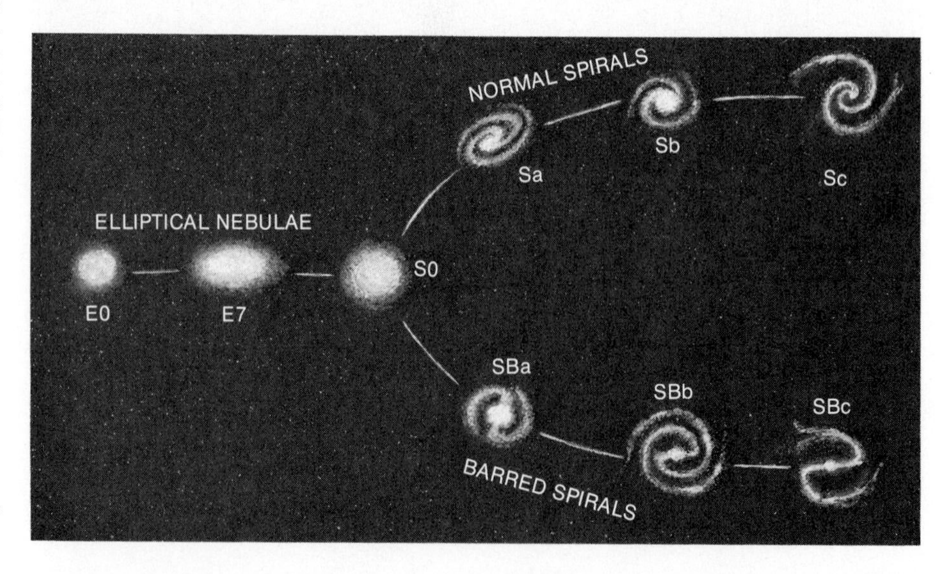

Figure 23–11 The Hubble classification of galaxies.

gas and dust. The relative amount of gas increases from types Sa (or SBa) to Sc (or SBc). Irregular galaxies usually have even denser interstellar media than do spirals.

At first it was thought that the arrangement of galaxies in the Hubble classification, and the differing amounts of gas in different types, might indicate that one type of galaxy evolves into another, but that is no longer thought to be the case. We believe that the differences in gas content result from differing conditions at the time of formation of the galaxies, and may be the result of processes that led some of the galaxies to become spiral and others elliptical.

There is in any case a correlation between the gas content (and thus the Hubble type) and the spectral types of stars that we see in different galaxies. Only in the galaxies with a substantial gas and dust content—mostly the Sc, SBc, and Irr galaxies—are the O and B stars to be found. Since these stars have short lifetimes on a stellar scale, they must have been formed comparatively recently, within the last several million years. Elliptical galaxies contain only older, cooler stars; this difference from spirals can be detected even with systems of filters, as the elliptical galaxies appear redder than spirals. Nor do we find H II regions, which surround hot stars, in elliptical galaxies.

Figure 23–12 The Seyfert galaxy NGC 4151, showing its bright nucleus.

23.1f Seyfert and N Galaxies

In the years since Hubble's work, other types of galaxies have been found that, although they can be fit into the Hubble classification scheme, have certain characteristics that make them suitable for special mention.

For example, in 1943, Carl Seyfert discovered a type of galaxy with a nucleus that is small and especially bright with respect to the arms (Fig. 23–12). *Seyfert galaxies* are defined both by this trait and by the appearance of broad emission lines in their spectra. The presence of emission lines indicates that hot gas is present in the nucleus, perhaps heated by violent activity there. The fact that the lines are exceptionally broad in wavelength is interpreted to mean that there are rapid motions of mass in the cores of Seyfert galaxies, causing Doppler shifts. The existence of x-radiation from NGC 4151 and strong infrared emission from many Seyfert nuclei is consistent with the idea that the nuclei are sites of activity.

The difference between Seyfert and N galaxies is partly based on their appearance and partly based on spectroscopic observations. The Seyferts are spiral galaxies with bright (although not extremely bright) nuclei, and emission-line spectra of hot gases. The N galaxies are probably ellipticals with very bright blue nuclei that can completely dominate the galaxies. Emission lines in N galaxies are generally broader than emission lines in Seyferts, although this spectroscopic distinction is not always clean cut.

Two to five per cent of galaxies are Seyferts. We do not know whether this means that all galaxies spend two to five per cent of their lives in this stage, or whether a small percentage of galaxies spend most of their lives in this stage.

Another type of galaxy with a bright nucleus is called an *N galaxy*. The intensity of optical radiation from the nucleus of an N galaxy overwhelms the radiation from the rest of the galaxy.

The Seyfert galaxies and the idea that M82 appeared to be exploding were the first indications of the existence of *active galaxies*, galaxies with internal activity on a powerful scale. Now the existence of radio, infrared, and x-ray emission from a variety of galaxies has changed our preconceived idea that galaxies were quiet places. We shall see in Section 24.8 that Seyfert galaxies and N galaxies are coming under increasing consideration because of their possible relation to quasars.

23.2 THE ORIGIN OF GALACTIC STRUCTURE

The overall shapes of galaxies, ranging from spherical (as in an E0 elliptical) to disk-shaped with only a slight central bulge (as in an Sc spiral) can be understood as a consequence of the conservation of angular momentum. (Angular momentum was discussed in Section 10.2; similarly, consideration of the conservation of angular momentum is important for understanding the formation of the solar system, as discussed in Section 10.4.) Presumably the elliptical galaxies were formed out of masses of gas that had only small overall angular momentum. For the spiral galaxies, on the other hand, the higher angular momenta provided a contribution from what we often call "centrifugal force" that tended to balance the inward force of gravity in one plane. Perpendicular to this plane, there was no such force acting to oppose gravity, and the gas collapsed into a disk.

23.2a Density Waves

We have discussed the density-wave explanation for spiral structure in Section 20.5c. We showed there that the pattern that we see as a spiral may be the result of increased density in a spiral form, resulting in increased compression of the gas and thus star formation. But the individual stars, once formed, usually orbit the galactic center at rates very different from that of the overall pattern.

23.2b Gravitational and Magnetic Effects

In the three hundred years since Newton presented the basic laws of gravity, many solutions have been found to equations that show the gravitational workings of a system containing one large mass near its center and another, lesser mass in orbit around the larger mass. To describe this system, we need only study the gravitational interaction of two bodies; this is thus called the *two-body problem.* In our solar system, we understand the orbit of a planet as basically a two-body problem involving the planet and the sun. The effects that the planets have on each other are much less than the effect that the sun has on each, and the mutual interactions of the planets are treated as small deviations (called *perturbations*) from the situation that would be present if only the sun-planet two-body problem had to be solved.

Even a situation that is as easy to state as a three-body problem, where the mutual gravitational interactions of three bodies must be studied, has no known general solution. Only certain limited cases of the three-body problem can be solved exactly.

One way of treating complicated gravitational problems is to use a com-

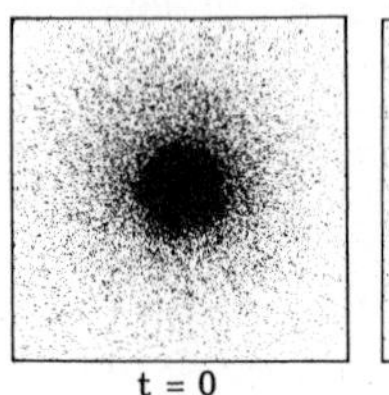
t = 0

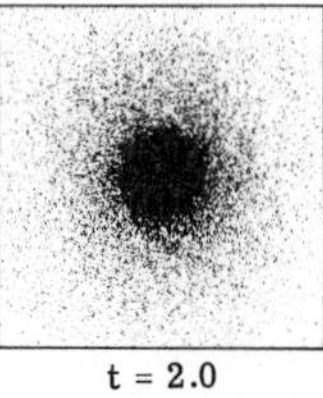
t = 2.0

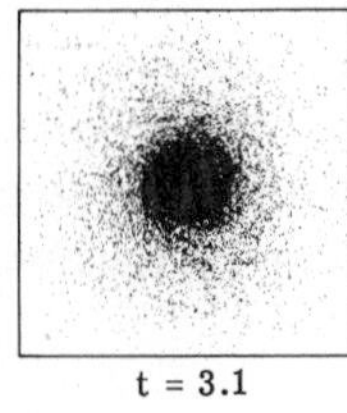
t = 3.1

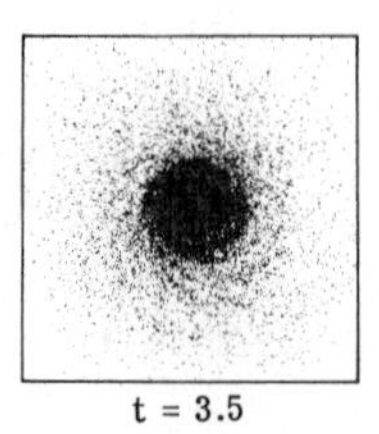
t = 3.5

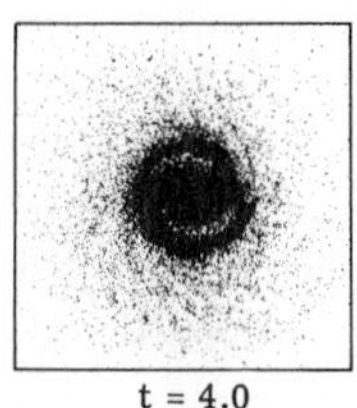
t = 4.0

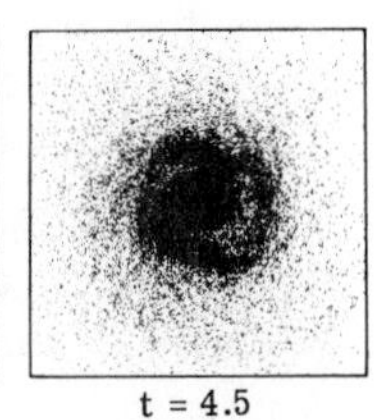
t = 4.5

puter to simulate the evolution of a system over time. One can consider a system of any number of masses at an instant in time, have the computer calculate all the independent gravitational interactions between each pair of masses, and use this information to see what the system would look like a short time later. In this way, one can generate a series of "snapshots" of the system, and thus follow its evolution (Fig. 23–13).

An alternate class of theories to explain the persistence of spiral structure involves the magnetic fields that are known to be present in galaxies. The field—which for our own galaxy is only one millionth the strength of the earth's magnetic field—is aligned along the spiral arms, as is known from observations of polarization of starlight. Magnetic and other theories cannot yet be ruled out as possible contributors to the spiral structure.

23.2c Interacting Galaxies

For a few of the peculiar galaxies, evidence exists that gravity is the dominant force in forming "arms" (often called "tails"). The situation depends on the interaction of two galaxies, and current calculations indicate that such interactions are too rare to have caused the many spirals that we see. Nonetheless, in particular cases, the effects of gravitational interactions can probably be dominant.

Two American astronomers, Juri Toomre and Alar Toomre, have used computer methods to follow the gravitational interaction between two galaxies that come very close together. In order to simplify the calculations, they made the assumption that the mass of each of the galaxies is concentrated at its center, and that the interactions of the individual pairs of particles they considered did not have to be taken into account. Figure 23–14 shows a time series they have calculated for the interaction of two massive galaxies. Long arms are drawn out by tidal forces; the original angular momentum of the galaxies contributes to the graceful curvature. At bottom we see the appearance of the galaxies at the time when they have moved sufficiently far away that they no longer affect each other to further change their form. The result of the computer simulation is very similar in appearance to that of a pair of galaxies known as "the Antennae," shown in Figure 23–15.

Figure 23–16 shows the result we saw in Figure 23–14 from three different points of view. We are fortunate in being able to see this beautiful pair of galaxies broadside, for it is the prettiest view.

The above example involved two massive galaxies. But the interaction of two galaxies of very unequal mass can lead to the more massive one developing forms, at least in its outer parts, that resemble spiral structure. Note that the Milky Way has close galactic companions—the Magellanic Clouds—so it is possible that there was a gravitational contribution to the origin of the spiral arms of our own galaxy.

Figure 23–13 A computer simulation by Frank Hohl of NASA's Langley Research Center in Virginia, showing the evolution of spiral structure.

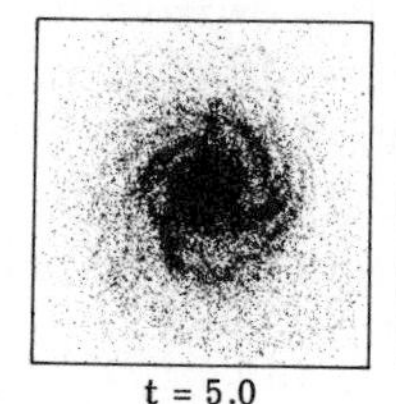
t = 5.0

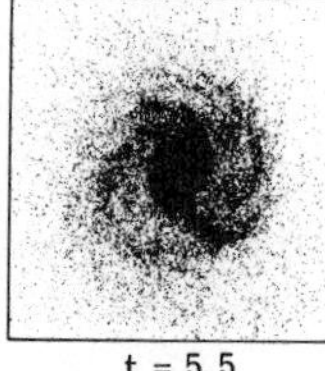
t = 5.5

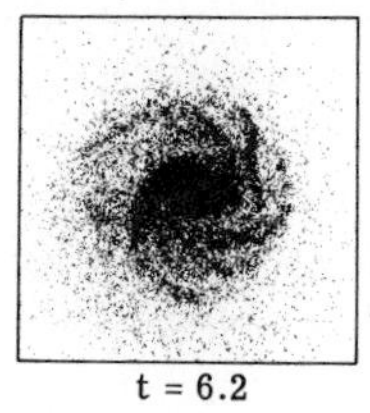
t = 6.2

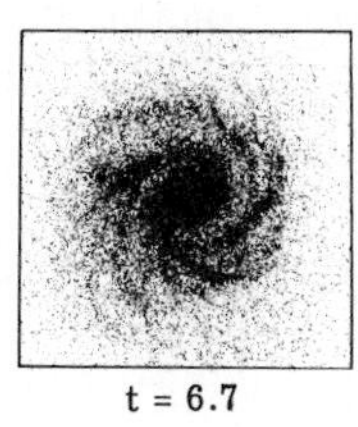
t = 6.7

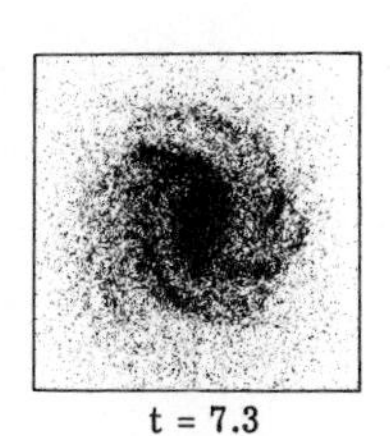
t = 7.3

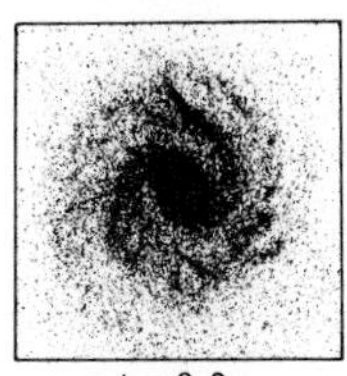
t = 8.0

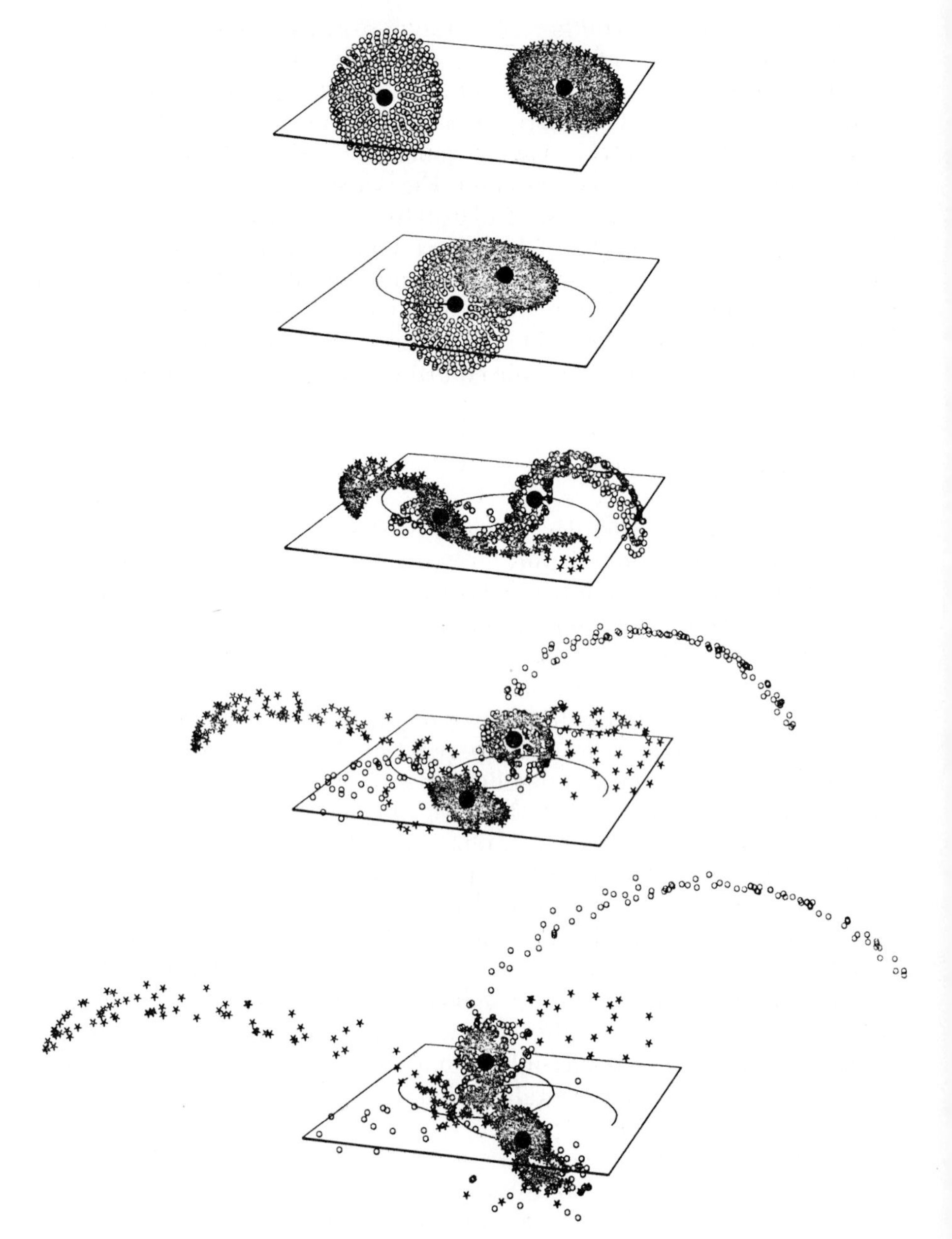

Figure 23–14 A sequence of drawings made on a computer by Alar and Juri Toomre, showing the very close encounter of two identical model galaxies. The large dot represents the center of each galaxy; the mass and central force of each galaxy is concentrated at its center for the purposes of this calculation. A disk of 350 particles is associated with each galaxy. The drawings are separated by an interval of 200 million years in the evolution of the galaxies.

23.3 CLUSTERS OF GALAXIES

Careful study of the positions of galaxies and their distances from us have revealed that most galaxies are part of clusters. These *clusters of galaxies* may have just a handful of members or may have hundreds of members; in any case, they are very large scale phenomena. (Obviously, they are on a completely different scale from that of the galactic clusters, a type of star cluster that we discussed in Sections 4.5 and 20.1.)

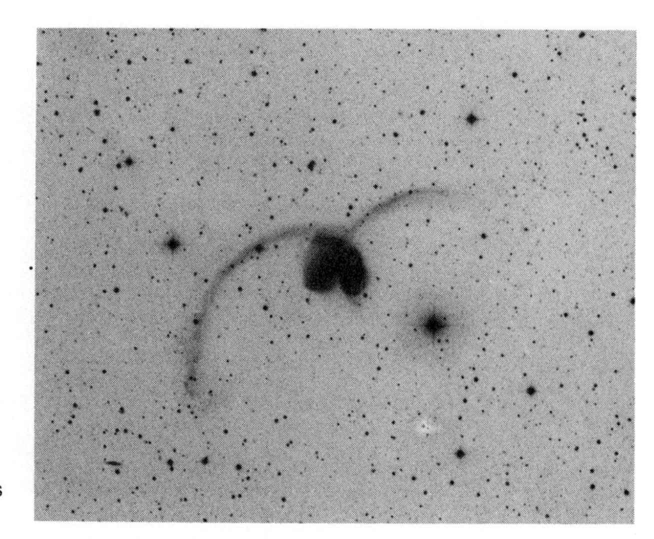

Figure 23–15 A negative print of the pair of galaxies NGC 4038 and NGC 4039, known as the Antennae.

23.3a The Local Group

The two dozen or so galaxies nearest us form the *Local Group*. The Local Group contains a typical distribution of types of galaxies and extends over a volume 1 megaparsec in diameter. It contains three spiral galaxies, each 15 to 50 kiloparsecs across—the Milky Way, Andromeda, and M33. There are four irregular galaxies, each 3 to 10 kiloparsecs across, including the Large and Small Magellanic Clouds. The rest of the galaxies are ellipticals, including 4 regular ellipticals, each 2 to 5 kiloparsecs across, two of which are the companions to the Andromeda Galaxy. The others are dwarf ellipticals, mostly less than 2 kiloparsecs across.

Previously unknown candidates for membership in the Local Group are occasionally found. Some of these newly discovered galaxies have been difficult to discover even though they are so close because they lie in the plane of our galaxy and are thus hidden from our view by dust. Others are simply not very prominent in the sky and were not noticed; three such diffuse dwarf ellipticals were discovered in 1972.

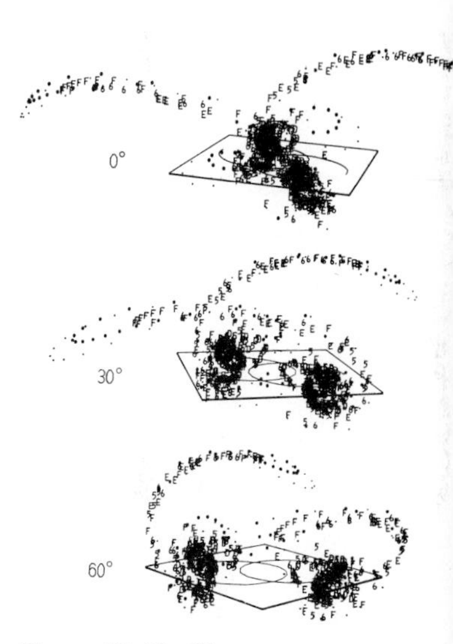

Figure 23–16 The numerical simulation shown in Fig. 23–15 closely resembles the Antennae when observed from a certain angle, as shown at top. From other angles, however, the resemblance is not obvious; this makes us realize that problems of the third-dimension and perspective limit our ability to recognize the true structure of galaxies.

In 1967 and 1968, an Italian astronomer, Paolo Maffei, examined photographs he had taken with a Schmidt camera and discovered two objects, now called Maffei I and II, which were brighter in the near infrared spectral region from 6800 Å to 8800 Å than they were at shorter wavelengths (Fig. 23–17). The detection of radio radiation from Maffei II, and optical and infrared observations made at the Leuschner Observatory at Berkeley, at Lick, and at Palomar have determined that these two objects are nearby galaxies. Maffei I is a giant elliptical, and is close enough that it would be visible to the naked eye if so much interstellar dust were not in the way. The dust provides a large amount of interstellar reddening, making the galaxy readily visible only in the infrared.

The early impressions were that Maffei I was only 1 megaparsec away from the Milky Way, not much farther away from us than is the Andromeda Galaxy. This would place it in the outskirts of the Local Group, though even if it were there, its velocity is sufficiently great that Maffei I would be only passing through the Local Group rather than being a permanent member. Radio observations at 21 cm indicate that Maffei II is rotating and so is a spiral. Assuming that its H II regions are normal in size leads to the conclusion that it is perhaps 5 megaparsecs away. This would indicate that Maffei

We discussed H II regions, regions of ionized hydrogen, in Section 20.2.

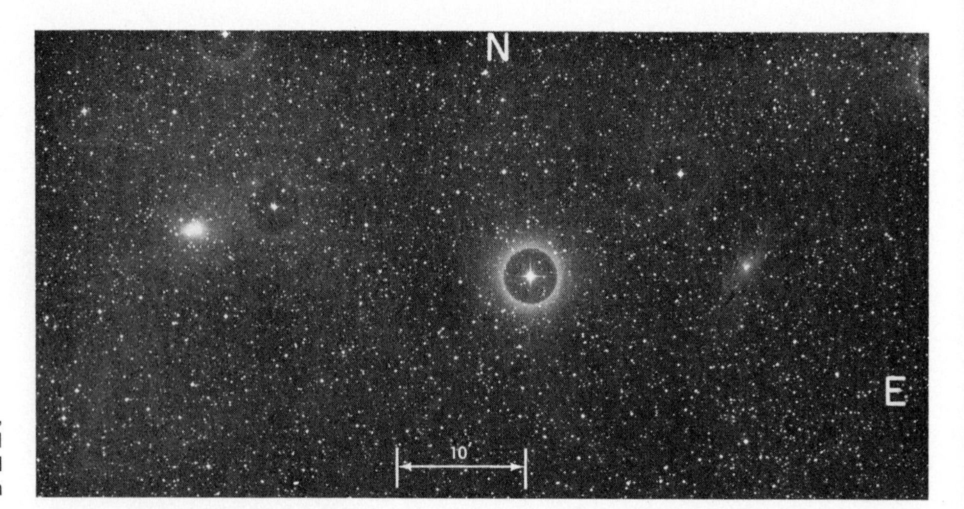

Figure 23–17 The fuzzy, non-stellar objects Maffei I and Maffei II, photographed in the infrared by Hyron Spinrad of Berkeley.

II would not be in the Local Group but rather in a nearby grouping of galaxies called the Ursa Major-Camelopardalis cloud. Since Maffei I has the same radial velocity as Maffei II, and the two galaxies are close together in the sky, it seems reasonable that they are located close to each other in space, which would also make Maffei I a member of this group of galaxies. Perhaps Maffei I and II dominate this small cluster of galaxies not far from our own.

The possible detection of another new member of the Local Group was reported in 1975 by S. Christian Simonson, then of the University of Maryland. Simonson interprets maps of the 21-cm structure in the direction of the constellation Gemini to indicate that a dwarf galaxy is passing nearby and is disrupting otherwise regular spiral structure there. No optical traces of the galaxy have been detected.

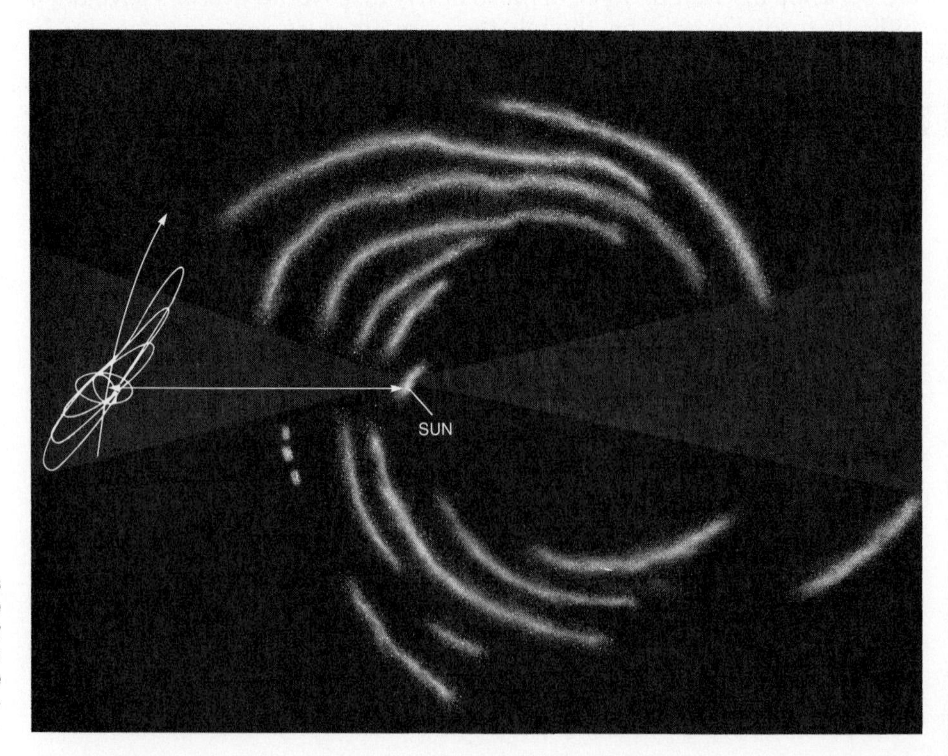

Figure 23–18 An artist's conception of the close companion of our galaxy that S. Christian Simonson reported to exist on the basis of 21-cm observations.

If this is a new galaxy, it would be only 20 kiloparsecs or so farther out from the galactic center than is the sun (Fig. 23–18). The new galaxy has been formally called Snickers, because its mass is only 10^8 $M_\odot$, peanuts compared with that of the Milky Way. It would be almost touching our galaxy, and some astronomers prefer to think that the 21-cm maps show a deformed part of the spiral structure of our galaxy rather than an independent galaxy.

23.3b Farther Clusters of Galaxies

There are apparently other small groups of galaxies in the vicinity of the Local Group, each containing only a dozen or so members. The nearest cluster of many galaxies (a rich cluster as opposed to a poor cluster) can be observed in the constellation Virgo (the Virgin) and is called the Virgo Cluster. It covers a region in the sky over 6° in diameter, 12 times greater than the angular diameter of the moon. The Virgo Cluster contains thousands of galaxies of all types. It is about 2 million parsecs across, and is located about 20 million parsecs away from us.

Other rich clusters are known at greater distances, including the Coma Cluster (Fig. 23–19) in the constellation Coma Berenices (Berenice's Hair).

Figure 23–19 A cluster of galaxies in the constellation Coma Berenices. A close look will reveal that many of the small dots are not round and sharp, and are thus galaxies rather than stars.

The Coma Cluster has spherical symmetry and a concentration of galaxies toward its center, not unlike the distribution of stars in a globular cluster. Thus it is a *regular cluster* as opposed to an *irregular cluster.*

X-rays have been discovered from the three rich clusters closest to us, the Virgo, Coma, and Perseus Clusters, so it seems reasonable that rich clusters of galaxies are generally x-ray sources. There are at least two possibilities for the origin of the observed x-rays. First, it could be the combination of emission from a few discrete x-ray sources.* Second, it could be radiation from a very hot (10^8 K) gas filling intergalactic space.

There may be as many as 10,000 galaxies in a cluster, and the density of galaxies near the center of a rich cluster may be higher than that near the Milky Way by a factor of one thousand to one million. Thousands of clusters of galaxies are known.

One interesting question to ask is whether these clusters of galaxies are in turn grouped into clusters of clusters—*superclusters.* Current evidence is that this is indeed the case. The Local Group, the several similar groupings nearby, and the Virgo Cluster form one such cluster of clusters, the *Local Supercluster.* This cluster of clusters contains 100 member clusters and is on the order of 100 megaparsecs across.

Does the clustering continue in scope? Are there clusters of clusters of clusters, and clusters of clusters of clusters of clusters, and so on in a hierarchical sequence? We do not know, and we are not able to investigate on this scale. In Section 25.5 we shall see that the answer to this question can have a profound impact on our cosmological ideas, and that we assume for the present that clusters of clusters, such as the Local Supercluster, are the largest scale of inhomogeneity in the universe.

23.3c The Missing Mass Problem

The missing mass problem *is the discrepancy found when the mass derived from consideration of the motions of galaxies in clusters is compared with the mass that we can observe.*

Clusters of galaxies appear to be stable configurations that have lasted for many years. But since galaxies have random velocities in various directions with respect to the center of mass of the cluster, why don't the galaxies escape? If we assume that the clusters of galaxies are actually stable, we can calculate the amount of gravity that must be present to keep the galaxies bound. Knowing the amount of gravity in turn allows us to calculate the mass. When this supposedly simple calculation is made for the Virgo Cluster, it turns out there should be fifty times more mass present than is observed. 98 per cent of the mass expected is not found. This is called the *missing mass problem.*

We know the missing mass is not simply in the form of neutral hydrogen, for if it were we would detect it through its 21-cm radiation. Now that we can survey the distribution of molecular hydrogen in our own galaxy by ultraviolet observations with the Copernicus satellite, we know that there is not enough H_2 in our own galaxy to provide the deficient amount of mass, and there is presumably not enough H_2 in the Virgo Cluster either. One leading thought is that the missing mass could be present in black holes. Another idea is that even if there were a very dense intergalactic medium, it would not radiate 21-cm emission if it were hot enough and thus almost

*Each source radiates by a mechanism technically known as "inverse Compton scattering," which can add energy to low-energy photons, converting them to x-rays. It does this through the interaction of the photons with electrons that are moving at relativistic velocities. This process appears to be very important to the understanding of x-ray astronomy.

entirely ionized. A hot gas, however, would radiate x-rays, so through x-ray observations of clusters of galaxies, we set limits on the amount of hot intergalactic gas that can be present. Thus we must continue to try to identify the sources of extragalactic x-radiation, both the component that comes from clusters of galaxies and also a diffuse component that comes isotropically (from all directions).

For any given cluster of galaxies, we can find an ad hoc way out of the missing mass problem. For example, we can assume that the Virgo Cluster is exploding rather than being stable, which would mean that we could no longer assume that the outward forces are in balance against gravity. But similar calculations have also been carried out for other clusters. For the Hercules Cluster (shown in the photograph opening this chapter), 80 per cent of the mass seems to be "missing." The missing mass problem seems to be widespread. Still it may simply be the case that when clusters of galaxies form, their members generally have higher velocities than they will have when the galaxy settles down (we say, technically, *relaxes*) to its final state. Thus the missing mass problem does not occur if clusters of galaxies are not old enough to have "relaxed."

In Section 25.7, we shall see how another method of measuring the density that also does not depend on whether mass is visible or invisible has been brought to bear on calculating the density of the universe. The results of this other, very indirect, method do not seem to show the discrepancy that we call the missing mass problem.

23.4 THE EXPANSION OF THE UNIVERSE

In the decade before the problem of the location of the "spiral nebulae" was settled, Vesto M. Slipher of the Lowell Observatory took many spectra that indicated that the spirals had large redshifts. This work was to lead to a profound generalization. In 1929 Hubble announced that galaxies in all directions are moving away from us and that there is a direct proportionality between the distance of a galaxy and its redshift. Hubble, in collaboration at Mount Wilson with Milton L. Humason, went on during the 1930's to establish the relation more fully. It is known as *Hubble's law*. The redshift is presumably caused by the Doppler effect (which was described in Section 2.15), and the law is usually stated in terms of the velocity that corresponds to the measured wavelength rather than in terms of the redshift itself.

Hubble's law states that the velocity of recession of a galaxy is proportional to its distance and is written

$$v = H_0 d,$$

where v is the velocity, d is the distance, and H_0 is the present-day value of the constant of proportionality, which is known as *Hubble's constant* (Figs. 23–20 and 23–21).

The best current value of Hubble's constant, measured by Allan Sandage and Gustav Tammann at the Hale Observatories, is 55 km/sec/Mpc. This is almost a factor of 10 lower than the value that Hubble originally announced, but Sandage and Tammann have been able to make use of new developments in finding the distance to distant galaxies to derive the improved value, incorporating as well earlier corrections to the distance scale.

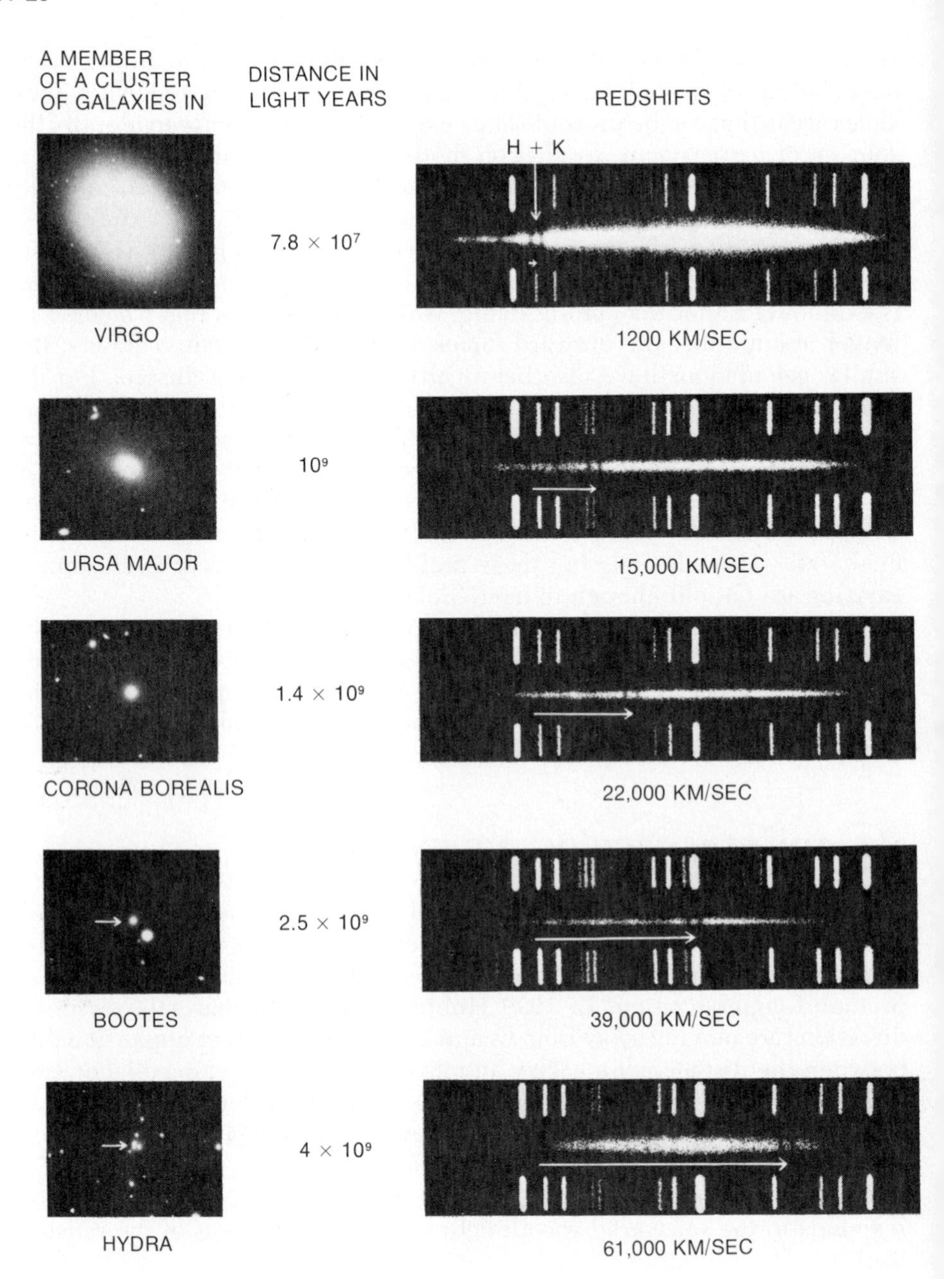

Figure 23–20 Spectra are shown at right for the galaxies at left.

Hubble's constant is given in units that appear strange, but it merely states that for each megaparsec (3.3×10^6 light years) of distance from the sun, the velocity increases by 55 km/sec (thus the units are 55 km/sec per Mpc). From Hubble's law, we see that a galaxy at 10 Mpc would have a redshift corresponding to 550 km/sec; at 20 Mpc the redshift of a galaxy would correspond to 1100 km/sec; and so on.

We use a subscript "o" on H_0 in order to retain the letter H for describing how Hubble's constant might vary over time. Obviously, if Hubble's constant varies over time, it isn't really a constant.

The major philosophical import of Hubble's law is that objects in the universe in all directions are moving away from us; the universe is expanding. Since the time when Copernicus moved the earth out of the center of the universe (and the time when Shapley moved the earth and sun out of even the center of the Milky Way Galaxy), we have not liked to think that we could be at the center of the universe. Fortunately, Hubble's law can be accounted for without our having to be at any such favored location, as we see below.

Imagine a raisin cake (Fig. 23–22) about to go into the oven. The raisins are spaced a certain distance away from each other. Then, as the cake rises, the raisins spread apart from each other. If we were able to sit on one of those raisins, we would see our neighboring raisins move away from us at a certain speed. It is important to realize that raisins farther away from us would be moving away faster: not only would the distance from us to the neighboring raisin have increased but also the additional distance beyond the neighbor to the farther raisin would have increased. Thus no matter in what direction we looked, the raisins would be receding from us, with the velocity of recession proportional to the distance.

The next important point to realize is that it doesn't matter which raisin we sit on; all the other raisins would always seem to be receding. Of course, the raisin cake is finite in size and the universe may have no limit, but other than that the analogy is exact. The fact that all the galaxies appear to be receding from us does not put us in a unique spot in the universe; there is no center to the universe. Each observer at each location would observe the same picture.

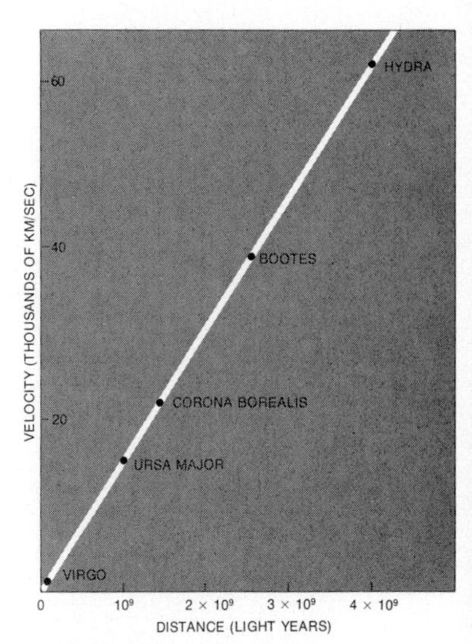

Figure 23–21 The Hubble diagram for the galaxies shown in Fig. 23–20.

Note also that the raisin cake—and the universe—must be expanding with respect to something. We are assuming that we are still able to measure lengths on an unchanging scale. In our analogy, the size of the raisins themselves is not changing; only the separations are changing. In the universe, the galaxies themselves and the clusters of galaxies are not expanding; only the distances between the clusters are increasing.

Note also that individual stars in our galaxy can appear to have small redshifts or blueshifts, caused either by their peculiar velocities or by the differential galactic rotation. Also, even some of the nearer galaxies have random velocities of sufficient size, or velocities less than our rotational velocity in our galaxy, so that they are approaching us. But except for these few nearby cases, all the galaxies are receding.

The major problem for setting the Hubble law on the firmest footing is finding the distances to the galaxies for which redshifts are measured. Only for the nearest galaxies can we detect Cepheid or RR Lyrae variable stars. Beyond those galaxies we use such methods as assuming that the magnitudes of supergiant stars or sizes of H II regions are more or less the same as they are in our own galaxy, and calculating, respectively, spectroscopic

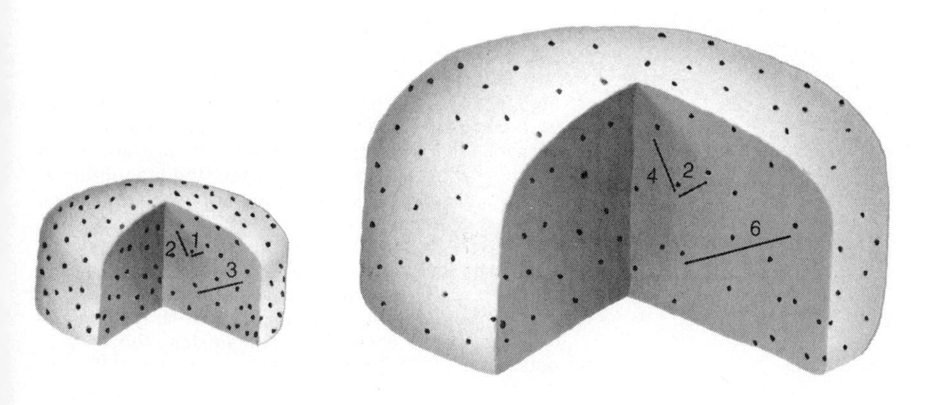

Figure 23–22 From every raisin in a raisin cake, every other raisin seems to be moving away from you at a speed that depends on its distance from you. This leads to a relation like the Hubble law between the velocity and the distance. Note also that each raisin would be at the center of the expansion measured from its own position, yet the cake is expanding uniformly. For a better analogy with the universe, consider an infinite cake; clearly, there is no center to its expansion.

parallaxes or the distances that would make the observed angular dimensions of the H II regions correspond with the linear dimensions we know for those objects in the Milky Way. At still greater distances, we assume that the brightest member in a cluster of galaxies has the same absolute magnitude as all other brightest members of other clusters; sometimes to limit the difficulty that one or two members might be exceptionally bright we rather consider the third-brightest member of a cluster. (The distribution of magnitudes of galaxies in clusters has also been considered, though this is much more difficult to do.) Certainly the methods we have to use grow less and less precise as we get farther and farther away from the sun. In particular, we are seeing galaxies that emitted their light so long ago that they then were different magnitudes than similar galaxies in our neighborhood are today. Also, we are observing radiation that was emitted in the ultraviolet, even though it is now redshifted into the visible, and we do not yet know much about ultraviolet spectra of galaxies. We may thus be making a systematic error in assessing their distances. In any case, all the current evidence indicates that when we correct for such effects as best we can the Hubble relation continues to be a straight-line proportionality deep into space (Fig. 23–23).

Beyond a certain range, we can no longer independently measure distances, and our only method of assessing distance is application of the Hubble law to the observed redshifts. The galaxy thus far detected with the greatest redshift, which is presumably also thus the most distant galaxy, is called 3C 123 (Fig. 23–24). It appears to be an elliptical galaxy in a compact cluster of galaxies, and is a strong source of radio waves. It has a redshift of 0.637, that is, the three spectral lines that are observed are shifted to the red by 63.7 per cent of their original values when they were emitted. (The H and K lines of ionized calcium, for example, appear in the red at 6496 Å and 6439 Å, respectively, instead of in the ultraviolet at 3968 Å and 3933 Å, their rest wavelengths. The third line, from ionized oxygen, is similarly redshifted.) The extent of the redshift indicates that the object is 8 billion light years away. 3C 123 is so faint, only magnitude 21.7, that only use of the most modern electronic devices allowed its spectrum to be obtained. Since the look-back time for this galaxy is 8 billion years, further studies might tell us how the overall properties of galaxies have evolved over time.

We discuss the redshift formula and its applications in Section 24.2, since more quasars than galaxies have such high redshifts. The redshift of 3C 123 is so great, 64 per cent of the speed of light, that the formula used for the Doppler shift must be the version that takes the special theory of relativity into account.

The time that it took the light to reach us is called the look-back time.

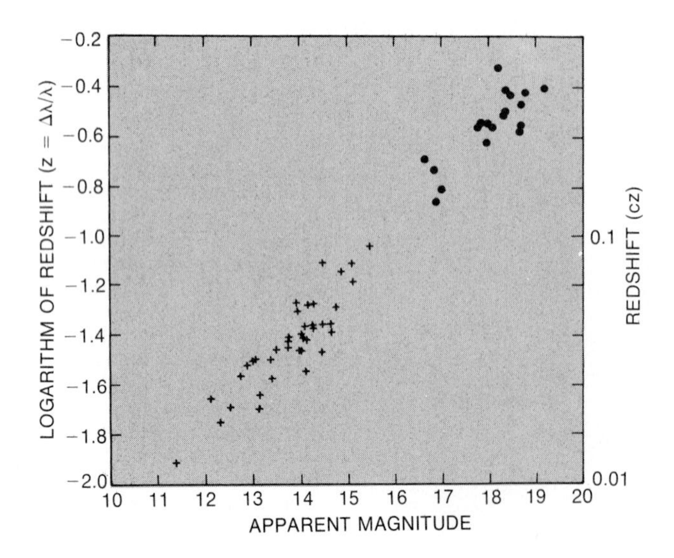

Figure 23–23 A recent Hubble diagram, put together by James Gunn and Beverley Oke at the Hale Observatories. Instead of distance and redshift, magnitude and redshift are graphed. (In this case, the magnitude at 5560 Å is used.) Thus the assumption is made that the galaxies graphed have essentially the same absolute magnitude, for only then can the inverse square law be applied to link magnitude and distance. Conventional photographic techniques gave the crosses at the left side of the diagram. New electronic techniques were used to measure the faintest galaxies, shown with dots.

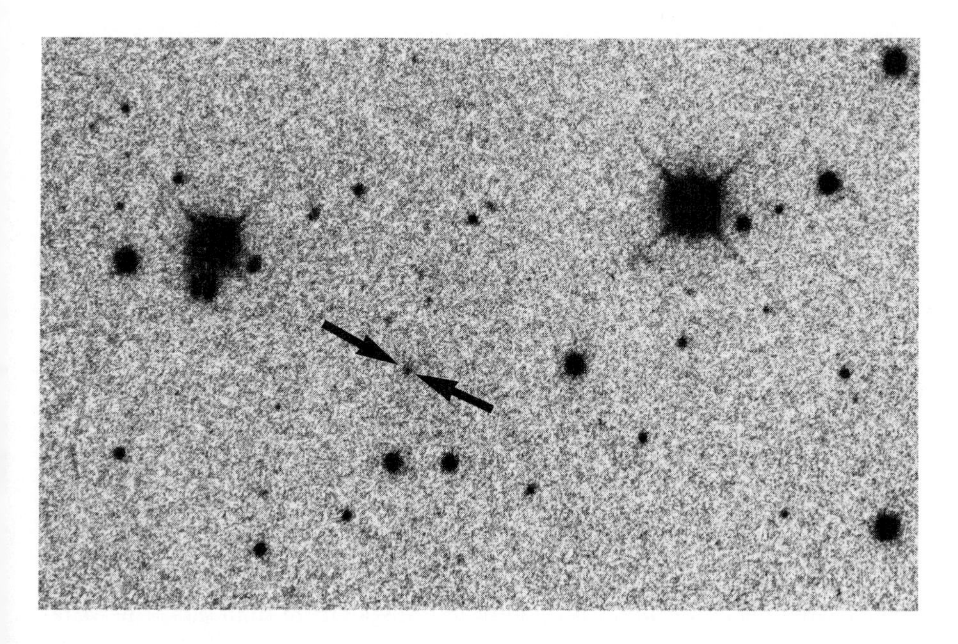

Figure 23–24 The farthest known galaxy, 3C 123, a galaxy fainter than 21st magnitude with a redshift of 0.637, can barely be observed. The photograph was taken by Hyron Spinrad with the 4-m Mayall reflector at the Kitt Peak National Observatory. The print is a negative.

23.5 RADIO GALAXIES

In Sections 20.2 and 20.3, we saw how a radio eye has been opened on the universe, and how observations made in the radio part of the spectrum have helped us understand the structure of our own Milky Way and of objects such as supernova remnants in it. However, most of the objects that we detect in the radio sky (that is, locations from which more than the background level of radio radiation can be detected) turn out not to be located in our galaxy. The study of these *extragalactic radio sources* is the subject of this section.

23.5a Low-Resolution Observations of Radio Galaxies

The core of our galaxy, the radio source we call Sagittarius A, is one of the strongest radio sources that we can observe in our galaxy. But if the Milky Way Galaxy were at the distance of other galaxies, its radio emission would be very weak.

We are able to detect a small amount of radio emission from many spiral galaxies. Among ellipticals, some seem to have quite strong radio emission while others at equal distances seem radio quiet. The radio radiation is presumably created by such processes as synchrotron radiation (which was described in Section 21.3), but the strength of the radiation is not exceptional and we shall not further deal with such "normal" radio galaxies.

Since the earliest work on radio astronomy, it has been clear that there is a class of galaxies whose members emit quite a lot of radio radiation, many orders of magnitude (that is, many powers of ten) more than "normal" radio galaxies. These have often been called "peculiar" radio galaxies; we shall limit the use of the term *radio galaxy* to mean these relatively powerful radio sources. They often appear optically as peculiar giant elliptical galaxies.

> **Box 23.1 X-ray Galaxies and Extragalactic Infrared Sources**
>
> The strength of the radio emission from a radio galaxy is non-thermal, that is, it does not follow Planck's law. Radio galaxies are usually also strong non-thermal emitters of x-rays and of infrared radiation. By analogy with the term "radio galaxy," we speak of an *x-ray galaxy* for such a strong x-ray emitter. Similarly to the way in which the brightest radio sources were assigned letters on discovery, the brightest x-ray sources were assigned letters. Thus the strongest x-ray source in the constellation Cygnus, the Swan, for example, is called Cygnus X-1.

The first radio galaxy to be detected, Cygnus A, radiates about a million times more energy in the radio region of the spectrum than does the Milky Way Galaxy. The most recent map is shown in Figure 23–25, with an optical image superimposed. As has been known for decades, Cygnus A, and dozens of other radio galaxies, emit radio radiation mostly from two zones, called *lobes,* located far to either side of the optical object that is visible. The optical object—two fuzzy blobs—has been the subject of much analysis, but its makeup is still not understood. At first, it was thought to be two galaxies in collision, but that idea was discarded when it was realized that there would not be enough collisions of galaxies to explain the many similar radio sources that had by then been discovered. Perhaps Cygnus A is a single galactic nucleus in the process of splitting, or perhaps opaque gas is blocking our view of part of a single object.

Such *double-lobed structure* is typical of many radio galaxies. Some have a third, central location of radio emission as well. Others did not seem to have such emission from the center as well as emission from the two lobes. But recent work at higher frequencies than had been customarily

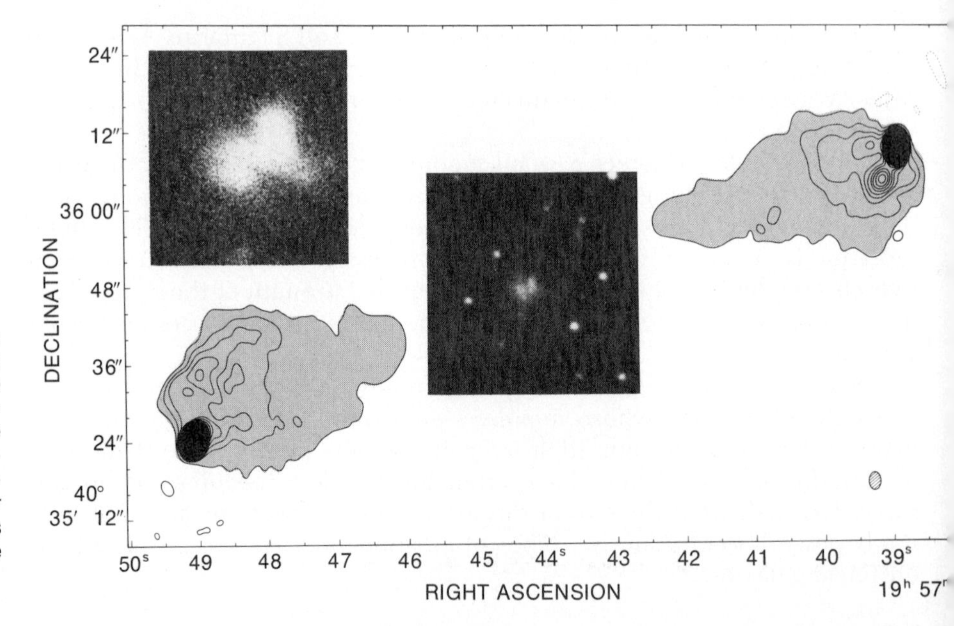

Figure 23–25 A radio map of Cygnus A, with shading and contours indicating the intensity of the radio emission. A photograph of the faint optical object or objects observable is superimposed at the proper scale. The beam size is shown by the ellipse at the lower right.

used for observing indicates that galactic nuclei probably always or almost always join the lobes in being radio sources, though the amount of emission from the nuclei may be relatively faint at the lower frequencies at which most observations have been made. It has long been thought that perhaps the gas that is emitting radiation in the lobes was ejected from the nucleus, and perhaps the relative strengths of the nucleus and the lobes represents evolution of the radio source.

Still other radio galaxies seem to emit radiation only from a single central volume, although that volume is usually greater in extent than the size of the optical image.

Often there are peculiarities in the optical images that correspond to radio sources of all kinds. For example, a small jet of gas can be seen on short exposures of M87, which corresponds to the powerful radio galaxy

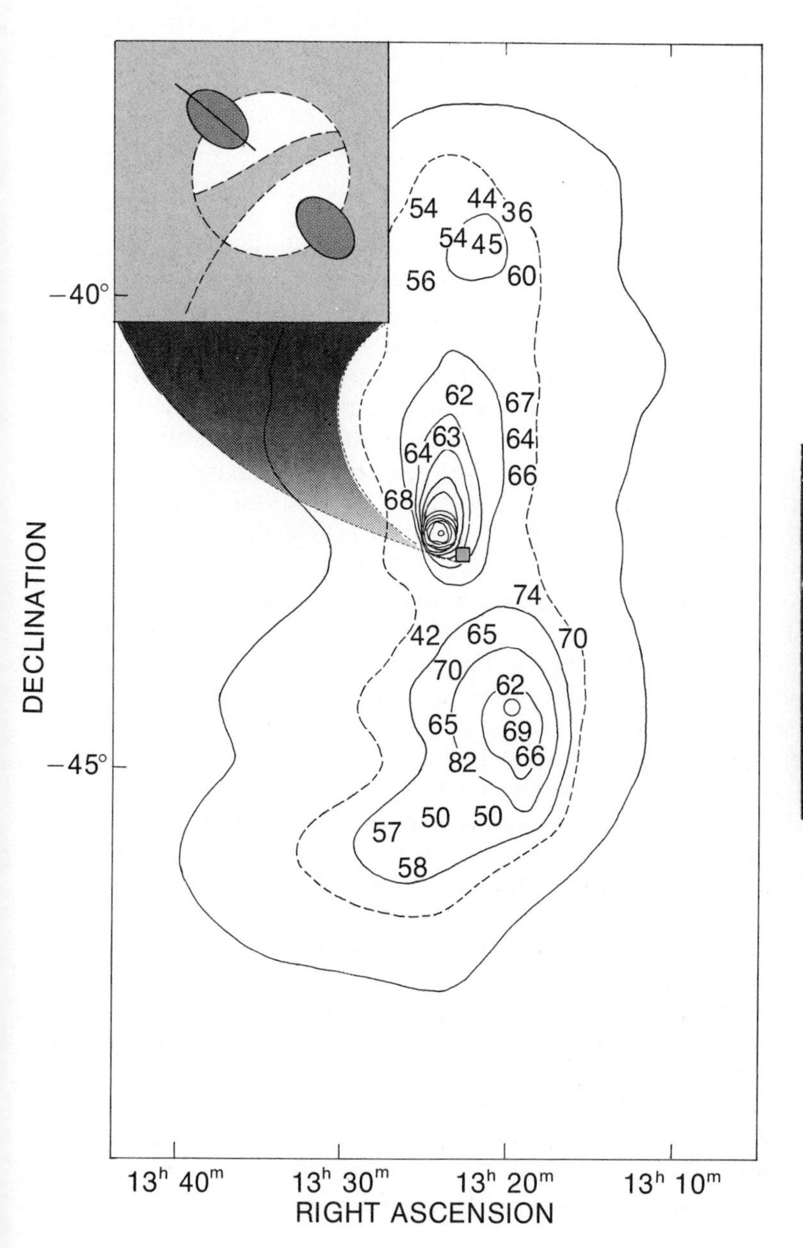

Figure 23–26 A radio map (*left*) of Centaurus A, with shading and contours indicating the intensity of the radio emission. In addition to the two large lobes, two smaller regions of radio emission are superimposed on the optical galaxy, NGC 5128, which is shown at right. The optical image is very peculiar, and resembles an elliptical galaxy with thick dust lanes wrapped around it.

1 ARC SEC

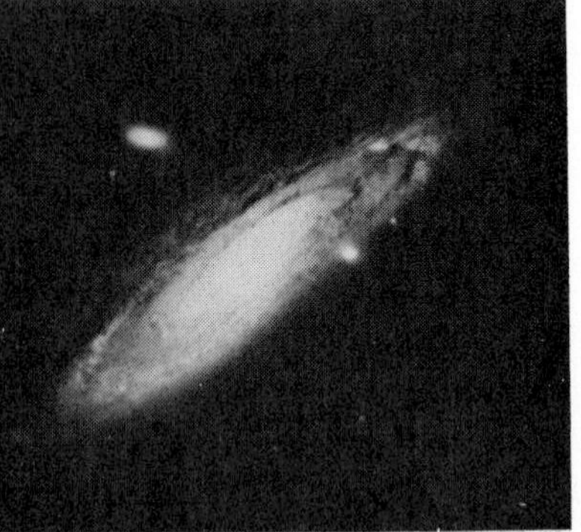

1 ARC MIN

10 ARC MIN

30 ARC MIN

Figure 23–27 The Andromeda Galaxy as it would appear at different resolutions.

Virgo A. Light from the jet is polarized, which confirms that the synchrotron process is at work here. M82, the galaxy that has the appearance of being in a state of explosion (Fig. 23–7), is also a radio galaxy.

Another strange optical image is shown in Figure 23–26 and Color Plate 48. This is the radio source Centaurus A, whose optical notation is NGC 5128. In this case, there are two different pairs of radio sources, a pair of lobes located outside the optical image and on each side of it, covering 5° to 8° of sky as seen from the earth, and a second pair of radio emitters located symmetrically to either side of the central dark band but close enough in to fall on the optical image, covering 7 arc min of sky. A weak radio source has been discovered at the center of this galaxy. The x-ray emission that has been detected varies over time by a factor of at least 4, and appears to be originating in this central "point source."

Observations of this galaxy with the new 4-m reflector at Cerro Tololo in the southern hemisphere have revealed the presence of a faint jet of gas extending over 40,000 parsecs from the center of the galaxy. This endorses the picture that an explosion took place, both expelling the jet and providing energy for the radio radiation.

23.5b Radio Interferometers

We have discussed in Section 21.2 that the resolution of single radio telescopes is very low, because of the long wavelength of radio radiation. Single radio telescopes may only be able to resolve structure a few minutes of arc or even a degree or so across, depending both on the wavelength of observation and on the size of the dish. For the past few years, the techniques of interferometry have been applied to radio astronomy, with the result that arrays of radio telescopes now in existence can map the sky with resolutions almost as high as the 1 arc sec or so that we can get with optical telescopes (Fig. 23–27). In the next sections, we describe first how these radio interferometers work and then discuss some of the high resolution results that have been obtained.

The wave structure of electromagnetic radiation manifests itself in the existence of *interference* phenomena, in which two beams of radiation—both of which are non-zero in strength (that is, have some intensity)—can be superimposed on each other to yield alternating bands, called *fringes,* of zero and non-zero radiation. For visible light, these fringes correspond to light and dark bands. The existence of interference cannot be understood when light is treated as a collection of particles (the photons), but can easily be understood on the basis of the wave theory (Fig. 23–28). (According to the laws of quantum mechanics, electromagnetic radiation has a dual nature in that it sometimes has properties of waves and sometimes has properties of photons.)

When we have a telescope—say the single dish of a radio telescope or the single mirror of an optical telescope—it receives radiation from a point in space as wavefront after wavefront of parallel radiation. The radiation is *coherent,* which means that all the waves are in step, with peaks arriving together, then troughs, then peaks, and so on.

The resolution of a single dish radio telescope depends on the diameter of the telescope (let us consider, for the moment, radiation of one frequency so that we do not have to discuss the variation of resolution with frequency). If, as we saw in Figure 4–11, we could somehow retain only the outer zone

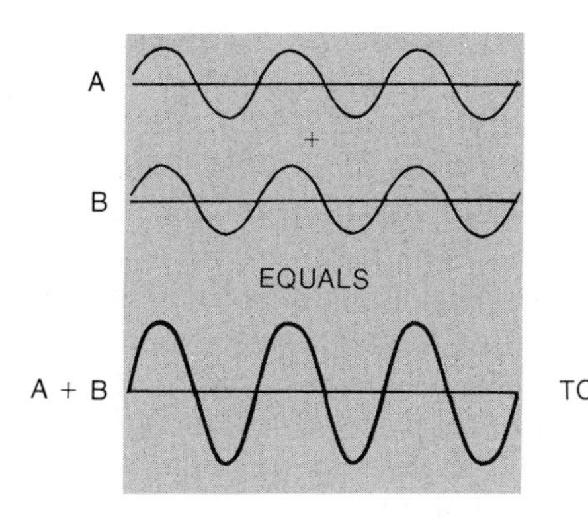

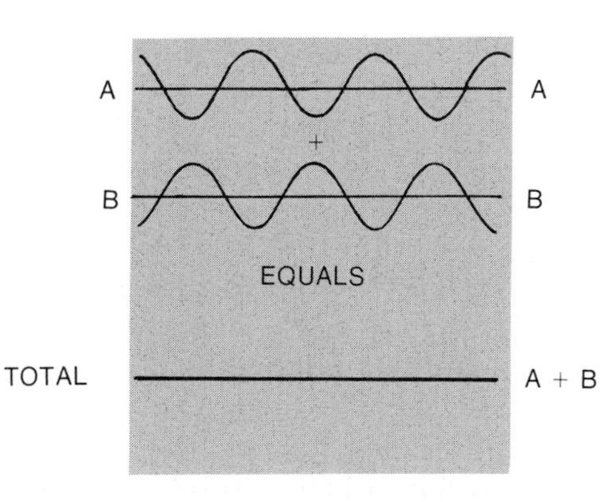

Figure 23–28 If electromagnetic waves—light or radio waves, for example—are *in phase* (i.e., oscillating in step), as at left, then the waves add in strength. If the waves are out of phase, as at right, then they can *interfere* with each other and cancel out. The phenomenon of interference cannot be understood if we think only in terms of particles (namely, photons).

of the dish, the resolution would remain the same (though the collecting area would be decreased, so we would have to collect the signal for a longer time to get the same intensity).

If we can maintain the coherence of the incoming radiation at two different dishes–that is, if we can maintain our knowledge of the relative arrival times of the wavefront at each of two dishes–we can retain the same resolution as though we had one whole dish of the spacing of the two small dishes, as shown in Figure 4–11. Whereas for a single dish, we would call this maximum spacing of the two most distant points from which we can detect radiation by the name "diameter," for a two-dish interferometer, we call it the *baseline.*

The main problem in building an interferometer is to maintain knowledge of the coherence of radiation from one dish to the other, that is, the distance that a given wavefront has to travel after it hits one dish until it hits the other (Fig. 23–29). Since wavefronts always travel at the same

Figure 23–29 In the top half of the figure, a given wave peak reaches both dishes simultaneously, so the amplitudes of the waves add. In the bottom half, the wave peak reaches one dish while a minimum of the wave reaches the other; the amplitudes subtract and zero total intensity results. Thus the interference phenomenon is set up, as long as we maintain a constant difference in the travel time of the electrical signals from the dishes where they are received to the control room where they are put together.

speed—the speed of light—this is equivalent to measuring the time of arrival of the waves of radiation to a high degree of accuracy.

Since the delay in arrival time between the waves depends on the angular position of an object in the sky with respect to the baseline, by studying the time delay one can figure out angular information about the object.

The first interferometers involved dishes that were linked with wires in order to make sure that the length of time remains constant between the incidence of the wavefront on the first dish to its incidence on the second. Thus the signals passed through lengths of wires that did not change, and in the control room where the signals were sent, the signals from the two dishes could be allowed to interfere with each other.

These interferometers, like the one at the National Radio Astronomy Observatory (which was shown in Figure 21–26A), enabled maps of radio sources to be made at much higher resolutions than had previously been possible.

In principle, however, if the dishes that make up an interferometer could be placed still farther away from each other, the resolution could be even better. A radio link between the dishes enabled an expansion of the baseline, but irregularities in the distance that the radio waves travel through the air limit this method to use within about 100 kilometers.

The breakthrough came with the invention of atomic clocks, which can keep time to an accuracy of one part in 10^{15} for the latest hydrogen masers (which is equivalent to a drift of three hundred-billionths of a second in a year). A time signal from an atomic clock can be recorded on a tape recorder on one tape band while the radio signal is put on an adjacent tape band. This signal can be compared at any later time with the signal from the other dish, synchronized accurately through comparison of the clock signals. The ability to have the time recorded so accurately freed radio astronomers of the need to have the dishes in direct contact with each other during the period of observation. Now all that is necessary is that the two telescopes observe the same object at the same period of time; the comparison of the signals can take place in a computer weeks later. With this ability, astronomers can make up an interferometer of two or more dishes very far apart, even thousands of kilometers. This technique is called *very-long-baseline interferometry (VLB* or *VLBI).*

Figure 23–30 VLB techniques with a baseline that is the diameter of the earth allow radio sources to be studied with extremely high resolution.

The maximum baseline that can be used in VLB research is approximately that of the diameter of the earth (Fig. 23–30), 12,700 km. This corresponds at wavelengths of a few centimeters to resolutions as small as 0.0001 arc sec, far better than resolutions that can be gotten with optical telescopes. So radio astronomy has moved in recent years from a situation of providing inferior resolution to a situation of providing superior resolution.

VLB techniques are difficult and time-consuming to apply. Also, when we work at high resolution, we sample only a small area of the sky at any time, and it therefore takes longer to study a region at high resolution than it does at low resolution. Therefore, VLB techniques can be applied only to very small areas of sky. But for those few areas, chosen for their special interest, our knowledge of the structure of radio sources has been fantastically improved.

We have already mentioned (in Section 5.11) that VLB techniques have provided an accurate measurement for the deflection of electromagnetic waves by the mass of the sun, providing a confirmation of Einstein's general theory of relativity.

Color Plate 45: The Eta Carinae Nebula, NGC 3372, in the southern constellation Carina. Visually, this is the brightest part of the Milky Way. The central dark cloud superimposed on the brightest gas is the Keyhole Nebula. η Carinae is the brightest star left of the Keyhole. (Cerro Tololo Inter-American Observatory photo)

Color Plate 46: The Andromeda Galaxy, also known as M31 and NGC 224, the nearest spiral galaxy to the Milky Way. It is accompanied by two elliptical galaxies, NGC 205 (below) and M32 (above). (Hale Observatories photo with the 1.2-m Schmidt camera)

Color Plate 47: M33, a spiral galaxy (class Sc) in the constellation Triangulum. Hot, blue stars outline the spiral arms while the central regions show cooler yellow and red stars. (Hale Observatories photo)

Color Plate 48: Centaurus A, a powerful radio source whose optical image, NGC 5128, shows an elliptical galaxy around which a heavy zone of dust appears to be wrapped. The galaxy rotates perpendicularly to the dust lane. (Cerro Tololo Inter-American Observatory photo with the 4-m telescope)

Color Plate 49 (top): NGC 7331, a spiral galaxy in Pegasus. This galaxy may be linked with the group of galaxies known as Stephan's Quintet. (Hale Observatories photo with the 5-m telescope)

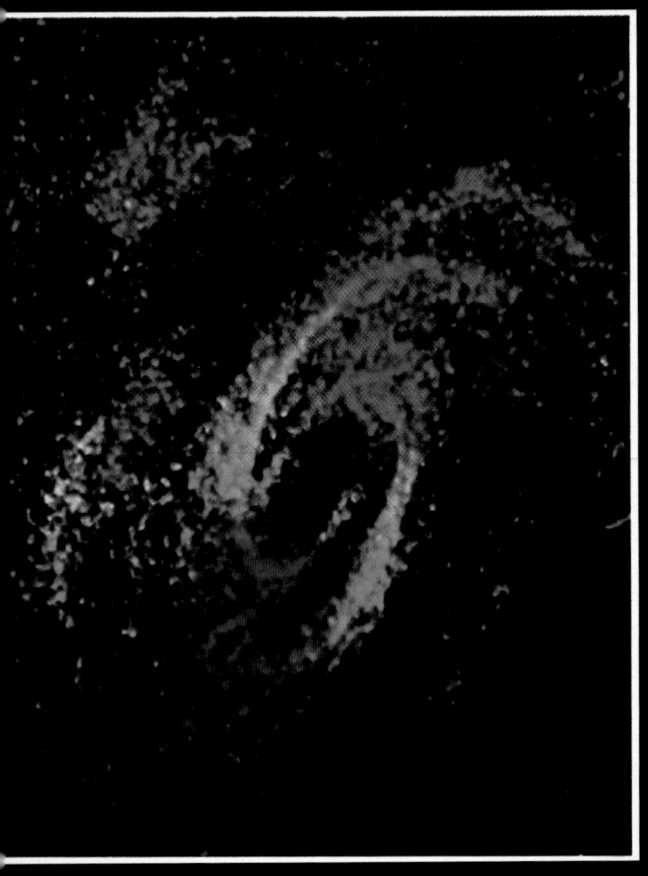

Color Plate 50 (bottom): Left is an image of 21-cm radiation from M81 in Ursa Major. Brightness represents intensity; color represents the Doppler shift: red = recession, violet = approach, green = no shift. (Westerbork data from A. H. Rots and W. W. Shane; imaged at NRAO.) Right is M81 observed optically at Palomar.

Also made possible by this extremely high resolution were observations of the nucleus of the radio galaxy M87; the nucleus turns out to be only 0.001 arc sec across, which tells us in how small a volume the large amount of energy emitted by M87 is generated (assuming that all the energy radiated by the galaxy is somehow generated in the nucleus).

The clouds of gas that give off spectral lines because of water vapor undergoing maser action (Section 21.6) have also been studied. The high resolution of the VLB system using the Haystack telescope in Westford, Massachusetts, together with a telescope at the Crimean Astrophysical Observatory in the U.S.S.R. was necessary to resolve the tiny clouds, which proved to be only about the size of our solar system. This endorses the idea that the clouds may be condensing into stars.

Other VLB observations have not only revealed the presence of a few small components in far-off radio sources like quasars, but also have shown that in some cases these components are separating at angular velocities that seem to correspond (at the distances of these objects based on Hubble's law) to velocities greater than the speed of light. Since this result is thought to be impossible, other theoretical explanations of the data have been sought. One model is the *Christmas tree model*, or *marquee model* (as in a theater marquee), in which there are really several components flashing on and off. Thus when we see two components apparently separated by a very much greater distance than they were earlier, we would really be seeing two different components from the previous set of observations (or at least one of the components is different). Thus the object need not be actually expanding at all. Alternatively, when one applies the special theory of relativity, one can find situations where rapidly moving objects, at speeds near to but less than the speed of light, can give the appearance of moving at speeds greater than the speed of light. If two objects are moving rapidly apart from each other, one toward and the other away from us, the light from the farther object takes longer to reach us and thus originated earlier than the light we receive at the same time from the nearer object. Under these circumstances, the movement at speeds greater than light could be only apparent and not actual. Still another model suggests that we are only seeing the delayed effects of events that happened previously. Observations of these radio sources are being continued to see whether a choice can be made among the above possibilities, or whether still other models are better. At present, many observations have been accumulated, and they always show expansion of the radio sources. The marquee model is therefore seeming less likely, since it would predict equal numbers of apparent expansions or contractions.

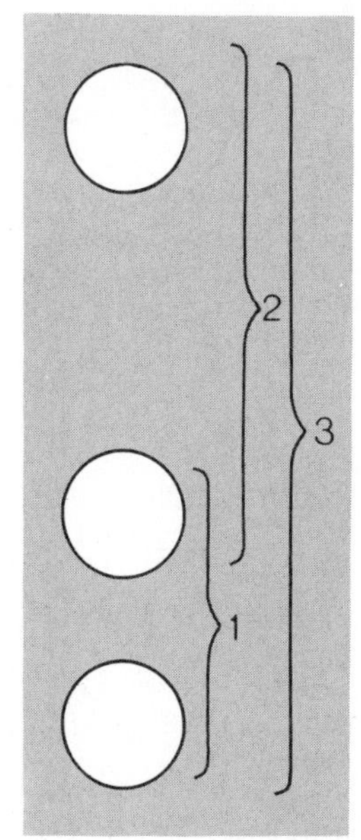

Figure 23–31 The lines joining the several dishes making up an interferometer represent pairs of dishes that give results equivalent to having several baselines simultaneously. The same results could be gathered with only two dishes, one of which was moved around, but it would take longer. With the three dishes shown, one can simultaneously make measurements with three different baselines.

23.5c *Aperture Synthesis Techniques*

By suitably arranging a set of radio telescopes across a landscape, one can simultaneously make measurements over a variety of baselines, because each two telescopes in the set can be considered to be a pair with a different baseline from each other pair (Fig. 23–31). Thus with such an arrangement one can more rapidly put together a map of a radio source than one can from two-dish interferometers. Also, the existence of several dishes instead of just two means that there is that much more collecting area.

Note that the resolution of an interferometer depends on the baseline, so that at any one time the resolution is quite good along the line in which the telescopes lie but is only the resolution of a single dish in the per-

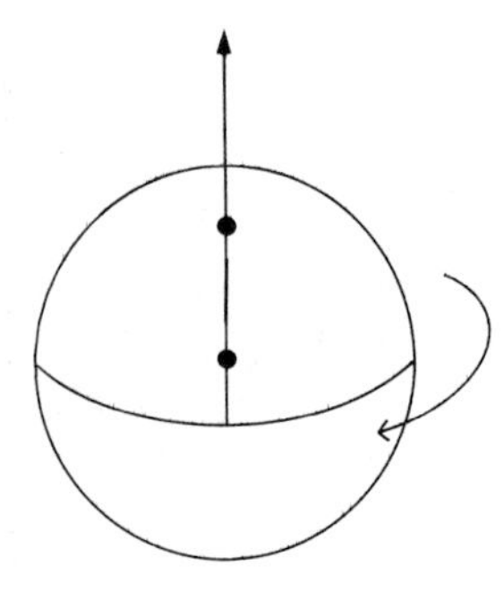

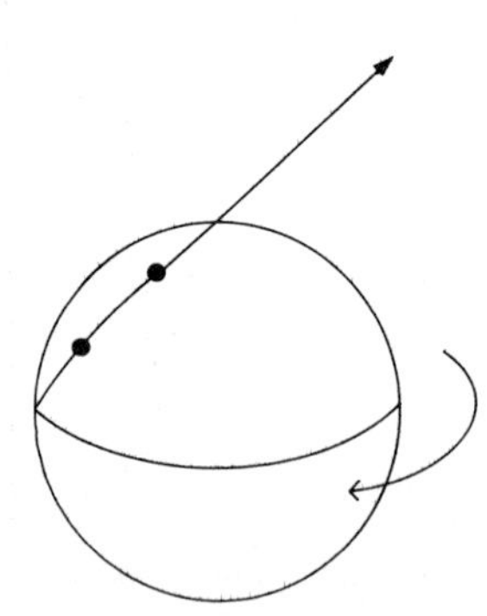

Figure 23–32 The earth's rotation carries around a straight line of dishes so that it maps out an ellipse in the sky. The figure shows the view of a line of dishes on the earth that we would see from the direction of the source being observed.

pendicular direction. Fortunately, we can take advantage of the fact that the earth's rotation changes the orientation of the telescopes with respect to the stars (Fig. 23–32). Thus by observing over a 12-hour period of time, one can improve the resolution in all directions.

In effect, we have synthesized the existence of a large telescope covering an elliptical area whose longest diameter is the same as the maximum separation of the outermost telescopes. This type of interferometric technique is known as *aperture synthesis*, and was pioneered by Martin Ryle in England. The interferometers that he built at Cambridge have been responsible for many high-resolution maps, including the map of Cygnus A shown in Figure 23–25.

The largest aperture synthesis telescope for the last few years has been at Westerbork in the Netherlands (Fig. 23–33). Twelve telescopes, each 25 m (82 ft) in diameter, are spaced over a 1.6 km (1 mi) baseline there. Observations made with the Westerbork array are discussed in Section 23.5d.

The most fantastic aperture synthesis radio telescope is now gradually being constructed in New Mexico. It is composed of 27 dishes, each 26 m (85 ft) in diameter, arranged in the shape of a "Y" over a flat area 27 km (17 mi) in diameter (Fig. 23–34). The control room is at the center of the "Y," and contains a large computer to analyze all the signals. This $76 million project is the major ground-based effort of U.S. astronomy. The system is prosaically called the ***Very Large Array*** (***VLA***).

When fully operational, which should be in 1981, the VLA will make pictures of a field of view a few minutes of arc across, with resolutions comparable to the 1 arc sec of optical observations from large telescopes, in about 10 hours. Though this resolution is lower than the resolution obtainable for a limited number of sources with VLB techniques, the ability to

Figure 23–33 The Westerbork array of telescopes near Groningen, The Netherlands.

A

make pictures in such a short time will be invaluable. And we have a lot to learn at a resolution of 1 arc sec (actually 0.6 arc sec at an observing wavelength of 6 cm and 2.1 arc sec at 21 cm).

Because the array is in the shape of a "Y" rather than a straight line, it does not need to wait for the earth to rotate in order to synthesize the aperture. The "Y" will be delineated by railroad tracks, on which the telescopes will be transported to 72 possible observing sites. The 27 dishes are coming into operation a few at a time, and even the first few telescopes on the first sections of track will be used to make preliminary observations over the next few years.

The techniques of aperture synthesis, developed for radio astronomy, are now also being developed for optical observations, as discussed in Section 4.3.

B

Figure 23–34 (*A*) An artist's conception of the VLA, currently being erected near Socorro, New Mexico. (*B*) One of the first completed antennas of the VLA being transported along the railroad track at the Socorro site.

23.5d Aperture Synthesis Observations

An aperture synthesis map of the spiral galaxy M51 made at Westerbork is shown in Figure 23–35. The array was operating at a wavelength of 21 cm in order to study the hydrogen distribution.

Westerbork has discovered several giant double radio sources (Fig. 23–36), much larger than any of the double-lobed sources previously known. Some are hundreds of times larger than our own galaxy (Fig. 23–37). These are the largest single objects currently known in the universe.

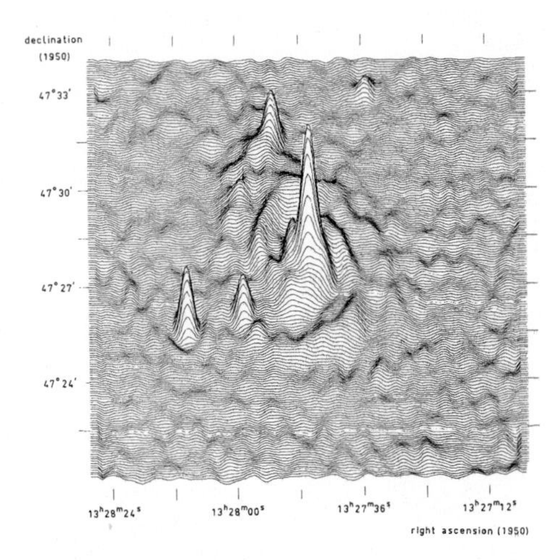

Figure 23–35 A Westerbork high-resolution radio map of the Whirlpool Galaxy, M51, superimposed on an optical photograph (*left*). At right is a computer view of the radio map, with the third dimension representing the intensity of the radio source.

Observations first with the Cambridge One Mile (1.6 km) Interferometer and then with the Westerbork array have revealed the existence of a class of galaxies with tails extending out from one side. They are called *head-tail galaxies,* and resemble tadpoles in appearance (Fig. 23–38). As these galaxies move through intergalactic space, it is thought that they expel the clouds of gas that we see as the tails.

The discrete blobs that we can see in the tails indicate that the galaxies give off puffs of ionized gas every few million years as they chug through intergalactic space. Perhaps by studying these puffs, we can learn about the main galaxies themselves as they were at earlier stages in their lives. Head-tail galaxies seem to be a common although hitherto unknown type; most are found in rich clusters of galaxies.

Figure 23–36 A Westerbork synthesis map of DA 240, a giant radio galaxy with two lobes of emission. It is 34 arc min across in our sky, larger than the full moon. Its central radio component coincides with a distant galaxy observed in visible light.

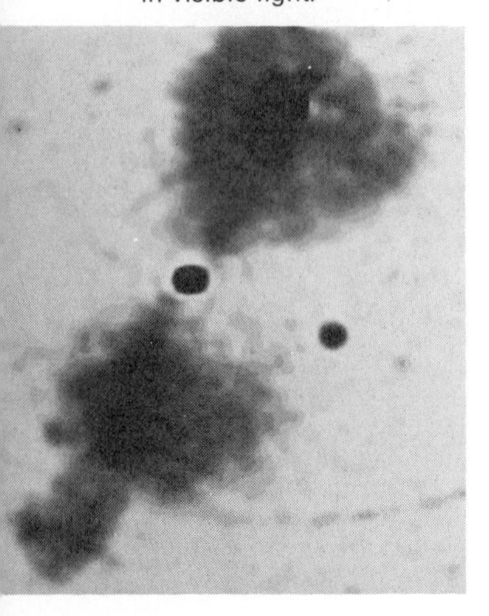

23.6 THE FORMATION OF GALAXIES

The idea that galaxies are important building blocks of the universe was not proved until 1924, so the separation of the problem of how the galaxies were formed from the cosmogonical problem we discussed in Section 10.5 did not become necessary until then. Since that time and to the present, most astronomers have believed that galaxies originated through some sort of *gravitational instability*. In this theory, a fluctuation in density either developed or pre-existed in the gas from which the galaxy was to form. This fluctuation grew in mass and then collapsed and cooled until the galaxy was formed. A fluctuation containing a large amount of mass would form a cluster of galaxies. After the galaxy or galaxies were formed, stars and indeed planets could form in a similar manner.

When Hubble's law was discovered, the theoretical problem of galaxy formation became more difficult, for one had to predict what would happen to gravitational instabilities in an expanding universe. When the primeval black body radiation (which we discuss in Section 25.4) was discovered in 1965, its existence became one of the major considerations, for it would

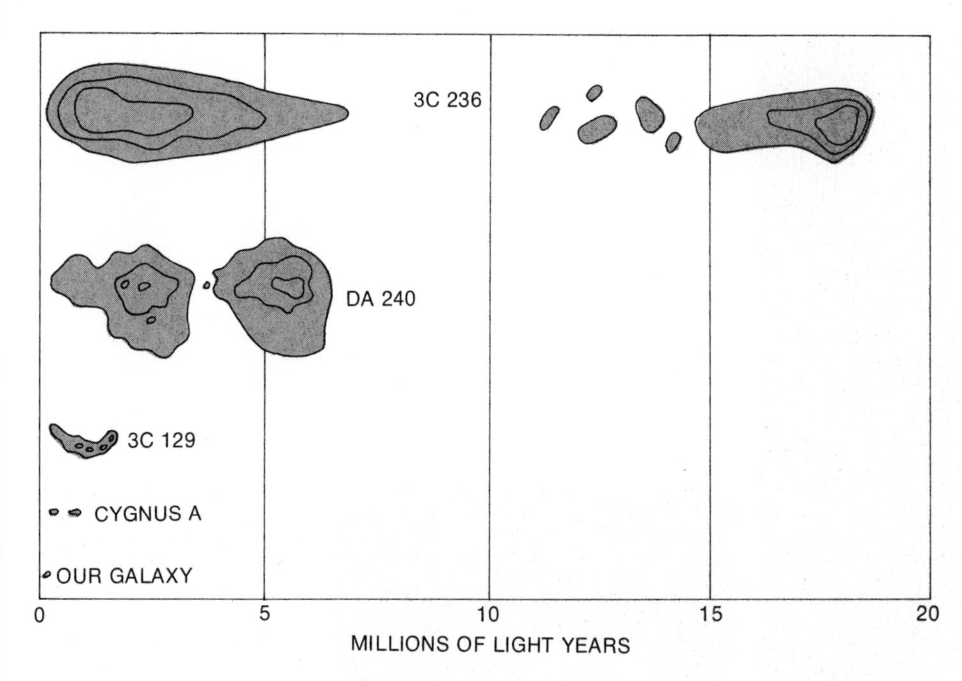

Figure 23–37 A comparison of the sizes of giant double-lobed radio galaxies with our own galaxy.

have been of overwhelming strength long ago. One problem all along with the gravitational instability theories has been the fact that calculations seem to indicate that fluctuations would not grow in size sufficiently rapidly to have given rise to objects on the scale of galaxies or clusters of galaxies. And where did the fluctuations come from? Were they present during the initial instants after the formation of the universe, or did they arise as chance statistical variations in density? Further, why do galaxies spin? Where does their angular momentum come from? Though the gravitational instability theory remains dominant, the above difficulties have not been entirely overcome.

In one sense, the gravitational instability theories hark back to the Aristotelian view that the universe was fundamentally simple, and that order cannot follow from basic disorder. The contrary view is that the universe was very complex at first and has evolved to its current relatively simple stage (assuming that the regularities we see when we study galaxies indicate a basic simplicity). The latest version of these alternative theories involves a fundamental *cosmic turbulence* that existed since the origin of the universe. At a certain stage in the expansion of the universe, amounts of mass suitable to become galaxies or clusters of galaxies would tend to separate out from the overall distribution of matter, carrying an intrinsic spin with them from the turbulence.

Turbulence *in a material is a disturbed state with swirls and eddies in motion.*

Such theories had fallen into disrepute until the last ten years because theoretical calculations had indicated that pre-existing turbulence would have disappeared in the early stages of the universe. But Leonid Ozernoi and his collaborators at the Lebedev Physics Institute in Moscow have found reasons why this need not have been so, and have elaborated on cosmic turbulence theories. With pre-existing turbulence, it is easier to understand why galaxies spin than it is on the basis of theories of gravitational instability. But it can be objected that we have merely changed the question from "where does the spin of the galaxies come from?" to "where does the turbulence come from?" without providing a fundamental answer.

Much theoretical work is currently being devoted to increasing our understanding of the dynamics of gases and to how the galaxies may have formed.

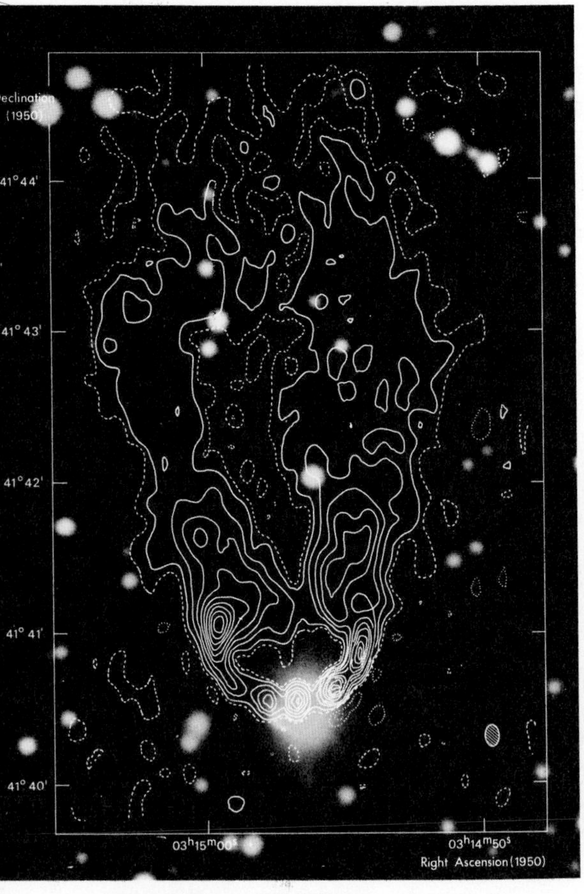

Figure 23–38 (*A*) A Westerbork radio map of the head-tail galaxy NGC 1265, superimposed on an optical photograph from the Palomar Sky Survey. The effective size of the radio beam at the wavelength of 6 cm used for these observations is shown as the shaded ellipse at lower right.

Figure 23–38 (*B*) A Westerbork radio map of the head-tail radio source 3C 129, converted to a radiophotograph in which the brightness of the image corresponds to the intensity of radio emission at a wavelength of 21 cm (near but not at the hydrogen wavelength). The resolution is about 30 arc sec. The front end of the head of the radio galaxy corresponds to the position of a faint optical galaxy, a member of a rich cluster of galaxies.

SUMMARY AND OUTLINE OF CHAPTER 23

Aims: To study the different types of galaxies that are observed in various parts of the spectrum, to see that galaxies are fundamental units of the universe, to study the expansion of the universe, to consider how interferometry is now allowing radio astronomy to make significant advances in the study of galaxies, and to consider how and why galaxies may have formed

Observations and catalogues of non-stellar objects
- Lord Rosse's early observations of spiral forms
- Messier's catalogue, General Catalogues by the Herschels, New General Catalogue (NGC) and Index Catalogues (IC) by Dreyer
- Galaxies as "island universes"
- Shapley-Curtis debate, 1920, dwelt on size of our own galaxy
- Incorrect measurements of proper motion and ignorance of the distinction between novae and supernovae led to the lack of knowledge of what galaxies are
- Matter was settled when Hubble observed Cepheids in galaxies
- Hubble classification (Section 23.1)
- Elliptical galaxies (E0–E7)
- Spiral galaxies (Sa–Sc) and barred spiral galaxies (SBa–SBc)
- Irregular galaxies (Irr I, Irr II)
- Peculiar galaxies (pec)
- Seyfert and N galaxies—bright cores

Origin of galactic structure (Section 23.2)
- Density waves lead to spiral appearance, though the actual distribution of matter may not be spiral
- Tidal effects create tails in interacting galaxies

Clusters of galaxies (Section 23.3)
- Local Group
- Rich clusters, such as Virgo and Coma, are x-ray sources
- Local Supercluster exists; is there high order clustering?
- Missing mass problem: some gravitational effects indicate the presence in clusters of galaxies of unobserved mass

The universe is expanding (Section 23.4)
- Hubble's law, $v = H_0 d$, expresses how the velocity of expansion increases with increasing distance
- Best current value for Hubble's constant is $H_0 = 55$ km/sec/Mpc
- The expansion is universal, and has no center

Radio galaxies (Section 23.5)
- Some objects are powerful sources of non-thermal radiation in the radio, x-ray, and infrared spectral regions
- Double-lobed shape is typical of radio galaxies; sometimes a peculiar optical object is present at the center
- Single-dish radio telescopes have low resolution; interferometers give resolution as high or higher than optical observations
- VLB, VLBI (very-long-baseline interferometry) uses widely separated dishes to provide the highest resolution now possible
- Aperture synthesis arrays provide quicker maps, still retain high resolution
- Current interferometers at Cambridge (England) and Westerbork (The Netherlands)
- VLA (Very Large Array) now under construction in U.S.
- Giant radio galaxies and head-tail galaxies studied

Galaxy formation (Section 23.6)
- Gravitational instability is leading model
- Some astronomers believe in pre-existing cosmic turbulence

QUESTIONS

1. Shapley and Curtis both argued on the basis of incomplete knowledge. In view of what we now know, evaluate the points that they made.
2. The sense of rotation of galaxies is determined spectroscopically (Section 23.1b). How is this done?
3. Which classes of galaxies are the most likely to have new stars forming? What evidence supports this?
4. What makes us think that the nuclei of Seyfert galaxies are very active?
5. How is it possible that galaxies could exist close to our own and not have been discovered before?
6. When we say that mass is missing, why do we expect it to be there in the first place?
7. How do we know that the missing mass is not neutral or molecular hydrogen?
8. To measure the Hubble constant, you must have a means (other than the redshift) to determine the distances to galaxies. What are two methods that are used?
9. (a) At what velocity in km/sec is a galaxy 5 million parsecs away receding from us? (b) Express this velocity in miles/sec and miles/hr.
10. Does α Centauri, the nearest set of stars to us, show a redshift that follows Hubble's law? Explain.

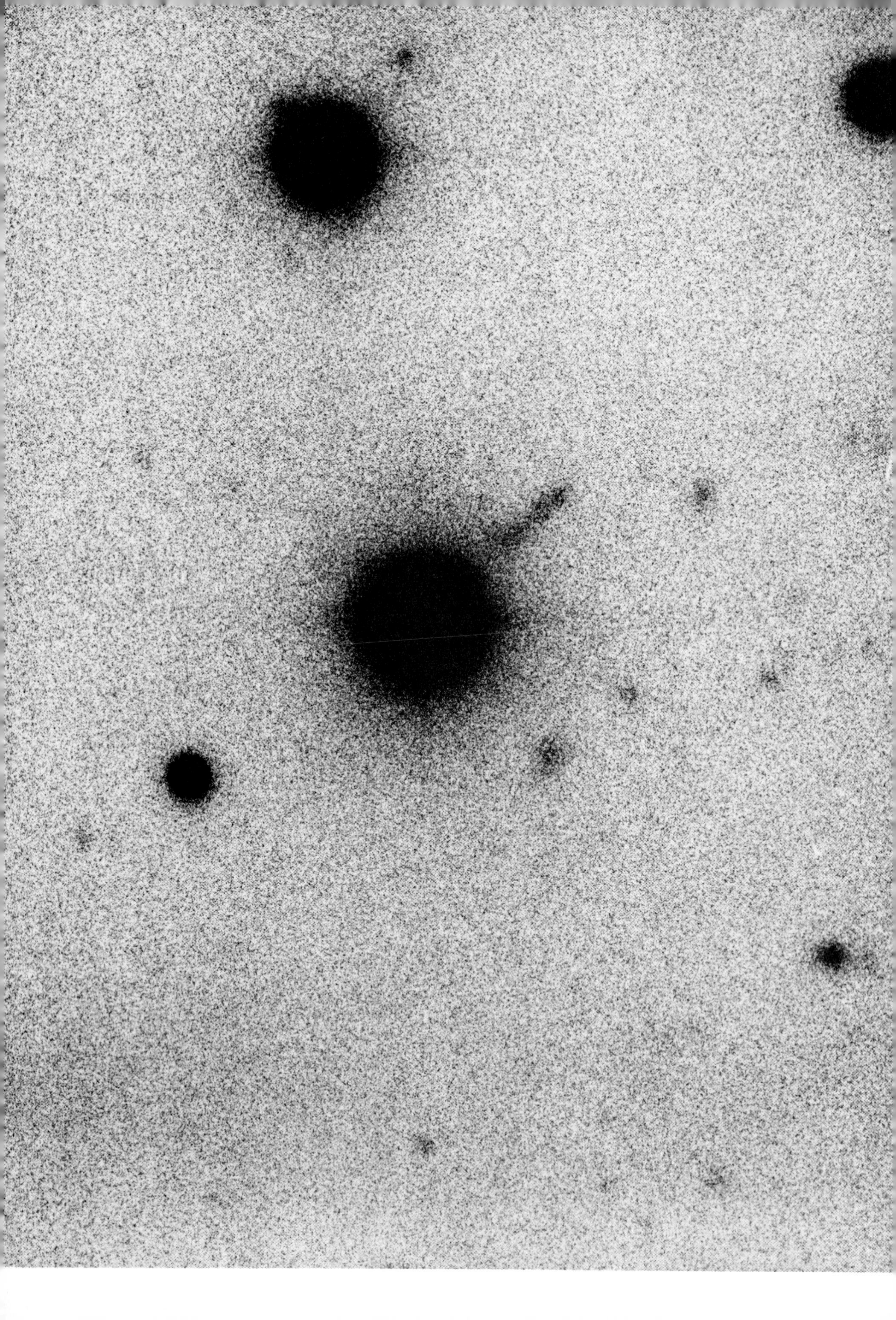

24

Quasars

Quasars are enigmatic objects that, like stars, appear as points of light in the sky. But unlike the stars that we see, the quasars occur in the farthest reaches of the universe and thus may be the key to our understanding the history and structure of space. We study quasars in both the optical and radio parts of the spectrum. Although they are relatively faint in visible light, they are among the strongest radio sources in our sky, and therefore must be prodigious radiators of energy. Astronomers have so far been unable to determine how they give off so much energy. Consequently the quasars raise fundamental questions about our basic physical laws and how energy is generated in the universe.

The word *quasar* originated from QSR, a contraction of "quasi-stellar radio source." However, the most important characteristic of quasars is not that they are emitting radio radiation but that they are traveling away from earth at tremendous speeds. Even certain "quasi-stellar objects" that are not radio sources have been included among the quasars because they too have tremendous velocities. Astronomers can deduce distances by observing the quasar spectra, measuring the Doppler shifts, and applying Hubble's law (Section 23.4).

Most astronomers accept the idea that quasars are the objects most distant from us, but a small minority contends that they are in fact relatively close to us. The reasoning that leads to these opposite conclusions will be discussed later in this chapter. Even if the quasars are relatively near, they are still important to astronomy since they would show that our present methods of measuring distances to faraway objects are not always valid. Whether nearby or fantastically distant, quasars have strange properties that raise basic questions.

24.1 THE DISCOVERY OF QUASARS

Quasars are a discovery resulting from the interaction of optical astronomy and radio astronomy. When maps of the radio sky turned out to be

---

An enlarged portion of a photograph of the quasi-stellar radio source 3C 273, taken with the 5-m Hale telescope. The object looks like any 13th magnitude star, except for the faint narrow jet that is visible out to about 20 seconds of arc from the quasi-stellar object.

very different in appearance from maps of the optical sky, many astronomers set out to correlate the radio objects with visible ones. The sun was easy to identify, and some of the brighter galaxies also proved to be radio sources (as discussed in Section 23.5).

Single-dish radio telescopes did not give sufficiently accurate positions of objects in the sky to allow identifications to be made, so interferometers (Section 23.5b) had to be used. Astronomers at the University of Cambridge (England), in particular, used interferometry to compile a series of catalogues. Most of the radio objects in these catalogues could be identified with optical objects, but a few had no clear identifications. Allan Sandage of the Mt. Wilson and Palomar Observatories and Thomas Matthews of Caltech's Owens Valley Radio Observatory (Matthews is now at the University of Maryland) used the Owens Valley interferometer to improve the accuracy of the positions and the 5-meter (200-inch) telescope to photograph the suspected regions. At least one of the strong radio sources, 3C 48 (the 48th source in the third Cambridge catalogue), seemed to be suspiciously near a faint (16th magnitude), bluish star, and Sandage and Matthews reported this identification of a "radio star" to the American Astronomical Society in 1960. At that time, no stars had been found to emit radio waves, with the sole exception of the sun, whose weak radio radiation we can detect only because of the sun's proximity to the earth.

Accurate positions can be measured for radio sources either by interferometry or by lunar occultation.

In Australia, Cyril Hazard used the 64-meter (210-foot) radio telescope at Parkes (Color Plate 2) to observe the passage of the moon across the position in the sky of another bright radio source, 3C 273 (the 273rd source in the 3rd Cambridge catalogue). Since we know the position of the moon in the sky very accurately, if it occults—hides—a radio source then we know that at the moment the signal strength decreases the source must have passed behind the advancing limb of the moon. (Actually, there are diffraction effects around the edge of the moon that make the intensity fall off in an oscillatory manner instead of sharply.) Later on, the instant the radio source emerges—for 3C 273 it was nearly an hour later—the position of the lunar limb marks another set of possible positions. Where these two sets of possible positions overlap is the actual position of the source.

Hazard and his colleagues were fortunate to have three lunar occultations of 3C 273 within a few months, and from the data they gathered, they derived a very accurate position for the source and a map of its structure. The optical object, which is 13th magnitude, is not completely starlike in appearance, for a luminous jet appears to be connected to the point nucleus (as shown in the photograph opening this chapter).

Figure 24–1 The first quasar, 3C 48, photographed with the 5-m Hale telescope.

There seem to be two components of the radio emission from 3C 273. One coincides with the jet, and the other coincides with the bluish stellar object, clinching the identification as a quasar. Maarten Schmidt of Mt. Wilson and Palomar photographed the spectrum of this "quasi-stellar radio source" with the 5-meter Palomar telescope, a telescope that has played a key role in the study of such sources because it was one of only a few telescopes large enough to allow optical spectra of these faint objects to be photographed in "reasonable" observing times, that is, with exposures of "only" a few hours.

Some quasars seem to be surrounded by faint luminosity, but on the whole their optical images are those of faint stars (Fig. 24–1), even though one might have expected such strong radio sources to be associated with more distinctive objects.

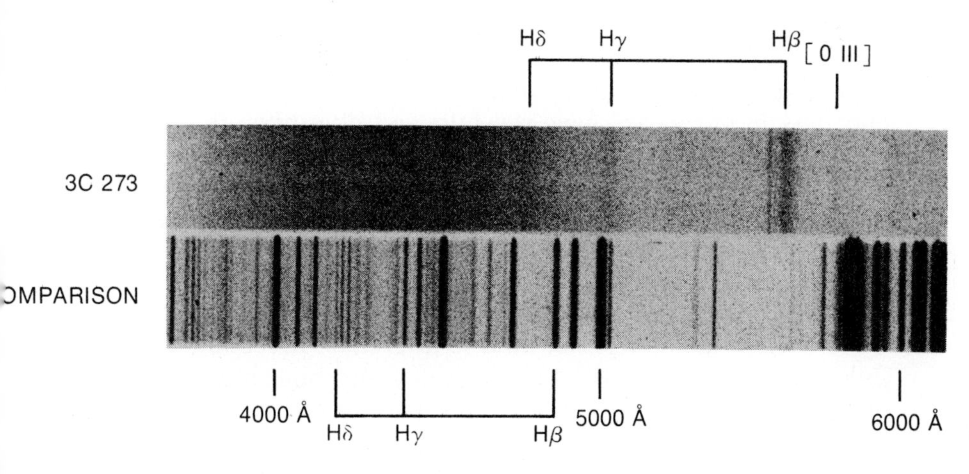

Figure 24–2 Spectrum of the quasar 3C 273. The lower spectrum consists of hydrogen and helium lines; it serves to establish the scale of wavelength. The upper part is the spectrum of the quasar, an object of 13th magnitude. The Balmer lines Hβ, Hγ, and Hδ in the quasar spectrum are at longer wavelengths than in the comparison spectrum. The redshift of 16 per cent corresponds, according to Hubble's law, to a distance of three billion light years. Note that the comparison spectrum represents hydrogen and helium sources on earth (more particularly, located inside the Palomar dome).

The spectra of the bluish stellar objects showed emission lines, but the lines were not of any known element. The lines had the general appearance of lines emitted by a gas of medium temperature, but the wavelengths of the lines observed were different from those that known gases emit.

The breakthrough in understanding came in 1963. At that time, Schmidt noted that the spectral lines of 3C 273 (Fig. 24–2) seemed to have the same pattern as lines of hydrogen under normal terrestrial conditions—that is, they had the characteristic pattern of lines that the hydrogen spectrum has. Schmidt then made a major scientific discovery: he asked himself whether he could simply be observing a hydrogen spectrum that had been greatly shifted in wavelength by the Doppler effect. The Doppler shift required would be huge: each wavelength would have to be shifted by 16 per cent toward the red to account for the spectrum of 3C 273. This would mean that 3C 273 is receding from us at approximately 16 per cent of the speed of light. Immediately, Schmidt's colleague Jesse Greenstein recognized that the spectrum of 3C 48, another of the bluish stellar objects associated with radio sources, could be similarly explained. All the lines in the spectrum of 3C 48 were shifted by 37 per cent, a still more astounding redshift.

Later, absorption lines were discovered in quasars as well as the emission lines just discussed. Many quasars have several systems of emission and absorption lines of differing redshifts. Some of these lines may be formed in clouds of gas of differing velocities surrounding the quasars.

24.2 THE DOPPLER SHIFT IN QUASARS

Most objects in the universe show some Doppler shift with respect to the earth, for they are either approaching us or receding from us. (This is normally the only component of their motion that we can measure.) For velocities that are small compared to the velocity of light, the amount of shift in the spectrum is written simply, for rest wavelength λ, velocity v, and speed of light c, as

$$\frac{\Delta\lambda}{\lambda} = \frac{v}{c} \qquad \text{(as discussed in Section 2.15).}$$

Astronomers often use the symbol z to stand for $\Delta\lambda/\lambda$, the amount of the redshift. Since the distances to quasars can be assessed only from the redshifts, let us discuss here the use of z.

The stars in our galaxy have redshifts and blueshifts; these Doppler shifts are small (they occur from random motions of stars or from organized motions around the center of our galaxy) but are quite large enough for us to measure. The spectrum of the star that has the largest radial velocity with respect to the earth has a Doppler shift of two-tenths of one per cent, that is s, $z = 0.2$ per cent $= 0.002$. Thus this star is moving at two-tenths of one per cent of the speed of light $= 0.002 \times 3 \times 10^5$ km/sec $= 600$ km/sec, about two million kilometers per hour in the radial direction.

Unlike stars, which have slight redshifts and blueshifts, all galaxies except the nearest are redshifted with respect to the sun. The redshifts follow Hubble's law, which is discussed in Section 23.5.

The Doppler shifts of quasars are much greater than those of most galaxies. Even for the nearest quasar, 3C 273, $z = 0.16$ (read "a redshift of 16 per cent"), and so the spectrum has been shifted by a substantial amount. Thus Hubble's law implies that 3C 273 is as far away from us as are distant galaxies. Only a handful of galaxies are known to have greater redshifts.

Let us consider a redshift of 0.2, to pick a round number, and calculate its effect on a spectrum (Fig. 24–3). $z = 0.2$ means that $v/c = 0.2$, and therefore $v = 0.2\,c = 0.2 \times 3 \times 10^5$ km/sec $= 6 \times 10^4$ km/sec.

If a spectral line were emitted at 4000 Å in the quasar, at what wavelength would we record it on our photographic plate on earth? (We would actually be observing particular spectral lines such as that of hydrogen at 1216 Å, but let us continue to use round numbers for this example.) $\Delta\lambda/\lambda = \Delta\lambda/4000 = 0.2$. Therefore $\Delta\lambda = 0.2 \times 4000 = 800$. Note that we have only calculated $\Delta\lambda$, the shift in wavelength. The **new** wavelength is equal to the old wavelength plus the shift in wavelength, $\lambda_{new} = \lambda + \Delta\lambda = 4000 + 800 = 4800$Å. Thus the line that was emitted at 4000 Å in the quasar was recorded at 4800 Å on earth.

Similarly, a line that was emitted at 5000 Å is shifted $\Delta\lambda/\lambda = \Delta\lambda/5000 = 0.2$. $\lambda + \Delta\lambda = 5000 + 1000 = 6000$ Å. The spectrum is thus not merely displaced by a constant number of angstroms, but is stretched more and more toward the higher wavelengths.

Remember, however, that this formula is valid only for velocities much less than c, the speed of light. For speeds closer to the speed of light, we must use a formula from the special theory of relativity:

$$\frac{\Delta\lambda}{\lambda} = \sqrt{\frac{1 + v/c}{1 - v/c}} - 1.$$

In this formula, positive values of v correspond to receding objects. When v is much less than c, then the relativistic formula approximates the same non-relativistic formula we used earlier $\Delta\lambda/\lambda = v/c$). But in the relativistic formula, when v gets very close to c, $\Delta\lambda/\lambda$ gets greater than 1 even though v is still less than c. We still use the letter z to stand for $\Delta\lambda/\lambda$. (Actually, for objects at cosmological distances we should really use an equation from the general theory of relativity, but the equation is not easily stated and the values are similar to the above.)

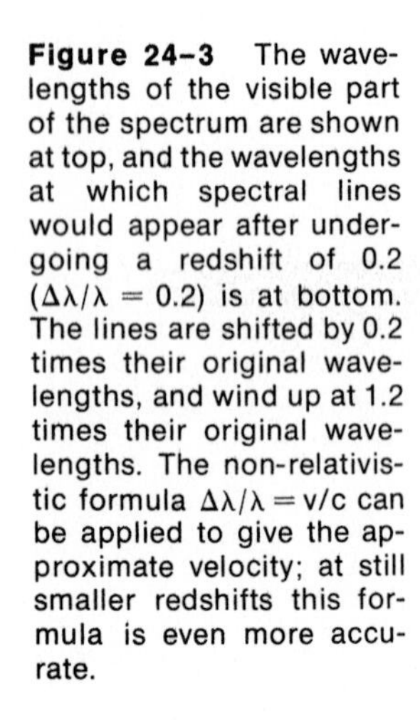

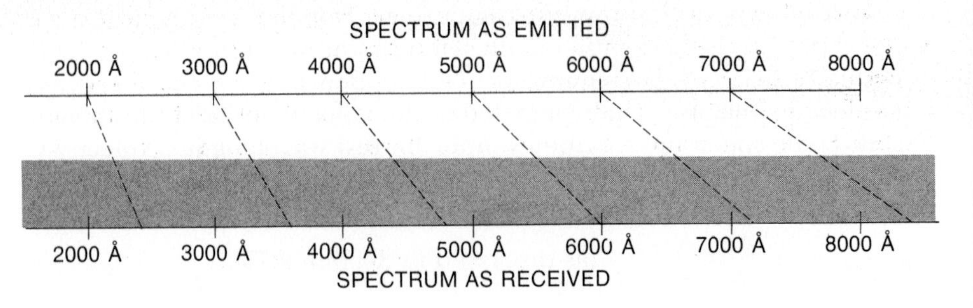

Figure 24–3 The wavelengths of the visible part of the spectrum are shown at top, and the wavelengths at which spectral lines would appear after undergoing a redshift of 0.2 ($\Delta\lambda/\lambda = 0.2$) is at bottom. The lines are shifted by 0.2 times their original wavelengths, and wind up at 1.2 times their original wavelengths. The non-relativistic formula $\Delta\lambda/\lambda = v/c$ can be applied to give the approximate velocity; at still smaller redshifts this formula is even more accurate.

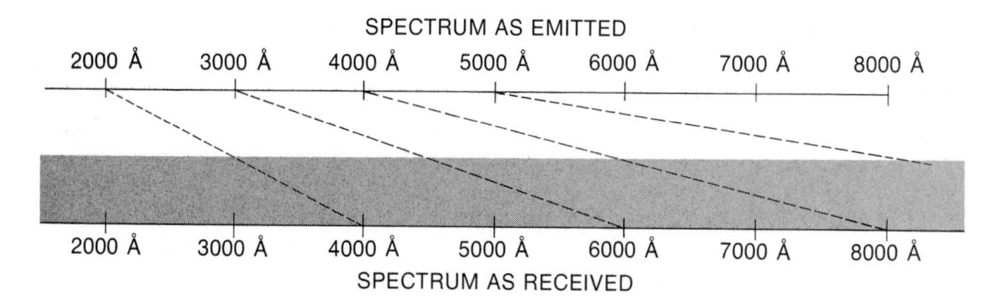

Figure 24–4 For a redshift of 1 ($\Delta\lambda/\lambda = 1$), as shown here, the lines are shifted (*bottom*) by an amount equal to their original wavelengths, and wind up at twice their original wavelengths. The velocity is a significant fraction of the speed of light and the simple non-relativistic formula for the Doppler shift cannot be applied.

As quasars are found that have larger and larger velocities of recession, the redshifts will be greater and greater, and z can even be larger than 1 without violating the special theory of relativity. For $z = 1$, we merely have the shift, $\Delta\lambda$, equaling the original wavelength, λ (Fig. 24–4). The velocity is still less than the speed of light.

24.3 NON-SPECTRAL METHODS OF DETECTING QUASARS

Candidates for quasar status can be found without going through the tedious job of randomly taking spectra. Quasars are slightly bluer in color than normal stars since most of their radiation is given off in the ultraviolet part of the spectrum that is normally cut off by the earth's atmosphere. When the strong ultraviolet light is Doppler-shifted toward the red, it winds up as a strong blue (which is the next color to longer wavelengths from ultraviolet). Martin Ryle and Allan Sandage photographed an area of the sky through an ultraviolet filter and then moved the photographic plate over a millimeter or so in its holder and took a second exposure on it. The second exposure was taken through a blue filter. Thus each star has two adjacent images that are easy to compare. When the ultraviolet image is stronger, we say that the object has an *ultraviolet excess.* Objects with strong ultraviolet excesses are likely to be quasars.

To confirm that the objects with large ultraviolet excesses are indeed quasars, we must take spectra. But it takes a long time to photograph the spectra of such faint objects. Thanks to the screening method of Ryle and Sandage, we have to take the spectra of a smaller number of objects (only those with large ultraviolet excesses) than we would if we had to examine the spectrum of every object in the sky. In this way, over a hundred quasars have been discovered, with Doppler shifts ranging from about 16 per cent to 353 per cent ($z = 0.16$ to $z = 3.53$). E. Margaret Burbidge and T. D. Kinman at Lick, Roger Lynds at Kitt Peak, and others joined Schmidt in measuring the redshifts, turning many suspected objects into known quasars.

Figure 24–5 This object appears stellar but is known to be not a star but a quasar.

Not only did quasi-stellar radio sources show up on the photographic plates as having large ultraviolet excesses, but another class of objects also appeared. Sources in this new class resembled the quasi-stellar radio sources in being quasi-stellar and in having large redshifts but did not emit any detectable radio radiation. These were called *quasi-stellar objects* (QSO's) (Fig. 24–5). We now use the term *quasar* to include both quasi-stellar radio sources and quasi-stellar objects. Although only a hundred spectra have been measured, astronomers estimate that there have been perhaps one million quasars in the universe. By now, however, most of them have probably lived out their lifetimes; that is, they have given off so much energy that they are no longer quasars. Perhaps about 35,000 quasars now exist, in the sense that they would be included if we could somehow take an instantaneous picture of the universe as it is at present, without consideration of the fact that because of the finite speed of light, we see distant sources by the radiation that they emitted long ago.

24.4 THE IMPORTANCE OF QUASARS

If we accept that the spectra of quasars are redshifted by tremendous amounts and apply Hubble's law, we realize that the quasars are the farthest known objects in the universe. At the moment, we have no independent way of measuring distances to objects so far away except by Hubble's law, and so it is important to test Hubble's law. Any deviations from the law would no doubt show up in the farthest objects, namely the quasars.

At present, Hubble's law is our only method of finding the distances to any objects with substantial redshifts. If the quasars were found not to satisfy Hubble's law, however, then doubt would be cast on all distances derived by Hubble's law, for we would never know whether we were observing an object that satisfied the law or one that did not.

If we do accept the quasars as the farthest objects, then they are billions of light years away from us and their light has taken billions of years to reach us. Thus we are looking back in time when we observe the quasars, and we can hope that they will help us understand the early phases of our universe.

24.5 THE ENERGY PROBLEM

If the quasars are as far away as this orthodox view, of which Maarten Schmidt is a leading proponent, holds, then they must be intrinsically very luminous to appear to us at their observed intensities. When quasars were discovered, scientists went back to the historical files of photographic plates and, in particular, to the extensive library of plates at the Harvard College Observatory. Harlan Smith, then of Yale and now of the University of Texas, and Dorrit Hoffleit of Yale and the Maria Mitchell Observatory on Nantucket, went back through the plate files looking at the brighter quasars. They could trace some for many years—plates of the sky region containing 3C 273 went back to 1887. They measured the optical brightness of the objects on many photographic plates and found that the brightness varied with time—scales of weeks or months. Of course, particular quasars were chosen for detailed study. It has since been found that the absolute magnitude of 3C 279 has been as high as –31, 10 magnitudes (and therefore 10,000 times) brighter than the Andromeda Galaxy.

If you want to pack as much energy into a quasar as they seem to radiate, you must have a certain amount of space in which to pack it. A quasar cannot be too small and still contain the 100 million solar masses that many astronomers think is necessary to generate that much energy. If something is, say, a tenth of a light year across, you would expect that it could not vary in brightness in less than a tenth of a year because one side cannot signal to the other side, so to speak, to join in the variation (Fig. 24–6).

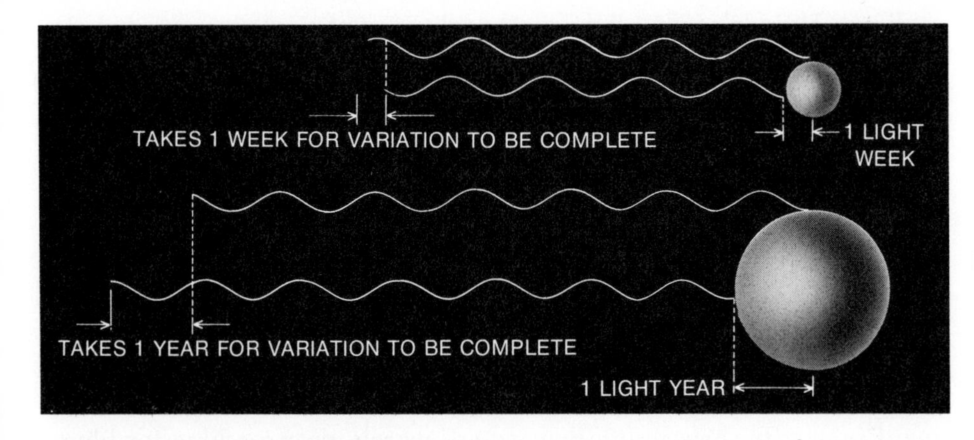

Figure 24–6 A smaller object can give off radiation that fluctuates more rapidly than a larger object can. If enough energy is given off so that we assume that the entire object is changing in some way, then the rapidity of the changing radiation can give us a limit to the size of the object.

Somehow the whole object has to know at the same time to do something, and that ability is limited by the speed of light.

So the variations in intensity mean that the quasars are probably fairly small. But the mechanism that makes them vary with these very short time scales is not known.

One possible model for the energy generation in a quasar is nuclear fusion in an extremely massive body (containing thousands of solar masses) known as a *supermassive star*. There is no confirming observation that such supermassive stars exist.

A model of many exploding supernovae seemed attractive for a time because one could explain the observed sporadic variations in the intensity by variations in the number going off from one moment to another. But the notion of the immense number of supernovae boggles the mind and does not seem acceptable in terms of the amount of mass required. Philip Morrison of M.I.T. has asked whether a quasar is not just some form of a giant pulsar. The pulsars are only little stars 10 km across, but perhaps they do have some characteristics in common with the giant, far-away quasars. Perhaps quasars are also rotating. We simply do not know. The pulsars pulse with very short time scales of a second or so and the quasars pulse, if they pulse, with time scales of a month or so. The amounts of energy they produce are quite different. We would really like to be able to monitor the intensities of quasars over a much greater period of time before we talk about their possible rotation.

Many other suggestions have been made to explain the source of energy in quasars, none of which is convincing. Perhaps matter and antimatter are annihilating each other in quasars. It may be a long time before we know, or agree on, the energy sources involved. Perhaps the answer will prove to be a new method, yet unknown, that can be applied to help our energy problems here on earth.

24.6 OTHER EXPLANATIONS OF REDSHIFTS

Some scientists say that it would be too difficult for quasars to generate the amount of energy that is necessary to make them as bright as they would need to be if they were very far from us. They feel that this problem of providing enough energy is so serious that they conclude that quasars must be close to us. If the quasars are close, we must think of some other way of accounting for their great redshifts. There are at least three other ways. They are all unattractive to many astronomers because they

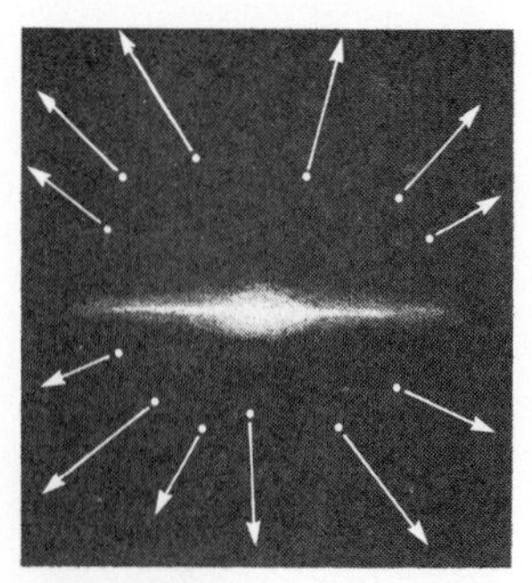

Figure 24–7 If quasars were local, and were ejected from our galaxy, that could explain why they all have high redshifts. But a tremendous amount of energy is necessary to make the ejection. So we do not find ourselves any closer to a solution of the energy problem—where all this energy comes from.

challenge Hubble's law and thus cast doubt on our knowledge of the whole scale of the universe.

The first way also relies on velocity and assumes that quasars are close but are going away from us very rapidly. Quasars could be relatively close to us yet show high redshifts if they had been ejected from the center of a galaxy at very high velocities (Fig. 24–7). If such quasars occurred in another galaxy, we would expect to see some of them going away from us and some coming toward us. Some of them would show tremendous redshifts signifying high velocities of recession; others would have tremendous blueshifts, signifying high velocities of approach. But all the quasars have redshifts; none of them has a blueshift, so astronomers think that this idea is unlikely. But one day if we find a blueshifted quasar, much of astronomy will have to be rewritten. As part of our search, we must guard against the possibility that we are merely not searching for blueshifted objects in the proper way. Note that the method of screening by studying objects with ultraviolet excesses would not work for blueshifted objects, since the peak in their spectra would no longer be shifted into the near ultraviolet for us to observe.

Since there are no blueshifted quasars, and if the quasars exploded from the center of our galaxy or a nearby galaxy, then the explosion took place so long ago that all quasars have expanded past us and are receding in all directions. Such tremendous and perhaps repeated explosions require a source of energy so great that it would be no easier to explain than it was with the redshifts following Hubble's law.

One of the leading proponents of the "local" theory of quasars, James Terrell of the Los Alamos Scientific Laboratory, points out that objects recently discovered near the radio source Centaurus A could be blueshifted quasars after all. That is, they look like faint blue stars from our vantage point in the Milky Way Galaxy, but from a star inside Centaurus A they could look redshifted and could appear as bright as 3C 273 appears to us. These faint objects, newly discovered with the 4-meter telescope at Cerro Tololo, are fainter than 23rd magnitude, which is too faint for us to be able to obtain their spectra in order to measure their Doppler shifts. If they are indeed traveling at a few tenths of the speed of light, however, we might hope to see them show proper motions within a few years.

A second kind of redshift is "gravitational." According to Einstein's theory of relativity, spectral lines are redshifted very slightly when they leave a star or any object that has a large mass and hence a large gravitational pull. The sun slightly redshifts the lines in the spectrum that leave it; a white dwarf star, which is much denser than the sun, redshifts the lines much more (see Section 7.5 for a discussion of this). Several scientists, including Fred Hoyle, then of Cambridge University, and William A. Fowler of Caltech, tried to find a way to allow for a sufficient redshift from gravity to match the huge redshifts in quasars. But the spectra of the quasars are not the spectra of very dense stars; they resemble in part that of a thin gas. So the scientists considered that dense objects were present to provide the gravity in addition to the presence of the gas that provides the spectral lines. Perhaps the dense objects are surrounded by the gas. At any rate, Fowler and Hoyle were not able to work out such a system that would be free of other, unobserved consequences.

A third possibility is that the quasars are acting on principles that we do not yet understand—some new kind of physics. Some specific modifications of physical laws have even been suggested. But most scientists feel (as discussed in Section 19.5) that the basic philosophy of science forbids us from inventing new physical laws as long as we can satisfactorily use the existing ones to explain all our data.

24.7 PECULIAR GALAXIES AND QUASARS

A principal attack on the theory that quasars are at great distances has come from Halton Arp of the Hale Observatories in California. He has been making observations of "peculiar galaxies" (discussed in Section 23.1d), objects in the sky whose shapes deviate from the regularity of the shapes of ordinary galaxies. These observations have led him to conclude that quasars might be linked to galaxies, both the peculiar and the ordinary types. He contends that he has found many examples in which two quasars are located on opposite sides of a galaxy from each other. He argues that the quasars and the galaxy must therefore be linked. We know the distances to the galaxies from Hubble's law; if the quasars and galaxies are indeed physically linked, then the quasars are at the same distance from us as the galaxies. These distances are much smaller than the quasars' velocities of recession indicate.

A similar argument that Doppler distances cannot be trusted is made with Stephan's Quintet (Fig. 24–8), five apparently linked galaxies, one of which has a redshift very different from the others. In another case, near the radio source 4C 11.50, two quasars with very different redshifts (0.44 and 0.90), lie within only 5 arc sec of each other. One of the quasars has the same velocity as a faint cluster of galaxies that was later discovered to be in the same direction. Both these cases can be explained without need to reject the Hubble law if the object with a discordant redshift only happens to appear in almost the same line of sight as do the other objects. As-

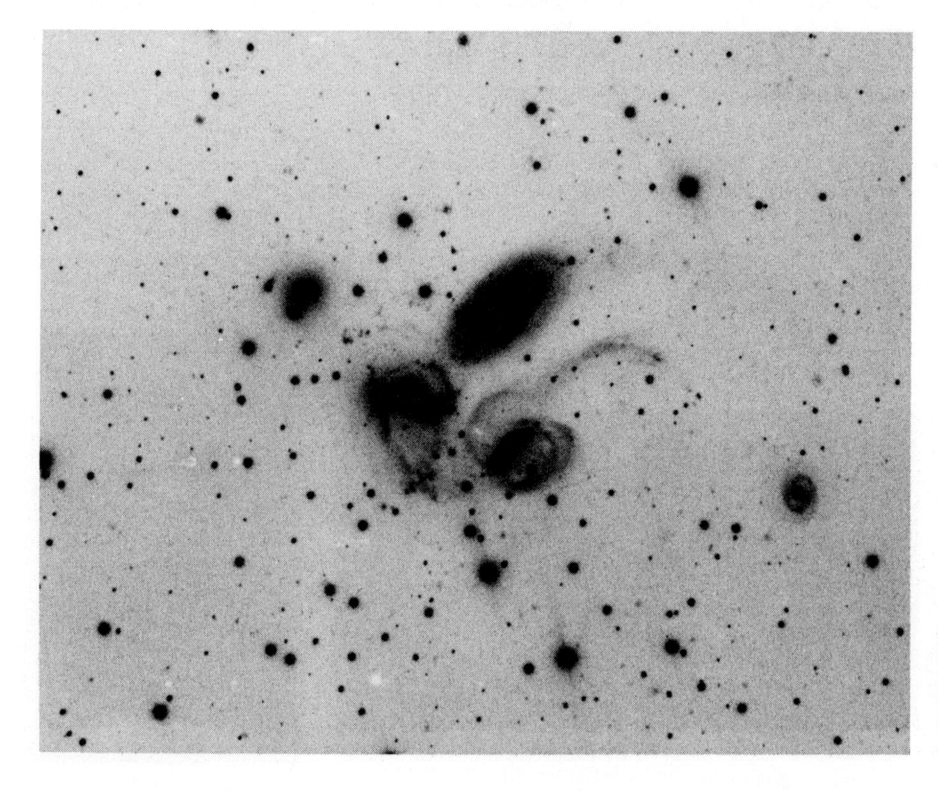

Figure 24–8 Stephan's Quintet, in a negative print of a photograph taken by Halton Arp with the 5-m telescope at the Hale Observatories.

"Any coincidence," said Miss Marple to herself, "is always worth noticing. You can throw it away later if it is only a coincidence."
Agatha Christie
NEMESIS
(London: Collins, 1971, p. 58)

Figure 24–9 One example of how even improbable things can happen. The galaxies shown at center, called 145-IG-03, appear to be interacting on this Cerro Tololo plate: But are two galaxies interacting or three? Surprisingly, the stellar-appearing object at the end of the luminous bridge—the round dot at the lower right end of the apparently connected objects—turns out (from its spectrum) to be a normal star of spectral type K and with a redshift very close to 0. This star is superimposed on the more distant galaxies to an accuracy of less than 0.1 arc sec. The example shown here points up the difficulty of concluding on the basis of probability arguments that objects are physically associated.

sociations of a quasar with a distant group of galaxies with redshifts that do agree with each other have been found.

Arp's data are subject to several objections, the main one being statistical. Arp feels that the associations he finds of objects of differing redshifts are genuine and not the result of chance. He points out that the probability of finding these objects so close together in the sky is very small. But most astronomers feel that the fact that a quasar and a galaxy are seen in the same directions from the earth does not prove that they are physically linked; they could be at different distances (Fig. 24–9). The associations would be harder to explain in this way if we found a group of galaxies in which not just one but two discrepant redshifts were found. No such group is known. So far, most astronomers do not accept Arp's arguments and feel that quasars are in fact very far away.

Figure 24–10 (*A*) Increasing exposure time reveals more and more galactic structure around the Seyfert galaxy NGC 4151, which appears almost stellar on the least exposed photograph. (*B*) The nucleus of the Seyfert galaxy photographed from an altitude of 25,000 m (80,000 ft) with the 90-cm (36-in) telescope carried aloft in 1970 by the Stratoscope II balloon.

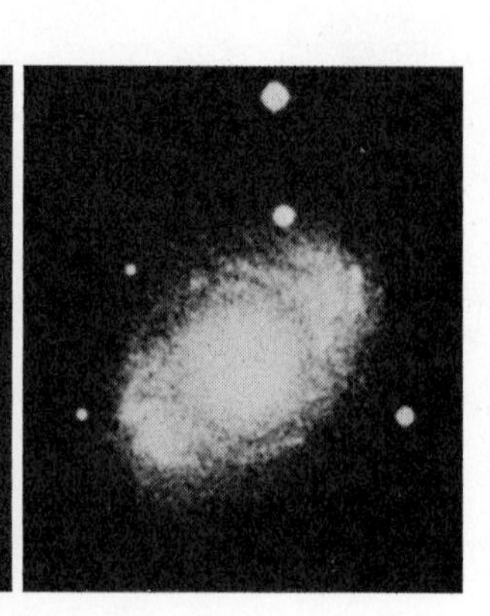

A

B

24.8 QUASARS AND GALAXIES

Another question is whether quasars are truly a new phenomenon or just unfamiliar types of galaxies. Some spiral galaxies have especially bright nuclei while others have fainter centers. One type of galaxy, discovered by Carl Seyfert of the Mount Wilson Observatory in 1943 and named after him, has a very bright nucleus indeed (Fig. 24–10). Can the quasars that we see be only the bright nuclei of galaxies? It has been recently calculated that although we can see such structures as the nucleus and the spiral arms of close Seyferts, only the nuclei would be visible if such galaxies were very far away. Another type of galaxy, an N galaxy, also has an especially bright core (the differences between the spectra of Seyfert and N galaxies are discussed in Section 23.1f). The brightest quasar (3C 273) and two Seyfert galaxies are among the few extra-galactic objects from which the Uhuru satellite has detected x-radiation.

Jerome Kristian of the Hale Observatories has studied a sample of about 25 quasars to see whether they could be galaxies with bright cores. He attempted to predict which would be close enough to us to reveal some structure around the core, which would appear as points like quasars, and which would be in between and thus might or might not show structure. He then studied the best photographs of these objects, and found that almost all objects in which one would expect to see structure did in fact show structure, almost all objects that would not show structure because they were too far away did not show structure, and that the middle group was mixed. It thus seems reasonable to believe that quasars are a class of objects somehow related to the cores of galaxies. Perhaps quasars and galaxies are the same, but are simply in different stages of evolution.

Now there is even stronger evidence that the redshift is a valid estimator of distance for quasars. A strange object named BL Lacertae has been known for years. At first, it was thought to be merely a variable star, but in recent years it was suspected to be stranger than that, partly because of its rapidly variable radio emission. Its spectrum showed only a featureless continuum, with no absorption or emission lines, so little could be learned about it.

J. Beverley Oke and James Gunn of Caltech had the idea of blocking out the bright central part of the star with a disk held up in the focus of their telescope, thereby obtaining the spectrum of the haze of gas that seems to surround the star. On one of the several nights that they observed, the central source was relatively faint and the spectrum of the surrounding haze of gas could be observed with an electronic device. The spectrum turns out to be a faint continuum with absorption lines that are redshifted by 7 per cent. Thus BL Lacertae is the nearest quasar!

If BL Lacertae is at the distance that corresponds to this redshift, it has several properties similar to other quasars if they are at their distances assigned from the Hubble law. This strengthens the notion that the quasars are indeed at those distances and that it is valid to use the Hubble relation. And the quasar again seems to be a special "event" in the center of a mass of gas, just as quasars may be similar to bright "events" in the centers of Seyfert galaxies or N galaxies. These observations may indicate that such "events" can happen late in the lifetime of a galaxy. Since most quasars are at large redshifts, the light we see now was emitted far back in time, so it had been thought that such events might only be linked to young galaxies.

The observation of the region surrounding BL Lacertae was very difficult to make, and must be thoroughly confirmed by other observations. One attempt to make such a confirmation with another large telescope did not succeed, but a follow-up with the 5-m telescope did confirm the earlier detection of a 7 per cent redshift. Work is continuing on BL Lacertae itself and on other, similar objects, which are known as *Lacertids*. At least one other Lacertid also has a similar redshift, 7 per cent, for its haze of gas. This object was observed with the new 3.9-meter Anglo-Australian telescope in the southern hemisphere.

In sum, there seems to be a strong relation between quasars and the cores of galaxies. This linkage may make the quasars seem somewhat less strange, but at the same time causes galaxies to seem more exotic.

SUMMARY AND OUTLINE OF CHAPTER 24

Aims: To understand the history and significance of quasars

Discovery of quasars (Section 24.1)
- QSR – quasi-stellar radio sources
- Interaction of optical and radio astronomy
- Positions from lunar occultations or from interferometry
- Huge Doppler shifts in spectra

Doppler shift in quasars (Section 24.2)

Non-spectral methods of detecting quasars (Section 24.3)
- Ultraviolet excesses
  - Quasi-stellar objects (QSO's) – non-radio sources

Importance of quasars (Section 24.4)
- Test of Hubble's law, which is thus a test of the accuracy of the distance scale

Energy problem (Section 24.5)
- Too much energy required to be accounted for
- Possible explanations
  - Supermassive star
  - Immense number of supernovae
  - Pulsarlike behavior
  - Annihilation of matter and anti-matter
  - Presently unknown method

Other explanations of redshifts (Section 24.6)
- Quasars ejected from cores of galaxies at very high velocities; no evidence of blueshifted quasars, however
- Gravitational redshifts
- Quasars operating under new physical laws?
- All these alternative explanations challenge Hubble's law

Peculiar galaxies and quasars (Section 24.7)
- Question of physical link between galaxies and quasars
- Stephan's Quintet and other sources with discrepant redshifts might be explained as chance alignments

Quasars and galaxies (Section 24.8)
- Are quasars truly a new phenomenon?
- Seyfert and N galaxies have bright nuclei that may be related to quasars
- BL Lacertae – the nearest quasar

QUESTIONS

1. Why was it important to know accurately the positions in the sky of radio sources for which there was at first no optical counterpart?

2. (a) You are heading toward a red traffic light so fast that it appears green. How fast are you going? (b) At $1 per mph over the speed limit of 55 mph, what would your fine be in court?

3. A quasar is receding at half the speed of light. (a) If its distance is given by the Hubble relation, how far away is it? (b) At what wavelength would the 21-cm line appear?

4. A quasar is observed and it is found that a line whose rest wavelength is 3000 Å is observed at 6000 Å. (a) How fast is the quasar receding? (b) How far away is it if its distance is given by the Hubble relation?

5. What is the difference between a QSO and a QSR?

6. Why does the rapid time variation in some quasars make the "energy problem" even more difficult to solve?

7. What are three differences between quasars and pulsars?

8. Briefly list the objections to each of the following "local" explanations of quasars:
   (a) They are local objects flying around at large velocities.
   (b) The redshift is gravitational.

9. If quasars were proved to be local objects, how would this help solve the "energy problem"?

10. What features of some quasars suggest that quasars may be some type of galaxy?

25

Cosmology

In Armagh, Ireland, in the mid-seventeenth century, Bishop Ussher declared that the universe was created at 9 a.m. on Sunday, October 23rd, in the year 4004 B.C. Nowadays we are less certain of the details of our origin.

We study the origin of the universe as part of the study of the universe as a whole. This larger field is called *cosmology.* Our discussion of astronomy has brought us to consider many different types of celestial objects and many different ways of observing them. All this observational knowledge and all these techniques, in conjunction with theoretical calculation, must be brought to bear on cosmological problems in order to understand the most fundamental questions about our universe. Still, there is a unifying theme in the study of cosmology. After all, we have only one universe, and its study is thus an investigation of just one thing. Yet the universe's uniqueness is also a fundamental limitation, for when we consider a class represented by just one member, we cannot tell whether conclusions are fundamental to the class (that is, would be true of other members, if there were any) or are restricted in applicability to that single member.

The part of cosmology devoted to the study of the origin of the universe (if, indeed, it had an origin) is called *cosmogony.* (Cosmogony as used in a more limited sense of the origin of the solar system was discussed in Section 10.5.) The study of where we have come from and of what the universe is like now leads us to consider where we are going. Is the universe now in its infancy, in its prime of life, or in its old age? Will it die? It is difficult for us, who, after all, spend most of our time thinking of topics like "which TV program shall I watch tonight" or "what's for dinner," to realize that we can seriously think about the structure of space around us, and to conclude what the future of the universe will be. One must take a little time every day, as Alice was told when she was in Wonderland, to think of impossible things. By and by we become accustomed to concepts that have at first overwhelmed us. You must sit back and ponder when studying cosmology; only in time will many of the ideas that we shall discuss take shape and form in your mind.

Albert Einstein in the Swiss Patent Office in Berne in 1905.

25.1 OLBERS'S PARADOX

Many of the deepest questions of cosmology can be very simply phrased. Why is the sky dark at night? Analysis of this simple observation leads to profound conclusions about the universe. We certainly know that the night sky is basically dark, with light from stars and planets scattered about on a dark background. But a bit of thought will show that if there is a uniform distribution of stars in space, it shouldn't be dark anywhere in the sky. If we look in any direction at all we will eventually see a star (Fig. 25–1), so the sky should appear uniformly bright. We can make the analogy to our standing in a forest. There, we would see some trees that are closer to us and some trees that are farther away, but if the forest is big enough our line of vision will always eventually stop at the surface of a tree. If all the trees were painted white, we would see a white expanse all around us. Similarly, when looking up at the night sky we would expect the sky to have the uniform brightness of the surface of a star.

Olbers phrased his paradox in terms of stars, whereas we know that the stars are actually grouped into galaxies. But we can carry on the same argument with galaxies, and deduce that we must see the average surface brightness of galaxies everywhere. This is patently not what we see.

The fact that this argument doesn't work, and that the sky **is** dark at night, is called *Olbers's paradox*. Heinrich M. W. Olbers phrased it in 1826, although the question had been discussed at least a hundred years earlier. The solution of Olbers's paradox leads us into basic considerations of the structure of the universe. This solution could not have been advanced in Olbers's time because it depends on astronomical discoveries that have been made more recently.

In explaining Olbers's paradox, it is best to clear up one possible point of confusion right at the outset. Remember—the fact that the total amount of light we get from a star decreases as the star gets farther away is not relevant to this discussion. In other words, we cannot explain the paradox merely by claiming that the farthest stars are so distant that they look relatively faint and therefore the sky looks black in their directions. We must realize that since each of our lines of sight would seem to end at the surface of a star, we would see the brightness of that star per unit area of its surface and not the total amount of light. As we consider bright surfaces that are farther and farther from us, even though the total amount of light we receive gets less, the brightness of each bit of surface stays the same. This is a difficult concept that engenders much confusion, and even astronomers sometimes find themselves momentarily making mistakes in calculations by forgetting that the surface brightness of an extended object like a nebula or galaxy (as opposed to a point source) is a constant. The surface brightness is constant because as the area of the object decreases, the total amount of light (the *flux*) we receive from the object decreases at the same rate, namely with the inverse square of the distance (see Fig. 2–42). For the purpose of this discussion, we must consider the actual sizes of the disks of the stars, even though the disks are so small when seen from the earth that the stars appear as points of light.

Figure 25–1 If we look far enough in any direction in an infinite universe, our line of sight will hit the surface of a star. This leads to Olbers's paradox.

You may be familiar with an example from photography that results from the same effect of constant surface brightness. Suppose that we set the aperture and exposure time of our camera by coming up close to a subject and making measurements. When we go 15 feet back to take our picture, these settings don't change from what they were when we were up close. This is simply because the surface brightness of the objects remains the same no matter how far we are away from them.

One might think that the easiest way out of the paradox, as was realized in Olbers's time, is simply to say that there is dust or other interstellar matter absorbing the light from the distant stars and galaxies. But this doesn't solve the problem, because the dust would soon absorb so much energy that it would heat up and begin glowing. So given a long enough time, all the matter in the universe would begin glowing with the same brightness, and we would have our paradox all over again.

One solution to Olbers's paradox lies in part in the existence of the redshift, and thus in the expansion of the universe (Section 23.4). This does not mean that the answer to the paradox is just that visible light from distant galaxies is redshifted out of the visible, for at the same time ultraviolet light is continually being redshifted out of the ultraviolet into the visible. The point is rather that each quantum of light undergoes a real diminution of energy as it is redshifted. Thus we do not see the level of brightness emitted at the surface of a faraway star or galaxy because the energy that was emitted has been diminished by this redshift effect before it reaches us. Since the energy, E, of a quantum at a given wavelength, λ, corresponding to a certain frequency, ν, is $E = h\nu = hc/\lambda$, the quantum has at first an energy corresponding to its original wavelength. As its wavelength gets longer (its frequency decreases) because of the Doppler effect and Hubble's law, the formula $E = hc/\lambda$ shows that its energy diminishes. This energy is actually lost from the quantum.

Of course, in Olbers's time, the large telescopes that later made the discovery of Hubble's law possible had not even been twinkles in the eyes of their builders, who were not yet born. The discovery that one resolution of the paradox depended on the expansion of the universe had to await later generations.

There are some complications to this major point that we could mention. As we look out in space, we are looking back in time, because the light has taken a finite amount of time to travel. If we could see far enough out we would possibly see back to a time before the formation of the stars. E. R. Harrison of the University of Massachusetts recently pointed out that this is an equally valid way out of the dilemma. In fact, Harrison calculates that this explanation is a more important contribution to the solution of the paradox than is the existence of the redshift. Most of the radiation, the absence of which contributes to the existence of Olbers's paradox, would have to come from stars that were very distant and that consequently emitted their radiation a long time ago—10^{24} years. The universe is simply not that old; on the basis of Hubble's law an age of 10^{10} years or so seems appropriate.

The fact that we have to know about the expansion and the age of the universe to answer Olbers's question—why is the sky dark at night?—shows how the most straightforward questions in astronomy can lead to important conclusions. In this case, we conclude that either (1) the universe is expanding, or (2) it is too young (assuming that if it had indeed been sufficiently old, earlier generations of stars would have been present to radiate), or (3) both.

25.2 THE BIG BANG THEORY

Astronomers looking out into space have noticed that on the large scale things look about the same in all directions. That is, ignoring the presence of local effects such as our being in the plane of a particular galaxy and thus seeing a Milky Way across the sky, the universe has no direction that is special. Further, it seems that there is no change with distance either, except insofar as time and distance are linked.

This notion has been codified as the *cosmological principle*: the universe is homogeneous and isotropic throughout space. The assumption of homogeneity says that the distribution of matter doesn't vary with distance from the sun, and the assumption of isotropy (Fig. 25–2) says that it is about the same no matter in which direction one looks. Actually, of course, we know that we have to look in certain directions to see out of our galaxy without having our vision ended by the interstellar dust, but remember that we are ignoring inhomogeneities or lack of isotropy on this small scale.

In fact, in most studies of cosmology, astronomers treat the universe as though all the matter in it were spread out uniformly in space. Of course we know that that isn't true; we have only to look at the stars and galaxies to see the lack of uniformity. But it is a great convenience in many theoretical calculations not to have to worry about "details" like galaxies.

"Big bang" is the technical as well as the popular name for these cosmologies; professionals write about the big bang in the scientific journals.

A set of theories of the universe known as *big bang* theories satisfies the cosmological principle. Basically, these cosmologies say that once upon a time there was a great big bang that began the universe. From that moment on, the universe expanded, and as the galaxies formed they shared in the expansion.

Remember that the galaxies themselves are not expanding; the stars in a galaxy don't tend to move away from each other.

Many students ask whether the fact that there was a big bang means that there was a center of the universe from which everything expanded. The answer is no; first of all, the big bang may have been the creation of space itself. Furthermore, the matter of this primordial cosmic egg was everywhere at once. There may be an infinite amount of matter in the universe, so it is possible that at the big bang an infinite amount of matter was compressed to an infinite density while taking up all space.

What was present before the big bang? Versions of the big bang theory (Fig. 25–3) have been put forward to deal with this question, but there is no

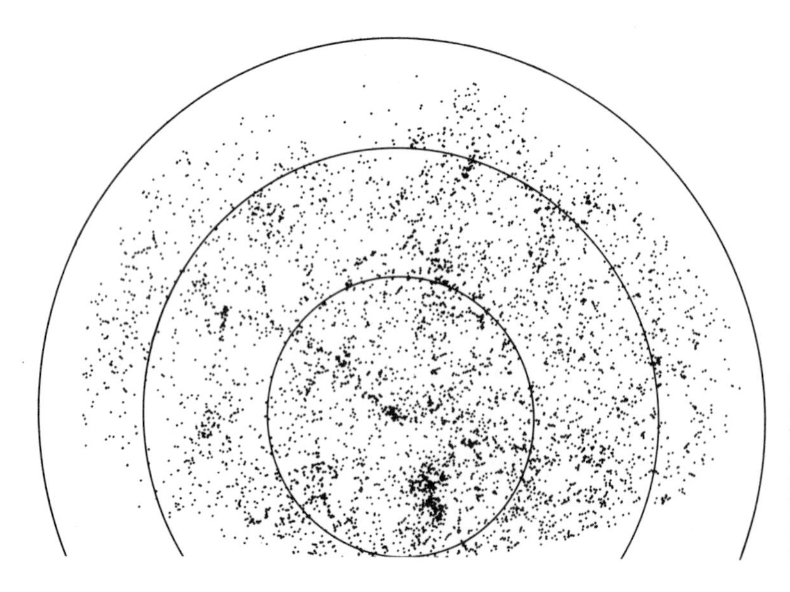

Figure 25–2 The distribution of 5634 galaxies on a plot that shows half the sky, with the center of the plot corresponding to the north pole of our galaxy and the circumference to the galactic equator. 60° and 30° lines of galactic latitude are also shown. The plot was made by P. J. E. Peebles of Princeton.

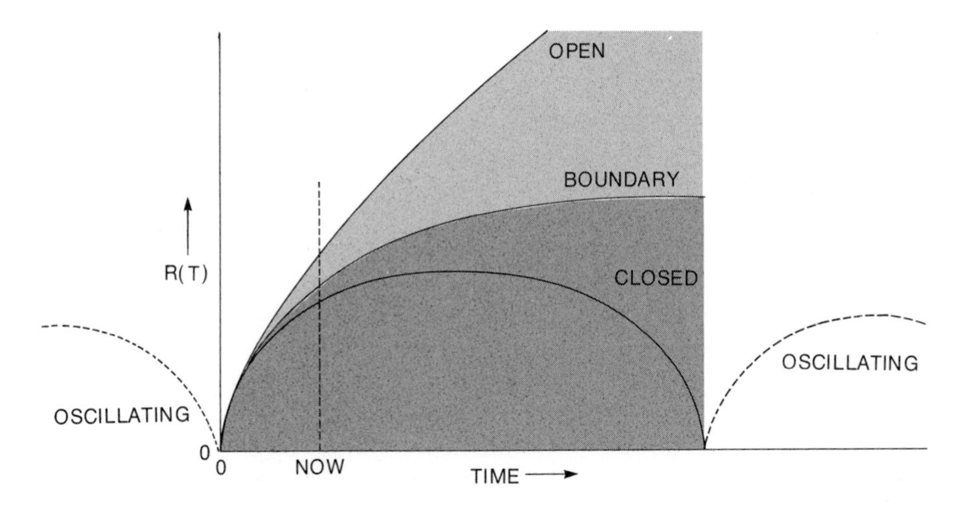

Figure 25–3 To trace back the growth of our universe, we would like to know the rate at which its rate of expansion is changing. Big bang models of the universe are shown; the vertical axis represents a scale factor, R, that represents some measure of distances and how they change as a function of time, T. The universe could be open and expand forever or closed and begin to contract again. If it is closed, we do not know whether it would oscillate or whether we are in the only cycle of expansion or contraction it will ever undergo. We will see later that current evidence favors the open model.

Astronomers use a quantity called the *deceleration parameter,* which is given the symbol q_0, to describe how fast the expansion is slowing down. $q_0 = 1/2$ marks the dividing line between an open universe (q_0 less than $1/2$, including 0) and a closed universe (q_0 greater than $1/2$, including 1).

real way to answer it. For one thing, we can say that time began at the big bang, and that it is meaningless to talk about "before" the big bang because time didn't exist. We do not now think that the universe can remain in a static condition, so it seems unlikely that the universe was always just sitting there in an infinitely compressed state, whatever "always" means. Of course, these possibilities don't answer the question of why the big bang happened. Another possibility is that there had been a prior big bang and then a recollapse, and that this most recent big bang was one in an infinite series of bangs. This version of the big bang is called the *oscillating universe.* It solves some questions, for if the universe is oscillating then we know that there were other cycles before the most recent big bang. However, the oscillating universe theory doesn't really tell us anything about a real origin of the universe because in this case there was no origin. The concepts discussed in this paragraph will not be easy to digest. They may take hours, years, or a lifetime to come to terms with.

If we consider Hubble's law with the current value for Hubble's constant—and the cosmologists Allan Sandage and Gustav Tammann have evidence that Hubble's constant is the same for "local" and "distant" galaxies—we can extrapolate backward in time. We simply calculate when the big bang would have had to take place for the universe to have reached its current state at its current rate of expansion. This calculation indicates that the universe is somewhere between 13 billion and 20 billion years old (Fig. 25–4). The range of values indicates in part the uncertainty in Hub-

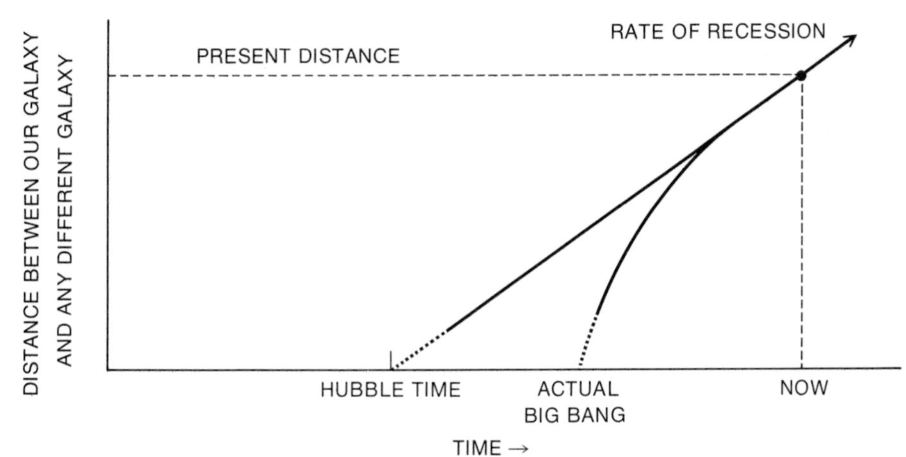

Figure 25–4 If we could ignore the effect of gravity, then we could trace back in time very simply; the Hubble time corresponds to the inverse of the Hubble constant ($1/H_0$). Actually, gravity has been slowing down the expansion. The vertical axis again represents some scale factor.

ble's constant and in part the uncertainty in the calculation of how much the effect of gravity would have by now slowed down the expansion of the universe.

The universe will last more than ten billion years, so we have nothing to worry about for the immediate future, but the study of the future of the universe is an exceptionally interesting investigation. Some of the methods of tackling this question involve measuring the amount of mass in the universe and thus the amount of gravity (as we will discuss in Section 25.7). Other methods of determining our destiny involve looking at objects as distant as possible to see if any deviation from the Hubble law can be determined. This investigation of distant bodies has been going on for many years, but a deviation from the straight-line relation between velocity and distance known as Hubble's law has not been found conclusively. Sometimes one hears that a slight deviation from a straight line has been found at the outermost end of the graph, but as yet there are no definitive results (Fig. 25–5). The deviations have always been within the margin of error of the statistics.

Since optical measurements are hindered because of the absorption of light by interstellar or even intergalactic dust, for a while there were high hopes that radio astronomy would provide a method for resolving the problem of the structure of the outermost parts of the universe. Since one cannot usually measure spectral lines, and therefore redshifts, in the radio spectra of these very distant galaxies, the method that has been used is that of *source counting.*

In radio source counting, one makes as complete a catalogue as possible of all the radio objects in the sky, or at least in a selected area of the sky, and lists the apparent brightnesses of all these objects. If all radio objects have approximately the same intrinsic brightness, or if at least the distribution of brightness does not change with time or from place to place, we

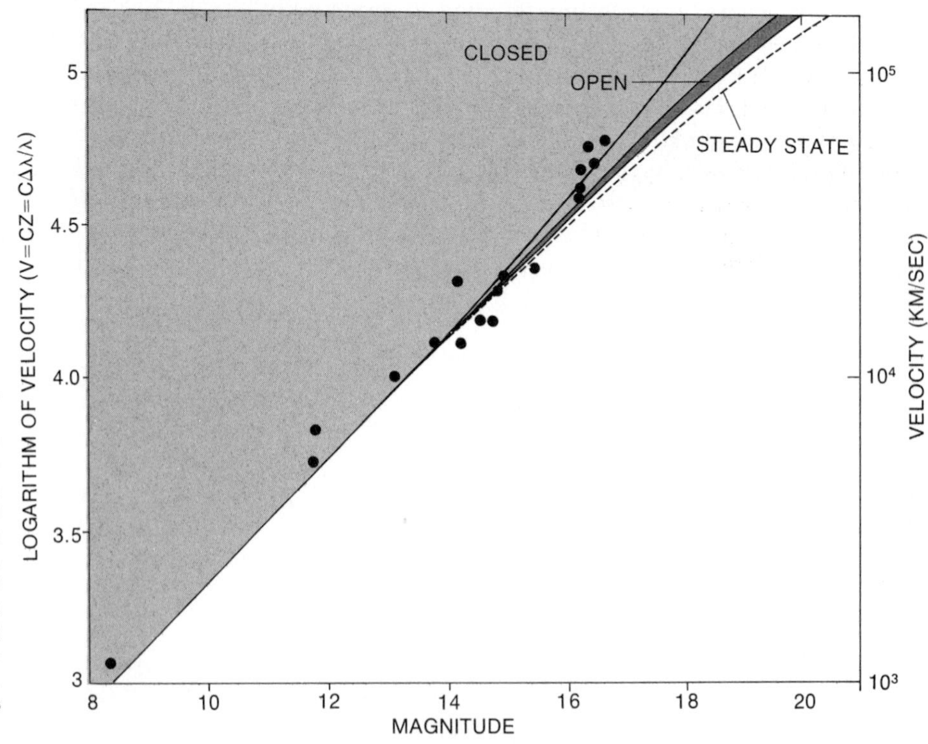

Figure 25–5 This Hubble diagram, plotted in terms of redshift and magnitude, shows faint clusters of galaxies (dots). In principle, we can determine the future of the universe from the slight deviations at upper right of the curves from a straight line. If the curve appears in any of the shaded regions, the expansion is slowing down. If it falls in the heavily shaded region, the universe is open, infinite, and will expand forever. If the curve falls farther to the left in the lightly shaded region (a possibility that is illustrated with a sample curve), then the universe is closed, finite, and will eventually begin to contract. The data are currently not sufficient to definitely conclude between these possibilities. The dashed line represents the steady state universe.

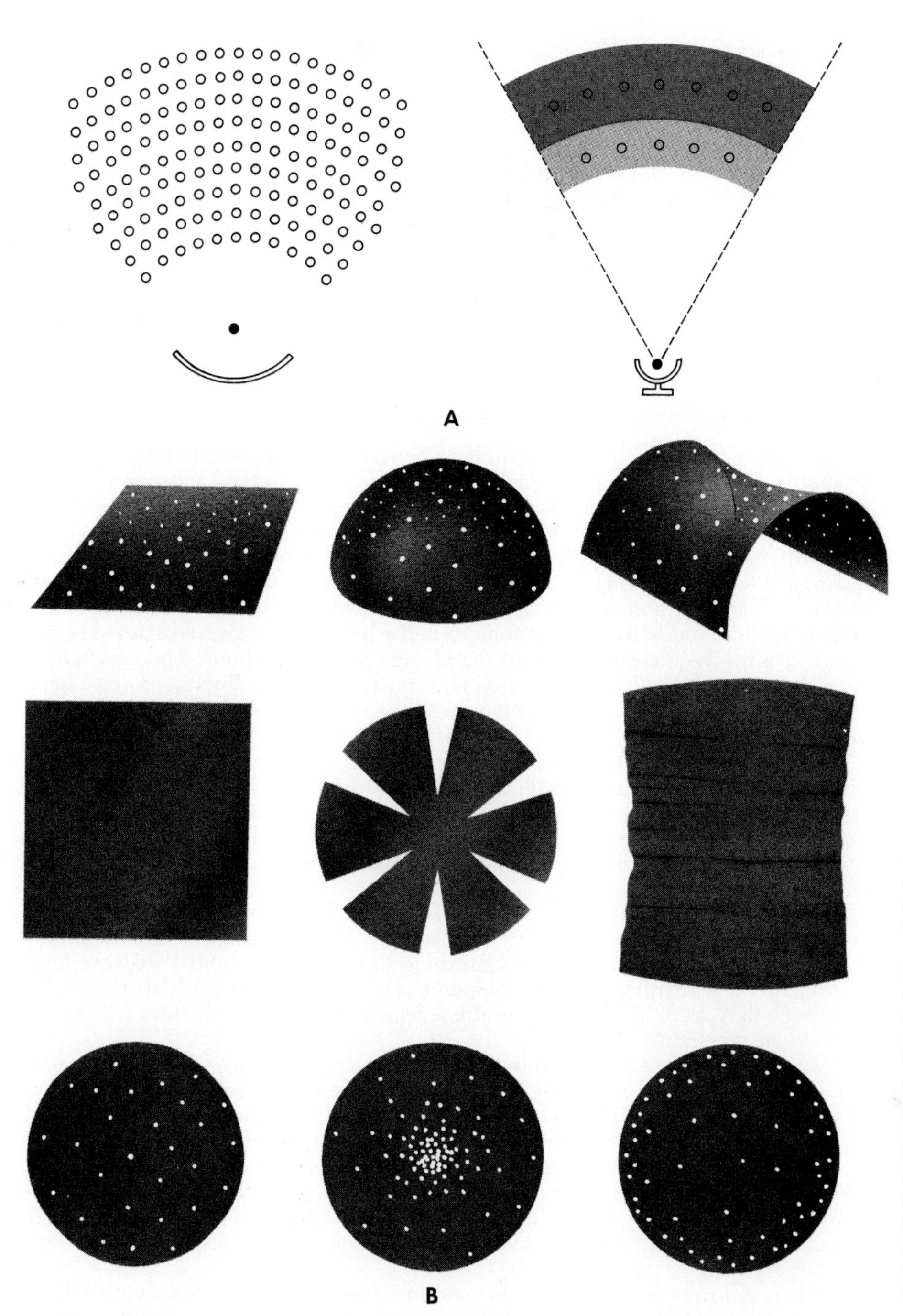

Figure 25–6(*A*) For a circular theatre, we can predict the number of seats at each distance from us if we assume that the seats are distributed uniformly. (*B*) We show here two-dimensional analogues to curved space. If a positively curved surface—a hemisphere, for example—is flattened, objects appear concentrated at the center. If a negatively curved surface—a saddle shape—is flattened, objects are relatively absent from the center. We try to use concepts like this to interpret counts of radio sources at different distances from us in terms of the curvature of space. It now seems that such arguments tell us more about how radio sources evolve over time than they do about cosmology.

would expect the number of sources at fainter and fainter magnitudes to increase in a known way.

We can illustrate this situation by considering the number of seats in a circular theater (Fig. 25–6*A*). Suppose that we are in the center of the stage, looking out at the audience. The last row has many more seats in it than the first row, because it is on a larger circle. If we simply know the distribution of seats in a row (that is, the width of the seats) and the distance from center stage to the various rows, we could calculate how many seats there are in each row without actually counting.

In radio source counting, we see if the number of sources at fainter magnitudes, which presumably corresponds to sources at greater distances, increases according to the way our calculations say it should. If this is not the case, we can interpret the result in terms of deviations from Hubble's law, or in terms of a curvature of space (Fig. 25–6*B*) that we can analyze according to Einstein's general theory of relativity. (See Sections 5.11 on the sun and 9.6 on black holes for discussions of general relativity and Section 1.1 for a discussion of the curvature of space.) The latest radio source counts do in fact indicate some deviation from the way in which we might expect for flat space, but it is now realized that the deviation is mainly a result of the fact that as we look out in space we are looking back in time. We are seeing the radio sources as they were long ago, when they undoubtedly had different brightnesses in all parts of the spectrum than they do now. Thus radio source counting tells us more about the evolution of radio sources than it does about cosmology.

Although we have discussed the big bang theories in words rather than in equations, these theories are actually discussed by theoretical astrophysicists as solutions to a set of equations that Einstein advanced as part of his general theory of relativity. Einstein tried to find solutions to his equations and the solutions he found in 1917 corresponded to an expanding or to a contracting universe. But at that time the observational evidence for the expansion of the universe was not known, and it was thought that the universe was static. So Einstein deliberately and artificially modified his equations to make his model universe static. He did this by arbitrarily adding a constant to one side of the equations. This constant is called the *cosmological constant,* and one can calculate the value that it must have to balance the expansion of the universe that would otherwise be implied by the equations and thus make the universe static.

A solution to Einstein's original equations was soon worked out by the Dutch scientist Willem de Sitter in 1917. But his solution had a deficiency too—it corresponded to a universe without any matter in it. So de Sitter's solution is also unsatisfactory, though it may wind up being approximately valid in our distant future when the universe has expanded so much that it has negligible average density.

Note that even though Einstein's and de Sitter's particular solutions give answers that are not in agreement with observational evidence, the investigations were still worth carrying out. Work on them developed methods that were later useful, and these extreme solutions helped us to understand the more realistic solutions once they were found. Much of the research done in science does not lead to ultimate answers at the first try.

More interesting sets of solutions that correspond to more physically reasonable universes were worked out in the early 1920's by the Soviet mathematician Alexander Friedmann. Friedmann's solutions are the mathematical formulations of the big bang cosmologies that we now accept. The Belgian abbé Georges Lemaître in 1927 also discovered interesting solutions of the equations and, in particular, Lemaître recognized the applicability of his solutions to the observational evidence. He suggested that the original condition, the "cosmic egg," from which the universe was expanding was hot and dense. He thus deserves credit for the theory of a *hot big bang.*

When Einstein realized that observations showed that the universe really is expanding, he withdrew his artificial modifications of the equations and accepted the newer solutions of his original equations. Still, some modern theoreticians keep trying to find a physical interpretation of the cosmological constant, even though most astronomers feel that it was never necessary to add the constant in the first place (which is equivalent to saying that the cosmological constant is simply equal to 0).

25.3 THE STEADY STATE THEORY

The cosmological principle, that the universe is homogeneous and isotropic, is very general in scope, but starting in the late 1940's, three British scientists began investigating a principle that is even more general.

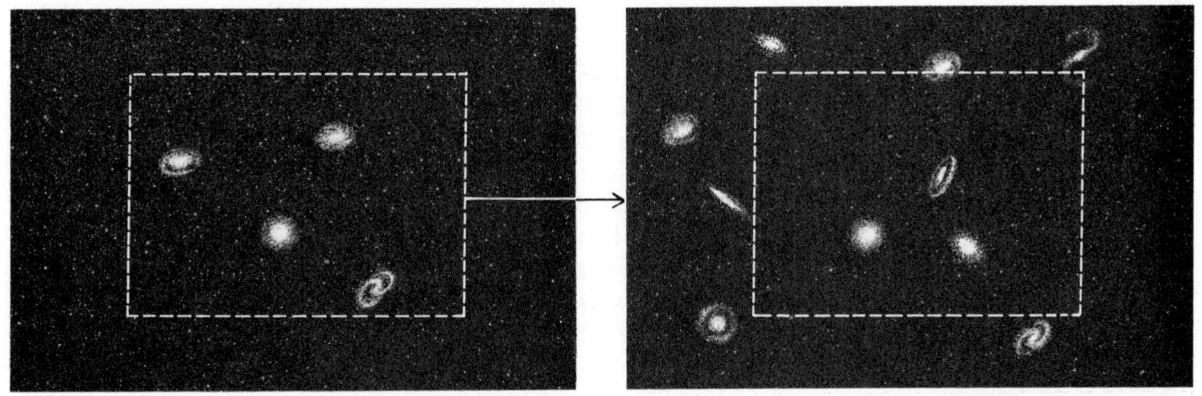

Figure 25–7 In the steady state theory, as the dotted box at left expands to fill the full box at right, new matter is created to keep the density constant. In the picture, the four galaxies shown at left can all still be seen at right, but new galaxies have been added so that the number of galaxies inside the dotted box is about the same as it was before.

Hermann Bondi, Thomas Gold, and Fred Hoyle considered what they called the *perfect cosmological principle*, that the universe is not only homogeneous and isotropic in space but also unchanging in time. According to Occam's Razor (Section 19.5), we must accept the simplest theory that is in accord with all the observations, but it is a matter of personal preference whether the cosmological principle or the perfect cosmological principle is more simple.

The theory that follows from the perfect cosmological principle is called the *steady state theory* (Fig. 25–7); it has certain philosophical differences from the big bang cosmologies. For one thing, according to the steady state theory, the universe never had a beginning and will never have an end. It always looked just about the way it does now and always will look that way. Some scientists are glad to see the question of what happened before the big bang, or the need for finding a cause for the big bang, eliminated in the steady state theory.

But the steady state theory must be squared with the one major fact that we know about cosmology, that the universe is expanding. How can the universe expand continually but not change in its overall appearance? Remember that in the basic formulation of the theory we consider the matter to be smoothed out throughout all space to an average density. It follows that for the density to remain constant, new matter must be created at the same rate that the expansion would decrease the density. Thus the density remains the same. Now the matter that is created in the steady state theory is not simply matter that is being converted from energy according to Einstein's famous formula, $E = mc^2$. No, this is matter that is appearing out of nothing, and is thus equivalent to energy appearing out of nothing.

Many scientists objected very strenuously to the idea that matter could appear out of nothing. After all, the "law" that matter is conserved (i.e., cannot be created or destroyed), generalized by Einstein into a law of conservation of matter and energy, seems very basic. Nevertheless, since we know the rate at which the universe is expanding, we can calculate how much new matter would have to be created. It works out to be only one hydrogen atom per cubic centimeter of space every 10^{15} years, which is equivalent to one thousand atoms of hydrogen per year in a volume the size of the Astrodome in Houston. This is the rate necessary to maintain the overall value observed for the density of matter in space. This value is far too small for us to be able to measure on earth, or even for us to be able to hope to measure on earth in the foreseeable future. So any deviation that the steady state theory would require from the "law" of conservation of matter and energy is certainly too small to be ruled out on observational grounds.

No one knows, of course, in what form the matter would appear. It could appear uniformly distributed throughout space, or perhaps it could preferentially appear in places where there was no matter already. The latter supposition would seem helpful in supplying the matter to places where it would be needed to form new galaxies to take up the space vacated by the receding old ones. But it has also been suggested that perhaps the matter appears preferentially at such places as the centers of galaxies or in quasars, possibly even as matter and anti-matter, in order to provide the powerful energy sources that seem to exist at those locations.

For many years a debate raged between proponents of the big bang theories and proponents (largely the inventors) of the steady state theory. The evidence that came in—optical studies of deviations from Hubble's law in distant galaxies, or radio source counts—consistently seemed to favor big bang cosmologies over the steady state theory. But it could be fairly pointed out that none of this evidence against the steady state theory was conclusive, and in several cases either alternative explanations for the data were conceivable or the steady state theory could be modified, sometimes extensively, to be consistent with the recent discoveries. Since the time of the discovery of quasars, many astronomers have felt that the distribution of quasars—all far away from us in space, which corresponds to their having been more numerous at an earlier time—is strong evidence against the steady state cosmology. In the next section we shall discuss still stronger evidence against the steady state theory.

25.4 THE PRIMORDIAL BACKGROUND RADIATION

In 1965, a discovery of the greatest importance was made: radiation was discovered that is most readily explained as a remnant of the big bang itself.

That much is easy to state, and if we have indeed discovered radiation from the big bang itself then clearly the steady state theory is discredited. But the explanation of just how the observed radiation is interpreted to come from the big bang is fairly technical and requires an understanding of abstract concepts. However, the sheer importance of the discovery warrants our taking the time to consider this matter thoroughly.

The original discovery was made by Arno A. Penzias and Robert W. Wilson of the Bell Telephone Laboratories in New Jersey. They were testing a radio telescope and receiver system (Fig. 25–8) to try to track down all possible sources of static. The discovery, in this way, thus parallels Jansky's discovery of radio astronomy itself.

Penzias and Wilson were observing at a wavelength of 7 cm in the radio spectrum. After they had subtracted the contributions of all known sources to the static, they were left with a residual signal that they could not explain.

At the same time, a group at Princeton University including Robert Dicke, P. J. E. Peebles, David Roll, and David Wilkinson had made the prediction that there should be radiation from the big bang that corresponded to a certain signal strength at 7 cm. These astronomers were, in fact, in the process of building their own receiver to observe at another

Actually, the Princeton group was not the first to predict that such radiation might be present. Many years earlier, in 1946, George Gamow had made a similar prediction, but his work on this had long since been forgotten by almost everyone. His work was only recalled after all the new measurements had been made and interpretations advanced.

Figure 25-8 The large horn-shaped antenna at the Bell Telephone Laboratories' space communication station in Holmdel, New Jersey. 3° background radiation was discovered with this instrument.

radio wavelength. The findings of the observers from Bell Labs confirmed their prediction.

The basic prediction was that there should be radiation permeating the entire universe, and that this radiation would follow the spectrum of a black body at a certain, very low temperature (Fig. 25-9).

In addition to the black-body nature of the radiation spectrum, the fact that the radiation was observed to be highly isotropic was other evidence that it came from the big bang. Because the big bang took place simultaneously everywhere in the universe, or even "was" the universe, we

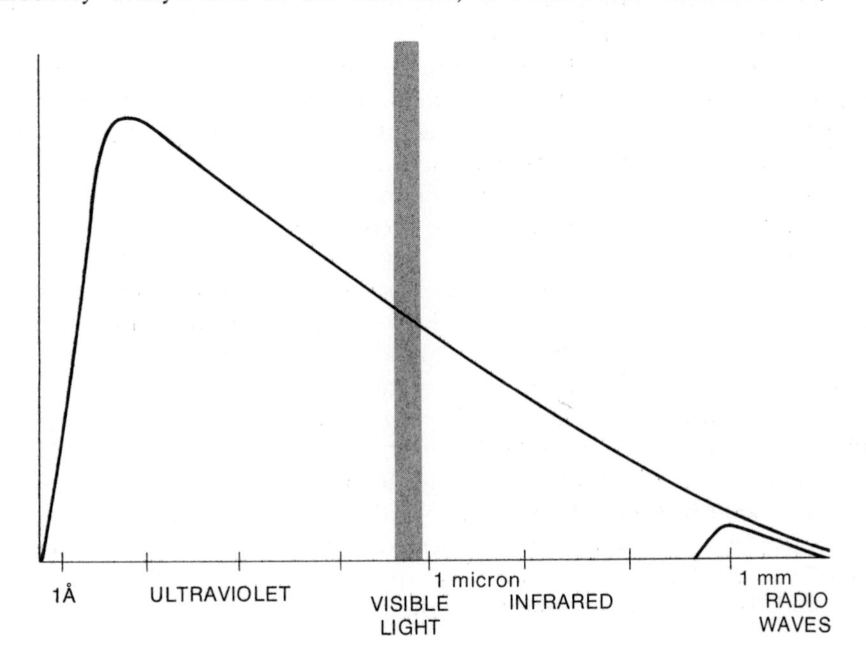

Figure 25-9 Planck curves for black bodies at different temperatures. Radiation from a 3° black body (the lower curve) peaks at very long wavelengths. On the other hand, radiation from a very hot black body (the upper curve) peaks at very short wavelengths.

would expect that radiation from it would fill the whole universe. It would thus have the property of being isotropic to a very high degree—that is, it would be the same in any direction that we observe. It was generated all through the universe at the same time, so its remnant must seem to come from all around us now.

Most mechanisms to account for such black-body radiation, other than the big bang, predict that the intensity of the radiation would vary slightly from direction to direction in the sky. For example, if the radiation originated in discrete sources, then we would expect the radiation to be most intense in the direction of the strongest sources. Of course, a model that provided an isotropic distribution of radiating bodies would not be subject to this objection.

We have dealt before with black bodies (Section 2.5) and have seen how the radiation from stars conforms in particular regions of the spectrum to emission from black bodies at certain temperatures. The sun, for example, corresponds in the visible spectrum to a black body of about 6000 K, though in other parts of the spectrum it does not conform to this black-body relation.

But this non-conformity in itself is enough to show that the sun **is not** a black body, even though its radiation agrees with the black-body relation in some part of its spectrum. A black body is a theoretical concept, an idealized version of the perfect radiating body. And while no individual body in existence can be a perfect black body, we can still describe the properties that a black body must have.

Basically, emission from a black body follows Planck's law of radiation (Fig. 2–12 and 2–14) in that for a given temperature there is an equation that links any wavelength we consider and the rate of radiation emitted at that wavelength. If radiation from anything follows Planck's law, then that thing is a black body; if the radiation deviates from Planck's law, even by the slightest amount, then that thing is not a black body.

The key fact to remember about Planck's law is that specifying just one number—the temperature—is enough to define the whole Planck curve. Given a particular temperature, we can predict the intensity of radiation at any given wavelength. Given another temperature, we predict a different wavelength-intensity relationship.

But still we tend to think of black bodies as physical rather than conceptualized objects, and that is a temptation that must be fought here in order to understand this universal radiation from the big bang. It is quite normal for us to try to visualize abstract concepts in concrete terms. Horatio, for example, speaks to Hamlet of "the morn, in russet mantle clad. . . ." Of course, the morn wasn't wearing any such thing; it has no particular shape to be clad.

Now that we have reviewed black bodies, let us discuss how radiation could have arisen in the big bang. The leading models of the big bang consider a hot big bang, with temperatures of billions of degrees in the fractions of a second following the beginning of time.

In the millennia right after the big bang, the universe was opaque. Photons did not travel very far before they were absorbed. But gradually the universe cooled, and after about 100,000 years, when the universe reached 3000 K, the hydrogen ions suddenly combined with electrons to become hydrogen atoms. With this new situation, there was less continuous opacity, since hydrogen has mainly a spectrum of lines. Thus the universe suddenly became transparent. Since the matter would no longer continually absorb and reradiate all the radiation, we are left with the remnant of this original radiation, which is traveling through space forever. As the universe continues to expand, the remnant radiation retains the shape of a Planck curve but the curve corresponds to cooler and cooler temperatures. Planck curves corresponding to cooler temperatures have lower intensities of radiation and have the peaks of their radiation shifted toward the red. Because of this, the sun, at 6000 K, peaks in the yellow-green; a cool star, at 3000 K, peaks in the infrared; and the universe, much cooler yet, peaks at still longer wavelengths, equivalent to the very shortest wavelength end of the radio region of the spectrum. In a moment, we shall see just how cool the universe is.

Now, we recall that Penzias and Wilson measured a particular flux at 7 cm. This flux corresponds to what would be emitted by a black body at a temperature of only 3 K. Thus we speak of the universe's 3° *background radiation*, or, sometimes, the *background radiation*, or the *primordial background radiation.** Following the original measurements at Bell Labs and at 3 cm by the Princeton group, various other groups measured values for the background radiation at different wavelengths (Fig. 25–10). Unfortunately, though, all of these measurements were obtained at wavelengths longer than that of the peak, so this left researchers with just the right-hand half of the curve defined. All the points corresponded to a temperature of approximately 3°, but still, a black-body curve does have a peak. It would have been more satisfying if some of the points measured lay on the left side of the peak graphed in Figure 25–10, or at least close enough to the peak on the right side to show that the curve in fact turns over. Points lying on the left side of the peak would prove conclusively that the radiation followed a black-body curve and was not caused by some other mechanism that could produce a straight line, or some other form that happened to mimic a 3° black-body curve in the centimeter region of the spectrum.

Though it would have been desirable to measure a few points on the short wavelength side of the black-body curve, astronomers were faced with a formidable opponent that frustrated their attempts to measure these

*Under the new temperature notation now in use, this would be called 3 K radiation, but we shall join the astronomical community in continuing to call it 3° radiation.

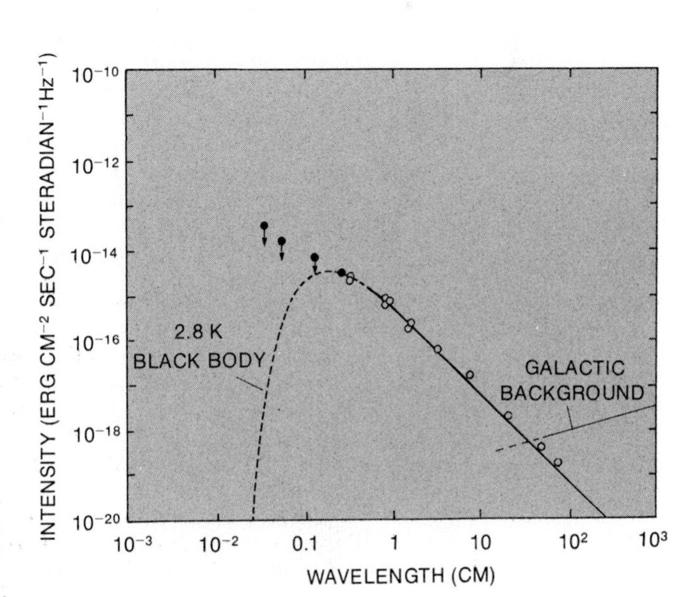

Figure 25–10 Direct measurements (dots) and upper limits from observations of the cyanogen radical (arrows) strongly indicated that the background radiation exists and is at a temperature of about 3°.

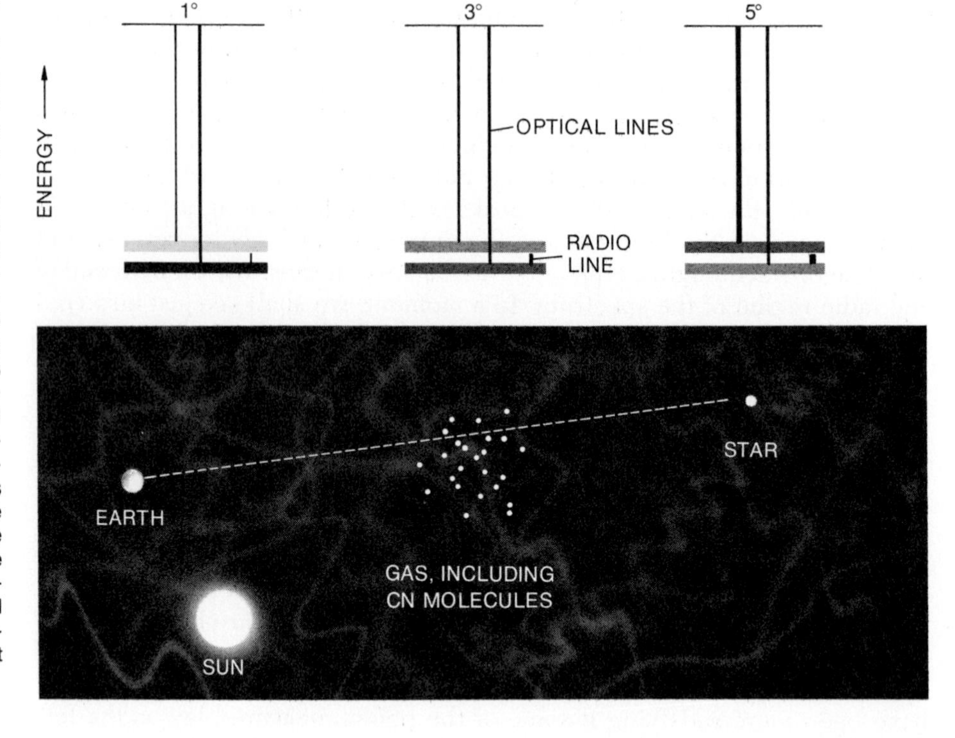

Figure 25–11 For a higher temperature of radiation, more CN molecules are excited to their second energy state. This changes the ratio of intensity of the two optical lines that connect these lowest two energy states with a higher state. Thus measuring this ratio gives the temperature of the background radiation at the radio wavelength that links the two lowest levels. The shading of energy levels shows schematically that the population of the middle level increases as the temperature increases. The thickness of the vertical lines shows schematically the resulting changes in the relative strengths of the spectral lines. The two optical CN lines were observed long before direct measurements could be made at the radio wavelength.

points: the earth's atmosphere. Our atmosphere absorbs radiation of wavelengths less than about 6 mm. This is a severe limitation as the turn-over in the 3° black-body curve occurs at about 2 mm, and very little radiation gets through at that wavelength.

For some years, we had only an indirect method of showing that the black-body curve indeed peaked at about 2 mm.

Absorption lines from interstellar molecules of cyanogen (CN) had long been known in the optical range of the spectrum. But the relative strengths of two of CN's spectral lines in the visible part of the spectrum depended on how many atoms were in the first excited level above the ground state (Fig. 25–11). And the easiest way to increase the number of atoms in this first upper level is to increase the amount of radiation at 2.6 mm, which corresponds to a transition between the two levels (the ground level and this first upper level). If there were no radiation at 2.6 mm, then essentially all the atoms would be in the lowest ground state. If there were a little bit of 2.6 mm radiation, corresponding to a 3° black body, there would be enough energy to excite a few of the atoms to a slightly higher state. If there were more 2.6 mm radiation, corresponding to a higher-temperature black body, there would be even more atoms in higher states. The visible lines of CN that were actually observed correspond to transitions between these levels and other, much higher, levels.

In this manner, the CN molecules acted as thermometers placed out in space. We would look with optical telescopes toward nearby stars to see the strengths of certain absorption lines in the optical part of the spectrum; these lines were formed by gas in interstellar space absorbing the radiation from the background stars (as in Fig. 21–20). The strengths of visible absorption lines that originated at the lowest ground state relative to other absorption lines tell us about the amount of radiation at 2.6 mm, near the peak of the 3° black-body curve, since the number of atoms in these energy levels was affected by the flux there. Thus we could learn about a wavelength that we could not observe directly. This complex chain of reasoning illustrates the indirect methods that astronomers must use when they are stymied by the atmosphere or by other limitations.

As a result of the indirect CN observations described above, we had a value at 2.6 mm that showed that the amount of background radiation did not keep increasing in a straight line toward shorter wavelengths. If the radiation spectrum near 2.6 mm was a straight line, then we would have expected the optical spectral line of CN to be stronger than it was observed to be. Values measured for even shorter radio wavelengths—obtained from another CN line and lines from CH and CH^+—were not accurate enough to match the 3° black-body curve exactly, but they did provide more evidence that the radiation being measured did not keep increasing and consequently probably turned over and followed a black-body law.

Actually, the fact that the CN observations implied a universal temperature of a few degrees above absolute zero had been noticed 15 years before the direct detection of the primordial radiation, by a Canadian scientist, Gerhard Herzberg, but since there was then no known theoretical reason for this to be the case, the observation was just briefly mentioned and set aside.

More direct observations of the region beyond the peak of the Planck curve were attempted, including observations from rockets, but the attempts failed because of observational difficulties.

In 1975, infrared observations were finally made that seem to have proved unequivocally that the radiation follows a black-body curve. A group from NASA's Institute for Space Studies in New York and the University of California at Berkeley launched instruments in a balloon that traveled to a height of 40 km (25 miles), high enough to get above most of the earth's atmosphere. The experiment was an especially difficult one to carry out because the instruments had to be kept very cold—only a few degrees above absolute zero—by immersing them in helium cooled so much that it had liquefied. It is a good trick to launch such complex equipment. The astronomers were able to measure the short wavelength side of the black-body curve down to 0.6 mm, and found that their measurements agreed with a black-body curve of 3° (Fig. 25–12).

The importance of the discovery of the background radiation cannot be overstressed. Dennis Sciama, the British cosmologist, put it succinctly by saying that up to 1965 we carried out all our cosmological calculations knowing just one fact: that the universe expanded according to Hubble's law. After 1965, he said, we had a second fact: the existence of the background radiation. That may be a bit oversimplified, but it is essentially true.

So at present almost all astronomers consider it settled that radiation has been detected that could only have been produced in a big bang. Accepting this clearly rules out the steady state theory as a viable cosmology, and means that some version of the big bang theory must hold. It doesn't settle the question of whether the universe will expand forever, or eventually contract, or oscillate.

It now seems that a very slight difference in temperature—about one part in a thousand—has been measured from one direction in space to the

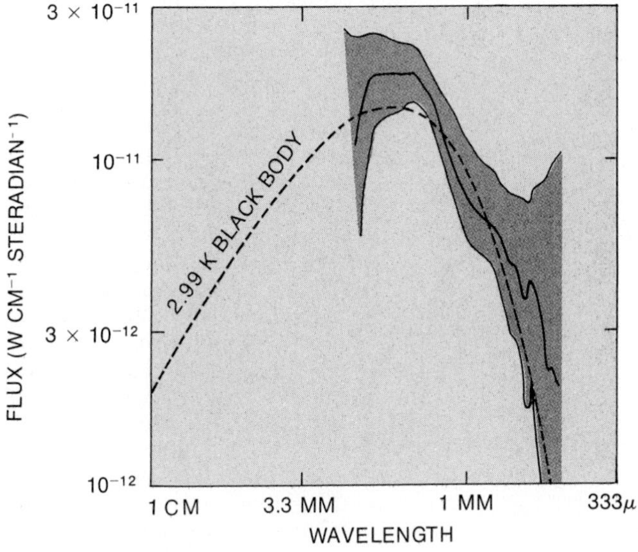

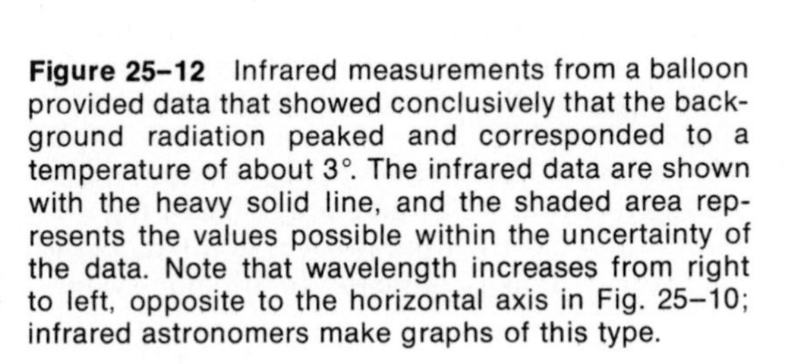
Figure 25–12 Infrared measurements from a balloon provided data that showed conclusively that the background radiation peaked and corresponded to a temperature of about 3°. The infrared data are shown with the heavy solid line, and the shaded area represents the values possible within the uncertainty of the data. Note that wavelength increases from right to left, opposite to the horizontal axis in Fig. 25–10; infrared astronomers make graphs of this type.

The notion that the universe as a whole might have a standard of rest is very exciting, and further studies of the anisotropy must be made.

opposite direction. This anisotropy is what would result from the Doppler effect if the earth had an absolute velocity of 330 km/sec with respect to the standard of rest of the universe as defined by the background radiation.

This section exemplifies a problem that exists in learning modern astronomy. Most of the readers of this book will not even have heard about 3° black-body radiation before, and here we are saying that it is one of the two most significant things that astronomers know about the universe. Students who take a course in modern history may expect to learn a lot about pre-war Europe, military strategy, and other background, but still, they have at least heard of World War II before they started the course. Such mind-boggling and non-intuitive conceptions as primordial radiation are some of the exciting things about modern astronomy. New developments continue to surprise us all. It takes time to become familiar with these concepts; their understanding requires a conceptualization of complex and abstract material.

25.5 THE HIERARCHICAL UNIVERSE

All the standard models of big bang and steady state cosmologies implicitly assume that we can treat the universe as though matter is spread out uniformly in it. This matter, to simplify calculation, is considered to fill all of space with a certain mean *cosmic density*. The density can be measured by adding up all the matter in a certain suitably large volume and dividing by the size of the volume. Stars and galaxies are then simply treated as fluctuations in this smoothed-out distribution of matter.

But is this really a fair assumption? Let us ask whether or not we really can define a cosmic density.

If we start with small volumes, the average density is fairly high. The average density of the earth, as we discussed, is about $5\frac{1}{2}$ grams/cm^3. But to consider the average density of the solar system, we add up the masses of the sun, planets, asteroids, comets, and interplanetary matter, and divide this sum by the volume of the solar system. Remember that there is a lot of empty space in the solar system, and thus the mean density of the solar system is much lower than the mean density of a single planet.

A general idea of the mean density of the solar system can be determined by the following calculations. (This procedure also serves as an example of the type of rough estimation that is common in astronomy.) Most of the solar system's mass is in the sun, which has an average density of approximately 1 gram/cm^3. The density of the solar system will be less than the density of the sun by the same factor that the volume of the solar system is greater than the volume of the sun. The sun has a radius of 0.7×10^6 km; let us say, for purposes of estimation, that it is 10^6 km. Its volume, $4\pi/3\,r^3$ (which is of the same order of magnitude as r^3 because $4\pi/3$ is not far from 4, which is of the same order of magnitude as 1), is thus about $(10^6)^3 = 10^{18}$ km^3. The solar system's radius is at least 50 A.U., because that is the maximum distance that Pluto can be from the sun. It is difficult to define the thickness of the solar system, but since Pluto has an orbit that is very inclined, we can get a lower limit by saying that the solar system is at least 10 A.U. thick. (If the solar system is really spherical, the volume would be even larger.) Thus the volume of the solar system is at least that of a volume of a cylinder: $\pi r^2 h$, which is approximately, very approximately, equal to r^2h. Now, one A.U. is 1.5×10^8 km, and the volume of the solar system is thus $(50 \times 50 \times 10)$ A.U.$^3 = (50)^2 \times 10 \times (1.5)^3 \times (10^8)^3$ km$^3 = 5^2 \times (1.5)^3 \times 10^3 \times 10^{24}$ km^3, which is roughly $10^2 \times 10^3 \times 10^{24}$ km$^3 = 10^{29}$ km^3. This volume is greater than the sun's volume by a factor of 10^{11}. (Note that this is a large enough factor that the smaller factors of 10 or so that we ignored in the estimating process don't affect the qualitative validity of this result.) The average density of matter in the solar system is thus approximately 10^{-11} gm/cm^3.

We have seen in the last two paragraphs that the density of the solar system is very much less than the density of a single planet because of all the empty space that is averaged in. Similarly, the density of a galaxy is very much less than the density of a solar system, and the density of a cluster of galaxies is very much less than the density of a single galaxy. When we add up all the material in the largest volume we can measure, that of the universe on the scale of the clusters of galaxies, we find a density of approximately 10^{-31} gm/cm^3, and call this the *cosmic density*. The presence of invisible matter may change this by a factor of 10 to 100, but that doesn't change the fact that the cosmic density of matter is very low by terrestrial, stellar, or even galactic standards.

A very few astronomers have pointed out that we cannot rule out the possibility that there are clusters of clusters of galaxies, and then clusters of clusters of clusters of galaxies, and so on. This model is called a *hierarchical universe* (Fig. 25–13). In this case, the mean density would keep going down as we evaluated larger and larger volumes. Thus if we want to say that the density is any very small, even negligible value, we need only consider a sufficiently large volume. Thus the density gets infinitely close to zero as we consider larger and larger volumes. Mathematicians would say that for an infinite volume the limit of the density is zero.

If this is the case, then many cosmological calculations are nonsense. If there is no such thing as a "mean density," then we cannot apply our simple ideas that the mass corresponding to a measured mean density exerts or even pulls it back in on itself. There is no real resolution of this problem and we shall put it aside for the rest of our discussion. Most astronomers ignore the question of the hierarchical universe and are satisfied to deal with the largest volumes that we can measure: clusters of galaxies. There are philosophical grounds for considering only volumes on which one can hope to make measurements, and it is largely a philosophical choice whether one chooses to believe that when one considers larger volumes the cosmic density remains at its value for clusters of galaxies, or whether a hierarchical universe exists and the cosmic density approaches a limit of zero.

Figure 25–13 In a hierarchical universe we would see not only galaxies (*left*) and clusters of galaxies (*center*) but also clusters of clusters of galaxies (*right*) and so on.

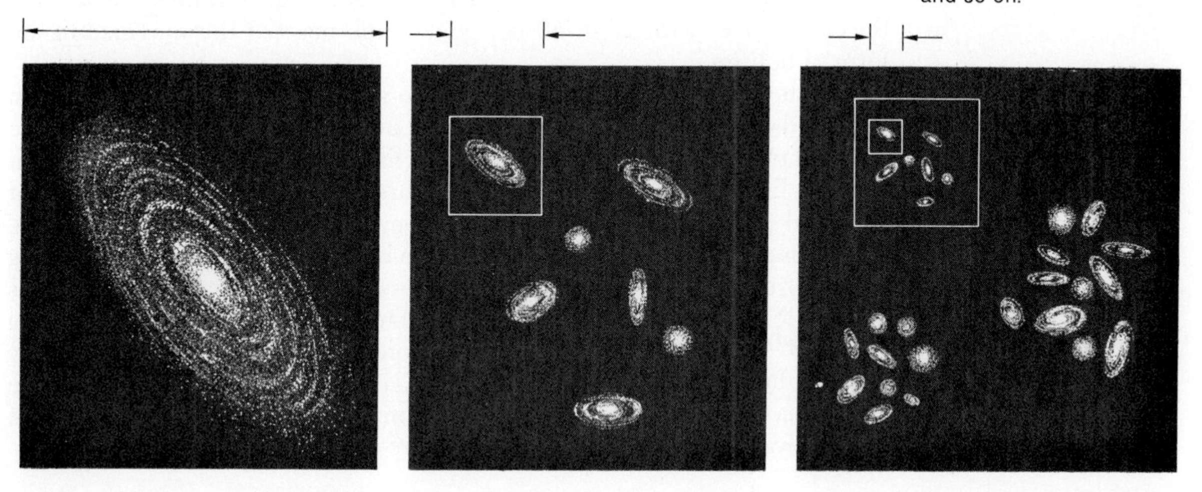

25.6 THE CREATION OF THE ELEMENTS

How did the tremendous explosion we call the big bang result in the universe we now know, with galaxies and stars and planets and people and flowers? Obviously, many complex stages of formation have taken place, and what was torn asunder at the beginning of time has now taken the form of an organized system.

We can trace back the Hubble law by following the expansion backward in time (see Fig. 25–4) and calculating how long ago all the matter we see would have been collapsed together. This calculation leads to the idea that the big bang took place some ten to twenty billion years ago—the best modern estimate is thirteen billion. It is difficult to comprehend that we can meaningfully talk about the first few **seconds** of that time so long ago. But we can indeed set up sets of equations that satisfy the physical laws we have derived, and make computer simulations and calculations that we think tell us a lot about what happened in the first few moments after the origin of the universe.

According to these calculations, as the big bang took place, the universe was compressed to a fantastically high density. At first the temperature was trillions of degrees, and only the simplest kinds of matter, such as protons, neutrons, electrons, and neutrinos, were present. Perhaps their antiparticles—antiprotons, antineutrons, positrons, and antineutrinos—were present as well. (Another model, presented earlier by George Gamow, had a slightly different scenario—at first all was neutrons. Gamow called this sea of neutrons at the beginning of the universe by the name *ylem*. Neutrons spontaneously decay in a few minutes into protons and electrons and antineutrinos, so the eventual effect was similar.)

Some of the earliest quantitative work on nucleosynthesis in the big bang was described in an article published under the names of Ralph Alpher, Hans Bethe, and Gamow in 1948. Actually Alpher and Gamow did the work, and just for fun included Bethe's name in the list of authors so that the names would sound like the first three letters of the Greek alphabet. These letters seem particularly appropriate for an article about the beginning of the universe.

After about a hundred seconds, the temperature dropped to a billion degrees (which is low enough for a deuterium nucleus to hold together), and the protons and neutrons began to combine into heavier assemblages, the nuclei of the heavier isotopes of hydrogen and elements like helium and lithium. The study of the formation of the elements is called *nucleosynthesis*, or sometimes *nucleogenesis*.

The first nuclear amalgam to form was simply a proton and neutron together. We call this a *deuteron*; it is the nucleus of deuterium, an isotope of hydrogen (see Fig.6–8). Then two protons and a neutron could combine to form the nucleus of a helium isotope, and then another neutron could join to form ordinary helium. Within minutes, the temperature dropped to 100 million degrees, too low for nuclear reactions to continue. Nucleosynthesis stopped, with 25 or 30 per cent of the mass of the universe in the form of helium.

To test this theory, we would like to study the distribution of helium around the universe, to see if it tends to be approximately this percentage everywhere. But helium is very difficult to observe; it has few convenient spectral lines for us to study, and those are usually observable only in hot stars and gaseous nebulae, where a lot of energy is available to excite them. Furthermore, even when we can observe helium in stars we are observing only the surface layers, which do not necessarily have the same abundances of the elements as the interiors of the stars. So even though "the helium problem" has been extensively studied, one cannot as yet conclusively say that the helium is uniformly distributed around the universe, though the evidence tends to support that conclusion. Another difficulty is that some helium was surely formed in the interiors of stars.

For the next million years, though the universe continued to cool gradually as it expanded, its temperature was still very high. Because of the high density, the universe was opaque. Then, as we have already described, the temperature dropped to about 3000 K, hydrogen atoms formed out of protons and electrons, and the universe became transparent. It was at this point that the black-body background radiation became able to circulate throughout space.

Did all the elements form in the big bang? Only the lightest elements could be formed in the big bang. The nuclear state of mass 5 and again the state of mass 8 are not stable and don't hold together long enough to form still heavier nuclei by the addition of another proton. In a stellar interior, processes such as the triple-alpha process (Section 6.4) can get past these mass gaps. So only hydrogen (and deuterium) and helium (and traces of one isotope of lithium and one of boron) would be formed in the big bang.

The heavier elements must, then, have been formed at times after the big bang. The theory of nucleosynthesis shows that these heavier elements are built up in the interiors of ordinary stars, in the explosions of supernovae, and possibly in explosions of more massive gaseous clouds. E. Margaret Burbidge, Geoffrey Burbidge, William A. Fowler, and Fred Hoyle had shown in 1957 how both heavier elements and additional amounts of the lighter elements can be synthesized in stars.

Thus heavy elements in our bodies do not date back to the earliest part of time. They were formed more recently, in the interiors of stars or in supernovae in our own galaxy.

It should be mentioned that work on element formation in stars was stimulated by the existence of the steady state theory. The lack of a big bang in that theory meant that all the elements had to be synthesized in ongoing processes. So even though the steady state theory has been dismissed, it was of great value as a stimulus of new ideas.

25.7 THE FUTURE OF THE UNIVERSE

How can we foretell how the universe will evolve in the distant future? We know that for the present it is expanding, but will that always continue? We have seen that the steady state theory now seems discredited, so let us discuss the alternatives that are predicted by different versions of big bang cosmologies.

Basically, the question is whether there is enough gravity in the universe to overcome the expansion (Fig. 25–3). If gravity is strong enough, then the expansion will gradually stop, and a contraction will begin. If gravity is not strong enough, then the rate of expansion might slow, but the universe would continue to expand forever, just as a rocket sent up from Cape Canaveral will never fall back to earth if it is launched with a high enough velocity.

To assess the amount of gravity, we must determine the average mass in a given volume of space, that is, the density. It might seem that to find the mass in a given volume we need only count up the objects in that volume: one hundred billion stars plus umpteen billion atoms of hydrogen plus so much interstellar dust, etc. But there are severe limitations to this method, for there are many kinds of mass that are invisible to us. How much matter is in black holes, for example? Until a few years ago, we couldn't measure the molecular hydrogen in space, and until a few decades ago we couldn't

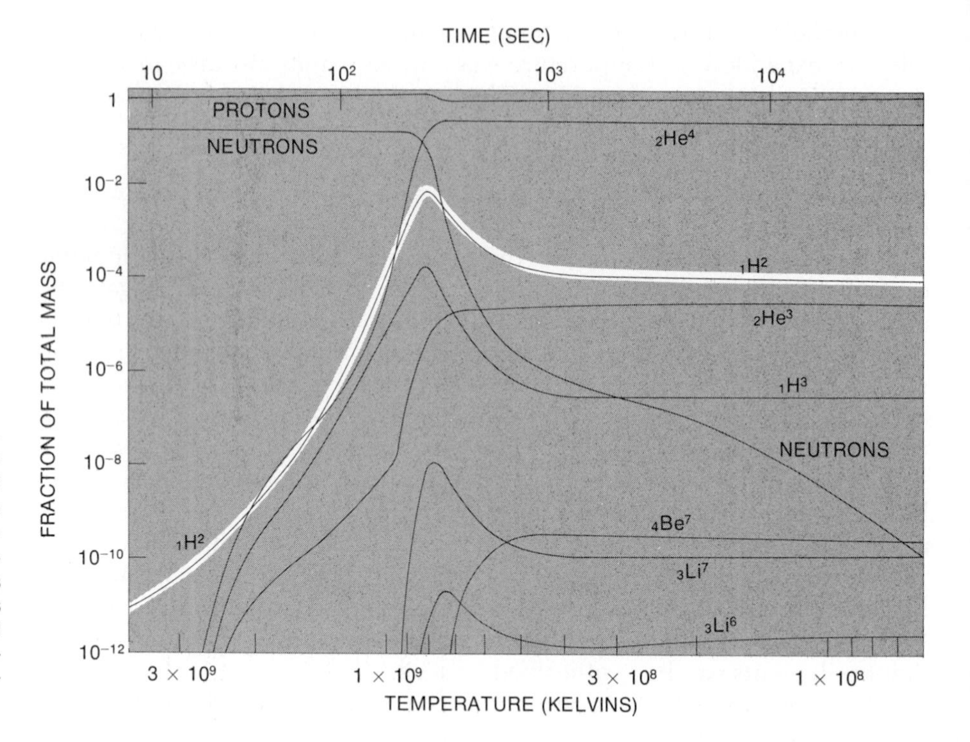

Figure 25–14 A theoretical model, worked out by Robert V. Wagoner of Stanford, that shows the changing relative abundances of the light elements in the first minutes after the big bang. Time is shown on the top axis and the corresponding temperature is shown on the bottom axis.

measure the atomic hydrogen in space either, although their abundances are now approximately known to us.

Moreover, we have seen in Section 23.3 that there are lines of evidence from dynamic studies of clusters of galaxies that indicate that much of the mass may be "missing," that is, invisible to us. We must find a method that doesn't depend on whether mass is visible or invisible.

One method that has recently attracted increasing interest concerns itself with the abundances of the light elements. Since these light elements were formed soon after the big bang (Fig. 25–14), they may tell us about conditions at the time of their formation.

But somehow we must distinguish between the amount of these elements that was formed in the big bang, and the amount that has been subsequently formed in stars. This consideration complicates the calculations for helium, for example, because the helium formed later in stellar interiors and then spewed out in supernovae has been added to the primordial helium formed in the big bang.

However, one isotope of hydrogen, deuterium, is free of this complication. We do not think that any is formed in stars, so all the deuterium now in existence was formed at the time of the big bang. Deuterium, "heavy hydrogen," contains one proton and one neutron in its nucleus instead of just the single proton of ordinary hydrogen. We are most familiar with deuterium as a constituent of "heavy water," HDO, which is used in atomic reactors. Deuterium is now being studied increasingly as an important constituent of the fusion process. We hope that we will be able to harness nuclear fusion, which would allow us to transform mass into energy here on earth to provide pollution-free new energy sources.

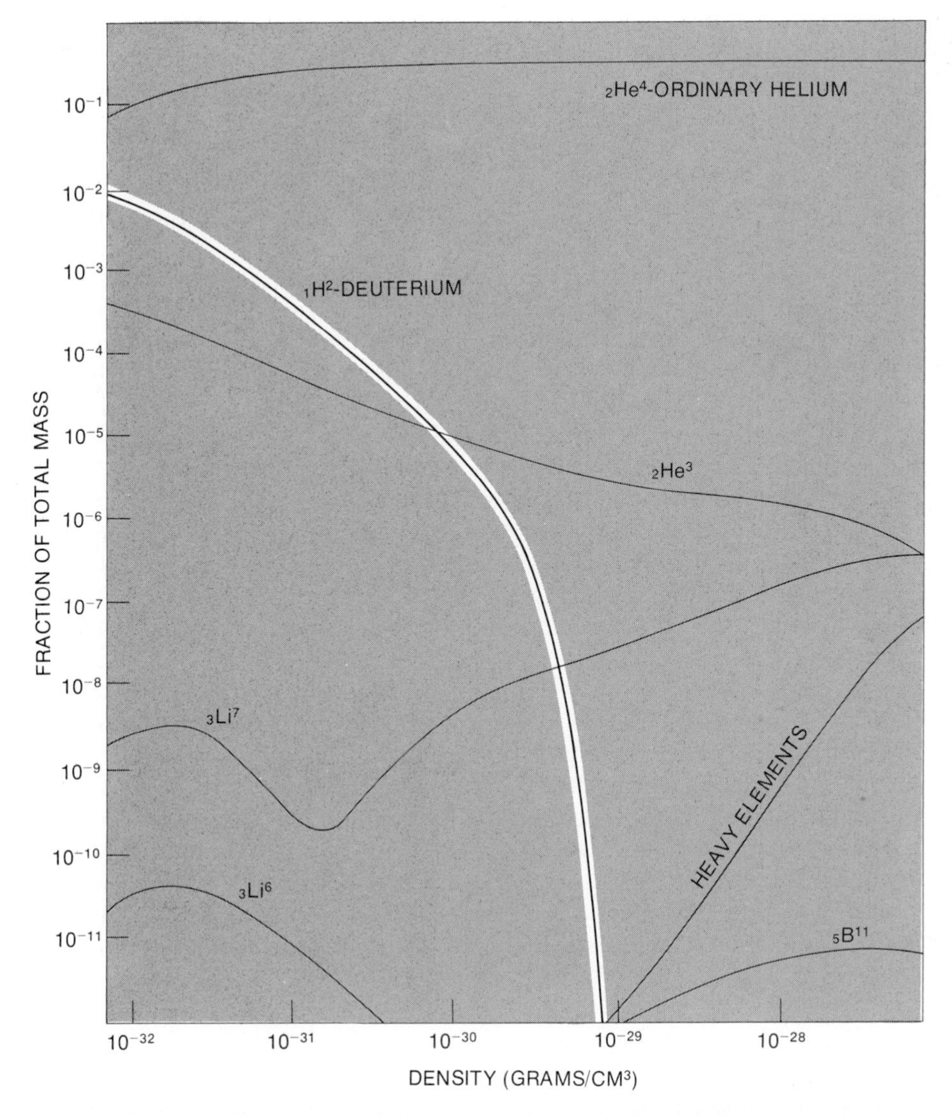

Figure 25–15 The horizontal axis shows the current day cosmic density of matter. From our knowledge of the approximate rate of expansion of the universe, we can deduce what the density was long ago. The abundance of deuterium, outlined in white, is particularly sensitive to the density at the time when the deuterium was formed, within the first 15 minutes after the big bang. If the density then was very high, then most of the deuterium quickly became helium and little deuterium survived. If the density, on the other hand, was low, then a relatively high abundance of deuterium would have remained without being "cooked" into helium. Thus present-day observations of the deuterium abundance tell us what the cosmic density is and was.

Deuterium has two properties that make it an important probe of conditions that existed in the first fifteen minutes after the big bang, when the deuterium was formed. First, the amount of deuterium, relative to the amount of hydrogen, that is formed is particularly sensitive to the density of matter at the time of formation (Fig. 25–15); that is, a slight variation in the primordial density makes a much larger change in the deuterium abundance than it does in the abundance of other isotopes or elements. Second, the nuclear processes that we think take place in stars all destroy deuterium, rather than create it. So the amount of deuterium we measure today is not greater than the amount that would have existed shortly after the big bang. (Indeed, we can estimate that the factor by which the amount of deuterium is depleted is two to four.)

Why is the ratio of deuterium to (ordinary) hydrogen sensitive to density? Deuterium very easily combines with an additional neutron, briefly forming another hydrogen isotope (tritium), and the new neutron quickly

Figure 25–16 The 40-m telescope at the Owens Valley Radio Observatory.

decays into a proton, leaving us with an isotope of helium. If the universe was very dense in its first few minutes, then it was easy for the deuterium to meet up with neutrons, and almost all the deuterium "cooked" into helium. If, on the other hand, the density of the universe was low, then most of the deuterium that was formed still survives. The theoretical calculations shown in Figure 25–15 can be used to indicate the density that would have been present soon after the big bang, if the ratio of deuterium to hydrogen can be determined observationally.

There have been a number of attempts to detect deuterium in interstellar space, but it was not until 1972 that any succeeded. My colleagues Diego A. Cesarsky of the Observatoire de Paris and Alan T. Moffet of Caltech, and I, observed a faint absorption feature that is probably an absorption line caused by the spin-flip transition of deuterium (see Section 21.3). We used the 40-meter (130-foot) radio telescope at Caltech's Owens Valley Radio Observatory (Fig. 25–16) and pointed it in the direction of the center of our galaxy, toward the radio source Sagittarius A. Sagittarius A is a strong continuous background source, and we observed the absorption line caused by deuterium located in space between the galactic center, 10 kiloparsecs away, and the earth (Fig. 25–17). The line is so weak (Fig. 25–18) that we would like to have more definitive results, and observations are continuing.

At the same time as our original observations, a group of astronomers from the Bell Telephone Laboratories was discovering deuterium as one of the constituents of a "deuterated" molecule, DCN, in the molecular cloud in the constellation Orion (see Section 21.7). The fact that deuterium is present is not ambiguous, for the DCN emission line is clearly present, but the interpretation of their data to give an abundance for deuterium is very complex because the processes of interstellar molecular formation are not well understood. The astronomers from Bell Labs have since mapped the strength of this DCN emission line in several molecular clouds in our galaxy.

Figure 25–17 We were trying to observe deuterium in interstellar space between the center of our galaxy and the earth. The deuterium would be in regions of neutral hydrogen, shown schematically as a dark cloud. The deuterium, and other gas in the cloud, could be detected from the absorption lines they form in the continuous spectrum emitted by the center of our galaxy.

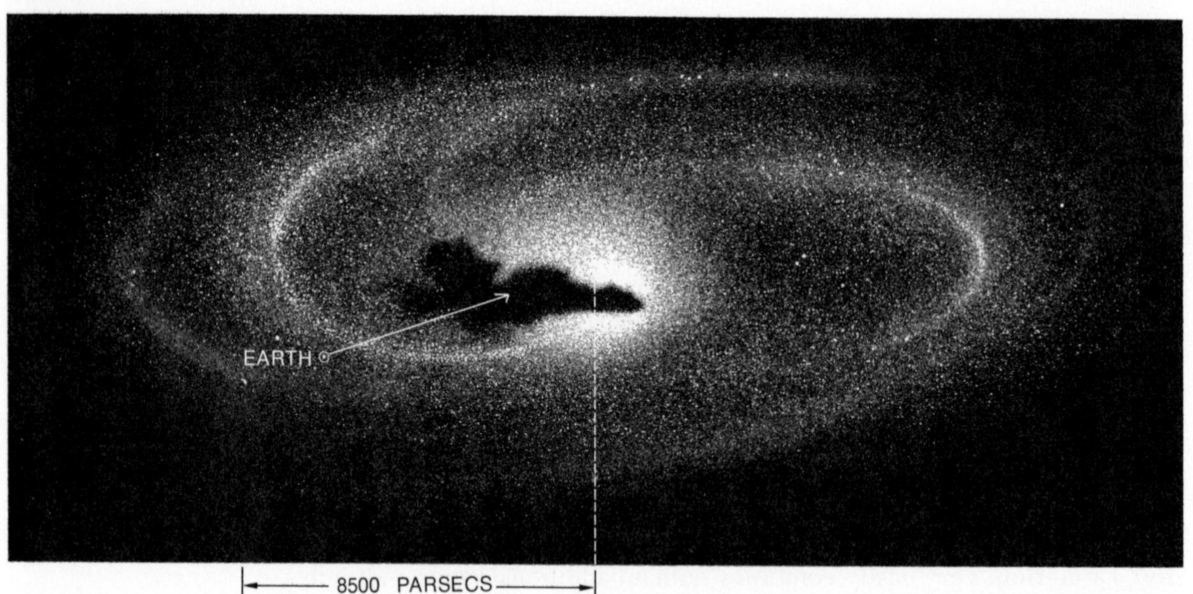

Since these discoveries, some of the Lyman lines of deuterium have been detected in absorption by the Princeton experiment aboard the Copernicus Orbiting Astronomical Observatory (Fig. 25–19). The Princeton group looked in the direction of a nearby star in order to try to detect an absorption line caused by deuterium in space between the star and the earth. Their observations not only show the presence of a deuterium absorption line more clearly than do the spin-flip observations of my own group but also can be interpreted with fewer complications than the molecular observations. The Copernicus satellite can study, however, only the nearest few hundred parsecs to us. The latest Copernicus results include stars up to 5 per cent of the distance to the galactic center, and establish a fraction of deuterium to hydrogen of 14 parts per million.

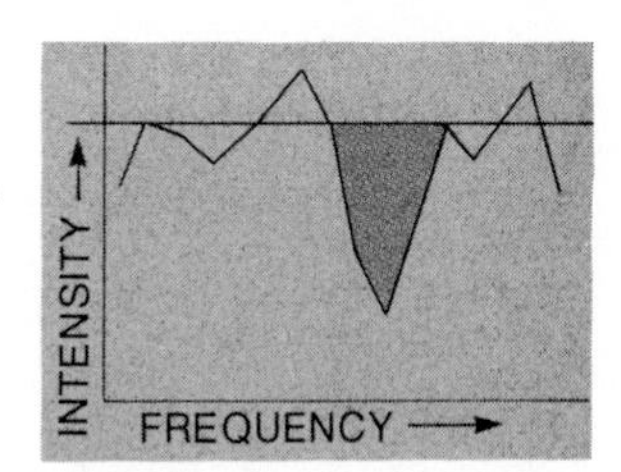

Figure 25–18 The data from nearly 16 weeks of observing in the direction of the strong radio source in the galactic center (which was shown in Fig. 22–4). The shaded area represents the deuterium absorption line at a wavelength of 92 cm.

Deuterium has also been discovered in molecules in the atmosphere of Jupiter. Just as the interstellar molecular observations, these observations must be interpreted in terms of chemical processes before cosmological use can be made of the data. Also, we don't know how the deuterium fared in the early stage of the formation of the solar system.

New studies of the Copernicus data by a French group of scientists in collaboration with the Princeton researchers have extended—and beclouded—the issue. Sophisticated analysis of the data indicates that there may be variations in the deuterium abundance from place to place to place, though we expect that primordial nucleosynthesis would have led to a homogeneous and isotropic distribution of deuterium. This is being further studied, as is the amount of variation we expect from place to place in our galaxy because of the deuterium consumption by stars. Nothing in astronomy ever seems to wind up as simple as it may seem at first.

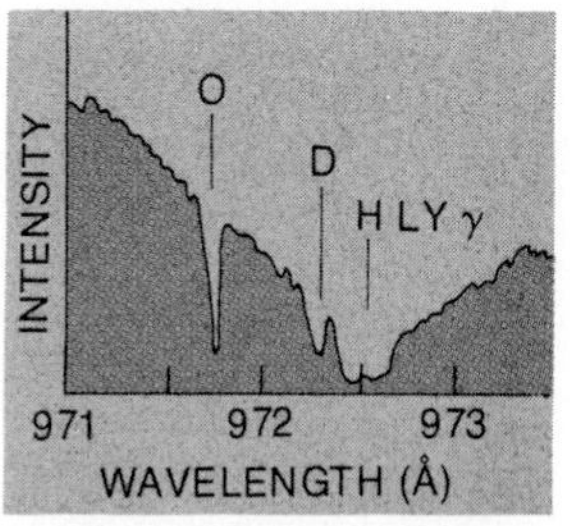

Figure 25–19 Observations from the Copernicus satellite of the Lyman lines of interstellar deuterium and normal hydrogen. The broad absorption that takes up the whole graph is from hydrogen; the corresponding deuterium line is the marked dip.

Notwithstanding all these uncertainties, all the studies of deuterium thus far agree that the amount of deuterium is such that there is not, and was not, enough mass to reverse the expansion of the universe. We say that the universe is *open*. (Alternatively, if the universe were oscillating or even to fall back on itself only once, it would be *closed*.) Recent new measurements of galaxies by Sandage and Tammann at Palomar now agree with this prediction that the universe is open. Sandage and Tammann measured the *deceleration parameter* (see Fig. 25–3), a measure of how much the expansion of the universe is slowing down at the outer limits of our ability to observe, taking into account that the galaxies they are observing are being studied far back in time.

It is particularly interesting that all these measurements tend to suggest the same result, that the universe will not stop its expansion, and that the galaxies will continue to fly away from each other. And it is a good example of modern astronomical research to note that these results were obtained from a wide variety of methods of observation: optical observation with large telescopes on earth, radio observations using a variety of techniques, and ultraviolet observations using telescopes in space. We are making use of all the methods we know to try to foretell our future.

SUMMARY AND OUTLINE OF CHAPTER 25

Aims: To study the origin and evolution of the universe as a whole, to discuss the origin of the elements, and to consider observations that indicate the future of the universe

- Olbers's paradox and its solutions (Section 25.1)
- Big bang theory (Section 25.2)
  - Cosmological principle: universe homogeneous and isotropic
  - Hubble's constant and the "age" of the universe – 13 to 20 billion years
  - Source counting techniques as test for Hubble's law
  - The solutions to Einstein's equations and the cosmological constant
- Steady state theory (Section 25.3)
  - Perfect cosmological principle: universe homogeneous, isotropic, and unchanging in time
  - Continuous creation of matter
  - Demise of the steady state theory with discovery of quasars and 3° black-body radiation
- Primordial background radiation (Section 25.4)
  - Provides strong support for the big bang theory
  - Evidence for 3° black-body radiation
    - direct radio measurements on one side of the peak of the black-body curve
    - optical measurements of CN molecular lines indirectly confirm that the radiation follows a Planck curve
    - new infrared observations provide convincing evidence for the existence of the 3° radiation because they confirm the peaked nature of the radiation
- Hierarchical universe (Section 25.5)
  - Cosmic density could be zero
  - We choose to ignore this possibility and deal with the largest volumes that we can measure: clusters of galaxies
- Creation of the elements (Section 25.6)
  - Origin of the light elements in the first minutes after the big bang
  - Heavier elements formed in the interiors of stars
- Future of the universe (Section 25.7)
  - Evidence from the study of deuterium
  - Deuterium-to-hydrogen ratio depends sensitively on cosmic density
  - All deuterium observed was formed in the first minutes after the big bang
  - Current evidence is that universe is open, i.e., expansion will not stop

QUESTIONS

1. Hindsight has allowed solutions to be found to Olbers's paradox. For example, knowing that the universe is expanding, we can come up with a solution. However, based on the reasoning in this chapter, do you think it is possible that scientists might have used Olbers's paradox to reach the conclusion that the universe is expanding before it was determined observationally? Explain.

2. We say in the chapter that the universe is homogeneous on a large scale. What is the largest scale on which the universe does not seem to appear homogeneous?

3. List the observational evidence favoring each of the following theories: (a) the big bang, and (b) the steady state theory.

4. What is "perfect" about the perfect cosmological principle?

5. If galaxies and radio sources increase in luminosity as they age, when we look to great distances then we are seeing them when they were less intense than similar nearby galaxies or radio sources are now. How does this affect the

method of radio source counting? Would it tend to make the universe decelerate at a greater or lesser rate than the actual rate of deceleration?

6. For a Hubble constant of 55 km/sec/Mpc, show how you calculate the Hubble time, the age of the universe ignoring the effect of gravity. (Hint: Take $1/H_0$, and simplify units so that only units of time are left.)

7. In actuality, if current interpretations are correct, the 3° background radiation is only indirectly the remnant of the big bang, but is directly the remnant of an "event" in the early universe. What event was that?

8. Why was the use of CN in studying the background radiation so important at the time?

9. Apply Wien's displacement law to the solar photospheric temperature and spectral peak in order to show where the spectrum of a 3° black body peaks.

10. Why must we resort to indirect methods to find out if the universe is open or closed?

11. Why is determining the abundance of deuterium so important?

12. What are two pieces of evidence that suggest that the universe is open?

TOPICS FOR DISCUSSION

1. Which is more appealing to you: the cosmological principle or the perfect cosmological principle? Discuss why we should adopt one or the other as the basis of our cosmological theory.

2. What effect would finding out whether the universe will expand forever, will begin to contract into a final black hole, or will contract and then oscillate with cycles extending into infinite time, have on the conduct of one's life or thoughts?

26

Astronomy Now

We have seen in the preceding chapters that astronomy encompasses not only a wide range of subjects but also a wide range of techniques. No longer can a single astronomer hope to be expert in all phases of the science. We have consistently seen how new discoveries often open more questions than they answer. The advance of science seems to leave us farther and farther behind!

Up to this point, this book has been organized by type and size of object, from stars (and their planets) at the beginning up to the universe as a whole at the end. Let us take a brief tour of topics that are now under active consideration, organized, for a change, by wavelength in the electromagnetic spectrum. The discussion should give us an idea of new discoveries that are expected in the next few years.

26.1 CURRENT PLANS

The shortest wavelengths in the electromagnetic spectrum represent gamma-rays and x-rays. Of course, we are dependent on satellites above the earth's atmosphere to study these radiations efficiently. Study of this region of the spectrum, in which each individual photon has a high energy, is a NASA priority for the next few years, and we can confidently look forward to discoveries in this field that will force us to revamp our ideas of many processes in astronomy. The two High Energy Astronomy Observatories that are scheduled for launch will provide a capability for research in this region of the spectrum that is a significant advance. In the meantime, other satellites continue in orbit. X-ray data will be used for further mapping of the x-ray sky and for further detailed studies of particularly interesting or suspicious objects. The Vela series of satellites, which mostly look for gamma-ray bursts in order to detect explosions of hydrogen bombs, are equally ready to report gamma-ray bursts from outer space. In the meantime, theoreticians continue to make progress in understanding the high-

---

Two of the first VLA antennas to be completed.

energy processes that dominate emission in this region of the spectrum. And most people want to know whether we find more (or any) black holes. We should also see whether the current tendency to invoke the presence of black holes to explain puzzling phenomena is a panacea or a fad.

In the ultraviolet, the Copernicus satellite continues to study the stars and interstellar space. Two of its recent observations, of deuterium in interstellar space and of the spectrum of a white dwarf, illustrate the unique and important research that can be carried out with this Orbiting Astronomical Observatory. The European TD-1 continues its ultraviolet studies of the stars with high spectral resolution. In solar research, the data from Skylab continue to be studied and together with the new data from the eighth Orbiting Solar Observatory may significantly advance our understanding of the processes by which energy is transferred from the solar interior out to the corona and beyond.

In visible light, a number of new giant telescopes are just beginning their operation or are about to be completed. Our ability to observe the optical sky will be increased manyfold. Electronic devices to receive and record the photons that are focused by the huge mirrors increase the capability of even the older telescopes. But we are still faced with a limitation in the size of telescopes with single mirrors. The completion of the multi-mirror telescopes now under construction will tell us whether this kind of instrument will allow us to study fainter objects than we have previously been able to study. If we can, we will be able to observe regions more distant from us, which could not help but improve the observations on which cosmological theories are based. Also in the optical spectrum, we can expect improvements in our knowledge of the outer planets. The images that will be sent to us from Saturn in 1979 and possibly images of Uranus from a subsequent spacecraft should increase manyfold our understanding of the planets, their atmospheres, and their origins.

The infrared is a burgeoning area of research. We are seeing stars in formation and so are getting to understand better the processes of stellar birth. We can study dust wherever it is in space—in the solar system or between the stars. New telescopes devoted exclusively to infrared studies are becoming operational, and electronic instrumentation used to study this part of the spectrum is improving rapidly.

In radio astronomy, there will continue to be tremendous strides in making high-resolution pictures of astronomical objects. The VLA (Very Large Array) will shortly become operational and will enable large-scale mapping of the sky in a manner never before possible (Fig. 26–1). In spectral-line radio astronomy, scientists will continue to shift their interest away from mere discovery of new molecules toward using the molecular studies to analyze the physical conditions in regions in space, especially those that are also under observation with infrared methods. Further studies of the ultraviolet and radio deuterium lines should help us decide whether or not deuterium was formed in the big bang, and what the future of the universe will be.

This is but the beginning of a long list of topics of active research. Different astronomers put high priority on very different areas. I suppose that we are fortunate that we don't all want to work on the same thing at the same time! There is surely no shortage of research topics available for astronomers.

Perhaps the most significant gains will come when observations from several different parts of the spectrum are put together. The sun, for ex-

Figure 26–1 The VLA, under construction near Socorro, New Mexico.

ample, is under careful scrutiny from wavelengths down below 0.01 angstroms up to wavelengths above 3 meters, a factor of more than a trillion (10^{12}).

In a textbook of astronomy, it is relatively easy to get across a sense of progress in observational results compared to transmitting a sense of the importance of theoreticians in the whole enterprise. The computer, especially, has enhanced the ability of theoreticians to make detailed models of what is occurring in a stellar atmosphere, in a stellar interior, or in a disk of material around a black hole. But even without the computer, our understanding of basic properties of the interaction of matter and a magnetic field—a field tongue-twistingly named *magnetohydrodynamics* and often abbreviated MHD—is not only improving rapidly but also is finding immediate application on earth. The ability of a scientist to sit quietly with a pencil and paper, confidently calculating how bodies 10^{26} (100,000,000,000,000,000,000,000,000) times the astronomer's size will act, remains at the heart of modern research.

"No one has yet programmed a computer to be of two minds about a hard problem, or to burst out laughing. . . ."

Lewis Thomas, THE LIVES OF A CELL (New York: Viking Press, 1974)

26.2 THE SCIENTIFIC ENTERPRISE

Science of all kinds, including astronomy, grows increasingly dependent on large instrumentation. It becomes more and more difficult—and is really nigh unto impossible—to carry out significant research as a lone individual. Even theoreticians are often dependent on the availability of large computers, and their calculations may be limited in accuracy mainly by the amount of computer time that is available to them.

All this points to the importance of funding to provide the equipment and the financial backing for the scientists' time and effort. Although many smaller sources of funds exist, the role of government funding has increased to the point where it is clearly the dominant source. A government supports scientific research for a variety of reasons. Its interest in supporting science for the sake of knowledge itself is one of them, just as governments often support museums of art, or opera companies.

But this is surely not the major reason why governments support science. Usually there is a desire for practical results. Astronomers do not ordinarily undertake their investigations for practical ends, but over the past centuries a number of basic ideas and processes of practical importance have been discovered in an astronomical context. We can confidently expect that today's astronomical research will pay off similarly in the long run.

Often, astronomers benefit from apparatus or projects that were set up mainly for other purposes. The whole Apollo program to land astronauts on the moon by the end of the decade of the 1960's had important political implications and implications for national pride and goals; these led the United States government to spend on the project sums previously unparallelled in the history of scientific support. But for whatever reasons the astronauts were sent to the moon, or for whatever reasons Skylab was sent into orbit, the astronomers were at the ready to make whatever observations they could with the new techniques available.

In the United States, the National Science Foundation is the organization that is the primary source of support for basic research in astronomy. The astronomy section of the NSF has been organized into two divisions. One, the Astronomy Centers Section, has responsibility for the national observatories that are wholly supported by the NSF: the Kitt Peak National Observatory (for optical stellar and solar work), the National Radio Astronomy Observatory (including telescopes at Green Bank and on Kitt Peak, and now the VLA as well), the Arecibo Ionospheric Observatory (where the giant radio telescope is used to study the earth's atmosphere as well as objects in outer space), and the National Center for Atmospheric Research (where studies of the atmospheres of stars as well as the atmosphere of the earth are carried out). Second, the Astronomy Research Section gives out a number of smaller grants to scientists at colleges and universities around the country, sometimes to allow them to maintain facilities or projects at their home institutions and sometimes to allow them to use the shared national facilities. The NSF might spend 30 million dollars or so on astronomy in a given year, a tiny fraction of a per cent of the national budget.

In Canada, in most European countries, in Australia, in the Soviet Union, and elsewhere, government organizations also have important influences on the state of astronomical research.

Astronomy has also certainly benefited by the interest of NASA in astronomical problems. The cost of space experiments is many times the cost of ground-based observing, so one must be prepared to think on another

Figure 26–2 The control room of the 100-meter radio telescope near Bonn of the Max Planck Institute for Radio Astronomy. This instrument is the largest fully-steerable radio telescope in the world.

scale of expenses when one considers using a satellite to make observations. Of course, most of the costs of experimenting in space are not really from the astronomical experiments but are rather engineering costs. And the tremendous amounts of money that are spent do not just go up into outer space; the money is spent here on earth, largely to pay contractors and subcontractors to build, test, and maintain the equipment.

Other organizations, non-governmental in type, also sometimes fund astronomical research. For example, the National Geographic Society has supported a series of eclipse expeditions and financed the fundamentally important National Geographic Society-Palomar Observatory Sky Survey with the Schmidt camera on Palomar. The Research Corporation, a private corporation that derives its income from patents assigned to it, sometimes gives grants for astronomical projects. The Alfred P. Sloan Foundation numbers several astronomers among the scientists to whom they give career development grants.

26.3 GRANTS

How does a scientist, say a professor at a university, go about getting a grant from the National Science Foundation? First, he or she would have to have a clear idea of the project under consideration, which would normally require some preliminary research. This preliminary research must be supported in some other way, perhaps by funds from the university. Eventually the project would be developed to a point where a proposal for a grant could be written (Fig. 26–3).

The proposal might be some two dozen pages in length. It would contain a clear exposition of the proposed project, its importance, and why it should be undertaken in this specific manner. The scientist's ability to undertake the project would have to be justified; his or her research ability as demonstrated by prior published papers would be an important factor in

HOPKINS OBSERVATORY
WILLIAMS COLLEGE
WILLIAMSTOWN, MASS. 01267
To: Universal Creation Foundation
REQUEST FOR SUPPLEMENT TO U.C.F. GRANT
#000-00-00000-001
"CREATION OF THE UNIVERSE"

This report is intended only for the internal uses of the contractor.

Period: Present to Last Judgment
Principal Creator: Creator
Proposal Writers and Contract Monitors:
Jay M. Pasachoff and Spencer R. Weart

BACKGROUND

Under a previous grant (U.C.F. Grant #000-00-00000-001), the Universe was created. It was expected that this project would have lasting benefits and considerable spinoffs, and this has indeed been the case. Darkness and light, good and evil, and Swiss Army knives were only a few of the useful concepts developed in the course of the Creation. It was estimated that the project would be completed within four days (not including a mandated Day of Rest, with full pay), and the 50% overrun on this estimate is entirely reasonable, given the unusual difficulties encountered. Infinite funding for this project was requested from the Foundation and granted. Unfortunately, this has not proved sufficient. Certain faults in the original creation have become apparent, which it will be necessary to correct by means of miracles. Let it not be said, however, that we are merely correcting past errors; the final state of the Universe, if this supplemental request is granted, will have many useful features not included in the original proposal.

PROGRESS TO DATE

Interim progress reports have already been submitted ("The Bible," "The Koran," "The Handbook of Chemistry and Physics," etc.). The millennial report is currently in preparation, and a variety of publishers for the text (tentatively entitled, "Oh, Genesis!") will be created. The Gideon Society has applied for the distribution rights. Full credit will be given to the Foundation.

Materials for the Universe and for the Creation of Man were created out of the Void at no charge to the grant. A substantial savings was generated when it was found that materials for Woman could be created out of Man, since the establishment of Anti-Vivisection Societies was held until Phase Three. Given the limitations of current eschatological technology, it can scarcely be denied that the Contractor has done His work at a most reasonable price.

SUPPLEMENT

We cannot overlook a certain tone of dissatisfaction with the Creation which has been expressed by the Foundation, not to mention by certain of the Created. Let us state outright that this was to be expected, in view of the completely unprecedented nature of the project. The need for a supplement is to be ascribed solely to inflation (not to be confused with expansion of the Universe, which was anticipated). Union requests for the accrual of Days of Rest at the rate of one additional Day per week per millenium ($Dw^{-1}m^{-1}$) must also be met. Concerning the problem of Sin, we can assure you that extensive experimentation is under way. Considerable experience is being accumulated and we expect a breakthrough before long. When we are satiated with Sin, we shall go on to consider Universal Peace.

We cannot deny—in view of the cleverness of the Foundation's auditors—that the bulk of the supplemental funds will go to pay off old bills. Nevertheless we do not anticipate the need for future budget requests, barring unforeseen circumstances. If this project is continued successfully, additional Universe—anti-Universe pairs can be created without increasing the baryon number, and we would keep them out of the light cone of the Original. By the simple grant of an additional Infinity of funds (and note that this proposal is merely for Aleph Null), the officers of the Foundation will be able to present their Board of Directors with the accomplished Creation of one or more successful Universes, instead of the current incomplete one.

We will attempt to minimize the additional delays that may temporarily exist during the changeover from fossil fuels to fusion for some minor locations in the Universe. For the time being, elements with odd masses (hydrogen, lithium, etc.) will be created only on odd days of the month and those with even masses (helium, beryllium, etc.) on the even days, except, of course, for Sundays. The 31st of each month will be devoted to creation of trans-uranic elements. The Universe has been depleted of deuterium; new creation of this will take place only on February 29th in leap years.

PROSPECTIVE BUDGET

Remedial miracles on fish of the sea	∞
Remedial miracles on fowl of the air	∞
Creeping things that creep upon the earth, etc.	∞
Hydrogen	n/c (created)
Heavier elements	n/c (nucleosynthesis)
(Note: The carbon will be reclaimed and ecologically recycled)	
Mountains (Sinai, Ararat, Palomar)	∞
Extra quasars, neutron stars	∞
Black holes (no-return containers)	∞
Miscellaneous, secretarial, office supplies, etc.	∞
Telephone installation (Princess model, white, one-time charge, tax included)	$16.50

SALARIES

Creator (1/4 time)	at His own expense
Archangels	
Gabriel	1 trumpet (Phase 5)
Beelzebub	misc. extra brimstone (low sulfur)
Others	assorted halos
Prophets	
Moses	stone tablets (to replace breakage)
Geniuses	finite

N.B. Due to the Foundation's regulations and changes in Exchange Rates, we have not yet been able to reimburse Euclid (drachmae), Leonardo (lire), Newton (pounds sterling), Descartes (francs), or Catfish Hunter (dollars). Future geniuses will be remunerated indirectly via the Alfred Nobel Foundation.

Graduate students (2 at 2/5 time)	reflected glory

MONITORING EQUIPMENT & MISC.

1 5-meter telescope (maintenance)	finite
Misc. other instruments, particle accelerators, etc.	large but finite
Travel to meetings	∞
Pollution control equipment	+40%
Total	$\infty + 40\% = \infty$
Overhead (51.97%)	∞
Total funds requested	∞

Starting date requested:
Immediate. Pending receipt of supplemental funds, layoffs are anticipated to reach the 19.5% level in the ranks of angels this quarter.

Figure 26–3 *See opposite page for legend*

assessing this. If staff or students will be included in the project, their tasks and abilities must be explained.

The budget is an important part of the proposal. It would contain funds for any equipment necessary for the experiment, and for expendable supplies like film. It would contain funds for salaries for technicians and possibly for students during the summer and even during the school term. It may even contain some salary for the professor in charge; most universities pay professors' salaries only during the academic year and any salary a professor might receive for summer work would come from a grant. The grant may also contain funds to travel to a telescope, or to travel to a scientific meeting to present the results of the research in either preliminary or final form. And a major item in the budget is the overhead of 50 per cent or more, as described in Section 21.8.

At the astronomy section of the NSF, the program managers would send out copies of the proposal to perhaps half a dozen scientists across the country who are knowledgeable in the specific field of research under discussion. These scientists are asked to act as anonymous *referees* of the proposal in a process known as *peer review*. The referees donate their time to this purpose as part of every scientist's role in keeping the national scientific enterprise running.

The program managers try to assess the referees' reports and assign ratings and priorities to each proposal. Then they try to divide their limited funds appropriately. There are always many more acceptable proposals than can be funded. And each year, especially recently, the number of proposals submitted has grown at a much higher rate than has the amount of available funds.

The process of review can take six months or more, so prospective proposers have to decide long in advance what they want to do. One day, successful proposers receive in the mail copies of letters announcing the awarding of grants. The originals of the letters are sent to the presidents of their respective colleges or universities. Checks from NSF find their way to the institutions' business offices, and accounts are set up against which the professors can write vouchers. The projects can then get under way.

26.4 SCIENTIFIC PUBLICATION

There is a great overlap between the systems by which grants are awarded and the systems by which scientific papers are published. The peer review process, in which the paper submitted is read by a number of scientists who comment anonymously on its quality, is often common to both processes.

It may be two to six months before a scientist hears whether or not a paper submitted to a journal has been accepted for publication, either in the form submitted or pending modifications suggested by the referees.

Once a paper is accepted, the mechanical part of the project begins. Many months after receiving the letter of acceptance, the scientist receives galley proofs of his article to read for typographical errors. The *Astrophysi-*

Figure 26–3 A mock proposal, including most of the kinds of descriptions and budgeting stages that a real proposal would have. Reprinted from the *Journal of Irreproducible Results* and *Theology Today.*

Figure 26–4 Some of the major technical astronomical journals.

cal Journal, in particular, the major astronomical journal of general distribution in the United States, always stamps on their proofs "Read and return within 48 hours; delay in return of these proofs may result in delay in publication of the article." Few scientists can resist giving their proofs the highest priority.

Many American scientific journals have a system called "page charges" for supporting themselves. The author's institution is assessed a certain fee per published page; the *Astrophysical Journal* currently assesses $50 per page. Thus a 10-page article may cost $500 to have published. This is usually justified as part of the ordinary costs of doing the research; after all, thousands of dollars of time and equipment went into the research. Page charges are often included in grant proposals, so that it is often the NSF rather than the college or university that pays them.

The rationale behind page charges is often that they cover the cost of setting the article in type. Then the other expenses, including printing and mailing, go into the cost of a subscription. This enables the subscription costs to be kept sufficiently low so that the circulations are much larger than they would be if page charges did not exist. Still, the *Astrophysical Journal* has a total circulation of only 3500 copies.

It must be stressed that page charges are the normal and ordinary way of publishing major American journals and should not be confused with vanity press charges for the publication of books in other fields. Some journals, both in the United States and abroad, do not assess page charges and often make up for the lack of income by charging very high rates for subscriptions to libraries. So the academic organization winds up paying in any case.

Finally, perhaps a year or so after an article was submitted, it appears in print. Libraries and individuals have subscriptions. Also, the scientists who wrote the articles usually arrange to have a number of "reprints" made of the specific pages on which their articles appear. Researchers in specific fields often write to the authors for reprints in order to have their own copies of the articles for ready reference.

Unlike the humanities, astronomy is not a field in which research contributions are ordinarily published in book form. Articles in the scientific journals are a more usual form of dissemination of research results. Of course, a number of important books do appear that describe research results, particularly in one of several excellent series of astronomical volumes. Sometimes these books are monographs and often they contain contributions from many authors.

The astronomical community is always interested in new ideas and new explanations of phenomena that have puzzled them for years. There are so many journals that almost any piece of reasoned substantial research will be published by one of them. And any member of the American Astronomical

Figure 26–5 Some of the major popular astronomical journals.

Society can read a paper on his research at one of the meetings that are held each year.

Of course it is rare that a real breakthrough is presented in one particular paper. Astronomers, as do all scientists, mainly advance step by step, searching for pieces to fill in the jigsaw puzzle of the universe.

26.5 THE NEW ASTRONOMY

In the decade of the seventies, astronomy seems to be maintaining its momentum as one of the most active sciences. It is exploding with new results.

The discussions in this book have necessarily been limited to a fraction of the topics with which contemporary astronomers concern themselves. Yet it is to be hoped that the reader has become aware of topics and ongoing investigations that will provide a basis for the astronomy of the 1980's and afterward.

Close to earth, the planets will continue to be explored, and we hope to gain a fundamental understanding of all the planets in the solar system. We are better prepared now to study comets than we have ever been, and the reappearance of Halley's comet in 1985 or of another bright comet will lead to much excellent data.

The sun remains the center of active investigation, not only for its fundamental properties as the nearest example to earth of a star but also because of its effect on our environment in space. From neutrinos and particles in the solar wind to the entire electromagnetic spectrum, we can study the sun in more ways than we can study any other object in the universe.

The stars and galaxies are also under scrutiny across the electromagnetic spectrum. They are all much fainter than the sun, of course, and so are correspondingly more difficult to study. But we seem to be approaching the point where we begin to understand the fundamental processes that govern stars and galaxies (although I would hate to read these words from the perspective of a hundred years hence, when it will be seen that we really now know nothing at all). Will quasars turn out to be something completely new?

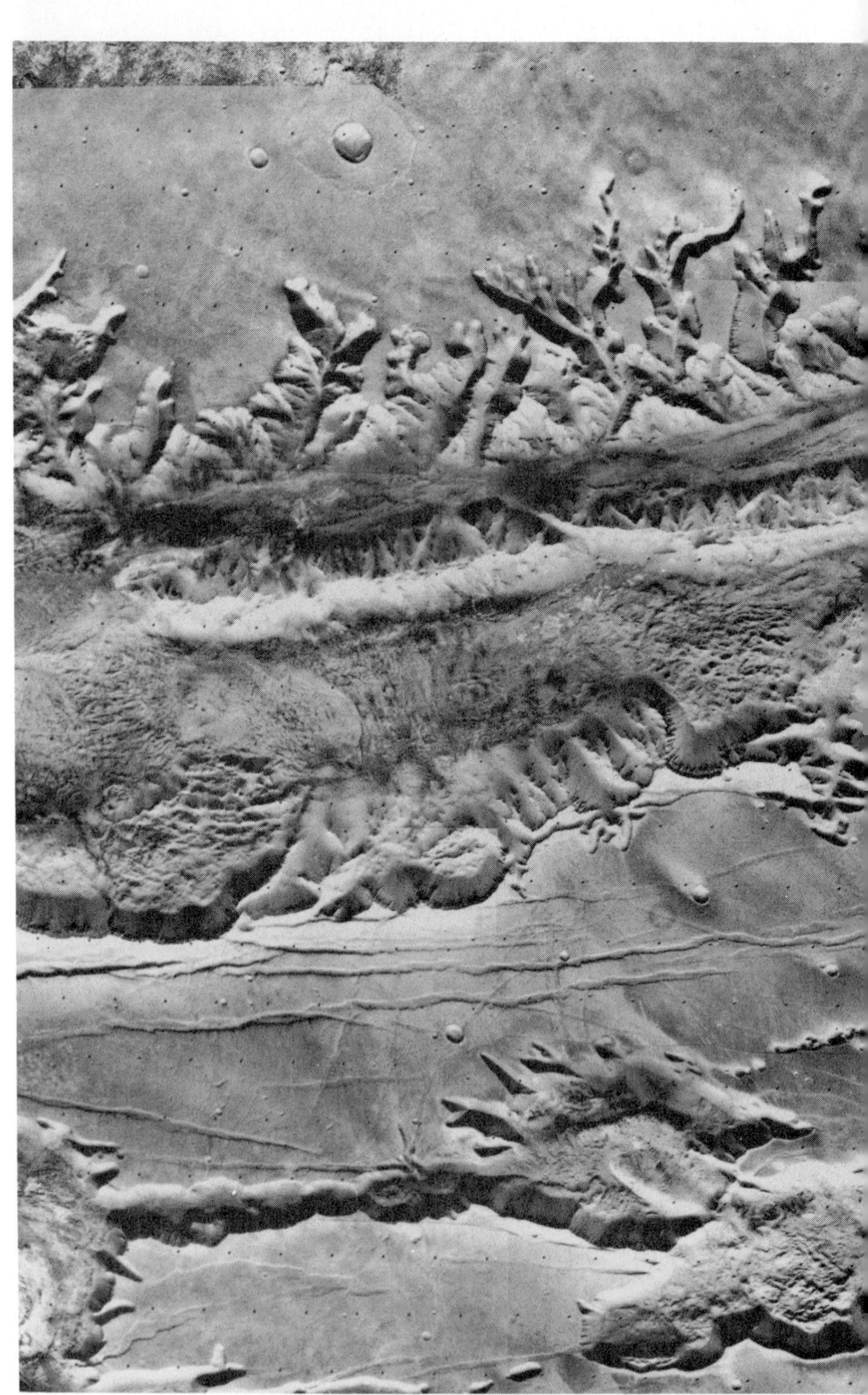

Figure 26–6 A mosaic of Viking Orbiter 1 pictures, showing part of the huge equatorial canyon on Mars.

Figure 26–7 The spiral galaxy M81, photographed with the new 4-meter telescope at the Kitt Peak National Observatory.

We are also making progress on our understanding of cosmology, a subject that has been at the center of human thought for over two thousand years but about which we have a lot to learn. The detection of the primordial radiation has brought us tidings of the big bang itself, and we hope that within a few years careful measurements of this radiation in different directions will tell us about the inhomogeneous nature of the universe in the first few instants of time some 13 billion years ago.

Astronomy has come a long way since early humans first noticed the objects in the sky and their motions. The pace of progress is dramatically increasing, and it is exciting to see the discoveries made in such quick succession. Still, we have at least as far to go.

SUMMARY AND OUTLINE OF CHAPTER 26

Aims: To summarize some of the future developments that we may expect in astronomical research, and to discuss the funding and publication of astronomical research

Some next steps (Sections 26.1 and 26.5)
The scientific enterprise (Section 26.2)
- Importance of funding
- National Science Foundation (NSF) oversees most funding for astronomy in the U.S.

Grants (Section 26.3)
- How a proposal is prepared and submitted
- Peer review

Scientific publication (Section 26.4)
- Peer review active in this area also
- Financing of publication through page charges
- Openness to new ideas

TOPICS FOR DISCUSSION

1. What are the astronomical projects **you** think are the most important for the next decade?

2. What is the value of astronomy to **you**?

Appendices

Appendix 1

Greek Alphabet

Upper Case	Lower Case	Pronunciation
A	α	alpha
B	β	beta
Γ	γ	gamma
Δ	δ	delta
E	ϵ	epsilon
Z	ζ	zeta
H	η	eta
Θ	θ	theta
I	ι	iota
K	κ	kappa
Λ	λ	lambda
M	μ	mu
N	ν	nu
Ξ	ξ	xi
O	o	omicron
Π	π	pi
P	ρ	rho
Σ	σ	sigma
T	τ	tau
Υ	υ	upsilon
Φ	ϕ	phi
X	χ	chi
Ψ	ψ	psi
Ω	ω	omega

Appendix 2

Measurement Systems

Metric units

Basic units

length	meter (m)
volume	liter (l)
mass	gram (gm or g)
time	second (sec or s)

Other metric units

1 micron (μ) = 10^{-6} meter
1 Ångstrom (Å or A) = 10^{-10} meter
= 10^{-8} cm

Prefixes for use with basic units of metric system

Prefix	Symbol	Power		Equivalent
tera	T	10^{12} =	1,000,000,000,000	Trillion
giga	G	10^{9} =	1,000,000,000	Billion
mega	M	10^{6} =	1,000,000	Million
kilo	k	10^{3} =	1,000	Thousand
hecto	h	10^{2} =	100	Hundred
deca	da	10^{1} =	10	Ten
– – –	–	10^{0} =	1	One
deci	d	10^{-1} =	.1	Tenth
centi	c	10^{-2} =	.01	Hundredth
milli	m	10^{-3} =	.001	Thousandth
micro	μ	10^{-6} =	.000001	Millionth
nano	n	10^{-9} =	.000000001	Billionth
pico	p	10^{-12} =	.000000000001	Trillionth
femto	f	10^{-15} =	.000000000000001	
atto	a	10^{-18} =	.000000000000000001	

Examples: 1000 meters = 1 kilometer = 1 km
10^{6} hertz = 1 megahertz = 1 MHz
10^{-3} sec = 1 millisecond = 1 msec

Some other units used in astronomy

Energy: 1 joule (J) = 10^{7} ergs
1 electron volt (eV) = 1.60207×10^{-12} ergs
Power: 1 watt (W) = 1 joule per sec (J/sec)
Frequency: hertz (Hz) = cycles per sec (sec^{-1})
1 megahertz (MHz) = 10^{6} Hz

Conversion factors

1 cm = 0.3937 in
1 m = 1.0936 yd
1 km = 0.6214 mi
= 5/8 mi
1 kg = 2.2046 lb
1 gm = 0.0353 oz

1 in = 25.400 mm
= 2.54 cm
1 ft = 0.3048 m
1 yd = 0.9144 m
1 mi = 1.6093 km
= 8/5 km
1 lb = 0.4536 kg

Appendix 3

Basic Constants

Physical constants

Speed of light*	c	$= 299\ 792\ 458$ m/sec
Constant of gravitation*	G	$= 6.672 \times 10^{-11}$ m³/kg·sec²
Planck's constant	h	$= 6.6262 \times 10^{-27}$ erg·sec
Boltzmann's constant	k	$= 1.3806 \times 10^{-16}$ erg/kelvin
Stefan-Boltzmann constant	σ	$= 5.66956 \times 10^{-5}$ erg/cm²·deg⁴·sec
Wien displacement constant	$\lambda_{max}T$	$= 0.289789$ cm·K $= 28.9789 \times 10^{6}$ Å·K
Mass of hydrogen atom	m_H	$= 1.6735 \times 10^{-24}$ gm
Mass of neutron	m_n	$= 1.6749 \times 10^{-24}$ gm
Mass of proton	m_p	$= 1.6726 \times 10^{-24}$ gm
Mass of electron	m_e	$= 9.1096 \times 10^{-28}$ gm
Rydberg's constant	R	$= 1.09677 \times 10^{5}$/cm

Mathematical constants

$\pi = 3.1415926536$
$e = 2.7182818285$

Astronomical constants

Astronomical unit*	1 A.U.	$= 1.495\ 978\ 70 \times 10^{11}$ m
Solar parallax*	$\pi_{\odot}$	$= 8.794148$ arc sec
Parsec	pc	$= 206\ 264.806$ A.U.
		$= 3.261633$ light years
		$= 3.085678 \times 10^{18}$ cm
Light year	ly	$= 9.460530 \times 10^{17}$ cm
		$= 6.324 \times 10^{4}$ A.U.
Tropical year (1900)* – equinox to equinox		$= 365.24219878$ ephemeris days
1 Julian century*		$= 36525$ days
1 day*		$= 86400$ sec
Sidereal year		$= 365.256366$ ephemeris days
		$= 3.155815 \times 10^{7}$ sec
Mass of sun*	$M_{\odot}$	$= 1.9891 \times 10^{33}$ gm
Radius of sun*	$R_{\odot}$	$= 696000$ km
Luminosity of sun	$L_{\odot}$	$= 3.827 \times 10^{33}$ erg/sec
Mass of earth*	$M_{\oplus}$	$= 5.9742 \times 10^{27}$ gm
Equatorial radius of earth*	$R_{\oplus}$	$= 6\ 378.140$ km
Mean distance center of earth to center of moon		$= 384\ 403$ km
Radius of moon*	R_M	$= 1\ 738$ km
Mass of moon*	M_M	$= 7.35 \times 10^{25}$ gm
Solar constant	S	$= 135.3$ mW/cm²
Direction of galactic center (precessed for 1950)	α	$= 17^{h}42.4^{m}$
	δ	$= -28° 55'$

*Adopted as "IAU (1976) system of astronomical constants" at the General Assembly of the International Astronomical Union.

Appendix 4

Elements and Solar Abundances*

Atomic Number	Element	Name	Atomic Weight	Solar Abundance (Relative to Hydrogen = 10^{12})
1	H	hydrogen	1.01	10^{12}
2	He	helium	4.00	6.3×10^{10}
3	Li	lithium	6.94	10×10^{1}
4	Be	beryllium	9.01	1.4×10^{1}
5	B	boron	10.81	1.3×10^{2}
6	C	carbon	12.01	4.2×10^{8}
7	N	nitrogen	14.01	8.7×10^{7}
8	O	oxygen	16.00	6.9×10^{8}
9	F	fluorine	19.00	3.6×10^{4}
10	Ne	neon	20.18	3.7×10^{7}
11	Na	sodium	22.99	1.9×10^{6}
12	Mg	magnesium	24.31	4.0×10^{7}
13	Al	aluminum	26.98	3.3×10^{6}
14	Si	silicon	28.09	4.5×10^{7}
15	P	phosphorus	30.97	3.2×10^{5}
16	S	sulphur	32.06	1.6×10^{7}
17	Cl	chlorine	35.45	3.2×10^{5}
18	Ar	argon	39.95	1.0×10^{6}
19	K	potassium	39.10	1.4×10^{5}
20	Ca	calcium	40.08	2.2×10^{6}
21	Sc	scandium	44.96	1.1×10^{3}
22	Ti	titanium	47.90	1.1×10^{5}
23	V	vanadium	50.94	1.0×10^{4}
24	Cr	chromium	52.00	5.1×10^{5}
25	Mn	manganese	54.94	2.6×10^{5}
26	Fe	iron	55.85	3.2×10^{7}
27	Co	cobalt	58.93	7.9×10^{4}
28	Ni	nickel	58.71	1.9×10^{6}
29	Cu	copper	63.55	1.1×10^{4}
30	Zn	zinc	65.37	2.8×10^{4}
31	Ga	gallium	69.72	6.3×10^{2}
32	Ge	germanium	72.59	3.2×10^{3}
33	As	arsenic	74.92	
34	Se	selenium	78.96	
35	Br	bromine	79.90	
36	Kr	krypton	83.80	
37	Rb	rubidium	85.47	4.0×10^{2}
38	Sr	strontium	87.62	7.9×10^{2}
39	Y	yttrium	88.91	1.3×10^{2}
40	Zr	zirconium	91.22	5.6×10^{2}
41	Nb	niobium	92.91	7.9×10^{1}
42	Mo	molybdenum	95.94	1.4×10^{2}
43	Tc	technetium	98.91	
44	Ru	ruthenium	101.07	6.8×10^{1}
45	Rh	rhodium	102.91	2.5×10^{1}
46	Pd	palladium	106.4	3.2×10^{1}
47	Ag	silver	107.87	7.1
48	Cd	cadmium	112.40	7.1×10^{1}
49	In	indium	114.82	4.5×10^{1}
50	Sn	tin	118.69	1.0×10^{2}
51	Sb	antimony	121.75	1.0×10^{1}
52	Te	tellurium	127.60	
53	I	iodine	126.90	
54	Xe	xenon	131.30	
55	Cs	caesium	132.91	$<7.9 \times 10^{1}$

Appendix 4 continued on the opposite page

Appendix 4 *Continued*

Atomic Number	Element	Name	Atomic Weight	Solar Abundance (Relative to Hydrogen = 10^{12})
56	Ba	barium	137.34	1.2×10^2
57	La	lanthanum	138.91	1.3×10^1
58	Ce	cerium	140.12	3.5×10^1
59	Pr	praseodymium	140.91	4.6
60	Nd	neodymium	144.24	1.7×10^1
61	Pm	promethium	146	
62	Sm	samarium	150.4	5.2
63	Eu	europium	151.96	5.0
64	Gd	gadolinium	157.25	1.3×10^1
65	Tb	terbium	158.93	
66	Dy	dysprosium	162.50	1.1×10^1
67	Ho	holmium	164.93	
68	Er	erbium	167.26	5.8
69	Tm	thulium	168.93	1.8
70	Yb	ytterbium	170.04	7.9
71	Lu	lutetium	174.97	5.8
72	Hf	hafnium	178.49	6.3
73	Ta	tantalum	180.95	
74	W	tungsten	183.85	5.0×10^1
75	Re	rhenium	186.2	≤ 0.5
76	Os	osmium	190.2	5.0
77	Ir	iridium	192.2	7.1
78	Pt	platinum	195.09	5.6×10^1
79	Au	gold	196.97	5.6
80	Hg	mercury	200.59	$< 1.3 \times 10^2$
81	Tl	thallium	204.37	7.9
82	Pb	lead	207.19	8.5×10^1
83	Bi	bismuth	208.98	$< 7.9 \times 10^1$
84	Po	polonium	210	
85	At	astatine	210	
86	Rn	radon	222	
87	Fr	francium	223	
88	Ra	radium	226.03	
89	Ac	actinium	227	
90	Th	thorium	232.04	1.6
91	Pa	protactinium	230.04	
92	U	uranium	238.03	< 4.0
93	Np	neptunium	237.05	
94	Pu	plutonium	242	
95	Am	americium	242	
96	Cm	curium	245	
97	Bk	berkelium	248	
98	Cf	californium	252	
99	Es	einsteinium	253	
100	Fm	fermium	257	
101	Md	mendelevium	257	
102	No	nobelium	255	
103	Lr	lawrencium	256	
104	Rf	rutherfordium	261	
105	Ha	hahnium	262	
106		(Reported 1974)	263	

*The solar abundance for hydrogen has been arbitrarily set at 10^{12}.

Atomic weights are averages for terrestrial abundances.

Solar abundances from John E. Ross and Lawrence H. Aller, *Science, 191*, 1223, 1976. Gaps indicate that the element has not been observed on the sun.

Appendix 5

The Planets

Appendix 5a. *Intrinsic and Rotational Properties*

Name	Equatorial Radius km	÷Earth's	Mass ÷Earth's	Mean Density (gm/cm^3)	Oblate-ness	Surface Gravity (Earth = 1)	Sidereal Rotation Period	Inclination of Equator to Orbit	Appare Magnitud Opposit
Mercury	2,439	0.3824	0.0553	5.44	0.0	0.378	59^d	<28°	−1.8
Venus	6,052	0.9489	0.8150	5.24	0.0	0.894	244.3^d R	177°	−4.3
Earth	6,378.140	1	1	5.497	0.0034	1	$23^h56^m04.1^s$	23°27′	–
Mars	3,397.2	0.5326	0.1074	3.9	0.009	0.379	$24^h37^m22.6^s$	23°59′	−2.0
Jupiter	71,398	11.194	317.89	1.3	0.063	2.54	9^h50^m to $>9^h55^m$	3°05′	−2.5
Saturn	60,000	9.41	95.17	0.7	0.098	1.07	10^h14^m to $>10^h38^m$	26°44′	+0.6
Uranus	25,400	3.98	14.56	1.2	0.06	0.919	12^h R	97°55′	+5.5
Neptune	24,300	3.81	17.24	1.7	0.021	1.19	15^h48^m	28°48′	+7.8
Pluto	2,500	0.4	0.11?	?	?	?	$6^d9^h17^m$	?	+14.9

R signifies retrograde rotation.

The masses and diameters are the values recommended by the International Astronomical Union in 1976; densities and su gravities were calculated from these values.

Appendix 5b. *Orbital Properties*

Name	Semimajor Axis A.U.	10^6 km	Sidereal Period Years	Days	Synodic Period (days)	Eccentricity	Inclina to Ecli
Mercury	0.3871	57.9	0.24084	87.96	115.9	0.2056	7°00′2
Venus	0.7233	108.2	0.61515	224.68	584.0	0.0068	3°23′4
Earth	1	149.6	1.00004	365.26	–	0.0167	0°00′1
Mars	1.5237	227.9	1.8808	686.95	779.9	0.0934	1°51′0
Jupiter	5.2028	778.3	11.867	4334.3	398.9	0.0483	1°18′2
Saturn	9.5388	1427.0	29.461	10,760	378.1	0.0560	2°29′1
Uranus	19.1914	2871.0	84.074	30,707	369.7	0.0461	0°48′2
Neptune	30.0611	4497.1	164.82	60,199	367.5	0.0100	1°46′2
Pluto	39.5294	5913.5	248.53	90,774	366.7	0.2484	17°09′0

Mean elements of planetary orbits for 1980, referred to the mean ecliptic and equinox of 1950 (Seidelmann, P. K., L. E. Dogg and M. R. DeLuccia, Astronomical Journal 79, 57, 1974).

Appendix 6

Planetary Satellites

atellite	Semimajor Axis of Orbit (km)	Sidereal Period (days)	Orbital Eccentricity	Orbital Inclination (°)	Radius (km)	Mass ÷ Mass of Planet	Discoverer	Apparent Magnitude at Opposition
ATELLITE OF THE EARTH								
he Moon	385,000	27.322	0.055	18–29	1738	0.01230002*	–	–12.7
ATELLITES OF MARS								
hobos	9,380	0.3189	0.018	1.0	14	2.7×10^{-8}	Hall (1877)	11.6
eimos	23,500	1.262	0.002	1.3	8	4.8×10^{-9}	Hall (1877)	12.7
ATELLITES OF JUPITER								
V Amalthea	181,000	0.4982	0.003	0.4	80	18×10^{-10}	Barnard (1892)	13.0
I Io	422,000	1.769	0.000	0	1830	4.70×10^{-5}*	Galileo (1610)	5.0
II Europa	671,000	3.551	0.000	0	1550	2.56×10^{-5}*	Galileo (1610)	5.3
III Ganymede	1,070,000	7.155	0.001	0	2640	7.84×10^{-5}*	Galileo (1610)	4.6
IV Callisto	1,880,000	16.69	0.01	0	2500	5.6×10^{-5}*	Galileo (1610)	5.6
III Leda	11,110,000	239	0.147	26.7	8	5×10^{-13}	Kowal (1974)	20
VI Himalia	11,500,000	250.6	0.158	27.6	60	8.5×10^{-10}	Perrine (1904)	14.7
VII Elara	11,700,000	259.7	0.207	24.8	20	0.35×10^{-10}	Perrine (1905)	16.0
X Lysithea	11,900,000	263.6	0.130	29.0	7	0.010×10^{-10}	Nicholson (1938)	18.8
XII Ananke	21,200,000	631.1 R	0.169	147	6	0.007×10^{-10}	Nicholson (1951)	18.3
XI Carme	22,600,000	692.5 R	0.207	164	7	0.020×10^{-10}	Nicholson (1938)	18.6
III Pasiphae	23,500,000	738.9 R	0.378	145	6	0.077×10^{-10}	Melotte (1908)	18.1
IX Sinope	23,700,000	758 R	0.275	153	7	0.015×10^{-10}	Nicholson (1914)	18.8
IV							Kowal (1975)	20
ATELLITES OF SATURN								
0 Janus	169,500	0.749	0.0	0.0	100	3×10^{-8}	Dollfus (1966)	14
1 Mimas	186,000	0.942	0.020	1.5	200	6.59×10^{-8}	W. Herschel (1789)	12.1
2 Enceladus	238,000	1.370	0.004	0.0	300	1.48×10^{-7}	W. Herschel (1789)	11.8
3 Tethys	295,000	1.888	0.000	1.1	500	1.10×10^{-6}	Cassini (1684)	10.3
4 Dione	377,000	2.737	0.002	0.0	400	2.04×10^{-6}	Cassini (1684)	10.4
5 Rhea	527,000	4.518	0.001	0.4	750	3.2×10^{-6}	Cassini (1672)	9.7
6 Titan	1,220,000	15.95	0.029	0.3	2900	2.41×10^{-4}*	Huygens (1655)	8.4
7 Hyperion	1,480,000	21.28	0.104	0.4	200	2×10^{-7}	Bond (1848)	14.2
8 Iapetus	3,560,000	79.33	0.028	14.7	750	3.94×10^{-6}	Cassini (1671)	11.0
9 Phoebe	13,000,000	550.5 R	0.163	150	100	5.2×10^{-8}	W. Pickering (1898)	16.5
ATELLITES OF URANUS								
Miranda	130,000	1.414	0.017	0	200	1×10^{-6}	Kuiper (1948)	16.5
Ariel	191,000	2.520 R	0.003	0	700	15×10^{-6}	Lassell (1851)	14.4
Umbriel	260,000	4.144 R	0.004	0	500	6×10^{-6}	Lassell (1851)	15.3
Titania	436,000	8.706 R	0.002	0	900	50×10^{-6}	W. Herschel (1787)	14.0
Oberon	583,000	13.46 R	0.001	0	800	29×10^{-6}	W. Herschel (1787)	14.2
ATELLITES OF NEPTUNE								
riton	354,000	5.877 R	0.000	160.0	1900	2×10^{-3}*	Lassell (1846)	13.6
ereid	5,570,000	359.4	0.76	27.4	300	10^{-6}	Kuiper (1949)	18.7

*Values recommended by the International Union in 1976.

Most other values from Wilkins, G. A., and A. T. Sinclair (Proc. Royal Society London A *336*, 85, 1974), with diameters for 5 of Sat-rn's satellites from Veverka, J., J. Elliot, and J. Goguen (Sky and Telescope, December 1975, p. 356.)

Apparent magnitudes at discovery given for Jupiter XIII and XIV.

R signifies retrograde revolution.

Appendix 7

Brightest Stars

Star	Name	Position 1980 *R.A.*	*Dec.*	Apparent Magnitude (V)	Spectral Type	Absolute Magnitude	Distance (ly)	Proper Motion (sec)
1. α CMa A	Sirius	06 44.2	−16 42	−1.47	Al V	+1.45	8.7	1.324
2. α Car	Canopus	06 23.5	−52 41	−0.72	F0 Ib-II	−3.1	98	0.025
3. α Boo	Arcturus	14 14.8	+19 17	−0.06	K2 IIIp	−0.3	36	2.284
4. α Cen A	Rigil Kentaurus	14 38.4	−60 46	0.01	G2 V	+4.39	4.3	3.676
5. α Lyr	Vega	18 36.2	+38 46	0.04	A0 V	+0.5	26.5	0.345
6. α Aur	Capella	05 15.2	+45 59	0.05	G8 III:+F	−0.6	45	0.435
7. β Ori A	Rigel	05 13.6	−08 13	0.14v	B8 Ia	−7.1	900	0.001
8. α CMi A	Procyon	07 38.2	+05 17	0.37	F5 IV-V	+2.7	11.3	1.250
9. α Ori	Betelgeuse	05 54.0	+07 24	0.41v	M2 Iab	−5.6	520	0.028
10. α Eri	Achernar	01 37.0	−57 20	0.51	B3 Vp	−2.3	118	0.098
11. β Cen AB	Hadar	14 02.4	−60 16	0.63v	B1 III	−5.2	490	0.035
12. α Aql	Altair	19 49.8	+08 49	0.77	A7 IV-V	+2.2	16.5	0.658
13. α Tau A	Aldebaran	04 34.8	+16 28	0.86v	K5 III	−0.7	68	0.202
14. α Vir	Spica	13 24.1	−11 03	0.91v	B1 V	−3.3	220	0.054
15. α Sco A	Antares	16 28.2	−26 23	0.92v	M1 Ib+B	−5.1	520	0.029
16. α PsA	Fomalhaut	22 56.5	−29 44	1.15	A3 V	+2.0	22.6	0.367
17. β Gem	Pollux	07 44.1	+28 05	1.16	K0 III	+1.0	35	0.625
18. α Cyg	Deneb	20 40.7	+45 12	1.26	A2 Ia	−7.1	1600	0.003
19. β Cru	Beta Crucis	12 46.6	−59 35	1.28v	B0.5 III	−4.6	490	0.049
20. α Leo A	Regulus	10 07.3	+12 04	1.36	B7 V	−0.7	84	0.248
21. α Cru A	Acrux	12 25.4	−62 59	1.39	B0.5 IV	−3.9	370	0.042
22. ϵ CMa A	Adhara	06 57.8	−28 57	1.48	B2 II	−5.1	680	0.004
23. λ Sco	Shaula	17 32.3	−37 05	1.60v	B1 V	−3.3	310	0.031
24. γ Ori	Bellatrix	05 24.0	+06 20	1.64	B2 III	−4.2	470	0.015
25. β Tau	Elnath	05 25.0	+28 36	1.65	B7 III	−3.2	300	0.178

Based on a table compiled by Donald A. MacRae in the *Observer's Handbook 1976* of the Royal Astronomical Society of Canada.

Appendix 8

The Nearest Stars

Name	1980 α	δ	Parallax π	Distance	Spectral Type	Proper Motion μ	Visual Magnitude m	Luminosity
	h m	° ′	″	ly		″		
Sun					G2		−26.8	1.0
α Cen A	14 38	−60 46	0.760	4.3	G2	3.68	0.1	1.3
B					K5		1.5	0.36
C	14 28	−62 36			M5*e*		11.0	0.00006
Barnard's	17 56	+04 36	.552	5.9	M5	10.30	9.5	0.00044
Wolf 359	10 56	+07 10	.431	7.6	M6*e*	4.84	13.5	0.00002
Lalande 21185	11 03	+36 07	.402	8.1	M2	4.78	7.5	0.0052
Sirius A	6 44	−16 42	.377	8.6	A1	1.32	−1.5	23.
B					*wd*		7.2	0.008
Luyten 726-8A	1 37	−18 04	.365	8.9	M6*e*	3.35	12.5	0.00006
B (UV Ceti)					M6*e*		13.0	0.00004
Ross 154	18 49	−23 50	.345	9.4	M5*e*	0.74	10.6	0.0004
Ross 248	23 40	+44 04	.317	10.3	M6*e*	1.82	12.2	0.00011
ε Eri	03 32	−09 32	.305	10.7	K2	0.97	3.7	0.30
Luyten 789-6	22 38	−15 28	.302	10.8	M6	3.27	12.2	0.00012
Ross 128	11 47	+00 58	.301	10.8	M5	1.40	11.1	0.00033
61 Cyg A	21 06	+38 38	.292	11.2	K5	5.22	5.2	0.083
B					K7		6.0	0.040
ε Ind	22 03	−56 52	.291	11.2	K5	4.67	4.7	0.13
Procyon A	07 39	+05 17	.287	11.4	F5	1.25	0.3	7.6
B					*wd*		10.8	0.0005
Σ 2398 A	18 42	+59 36	.284	11.5	M3.5	2.29	8.9	0.0028
B					M4		9.7	0.0013
Groombridge 34 A	00 18	+43 54	.282	11.6	M1	2.91	8.1	0.0058
B					M6		11.0	0.00040
Lacaille 9352	23 05	−35 59	.279	11.7	M2	6.87	7.4	0.012
τ Ceti	01 43	−16 03	.273	11.9	G8	1.92	3.5	0.44
BD+5°1668	07 27	+05 27	.266	12.2	M4	3.73	9.8	0.0014
L725-32	01 11	−17 06	.262	12.4	M5*e*	1.31	11.5	0.0003
Lacaille 8760	21 16	−38 58	.260	12.5	M1	3.46	6.7	0.025
Kapteyn's	05 11	−44 59	.256	12.7	M0	8.79	8.8	0.0040
Kruger 60 A	22 27	+57 36	.254	12.8	M4	0.87	9.7	0.0017
B					M6		11.2	0.00044
Ross 614 A	06 28	−02 48	.249	13.1	M5*e*	0.97	11.3	0.0004
B					?		14.8	0.00002
BD−12°4523	16 30	−12 36	.249	13.1	M5	1.18	10.0	0.0013
van Maanen's	00 48	+05 19	.234	13.9	*wd*F	2.98	12.4	0.00017
Wolf 424 A	12 33	+09 09	.229	14.2	M6*e*	1.87	12.6	0.00014
B					M6*e*		12.6	0.00014
CD−37°15492	00 04	−37 27	.225	14.5	M3	6.09	8.6	0.0058
G158 27	00 06	−07 38	.224	14.6		2.1	13.8	0.00005
Groombridge 1618	10 10	+49 33	.217	15.0	M0	1.45	6.6	0.040
CD−46°11540	17 28	−46 53	.216	15.1	M4	1.15	9.4	0.0030
CD−49°13515	21 32	−49 11	.214	15.2	M3	0.78	8.7	0.0058
CD−44°11909	17 37	−44 17	.213	15.3	M5	1.14	11.2	0.00063
Luyten 1159-16	01 59	+13 00	.212	15.4	(M7)	2.08	12.3	0.00023
Lalande 25372	13 44	+15 01	.208	15.7	M3.5	2.30	8.5	0.0076
AOe 17415-6	17 37	+68 22	.207	15.7	M3.5	1.31	9.1	0.0044
CC 658	11 44	−64 42	.206	15.8	*wd*	2.69	11.0	0.0008
Ross 780	22 52	−14 22	.206	15.8	M5	1.17	10.2	0.0016
40 Eri A	04 14	−07 41	.205	15.9	K0	4.08	4.4	0.33
B					*wd*A		9.9	0.0027
C					M4*e*		11.2	0.00063

Based on a table compiled by Alan H. Batten in the *Observer's Handbook 1976* of the Royal Astronomical Society of Canada.

Appendix 9
Messier Catalogue

M	NGC	α 1980 h m	δ ° ′	m_v	Description
1	1952	5 33.3	+22 01	11.3	Crab Nebula in Taurus
2	7089	21 32.4	−00 54	6.3	Globular cluster in Aquarius
3	5272	13 41.3	+28 29	6.2	Globular cluster in Canes Venatici
4	6121	16 22.4	−26 27	6.1	Globular cluster in Scorpio
5	5904	15 17.5	+02 11	6	Globular cluster in Serpens
6	6405	17 38.9	−32 11	6	Open cluster in Scorpio
7	6475	17 52.6	−34 48	5	Open cluster in Scorpio
8	6523	18 02.4	−24 23		Lagoon Nebula in Sagittarius
9	6333	17 18.1	−18 30	7.6	Globular cluster in Ophiuchus
10	6254	16 56.0	−04 05	6.4	Globular cluster in Ophiuchus
11	6705	18 50.0	−06 18	7	Open cluster in Scutum
12	6218	16 46.1	−01 55	6.7	Globular cluster in Ophiuchus
13	6205	16 41.0	+36 30	5.8	Globular cluster in Hercules
14	6402	17 36.5	−03 14	7.8	Globular cluster in Ophiuchus
15	7078	21 29.1	+12 05	6.3	Globular cluster in Pegasus
16	6611	18 17.8	−13 48	7	Open cluster in Serpens
17	6618	18 19.7	−16 12	7	Omega Nebula in Sagittarius
18	6613	18 18.8	−17 09	7	Open cluster in Sagittarius
19	6273	17 01.3	−26 14	6.9	Globular cluster in Ophiuchus
20	6514	18 01.2	−23 02		Trifid Nebula in Sagittarius
21	6531	18 03.4	−22 30	7	Open cluster in Sagittarius
22	6656	18 35.2	−23 55	5.2	Globular cluster in Sagittarius
23	6494	17 55.7	−19 00	6	Open cluster in Sagittarius
24	6603	18 17.3	−18 27	6	Open cluster in Sagittarius
25	IC4725	18 30.5	−19 16	6	Open cluster in Sagittarius
26	6694	18 44.1	−09 25	9	Open cluster in Scutum
27	6853	19 58.8	+22 40	8.2	Dumbbell Nebula; planetary nebula in Vulpecula
28	6626	18 23.2	−24 52	7.1	Globular cluster in Sagittarius
29	6913	20 23.3	+38 27	8	Open cluster in Cygnus
30	7099	21 39.2	−23 15	7.6	Globular cluster in Capricornus
31	224	0 41.6	+41 09	3.7	Andromeda Galaxy (Sb)
32	221	0 41.6	+40 45	8.5	Elliptical galaxy in Andromeda; companion to M31
33	598	1 32.8	+30 33	5.9	Spiral galaxy (Sc) in Triangulum
34	1039	2 40.7	+42 43	6	Open cluster in Perseus
35	2168	6 07.6	+24 21	6	Open cluster in Gemini
36	1960	5 35.0	+34 05	6	Open cluster in Auriga
37	2099	5 51.5	+32 33	6	Open cluster in Auriga
38	1912	5 27.3	+35 48	6	Open cluster in Auriga
39	7092	21 31.5	+48 21	6	Open cluster in Cygnus
40	–	–	–		Double star in Ursa Major
41	2287	6 46.2	−20 43	6	Open cluster in Canis Major
42	1976	5 34.4	−05 24		Orion Nebula
43	1982	5 34.6	−05 18		Orion Nebula; smaller part
44	2632	8 38.8	+20 04	4	Praesepe; open cluster in Cancer
45	–	3 46.3	+24 03	2	The Pleiades; open cluster in Taurus
46	2437	7 40.9	−14 46	7	Open cluster in Puppis
47	2422	7 35.6	−14 27	5	Open cluster in Puppis
48	2548	8 12.5	−05 43	6	Open cluster in Hydra
49	4472	12 28.8	+08 07	8.9	Elliptical galaxy in Virgo
50	2323	7 02.0	−08 19	7	Open cluster in Monoceros

Appendix 9 continued on the opposite page

M	NGC	α 1980 h m	δ ° ′	m_v	Description
51	5194	13 29.0	+47 18	8.4	Whirlpool Galaxy; spiral galaxy (Sc) in Canes Venatici
52	7654	23 23.3	+61 29	7	Open cluster in Cassiopeia
53	5024	13 12.0	+18 17	7.7	Globular cluster in Coma Berenices
54	6715	18 53.8	−30 30	7.7	Globular cluster in Sagittarius
55	6809	19 38.7	−31 00	6.1	Globular cluster in Sagittarius
56	6779	19 15.8	+30 08	8.3	Globular cluster in Lyra
57	6720	18 52.9	+33 01	9.0	Ring Nebula; planetary nebula in Lyra
58	4579	12 36.7	+11 56	9.9	Spiral galaxy (SBb) in Virgo
59	4621	12 41.0	+11 47	10.3	Elliptical galaxy in Virgo
60	4649	12 42.6	+11 41	9.3	Elliptical galaxy in Virgo
61	4303	12 20.8	+04 36	9.7	Spiral galaxy (Sc) in Virgo
62	6266	16 59.9	−30 05	7.2	Globular cluster in Scorpio
63	5055	13 14.8	+42 08	8.8	Spiral galaxy (Sb) in Canes Venatici
64	4826	12 55.7	+21 48	8.7	Spiral galaxy (Sb) in Coma Verenices
65	3623	11 17.8	+13 13	9.6	Spiral galaxy (Sa) in Leo
66	3627	11 19.1	+13 07	9.2	Spiral galaxy (Sb) in Leo; companion to M65
67	2682	8 50.0	+11 54	7	Open cluster in Cancer
68	4590	12 38.3	−26 38	8	Globular cluster in Hydra
69	6637	18 30.1	−32 23	7.7	Globular cluster in Sagittarius
70	6681	18 42.0	−32 18	8.2	Globular cluster in Sagittarius
71	6838	19 52.8	+18 44	6.9	Globular cluster in Sagitta
72	6981	20 52.3	−12 39	9.2	Globular cluster in Aquarius
73	6994	20 57.8	−12 44		Open cluster in Aquarius
74	628	1 35.6	+15 41	9.5	Spiral galaxy (Sc) in Pisces
75	6864	20 04.9	−21 59	8.3	Globular cluster in Sagittarius
76	650	1 40.9	+51 28	11.4	Planetary nebula in Perseus
77	1068	2 41.6	−00 04	9.1	Spiral galaxy (Sb) in Cetus
78	2068	5 45.8	+00 02		Small emission nebula in Orion
79	1904	5 23.3	−24 32	7.3	Globular cluster in Lepus
80	6093	16 15.8	−22 56	7.2	Globular cluster in Scorpio
81	3031	9 54.2	+69 09	6.9	Spiral galaxy (Sb) in Ursa Major
82	3034	9 54.4	+69 47	8.7	Irregular galaxy (Irr) in Ursa Major
83	5236	13 35.9	−29 46	7.5	Spiral galaxy (Sc) in Hydra
84	4374	12 24.1	+13 00	9.8	Elliptical galaxy in Virgo
85	4382	12 24.3	+18 18	9.5	Elliptical galaxy (SO) in Coma Berenices
86	4406	12 25.1	+13 03	9.8	Elliptical galaxy in Virgo
87	4486	12 29.7	+12 30	9.3	Elliptical galaxy (Ep) in Virgo
88	4501	12 30.9	+14 32	9.7	Spiral galaxy (Sb) in Coma Berenices
89	4552	12 34.6	+12 40	10.3	Elliptical galaxy in Virgo
90	4569	12 35.8	+13 16	9.7	Spiral galaxy (Sb) in Virgo
91	–	–	–		M58 ?
92	6341	17 16.5	+43 10	6.3	Globular cluster in Hercules
93	2447	7 43.6	−23 49	6	Open cluster in Puppis
94	4736	12 50.1	+41 14	8.1	Spiral galaxy (Sb) in Canes Venatici
95	3351	10 42.8	+11 49	9.9	Barred spiral galaxy (SBb) in Leo
96	3368	10 45.6	+11 56	9.4	Spiral galaxy (Sa) in Leo
97	3587	11 13.7	+55 08	11.1	Owl Nebula; planetary nebula in Ursa Major
98	4192	12 12.7	+15 01	10.4	Spiral galaxy (Sb) in Coma Berenices
99	4254	12 17.8	+14 32	9.9	Spiral galaxy (Sc) in Coma Berenices
100	4321	12 21.9	+15 56	9.6	Spiral galaxy (Sc) in Coma Berenices

Appendix 9 continued on the following page

Appendix 9 *Continued*

M	NGC	α 1980 h m	δ ° ′	m_v	Description
101	5457	14 02.5	+54 27	8.1	Spiral galaxy (Sc) in Ursa Major
102	–	–	–		M101 ?
103	581	1 31.9	+60 35	7	Open cluster in Cassiopeia
104	4594	12 39.0	−11 35	8	Sombrero Nebula; spiral galaxy (Sa) in Virgo
105	3379	10 46.8	+12 51	9.5	Elliptical galaxy in Leo
106	4258	12 18.0	+47 25	9	Spiral galaxy in (Sb) Canes Venatici
107	6171	16 31.8	−13 01	9	Globular cluster in Ophiuchus
108	3556	11 10.5	+55 47	10.5	Spiral galaxy (Sb) in Ursa Major
109	3992	11 56.6	+53 29	10.6	Barred spiral galaxy (SBc) in Ursa Major

Based in part on a table in the *Observer's Handbook 1976* of the Royal Astronomical Society of Canada.

Appendix 10

The Constellations

Latin Name	Genitive	Abbreviation	Translation
Andromeda	Andromedae	And	Andromeda*
Antlia	Antliae	Ant	Pump
Apus	Apodis	Aps	Bird of Paradise
Aquarius	Aquarii	Aqr	Water Bearer
Aquila	Aquilae	Aql	Eagle
Ara	Arae	Ara	Altar
Aries	Arietis	Ari	Ram
Auriga	Aurigae	Aur	Charioteer
Boötes	Boötis	Boo	Herdsman
Caelum	Caeli	Cae	Chisel
Camelopardalis	Camelopardalis	Cam	Giraffe
Cancer	Cancri	Cnc	Crab
Canes Venatici	Canum Venaticorum	CVn	Hunting Dogs
Canis Major	Canis Majoris	CMa	Big Dog
Canis Minor	Canis Minoris	CMi	Little Dog
Capricornus	Capricorni	Cap	Goat
Carina	Carinae	Car	Ship's Keel**
Cassiopeia	Cassiopeiae	Cas	Cassiopeia*
Centaurus	Centauri	Cen	Centaur*
Cepheus	Cephei	Cep	Cepheus*
Cetus	Ceti	Cet	Whale
Chamaeleon	Chamaeleonis	Cha	Chameleon
Circinus	Circini	Cir	Compass
Columba	Columbae	Col	Dove
Coma Berenices	Comae Berenices	Com	Berenice's Hair*
Corona Australis	Coronae Australis	CrA	Southern Crown
Corona Borealis	Coronae Borealis	CrB	Northern Crown
Corvus	Corvi	Crv	Crow
Crater	Crateris	Crt	Cup
Crux	Crucis	Cru	Southern Cross
Cygnus	Cygni	Cyg	Swan

Appendix 10 continued on the opposite page

Appendix 10 *Continued*

Latin Name	Genitive	Abbreviation	Translation
Delphinus	Delphini	Del	Dolphin
Dorado	Doradus	Dor	Swordfish
Draco	Draconis	Dra	Dragon
Equuleus	Equulei	Equ	Little Horse
Eridanus	Eridani	Eri	River Eridanus*
Fornax	Fornacis	For	Furnace
Gemini	Geminorum	Gem	Twins
Grus	Gruis	Gru	Crane
Hercules	Herculis	Her	Hercules*
Horologium	Horologii	Hor	Clock
Hydra	Hydrae	Hya	Hydra* (water monster)
Hydrus	Hydri	Hyi	Sea serpent
Indus	Indi	Ind	Indian
Lacerta	Lacertae	Lac	Lizard
Leo	Leonis	Leo	Lion
Leo Minor	Leonis Minoris	LMi	Little Lion
Lepus	Leporis	Lep	Hare
Libra	Librae	Lib	Scales
Lupus	Lupi	Lup	Wolf
Lynx	Lyncis	Lyn	Lynx
Lyra	Lyrae	Lyr	Harp
Mensa	Mensae	Men	Table (mountain)
Microscopium	Microscopii	Mic	Microscope
Monoceros	Monocerotis	Mon	Unicorn
Musca	Muscae	Mus	Fly
Norma	Normae	Nor	Level (square)
Octans	Octantis	Oct	Octant
Ophiuchus	Ophiuchi	Oph	Ophiuchus* (serpent bearer)
Orion	Orionis	Ori	Orion*
Pavo	Pavonis	Pav	Peacock
Pegasus	Pegasi	Peg	Pegasus* (winged horse)
Perseus	Persei	Per	Perseus*
Phoenix	Phoenicis	Phe	Phoenix
Pictor	Pictoris	Pic	Easel
Pisces	Piscium	Psc	Fish
Piscis Austrinus	Piscis Austrini	PsA	Southern Fish
Puppis	Puppis	Pup	Ship's Stern**
Pyxis	Pyxidis	Pyx	Ship's Compass**
Reticulum	Reticuli	Ret	Net
Sagitta	Sagittae	Sge	Arrow
Sagittarius	Sagittarii	Sgr	Archer
Scorpius	Scorpii	Sco	Scorpion
Sculptor	Sculptoris	Scl	Sculptor
Scutum	Scuti	Sct	Shield
Serpens	Serpentis	Ser	Serpent
Sextans	Sextantis	Sex	Sextant
Taurus	Tauri	Tau	Bull
Telescopium	Telescopii	Tel	Telescope
Triangulum	Trianguli	Tri	Triangle
Triangulum Australe	Trianguli Australis	TrA	Southern Triangle
Tucana	Tucanae	Tuc	Toucan
Ursa Major	Ursae Majoris	UMa	Big Bear
Ursa Minor	Ursae Minoris	UMi	Little Bear
Vela	Velorum	Vel	Ship's Sails**
Virgo	Virginis	Vir	Virgin
Volans	Volantis	Vol	Flying Fish
Vulpecula	Vulpeculae	Vul	Little Fox

*Proper names.
**Formerly formed the constellation Argo Navis, the Argonaut's ship.

Selected Readings

(A more detailed list of readings appears in the Student Study Guide.)

Non-Technical Books on Astronomy

Herbert Friedman, *The Amazing Universe* (Washington: National Geographic Society, 1975). *National Geographic's* survey of astronomy, written by a pioneer in space science and profusely illustrated in color.

Donald H. Menzel, *Astronomy* (New York: Random House, Inc., 1970). A well-illustrated large-format popular description of the universe.

Isaac Asimov, *The Universe* (New York: Avon Books, 1966). A popular survey.

Kees Boecke, *Cosmic View: The Universe in 40 Jumps* (New York: John Day Co., Inc., 1957.) Views of objects in the universe increasing in scale by factors of 10, ranging up to clusters of galaxies and down to the nucleus of an atom.

Lawrence H. Aller, *Atoms, Stars, and Nebulae*, rev. ed. (Cambridge, Mass.: Harvard University Press, 1971). Readable discussions of the material in Chapters 2 and 4.

Bart J. Bok and Priscilla F. Bok, *The Milky Way*, 4th ed. (Cambridge, Mass.: Harvard University Press, 1974). A readable and well-illustrated survey of the Milky Way; includes good discussions of star clusters and H-R diagrams.

John C. Brandt, *The Sun and Stars* (New York: McGraw-Hill Book Co., 1966). A general discussion.

Donald H. Menzel, *Our Sun*, rev. ed. (Cambridge, Mass.: Harvard University Press, 1959). In need of updating, but still the best available popular discussion of the sun.

Robert Jastrow, *Red Giants and White Dwarfs* (New York: Harper & Row Pubs., Inc., 1967; New American Library, 1969). "The evolution of stars, planets and life."

Harry L. Shipman, *Black Holes, Quasars, and the Universe* (Boston: Houghton Mifflin Co., 1976). A careful presentation to a non-specialist audience.

Frederick Golden, *Quasars, Pulsars, and Black Holes* (New York: Charles Scribner's Sons, 1976). A popular discussion.

William J. Kaufmann III, *Relativity and Cosmology* (New York: Harper & Row Pubs., Inc., 1973). A brief and easy treatment of black holes, curved space, quasars, and other strange concepts.

Fred L. Whipple, *Earth, Moon, and Planets*, 3rd ed. (Cambridge, Mass.: Harvard University Press, 1968). A non-technical discussion of the solar system.

Richard S. Lewis, *The Voyages of Apollo* (New York, Quadrangle, 1974). A general discussion of the moon-landing program written for non-scientists.

Walter Sullivan, *Continents in Motion* (New York: McGraw-Hill Book Co., 1974). A popular discussion of the earth and plate tectonics.

Carl Sagan, *The Cosmic Connection: An Extraterrestrial Perspective* (Dell Publishing Co., Inc., 1975). Anecdotes and reminiscences about the search for extraterrestrial life.

Ronald Bracewell, *The Galactic Club—Intelligent Life in Outer Space* (San Francisco: W. H. Freeman & Co., 1974). On extraterrestrial life.

Richard Berendzen, Richard Hart, and Daniel Seeley, *Man Discovers the Galaxies* (New York: Neale Watson Academic Publications, Inc., 1976). A historical view of studies of the Milky Way and other galaxies.

Charles A. Whitney, *The Discovery of Our Galaxy* (New York, Alfred A. Knopf, Inc., 1971). A readable history.

Harlow Shapley, *Galaxies*, 3rd ed., revised by Paul W. Hodge (Cambridge, Mass.: Harvard University Press, 1972). A non-technical study of galaxies, written by the master.

Gerrit Verschuur, *The Invisible Universe* (New York: Springer-Verlag New York, Inc., 1974). A popular survey of radio astronomy.

George Gamow, *One, Two, Three . . . Infinity* (New York: Bantam Books, Inc., 1971). A reprinting of a wonderful description of the structure of space that has introduced many a contemporary astronomer to his or her profession.

Evry Schatzman, *The Structure of the Universe* (New York: McGraw-Hill Book Co., 1968). A popular discussion.

Fred Hoyle, *Highlights in Astronomy* (San Francisco: W. H. Freeman & Co., 1975). A well-illustrated personal tour through the heavens.

Lloyd Motz, *The Universe: Its Beginning and End* (New York: Charles Scribner's Sons, 1975). A non-technical survey of the universe, stars, and the solar system.

More Advanced Books:

Eugene H. Avrett, ed., *Frontiers in Astrophysics* (Cambridge, Mass.: Harvard University Press, 1976). A series of chapters, each written by an expert, on contemporary research.

Owen Gingerich, ed., *Frontiers in Astronomy, New Frontiers in Astronomy* (San Francisco: W. H. Freeman & Co., 1971 and 1976, respectively). Two collections of reprints from Scientific American.

The Solar System (San Francisco: W. H. Freeman & Co., 1975). A reprint of the special Scientific American issue of September 1975.

Allen Sandage, *The Hubble Atlas* (Washington, D.C.: Carnegie Institution of Washington, 1961). Beautiful photographs of galaxies and their descriptions.

Elske v. P. Smith and Kenneth C. Jacobs, *Introductory Astronomy and Astrophysics* (Philadelphia, W. B. Saunders Co., 1973). A mathematical introductory textbook.

Monthly Non-technical Magazines:

Sky and Telescope, 49-50-51 Bay State Road, Cambridge, MA 02138.

Astronomy, 411 East Mason Street, Milwaukee, WI 53202.

Popular Astronomy, 270 Madison Avenue, New York, NY 10016.

Mercury, Astronomical Society of the Pacific, 1244 Noriega Street, San Francisco, CA 94122.

The Griffith Observer, Griffith Observatory, 2800 East Observatory Road, Los Angeles, CA 90027.

Science News, 1719 N Street, N.W. Washington, DC 20036. Published weekly.

Magazines and Annuals Carrying Articles on Astronomy:

Scientific American, 415 Madison Avenue, New York, NY 10017.

National Geographic, Washington, DC 20036.

Natural History, Membership Services, Box 6000, Des Moines, IA 50340.

Science Year (Chicago, IL, Field Enterprises Educational Corp.). The World Book Science Annual.

Observing Reference Books:

Donald H. Menzel, *A Field Guide to the Stars and Planets* (Boston: Houghton Mifflin Co., 1975).

Charles A. Whitney, *Whitney's Star Finder* (New York, Alfred A. Knopf, Inc., 1974).

Arthur P. Norton, *Norton's Star Atlas and Reference Handbook*, 16th ed., revised by the successors of the late Mr. Norton (Cambridge, Mass.: Sky Publishing Corp., 1973).

The Observer's Handbook (yearly), Royal Astronomical Society of Canada, 252 College Street, Toronto M5T 1R7, Canada.

The American Ephemeris and Nautical Almanac (yearly), U.S. Government Printing Office, Washington, DC 20402.

Hans Vehrenberg, *Atlas of Deep Sky Splendors* (Cambridge, Mass.: Sky Publishing Corp., 1971). Photographs, descriptions, and finding charts for hundreds of beautiful objects.

General Reference Books:

C. W. Allen, *Astrophysical Quantities*, 3rd ed. (London: The Athlone Press of the University of London, 1973). Tables and lists of almost every conceivable kind.

Kenneth R. Lang, *Astrophysical Formulae* (New York: Springer-Verlag New York, Inc., 1974). Formulas of all kinds, plus many tables.

Van Nostrand's Scientific Encyclopedia, 5th ed. (New York: Van Nostrand Reinhold Co., 1976). A one-volume encyclopedia.

Jeanne Hopkins, *Glossary of Astronomy and Astrophysics* (Chicago: University of Chicago Press, 1976). A technical glossary.

For Information about Amateur Societies:

American Association of Variable Star Observers (AAVSO), 187 Garden Street, Cambridge, MA 02138.

Association of Lunar and Planetary Observers (ALPO), 8930 Raven Drive, Waco, TX 76710.

Careers in Astronomy:

The Executive Officer, American Astronomical Society, 211 FitzRandolph Road, Princeton, NJ 08540. A free booklet is available on request.

Acknowledgments

Color Plates

Plate 1 Copyright by The Association of Universities for Research in Astronomy, Inc. The Kitt Peak National Observatory;
Plates 2, 13, 14, and 15 Jay M. Pasachoff;
Plate 3 Bausch and Lomb;
Plates 4, 45, and 48 Copyright by the Association of Universities for Research in Astronomy, Inc. The Cerro Tololo Inter-American Observatory;
Plates, 5, 6, 7, 8, 10, 11, 12, 41, 42, 43, 44, 46, 47, and 49 Copyright California Institute of Technology and Carnegie Institution of Washington;
Plate 9 Richard J. Borken and Saul A. Rappaport;
Plate 16 William C. Atkinson;
Plates 17 and 18 Naval Research Laboratory/NASA, courtesy J. David Bohlin;
Plates 19, 20, and 21 High Altitude Observatory/NASA, Courtesy of Robert M. MacQueen and Charles L. Ross;
Plates 22, 23, and 24 Harvard College Observatory/NASA, courtesy of Peter V. Foukal, Edward J. Schmahl, and Edmond M. Reeves;
Plates 25, 26, 27, 28, 29, 30, and 31 NASA;
Plate 32 International Planetary Patrol Photograph furnished by Lowell Observatory;
Plate 33 Gustav Lamprecht;
Plates 34, 35, 36, and 37 NASA;
Plate 38 NASA/Lunar and Planetary Laboratory, University of Arizona;
Plate 39 Lunar and Planetary Laboratory, University of Arizona;
Plate 40 © 1976 Saul G. Levy;
Plate 50 Netherlands Foundation for Radio Astronomy, National Radio Astronomy Observatory, Arnold H. Rots and William W. Shane. Optical image from the Hale Observatories.

Illustrations

PART I—Opener Lick Observatory;
Figs. 1–2 and 1–3 Jay M. Pasachoff;
Fig. 1–4 Skyviews Survey Inc.;
Figs. 1–5 and 1–6 NASA;
Fig. 1–12 Harvard College Observatory;
Fig. 1–14 Hale Observatories.

PART II—Opener (*left*) Hale Observatories, (*right*) Big Bear Solar Observatory.

PART II Introduction NASA and TRW Systems Group;

CHAPTER 2–Opener By permission of The Houghton Library, Harvard University;
Fig. 2–1 From THE LITTLE PRINCE by Antoine de Saint-Exupéry, copyright 1943, by Harcourt Brace Jovanovich, Inc. in America and William Heinemann Ltd. in England; copyright, 1971, by Consuelo de Saint-Expéry. Reproduced by permission of the publishers;
Fig. 2–8 From Leo Goldberg, "Ultraviolet Astronomy," *Scientific American*, June 1969, and *Frontiers in Astronomy*, reprinted courtesy of *Scientific American;*
Fig. 2–15 © National Geographic Society–Palomar Observatory Sky Survey. Reproduced by permission from the Hale Observatories;
Fig. 2–16 American Institute of Physics, Meggers Gallery of Nobel Laureates;
Fig. 2–17 Harvard College Observatory;
Fig. 2–18 The Kitt Peak National Observatory;
Fig. 2–28 American Institute of Physics, Niels Bohr Library, Margrethe Bohr Collection;
Fig. 2–30 Jay M. Pasachoff;
Fig. 2–31 Jay M. Pasachoff/Hale Observatories;
Fig. 2–34 After Bok and Bok, *The Milky Way*, courtesy Harvard University Press;
Fig. 2–35 Hale Observatories;
Fig. 2–38 © 1971 by *Playboy Magazine;*
Figs. 2–44 and 2–45 Courtesy of Jean Pierre Swings;
Fig. 2–49 Peter van de Kamp, Sproul Observatory.

CHAPTER 3–Opener The Kitt Peak National Observatory;
Fig. 3–2 Jay M. Pasachoff;
Fig. 3–6 Lick Observatory;
Fig. 3–7 American Institute of Physics, Niels Bohr Library;
Fig. 3–13 Copyright Royal Greenwich Observatory;
Figs. 3–15 and 3–16 Hale Observatories;
Fig. 3–17A Dale P. Cruikshank, University of Hawaii;
Figs. 3–17B and 3–17C Sovfoto;
Fig. 3–18 NASA;
Fig. 3–19 The Kitt Peak National Observatory;
Fig. 3–20 The Cerro Tololo Inter-American Observatory;
Fig. 3–21 The Kitt Peak National Observatory and The Cerro Tololo Inter-American Observatory/Arthur A. Hoag;
Fig. 3–22 Jay M. Pasachoff;
Fig. 3–24 Center for Astrophysics [of the Harvard College Observatory and the Smithsonian Astrophysical Observatory];
Fig. 3–25 Hale Observatories;
Fig. 3–27 © National Geographic Society–Palomar Observatory Sky Survey. Reproduced by permission from the Hale Observatories;
Fig. 3–29 (*left*) Jay M. Pasachoff, (*middle*) Harvard College Observatory, (*right*) Hale Observatories;
Fig. 3–30 Thomas B. McCord, Massachusetts Institute of Technology;
Fig. 3–31 Hale Observatories;
Fig. 3–32 Giovanni Fazio, Center for Astrophysics;
Fig. 3–34A Herbert Gursky, Center for Astrophysics;
Fig. 3–34B Gordon P. Garmire, California Institute of Technology;
Figs. 3–35, 3–36, and 3–37 Hale Observatories;
Fig. 3–38 Jay M. Pasachoff;
Fig. 3–43 Chapin Library, Williams College;
Fig. 3–46 Emil Schulthess, Black Star;
Fig. 3–47 By permission of The Houghton Library, Harvard University;
Fig. 3–48 Dale P. Cruikshank, University of Hawaii;
Fig. 3–49 Yerkes Observatory;
Figs. 3–52A and 3–52B © 1970 United Features Syndicate, Inc.

CHAPTER 4–Opener Hale Observatories;
Fig. 4–2 Hale Observatories;
Fig. 4–3 Lick Observatory;
Fig. 4–6 Peter van de Kamp;
Fig. 4–7 Based on data from D. L. Harris III, K. Aa. Strand, and C. E. Worley in *Basic Astronomical Data,* K. Aa. Strand, ed., University of Chicago Press, copyright 1963 by The University of Chicago;
Fig. 4–8 The Kitt Peak National Observatory;
Fig. 4–9 R. Hanbury Brown;
Fig. 4–10 Peter van de Kamp and Sarah Lee Lippincott, from *Vistas in Astronomy, 19,* 231, (1975), courtesy Pergamon Press;
Fig. 4–12 Courtesy of *La Recherche;*
Fig. 4–13 After Lawrence H. Aller, *Atoms, Stars and Nebulae,* courtesy Harvard University Press;
Fig. 4–14 American Association of Variable Star Observers/Janet Mattei;
Fig. 4–15A After Bok and Bok, *The Milky Way,* courtesy Harvard University Press;
Fig. 4–15B American Association of Variable Star Observers/Janet Mattei;
Fig. 4–17 Harvard College Observatory;
Fig. 4–18 Shigetsugu Fujinami;
Fig. 4–19 After Bok and Bok, *The Milky Way,* courtesy Harvard University Press;
Fig. 4–21 After Bok and Bok, *The Milky Way,* courtesy Harvard University Press;
Fig. 4–22 Hale Observatories;
Fig. 4–23 By permission of The Houghton Library, Harvard University;
Fig. 4–24 Lick Observatory;
Fig. 4–25 Dept. of Astronomy, University of Michigan, courtesy of Freeman D. Miller;
Fig. 4–26 After Bok and Bok, *The Milky Way,* courtesy Harvard University Press;
Fig. 4–29 Lick Observatory;
Fig. 4–30 After Bok and Bok, *The Milky Way,* courtesy Harvard University Press, and a graph of Harold L. Johnson and Allan R. Sandage in the *Astrophysical Journal 124,* 379. Reprinted by permission of the University of Chicago Press © 1956 by the American Astronomical Society;
Fig. 4–31 Photograph courtesy of Harvard College Observatory. X-ray data courtesy of George W. Clark, Massachusetts Institute of Technology;
Fig. 4–32 Astronomy Netherlands Satellite data/Space Research Laboratory, Utrecht.

CHAPTER 5–Opener The Kitt Peak National Observatory/William C. Livingston;
Fig. 5–2 Sacramento Peak Observatory;
Figs. 5–3 and 5–4 Hale Observatories;
Fig. 5–5 Jay M. Pasachoff;
Fig. 5–8 Sacramento Peak Observatory;
Fig. 5–9 A. Keith Pierce, The Kitt Peak National Observatory;
Fig. 5–10 From John Kohl and W. H. Parkinson, Center for Astrophysics. Reprinted from the *Astrophysical Journal 205,* 599 with permission of the University of Chicago Press © 1976 by the American Astronomical Society;
Figs. 5–12, 5–13, and 5–14 Jay M. Pasachoff;
Fig. 5–15 Hale Observatories;
Fig. 5–16 Sacramento Peak Observatory;
Fig. 5–17 The Kitt Peak National Observatory;
Fig. 5–18 Big Bear Solar Observatory;
Fig. 5–19 The Aerospace Corp./David K. Lynch;
Fig. 5–20 From Olin C. Wilson, Hale Observatories. Reprinted from the *Astrophysical Journal 130,* 499 with permission of the University of Chicago Press © 1959 by the American Astronomical Society;
Fig. 5–21 Eugene H. Avrett, Center for Astrophysics;
Fig. 5–22 M. Kanno, Hida Observatory, University of Kyoto;
Fig. 5–23 Harvard College Observatory, Imperial College (London), York University (Toronto), and Culham Laboratory (England), courtesy of Robert Speer;
Figs. 5–24 and 5–25 High Altitude Observatory/Robert M. MacQueen and Charles L. Ross;

Fig. 5–28 Based on a diagram by H. C. van de Hulst;
Fig. 5–29 University of Hawaii;
Fig. 5–31 NASA;
Fig. 5–32 Naval Research Laboratory/NASA;
Fig. 5–33 American Science and Engineering, Inc./NASA;
Fig. 5–34A Courtesy G. Nikolskii, Institute of Terrestrial Magnetism, Moscow;
Fig. 5–34B Wide World Photos, Inc.;
Fig. 5–35 Jay M. Pasachoff;
Fig. 5–37 NASA;
Fig. 5–38 Big Bear Solar Observatory;
Figs. 5–39 and 5–40 Jay M. Pasachoff;
Fig. 5–41A Jay M. Pasachoff and Donald H. Menzel;
Fig. 5–41B NASA;
Figs. 5–42, 5–43, 5–46, 5–47, and 5–48 Jay M. Pasachoff;
Fig. 5–49 Dale R. Corson;
Fig. 5–50 M. Kanno, Hida Observatory, University of Kyoto;
Fig. 5–51 Sacramento Peak Observatory/Air Force Cambridge Research Laboratories;
Fig. 5–52 M. Waldmeier, Swiss Federal Observatory;
Fig. 5–53 Hale Observatories/Robert F. Howard and John M. Adkins;
Fig. 5–55 Randolph H. Levine, Center for Astrophysics and American Science and Engineering, Inc.;
Fig. 5–56 The Kitt Peak National Observatory/William C. Livingston;
Fig. 5–57 The Aerospace Corp./Dale Vrabec and Thomas Janssens;
Fig. 5–60 Max-Planck Institut für Radioastronomie, Bonn, G.F.R.;
Fig. 5–61 Big Bear Solar Observatory/Harold Zirin;
Fig. 5–62 Gustav Lamprecht;
Fig. 5–63 Big Bear Solar Observatory/Harold Zirin;
Fig. 5–64 Sacramento Peak Observatory;
Figs. 5–65 and 5–66 Paul Wild, Commonwealth Scientific and Industrial Research Organization, Australia;
Fig. 5–67 The Kitt Peak National Observatory;
Fig. 5–68 National Oceanic and Atmospheric Administration/Patrick S. McIntosh;
Fig. 5–69 Hale Observatories;
Fig. 5–70 Harvard College Observatory/NASA;
Fig. 5–71 Naval Research Laboratory/NASA;
Fig. 5–72 Stephen H. Schneider and Clifford Mass, *Science 190,* 741, copyright 1975 by The American Association for the Advancement of Science;
Fig. 5–73 Hale Observatories;
Fig. 5–74 Naval Research Laboratory/NASA;
Fig. 5–75 NASA;
Fig. 5–76 Laboratory for Atmospheric and Space Physics/NASA;
Fig. 5–77 Matthew P. Thekaekara/NASA Goddard Space Flight Center;
Fig. 5–78 NASA;
Fig. 5–79 © 1975 Field Enterprises Education Corp.;
Fig. 5–80 Courtesy of The Archives, California Institute of Technology;
Fig. 5–81 Courtesy of Hale Observatories and Otto Nathan;
Fig. 5–85 Einstein Archives/American Institute of Physics;
Fig. 5–86 Harriet H. Malitson/NASA Goddard Space Flight Center;
Fig. 5–87 American Science and Engineering, Inc./Harvard College Observatory.

CHAPTER 6—Opener Theodore R. Gull/The Kitt Peak National Observatory;
Fig. 6–2 George H. Herbig, Lick Observatory;
Fig. 6–3 Lick Observatory;
Fig. 6–4 George H. Herbig, Lick Observatory;
Fig. 6–10 American Institute of Physics, Niels Bohr Library, Segrè Collection;
Fig. 6–14 Lund Observatory, Sweden;
Figs. 6–18 and 6–19 Raymond Davis, Jr., Brookhaven National Laboratory.

CHAPTER 7—Opener Lick Observatory;
Fig. 7–3 Hale Observatories;
Fig. 7–4 Lick Observatory;

Fig. 7–5 Hale Observatories;
Figs. 7–7 and 7–9 From C. R. O'Dell, in IAU Symposium No. 34, "Planetary Nebulae," eds. D. E. Osterbrock and C. R. O'Dell. Reprinted by permission of the International Astronomical Union;
Fig. 7–11 Peter van de Kamp;
Figs. 7–12 and 7–13 Lick Observatory;
Figs. 7–14A and 7–14B Hale Observatories;
Fig. 7–15 © Ben Mayer, Los Angeles;
Figs. 7–16 and 7–18 Bruce E. Bohannan and Gary Emerson, University of Colorado, Boulder;
Fig. 7–17 Luigi G. Jacchia, Center for Astrophysics.

CHAPTER 8—Opener Lick Observatory;
Fig. 8–2 By permission of The Houghton Library, Harvard University;
Figs. 8–3 and 8–4 Hale Observatories;
Fig. 8–5 Hale Observatories, courtesy of Sidney van den Bergh. Photograph by R. Minkowski;
Fig. 8–6 The Cerro Tololo Inter-American Observatory;
Fig. 8–7 National Radio Astronomy Observatory/Hale Observatories;
Fig. 8–8 William Miller and Museum of Northern Arizona;
Fig. 8–9 Sidney van den Bergh/Hale Observatories;
Fig. 8–11 Jocelyn Bell Burnell;
Fig. 8–12A National Radio Astronomy Observatory;
Fig. 8–13 Nathaniel P. Carleton, Center for Astrophysics;
Fig. 8–14 Joseph H. Taylor, University of Massachusetts-Amherst;
Figs. 8–18 and 8–19 Lick Observatory;
Fig. 8–22 John C. Brandt, Robert G. Roosen, J. Thompson, and D. J. Ludden/NASA Goddard Space Flight Center;
Fig. 8–23 Richard N. Manchester, CSIRO;
Fig. 8–25 Herbert Friedman and Seth Shulman, Naval Research Laboratory.

CHAPTER 9—Opener Hale Observatories/Jerome Kristian;
Fig. 9–5 Drawing by Charles Addams; reprinted with permission of *The New Yorker Magazine, Inc.* © 1974;
Fig. 9–8 Chapin Library, Williams College;
Fig. 9–10 Herbert Gursky, Center for Astrophysics;
Fig. 9–11 Lois Cohen—Griffith Observatory;
Fig. 9–12 NASA and TRW Systems Group;
Fig. 9–13 Joseph Weber, University of Maryland.

PART III—Opener (*right*) Lunar and Planetary Laboratory, University of Arizona; (*left*) NASA;
Fig. 10–2 Lick Observatory;
Fig. 10–6B Harvard College Observatory;
Fig. 10–8 George Lovi, Vanderbilt Planetarium;
Fig. 10–9 Burndy Library, photograph by Owen Gingerich;
Fig. 10–10 Chapin Library, Williams College;
Fig. 10–12 Courtesy of Owen Gingerich;
Fig. 10–13 Chapin Library, Williams College;
Fig. 10–14 Uppsala University Library, photographed by Charles Eames, reproduced courtesy of Owen Gingerich;
Fig. 10–15 Chapin Library, Williams College;
Fig. 10–17 University of Michigan Library, Dept. of Rare Books and Special Collections, translation by Stillman Drake, reprinted courtesy of *Scientific American;*
Fig. 10–18 New Mexico State University Observatory;
Figs. 10–19, 10–20, and 10–25 Chapin Library, Williams College;
Fig. 10–27 B.C. by permission of John Hart and Field Enterprises, Inc.;
Fig. 10–29 United Press International photograph.

CHAPTER 11—Opener NASA;
Fig. 11–1 Base photograph: Lick Observatory;
Fig. 11–2 Hale Observatories;

Fig. 11–4 Lick Observatories;
Figs. 11–5 and 11–6A NASA;
Fig. 11–6B Drawing by Alan Dunn; reprinted with permission of *The New Yorker Magazine, Inc.* © 1971;
Fig. 11–7 General Electric Research and Development Center;
Fig. 11–8 NASA;
Fig. 11–9 Lick Observatory;
Figs. 11–10 and 11–11 NASA;
Fig. 11–12 Alfred J. Ferrari, California Institute of Technology;
Fig. 11–13 Judith W. Frondel, Harvard University;
Fig. 11–15 NASA;
Fig. 11–16 Data from Gerald J. Wasserburg and D. A. Papanastassiou, courtesy of Gerald J. Wasserburg, California Institute of Technology.

CHAPTER 12–Opener NASA;
Fig. 12–2 New Mexico State University Observatory;
Figs. 12–7, 12–8, 12–9, 12–10, 12–11, 12–12, 12–13, 12–14, 12–15, 12–16, 12–17, and 12–18 NASA.

CHAPTER 13–Opener NASA;
Fig. 13–1 Hale Observatories;
Fig. 13–4 Jet Propulsion Laboratories/R. M. Goldstein;
Fig. 13–6 Leonard Linton;
Fig. 13–8 Lick Observatory;
Figs. 13–9 and 13–10 NASA;
Fig. 13–11 Lick Observatory;
Fig. 13–12 Sovfoto;
Fig. 13–13 NASA;
Figs. 13–14 and 13–15 Science Service;
Fig. 13–16 Sovfoto.

CHAPTER 14–Opener NASA;
Fig. 14–1 U.S. Geological Survey, EROS Data Center;
Figs. 14–2 and 14–3 NASA;
Figs. 14–5, 14–6, 14–7, and 14–8 Jay M. Pasachoff;
Quotation from Paul A. Weiss (abridged from Rockefeller Institute Review 2, no. 6, pp 8–14) with permission of Dr. Weiss and the Rockefeller University Press. Reprinted in *A Random Walk in Science,* R. W. Weber, ed. (London: The Institute of Physics, 1973).

CHAPTER 15–Opener NASA;
Figs. 15–1 and 15–2 Lick Observatory;
Fig. 15–4 U.S. Geological Survey;
Figs. 15–5 through 15–20 NASA.

CHAPTER 16–Opener NASA/Lunar and Planetary Laboratory, University of Arizona;
Fig. 16–1 Lunar and Planetary Laboratory, University of Arizona;
Fig. 16–5A *Sky and Telescope;*
Fig. 16–5B Yerkes Observatory;
Fig. 16–6 Hale Observatories/Charles T. Kowal;
Figs. 16–7 and 16–9, and 16–11 NASA;
Fig. 16–12 NASA/Hale Observatories, Guido Munch;
Fig. 16–13 NASA/Lunar and Planetary Laboratory, University of Arizona;
Figs. 16–14 and 16–15 NASA;
Fig. 16–16 NASA/B. Roy Frieden, University of Arizona.

CHAPTER 17–Opener New Mexico State University Observatory;
Fig. 17–2 New Mexico State University Observatory;
Fig. 17–3 Lowell Observatory photograph;
Fig. 17–5 Smithsonian Astrophysical Observatory/Kaare Aksnes;
Fig. 17–6 Pic du Midi Observatory/Audouin Dollfus, Observatoire de Paris;
Fig. 17–8 NASA;

Figs. 17–9 and 17–12 Lunar and Planetary Laboratory, University of Arizona;
Fig. 17–13 Master and Fellows of St. John's College of Cambridge;
Fig. 17–14 Lick Observatory;
Fig. 17–16 Lowell Observatory photograph;
Fig. 17–17 Hale Observatories.

CHAPTER 18—Opener Institut d'Astrophysique de Liège, Belgium with the Schmidt camera at the Observatoire de Haute Provence, France;
Fig. 18–1 William Liller, Harvard College Observatory;
Fig. 18–2 Hale Observatories;
Fig. 18–3 European Southern Observatory/Jean Pierre Swings;
Fig. 18–4 Hale Observatories;
Fig. 18–5 NASA;
Fig. 18–6 National Portrait Gallery, London, painted by R. Phillips prior to 1721;
Fig. 18–7 Hale Observatories;
Fig. 18–8A Leon Tadrick in *The Sciences*, Vol. 15, No. 8, Nov. 1975 © 1975 The New York Academy of Sciences;
Figs. 18–8B and 18–8C *Sky and Telescope* by Roger Sinnott from orbital elements by Joseph Brady and Edna Carpenter, Lawrence Radiation Laboratory, University of California;
Fig. 18–9 NASA;
Fig. 18–10 Hale Observatories;
Fig. 18–11 Smithsonian Astrophysical Observatory;
Fig. 18–12 Peter Bloomer, *Horizons West*, by permission of Meteor Crater Enterprises, Inc.;
Fig. 18–13 Wide World Photos;
Fig. 18–14 Hale Observatories/John Huchra;
Fig. 18–15 Harvard College Observatory.

CHAPTER 19—Opener Museum of Modern Art/Twentieth-Century Fox, Inc.;
Fig. 19–1 © 1974 United Features Syndicate, Inc.;
Fig. 19–2 Peter van de Kamp, Sproul Observatory;
Fig. 19–3 The Kitt Peak National Observatory;
Fig. 19–4 © 1975 Field Enterprises Educational Corp.;
Figs. 19–5 and 19–6 NASA Ames Research Center;
Figs. 19–7 and 19–8 Cornell University photograph;
Fig. 19–9 Museum of Modern Art/Twentieth-Century Fox, Inc.;
Figs. 19–10 and 19–11 Robert M. Sheaffer;
Fig. 19–12 © 1975 by United Features Syndicate, Inc.;
Fig. 19–13 Royal Library of Copenhagen;
Marginal Note, page 435 Written by Larry Rhine and Mel Tolkin © copyright 1975 Tandem Productions, Inc. All Rights Reserved.

PART IV—Opener (*right*) Lick Observatory; (*left*) Hale Observatories;

PART IV Introduction Reproduced by courtesy of the Trustees, The National Gallery, London, painted circa 1578;

CHAPTER 20—Opener Hale Observatories;
Fig. 20–1 Yerkes Observatory;
Fig. 20–2 Harvard College Observatory;
Figs. 20–4 and 20–5 Hale Observatories;
Figs. 20–6 and 20–7 Lick Observatory;
Fig. 20–9 Hale Observatories;
Figs. 20–10 and 20–11 Lick Observatory;
Fig. 20–12 Data from W. Becker, photograph from Hale Observatories;
Fig. 20–14 Agris Kalnajs, Mt. Stromlo Observatory.

CHAPTER 21—Opener Jay M. Pasachoff;
Figs. 21–1 and 21–2 Courtesy of Bell Laboratories;
Fig. 21–4 Jay M. Pasachoff;

Fig. 21–9 Hale Observatories;
Fig. 21–15 Harvard University/E. M. Purcell;
Fig. 21–18A Gart Westerhout, University of Maryland;
Fig. 21–18B Gerrit Verschuur, University of Colorado;
Fig. 21–19 Hale Observatories;
Fig. 21–21 National Radio Astronomy Observatory;
Fig. 21–23 Graphic Films, Inc., Hollywood;
Fig. 21–24 Data from Marc L. Kutner, Rensselaer Polytechnic Institute, photograph from Lick Observatory;
Fig. 21–27 National Radio Astronomy Observatory;
Figs. 21–28 and 21–29 Jay M. Pasachoff.

CHAPTER 22 – Opener Eric Becklin, California Institute of Technology;
Fig. 22–1 Institute d'Astrophysique de Liège, Belgium;
Fig. 22–2 Hale Observatories;
Fig. 22–3 Photograph from Lick Observatory;
Fig. 22–4A Dennis Downes, Max Planck Institut für Radioastronomie, Bonn, G.F.R.;
Fig. 22–4B Eric Becklin, California Institute of Technology;
Fig. 22–5 Lund Observatory, Sweden;
Fig. 22–6 Harvard-Smithsonian Center for Astrophysics/Riccardo Giacconi and Herbert Gursky; with permission of D. Reidel Publishing Co.;
Fig. 22–7 NASA;
Fig. 22–8 Smithsonian Astrophysical Observatory.

PART V – Opener Hale Observatories;

PART V – Introduction Lick Observatory Archives;

CHAPTER 23 – Opener Hale Observatories;
Fig. 23–1 Harvard College Observatory;
Figs. 23–2 to 23–5 Hale Observatories;
Fig. 23–6 Courtesy of the U.K. Schmidt Telescope Unit, Royal Observatory, Edinburgh;
Figs. 23–7 to 23–9 Hale Observatories;
Fig. 23–10 European Southern Observatory;
Fig. 23–12 Hale Observatories;
Fig. 23–13 NASA Langley Research Center/Frank Hohl;
Fig. 23–14 Alar Toomre and Juri Toomre;
Fig. 23–17 Hyron Spinrad, University of California;
Fig. 23–19 Lick Observatory;
Fig. 23–20 Hale Observatories;
Fig. 23–23 Hale Observatories/James E. Gunn and J. Beverley Oke, from J. E. Gunn, Seventh Texas Symposium of Relativistic Astrophysics, copyright 1975 by The New York Academy of Sciences;
Fig. 23–24 The Kitt Peak National Observatory/Hyron Spinrad;
Fig. 23–25 Middle photograph from Lick Observatory. Top left photograph from Hale Observatories. Radio map based on Phillip J. Hargrave and Martin Ryle, *Monthly Notices of the Royal Astronomical Society 166,* 305, copyright 1974 by the Royal Astronomical Society;
Fig. 23–26 Radio map based on B. F. C. Cooper, R. M. Price, and D. J. Cole, *Australian Journal of Physics 18,* 589, copyright 1965. Photograph from Hale Observatories;
Fig. 23–27 Top photograph from Hale Observatories;
Fig. 23–33 Westerbork Synthesis Telescope/H. C. Kahlmann;
Fig. 23–34 National Radio Astronomy Observatory and E-Systems, Inc.;
Fig. 23–35 Radio data from Westerbork, D. S. Mathewson, P. C. van der Kruit, and W. N. Brouw, *Astronomy and Astrophysics 17,* 468, copyright 1972 by European Southern Observatory. Optical data from Hale Observatories;
Fig. 23–36 Westerbork Synthesis Telescope, Wilkes, Strom, and Wilson, *Nature 250,* 625, copyright 1974 by Macmillan Journals Limited, London;

Fig. 23–37 Courtesy of Richard G. Strom, George K. Miley, Jan Oort, and *Scientific American;*
Fig. 23–38A Westerbork Synthesis Telescope, K. J. Wellington, G. K. Miley, and H. van der Laan, *Nature 244,* 502, copyright 1973 by Macmillan Journals Limited, London; photograph © National Geographic Society—Palomar Observatory Sky Survey, reproduced by permission of the Hale Observatories.
Fig. 23–38B Westerbork Synthesis Telescope, G. K. Miley, G. C. Perola, P. C. van der Kruit, and H. van der Laan, *Nature 237,* 269, copyright 1972 by Macmillan Journals Limited, London.

CHAPTER 24—Opener Hale Observatories/Maarten Schmidt;
Fig. 24–1 Hale Observatories;
Fig. 24–2 Hale Observatories/Maarten Schmidt;
Fig. 24–5 Hale Observatories;
Fig. 24–8 Hale Observatories/Halton Arp;
Fig. 24–9 The Cerro Tololo Inter-American Observatory;
Fig. 24–10A Hale Observatories/montage by W. W. Morgan, Yerkes Observatory;
Fig. 24–10B Princeton University Observatory.

CHAPTER 25—Opener Lotte Jacobi;
Fig. 25–2 P. J. E. Peebles, Princeton University;
Fig. 25–8 Courtesy of Bell Laboratories;
Fig. 25–10 Based on Patrick Thaddeus, *Annual Review of Astronomy and Astrophysics 10,* 305, copyright 1972 by Annual Review Inc.;
Fig. 25–12 Based on D. P. Woody, J. C. Mather, N. S. Nishioka, and P. L. Richards, *Physical Review Letters 34,* 1036, copyright 1975 by The American Physical Society;
Figs. 25–14 and 25–15 Robert V. Wagoner, Stanford University;
Fig. 25–16 Jay M. Pasachoff and Diego A. Cesarsky, reprinted from the *Astrophysical Journal 193,* 65 with permission of the University of Chicago Press © 1973 by the American Astronomical Society;
Fig. 25–19 John B. Rogerson, Jr. and Donald H. York, Princeton University Observatory. Reprinted from the *Astrophysical Journal 186,* L95, with permission of the University of Chicago Press © 1973 by the American Astronomical Society.

CHAPTER 26—Opener and Fig. 26–1 National Radio Astronomy and E-Systems, Inc.;
Fig. 26–2 Jay M. Pasachoff;
Fig. 26–3 © 1973 *The Journal of Irreproducible Results, Inc.*;
Figs. 26–4 and 26–5 Jay M. Pasachoff;
Fig. 26–6 NASA;
Fig. 26–7 The Kitt Peak National Observatory;
Marginal Note, Page 579 Copyright © 1973 by The Massachusetts Medical Society. Reprinted by permission of The Viking Press and Lewis Thomas.

I also thank Frederick P. Woodson and the Photo Laboratory of the Hale Observatories, Dennis di Cicco at *Sky and Telescope*, Martha Liller at the Harvard College Observatory, Mildred Shapley Matthews at the Lunar and Planetary Laboratory, Rita Beebe at New Mexico State University Observatory, Spencer R. Weart and Joan N. Warnow at the Center for the History of Physics at the American Institute of Physics, and John Lancaster at the VLA of the National Radio Astronomy Observatory for their assistance in obtaining photographs.

Index

Principal references are in **boldface** type.
References to figures or marginal notes are in *italic* type.
References to tables are followed by t.
References to Color Plates are prefaced by CP.

Entries are alphabetized by letter, ignoring spaces, rather than by word. For example, B stars immediately precedes Bubble Nebula rather than appearing at the beginning of the B's. Significant initial numbers are alphabetized under their spellings; for example, 21 cm is alphabetized as twenty-one. Less important initial numbers and subscripts are ignored in alphabetizing; for example, 3C 273 is at the beginning of the C's, M1 appears at the beginning of the M's, and CO_2 appears next to CO. Greek letters are alphabetized under their English spellings.

Summer

June 1, 1:00 a.m. Local Daylight Time
July 1, 11:00 p.m. Local Daylight Time
August 1, 9:00 p.m. Local Daylight Time

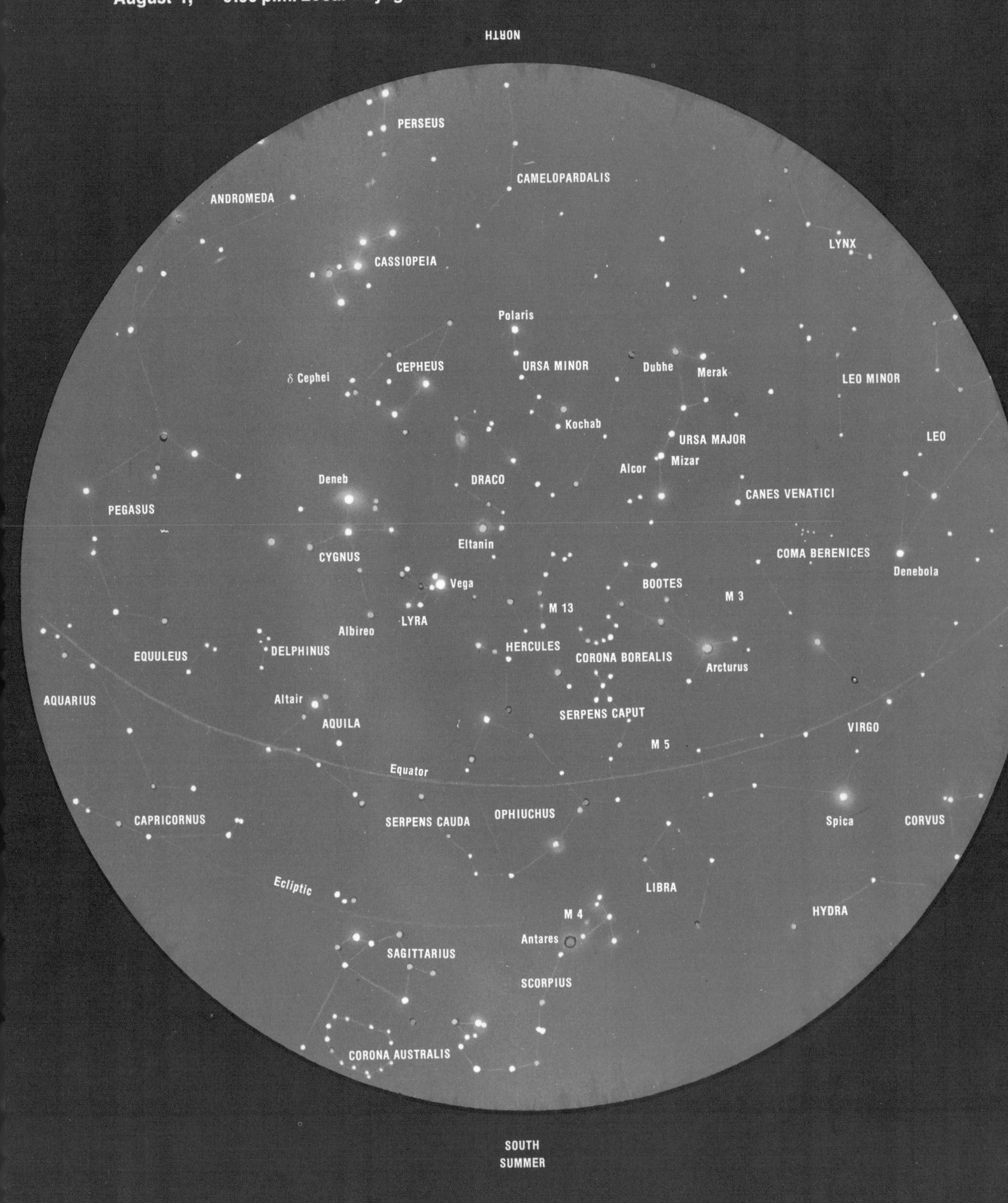

ASTRONOMY Magazine Star Dome Map by Adolf Schaller.